3/34

THE BIOLOGY OF THE ACTINOMYCETES

THE BIOLOGY OF THE ACTINOMYCETES

EDITED BY

M. GOODFELLOW
*Department of Microbiology,
The University,
Newcastle upon Tyne, UK*

M. MORDARSKI
*Department of Biosynthesis,
Institute of Immunology and
Experimental Therapy,
Wrocław, Poland*

S. T. WILLIAMS
*Department of Botany,
University of Liverpool,
Liverpool, UK*

1984

ACADEMIC PRESS

(*Harcourt Brace Jovanovich, Publishers*)

London   Orlando   San Diego   San Francisco   New York
Toronto   Montreal   Sydney   Tokyo   São Paulo

ACADEMIC PRESS INC. (LONDON) LTD.
24–28 Oval Road
London NW1 7DX

U.S. Edition published by
ACADEMIC PRESS INC.
(Harcourt Brace Jovanovich Inc)
Orlando, Florida 32887

Copyright © 1983 by
ACADEMIC PRESS INC. (LONDON) LTD.

All Rights Reserved

No part of this book may be reproduced in any form by photostat, microfilm, or any other means, without written permission from the publishers

British Library Cataloguing in Publication Data

The biology of actinomycetes.
1. Actinomycetes
I. Goodfellow, M. II. Mordarski, M.
III. Williams, S. T.
589.9′2   QR82.A35

ISBN 0-12-289670-X
LCCCN 83-71680

Printed in Great Britain by J. W. Arrowsmith Ltd., Bristol

Contributors

B. L. Beaman Department of Medical Microbiology, School of Medicine, University of California, Davis, California 95616, USA.

G. H. Brownell Department of Cell and Molecular Biology, Medical College of Georgia, 1459 Laney Walker Boulevard, Georgia 30912, USA.

K. F. Chater John Innes Institute, Norwich NR4 7UH, UK.

T. Cross Postgraduate School of Studies in Biological Sciences, University of Bradford, Bradford, West Yorkshire BD7 1DP, UK.

Katherine Denniston Division of Molecular Virology and Immunology, Georgetown University, Rockville, Maryland 20852, USA.

M. Goodfellow Department of Microbiology, The Medical School, The University, Newcastle upon Tyne NE1 7RU, UK.

D. A. Hopwood John Innes Institute, Norwich NR4 7UH, UK.

Shirley Lanning Department of Botany, University of Liverpool, PO Box 147, Liverpool L69 3BX, UK.

R. Locci Mycology Laboratory, Plant Pathology Department, University of Milan, Via Celoria 2, 20133 Milan, Italy.

D. E. Minnikin Department of Organic Chemistry, The University, Newcastle upon Tyne NE1 7RU, UK.

M. Mordarski Institute of Immunology and Experimental Therapy, Polish Academy of Sciences, ul Czerska 12, Wrocław 53-114, Poland.

A. G. O'Donnell Department of Organic Chemistry, The University, Newcastle upon Tyne NE1 7RU, UK.

Wanda Peczyńska-Czoch Institute of Immunology and Experimental Therapy, Polish Academy of Sciences, ul Czerska 12, Wrocław 53-114, Poland.

K. P. Schaal Institute of Hygiene, University of Cologne, Goldenfelstrasse 21, D-5000 Cologne 41, West Germany.

G. P. Sharples Department of Biology, Liverpool Polytechnic, Liverpool L3 3AF, UK.

Elizabeth M. H. Wellington Department of Biology, Liverpool Polytechnic, Liverpool L3 3AF, UK.

S. T. Williams Department of Botany, University of Liverpool, PO Box 147, Liverpool L69 3BX, UK.

Preface

Actinomycetes are a successful and widely distributed group of bacteria which have a number of properties that favour them in competition with other saprophytic microorganisms. They are best known for their economic importance as producers of antibiotics, vitamins and enzymes and are certain to have a significant role in the future of biotechnology. Some are causal agents of important human and animal diseases, others are plant pathogens but most are involved in turnover of organic matter. Given their industrial, ecological and medical importance it is not surprising that actinomycetes have been the subject of several recent books. Most of these have originated from the proceedings of symposia and, while useful to the specialist, they do not, by their very nature, always present a comprehensive view of the properties and biological importance of actinomycetes. The purpose of the present book is to give readers a balanced survey of the current knowledge of actinomycete biology.

All of the chapters have been written by specialists so that the book as a whole constitutes a unique collection of information on an interesting, but all too often neglected, group of bacteria. There are valuable and detailed reviews on ecology, genetics, morphology, pathogenicity, systematics, wall envelope composition and on the clinically significant actinomycetes. Specialist texts should be consulted for additional consideration of important pathogens, such as *Corynebacterium diphtheriae*, *Mycobacterium leprae* and *Mycobacterium tuberculosis*, and on the principles and methods of antibiotic production. We hope that the book will find a place in advanced courses of microbiology, will provide a useful general background to all those who work with, and try to unravel the nature of, actinomycetes, whether this be in industry, the health service or institutes of higher education.

We would like to take this opportunity to thank all of the contributors for bearing with us and for providing such excellent and well researched chapters. We would also like to thank many colleagues who have assisted us in a variety of ways and last but not least we extend our gratitude to Judith Gaigy and Dorothy Lewis whose help in typing was invaluable. Throughout, we have been greatly encouraged and assisted by Academic Press.

July, 1983 M. G.
  M. M.
  S. T. W.

Contents

List of contributors ... v

Preface ... vii

1. **Introduction to and Importance of Actinomycetes**   M. GOODFELLOW, M. MORDARSKI and S. T. WILLIAMS ... 1

2. **Classification**   M. GOODFELLOW and T. CROSS ... 7

3. **Morphology**   R. LOCCI and G. P. SHARPLES ... 165

4. **Genetics of the Nocardioform Bacteria**   G. H. BROWNELL and KATHERINE DENNISTON ... 201

5. ***Streptomyces* Genetics**   K. F. CHATER and D. A. HOPWOOD ... 229

6. **Transformation of Xenobiotics**   WANDA PECZYŃSKA-CZOCH and M. MORDARSKI ... 287

7. **Actinomycete Envelope Lipid and Peptidoglycan Composition**   D. E. MINNIKIN and A. G. O'DONNELL ... 337

8. **Clinical Significance of Actinomycetes**   K. P. SCHAAL and B. L. BEAMAN ... 389

9. **Laboratory Diagnosis of Actinomycete Diseases**   K. P. SCHAAL ... 425

10. **Actinomycete Pathogenesis**   B. L. BEAMAN ... 457

11. **Ecology of Actinomycetes**   S. T. WILLIAMS, S. LANNING and E. M. H. WELLINGTON ... 481

Index ... 529

1
Introduction to and Importance of Actinomycetes

M. GOODFELLOW*, S. T. WILLIAMS†
and M. MORDARSKI‡

*Department of Microbiology, The University, Newcastle upon Tyne, UK,
† Department of Botany, University of Liverpool, Liverpool, UK and ‡ Institute of Immunology and Experimental Therapy, Polish Academy of Sciences, Wrocław, Poland

1. Introduction 1
2. Causal agents of disease (from 1875) 2
3. Ecology and physiology of saprophytes (from 1900) . . 3
4. Production of antibiotics and other useful secondary metabolites (from 1940) 3
5. Genetics (from 1955) 4
   References 5

1. Introduction

The existence of the actinomycetes has been recognized for over a hundred years. For much of this time they were regarded as an exotic group of organisms with affinities to both bacteria and fungi. However, determinations of their fine structure and chemical composition, initiated in the 1950s, confirmed their prokaryotic nature (see Chapters 3 and 7). They now constitute the order *Actinomycetales* (Buchanan, 1917) and their removal from the mycologist's sphere of influence has been completed. Their change of status paralleled that of the blue-green algae to the cyanobacteria but it was accepted more rapidly and less acrimoniously. It is not easy to give a short, accurate definition of an actinomycete. They are frequently described as bacteria which have the ability to form branching hyphae at some stage of their development. However, this attribute is tenuous and it often requires imagination to believe in it (Gottlieb, 1973). The exact composition and boundaries of the order *Actinomycetales* are still open to question and modification by the application of new taxonomic techniques which have also led to improvements in the classification and identification of actinomycete genera and species (see Chapter 2). Despite their relegation to a single order of the kingdom *Prokaryotes*, their biological attributes, their importance to man and their history have ensured that actinomycetes are still generally studied as a group distinct from other bacteria.

Since Waksman's volumes on the actinomycetes (Waksman, 1950, 1959, 1961, 1967; Waksman and Lechevalier, 1962), most books on their biology have originated from the proceedings of symposia (e.g. Prauser, 1970; Arai, 1976; Goodfellow *et al.*, 1976; Lloyd and Sellers, 1976; Freerksen *et al.*, 1978; Mordarski *et al.*, 1978; Schaal and Pulverer, 1981). Although such publications are particularly useful to the specialist, they do not, by their very nature, always present a balanced coverage of the subject. There have been many advances in actinomycete biology since Waksman's publications, so that it is now appropriate to stand back and present a balanced survey of current knowledge. Although the actinomycetes have been the subject of an extensive literature in recent years, many aspects of their nature, physiology, and especially their role in natural ecosystems, have still to be understood. Clearly much remains to be done but an exciting array of powerful molecular and microbiological methods are available to those with the responsibility for developing actinomycete biology further.

The development of knowledge on the biology of the actinomycetes over the last hundred years is summarized below. Detailed accounts of early studies of actinomycetes can be found elsewhere (Lieske, 1921; Henrici, 1930; Krasilnikov, 1941; Waksman, 1959).

2. Causal Agents of Disease (from 1875)

As with many microbes, the study of actinomycetes was initiated in the late 19th century by workers examining diseased material from humans, animals or plants. The first unambiguous description of an actinomycete was probably that of Cohn (1875) who observed filamentous growth in concretions from lachrymal ducts and named the organism *Streptomyces foersteri*, but this generic name had been used by Corda (1839) for a group of fungi and was invalid. Shortly after this, an organism seen in a specimen of 'lumpy jaw' of cattle was described as *Actinomyces bovis* (Harz, 1877). Other observations of actinomycete-like microbes associated with human or animal infections soon followed, but their taxonomy and pathogenicity were confused due to the lack of pure cultures. The first plant pathogen *Streptomyces* (née *Oospora*) *scabies*, was isolated from potato scab by Thaxter (1891).

Subsequently, actinomycetes have proved to be causal agents of many human and animal infections. These include some widespread and intensively studied diseases, such as diphtheria (*Corynebacterium diphtheriae*), tuberculosis (*Mycobacterium tuberculosis*) and leprosy (*Mycobacterium leprae*), but it must be noted that the inclusion of such microbes in the actinomycetes has been a matter for debate (see Chapter 2). Consideration of the medical aspects of such diseases is beyond the scope of this book, but the relevant information can be found in recent detailed reviews (e.g. Saragea *et al.*, 1979; Ratledge and Stanford, 1982; Kubica and Wayne, 1983).

However, there is a wide range of actinomycete infections which are less widely known (Slack and Gerencser, 1975; Lloyd and Sellers, 1976). Many of these are proving to be more clinically significant than previously thought, partly due to improvements in procedures for their diagnosis (see Chapters 8, 9 and 10). It is also becoming increasingly evident that *Actinomyces* sp. can play a role in the aetiology of caries and periodontal disease.

3. Ecology and Physiology of Saprophytes (from 1900)

One of the first truly saprophytic actinomycetes to be detected was *Streptothrix chromogena* which was isolated from soil by Beijerinck (1900). The widespread occurrence of actinomycetes (particularly streptomycetes) in soil was demonstrated by Krainsky (1914) and Waksman and Curtis (1916, 1918). Over the next 20 years, knowledge of the ecology of actinomycetes in soil, composts and other habitats was considerably extended by Waksman, Jensen and other soil microbiologists. These studies also provided basic information on the isolation, cultivation, identification and physiology of saprophytic actinomycetes; indeed, many of the concepts and techniques originated in this period are still accepted today. This work also proved to be a valuable prelude to the exploitation of the actinomycetes as producers of antibiotics.

Subsequent studies of actinomycete ecology have sometimes consisted of little more than elaboration or repetition of the results of these pioneering workers. However, there has also been increasing emphasis on their roles in extreme environments and in many natural processes such as nitrogen fixation, decomposition of ligno-celluloses and the control of root pathogens (see Chapter 11). An impressive array of procedures are now available for the selective isolation of specific actinomycete taxa from natural habitats (Williams and Wellington, 1982; Cross, 1982).

4. Production of Antibiotics and Other Useful Secondary Metabolites (from 1940)

The first purified antibiotic to be obtained from an actinomycete was actinomycin (Waksman and Woodruff, 1940). This heralded a new era in which the potential medical and commercial value of these microbes was realized; this in turn influenced all aspects of research on the group. The discovery of actinomycin was soon superceded by that of streptomycin (Schatz et al., 1944) which is probably best known for its use in the control of tuberculosis. Many other medically useful antibiotics followed, most of which are still in use. Although most of these antibiotics originated from streptomycetes, other genera such as *Actinoplanes, Actinomadura* and *Micromonospora*, also produce useful or potentially useful antibiotics. Although the rate of return has decreased in recent years, new antibiotics and other useful metabolites from actinomycetes are still being discovered. This is well illustrated by β-lactamase inhibitors, such as clavulanic acid from *Streptomyces clavuligerus* (Reading and Cole, 1977), which have recently been commercially produced to overcome bacterial resistance to existing β-lactam antibiotics. The ability of actinomycetes to produce useful secondary metabolites remains unsurpassed, although the reasons for this and the biological significance of such products are still not clear (see Chapter 11).

The importance of actinomycetes in industrial biosynthesis has undoubtedly stimulated many aspects of basic research on these microbes. However, it inevitably led to concentration on the detection and selection of potentially useful strains (mainly streptomycetes) and the optimization of their fermentations. Therefore, research on some basic aspects of the biology of actinomycetes was somewhat retarded and overshadowed by their practical exploitation. Thus, most research from 1940–1970 was concentrated on the streptomycetes, to the comparative neglect of other saprophytic actinomycetes. In recent years, however, there has been an increasing recognition of the mutual interests of pure and applied microbiologists working with actinomycetes.

The principles and methods of antibiotic production have been extensively reviewed and will not be covered here. However, the use of microbes to transform known organic compounds into novel, useful agents is increasing, and the potential of actinomycetes in this field is also considerable (see Chapter 6). In addition, the genetic (see Chapter 5) and ecological (Chapter 1) aspects of antibiotic production are of theoretical and practical importance.

5. Genetics (from 1955)

Until the prokaryotic nature of actinomycetes was recognized, knowledge of their genetics was extremely limited. Since the 1950s, research on actinomycete genetics has generally paralleled that on other bacteria, most attention being paid to *Streptomyces coelicolor* and other *Streptomyces* species (see Chapter 5), and to *Nocardia* and *Rhodococcus* species (see Chapter 4).

The first report of recombination in streptomycetes (Sermonti and Spada Sermonti, 1955; Hopwood, 1957) initiated a period of study of the location of genes on the chromosome which facilitated study of its behaviour during conjugation. In the 1970s, the role of plasmids in controlling fertility and the genetic control of differentiation, primary metabolism and antibiotic biosynthesis in streptomycetes were elucidated by Hopwood and his co-workers (see Chapter 5).

These studies and others provided a basic model of actinomycete genetics, which has and will facilitate the application of the techniques of 'genetic engineering', such as gene cloning and protoplast fusion. Hopefully these will lead to still further exploitation of actinomycetes for their useful metabolites. Such developments, allied to their natural biochemical, physiological and ecological diversity, should ensure that actinomycetes have a role in the future of biotechnology which is at least as significant as their current one.

REFERENCES

Arai, T. (ed) (1976). *Actinomycetes: The Boundary Microorganisms.* Toppan Company Limited, Tokyo.
Beijerinck, M. W. (1900). Uber Chinonbildung durch *Streptothrix chromogena* und Lebensweise dieses Mikroben. *Centralblatt für Bakteriologie und Parasitenkunde. Abteilung II* **6**, 2–12.
Buchanan, R. E. (1917). Studies on the nomenclature and classification of the bacteria. II. The primary subdivisions of the Schizomycetes. *Journal of Bacteriology* **2**, 155–164.
Cohn, F. (1875). Untersuchungen über Bakterien. II. *Beiträge zur Biologie der Pflanzen* **1**, 141–207.
Corda, A. C. J. (1839). *Pracht-Flora Europaeischer Schimmelbildungen.* Gerhard Fleischer, Leipzig.
Cross, T. (1982). Actinomycetes: A continuing source of new metabolites. *Developments in Industrial Microbiology* **23**, 1–18.
Freerksen, E., Tarnok, I. and Thumin, J. H. (eds) (1978). *Genetics of the Actinomycetales.* Gustav Fischer Verlag, Stuttgart and New York.
Goodfellow, M., Brownell, G. H. and Serrano, J. A. (eds) (1976). *The Biology of the Nocardiae.* Academic Press, London and New York.
Gottlieb, D. (1973). General consideration and implications of the *Actinomycetales.* In: *Actinomycetales, Characteristics and Practical Importance*, (G. Sykes and F. A. Skinner, eds.) pp. 1–10. Academic Press, London and New York.
Harz, C. O. (1877). *Actinomyces bovis,* ein neuer Schimmel in den Geweben des Rindes. *Jahrebericht der Königlichen Centralen Thierarzneischule München für 1877/1878* **5**, 125–140.
Henrici, A. T. (1930). *Molds, Yeasts and Actinomycetes.* J. Wiley and Sons, New York.
Hopwood, D. A. (1957). Genetic recombination in *Streptomyces coelicolor. Journal of General Microbiology* **16**, ii–iii.
Krainsky, A. (1914). Die Actinomyceten und ihre Bedeutung in der Natur. *Centralblatt für Bakteriologie und Parasitenkunde. Abteilung II* **41**, 649–688.
Krasilnikov, N. A. (1941). *Guide to Actinomycetales.* Akademiya Nauk Soyuza SSR, Moscow.
[English translation; Israel Program for Scientific Translations Limited, Jerusalem].
Kubica, G. P. and Wayne, L. G. (eds.) (1983). *The Mycobacteria: A Sourcebook Volume 1. The Mycobacteria.* Marcel Dekker, New York.

Lieske, R. (1921). *Morphologie und Biologie der Strahlenpilze.* G. Borntraeger, Leipzig.
Lloyd, D. H. and Sellers, K. C. (eds.) (1976). *Dermatophilus Infection in Animals and Man.* Academic Press, London and New York.
Mordarski, M., Kurylowicz, W. and Jeljaszewicz, J. (eds) (1978). *Nocardia and Streptomyces.* Gustav Fischer Verlag, Stuttgart.
Prauser, H. (ed) (1970). *The Actinomycetales.* Gustav Fischer Verlag, Jena.
Ratledge, C. and Stanford, J. L. (eds) (1982). *The Biology of the Mycobacteria. Volume 1. Physiology, Identification and Classification.* Academic Press, London and New York.
Reading, C. and Cole, M. (1977). Clavulanic acid: a beta-lactamase-inhibiting beta-lactam from *Streptomyces clavuligerus. Antimicrobial Agents and Chemotherapy* **11**, 852–857.
Saragea, A., Maximescu, P. and Meitert, E. (1979). *Corynebacterium diphtheriae*: Microbiological methods used in clinical and epidemiological investigations. In: *Methods in Microbiology, Volume 13* (T. Bergan and J. R. Norris, eds.), pp. 61–176. Academic Press, London and New York.
Schaal, K. P. and Pulverer, G. (eds) (1981). *Actinomycetes.* Gustav Fischer Verlag, Stuttgart and New York.
Schatz, A., Bugie, E. and Waksman, S. A. (1944). Streptomycin, a substance exhibiting antibiotic activity against gram-positive and gram-negative bacteria. *Proceedings of the Society for Experimental Biology and Medicine* **55**, 66–69.
Sermonti, G. and Spada-Sermonti, J. (1955). Genetic recombination in *Streptomyces. Nature (London)* **176**, 121.
Slack, J. M. and Gerencser, M. A. (1975). *Actinomyces, Filamentous Bacteria. Biology and Pathogenicity.* Burgess Publishing Company, Minneapolis.
Thaxter, R. (1891). The potato scab. *Connecticut Agricultural Experimental Station Report 1890*, pp. 81–85.
Waksman, S. A. (1950). *The Actinomycetes. Their Nature, Occurrence, Activities and Importance.* Chronica Botanica Company, Waltham, Mass., U.S.A.
Waksman, S. A. (1959). *The Actinomycetes, Volume 1. Nature, Occurrence and Activities.* Williams and Wilkins, Baltimore.
Waksman, S. A. (1961). *The Actinomycetes, Volume 2. Classification, Identification and Description of Genera and Species.* Williams and Wilkins, Baltimore.
Waksman, S. A. (1967). *The Actinomycetes. A Summary of Current Knowledge.* The Ronald Press, New York.
Waksman, S. A. and Curtis, R. E. (1916). The actinomyces of the soil. *Soil Science* **1**, 99–134.
Waksman, S. A. and Curtis, R. E. (1918). The occurrence of actinomycetes in the soil. *Soil Science* **6**, 309–319.
Waksman, S. A. and Woodruff, H. B. (1940). Bacteriostatic and bactericidal substances produced by a soil actinomyces. *Proceedings of the Society for Experimental Biology and Medicine* **45**, 609–614.
Waksman, S. A. and Lechevalier, H. A. (1982). *The Actinomycetes, Volume 3. Antibiotics of Actinomycetes.* Baillière, Tindall and Cox, London.
Williams, S. T. and Wellington, E. M. H. (1982). Principles and problems of selective isolation of microbes. In: *Bioactive Microbial Products: Search and Discovery* (J. D. Bu'lock, L. J. Nisbet and D. J. Winstanley, eds.), pp. 9–26. Academic Press, London and New York.

2
Classification

M. GOODFELLOW* and T. CROSS†

Department of Microbiology, The University, Newcastle upon Tyne, UK, and † Postgraduate School of Studies in Biological Sciences, University of Bradford, Bradford, West Yorkshire, UK

1. Introduction 8
2. Chemotaxonomy 12
3. Numerical taxonomy 21
4. Towards a definition of actinomycetes 28
5. Actinobacteria 29
   - A. *Actinomyces* 32
   - B. *Agromyces* 35
   - C. *Arachnia* 37
   - D. *Arcanobacterium* 38
   - E. *Arthrobacter* 40
   - F. *Brevibacterium* 44
   - G. *Cellulomonas* 46
   - H. *Curtobacterium* 48
   - I. *Microbacterium* 50
   - J. *Oerskovia* 52
   - K. *Promicromonospora* 53
   - L. *Renibacterium* 54
   - M. *Rothia* 56
   - N. *Unassigned actinobacteria* 57
6. Actinoplanetes 59
   - A. *Actinoplanes* 60
   - B. *Dactylosporangium* 66
   - C. *Micromonospora* 67
7. Actinomycetes with multilocular sporangia . . . 69
   - A. *Dermatophilus* 70
   - B. *Frankia* 71
   - C. *Geodermatophilus* 73
8. Nocardioform actinomycetes 74
   - A. *Caseobacter* 75
   - B. *Corynebacterium* 77
   - C. *Mycobacterium* 80
   - D. *Nocardia* 88
   - E. *Rhodococcus* 91
9. Streptomycetes 94
   - A. *Intrasporangium* 98
   - B. *Kineosporia* 99

THE BIOLOGY OF THE ACTINOMYCETES
ISBN 0-12-289670-X

Copyright © 1983 by Academic Press, London
All rights of reproduction in any form reserved

	C. *Nocardioides*	100
	D. *Sporichthya*	101
	E. *Streptomyces*	102
	F. *Streptoverticillium*	104
10.	Maduromycetes	105
	A. *Actinomadura*	106
	B. *Excellospora*	108
	C. *Microbispora*	109
	D. *Microtetraspora*	110
	E. *Planobispora*	111
	F. *Planomonospora*	112
	G. *Spirillospora*	113
	H. *Streptosporangium*	114
11.	Thermomonosporas	115
	A. *Actinosynnema*	116
	B. *Nocardiopsis*	117
	C. *Streptoalloteichus*	118
	D. *Thermomonospora*	119
12.	Micropolysporas	121
	A. *Actinopolyspora*	122
	B. *Micropolyspora nomen conservandum*	123
	C. *Pseudonocardia*	125
	D. *Saccharomonospora*	126
	E. *Saccharopolyspora*	128
	F. Unassigned micropolysporas	128
13.	Thermoactinomyces	130
	A. *Thermoactinomyces*	130
	References	131

1. Introduction

The actinomycetes have been traditionally considered to be prokaryotic bacteria with elongated cells or filaments that usually showed some degree of true branching. Although the morphology of these organisms ranges from simple to complex, most strains of most species can be assigned to one of two broad morphological groups, nocardioform- and sporo-actinomycetes (Prauser, 1970, 1976a, 1978, 1981). Nocardioform bacteria form hyphae which eventually fragment into coccoid or rod-like elements that give rise to new mycelia (Locci, 1976, 1978, 1981). The genera *Caseobacter*, *Mycobacterium* and *Rhodococcus* are generally included in this group, though all of them contain strains that exhibit little, if any, branching, growing merely as rod or coccoid elements. The sporoactinomycetes encompass a greater morphological complexity that includes the formation of spores in or on definite parts of the mycelium (Locci, 1976; Williams et al., 1976; Williams and Wellington, 1980). A third level of organization is presented by *Dermatophilus* and *Geodermatophilus* which form a substrate mycelium that divides both transversely and longitudinally to give a primitive multilocular sporangium. Coccoid elements are released which may gain motility and eventually germinate into filaments or hyphae (Cross and Goodfellow, 1973). *Frankia* shows some of the morphological traits associated with dermatophili and geodermatophili (Callaham *et al.*, 1978; Becking, 1981).

Actinomycetes have for many years been grouped together solely on morphological grounds even though they have never been satisfactorily distinguished from coryneform bacteria on this basis (Bousfield and Goodfellow, 1976; Goodfellow and Minnikin, 1981a). It is now beyond dispute that innumerable morphological transitions exist between nocardioform actinomycetes like *Actinomyces, Nocardia* and *Rhodococcus* and coryneform bacteria, such as *Arthrobacter, Cellulomonas* and *Corynebacterium*, that have a tendency to form branched elements (Locci, 1976, 1978, 1981; Williams *et al.*, 1976; Prauser, 1978, 1981). Recent morphological studies help to explain why early workers (eg. Lehmann and Neumann, 1920, 1927; Lieske, 1921; Ørskov, 1923; Jensen, 1953) found it impossible to distinguish between corynebacteria, mycobacteria, nocardiae and related bacteria simply on morphological features. In the light of current knowledge it is clear that many of the earlier classifications of coryneform and nocardioform bacteria were artificial, had a narrow data base, and were consequently unreliable vehicles for the identification of unknown isolates (Goodfellow and Minnikin, 1977, 1981a, b, c, 1982). A tentative step towards remedying this situation was taken in the current edition of *Bergey's Manual of Determinative Bacteriology* (Buchanan and Gibbons, 1974) where actinomycetes and coryneform bacteria were considered together in a section entitled '*Actinomycetes and related organisms*'.

Bacterial systematics has undergone revolutionary change in the last 20 years. The application of new and reliable biochemical, chemical, genetical, numerical and molecular biological techniques have been responsible for rapidly changing views on how bacteria ought to be classified and identified (see Goodfellow and Board, 1980; Berkeley and Goodfellow, 1981). These techniques have generally been applied to greatest effect on taxa where dependence on form and function proved most unsatisfactory and they have provided a framework for revised classification of both coryneform (Bousfield and Callely, 1978; Stackebrandt *et al.*, 1980a, b; Goodfellow and Minnikin, 1981a; Keddie and Bousfield, 1980; Keddie and Jones, 1981; Döpfer *et al.*, 1982) and nocardioform bacteria (Bradley and Mordarski, 1976; Lechevalier, 1976; Minnikin and Goodfellow, 1976, 1980, 1981a; Goodfellow and Minnikin, 1977, 1978, 1981b, c, 1983; Mordarski *et al.*, 1977, 1978a, b, 1980a, b, 1981a, b; Stackebrandt and Woese, 1981a, b). The newer methods are now being applied to sporoactinomycetes with interesting results (Stackebrandt and Woese, 1981a, b; Stackebrandt *et al.*, 1981, 1982; Williams *et al.*, 1981, 1983a, b; Goodfellow and Pirouz, 1982).

Most of the new taxonomic techniques only serve to detect affinities between closely related taxa; they are of limited value in determining relationships among distantly related species, genera and families. Suprageneric and evolutionary relationships of bacteria can, however, be detected using the powerful techniques of DNA–(ribosomal r) RNA association and 16S rRNA cataloguing. Base sequences of rRNA cistrons are more highly conserved than most of the genes forming the bacterial genome (Doi and Igarashi, 1965; Dubnau et al., 1965; Moore and McCarthy, 1967), a fact that allows comparisons to be made between nucleotide sequences of rRNA preparations from representatives of diverse taxa (De Smedt and De Ley, 1977; De Smedt et al., 1980; Gillis and De Ley, 1980). Ribosomal RNA cistron similarity data show that the acid-fast fast actinomycetes are phylogenetically close (Mordarski et al., 1980a, 1981b) and indicate that the sporoactinomycetes fall into at least three major homology groups: *Actinoplanes, Amorphosporangium, Ampullariella* and *Micromonospora*; *Planobispora, Planomonospora* and *Streptosporangium*; and *Chainia, Elytrosporangium, Kitasatoa, Microellobosporia, Streptomyces* and *Streptoverticillium* (Stackebrandt et al., 1981). These groupings are in good agreement with the current trends in the taxonomy of these organisms.

Ribosomal RNA cataloguing provides an even more exacting way of detecting phylogenetic relationships amongst prokaryotes (Fox et al., 1977a, b; Woese and Fox, 1977; Stackebrandt and Woese, 1981a, b). In this method, purified RNA is digested by TI ribonuclease, the oligonucleotides separated by two-dimensional electrophoresis are sequenced by a combination of endonuclease digestion procedures which yield a catalogue of sequences characteristic of the strain under study. The oligonucleotide catalogues of any two strains are compared one with another and oligonucleotides, of six residues or larger, common to any two catalogues, are scored to produce a 'S_{AB} value' characteristic of that pair of organisms. The function S_{AB} is equivalent to twice the total number of residues in sequences common to a pair of catalogues, divided by the total number of residues in all of the sequences in the two catalogues. S_{AB} values are analysed using standard clustering algorithms and data presented as dendrograms or, more appropriately, as evolutionary trees (Stackebrandt and Woese, 1981a).

16S rRNA cataloguing data show that Gram-positive bacteria form a distinct phyletic line that can readily be divided into two branches on the basis of DNA base composition (Stackebrandt and Woese, 1981a). The actinomycete–coryneform line includes bacteria with a guanine (G) plus cytosine (C) content above about 55 mol% and can be separated from the low G+C content (below 50 mol%) *Clostridium-Bacillus-Streptococcus* branch. Several taxa previously associated with the actinomycetes clearly belong to this second evolutionary branch. The genus *Eubacterium* is phylogenetically related to *Clostridium*, *Kurthia* to the lactic acid bacteria and, perhaps most surprisingly of all, *Thermoactinomyces* to the *Bacillaceae* (Ludwig et al., 1981; Tanner et al., 1981; Stackebrandt and Woese, 1981a).

The thermophilic genus *Thermoactinomyces* has long been regarded as a 'good actinomycete' because of the gross appearance of its powdery white or yellow colonies on agar, and branching hyphae which carry lateral spores on both substrate and aerial hyphae. Several taxonomic studies have shown that the genus can be clearly distinguished from actinomycete genera with single spores, for example *Micromonospora*, *Saccharomonospora* and *Thermomonospora*, and the present authors earlier suggested that it should be placed in a distinct family because of its ability to produce endospores (Cross and Goodfellow, 1973). However, the reclassification of thermoactinomycetes in the *Bacillaceae* is a revolutionary move which requires some explanation and justification.

Wall analyses showed that the peptidoglycan of *Thermoactinomyces* species contained *meso*-diaminopimelic acid (*meso*-DAP) with no other characteristic amino acids or sugars (Becker et al., 1965) and the genus was therefore included in the actinomycete wall chemotype III classification of M. P. Lechevalier and Lechevalier (1970a, b). This peptidoglycan, namely the directly cross-linked *meso*-DAP-D-alanine type, is also found in the majority of *Bacillus* species (Schleifer and Kandler, 1972) and is not restricted to actinomycetes. The DNA base composition of *Thermoactinomyces* species show that they have a much lower mol% G+C content than is found in actinomycete genera (see Table 1). Values of 52.0 (Fritzsche, 1967), 53.4–54.4 (Craveri and Manachini, 1966), 54.1 and 54.8 (Craveri et al., 1966) are much nearer those found in some mesophilic and thermophilic *Bacillus* species (Priest, 1981). When detailed studies on the fine structure of *Thermoactinomyces* spores showed that they had the typical structure of bacterial endospores and contained dipicolinic acid (Cross, 1968; Cross et al., 1968a; Dorokhova et al., 1968) the relationship was even more apparent.

Further evidence for the relationship between *Bacillus* and *Thermoactinomyces* has recently come from a study of their isoprenoid quinones. Both genera typically contain major amounts of unsaturated menaquinones with seven or nine isoprene units (Collins and Jones, 1981a; Collins et al., 1982g; Minnikin and Goodfellow, 1981b) in contrast to the vast majority of actinomycetes which contain complex mixtures of partially saturated or hydrogenated menaquinones (Collins and Jones, 1981a). It is also interesting to note that the purified malate dehydrogenases from *Team. sacchari* and several *Bacillus* spp. were found to be tetramers which exhibited immunochemical homology and differed from the dimeric enzymes present in *Temo. fusca* and other non-endospore bacteria (Sundaram et al., 1980).

It is clear that the mere possession of branching hyphae should not automatically place a bacterium within the actinomycetes, and that the earlier observation, that the ability to produce endospores was confined to the *Bacillaceae* (Lechevalier and Lechevalier, 1967) may prove correct. Conversely, coryneform taxa such as *Arthrobacter, Brevibacterium, Cellulomonas, Corynebacterium, Curtobacterium* and *Microbacterium*, which rarely, if ever, form a primary mycelium, are phylogenetically intermixed with classical actinomycete taxa (Stackebrandt and Woese, 1981a, b) and must be classified accordingly. Indeed, the 16S data bear eloquent testimony to the inherent dangers in constructing taxonomies solely on the basis of morphological features (van Niel, 1946). This early warning to taxonomists was given even further substance with the discovery that the genus *Micrococcus* is phylogenetically close to the genus *Arthrobacter* (Stackebrandt *et al.*, 1980a; Stackebrandt and Woese, 1981a). These evolutionary data raise the important question as to what is, or what is not, an actinomycete, a point to which we will return.

It is not yet possible to compile a comprehensive phylogeny of the actinomycetes using data derived by comparative cataloguing of 16S rRNA but the outline of such a classification is emerging (Fig. 1). The four aggregate taxa recognized from the ribosomal RNA work (Mordarski *et al.*, 1980a, 1981b; Stackebrandt *et al.*, 1981) are again evident as is a fifth group containing the 'coryneform taxa' and the genera *Actinomyces, Agromyces, Oerskovia* and *Rothia*. The genera *Arachnia* and *Renibacterium* have properties which allow them to be added to this aggregate taxon. We propose the name Actinobacteria for actinomycetes assigned to this group. The composition of these five aggregate groups, which provide the core of the supragenetic classification recognized by the authors, is supported by criteria derived from other modern taxonomic methods notably chemotaxonomy and numerical taxonomy. The genera *Geodermatophilus* and *Thermomonospora* may form the nuclei of additional clusters (Fig. 1, Table 1) but the final aggregate taxon, which includes the genus *Micropolyspora*, is underpinned essentially by chemical and morphological features. It is premature to equate the aggregate groups with higher ranks in the taxonomic hierarchy especially since previous attempts to classify actinomycete genera into families yielded groups now known to be heterogeneous (Baldacci, 1958; Waksman, 1961; Baldacci and Locci, 1966; H. A. Lechevalier and Lechevalier, 1970; Prauser, 1970; Cross and Goodfellow, 1973; Buchanan and Gibbons, 1974).

2. Chemotaxonomy

Chemical systematics or chemotaxonomy is a rapidly expanding discipline in which information from chemical analyses of whole organisms or cell fractions is used not only for classification and identification but also for

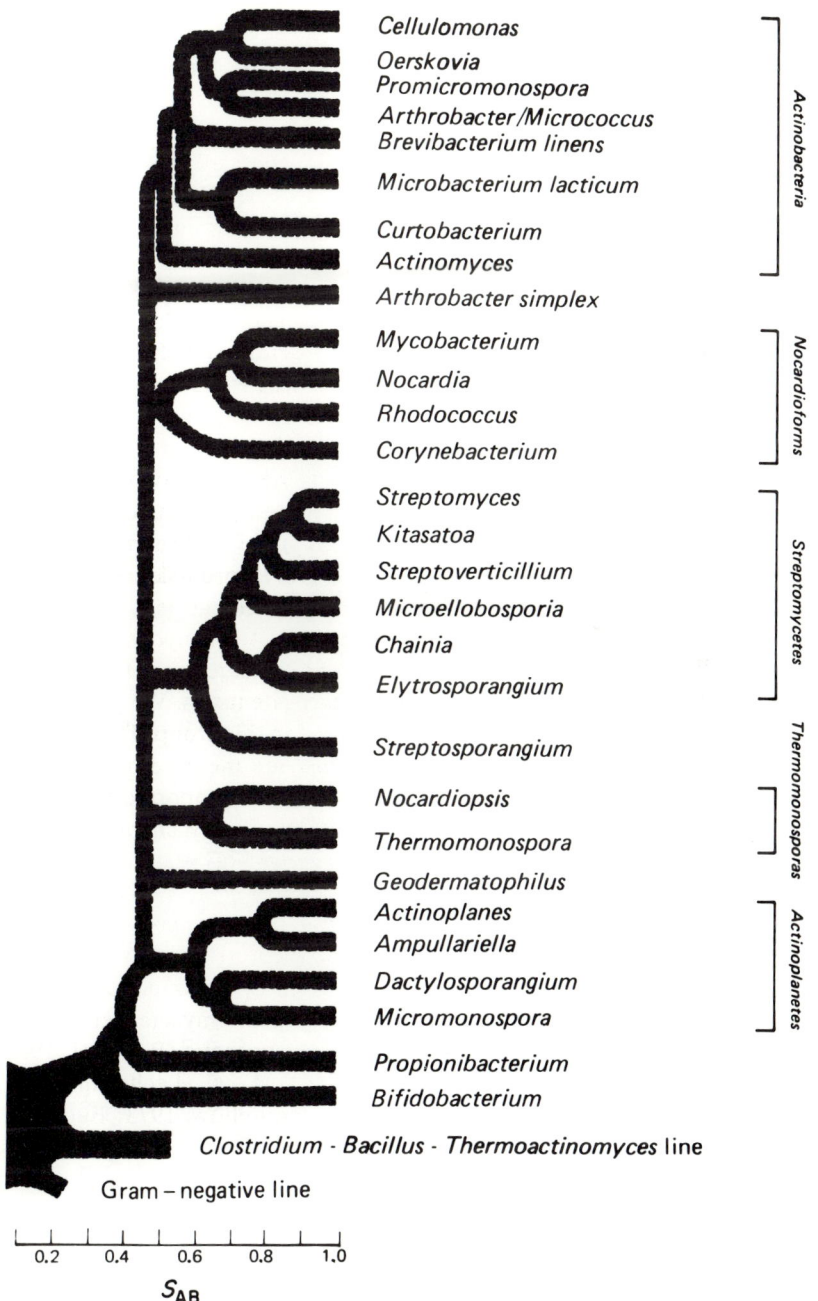

FIG. 1. Phylogenetic relationships based on 16S rRNA similarity (see Stackebrandt and Woese, 1981a).

TABLE 1. Major wall components, excluding lipids, and DNA base composition of actinomycetes

Aggregate group	Genus	Wall chemotype*	Major distinguishing wall constituents*[†]		Peptidoglycan type*	Muramic acid acyl group[‡]	Mole % G+C[§]
			diamino-acid	others			
Actinobacteria	Actinomyces	V, VI	lysine/ornithine	aspartic acid	A4α, A4β A5α	ND	60–73
	Agromyces	VII	DAB	glycine	B2γ	ND	71–72
	Arachnia	I	LL-DAP	galactose glycine	A3γ	ND	63–65
	Arcanobacterium	VI	lysine	rhamnose	A5α	ND	50–52
	Arthrobacter	VI	lysine	aspartic acid, galactose, glycine	A3α	acetyl	59–66
	Brevibacterium	III	meso-DAP	galactose, glucose, glycerol teichoic acid	A1γ	acetyl	60–64
	Cellulomonas	VIII	ornithine	rhamnose	A4β	acetyl	71–77
	Curtobacterium	VIII	ornithine	galactose	B2β	acetyl	66–73
	Microbacterium	VI	lysine	galactose, glycine, rhamnose	B1α	glycolyl	69–70
	Micrococcus	VI	lysine	none	A3α	acetyl	66–75
	Oerskovia	VI	lysine	galactose	A4α	acetyl	70–75

Group	Genus	Wall type	Diamino acid	Sugars	Mycolic acid type	Acyl type	%G+C
	Promicromonospora	VI	lysine	aspartic acid	A4α	ND	73–74
	Renibacterium	VI	lysine	glycine, glucose	ND	ND	53–54
	Rothia	VI	lysine	fructose, galactose, glucose, ribitol	A3α	ND	65–70
Actinoplanetes	*Actinoplanes*						72–73
	Amorphosporangium						71
	Ampullariella						72–73
	Dactylosporangium	II	*meso* or HO-DAP	glycine	A1γ	glycolyl	71–73
	Micromonospora						71–73
	Pilimelia						ND
Multilocular sporangia	*Dermatophilus*			madurose fucose,	A1γ	ND	57–59
	Frankia	III	*meso*-DAP	madurose, xylose	ND	ND	68–72
	Geodermatophilus			none	A1γ	ND	73–75
Nocardioforms	*Caseobacter*	IV	*meso*-DAP	arabinose, galactose	ND	ND	65–67
	Corynebacterium					acetyl	51–59
	Mycobacterium	IV	*meso*-DAP	arabinose, galactose	A1γ	glycolyl	62–70
	Nocardia					glycolyl	64–69
	Rhodococcus	IV	*meso*-DAP	arabinose, galactose	ND	glycolyl	59–69
	"*aurantiaca*" taxon					ND	ND

TABLE 1. (cont.)

Aggregate group	Genus	Wall chemotype*	Major distinguishing wall constituents*†		Peptidoglycan type*	Muramic acid acyl group‡	Mole % G+C§
			diamino-acid	others			
Streptomycetes	Intrasporangium	⎫				ND	ND
	Kitasatoa	⎪				ND	ND
	Nocardioides	⎪				ND	66–67
	Sporichthya	⎬ I	LL-DAP	glycine	A3γ	ND	ND
	Streptomyces	⎪				acetyl	69–78
	"Chainia"	⎪				acetyl	71–72
	"Elytrosporangium"	⎪				ND	ND
	"Microellobosporia"	⎪				ND	68–73
	Streptoverticillium	⎭				ND	69–73
	Kineosporia				ND	ND	ND
Maduromycetes	Actinomadura	⎫					65–77
	"Excellospora"	⎪					ND
	Microbispora	⎪					70–74
	Microtetraspora	⎬ III	meso-DAP	madurose	A1γ	acetyl	ND
	Planobispora	⎪					70–72
	Planomonospora	⎪					72
	Spirillospora	⎪					71–73
	Streptosporangium	⎭					69–71

Thermomonosporas	Actinosynnema	III	meso-DAP	none	ND	ND
	Nocardiopsis					70–76
	Streptoalloteichus					ND
	Thermomonospora					ND
Micropolysporas	Actinopolyspora	IV	meso-DAP	arabinose, galactose	ND	64
	Micropolyspora					ND
	Pseudonocardia					79
	Saccharomonospora	IV	meso-DAP	arabinose, galactose	ND	74–75
	Saccharopolyspora				Alγ	77
Thermo-actinomycetes	Thermoactinomyces	III	meso-DAP	none	Alγ	53–55

* Wall chemotypes and peptidoglycan type after M. P. Lechevalier and H. A. Lechevalier (1970a) and Schleifer and Kandler (1972), respectively; data also from Collins et al. (1982d), Döpfer et al. (1982), Goodfellow and Minnikin (1981a, c), Goodfellow and Pirouz (1982), Lechevalier and Lechevalier (1979), Minnikin et al. (1978a), Prauser (1978), Sanders and Fryer (1980), Seidl et al. (1980), Tomita et al. (1978), Yamada and Komagata (1970a), Yamaguchi (1965) and Weiss (personal communication).

† All wall preparations contain major amounts of alanine, glutamic acid, glucosamine and muramic acid; DAB, diaminobutyric acid and DAP, diaminopimelic acid.

‡ Date from Kawamoto et al. (1981), Uchida and Aida (1977, 1979), ND, not determined.

§ Taken from An et al. (1983), Baird-Parker (1974), Collins et al. (1982d), Craveri and Manachini (1966), Döpfer et al. (1982), Farina and Bradley (1970), Goodfellow and Minnikin (1981a, c), Johnson and Cummins (1972), Keddie and Bousfield (1980), Pine and Georg (1974), Sanders and Fryer (1980), Slack (1974a), Stackebrandt and Kandler (1979) and Yamaguchi (1967).

tracing evolutionary trends. A wide array of chemical methods are now commonly used to determine DNA base (Owen and Hill, 1979) and whole organism and wall sugar and amino acid composition of bacteria (Schleifer and Kandler, 1972; Keddie and Cure, 1977, 1978), with lipids being increasingly exploited as chemical markers (Asselineau, 1968; Goldfine, 1972; Shaw, 1974; Lechevalier, 1977; Lechevalier et al., 1977, 1981; Minnikin and Goodfellow, 1976, 1980). Chemical features have proved to be of particular importance in actinomycete systematics and have helped to resolve the taxonomic affinities of the genus *Thermoactinomyces*. Indeed, descriptions of actinomycete genera and species that do not include chemical properties can normally be considered to be incomplete.

Data from wall sugar and amino acid analyses prepared the ground for the reappraisal of the systematics of actinomycetes and coryneform bacteria (Lechevalier, 1976; Keddie and Bousfield, 1980). Simple qualitative analyses of the composition of walls and, later, whole organisms allowed actinomycetes and related organisms to be classified into eight large groups or wall chemotypes based on the limited distribution of a small number of wall components (Table 1; Becker et al., 1965; Lechevalier et al., 1966a; M. P. Lechevalier and Lechevalier, 1970a, b). The diamino acid of the peptidoglycan was found to be especially important, particularly in the classification of coryneform bacteria (Table 1; Cummins and Harris, 1956; Cummins, 1962; Keddie et al., 1966; Yamada and Komagata, 1970a, 1972a, b; Keddie and Cure, 1977, 1978; Minnikin et al., 1978a). Wall and whole organism analyses also provided the first unambiguous evidence of a relationship between *Corynebacterium sensu stricto*, which contains *Cnbc. diphtheriae*, *Cnbc. glutamicum* and a number of pathogenic and parasitic species, *Mycobacterium*, *Nocardia* and *Rhodococcus* (Goodfellow and Minnikin, 1981b, 1983), all of which contain major amounts of *meso*-DAP, arabinose and galactose, ie. they have a wall chemotype IV (*sensu* M. P. Lechevalier and Lechevalier, 1970a, b). These nocardioform taxa can be distinguished from *Streptomyces* and related strains, which have the L-isomer of DAP and glycine in their walls (chemotype I), but not from *Micropolyspora* and allied sporoactinomycetes which also possess a *meso*-DAP containing peptidoglycan associated with an arabinogalactan polymer (Table 1).

Determination of the primary structure of the wall peptidoglycan provides more precise data for classification and identification than that derived from simple qualitative studies. In 1967, Schleifer and Kandler introduced a quantitative procedure which was subsequently used to unravel the peptidoglycan structure of many Gram-positive organisms, not least the actinomycetes and coryneform bacteria (Schleifer and Kandler, 1970; Fiedler and Kandler, 1973a, b; Fiedler et al., 1973, 1981; Stackebrandt et al., 1978, 1980b; Seidl et al., 1980). In an excellent review of the contribution of wall analysis to the classification of Gram-positive bacteria, Schleifer and Kandler (1972) emphasized the value of the mode of cross-linkage and the composition of interpeptide bridges as an important source of taxonomic information. They also proposed a new classification of peptidoglycan types.

The variation in peptidoglycan structure has provided a useful framework for the reclassification of coryneform bacteria (Schleifer and Kandler, 1972; Minnikin et al., 1978a; Döpfer et al., 1982) and certain sporoactinomycetes (Alderson et al., 1981). Indeed, coryneform taxa account for almost half of the peptidoglycan types recognized (Table 1). True corynebacteria, like many other actinomycetes, belong to the directly cross-linked meso-DAP containing peptidoglycan type, namely the $A1_\gamma$ variation. Structural variations in the peptidoglycans of wall chemotype IV actinomycetes show that the muramic acid acyl groups of *Mycobacterium, Nocardia* and *Rhodococcus* are *N*-glycolated, not *N*-acetylated as in *Cornybacterium sensu stricto* and most other actinomycetes (Table 1; Adam et al., 1969; Azuma et al., 1970; Lederer et al., 1975; Uchida and Aida, 1977, 1979). It is also interesting that the muramic acid moeity of the rifamycin producing '*Nrda.*' *mediterranea* is *N*-acetylated (Bordet et al., 1972; Vacheron et al., 1972) and the peptide moeity is monoaminated on the diaminopimelic acid, not diaminated on the diaminopimelic acid and glutamic acid as in true nocardiae (Bordet et al., 1972; Alderson et al., 1981).

Actinomycetes are unusually rich in lipids and the structural variations that exist amongst the various lipid classes are being tapped for taxonomic purposes. An extensive literature is accumulating, and detailed reviews on the role of lipids as chemical markers are available for coryneform bacteria (Minnikin et al., 1978a), nocardioform actinomycetes (Minnikin and Goodfellow, 1976, 1978, 1980; Asselineau and Asselineau, 1978a, b) and, to a lesser extent, for sporoactinomycetes (Lechevalier et al., 1977, 1981; Minnikin et al., 1977a). Fatty acid, polar lipid and isoprenoid quinone analyses have been especially rewarding but a substantial body of data exist on other lipids such as prodiginine pigments, trehalose mycolates and complex mycobacterial lipids (Gerber and Lechevalier, 1976; Asselineau and Asselineau, 1978a, b; Minnikin and Goodfellow, 1980).

The long-chain fatty acids of actinomycetes fall into two broad groups, the hydroxylated and non-hydroxylated types. The latter are usually analysed by gas–liquid chromatography (GLC), whereas polar lipids are detected using two-dimensional thin-layer chromatography (TLC) with specific spray reagents being employed for the recognition of various functional groups. Polar lipid patterns are proving to be of value in the classification of actinomycetes (Lechevalier et al., 1977, 1981; Minnikin and Goodfellow, 1978) and coryneform bacteria (Komura et al., 1975; Minnikin et al., 1978a) including *Arthrobacter* (Collins et al., 1982a), *Curtobacterium* (Collins et al., 1980a), *Cellulomonas* and *Oerskovia* (Minnikin et al., 1979) and strains with a peptidoglycan based on 2,4-diaminobutyric acid (DAB) (Collins and

Jones, 1980). It is, however, important to examine polar lipid data with care as the properties of individual lipids can be influenced by subtle changes in growth environment (Minnikin and Goodfellow, 1981b). Menaquinones, 2-methyl-3-polyprenyl-1,4-naphthoquinones, the characteristic isoprenoid quinone type found in actinomycetes and coryneform bacteria (Minnikin et al., 1978a, b; Collins and Jones, 1981a) can be detected by reverse-phase TLC (Collins et al., 1980b). Their chemotaxonomic value lies in the variation of the number of isoprene units and hydrogenated double bonds.

Long-chain 2-alkyl-branched 3-hydroxy acids, collectively known as mycolic acids, were originally characterized from *Mybc. tuberculosis* (Stodola et al., 1938) and later found in extracts of a number of taxa that have an arabinogalactan in their walls and a peptidoglycan based on *meso*-DAP (Minnikin et al., 1978a; Minnikin and Goodfellow, 1980, 1981a; Collins et al., 1982b). Mycolic acids are found in some but not all wall chemotype IV actinomycete genera (Table 2) and can readily be detected by TLC (Minnikin et al., 1975, 1980). The separation of the wall chemotype IV actinomycetes into two groups is underlined by preliminary fatty acid and polar lipid data (Table 2). Thus, mycolic acid-containing strains characteristically contain straight-chain saturated and unsaturated fatty acids (Minnikin and Goodfellow, 1976, 1980; Minnikin et al., 1978a), diphos phatidylglycerol, phosphatidylinositol and phosphatidylinositol mannosides (Lechevalier et al., 1977, 1981; Minnikin and Goodfellow, 1978) whereas those that lack these characteristic compounds contain large proportions of branched chain *iso-* and *anteiso-* fatty acids (Guzeva et al., 1973; Kroppenstedt and Kutzner, 1976, 1978; Alderson et al., 1981) and a variety of polar lipids that include phosphatidylcholine and phosphatidylmonomethylethanolamine (Gochnauer et al., 1975; Pommier and Michel, 1973; Lechevalier et al., 1977).

Mycolic acid-containing taxa can be distinguished to some extent by differences in overall mycolic acid size and structure, and by their fatty acid, menaquinone and polar lipid content (Table 2; Minnikin and Goodfellow, 1980). At present, however, a clear distinction between caseobacters, corynebacteria and rhodococci cannot be made by mycolic acids alone. In contrast, representatives of the '*aurantiaca*' taxon are especially well defined as they have mycolic acids intermediate in size between those of mycobacteria and contain fully unsaturated menaquinones with nine isoprene units (Goodfellow et al., 1978). *Nocardia* strains, apart from *Nrda. amarae*, characteristically have tetrahydrogenated menaquinones with eight isoprene units as the major isoprenologue (Yamada et al., 1976, 1977a, b; Collins et al., 1977; Goodfellow et al., 1982a). *Corynebacterium* can also be distinguished from the other mycolic acid-containing taxa by its relatively low G+C composition (Table 1) and absence of phosphatidylethanolamine, though it has to be stressed that proportions of the latter can vary dramatically in extracts of mycolic acid-containing bacteria (Minnikin *et al.*, 1977b). Further, apart from *Cnbc. bovis*, corynebacteria may be distinguished from related strains by the absence of tuberculostearic acid (Lechevalier *et al.*, 1977; Collins *et al.*, 1982c).

The full impact of lipid analyses on the taxonomy of actinomycetes with a wall chemotype other than IV has yet to be realized but preliminary data (Table 3) are in line with other developments in the systematics of these organisms. The lipid data support the relationships already noted between *Streptomyces*, *Streptoverticillium* and certain other wall chemotype I taxa, and between the sporangia-forming taxa with a wall chemotype II. Fatty acid (Agre *et al.*, 1975), menaquinone (Collins *et al.*, 1977; Yamada *et al.*, 1977a, b) and polar lipid studies (Lechevalier *et al.*, 1977; Minnikin *et al.*, 1977b) also strongly support the creation of the genus *Nocardiopsis* (Meyer, 1976) for organisms previously classified as *Acmd. dassonvillei*. Similar lipid analyses highlight affinities that exist between the genera *Cellulomonas*, *Oerskovia* and *Promicromonospora* (Jones and Bradley, 1964; Lechevalier, 1972; Stackebrandt *et al.*, 1980b). It is also encouraging that lipid data lend some weight for assigning *Actinomadura* and the sporangia-forming actinomycetes with a wall chemotype III to a single aggregate taxon.

Chemical markers are being increasingly recommended for identification (Alderson and Goodfellow, 1979; Goodfellow and Schaal, 1979; Minnikin and Goodfellow, 1980), have proved effective in detecting heterogeneous taxa (Goodfellow and Minnikin, 1981c; Lechevalier *et al.*, 1981), and in spotting strains wrongly assigned to genera such as *Corynebacterium*, *Nocardia* and *Streptomyces*. The creation and subsequent endorsement of the genera *Actinomadura*, *Nocardiopsis*, *Oerskovia*, *Rhodococcus* and *Rothia* for strains previously classified as *Nocardia sensu* Waksman (1961) owes much to the labour of the chemical taxonomist. Similarly, organisms incorrectly classified as *Corynebacterium sensu lato* (Goodfellow and Minnikin, 1981a) have been transferred to the genera *Actinomyces*, *Arcanobacterium*, *Arthrobacter*, *Cellulomonas*, *Curtobacterium*, *Oerskovia* and *Rhodococcus* (Yamada and Komagata, 1970a, b, 1972a, b; Keddie and Cure, 1977, 1978; Minnikin *et al.*, 1978a, 1979; Keddie and Bousfield, 1980; Collins *et al.*, 1982d, e).

3. Numerical Taxonomy

Chemical features are frequently used to describe and distinguish between actinomycete genera (Minnikin *et al.*, 1978a; Goodfellow and Schaal, 1979; Minnikin and Goodfellow, 1980; Goodfellow and Wayne, 1982) but conventional numerical taxonomy has been the most effective modern method used to establish relationships between actinomycetes at the subgeneric

TABLE 2. Lipids of actinomycetes with a wall chemotype IV*

Taxon	Long chain fatty acids[†]	Mycolic acids			Predominant menaquinone[§]	Diagnostic phospholipids[¶]
		No. of carbons	No. of double bonds	Acid released on pyrolysis GC of mycolate[‡]		
Caseobacter	S, U, T	30–36	0–2	14:0–18:0	MK-9(H$_2$)	ND
Corynebacterium	S, U(T)	22–38	0–2	8:0–18:0 (14:1–18:1)	MK-8, 9(H$_2$)	PI, PIM
Mycobacterium	S, U, T	60–90	1–2	22:0–26:0	MK-9(H$_2$)	PE, PI, PIM
Nocardia	S, U, T	46–60	0–3	12:0–18:0 16:1–18:1[∥]	MK-8(H$_4$), 9(H$_2$)	
Rhodococcus	S, U, T	34–66	0–4	12:0–18:0	MK-8(H$_2$), 9(H$_2$)	
'aurantiaca' taxon	S, U, T	68–74	1–5	20:0–20:1 22:0–22:1	MK-9	
Actinopolyspora	S, I, A	—	—	—	MK-9(H$_4$)	PC, PG
Micropolyspora (faeni type)	S, I, A	—	—	—	MK-9(H$_4$, H$_6$)	PC, PG, PI, PME
Pseudonocardia	S, I, A(U, T)	—	—	—	MK-9(H$_4$)	PC, PE, PI, PME, (PIM)
Saccharomonospora	S, U, I, A	—	—	—	MK-9(H$_4$)	PE, PI, PIM (APG)
Saccharopolyspora	ND	—	—	—	ND	PC, PI, PME, (APG, PE, PG, PIM)

ND, not determined.
* Data from Lechevalier et al. (1977, 1981), Minnikin and Goodfellow (1980, 1981a), Collins et al. (1981a, 1982b, c) and Goodfellow and Minnikin (1982).
[†] Abbreviations: S, straight-chain; U, monounsaturated; T, tuberculostearic (10-methyloctadecenoic); A, anteiso; I, iso; bracket indicates variable occurrence.
[‡] Abbreviations exemplified by: 14:0, tetradecanoate; 14:1, tetradecanoate.
[§] Abbreviations exemplified by MK-8(H$_4$); menaquinone with two of the eight isoprene units hydrogenated.
[¶] Diagnostic phospholipids: DPG, diphosphatidylglycerol; PC, phosphatidylcholine; PG, phosphatidylglycerol; PE, phosphatidylethanolamine; PI, phosphatidylinositol; PIM, phosphatidylinositol mannosides and PME, phosphatidylmonomethylethanolamine.
[∥] Restricted to Nocardia amarae.

level. The principles and practice of numerical taxonomy are described in detail elsewhere (Sneath and Sokal, 1973; Sneath, 1978a, b; Jones and Sackin, 1980). In essence, numerical taxonomy involves the construction of a large data base for many strains which are grouped into clusters on the basis of shared similarities. Initially, all characters are given equal weight but, once a classification has been erected, cluster specific characters can be chosen and, after reproducibility studies, used to construct dichotomous keys, diagnostic tables and/or probability matrices for the identification of fresh isolates. This procedure is in sharp contrast to traditional practice in actinomycete taxonomy as taxa are recognized and defined using many equally weighted features, not on a small number of subjectively chosen morphological and staining properties (Goodfellow and Minnikin, 1977, 1978; Goodfellow and Pirouz, 1982; Williams *et al.*, 1983a). It should be noted that numerical taxonomies are based on phenetic data so that affinities between strains, and the hierarchies built on them, are entirely phenetic though phylogenetic deductions can be drawn from numerical phenetic classifications.

Actinomycete taxonomists were amongst the first to apply numerical techniques and, beginning with the study of Bojalil *et al.* (1962), a lot of work has been devoted to constructing numerical taxonomies of both coryneform (Jones, 1978) and nocardioform bacteria (Goodfellow and Minnikin, 1981b, c; Goodfellow and Wayne, 1982). In particular, numerical phenetic analyses have provided frameworks for the modern taxonomy of *Corynebacterium* (Jones, 1975; Goodfellow *et al.*, 1982b), *Mycobacterium* (Wayne *et al.*, 1971, 1978, 1981; Kubica *et al.*, 1972; Saito *et al.*, 1977b), *Nocardia* (Tsukamura, 1969, 1977; Goodfellow, 1971; Schaal and Reutersberg, 1978; Orchard and Goodfellow, 1980; Goodfellow *et al.*, 1982a), have led to the reintroduction of the genus *Rhodococcus* (Tsukamura, 1971, 1974; Goodfellow and Alderson, 1977; Rowbotham and Cross, 1977; Goodfellow *et al.*, 1982b, c) and the '*aurantiaca*' taxon (Goodfellow *et al.*, 1978), and endorsed proposals for the recognition of the genera *Actinomadura* (Tsukamura, 1969; Alderson and Goodfellow, 1979; Goodfellow *et al.*, 1979; Goodfellow and Pirouz, 1982), *Nocardiopsis* (Goodfellow and Minnikin, 1981c; Goodfellow and Pirouz, 1982), *Oerskovia* (Goodfellow, 1971) and *Rothia* (Holmberg and Hallander, 1973; Schofield and Schaal, 1982). In addition, numerical taxonomy also shows that the mycolic acid-containing genera are closely related to one another but not to the wall chemotype IV taxa which lack these fatty acids (Orchard and Goodfellow, 1980; Alderson *et al.*, 1981; Goodfellow and Minnikin, 1981c; Goodfellow and Pirouz, 1982). Predictably, *Corynebacterium* strains show higher overall similarities with other mycolic acid-containing taxa than with representatives of the genera *Arthrobacter, Brevibacterium, Cellulomonas* and *Microbacterium* (Goodfellow *et al.*, 1982b).

TABLE 3. Lipids in actinomycetes excluding wall chemotype IV taxa*,†

Wall chemotype	Taxon	Long-chain fatty acids	Predominant menaquinones	Diagnostic phospholipids
I	"Chainia", "Microellobosporia", Streptomyces, Streptoverticillium	S, I, A	MK-9(H$_4$, H$_6$, H$_8$)	PE, PI, PIM
	Arachnia	S, U	ND	ND
	Intrasporangium	ND	ND	PE, PI, PIM?
	Kineosporia	ND	ND	PC, PI, PME
	Nocardioides	S, I, A[T]	MK-8(H$_4$)	APG, PG, PI[PIM]
II	Actinoplanes, Amorphosporangium, Ampullariella, Dactylosporangium	S, I, A[U]	ND	PE, PI, PIM[APG]
	Micromonospora	ND	MK-9,10(H$_4$)	PE, PI[PIM]
	Pilimelia	S, I, A[U]	ND	ND
IIIA [plus madurose]	Actinomadura	S, U, T[I, A]	MK-9[H$_4$, H$_6$, H$_8$]	PI, PIM
	Actinosynnema	ND	ND	PE, PI, PIM [APG, PG]
	Dermatophilus	S, U	ND	PI, PIM[APG, PG]
	Microbispora, Planobispora, Planomonospora, Streptosporangium	S, U, I, A, T	MK-9(H$_4$)	PE, PI, PIM
		S, U, I, A, T	ND	PE, PI, PIM
	Microtetraspora	S, U, T[I, A]	MK-9(H$_4$)	PI, PIM
	Spirillospora	S, U, I, A, T	ND	PI, PIM[APG, PE, PG]
IIIB [minus madurose]	Brevibacterium	S, U, I, A	MK-8(H$_2$)	PG, PI, PIM
	Geodermatophilus	ND	ND	PE, PI[APG, PG]
	Nocardiopsis	S, U, I, A, T	MK-10(H$_2$, H$_4$, H$_6$)	APG, PC, PG[PIM
	Thermomonospora	S, U, A	MK-10(H$_8$)	ND
IV	Actinomyces [israelii type]	S, U[C]	MK-10(H$_2$, H$_4$)	PC, PG, PIM

VI	Actinomyces [bovis type]	S, U[C]	ND	ND
	Arcanobacterium	S, U	MK-9(H$_4$)	ND
	Arthrobacter	S, U, I, A	MK-9(H$_2$)	DPG, PG, PI
	Microbacterium	ND	MK-10, 11	DPG, PG
	Oerskovia, Promicromonospora	S, I, A	MK-9(H$_4$)	PG, PI, PIM? [APG, PE]
	Renibacterium	S, I, A	MK-9	DPG
	Rothia	S, I, A[U]	MK-7	DPG, PG
VII	Agromyces	S, I, A	MK-12	DPG, PG
VIII	Cellulomonas	S, I, A	MK-9(H$_4$)	PI, PIM?
	Curtobacterium	S, U, I, A	MK-9	PG, PI, PIM
—	Thermoactinomyces	S, I, A	MK-7,9	ND

ND, not determined.
* Data from Athalye et al. (1983), Lechevalier et al. (1977, 1981), Minnikin et al. (1978a), Collins et al. (1981a, 1982a, d, e), Collins (1982), Minnikin and Goodfellow (1981a) and Embley et al. (1983 and personal communication).
† Abbreviations: See footnote to Table 2.

Numerical phenetic methods have been applied relatively sparingly in the classification of sporoactinomycetes as many investigators (Pridham et al., 1965; Hütter, 1967; Shirling and Gottlieb, 1968a, b, 1969, 1972; Kutzner et al., 1978) have continued to rely on a small number of traditional morphological, staining and physiological tests for the classification of these organisms. This neglect is surprising, especially after Silvestri et al. (1962) had successfully applied taxometric techniques to the classification of the genus Streptomyces. In a logical development from this early study, Williams et al. (1983a) recovered over 300 species of Streptomyces in eighteen centres of variation encompassed in an aggregate taxon which corresponded to the genus Streptomyces. Representatives of the genera Actinopycnidium, Actinosporangium, Chainia, Elytrosporangium and Microellobosporia, all of which were recognized on the basis of supposedly diagnostic morphological features, were recovered in the aggregate Streptomyces cluster and were considered to be synonyms of Streptomyces. In contrast, marker strains of Intrasporangium, Kitasatoa, Nocardioides and Streptoverticillium, which like streptomycetes have a wall chemotype I, were distinguished from Streptomyces and from one another.

Numerical phenetic data (Goodfellow and Pirouz, 1982) are in line with those from 16S rRNA cataloguing studies in showing that sporoactinomycetes containing meso-DAP but no characteristic wall sugars (wall chemotype III sensu Lechevalier and Lechevalier, 1970a, b) form a diverse group. In the numerical study the genera Dermatophilus, Geodermatophilus, Nocardiopsis, Planobispora, Planomonospora and Thermomonospora were found to be homogeneous, Actinomadura, Microbispora and Microtetraspora heterogeneous, and the wall chemotype IV taxon, Mips. brevicatena was transferred to the genus Nocardia as Nrda. brevicatena. Further, the sharp separation of the Dermatophilus and Geodermatophilus clusters provided further evidence of the heterogeneity of the family Dermatophilaceae (Austwick, 1958; Cross and Goodfellow, 1973); the numerical data also questioned the continued classification of all of the sporangia-containing taxa in the family Actinoplanaceae (Couch, 1955).

Numerical phenetic surveys (Melville, 1965; Holmberg and Hallander, 1973; Holmberg and Nord, 1975; Fillery et al., 1978; Schaal and Schofield, 1981a, b) show that the family Actinomycetaceae (Buchanan, 1918) is markedly heterogeneous. In the most comprehensive study to date, Schofield and Schaal (1982) found that species of Actinomyces, with the exception of Actn. bovis, formed a well-defined aggregate taxon that was sharply separated from phena corresponding to the genera Arachnia, Bifidobacterium, Eubacterium, Propionibacterium and Rothia. Corynebacterium pyogenes, which shows a low overall similarity with Corynebacterium sensu stricto (Jones, 1975), was found to be closely related to Actn. bovis.

Numerical classifications need to be interpreted with care as similarity values between strains can be distorted by factors such as test and sampling error, the statistics used, and failure to allow for differences in growth rates and metabolic activity (Sneath and Johnson, 1972; Goodfellow et al., 1979). Most confidence can usually be placed in the major centres of variation defined in numerical analyses, it is the relationships of strains lying towards the periphery of clusters that are not always clear. It is, therefore, important that numerical classifications be evaluated in light of affinities based on independent criteria such as those derived from chemical, genetical and serological analyses (Sneath, 1976; Jones and Sackin, 1980). Ideally, following numerical analyses, strains should be selected to represent the whole range of variation within clusters and examined using the more analytical techniques that cannot readily be applied to large numbers of strains. Chemical, nucleic acid pairing, serological and phage host range studies have all been successfully used to evaluate numerical classifications of actinomycetes (Prauser, 1976b; Sneath, 1976; Goodfellow and Minnikin, 1978, 1982; Keddie and Bousfield, 1980; Williams et al., 1980; Goodfellow and Wayne, 1982). DNA:DNA pairing studies provide especially high quality data but cannot be used to establish relationships beyond a relatively narrow range of bacterial variation. DNA homology values can fall to quite low levels, below 20%, for species that are only moderately different phenotypically (Mordarski et al., 1981a).

Relatively few DNA pairing studies have been carried out on actinomycetes and the importance of some of them has been offset by the use of unrepresentative reference strains. Genetic data show that *Nocardia* and *Rhodococcus* have relatively little DNA in common with one another or with representatives of the genera *Mycobacterium* and *Streptomyces* (Mordarski et al., 1978a, b) and underline the heterogeneity of the coryneform taxa (Suzuki et al., 1981; Döpfer et al., 1982). Very low DNA homology values were also recorded between DNA from *Nrdo. albus* and that from *Stmy. griseus* (Tille et al., 1978). Farina and Bradley (1970) examined representatives of the sporangia-forming actinomycetes by DNA:DNA pairing in order to determine the degree of relatedness among selected genera. They found that the sporangia-forming taxa fell into two broad and unrelated groups and considered that the genus *Microellobosporia* be reduced to a synonym of *Streptomyces*. The most comprehensive work, however, has been carried out on carefully chosen representatives of the genus *Rhodococcus* where good congruence has been found between numerical phenetic and genetic data (Mordarski et al., 1978a, 1980b, 1981b). The DNA pairing method is particularly useful in resolving important taxonomic dilemmas at the species level. Thus, DNA pairing is helping to resolve the speciation of the *Nrda. asteroides* (Mordarski et al., 1978b) and *Actn. naeslundii/viscosus* complexes (Coykendall and Munzenmaier, 1979) and has shown that *Mybc. bovis* must be considered a sub-species of *Mybc. tuberculosis* (Bradley, 1975).

4. Towards a Definition of Actinomycetes

Morphology has always featured prominently in the recognition and definition of actinomycetes and their classification into families and genera. In 1973, Gottlieb considered that the actinomycetes consisted 'of varied groups of bacteria whose common feature is the formation of hyphae at some stages of development' but he went on to say that, in some organisms, hyphal formation was tenuous and that it required imagination to believe in it. In the following year, in the current edition of *Bergey's Manual of Determinative Bacteriology* (Gottlieb, 1974), organisms classified in the order *Actinomycetales* were considered to be 'bacteria that tend to form branching filaments which in some families develop into a mycelium'. It was, however, conceded that the filaments might be short, as in members of the families *Actinomycetaceae* and *Mycobacteriaceae*, and that in certain taxa they underwent fragmentation and consequently could only be observed in some stages of the growth cycle. The relatively simple morphology of most mycobacteria partly explains why these organisms were sometimes omitted from classifications of the actinomycetes (Waksman, 1961, 1967). Other workers questioned the collocation of the actinomycetes as a natural group and preferred to regard them as a convenient but artificial taxon (Sneath, 1970; Prauser, 1970, 1978, 1981; Goodfellow and Cross, 1974) and the difficulty of distinguishing between nocardioform actinomycetes and coryneform bacteria was widely recognized (Williams *et al.*, 1976; Locci, 1981). It was also conceded that an overreliance on morphological criteria had blurred the boundaries between *Corynebacterium*, *Mycobacterium* and *Nocardia* so that sections of these genera were more or less interchangeable (Bousfield and Goodfellow, 1976; Goodfellow and Wayne, 1982).

The morphological concept of an actinomycete can now be considered critically in light of information derived from the application of genetic and chemical methods. It is already quite apparent from 16S rRNA cataloguing studies that morphological features are poor tokens of phylogenetic relationships and that the traditional morphological definition of an actinomycete cannot be sustained (Stackebrandt *et al.*, 1980a, b, 1983; Ludwig *et al.*, 1981; Stackebrandt and Woese, 1981a, b). It is perhaps not too surprising that the morphologically simple corynebacteria show a close evolutionary relationship with the more elaborate mycobacteria, nocardiae and rhodococci as this grouping is consistent with the results of a host of chemical (Minnikin and Goodfellow, 1980, 1981a), comparative immunodiffusion (Lind and Ridell, 1976, 1982) and numerical phenetic studies (Goodfellow and Minnikin, 1981b, c; Goodfellow and Wayne, 1982).

Indeed, the traditional practice of separating the more highly differentiated actinomycetes from the morphologically simple coryneform bacteria no longer holds as strains of *Actinomyces, Oerskovia* and *Promicromonospora* show a closer phylogenetic affinity to *Arthrobacter, Brevibacterium, Cellulomonas, Curtobacterium* and *Microbacterium* than to mycelial-forming organisms such as *Nocardia* and *Streptomyces*. Further, the mycelium-forming *Thermoactinomyces* must now be classified in the family *Bacillaceae* with the aerobic, endospore-forming bacilli and *Micrococcus* and *Arthrobacter* are phylogenetically indistinguishable (Stackebrandt and Woese, 1979, 1981a, b; Stackebrandt *et al.*, 1980a). It is evident from these findings that the possession of branched hyphae should not automatically place a strain in the actinomycetes, conversely, the inability of a bacterium to produce branching filaments does not necessarily exclude it from this group of bacteria.

The order *Actinomycetales* (Buchanan, 1917) has to be redefined to include *Micrococcus* and taxa, apart from *Kurthia*, considered in the current edition of *Bergey's Manual of Determinative Bacteriology* (Rogosa *et al.*, 1974) to belong to 'the coryneform group of bacteria'. These bacteria should be added to those currently classified in the order *Actinomycetales* (Gottlieb, 1974), except for *Bifidobacterium* and *Thermoactinomyces*. It is perhaps premature to specify the unifying characters for defining this expanded group of bacteria but it is a task that will be expected and shall be attempted, albeit with some trepidation knowing the fate of the many previous attempts. The Actinomycetes are Gram-positive bacteria with a high $G+C$ content in their DNA (above 55 mol%) which are phylogenetically related from the evidence of 16S rRNA oligonucleotide sequencing and nucleic acid hybridization studies (thus excluding the genera *Bifidobacterium, Kurthia* and *Propionibacterium*). The actinomycetes include genera exhibiting a very wide range of morphology extending from the coccus, e.g. *Micrococcus*, though fragmenting hyphal forms to genera with a permanent and highly differentiated branched mycelium, e.g. in *Streptoverticillium*. There are no clear encompassing chemotaxonomic characters but the genera are linked in a series on the evidence of wall structure and lipid composition. Some, but not all, genera form spores which include motile zoospores and specialized structures that resist desiccation and mild heat but do not display the organization and extreme resistance properties of endospores (so excluding the genus *Thermoactinomyces*).

5. Actinobacteria

In the current edition of *Bergey's Manual of Determinative Bacteriology* (Slack, 1974b) the genera *Actinomyces, Arachnia, Bacterionema, Bifidobacterium* and *Rothia* were classified in the family *Actinomycetaceae*

(Buchanan, 1918). The genus *Agromyces*, which has an oxidative type of carbohydrate metabolism but grows under microaerophilic conditions, was tentatively added to the family by Cross and Goodfellow (1973). Slack and his colleagues made a substantial contribution to the systematics of the fermentative actinomycetes (see Slack and Gerencser, 1975) but, despite their sterling efforts, the description of the family *Actinomycetaceae* is still based on properties that are essentially classical. The taxon accommodates Gram-positive, non-acid-fast, non-motile actinomycetes that form a substrate mycelium which fragments into rod- and coccoid-like elements, have relatively exacting nutritional requirements, possess a fermentative carbohydrate metabolism and grow as facultative to obligate anaerobes. Several detailed accounts are available on the systematics of the fermentative actinomycetes (Slack and Gerencser, 1975; Bowden and Hardie, 1978; Gerencser, 1979; Schaal and Pulverer, 1981).

The family *Actinomycetaceae* is markedly heterogeneous given the data derived from the application of genetical, chemical and numerical phenetic techniques. *Bacterionema matruchotii* was clearly shown to be closely related to true corynebacteria in wall (Pine, 1970), lipid (Minnikin and Goodfellow, 1980), DNA base composition (Gilmour, 1974) and numerical phenetic analyses (Goodfellow and Minnikin, 1981c) and has recently been transferred to *Corynebacterium sensu stricto* as *Corynebacterium matruchotii* comb. nov. (Collins, 1982c; Goodfellow et al., 1982b). The taxonomic affinities of the genera *Arachnia*, *Rothia* and *Bifidobacterium* have also been questioned, the latter has been considered to resemble *Lactobacillus* on the basis of wall composition and carbohydrate metabolism. Kandler (1970) proposed that *Bifidobacterium* be returned to the family *Lactobacillaceae*, but 16S rRNA homology data indicate that bifidobacteria can readily be distinguished from both lactobacilli and actinomycetes. Arachniae have properties in common both with *Actinomyces* and *Propionibacterium* but were more closely related with the former in an extensive numerical phenetic survey (Schofield and Schaal, 1982). Further comparative studies are required to determine the supra-generic affinities of both arachniae and rothiae; the latter possess a peptidoglycan unique amongst actinomycetes (Schleifer and Kandler, 1972).

Traditionally, morphological resemblances have played an important role in the classification of both fermentative actinomycetes and coryneform bacteria, although it was not possible to draw a line between these two broad groups on the basis of morphological criteria (Bousfield and Goodfellow, 1976; Goodfellow and Minnikin, 1981a). It now appears that less reliance should be placed on morphological properties in the classification of these organisms as taxa from the two broad groups are phylogenetically interrelated (Fig. 1). Indeed, the genus *Actinomyces* shows a closer evolutionary relationship with *Agromyces*, *Arthrobacter*, *Brevibacterium*, *Cellu-*

lomonas, Curtobacterium, Microbacterium, Oerskovia and *Promicromonospora* (Stackebrandt *et al.*, 1980a; Stackebrandt and Woese, 1981a, b; Döpfer *et al.*, 1982) than with erstwhile related taxa such as *Nocardia*. The inclusion of fermentative actinomycetes and corynebacteria in a single aggregate taxon, the Actinobacteria, has much to recommend it but clearly further comparative studies are required to determine to what extent these organisms form a distinct natural group. The genera *Arcanobacterium, Arachnia, Micrococcus* and *Renibacterium* have properties which allow them to be tentatively designated as Actinobacteria (Table 1).

The Actinobacteria include a seemingly bewildering variety of chemical, morphological and physiological types. The organisms fall into at least eight different peptidoglycan types but, with a single exception, they contain N-acetylated muramic acid (Table 1). Lysine is the most common diamino acid in the wall peptidoglycan (*Micrococcus, Oerskovia, Renibacterium, Rothia*), but ornithine (*Cellulomonas, Curtobacterium*), diaminobutyric acid (*Agromyces*) and both the L- (*Arachnia*) and *meso-* (*Brevibacterium*) forms of diaminopimelic acid also occur. Some Actinobacteria contain unsaturated menaquinones as the predominant isoprenologue, others have hydrogenated components. The first group includes *Agromyces* (MK-*12*), *Curtobacterium, Renibacterium* (MK-*9*) and *Microbacterium* (MK-*10, 11*), the second *Arthrobacter, Brevibacterium* (MK-*9*(H_2)) and *Arcanobacterium, Cellulomonas* and *Oerskovia* (MK-*9*(H_4)). The aggregate group accommodates taxa with fermentative (*Actinomyces, Arachnia*) and oxidative metabolisms (*Agromyces, Arthrobacter, Brevibacterium, Curtobacterium*) and includes both motile (*Cellulomonas, Curtobacterium, Oerskovia*) and non-motile organisms (*Actinomyces, Agromyces, Arachnia* and *Rothia*). It also incorporates taxa that form cocci (*Micrococcus*), short rods (*Renibacterium*) or pleomorphic elements (*Arthrobacter, Brevibacterium, Cellulomonas, Microbacterium*) and those that produce a mycelium that fragments (*Actinomyces, Oerskovia, Rothia*). It is too early to recognize any taxonomic structure within the Actinobacteria but there is some evidence of a relationship between *Cellulomonas, Oerskovia* and *Promicromonospora* (Jones and Bradley, 1964; Jones, 1975; Stackebrandt and Kandler, 1979; Minnikin *et al.*, 1979; Stackebrandt *et al.*, 1980b) and between *Agromyces, Curtobacterium* and *Microbacterium* (Döpfer *et al.*, 1982).

A. *Actinomyces* (Harz, 1977)

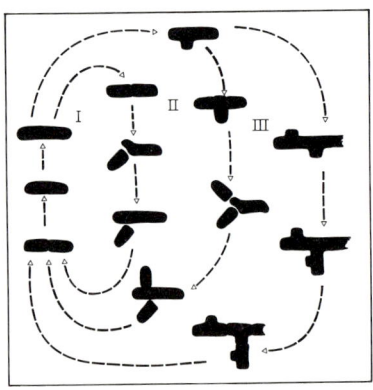

Gram-positive, asporogenous, non-acid-fast, non-motile actinomycetes which form a rudimentary primary mycelium particularly evident in undisturbed 18 to 48 hour microcolonies on agar or in broth culture. Branched rods which appear as V, Y and T forms are common. Long filaments, 1 μm or less in diameter, which vary in length and degree of branching occur in most strains. The organisms are facultatively anaerobic and, with the exception of one species which grows well under aerobic conditions, they are preferentially anaerobic. Growth is almost invariably enhanced by the addition of carbon dioxide. Catalase may or may not be formed. Carbohydrates are fermented with the formation of acid but not gas. The acid end products of glucose fermentation vary with cultivation conditions but usually include acetic, formic, lactic and succinic acids but not propionic acid.

The wall peptidoglycan contains alanine, glutamic acid and lysine with either aspartic acid or ornithine. The wall sugars show variation between species and include glucose, galactose, fucose, 6-deoxytalose, mannose, rhamnose but not arabinose. The DNA base composition also varies between species (Table 4). *Actinomyces* form an important part of the indigenous microflora of human, and possibly animal, mucous membranes, and are common in the oral cavity. Some strains are pathogenic for man, cattle, swine and other warm-blooded animals.

The type species is *Actinomyces bovis*, the type strain ATCC 13683.

Comprehensive information on the early taxonomic history and biology of *Actinomyces* can readily be obtained (Pine, 1970; Bowden and Hardie, 1973, 1978; Slack and Gerencser, 1975; Gerencser, 1979; Schaal and Pulverer, 1981). Members of the genus have been the subject of considerable interest in recent years due to their potential role as pathogens in periodontal disease (Gerencser and Slack, 1969; Socransky *et al.*, 1970; Jordan *et al.*,

TABLE 4. Chemical properties of *Actinomyces* species*

Actinomyces sp.	Major distinguishing wall constituents' amino acids	sugars	G+C (mol%)
Actm. bovis	aspartic acid, lysine	6-deoxytalose, fucose, rhamnose	57–63
Actm. humiferus	aspartic acid, lysine, ornithine	rhamnose	73
Actm. israelii	lysine, ornithine	galactose	57–65
Actm. naeslundii	lysine, ornithine	6-deoxytalose, glucose, mannose, rhamnose	63–69
Actm. odontolyticus	lysine, ornithine	galactose, glucose, mannose	62
Actm. pyogenes	lysine	glucose, rhamnose	58
Actm. "suis"	lysine, ornithine	rhamnose	ND
Actm. viscosus	lysine, ornithine	6-deoxytalose, galactose, glucose, mannose, rhamnose	59–70

* Data from Slack and Gerencser (1975) and Collins and Jones (1982a). All walls contain alanine, glutamic acid, glucosamine and muramic acid. ND, not determined.

1972, 1974) and root surface caries (Jordan and Hammond, 1972; Jordan and Sumney, 1973; Syed *et al.*, 1975). *Actinomyces naeslundii* and *Actn. viscosus* initiate periodontitis in hamsters and gnotobiotic rats (Jordan *et al.*, 1965; Socransky, 1970; Crawford *et al.*, 1978) and several new serotypes have been demonstrated for such organisms (Slack and Gerencser, 1975; Gerencser, 1979). *Actinomyces* strains are more usually associated with actinomycosis, a relatively uncommon disease of man and some other homoeothermic animals (see Chapter 8).

Actinomycotic lesions have been reported from many sites in man but are most common in the cervicofacial region (Pulverer and Schaal, 1978; Schaal, 1981). The principal agent, *Actn. israelii*, is usually accompanied in severe and chronic cases by *Actinobacillus actinomycetemcomitans* (Pulverer and Schaal, 1978; Schaal and Pulverer, 1981). *Actinomyces naeslundii*, *Actn. odontolyticus* and *Actn. viscosus* are also found in human infections (Georg and Coleman, 1970) but their aetiological role requires clarification. *Actinomyces bovis*, the major cause of bovine actinomycosis, has not been demonstrated in, or isolated from, man. In contrast, *Actn. israelii* has been isolated from animals (Brock and Georg, 1969; Slack *et al.*, 1969) and *Actn. viscosus* causes both natural and experimental disease in hamsters (Howell, 1963; Howell and Jordan, 1963; Keyes and Jordan, 1964), though similar strains have been recovered from dental calculus and other human sources (Snyder *et al.*, 1967; Gerencser and Slack, 1969; Georg *et al.*, 1969).

Actinomyces strains have relatively fastidious growth requirements but grow well on rich media at 37°C. Detailed procedures for the isolation of *Actinomyces* from clinical material are available (Slack and Gerencser, 1975; Schaal and Pulverer, 1981; see Chapter 9). Microcolony and cell morphology are used as the initial basis for isolation and primary differentiation. Many strains form branched filamentous or 'spider-like' microcolonies, others rough, heaped 'bread crumb' type colonies. The genus also contains facultative anaerobes that produce only pleomorphic cells and smooth colonies (Slack and Gerencser, 1975). The walls of the 'spider-like' forms of *Actn. bovis*, in contrast to those of the smooth forms, have little or no aspartic acid but a high proportion of hexosamine (Pine and Boone, 1967).

Actinomyces bovis, *Actn. israelii* and *Actn. odontolyticus* are good taxospecies (Melville, 1965; Holmberg and Nord, 1975; Schaal and Schofield, 1981a, b; Schofield and Schaal, 1982) but the status of *Actn. naeslundii* and *Actn. viscosus* is still not clear. These two species can be distinguished by their catalase reaction, wall carbohydrates and serological properties (Slack and Gerencser, 1975) but share many physiological features and, when compared by numerical taxonomic methods, representative strains share a high overall similarity (Holmberg and Hallander, 1973; Fillery *et al.*, 1978; Schaal and Schofield, 1981a, b; Schofield and Schaal, 1982). In addition, strains of *Actn. viscosus* of human origin differ serologically from the original hamster isolate and represent a second serotype (Gerencser and Slack, 1969; Bellack and Jordan, 1972). In a comprehensive study, Coykendall and Munzenmaier (1979) detected four DNA homology groups which corresponded to typical and atypical *Actn. naeslundii* strains and to *Actn. viscosus* serotypes 1 and 2, respectively. These workers also raised the prospect of equating the DNA homology groups with species. *Actinomyces viscosus* serotype 2 and *Actn. naeslundii* can also be distinguished by a number of biochemical and serological properties (Coykendall and Munzenmaier, 1979) and occupy different ecological niches.

Actinomyces humiferus, '*Actn. eriksonii*' and '*Actn. suis*' are listed as *species incertae sedis* in the current edition of *Bergey's Manual of Determinative Bacteriology* (Slack, 1974a), but only the former is cited on the *Approved Lists of Bacterial Names* (Skerman *et al.*, 1980). The original description of *Actn. humiferus* (Gledhill and Casida, 1969a) was sound but it is difficult to judge the affinities of this organism with established species in the absence of detailed comparative studies. The organism shares an affinity with *Actinomyces* strains with respect to morphology, wall composition and fermentation end products, but differs from members of this genus by growing well at 30°C, poorly, if at all, at 37°C, by its high G+C content (Table 4) and by its sensitivity to lysozyme. Further, *Actn. humiferus* has only been isolated from soil, whereas *Actinomyces* strains are primarily inhabitants of the mucous membranes of man and other animals. '*Actinomyces eriksonii*' (Georg *et al.*, 1965) has been implicated as an agent of actinomycosis but is probably more closely related to *Bifidobacterium* than to *Actinomyces* (Holmberg and Nord, 1975; Slack and Gerencser, 1975). It has been considered to be identical to *Bibc. adolescentis* (Mitsuoka *et al.*, 1974). However, '*Actn. eriksonii*' has been distinguished from typical strains of *Bibc. adolescentis* in DNA:DNA pairing experiments (Scardovi *et al.*, 1971) and shows a close genetic affinity to strains designated *Bibc. dentium* (Scardovi, 1980). Grässer (1957) isolated two groups of fermentative actinomycetes from granules in udder actinomycosis of swine, one group was identified as *Actn. israelii* and the other designated '*Actn. suis*'. More recent isolates from swine (Franke, 1973) are difficult to compare with '*Actn. suis*' given the poor original description of the latter and the fact that none of the original isolates are extant. The status of '*Actinobacterium meyerii*' also needs to be clarified as organisms bearing this name formed a distinct taxon related to species of *Actinomyces* in a numerical phenetic survey (Holmberg and Nord, 1975).

The taxonomic position of *Cnbc. pyogenes*, the causal agent of a variety of pyogenic infections in domestic animals and occasionally in man, has been the subject of some debate. The taxon was assigned to the genus *Corynebacterium* in the last edition of *Bergey's Manual of Determinative Bacteriology* (Cummins *et al.*, 1974) but it is clear that the organism has little in common with true corynebacteria (Cummins and Harris, 1956; Barksdale *et al.*, 1957; Cummins, 1962; Barksdale, 1970; Jones, 1975; Goodfellow *et al.*, 1976; Minnikin *et al.*, 1978a). Several workers have noted a relationship between *Cnbc. pyogenes* and *Actinomyces* (Schaal and Schofield, 1981a, b; Collins *et al.*, 1982d, e; Schofield and Schaal, 1982) and recently *Cnbc. pyogenes* was transferred to the genus *Actinomyces* as *Actn. pyogenes* (Reddy and Cornell, 1981; Collins and Jones, 1982a; Reddy *et al.*, 1982). A number of biochemical and physiological tests have been recommended for the identification of *Actinomyces* strains (see Chapter 9).

B. *Agromyces* (Gledhill and Casida, 1969b)

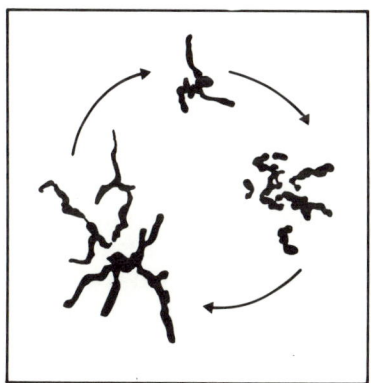

Gram-positive, asporogenous, non-motile, catalase negative actinomycetes which form a substrate mycelium that fragments into coccoid and branched rod-like elements. The organism is microaerophilic to anaerobic, does not require carbon dioxide for growth and grows poorly, if at all, under strictly anaerobic conditions. It is sensitive to lysozyme and has an oxidative metabolism.

The wall peptidoglycan contains alanine, glycine, glutamic acid and 2,4-diaminobutyric acid (DAB) and is of the uncommon B2γ type. Wall sugar composition varies, some strains contain galactose, mannose, rhamnose and xylose in major amounts, others have only glucose and rhamnose. *Agromyces* contain major amounts of *anteiso*-methyl branched chain acids, small proportions of *iso*-methyl branched, straight-chain saturated and monounsaturated acids, have unsaturated menaquinones with twelve isoprene units, MK-*12*, as the major isoprenologue and diphosphatidylglycerol, phosphatidylglycerol and two uncharacterized glycolipids. The organism is common in some soils.

The genus is monotypic, the type strain *Agromyces ramosus* ATCC 25173.

Agromyces ramosus was not recognized in the current edition of *Bergey's Manual of Determinative Bacteriology* (Buchanan and Gibbons, 1974) though it does appear on the *Approved Lists of Bacterial Names* (Skerman et al., 1980). The organism is nutritionally fastidious, requires organic nitrogen for growth, grows optimally on rich media such as brain heart infusion at 30°C, and has been isolated in very large numbers from a variety of soils using a modified dilution frequency procedure (Casida, 1965). Intraperitoneal injections do not appear to cause disease in mice. The type strain is not deficient in cytochromes, as was originally suggested; it produces small amounts of hydrogen peroxide (Jones et al., 1970). These latter workers also reported heavy growth of yellow colonies under full oxygen tension when the accumulating hydrogen peroxide was destroyed by the addition of catalase, fresh horse blood or manganese dioxide. The organism has recently been the subject of chemotaxonomic studies (Collins, 1982b).

Comment: The few Gram-positive bacteria with a peptidoglycan belonging to the B group are characterized by a cross-linkage between the α-carboxyl group of D-glutamic acid in position 2 of the peptide subunit and the C-terminal D-alanine of the adjacent subunit. *Curtobacterium* and *Microbacterium* also have this rare peptidoglycan type and together with *Agromyces* form a homogeneous cluster in DNA:rRNA pairing studies (Döpfer et al., 1982). Organisms currently labelled *Atbc. terregens*, '*Brev. helvolum*', *Brev. imperiale*, '*Brev. insectiphilum*', '*Cnbc. aquaticum*' and '*Cnbc. laevaniformis*' also have peptidoglycans belonging to the B group, as do a number of plant pathogenic corynebacteria (see page 57). It is also interesting that the lipids of *Agromyces* are similar to those of other actinomycetes that have a DAB-containing peptidoglycan (Collins, 1982a, b).

C. *Arachnia* (Pine and Georg, 1969)

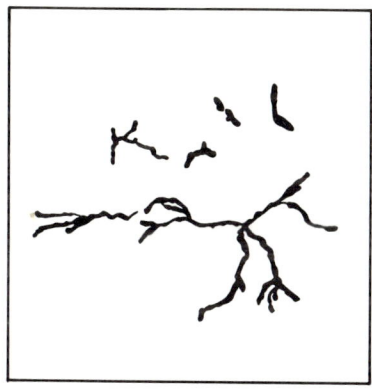

Gram-positive, non-acid-fast, non-motile, catalase negative actinomycetes which produce a primary mycelium on agar and fragment into irregular branched rods (0.2–0.3 × 3–5 μm) and filaments (5–20 μm). Swollen spherical cells of various sizes up to 5 μm are formed during stationary growth phase on certain media but neither aerial hyphae nor spores are produced. The organism is facultatively anaerobic but grows slowly aerobically; under neither of these gaseous regimes does it require carbon dioxide, nor is it stimulated by it. The end products from glucose fermentation are mainly propionic and acetic acids with traces of succinic, formic and lactic acids. A number of sugars, notably raffinose, are fermented and nitrate is reduced to nitrite.

The major amino acids of the wall peptidoglycan are alanine, aspartic acid, glycine, glutamic acid and LL-DAP, and the predominant wall sugar is galactose though glucose and mannose may also be present. Arachniae have a DNA base composition between 63 and 65 mol% G+C and form a tight DNA homology group. They cause human actinomycosis and lachrymal canaliculitis.

The genus is monotypic and the type strain *Arachnia propionica* ATCC 14157.

Arachniae are pathogenic both for man and experimental animals (Pine and Hardin, 1959; Gerencser and Slack, 1967; Brock *et al.*, 1973) and may form part of the oral flora (Rasmussen *et al.*, 1966; Gerencser and Slack, 1967; Holmberg and Forsum, 1973). Saline suspensions injected intraperitoneally into mice cause severe actinomycosis (Buchanan and Pine, 1962; Georg and Coleman, 1970). In man and experimental animals the disease state appears to be identical to that caused by *Actn. israelii*, an organism with which *Arac. propionica* can be confused (Brock *et al.*, 1973). Fluorescent antibody staining procedures provide an effective way of distinguishing between *Arachnia* and *Actinomyces* (Gerencser and Slack, 1967,

Holmberg and Forsum, 1973); biochemical tests are also recommended for this purpose (Holmberg and Nord, 1975; Gerencser, 1979).

Arachnia strains have fastidious growth requirements but grow well at 37°C on complex media such as brain heart infusion agar. The genus is heterogeneous, strains can be assigned to two serotypes (Gerencser and Slack, 1967; Brock *et al.*, 1973), which cross-react at low titre in fluorescent antibody tests, but can readily be made type specific by either dilution or absorption. The serotypes may represent two distinct species, a view that is not contradicted by numerical phenetic (Schaal and Schofield, 1981a, b; Schofield and Schaal, 1982), non-numerical (Kilian, 1978) and DNA:DNA pairing data (Johnson and Cummins, 1972).

Arachnia has had a short but eventful taxonomic history and its suprageneric affinities need to be resolved by the application of techniques such as DNA:rRNA pairing and 16S rRNA oligonucleotide sequencing. The organism was originally identified as *Actn. israelii* (Pine and Hardin, 1959), was renamed '*Actn. propionicus*' when found to produce propionic acid (Buchanan and Pine, 1962) and is similar to *Actinomyces* in morphology, pathogenicity and in its ability to cause typical actinomycosis with granules. It was renamed *Arac. propionica* (Pine and Georg, 1969) because it contained LL DAP, as opposed to lysine, in the wall peptidoglycan (Johnson and Cummins 1972; Schleifer and Kandler, 1972) and formed propionic acid as the major end product of glucose fermentation. These latter properties suggest a similarity with *Propionibacterium*. Arachniae differ, however, from propionibacteria by the consistent absence of catalase (Pine and Georg, 1974), their pathogenicity in man and animals; they also exhibit little genetic homology with representatives of the two major groups of propionic acid producing bacteria (Johnson and Cummins, 1972). In addition, numerical phenetic data show that *Arachnia* has more properties in common with *Actinomyces* than with *Propionibacterium* (Schaal and Schofield, 1981a, b; Schofield and Schaal, 1982).

D. *Arcanobacterium* (Collins *et al.*, 1982d)

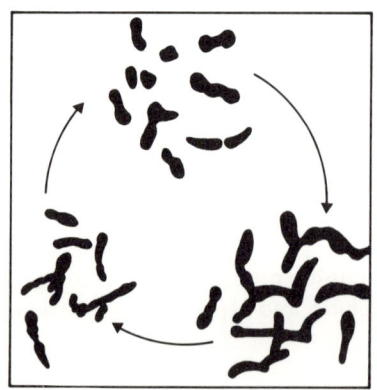

Gram-positive, asporogenous, non-acid-fast, non-motile actinomycetes in which irregular rods predominate during the first 18 hours on blood agar plates. V-forms are common. As growth proceeds, the bacillary elements become granular and segmented, and resemble small irregular cocci. Slender irregular forms are produced on Loeffler's medium but, after 48 hours, club and comma-shaped elements are produced. Arcanobacteria grow poorly on most laboratory media but growth is enhanced in an atmosphere of carbon dioxide and in media enriched with blood or serum. Surface colonies on blood agar plates are small (0.75 mm diameter) after 24 hours, becoming larger (1.5–2.5 mm diameter) on extended incubation. The organism is facultatively anaerobic, β haemolytic, produces acid from some sugars and can withstand heating at 60°C for 15 minutes.

The wall peptidoglycan contains lysine as diamino acid and rhamnose as characteristic wall sugar. The organisms have major amounts of straight chain and monounsaturated fatty acids, tetrahydrogenated menaquinones with nine isoprene units, MK-9 (H_4), as the major isoprenologue, and contain DNA within the range 50 to 52 mol% G+C.

The genus is monotypic, the type strain *Arcanobacterium haemolyticum* ATCC 9345.

The genus *Arcanobacterium* was proposed by Collins *et al.* (1982d) for bacteria previously classified as *Cnbc. haemolyticum*. The organism was originally isolated from infections amongst American soldiers, was assigned to the genus *Corynebacterium* by Maclean *et al.* (1946) but was considered to be a mutant of *Cnbc. pyogenes* by Barksdale *et al.* (1957). Neither *Cnbc. haemolyticum* nor *Cnbc. pyogenes* have much in common with true corynebacteria (Barksdale, 1970, 1981; Jones, 1975; Minnikin *et al.*, 1978a; Goodfellow and Minnikin, 1981a; Schofield and Schaal, 1981; Collins *et al.*, 1982d) but were considered to be closely related to streptococci on the basis of wall composition (Cummins and Harris, 1956). It is, however, now clear that '*Cnbc. haemolyticum*' and '*Cnbc.*' *pyogenes* can readily be distinguished from one another and from streptococci (Collins *et al.*, 1982d; Schofield and Schaal, 1982). In the 8th edition of *Bergey's Manual of Determinative Bacteriology*, *Cnbc. haemolyticum* was listed in an addendum to the genus *Corynebacterium* (Cummins *et al.*, 1974) and it did not appear on the *Approved Lists of Bacterial Names* (Skerman *et al.*, 1980).

E. *Arthrobacter* (Conn and Dimmick, 1947)

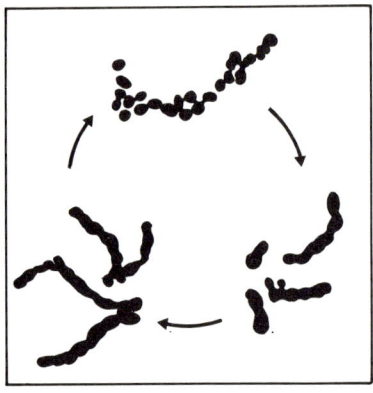

Gram-positive, asporogenous, non-acid-fast actinomycetes which, in complex media, undergo a marked change in form during the growth cycle. Stationary phase cultures (usually 2 to 7 days) are composed entirely or largely of coccoid cells which may be uniform in size and spherical, resembling micrococci, spherical to ovoid, or slightly elongate. On transfer to fresh complex medium, growth occurs by enlargement of the coccoid cells followed by the production of one, sometimes two or more, outgrowths to give rods that usually have a diameter less than that of the enlarged coccoid cell. Subsequent growth and division yield irregular rods which vary in size and shape and include straight, bent and curved, wedge-shaped and club-shaped forms. Cells may show primary branching but true mycelia are not produced. As exponential growth proceeds, the rods become shorter and are eventually replaced by coccoid forms characteristic of stationary phase cultures. The coccoid elements are formed either by gradual shortening of the rods at each successive division or, especially in richer media, by multiple fragmentation of larger rods. The rods are motile or non-motile by one subpolar or a few lateral flagella. The organism is an obligate aerobe and the mode of metabolism respiratory, never fermentative. Strains are catalase positive, form little or no acid from sugars in peptone water, grow in media containing soil extract and yeast extract but do not attack cellulose or survive heating at 63°C for 30 minutes in skimmed milk.

The wall peptidoglycan contains N-acetylated muramic acid, lysine in position 3 of the peptide subunit, and an interpeptide bridge linking the peptide subunits of adjacent glycan strands between L-lysine and the terminal D-alanine. Considerable variation is found in the number and nature of the amino acids in the interpeptide bridges. The predominant wall sugar is galactose; mannose and rhamnose are found in some strains. Arthrobacters contain dihydrogenated menaquinones with nine isoprene units, MK-

9(H_2), as the major isoprenologue, are rich in branched-chain fatty acids and contain major amounts of diphosphatidylglycerol, phosphatidylglycerol and phosphatidylinositol and a number of characteristic glycolipids. They have a DNA base composition of 59–66 mol% G+C. The organism is widely distributed and abundant in soil.

The type species is *Arthrobacter globiformis*, the type strain ATCC 8010.

The genus was introduced by Conn and Dimmick (1947) for highly aerobic, nutritionally non-exacting soil bacteria that liquefied gelatin slowly and exhibited a characteristic growth cycle. The taxon was later extended to include a number of nutritionally exacting species such as *Atbc. citreus*, *Atbc. duodecadis*, *Atbc. flavescens* and *Atbc. terregens* (Keddie and Jones, 1981). Thus, *Arthrobacter* became a depository for aerobic soil bacteria that characteristically exhibited a growth cycle in which the irregular rods in young cultures give rise to coccoid forms in older cultures. These coccoid forms, when transferred to fresh medium, produced outgrowths to give irregular rods again. The 'rod/coccus' growth cycle is not unique to *Arthrobacter* but occurs in other Actinobacteria, such as *Brevibacterium*, and in some members of the genus *Rhodococcus*. There is also evidence that the morphology of arthrobacters is markely influenced by the nutritional status of the medium (Luscombe and Gray, 1971; Clarke, 1972). In such circumstances it is hardly surprising that the heavy dependence on morphological features in the definition of *Arthrobacter* led to considerable confusion in the systematics of this taxon.

Modern taxonomic techniques showed that the genus *Arthrobacter* accommodated a number of very diverse taxa. As early as 1959, Cummins and Harris noted that arthrobacters were heterogeneous in wall composition. This heterogeneity was underlined in additional wall composition studies (Yamada and Komagata, 1972a; Keddie and Cure, 1978; Keddie and Bousfield, 1980), analyses of peptidoglycan structure (Schleifer and Kandler, 1972; Stackebrandt and Fiedler, 1979), numerical phenetic surveys (Jones, 1978), lipid analyses (Bowie *et al.*, 1972; Keddie and Cure, 1977; Minnikin *et al.*, 1978a; Collins *et al.*, 1979) and by determination of DNA base ratios (Skyring and Quadling, 1969; Skyring *et al.*, 1971; Suzuki *et al.*, 1981). These and many other studies led to the view that the genus *Arthrobacter* should be restricted to organisms that contain lysine in the wall peptidoglycan and have many phenotypic properties in common with the type species (Schleifer and Kandler, 1972; Yamada and Komagata, 1972a; Keddie, 1978; Keddie and Cure, 1978). *Arthrobacter* in this narrow sense is referred to as *Arthrobacter sensu stricto*. In contrast, *Arthrobacter sensu lato* also accommodates species traditionally associated with the genus (Table 5).

In *Bergey's Manual of Determinative Bacteriology* (Keddie, 1974a) several species of *Arthrobacter* were reduced to synonyms of *Atbc. globiformis*

TABLE 5. Some characteristics of *Arthrobacter globiformis* and certain species currently assigned to *Arthrobacter sensu lato**,[†]

	Atbc. globiformis	Atbc. nicotianae	Atbc. duodecadis	Atbc. flavescens	Atbc. terregens	Atbc. simplex	Atbc. tumescens
Wall diamino acid	lysine	lysine	meso-DAP	ornithine	ornithine	LL-DAP	LL-DAP
Peptidoglycan type	A3α	A4α	A4γ	ND	B2β	A3γ	A3γ
Wall sugars	galactose	galactose, glucose	ND	galactose, glucose, rhamnose	galactose, rhamnose, 6-deoxytalose	galactose, glucose, mannose	galactose, glucose, mannose
Major menaquinones	MK-9(H_2)	MK-8,9	ND	ND	ND	MK-8(H_4)	MK-8(H_4)
Fatty acids	S, U, I, A	S, U, I, A	ND	ND	ND	S, U, I, A, T	S, U, I, A, T
G+C (mol%)	59–66	60–66	ND	70	69–76	72–76	70–76
Growth factor requirements:							
Biotin	d	–	–	+	+	–	–
B_{12}	–	ND	+	–	–	–	–
Pantothenic acid	–	ND	–	–	+	–	–
Terregens factor	–	ND	–	+	+	–	–
Thiamine	–	ND	+	–	+	–	+

* Data from Schleifer and Kandler (1972), Keddie (1974a), Minnikin *et al.* (1978a), Keddie and Cure (1978), Goodfellow and Minnikin (1981a), Keddie and Jones (1981) and Döpfer *et al.* (1982). See also footnote to Table 2.
[†] Abbreviations: +, positive; –, negative; d, some strains positive, some negative.

primarily on the basis of wall and DNA base composition and a host of phenotypic properties. Other taxa for which the results were less complete were considered to be possible synonyms. Recent studies from DNA:DNA pairing, and wall composition analyses have shown that the *Atbc. globiformis* group accommodates a considerable degree of variation. It is now evident that taxa such as *Atbc. atrocyaneus*, *Atbc. aurescens*, *Atbc. crystallopoites*, *Atbc. histidinolovorans*, *Atbc. ilicis*, *Atbc. pascens*, *Atbc. polychromogenes*, *Atbc. ramosus* and *Atbc. ureafasciens* merit species status (Fiedler, 1973; Stackebrandt and Fiedler, 1979; Collins *et al.*, 1981b) while the high genetic similarity of '*Brev. sulphureum*' and '*Brev. protophormiae*' to some arthrobacters is consistent with their classification in *Arthrobacter sensu stricto*. In contrast, *Atbc. nicotianae* is probably in need of reclassification as it has a peptidoglycan based on lysine with the A4α variation (Schleifer and Kandler, 1972) and contains fully unsaturated menaquinones (Yamada *et al.*, 1976). The numerical predominance of arthrobacters in soils of various types, together with the nutritional versatility of the commonly occurring species (Keddie, 1974a; Hagedorn and Holt, 1975), indicate a possible role in mineralization.

Arthrobacter simplex and *Atbc. tumescens* have little in common with the *Atbc. globiformis* group. They have been sharply distinguished from the latter in numerical phenetic (Bousfield, 1972; Jones, 1975), chemotaxonomic (Minnikin *et al.*, 1978a; Goodfellow and Minnikin, 1981a; Collins *et al.*, 1982a) and in nucleic acid pairing and sequencing studies (Stackebrandt *et al.*, 1980a; Fiedler *et al.*, 1981). *Arthrobacter simplex* has been shown to be related to the genus *Nocardioides* in phage host range (Prauser, 1976b, 1981), wall peptidoglycan (Schleifer and Kandler, 1972; Prauser, 1978), lipid (O'Donnell *et al.*, 1983) and DNA:DNA pairing studies (Prauser, 1981). Representatives of both taxa have DNA rich in G+C, share a wall chemotype 1, have a peptidoglycan of the A3γ type, show cross-susceptibility to phages, contain major amounts of tetrahydrogenated menaquinones with eight isoprene units and have similar fatty acids and polar lipid profiles. Further, the morphological cycle in *Nrdo. albus* can be regarded as a development of the bending type of cell division shown by *Arthrobacter sensu stricto* (Prauser, 1981). On the basis of such data, O'Donnell and his colleagues proposed that *Atbc. simplex* be transferred to the genus *Nocardioides* as *Nrdo. simplex*. Additional work is required to determine the relationships between *Nrdo.* (*Atbc.*) *simplex*, *Atbc. tumescens*, the chemically related *Brev. lipolyticum* (Yamada and Komagata, 1970b; Minnikin *et al.*, 1978a) and LL-DAP containing isolates from soil, herbage and human skin (Keddie *et al.*, 1966; Pitcher, 1976; Prauser, 1976b; Keddie and Cure, 1977).

The two terregens factor-requiring species, *Atbc. flavescens* and *Atbc. terregens*, containing ornithine in the wall peptidoglycan (Schleifer and

Kandler, 1972; Keddie and Cure, 1977), have a DNA base composition 3 to 5% higher than *Atbc. globiformis* (Table 5) and can be distinguished from the latter in numerical phenetic surveys (Skyring *et al.*, 1971; Jones, 1975). These characteristics suggest an affinity to the genus *Curtobacterium* (Yamada and Komagata, 1972b), a relationship that is not contradicted by the peptidoglycan structure of *Atbc. terregens* (Schleifer and Kandler, 1972). The taxonomic position of *Atbc. duodecadis* remains uncertain (Skyring and Quadling, 1969; Keddie and Cure, 1979), for although this organism has lysine in its wall it differs in other respects from true arthrobacters (Keddie and Cure, 1977).

Comment: Extensive comparative studies should be undertaken to determine the relationships between *Arthrobacter* and *Micrococcus* given the difficulty of distinguishing between good representatives of these taxa in 16S rRNA cataloguing studies (Stackebrandt and Woese, 1979, 1981a; Stackebrandt *et al.*, 1980a). It is interesting that representative strains of *Arthrobacter* and *Micrococcus* have an A3α peptidoglycan (Schleifer and Kandler, 1970, 1972; Kloos *et al.*, 1974), major amounts of dihydrogenated menaquinones with nine isoprene units (Yamada *et al.*, 1976, Collins *et al.*, 1979) and have been found to share a high overall similarity in a numerical phenetic survey (Feltham, 1981).

F. *Brevibacterium* (Breed, 1953)

Gram-positive, asporogenous, non-acid-fast, non-motile actinomycetes which show a marked change in form in complex medium. Older cultures (*ca.* 3 to 7 days at 25°C) are usually composed entirely or largely of coccoid elements, which, on transfer to fresh complex medium, give rise to slender irregular rods typical of exponential phase cultures. V-forms are common and primary branching may occur. As growth continues the rods become shorter and are eventually replaced by coccoid cells characteristic of most stationary phase cultures. Yellow-orange to orange-red colonies are produced on suitable media, though incubation in light is usually required for pigment formation. The organism is an obligate aerobe, catalase positive, salt tolerant and the mode of metabolism is respiratory. Acids are not formed from glucose and other sugars in peptone water.

The wall peptidoglycan contains N-acetylated muramic acid and is of the directly cross-linked *meso*-DAP containing type (A1γ). The polysaccharide moiety of the wall contains galactose, glucose and glycerol teichoic acids; mannitol or ribitol teichoic acids may be present in some strains. Brevibacteria have dihydrogenated menaquinones with nine isoprene units, MK-9(H_2), as the major isoprenologue, are rich in branched-chain fatty acids, do not have mycolic acids but contain phosphatidylglycerol, phosphatidylinositol and phosphatidylinositol mannosides. The DNA base composition is in the range 60 to 64 mol% G+C. The organism forms a substantial fraction of the surface microflora of certain surface-ripened soft cheeses.

The type species is *Brevibacterium linens*, the type strain ATCC 9172.

The genus *Brevibacterium* was introduced by Breed (1953) for certain Gram-positive, non-spore forming rods previously assigned to the genus *Bacterium*. The poor generic description led to the taxon becoming a dumping ground for a diverse collection of ill-described coryneform bacteria. Modern taxonomic methods underlined the extreme heterogeneity of the genus but also resulted in many of the nomenclatural species being transferred to more appropriate niches such as *Arthrobacter, Corynebacterium, Curtobacterium, Oerskovia* and *Rhodococcus* (Yamada and Komagata, 1972a, b; Keddie and Cure, 1977, 1978; Minnikin *et al.*, 1978a; Lanéelle *et al.*, 1980; Suzuki *et al.*, 1981; Collins *et al.*, 1982b, c). These developments leave *Brev. linens* as a distinct taxon on the basis of chemical, genetical, molecular and numerical phenetic properties (Jones, 1978; Minnikin *et al.*, 1978a; Stackebrandt *et al.*, 1980c; Fiedler *et al.*, 1981; Suzuki *et al.*, 1981).

Brevibacterium linens can be isolated from cheeses by plating out suitable dilutions of surface material onto non-selective media such as Cheese Agar and Tryptic Soy Agar supplemented with sodium chloride (4%, w/v). Orange colonies picked off after 5 to 7 days at 25°C and shown to have a coryneform morphology are identified further (Keddie and Jones, 1981). The production of methanethiol by *Brev. linens* may contribute to the aroma and flavour of surface-ripened cheeses (Sharpe *et al.*, 1976, 1977). Pigmented coryneform bacteria considered to be similar to *Brev. linens* have been isolated from human skin, marine fish, pig manure slurry and poultry deep litter (Keddie and Jones, 1981).

Comment: *Brevibacterium* was listed as a genus *incertae sedis* in *Bergey's Manual of Determinative Bacteriology* (Rogosa and Keddie, 1974a) but it is now abundantly clear that it should be considered as the nucleus of a redefined genus as suggested by Yamada and Komagata (1972b). It seems likely that more detailed studies of *Brev. linens* and related strains from natural habitats will reveal the existence of additional species (Fiedler *et al.*, 1981; Keddie and Jones, 1981). Indeed, Collins *et al.* (1980c) have reclassified '*Chromobacterium iodinum*' in the redefined genus *Brevibacterium* as *Brev. iodinum* nom. rev.; comb. nov.

G. *Cellulomonas* (Bergey *et al.*, 1923)

Gram-positive, non-acid-fast, asporogenous actinomycetes which initially produce slender irregular rods (0.5–0.6 × 0.7–2.0 μm) that may be straight, angular, slightly curved and occasionally club-shaped; some of the rods show snapping division and appear as V-forms. In exponential growth, cultures are frequently filamentous and may show rudimentary branching but true mycelia are not formed and, as growth proceeds, rods become shorter and a small proportion of coccoid cells appear. Aerial hyphae are not formed and organisms are motile, by one subpolar or a few lateral flagella, or non-motile. Most strains are facultatively anaerobic but growth is markedly reduced under anaerobic conditions in glucose containing media. Acid is produced from glucose both oxidatively and fermentatively, nitrate is reduced to nitrate, starch is degraded and gelatin slowly hydrolysed. Thiamine and biotin are required for growth. Cellulose is degraded.

The wall peptidoglycan contains *N*-acetylated muramic acid, alanine, glutamic acid and ornithine while the sugar composition of the wall polysaccharides is complex and varies considerably between organisms. All strains contain an A3α peptidoglycan in which the cross-linkage is between L-ornithine in position 3 of one and D-alanine in position 4 of the other of the two peptide subunits, and a common feature of the interpeptide bridges is the presence of a dicarboxylic monoamino acid. Cellulomonads contain straight chain, *iso-* and *anteiso*-fatty acids, 12-methyltetradecanoic acid (*anteiso* 15 predominates), diphosphatidylglycerol, phosphatidylinositol, two unidentified phosphoglycolipids and tetrahydrogenated menaquinones with nine isoprene units, MK-$9(H_4)$, as the major isoprenologue. The G+C content of the DNA ranges from 71 to 77 mol%. Soil seems to be the primary reservoir of cellulomonads.

The type species is *Cellulomonas flavigena*, the type strain ATCC 482.

Cellulomonas strains grow well on peptone meat extract media at 30°C and usually form a yellow non-diffusible pigment on nutrient agar. The organisms are unambiguously Gram-positive (Stackebrandt and Kandler, 1979), but great care must be taken when treating them with alcohol in the Gram-staining procedure as they are readily decolourized and may be mistaken for Gram-negative rods. A method for the selective isolation of cellulomonads is still required, and little is known about their occurrence, distribution, number or activity in natural habitats. A common isolation method involves the preparation of enrichment cultures in yeast extract (0.05–0.1%, v/v) mineral media containing filter paper, followed by plating on cellulose agar (Keddie and Jones, 1981). Cellulolytic bacteria produce zones of clearing on cellulose agar and those with a coryneform morphology can be considered as cellulomonads. The latter may be of value for the production of protein from wood products (Han and Callihan, 1974; Thayer *et al.*, 1975) and in the decomposition of solid compost material (Kaufmann *et al.*, 1976).

The genus *Cellulomonas* has had a somewhat chequered history since it was proposed by Bergey *et al.* (1923) for a heterogeneous collection of strains whose main distinctive feature was their ability to decompose cellulose. The organisms received little attention until Clarke (1951, 1952) clearly demonstrated that they had a coryneform morphology and redescribed the taxon. In a subsequent study, Clarke (1953) reduced the 27 putative species to 10, distinguished primarily by a small number of biochemical properties. Only *Celm. flavigena* was recognized in the eighth edition of *Bergey's Manual of Determinative Bacteriology* (Keddie, 1974b).

The recognition of a separate genus for coryneform bacteria able to degrade cellulose was questioned by Jensen (1966) but subsequent studies based on modern taxonomic methods have demonstrated that the genus *Cellulomonas*, as revised by Clarke (1952), is well founded. Authentic *Cellulomonas* strains contain ornithine in the wall peptidoglycan (Yamada and Komagata, 1970a; Keddie and Cure, 1978; Keddie and Bousfield, 1980), the structure of which is unique (Schleifer and Kandler, 1972; Fiedler and Kandler, 1973b; Stackebrandt and Kandler, 1979). Further evidence of the integrity of the genus is provided by numerical phenetic (Jones, 1978), non-numerical (Yamada and Komagata, 1972a, b; Stackebrandt and

Kandler, 1974), nutritional (Keddie *et al.*, 1966; Owens and Keddie, 1969) and serological (Braden and Thayer, 1976) studies and from menaquinone (Yamada *et al.*, 1976; Collins *et al.*, 1979), fatty acid and polar lipid (Minnikin *et al.*, 1979), and DNA base composition determinations (Yamada and Komagata, 1970b). In an extensive comparative study, Stackebrandt and Kandler (1979) applied a variety of modern taxonomic methods to well documented strains of *Cellulomonas* and recognized seven distinct species: *Celm. biazotea, Celm. cellasea, Celm. flavigena, Celm. fimi, Celm. gelida, Celm. uda* and '*Celm. cartalyticum*'. The polysaccharide and amino-acid composition of the cell wall, the electrophoretic mobility of L-lactate dehydrogenase, and the utilization of a number of sugars and organic acids were found to be useful properties for species differentiation.

Cellulomonads can be distinguished from related actinomycetes using chemical criteria (see Table 3). A close similarity has been demonstrated between *Cellulomonas* and *Oerskovia* in numerical phenetic (Jones and Bradley, 1964; Jones, 1975), nucleic acid pairing and 16S rRNA sequencing (Stackebrandt and Kandler, 1979; Stackebrandt *et al.*, 1980b) studies and strains in both genera have similar menaquinone and polar lipid profiles (Collins *et al.*, 1979; Minnikin *et al.*, 1979). Further comparative studies are required to determine whether the differences in morphology, growth habits, wall composition and fatty acid profiles (Sukapure *et al.*, 1970; Schleifer and Kandler, 1972; Minnikin *et al.*, 1979) are sufficient to merit the continued separation of the genera *Cellulomonas* and *Oerskovia* or whether the latter should become a synonym of *Cellulomonas* as suggested by Stackebrandt *et al.* (1980b). Representatives of the genus *Promicromonospora* should be included in such studies as there is evidence of a relationship between the latter and *Oerskovia* (Lechevalier, 1972; Lechevalier *et al.*, 1977).

H. *Curtobacterium* (Yamada and Komagata, 1972b)

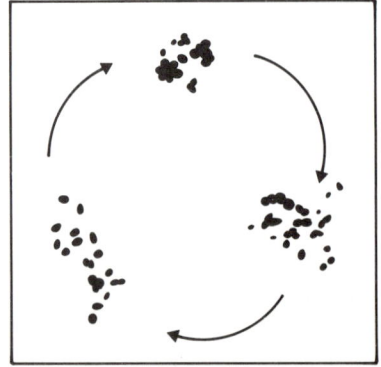

Gram-positive, non-acid-fast, asporogenous actinomycetes which produce irregular rods that may be straight, angular, club-shaped or appear as V-forms. Older cultures (*ca.* 7 days at 25°C) may show a proportion of coccoid cells. Aerial hyphae are not produced. The organisms are either motile, by a few lateral flagella, or non-motile. They are strict aerobes, acids are formed slowly and weakly from some carbohydrates and gelatin is slowly hydrolysed. Cellulose is not degraded.

The wall peptidoglycan contains *N*-acetylated muramic acid, ornithine as the diamino acid and, although the sugar composition of the wall polysaccharide varies between species, galactose is commonly present. All strains contain the unusual B2β peptidoglycan. Curtobacteria have straight-chain *iso-* and *anteiso*-fatty acids, with 12-methyltetradecanoic and 14-methylhexadecanoic predominating, major amounts of unsaturated menaquinones with nine isoprene units (MK-9), and have a characteristic polar lipid profile consisting of phosphatidylglycerol, diphosphatidylglycerol and a number of uncharacterized glycolipids. The G+C content of the DNA ranges from 66 to 73 mol%. Curtobacteria are widespread in plant matter and soil. Some are phytopathogens.

The type species is *Curtobacterium citreum*, the type strain ATCC 15828.

The genus *Curtobacterium* was introduced to accommodate motile brevibacteria, including *Brev. albidum*, *Brev. citreum*, *Brev. luteum*, *Brev. pusillum*, *Brev. saperdae* and *Brev. testaceum*, as well as *Cnbc. flaccumfaciens* and *Cnbc. poinsettiae*, the only two ornithine-containing plant pathogenic corynebacteria studied by Yamada and Komagata (1972b). *Curtobacterium citreum*, *Curt. flaccumfaciens*, *Curt. luteum* and *Curt. pusillum* form a well defined DNA homology group (Döpfer *et al.*, 1982). *Corynebacterium betae* and *Cnbc. oortii* were transferred to the genus *Curtobacterium* (Schleifer and Kandler, 1972; Jones, 1975; Starr *et al.*, 1975; Keddie and Cure, 1978; Collins *et al.*, 1980a) which was seen to be heterogeneous on the basis of chemical and nucleic acid pairing data. Thus, *Curt. saperdae* and *Curt. testaceum* can be distinguished from true curtobacteria by menaquinone, polar lipid (Yamada *et al.*, 1976; Collins *et al.*, 1980a) and detailed peptidoglycan composition (Schleifer and Kandler, 1972; Uchida and Aida, 1977), and in both DNA:DNA and DNA:rRNA pairing assays (Suzuki *et al.*, 1981; Döpfer *et al.*, 1982). *Corynebacterium nebraskense*, the cause of leaf freckle and wilt of maize and sweet corn (*Zea mays*) has polar lipid and menaquinone profiles in common with curtobacteria (Collins *et al.*, 1980a) but is in a DNA homology group with *Cnbc. michiganense* and *Cnbc. sepedonicum* (Döpfer *et al.*, 1982), organisms which contain DAB as the diamino-acid of the wall peptidoglycan (Keddie and Cure, 1978).

Curtobacteria grow well on media containing peptone, yeast extract and glucose (Yamada and Komagata, 1972a, b) and some of the economically important phytopathogenic strains can be grown on the CNS agar medium of Gross and Vidaver (1979). *Curtobacterium betae* causes vascular wilt and leaf spot of red beet (*Beta vulgaris*), *Curt. flaccumfaciens*, vascular wilt of beans (*Phaseolus* spp.), *Curt. oortii*, vascular disease with leaf and bulb lesions of tulip (*Tulipa* spp.), and *Curt. poinsettiae*, stem canker and leaf spot of poinsettiae (*Euphorbia pulcherrima*). In a numerical phenetic study, Dye and Kemp (1977) reduced *Curt. betae*, *Curt. oortii* and *Curt. poinsettiae* to pathovars of *Curt. flaccumfaciens* using numerical phenetic data and Döpfer *et al.* (1982) found that *Curt. betae* was genetically identical to *Curt. flaccumfaciens*.

Comment: The genus *Curtobacterium* should be restricted to strains containing D-ornithine in the wall peptidoglycan (variation B2β) and unsaturated menaquinones with nine isoprene units. The genus is phylogenetically close to *Microbacterium* (Stackebrandt *et al.*, 1980a) but can be distinguished from the latter in nucleic acid (Döpfer *et al.*, 1982) and wall composition studies (Schleifer and Kandler, 1972).

1. *Microbacterium* (Orla-Jensen, 1919)

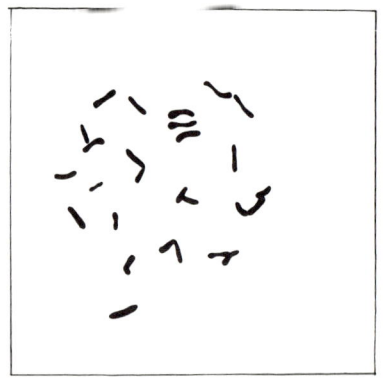

Gram-positive, non-acid-fast, asporogenous actinomycetes which produce small, slender, irregular rods (*ca*. 0.5 μm in diameter) which show V-formation in young cultures. In older cultures, the rods are shorter but a marked rod–coccus cycle does not occur. Aerial hyphae are not formed and the organism is non-motile. Colonies are 1 to 1.5 mm in diameter, circular, opaque, glistening and vary in colour from grey-white to pale greenish-yellow. All strains grow well under aerobic conditions, are catalase-positive, and produce acid weakly from glucose and other sugars in peptone water. L(+)-lactic acid is formed from glucose. The organism is nutritionally exacting, requires β-vitamins, and often amino acids, for growth but can survive heating at 63°C for 30 minutes in skim milk.

The wall peptidoglycan contains N-glycolylmuramic acid, lysine as diamino-acid, and the wall sugars are galactose, rhamnose and occasionally mannose. Strains have the rare B1α peptidoglycan and an interpeptide bridge containing one glycine and one L-lysine residue. Microbacteria have menaquinones with ten and eleven isoprene units, MK-*10,11*, as the major isoprenologues, and contain phosphatidylglycerol, diphosphatidylglycerol and dimannosyldiacylglycerol. The DNA base ratio is in the range 69–70 mol% G+C. The organism is found chiefly in milk, dairy products and on dairy equipment.

The genus is monotypic, and the type strain *Microbacterium lacticum* ATCC 8180.

The genus *Microbacterium* was proposed by Orla-Jensen (1919) for an ill-assorted group of Gram-positive, non-spore-forming rods that were heat resistant, and produced small amounts of L(+)-lactic acid from glucose. Four species were recognized, *Miba. flavum, Miba. lacticum, Miba. liquefaciens* and *Miba. mesentericum* and a fifth, *Miba. thermosphactum* (McLean and Sulzbacher, 1953), was proposed for strains that were neither heat resistant nor pleomorphic. *Microbacterium thermosphactum* was subsequently found to be quite different from the other microbacteria and the genus *Brochothrix* was established to contain the species *Brochothrix thermosphacta* (Sneath and Jones, 1976). Modern taxonomic methods have also shown that the genus *Microbacterium*, as envisaged by Orla-Jensen, was heterogeneous (Keddie and Jones, 1981). *Microbacterium flavum* has been reclassified as *Cnbc. flavescens* (Barksdale et al., 1979); *Miba. mesentericum* is no longer recognized, and *Miba. liquefaciens*, which has a β-type peptidoglycan but ornithine as the diamino-acid, shares an affinity with *Curtobacterium* (Keddie and Jones, 1981). These proposals leave *Miba. lacticum* as a distinct and recognizable taxon on the basis of chemical (Schleifer, 1970; Minnikin et al., 1978a; Collins et al., 1979), genetical (Döpfer et al., 1982), numerical phenetic (Jones, 1975) and non-numerical data (Jayne-Williams and Skerman, 1966).

Microbacterium lacticum grows on rich media such as yeast extract milk agar (Harrigan and McCance, 1976) and is generally isolated from dairy products by plating out pasteurized samples onto suitable, non-selective media. Nonthermoduric strains, indistinguishable in other respects from *Miba. lacticum*, have been reported from sources that have not been subjected to heat treatment (Jayne-Williams and Skerman, 1966). Microbacteria form a substantial part of the thermoduric count of raw and pasteurized milk and milk powder, and their occurrence in such products can be taken as reliable evidence of improperly cleaned dairy equipment (Thomas et al., 1967). This bacterial contamination is of particular importance to producers and processors in countries where statutory colony count standards are applied to spray-dried milk powder (Griffiths, 1977).

Comment: Although the genus *Microbacterium* is currently considered to be a genus *incertae sedis* (Rogosa and Keddie, 1974b) it is now clear that *Miba. lacticum* should be considered as the nucleus of a redefined genus as suggested by Jones (1975) and Keddie and Cure (1978). '*Brevibacterium' imperiale* and '*Cnbc. laevaniformis*' are similar to *Miba. lacticum* phenetically (Bousfield, 1972; Jones, 1975), in peptidoglycan type (Schleifer and Kandler, 1972) and on the basis of nucleic acid pairing (Döpfer *et al.*, 1982) and menaquinone data (Collins and Jones, 1981).

J. *Oerskovia* (Prauser *et al.*, 1970; Lechevalier, 1972)

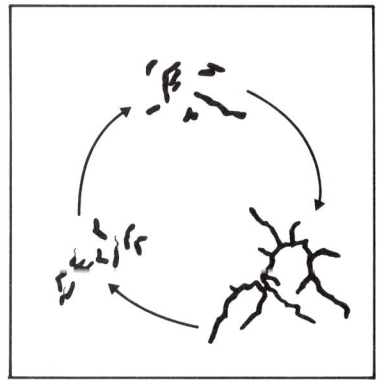

Gram-positive, non-acid-fast, asporogenous actinomycetes that form a primary mycelium which breaks up into motile rod-like elements. The motile rods are monotrichous when small (*ca.* 0.4×1.1 μm) but peritrichous when long and germinate to form extensively branched hyphae about 0.5 μm in diameter. Oerskoviae produce yellow colonies which lack aerial hyphae and are catalase positive when grown aerobically. The organism is facultatively anaerobic on Trypticase-soy medium, attacks glucose both oxidatively and fermentatively, hydrolyses casein, gelatin, starch and DNA, and reduces nitrate to nitrite.

Cell wall preparations contain *N*-acetylated muramic acid, major amounts of lysine and galactose with D-aspartic acid, D-glutamic acid, D-serine and L-threonine present in some strains. The wall peptidoglycan is of the A4α type. Oerskoviae contain *iso-*, *anteiso-* and straight-chain fatty acids, with 12-methyltetradecanoic acid predominating, diphosphatidylglycerol, phosphatidylinositol and two unidentified glycolipids and tetrahydrogenated menaquinones with nine isoprene units, MK-$9(H_4)$, as the major isoprenologue. The DNA base composition is within the range 70.5 to 75 mol% G+C. The organism has been isolated from soil, aluminium hydroxide gel antacid, dry grass cuttings and clinical material.

The type species is *Oerskovia turbata*, the type strain ATCC 25835.

Actinomycetes with the characteristic properties of the genus *Oerskovia* were first isolated from soil by Ørskov (1938), were designated 'motile nocardiae' but later classified as *Nrda. turbata* (Erikson, 1954). Several workers considered such strains to be quite different from typical nocardiae (Jones and Bradley, 1964; Prauser, 1967a; Sukapure *et al.*, 1970) and they were reclassified as *Orsk. turbata* by Prauser *et al.* (1970). Subsequently, a second species, *Orsk. xanthineolytica* (Lechevalier, 1972) was proposed for isolates from soil and grass cuttings. Similar organisms have been isolated from clinical material with one strain being recognized as the cause of endocarditis (Reller *et al.*, 1975; Sottnek *et al.*, 1977). *Oerskovia xanthineolytica* differs from *Orsk. turbata* in its ability to degrade xanthine and hypoxanthine, in the production of a phosphatase, and can also be separated by peptidoglycan structure and cytochrome pattern (Seidl *et al.*, 1980), and in DNA:DNA pairing assays (Stackebrandt *et al.*, 1980b).

The separation of *Oerskovia* from *Nocardia* is supported by numerical phenetic (Goodfellow, 1971; Jones 1975), chemical (Minnikin and Goodfellow, 1980) and phage host range studies (Prauser and Falta, 1968) but its relationship with *Cellulomonas* requires further study. Data from numerical phenetic (Goodfellow, 1971; Jones, 1975), wall (Seidl *et al.*, 1980), lipid (Yamada *et al.*, 1976; Collins *et al.*, 1977, 1979; Minnikin *et al.*, 1979) and genetical studies (Stackebrandt *et al.*, 1980b) show that *Brev. fermentans*, '*Cnbc. manihot*' and *Nrda. cellulans* have properties consistent with their transfer to the genus *Oerskovia*. Similarly, wall composition and molecular genetic data support the transfer of *Celm. carta* to *Oerskovia* (Stackebrandt *et al.*, 1978, 1980b; Seidl *et al.*, 1980). *Brevibacterium lyticum* has been transferred to the genus on the basis of wall and lipid composition (Collins and Jones, 1981c).

K. *Promicromonospora* (Krasilnikov *et al.*, 1961)

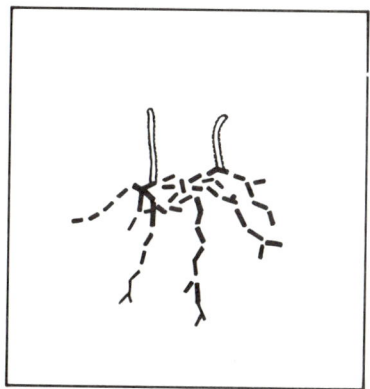

Gram-positive, asporogenous, non-acid-fast, non-motile actinomycetes which form a mycelium that fragments. At 37°C fragmentation is rapid and elements of varying length and shape are formed. The organism grows well between 10 and 48°C but poorly at extreme temperatures. The wall peptidoglycan contains lysine as the diamino acid and galactose is absent. Promicromonosporae contain tetrahydrogenated menaquinones with nine isoprene units, MK-9(H_4), as the predominant isoprenologue, major amounts of branched chain fatty acids and characteristic phospholipids which include phosphatidylglycerol, diphosphatidylglycerol and glucosamine containing phospholipids. The DNA base composition is 73 mol% G+C (Tsyganov et al., 1970).

The genus is monotypic, the type strain *Promicromonospora citrea* ATCC 15908.

The genus *Promicromonospora* was originally proposed for soil actinomycetes that formed yellow colonies, a sparse sterile aerial mycelium, and a substrate mycelium which not only fragmented into bacillary elements but was also reported to carry single spores. Subsequent studies on promicromonosporae, and on similar actinomycetes isolated from aluminium hydroxide antacid gels (Lechevalier, 1972), served to expand the original description of the genus but notably failed to confirm the ability to form spores. It now seems most unlikely that promicromonosporae can produce spores. The strong affinity of *Promicromonospora* to *Oerskovia* and non-motile *Oerskovia*-like strains deserves further consideration (Lechevalier, 1972; Collins et al., 1980b).

L. *Renibacterium* (Sanders and Fryer, 1980)

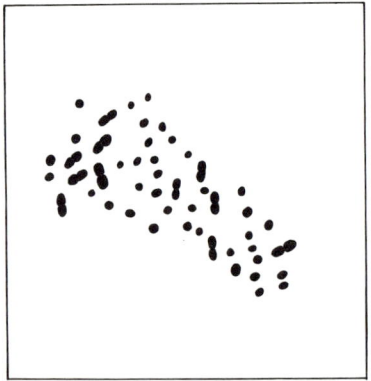

Gram-positive, asporogenous, non-acid-fast, non-motile actinomycetes which form short rods ($0.3–1.0 \times 1.0–1.5$ μm) that often occur in pairs. Cysteine is required for growth. Circular, convex, white colonies of various sizes are formed on cysteine serum agar. Optimal growth occurs at 15 to 18°C. The organism is aerobic, catalase positive, cytochrome oxidase negative, does not liquefy gelatin or form acid from glucose.

The wall peptidoglycan contains alanine, glycine, glutamic acid and lysine. The principal wall sugar is glucose but arabinose, mannose and rhamnose are also present. Renibacteria contain unsaturated menaquinones with nine isoprene units, MK-9, as the major isoprenologue, lack mycolic acids but are rich in methyl-branched chain fatty acids with 12-methyltetradecanoic, 13-methyltetradecanoic and 14-methylhexadecanoic predominating and have very characteristic polar lipid patterns consisting of diphosphatidylglycerol, two major and six or seven uncharacterized glycolipids and two unidentified minor phospholipids. The DNA base composition is within the range 53 to 54 mol% G+C. The organism causes Bacterial Kidney Disease (BKD, corynebacterial kidney disease, salmonid kidney disease), an important infectious disease of hatchery-reared salmonids.

The genus is monotypic, the type strain *Renibacterium salmoninarum* ATCC 33209.

Renibacteria can be cultivated on media such as cysteine serum agar, Loeffler coagulated serum and Dorset egg medium. Maximum growth is obtained after 3 to 4 weeks incubation at 15 to 18°C at pH 6.5 to 7.5. A semi-defined medium promotes good growth within 2 weeks (Embley *et al.*, 1982). The organism causes one of the most important infectious diseases of hatchery-reared salmonid fishes and can cause heavy mortalities over a wide range of temperatures (Fryer and Sanders, 1981).

Renibacteria were originally considered to be pleomorphic and were associated with the genus *Corynebacterium* (Ordal and Earp, 1956) and the coryneform bacteria (Bullock and Stuckey, 1975). They are now known to occur mainly as short rods (Young and Chapman, 1978) and can be distinguished both from true corynebacteria and other Actinobacteria by many chemical properties (Sanders and Fryer, 1980; Embley *et al.*, 1983). Isolates have been shown to cross-react by the use of antisera raised against whole cells and sonicated extracts; 20-fold differences in agglutination titres between some strains suggested differences in the degree of cross-reactivity (Bullock *et al.*, 1974). Further work is required to determine the relationship between '*Listeria denitrificans*' and *Renibacterium* for strains in these taxa have a number of chemical properties in common (Embley *et al.*, 1983).

M. *Rothia* (Georg and Brown, 1967)

Gram-positive, non-acid-fast asporogenous, non-motile, catalase-positive actinomycetes which form a rudimentary primary mycelium on agar that readily fragments into bacillary and coccoid elements. Branched filaments and hyphae are usually about 1 μm in diameter but may be up to 5 μm due to the presence of irregular swellings with club-shaped ends. Coccoid forms vary from 1 to 5 μm in diameter. Aerial hyphae are not produced. The organisms are facultatively anaerobic with a preference for aerobic growth, carbon dioxide is not stimulatory. A number of sugars are fermented with the production of acid but not gas, nitrate is reduced to nitrite, and hydrogen sulphide is formed on triple sugar iron agar. The main product of glucose fermentation is lactic acid but small amounts of acetic, formic and succinic acids are also found. Propionic acid is not produced.

The main amino acids in the peptidoglycan are alanine, lysine and glutamic acid and the major wall sugars fructose, galactose, glucose and ribitol. The peptidoglycan belongs to the L-lysine–L-alanine$_3$ type A3α which is common in bacteria but has not been found in any other actinomycetes. The DNA base composition is 65 to 70 mol% G+C. Rothiae are indigenous in the human oral cavity but are not known to cause natural infections in either man or animals.

The genus is monotypic and the type strain *Rothia dentocariosa* ATCC 17931.

Rothiae are fastidious in their growth needs, require organic nitrogen and grow poorly on nutrient agar. For primary isolation and maintenance, enriched media such as brain heart infusion (BHI) and tryptic soy are recommended with aerobic incubation at 37°C. Strains growing anaerobically on BHI plates produce spider-like microcolonies whereas macrocolonies are creamy white, and soft with smooth to rough surfaces. The inability of the organism to grow on Sabouraud's dextrose agar (Difco) provides a useful way of separating it from *Nocardia* species (Brown *et al.*, 1969). Rothiae have been isolated from blood and spinal fluids, and abscess formation demonstrated experimentally in mice (Roth and Flanagan, 1969). The organism can be identified in clinical samples using the fluorescent antibody technique (Hammond, 1970; Gerencser, 1979) and has been detected in animals (Dent *et al.*, 1976).

Rothia strains bear a superficial resemblance to *Actinomyces* and *Nocardia* and were originally classified as *Actn. dentocariosus* (Onisi, 1949; Onisi and Nuckolls, 1958), *Nrda. dentocariosus* (Roth, 1957; Roth and Thorn, 1962) and *Nrda. salivae* (Davis and Freer, 1960). They have, however, been sharply distinguished from nocardiae and fermentative actinomycetes in numerical phenetic (Goodfellow, 1971; Holmberg and Hallander, 1973; Schaal and Schofield, 1981a, b; Schofield and Schaal, 1982), chemotaxonomic (Georg and Brown, 1967; Sukapure *et al.*, 1970; Schleifer and Kandler, 1972; Pandhi and Hammond, 1975) and serological analyses (Holmberg and Forsum, 1973; Gerencser, 1979). Lipid studies have still to contribute decisively in *Rothia* systematics but the presence of a dimannosyl-diacylglycerol may prove to be a useful chemical marker (Pandhi and Hammond, 1975). The genus may be heterogeneous as organisms that differ biochemically and serologically from the type strain have been reported (Jordan and Hammond, 1972; Lesker *et al.*, 1974; Slack and Gerencser, 1975). Indeed, Lesker and his co-workers recognized four biotypes and three serotypes and suggested that one of the former merited the rank of species. Additional comparative studies are needed not only to find the sub-generic structure of *Rothia* but also to determine its supra-generic affinities.

N. UNASSIGNED ACTINOBACTERIA

A number of animal and plant species associated with the genera *Brevibacterium* and *Corynebacterium* need to be reclassified (Table 6; Keddie and Cure, 1978; Minnikin *et al.*, 1978a; Goodfellow and Minnikin, 1981a). These taxa include *Cnbc. beticola* and '*Cnbc. nephridii*' which are Gram-negative bacteria (Minnikin *et al.*, 1978a; Collins and Jones, 1982b) and phytopathogenic strains that are rich in G+C and contain 2,4-diaminobutyric acid as the dibasic acid of the wall peptidoglycan.

Schleifer and Kandler (1972) noted a relationship between the DAB-containing phytopathogens, ornithine-containing curtobacteria and the microbacteria that contain lysine in the wall peptidoglycan. Such strains, and probably *Agrm. ramosus*, have a peptidoglycan belonging to the rare B group, contain similar fatty acids and are closely related phylogenetically (Minnikin *et al.*, 1978a; Döpfer *et al.*, 1982). In addition, '*Cnbc.*' *insidiosum*, '*Cnbc*'. *michiganense*, '*Cnbc.*' *nebraskense* and '*Cnbc.*' *sepedonicum* form a

TABLE 6. Chemical properties of some unassigned animal and plant pathogenic actinobacteria*

	Wall diamino-acid	Peptidoglycan type	Fatty acids	Predominant menaquinone(s)	Phospholipids	Glycolipids	G+C (mol%)
'Brev. insectiphilum'	ornithine	B2β	ND	ND	ND	ND	73
Brev. protophormiae	lysine	ND	ND	MK-8, 9	ND	ND	ND
'Cnbc. aquaticum'	DAB	B2γ	S, U, I, A	MK-10, 11	DPG, PG (PI, PIM)	DiMaDAG	69–73
'Cnbc. barkeri'	ornithine	B2β	ND	MK-11, 12	ND	ND	ND
'Cnbc.' iranicum'	DAB	ND	S, U, I, A	MK-10	DPG, PG	+	68–71
'Cnbc.' insidiosum	DAB	B2γ	S, U, I, A	MK-9	DPG, PG	+	72–76
'Cnbc. mediolanum'	DAB	B2γ	S, U, I, A	MK-11, 12	DPG, PG	+	74
'Cnbc.' michiganense	DAB	B2γ	S, U, I, A	MK-9	DPG, PG	+	67–75
'Cnbc.' nebraskense	DAB	B2γ	S, U, I, A	MK-9	DPG, PG	+	77
'Cnbc. okanaganae'	DAB	ND	S, U, I, A	MK-11, 12	DPG, PG	+	64
'Cnbc.' sepedonicum	DAB	B2γ	S, U, I, A	MK-9	DPG, PG	+	70–76
'Cnbc. tritici'	DAB	B2γ	S, U, I, A	MK-10	DPG, PG	+	70–73

ND, not determined. * Data from Collins and Jones (1980, 1981a, d); Cummins et al. (1974); Döpfer et al. (1982); Keddie and Cure (1978); Luthy (1974); Minnikin et al. (1978a) and Starr et al. (1975).

loose DNA homology group (Starr *et al.*, 1975; Döpfer *et al.*, 1982), have unsaturated menaquinones with nine isoprene units, MK-9, as the major isoprenologue, large proportions of branched-chain fatty acids and very characteristic polar lipid patterns composed of diphosphatidylglycerol, phosphatidylglycerol and a number of uncharacterized glycolipids (Bowie *et al.*, 1972; Minnikin *et al.*, 1978a; Collins and Jones, 1980). The strains in these taxa also have many cultural and physiological properties in common (Lelliott, 1966; Dye and Kemp, 1977) and may form the nucleus of a new genus as suggested by Yamada and Komagata (1927b). '*Corynebacterium*' *iranicum* and '*Cnbc*'. *tritici* are closely related to this group (Table 6; Dye and Kemp, 1977) whereas '*Cnbc. aquaticum*' and '*Cnbc. mediolanum*' have been found to be loosely associated with one another and with *Agrm. ramosus* in DNA:DNA pairing experiments (Döpfer *et al.*, 1982). Further detailed comparative studies are required to determine the relationships between the DAB-containing taxa and the genera *Agromyces, Curtobacterium* and *Microbacterium*.

6. Actinoplanetes

The family *Actinoplanaceae* currently includes those hyphae-forming, Gram-positive, non-acid-fast members of the Order *Actinomycetales* in which the spores (motile or non-motile depending on the genus) are produced within spore vesicles or sporangia. About 15 genera have, at one time or another, been included in the family but this number has recently been reduced to the following 9 (Bland and Couch, 1981): *Actinoplanes, Amorphosporangium, Ampullariella, Dactylosporangium, Pilimelia, Planobispora, Planomonospora, Spirillospora* and *Streptosporangium*. This classification has been based almost entirely on the morphology of the spore vesicle or sporangium. Only one of the above genera has non-motile spores (*Streptosporangium*). The number of spores varies from one (in *Planomonospora*) to over a thousand (in *Pilimelia*). Spore formation in most genera occurs as a result of the septation of unbranched (*Streptosporangium*) or branched (*Actinoplanes*) and coiled sporogenous hyphae within an outer sheath which forms the boundary of the vesicle or sporangium. However, an alternative endogenous spore formation process would appear to be a feature of *Dactylosporangium, Planomonospora* and possibly *Planobispora*.

Recent studies of peptidoglycan composition and DNA:DNA reassociations (Farina and Bradley, 1970; Stackebrandt *et al.*, 1981) have shown that this classification is artificial and that at least two major DNA homology clusters with distinct peptidoglycans can be recognized. The genera in DNA homology cluster I, which we will now call the *Actinoplanetes*, have a wall chemotype II (Szaniszlo and Gooder, 1967). The peptidoglycan contains *meso-* or 3-hydroxy diaminopimelic acid and glycine, the latter amino acid would appear to replace L-alanine in the peptide attached to muramic acid which is glycolated (Kawamoto *et al.*, 1981). Whole-organism hydrolysates contain xylose and arabinose with variable amounts of other sugars. The same peptidoglycan structure and sugar composition has been found in the genus *Micromonospora* (Kawamoto *et al.*, 1981) which forms single spores not enclosed within a vesicle or sporangium on the substrate mycelium. Species of this genus also show high DNA homologies and ribosomal nucleotide similarities (Stackebrandt and Woese, 1981a,b) with other members of the DNA homology cluster I; *Actinoplanes, Amorphosporangium, Ampullariella, Pilimelia* and *Dactylosporangium* (Stackebrandt *et al.*, 1981), and *Micromonospora* must now be included in the Actinoplanetes.

The other genera at present included in the *Actinoplanaceae* have a wall chemotype III, belong to DNA homology cluster II and exhibit low ribosomal ribonucleic acid similarities with the Actinoplanetes. They have been considered separately and termed Maduromycetes in Section 10 (p. 105).

The genera *Actinoplanes, Amorphosporangium, Ampullariella* and *Pilimelia* have been separated mainly on the basis of spore vesicle morphology, spore shape and the arrangement of flagella on the zoospore, but this separation is not supported by the available chemotaxonomic and genetic evidence. When one considers that the shape of the spore vesicle is determined by the extent of the branching of the sporulating hyphae within a structureless sac and their eventual arrangement (e.g. in coils or parallel rows) and that this sac may split or be partially digested, it is not surprising that a variety of shapes are seen. Sporulation of hyphae within a sheath is characteristic of many other actinomycete genera, e.g. *Streptomyces*, where globose and cylindrical spores are formed and not accorded generic character status. If we persist in separating these genera then speciation will have to rely on traditional small differences in pigmentation, morphology and growth on numerous carbon sources (e.g. see Palleroni, 1979) and will be extremely time consuming. We must therefore support the earlier suggestions that the classification of these genera be simplified by accommodating them within the single genus *Actinoplanes* (Szaniszlo and Gooder, 1967; M. P. Lechevalier and Lechevalier, 1970b; Cross and Goodfellow, 1973) in the hope that this will prompt a more objective search for reliable specific characters. However, we feel that a useful classification would be one that still recognized the separate status of the morphologically distinct genera *Micromonospora* and *Dactylosporangium*.

A. *Actinoplanes* (Couch, 1950, 1955)

Actinomycetes with a very fine branching mycelium (0.2–1.5 μm diameter) which bear the characteristic spore vesicles (sporangia) and release motile spores. Spore vesicles of the type species, *Actp. philippinensis* are globose and contain irregularly coiled spore chains (Couch and Bland, 1974a). Other species may have asymmetric spore vesicles which vary in shape; they may be lobed, digitate, club-shaped or even cylindrical (Lechevalier and Lechevalier, 1975). When cultured on agar in light, the majority of species form yellow or orange colonies because of contained carotenoid pigments (Szaniszlo, 1968). *Actinoplanes italicus* (Beretta, 1973) has cherry-red colonies, those of '*Actp. coeruleus*' (Wagman et al., 1975) are blue, and brown to dark red colonies are characteristic of *Actp. ferrugineus* (Palleroni, 1979) but Parenti and Coronelli (1979) have also isolated purple-violet and green-pigmented species. Colonies normally lack aerial mycelium but bear a surface pallisade of hyphae terminating in distinct sporangiophores bearing the spore vesicles. Parenti and Coronelli (1979) examined several thousand fresh isolates of *Actinoplanes* and found only one, '*Actp. teichomyceticus*' (Parenti et al., 1978) with a well developed and powdery aerial mycelium. A very short aerial mycelium lacking spores was also reported to be present on *Actp. rectilineatus* which gave the colonies a brownish, powdery appearance (Lechevalier and Lechevalier, 1975). The wall of the spore vesicle is an extension of the sheath surrounding the pallisade hyphae and sporangiophores. It encloses branching hyphae which eventually become regularly subdivided into spores by annular ingrowths from the hyphal wall (Lechevalier and Holbert, 1965; Lechevalier et al., 1966b; Williams, 1970). The ability to produce spore vesicles may not be evident or be lost quite rapidly in culture but can sometimes be stimulated by the addition of humic acids (Willoughby et al., 1968) or tea extract (Parenti et al., 1978) to the growth medium. Spore vesicles would appear to survive desiccation for years in dry soil and leaf litter (Makkar and Cross, 1982), swell when rehydrated and release the contained spores through the burst wall (Higgins, 1967) or following its partial dissolution (Couch, 1963). Motility in spores requires an exogenous energy source (Higgins, 1967) and is actuated by a polar tuft of flagella (Higgins *et al.*, 1967; Locci and Petrolini-Baldan, 1971; Schäfer, 1973). The zoospores are globose to subglobose (1.0–1.5 μm diameter) and exhibit chemotaxis (Palleroni, 1976) although no single attractant or repellant was active for all strains.

Walls of actinoplanes were reported to contain *meso*-DAP and glycine, and were consequently placed in the chemotype II wall classification of H. A. Lechevalier and Lechevalier (1970). Yamaguchi (1965) and Szaniszlo and Gooder (1967) also reported that walls may contain 3-hydroxy diaminopimelic acid (HO-DAP), and glycine would appear to replace L-alanine in the peptide attached to muramic acid (Kawamoto *et al.*, 1981). It is interesting to note that some species (e.g. *Actp. philippinensis*) contain only *meso*-DAP, others (e.g. *Actp. utahensis*) contain only HO-DAP, and in strains of *Actp. missouriensis* some walls contain *meso*-DAP with a trace of HO-DAP, and other strains have only HO-DAP. It appears that morphologically similar actinoplanes sometimes have one or both of the diaminopimelic acids in their walls (Szaniszlo and Gooder, 1967). All species contain galactose in the wall polysaccharide with various amounts of glucose, mannose, arabinose, xylose and a deoxyhexose (Szaniszlo and Gooder, 1967). Actinoplanetes contain phosphatidylethanolamine (phospholipid pattern PI1) but the phospholipid fatty acid type differed in the two species examined (Lechevalier *et al.*, 1977).

The type species is *Actinoplanes philippinensis*, the type strain ATCC 12427.

Actinoplanes would appear to be present in most soils and are especially abundant in those with a neutral pH (Nonomura and Takagi, 1977). They have also been isolated from lake and river water, and from decaying plant material in streams and cast up on lake shores (Willoughby, 1966, 1968, 1969a; Makkar and Cross, 1982). Species were first isolated by Couch (1949) on baits of pollen grains and plant fragments floating on water above soil samples. They can now be more easily isolated from soil and litter by first drying the substrate and then rehydrating for 60 minutes to ensure release of zoospores before plating the suspension on colloidal chitin agar (Makkar and Cross, 1982). The recent observation that zoospores exhibit chemotaxis can be used to preferentially concentrate zoospores in capillaries for subsequent transfer to the isolation medium (Palleroni, 1980).

Actinoplanes are capable of producing a range of potentially useful secondary metabolites and consequently we have recently seen descriptions of many species able to synthesize new antibiotics (Parenti and Coronelli, 1979). Species have been distinguished on the basis of morphology, pigmentation and biochemical characters and there is a danger of overspeciation.

Actinoplanes armeniacus (Kalakoutskii and Kuznetsov, 1964) was originally reported to produce spherical spore vesicles (20–50 μm diameter) containing spherical to ovoid spores with peritrichous flagella. The colonies also had abundant aerial mycelium and chains of spores. Recent studies on the type strain ATCC 15676 failed to reveal the presence of spore vesicles; whole-organism hydrolysates contained LL-DAP and phage activity spectra showed extensive cross-reactions with streptomycetes. The species was accordingly transferred to the genus *Streptomyces* as *Stmy. armeniacus* (Kalakoutskii & Kuznetsov) comb. nov. (Wellington and Williams, 1981a, b). Kroppenstedt *et al.* (1981) independently reached the same conclusion on the basis of extensive biochemical and genetic evidence.

1. Amorphosporangium

Amorphosporangium (Couch, 1963) was separated from the genus *Actinoplanes* mainly on morphological grounds. The spore vesicles were said to be very irregular in shape (6–25 μm × 8–15 μm) with many lobes and contained rod-shaped spores arranged in coils which were not motile when released (Couch, 1963; Willoughby, 1969a). Hanton (1968) showed that the spores became motile in water after a short delay and were equipped with a polar tuft of flagella. The shape of the spores varied from spherical to rod-shaped and a few curved spores were seen. The majority, however, were short rods.

The type species is *Amorphosporangium auranticolor*, the type strain ATCC 15330.

The walls of the type species contain *meso*-DAP with some HO-DAP together with xylose, arabinose, galactose, glucose, mannose and deoxyhexose (Szaniszlo and Gooder, 1967). The correct classification of *Amfs. globisporus* (Thiemann, 1967), which was also included in the

Approved List of Bacterial Names (Skerman *et al.*, 1980), appears to be in doubt. Thiemann later opined that it belonged to the genus *Actinoplanes* (see Couch and Bland, 1974b) and this uncertainty exemplifies the problem of attempting to differentiate between the two genera.

2. *Ampullariella*

Ampullariella (Couch, 1964) was also described as morphologically distinct in having bottle- or flask-shaped, digitate or lobate spore vesicles (5–20 μm × 8–30 μm) with the spores arranged in parallel chains. The individual spores are rod-shaped and motile by a tuft of polar flagella (Higgins *et al.*, 1967; Schäfer, 1973). Walls contain either HO-DAP alone or a mixture of *meso*-DAP and HO-DAP as the diamino acid together with glycine. Wall polysaccharides always contain galactose with various amounts of glucose, mannose, arabinose, xylose and deoxyhexose (Szaniszlo and Gooder, 1967). Species have been isolated from soil (Couch, 1963; Willoughby, 1968; Nonomura *et al.*, 1979) and from leaf washings (Willoughby, 1969a). Isolates often give sterile colonies on agar but spore vesicles appear when portions of colonies are transferred to water (Willoughby, 1969b). The recent interest in actinoplanetes as a possible source of new antibiotics has resulted in the description of several new *Ampullariella* species (Nonomura *et al.*, 1979).

The type species is *Ampullariella regularis*, type strain DSM 43151.

3. *Pilimelia*

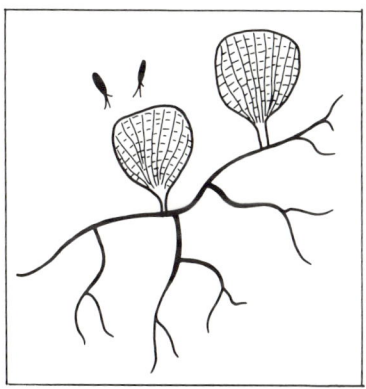

Pilimelia (Kane, 1966) was the generic name proposed for keratinophilic actinomycetes isolated from soil samples by baiting with hair. Spore vesicles are spherical or cylindrical (2–24 μm × 2–35 μm) and contain parallel chains of rod-shaped spores (0.3–0.7 μm × 0.9–1.3 μm). The motile zoospores were originally stated to have a single polar flagellum but un-named species were reported to have one to four sub-polar or lateral flagella (Schäfer, 1973; Hanton, 1974). The spore vesicles are typically seen on hairs but are rarely produced on the soft pale-yellow colonies growing on agar. Cultures in the laboratory soon loose their ability to grow on natural keratin. The wall peptidoglycan contains *meso*-DAP and glycine and no HO-DAP (Szaniszlo and Gooder, 1967) and wall polysaccharides contain minor amounts of galactose together with glucose, mannose, arabinose and xylose.

The type species is *Pilimelia terevasa*, the type strain ATCC 25603.

Keratinophilic actinomycetes with spherical spore vesicles releasing rod-shaped motile spores had been earlier isolated from soil on hair baits by Karling (1954). Similar organisms were isolated in mixed culture on hair baits by Gaertner (1955) and studied by Rothwell (1957) but these authors were unable to isolate their organisms in pure culture. An extensive study by Schäfer (1973) provides more information on the morphology of the spore vesicles and zoospores, and the growth of isolates in pure culture which was made possible by the use of complex media containing skimmed milk and cattle horn meal. He was also able to see the star- or fan-shaped chains of spores radiating from the tips of sporangiophores and lacking a surrounding vesicle that had been seen earlier by Gaertner. *Pilimelia* strains were found to be common in soil (Schäfer, 1973; Tribe and Abu El-Souod, 1979). An unusual species, named '*Pilm. columellifera*' by Schäfer (1973), was recognized by the presence of a characteristically nail-shaped columella projecting into the spore vesicle. It was the most common species encountered by Tribe and Abu El-Souod (1979) but the latter authors suggested that it should be transferred to the genus *Spirillospora* because the spore chains within the spore vesicle were spiral. This suggested transfer cannot be accepted until details of its wall composition are known.

B. *Dactylosporangium* (Thiemann et al., 1967a)

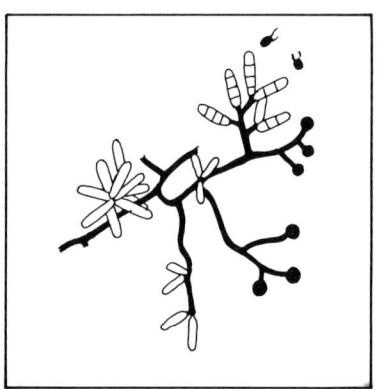

Mesophilic actinomycetes with a fine branching substrate mycelium which does not fragment and may appear twisted or even coiled in agar media. Colonies are tough and leathery and lack aerial mycelium but carry the characteristic clusters of finger-like sporangia (1.0–1.2 μm × 4.0–6.0 μm) containing a single row of three to five oval, pyriform or cylindrical spores furnished with a tuft of polar or peritrichous flagella (Higgins et al., 1967; Thiemann, 1970). The substrate mycelium also produces conspicuous, phase-bright, globose spores (Ensign, 1978) which are more frequently seen in old cultures on nutrient deficient media. They are heat sensitive and germinate when transferred to fresh nutrient solutions: they are probably resistant forms enabling the organism to survive in soil and lake sediments.

Dactylosporangia have a wall chemotype II and contain glycolyl muramic acid (Kawamoto et al., 1981) so resembling *Actinoplanes, Amorphosporangium, Ampullariella* and *Micromonospora*. These genera were also found to be closely related from the results of 16S rRNA oligonucleotide sequence studies (Stackebrandt and Woese, 1981a, b) but it would be premature at this stage to encompass them all within a single genus. Dactylosporangia have a distinctive morphology and can be immediately recognized on primary isolation plates. In contrast to sporulation in *Actinoplanes*, the spores of dactylosporangia are formed endogenously within the parent hypha which acts as a sporangium (Sharples et al., 1974). The resultant sporangia are comparatively regular in shape, release the spores at a defined apical point and retain their integrity after zoospore release.

Strains are widely distributed in soil (Thiemann, 1970) and have been isolated from lake sediments (Johnston and Cross, 1976). Two species were originally described, *Dcso. aurantiacum* and *Dcso. thailandense* (Thiemann *et al.*, 1967a), but subsequent studies on 140 isolates suggested that a spectrum of strains existed which overlapped species boundaries. Predictably, the search for new antibiotics has unearthed another species which was named '*Dcso. matsukiense*' (Shomura *et al.*, 1980).

The type species is *Dactylosporangium aurantiacum*, the type strain ATCC 23491.

c. *Micromonospora* (Ørskov, 1923)

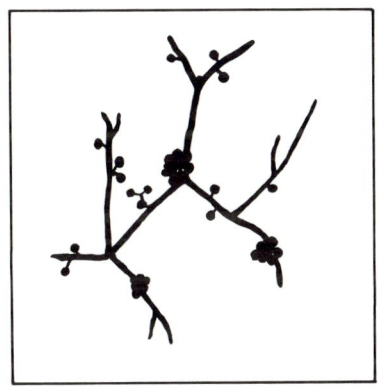

Mesophilic, sporoactinomycetes, with a fine, stable, branching, substrate mycelium (0.2–0.6 μm diameter) which bears the characteristic single spores on short or long sporophores. The non-motile spores often occur in dense clusters within the agar medium or embedded in the colony and, because they are darkly pigmented, eventually blacken the yellow-brown, red-brown, blue-green or purple colonies. Dormant spores show considerable resistance to heat (Suarez *et al.*, 1980) being unaffected at 75°C but sensitive to higher temperatures ($D_{80°C}$ = 12.5 minutes). The spores are also resistant to desiccation and can remain viable in soil or in lake and marine sediments for hundreds of years (Cross and Attwell, 1974). Aerial mycelium is normally absent or restricted to a bloom of sterile hyphae.

The wall peptidoglycan contain *meso*-DAP with a trace of LL-DAP, both *meso*- and 3-hydroxy-DAP or only 3-hydroxy-DAP as the diamino acid. Glycine is always present and appears to replace L-alanine as the first amino acid of the peptide chain attached to muramic acid which is normally glycollated (Kawamoto *et al.*, 1981). Xylose and arabinose were present in all wall preparations though the amounts varied; glucose, galactose, mannose and rhamnose contents differed from species to species and may even vary within strains of the same species (Kawamoto *et al.*, 1981). The characteristic phospholipids are phosphatidylethanolamine, phosphatidylinositol and phosphatidylinositol mannosides (phospholipid type II of Lechevalier *et al.*, 1977). The G +C content of the DNA is within the range 71 to 73 mol%.

The type species is *Micromonospora chalcea*, the type strain ATCC 12452.

The majority of species are aerobic and very common in soil (Cross, 1981a, b) and salt marshes (Hunter *et al.*, 1981), where they are probably active in the degradation of natural polymers such as cellulose, chitin and xylan (Erikson, 1941). Spores are washed from soil into lakes and the sea where they accumulate in sediments (Colmer and McCoy, 1943, 1950; Weyland, 1969, 1981; Johnston and Cross, 1976) but there is still some doubt as to whether they can grow in these habitats. They are relatively easy to isolate from soil or sediments by first heating suspensions at 70°C for 30 minutes and then spreading dilutions on agar media that discourage the outgrowth of *Bacillus* endospores. Alternatively, a selective medium containing novobiocin may be used (Orchard, 1979).

There are three obligately anaerobic species in the genus which can be separated by their ability to utilize cellulose and the nature of their fermentation products. '*Micromonospora acetoformici*' (Sebald and Prévot, 1962) is unable to utilize cellulose but will ferment glucose and starch to give a mixture of acetic, formic and lactic acids; '*Mims. propionici*' (Hungate, 1946) ferments glucose and cellulose to form propionic and acetic acids; and '*Mims. ruminantium*' (Małuszyńska and Janota-Bassalik, 1974) produced lactic and acetic acids from cellulose and cellobiose. The classification of these anaerobic species within the genus *Micromonospora*, previously based largely on morphological criteria, must now be confirmed by an analysis of their wall peptidoglycan and DNA:DNA homologies. '*Micromonospora ruminantium*' was reported to have a DNA base composition of 54.5 ± 1.6 mol % G +C which is much lower than that found in the aerobic members of the genus. The spores at the ends of branched filaments which 'absorbed malachite green less intensively than bacillus endospores' have not been studied with the precision applied to the spores of the aerobic species (Luedemann and Casmer, 1973; Hardisson and Suarez, 1979).

It is possible that some species may be implicated in diseases of man and animals but the descriptions of the isolates and proof that they were definitely responsible for the disease symptoms are very sketchy and inconclusive. *Micromonospora gallica* (Erikson, 1935) was isolated from blood cultures, '*Mims. caballi*' (Morquer and Comby, 1943) from a case of cutaneous actinomycosis in a horse and '*Mims. melanosporea*' (Baldacci and Locci, 1961) from a blood specimen of a woman suffering from mycetoma of the knee. '*Actinomonospora lusitanica*' was the name given to a single-spored organism isolated from cases of mycetoma in Portugal and Italy (Castellani *et al.*, 1959; Mungelluzi, 1966), but there is now considerable doubt that true spores were observed. The wall composition (chemotype III according to Luedeman, 1974) would support the suggestion that the organism was probably a strain of *Actinomadura madurae* (Lechevalier and Lechevalier, 1967) and that the vesicles or swellings on the hyphae were mistaken for spores.

A very unusual organism was recently isolated during the course of successive mutagenic treatments on *Streptomyces peuceticus* var. *caesius* (Grein *et al.*, 1980). The survivor lacked aerial mycelium, formed single and clusters of individual spores on the substrate mycelium, and its peptidoglycan contained *meso*- and hydroxy-DAP with xylose, arabinose, glucose, galactose and ribose – all characteristic of the genus *Micromonospora* and quite unlike the parent strain which had a wall chemotype I. Yet the mutant produced the same anthracycline and polyene antibiotics as the parent strain, so making one possible explanation, that the new strain was simply a contaminant, seem unlikely. The apparent transfiguration of one genus to another by treatment with nitrosoguanidine and heat is hard to accept and requires further study.

7. Actinomycetes With Multilocular Sporangia

This aggregate group contains three genera, *Dermatophilus*, *Frankia* and *Geodermatophilus*. The taxon is essentially morphological in concept as all of the strains it contains produce branching filaments that divide by longitudinal and transverse septa, eventually giving rise to large numbers of coccoid-like elements which may be motile (*Dermatophilus* and *Geodermatophilus*) or non-motile (*Frankia*). Organisms from all three taxa are Gram-positive, non-acid-fast and contain *meso*-DAP as the wall diaminoacid but it is now clear that dermatophili and geodermatophili are not closely related. Dermatophili contain DNA with a much lower $G+C$ content than geodermatophili (Samsonoff *et al.*, 1977) and, unlike the latter, contain madurose (H. A. Lechevalier and Lechevalier, 1970) and have a different polar lipid profile (Hasegawa *et al.*, 1979). Further, in an extensive numerical phenetic survey (Goodfellow and Pirouz, 1982), representative strains of *Dermatophilus* and *Geodermatophilus* were recovered in clusters that were sharply separated both from one another and from phena corresponding to other wall chemotype III taxa. The relationships between *Dermatophilus*, *Frankia* and *Geodermatophilus* need to be unravelled by the application of powerful techniques such as DNA–rRNA pairing and rRNA cataloguing.

A. *Dermatophilus* (Van Sacegham, 1915; Gordon, 1974)

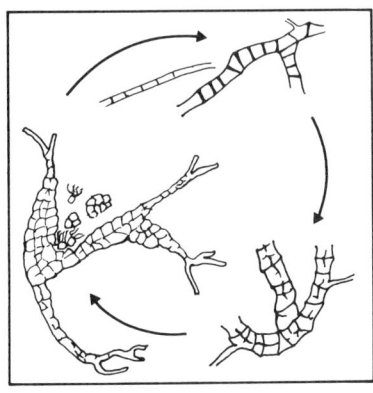

Gram-positive, non-acid-fast, aerobic or facultatively anaerobic actinomycetes that are obligate animal pathogens in nature. At the beginning of the growth cycle filaments less than a micron in diameter are formed. Each filament extends at its growth tip while transverse septation at the other end yields long tapering elements which branch laterally at right angles. The branching substrate filaments (0.5–5.0 μm diameter) are surrounded by a tough gelatinous capsule formed from the disintegrating cell wall. Further septation leads to the production of large numbers of coccoid-like bodies that mature as motile, isodiametric spores (zoospores) within the containing capsule. Branches divide in the same way as the original filament. On laboratory culture media, septa are formed in transverse and in horizontal and vertical longitudinal planes, giving rise to up to eight parallel rows of cells. Zoospores (0.5–1.0 μm) have a tuft of 5 to 50 flagella (8–9 nm diameter) and, after a period of motility, germinate to given filament initials. The capsule and contents can be considered as a primitive multilocular sporangium (Roberts, 1961).

Dermatophili are catalase positive, form acid from glucose and fructose, and degrade casein, elastin, gelatin, starch, tyrosine and urea. Initially colonies are white to grey but later turn orange to yellow. The capsular material is responsible for the moist, mucoid consistency of mature colonies. The peptidoglycan contains *meso*-DAP as the diamino acid and madurose has been detected in whole-organism hydrolysates. Preliminary polar lipid analyses show the presence of phosphatidylglycerol, diphosphatidylglycerol and phosphatidylinositol. The DNA base composition is 57–59 mol% G+C. The organism is pathogenic for mammals, invading only the uncornified epidermis.

The genus is monotypic and the type strain is *Dermatophilus congolensis* ATCC 14637.

Dermatophilus congolensis, the aetiological agent of streptothricosis, was first described by Van Saceghem, in 1915. The organism has never been isolated from any natural source other than diseased tissues of infected animals though infections have been reported from virtually every region of the world, mainly in herbivores including cattle and sheep (see Lloyd and Sellars, 1976; Hyslop, 1980). There have been a few cases of infection in man following the handling of cultures or infected animals. Active cases of the disease are probably required to initiate new infections given the poor ability of *Derm. congolensis* to survive in soil, or even in infected scab (Roberts, 1967). Zoospores that develop in large numbers in infected scab are the agents of infection. When the scab is wetted, the zoospores migrate to the surface and can be transmitted from one animal to another through contact with plants or insects. Experimental inoculations of abraded skin of guinea pigs, mice and other mammals results in acute ulcerative dermatitis within a week (Gordon and Perrin, 1971).

Dermatophili are exacting in their growth requirements but can readily be cultivated at 37°C on enriched media such as brain–heart infusion–horse blood agar on which they exhibit β-haemolysis. In an atmosphere of 10% carbon dioxide at 37°C growth is enhanced, aerial hyphae may be formed, but septation and spore formation are delayed. Several workers have given detailed accounts of the distinctive life cycle and properties of the organism (Roberts, 1970, 1981; Gordon, 1974, 1976; Hyslop, 1980; Goodfellow and Pirouz, 1982) which is susceptible to a wide range of antibiotics, including penicillin, streptomycin, erythromycin and the tetracyclines (Roberts, 1967). Strains from various hosts are relatively homogeneous in their serological properties (Richard *et al.*, 1976).

B. *Frankia* (Brunchorst, 1886)

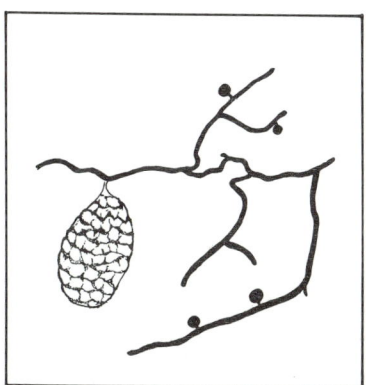

This genus was proposed by Brunchorst in 1886 and was redefined by Becking (1970, 1974) for symbiotic actinomycetes which lived in root nodules of non-leguminous plants. The organism was considered to be an obligate symbiont capable of fixing atmospheric nitrogen in root nodules. The root-nodule symbiosis occurs in seven orders and eight families of non-leguminous woody dicotyledonous plants (Becking, 1981). Ten species of *Frankia* were recognized on the basis of specific cross inoculation groups. Two morphologically distinct types of root nodules are formed. In *Casuarina* and *Myrica*, root nodules carry masses of upward growing rootlets whereas in species belonging to the families *Betulaceae*, *Coriariaceae*, *Datiscacieae*, *Elaeagnaceae*, *Rhamnaceae* and *Rosaceae*, nodules are formed from modified dichotomous branched roots and have a coralloid appearance. The endophyte is usually restricted to the cortical parenchyma of nodules and to nodular tissue just below the meristem where it produces new infections on young host cells.

Until recently, comparatively little was known about *Frankia* as attempts to isolate and grow the organism on various complex media were unsuccessful (Baker and Torrey, 1970). Pommer isolated *Frankia* from nodules of *Alnus glutinosa* (European black alder) in 1959 but his report aroused little interest. In 1978, Callaham *et al.* isolated a similar strain of *Frankia* from nodules of sweet fern (*Comptonia peregrina*). This strain grew well under microaerophilic but not anaerobic conditions, and produced effective nodules when the original host plant was reinoculated. In yeast extract and modified *Leptospira* broths it formed a branched septate mycelium and sporangenous bodies with transverse and longitudinal septation. Large numbers of ovoid or polyhedral spores (1.5–3.5 μm diameter) were formed in the 'sporangia'. The club-shaped vesicles, typical of effective nodules, were formed by an isolated strain in pure culture when grown on a nitrogen free medium (Tjepkema *et al.*, 1980). Nitrogenase activity was observed only in cultures with vesicles and was correlated with the number of vesicles. Since the seminal isolation studies frankiae have been isolated from several plant species (see Chapter 11) and their actinomycete nature confirmed. It has, however, been emphasized that the genus should not necessarily be restricted to endophytes but might include any free-living actinomycetes which may have the ability to produce non-motile spore-containing 'sporangia' in liquid culture (Lechevalier and Lechevalier, 1979).

In the host plant, frankiae form branching hyphae (0.3–2.8 μm diameter) which penetrate the cells. Depending on the host plant, the endophyte may also form spherical or club-shaped terminal vesicles (3–5 μm diameter) and 'sporangia' which may reach 25 × 35 μm. There is evidence that the host plants affect the morphology of the endophyte *in vivo* and that the incompatibility barriers between the representatives of the various plant taxa are not as strict as was previously thought (Becking, 1981). There is some evidence that the endophyte is taxonomically heterogeneous though all strains examined to date have a DNA base composition within the narrow range of 68–72 mol% G+C (An *et al.*, 1983). Frankiae contain *meso*-DAP as the diamino-acid of the wall peptidoglycan but, while strains from *Alnus rubra* and *Comptonia peregrina* contained xylose in their whole-organism hydrolysates, an *Elaeagnus* strain possessed a hexose sugar tentatively identified as fucose (Lechevalier and Lechevalier, 1979). Whole-organism hydrolysates of all three strains, however, contained an unknown polar lipid which stained blue with ninhydrin. In comparative immunodiffusion studies, frankiae were assigned to two serotypes (Baker *et al.*, 1981), the first included organisms isolated from root nodules of *Alnus*, *Comptonia* and *Myrica*, the second strains from nodules of *Elaeagnus*. It seems probable that additional centres of variation will be distinguished as more extensive comparative studies are carried out. A natural classification of *Frankia* must await the outcome of such comprehensive studies for we must agree with Lechevalier (1979) that speciation based on the host plant is unacceptable. At present, the genus contains only one species *Frankia alni*, a good representative of which is ATCC 33029.

c. *Geodermatophilus* (Luedemann, 1968)

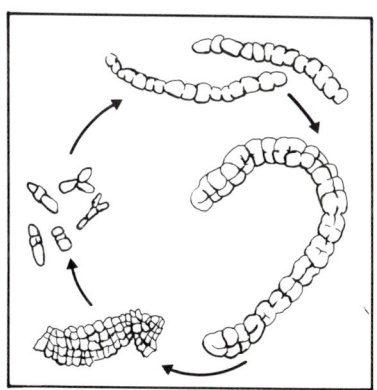

Gram-positive, non-acid-fast, aerobic actinomycetes that form a muriform, tuber-shaped, noncapsulated, holocarpic thallus containing masses of cuboid cells 0.5–2.0 μm in diameter. Under favourable conditions the thallus breaks up releasing cuboid or coccoid non-motile cells and elliptical to lanceolate zoospores. The latter have a tuft of one to four flagella and often retain their motility after giving rise to germ tubers or buds. A prominent feature of geodermatophili is their ability to form buds on both zoospores and young thalli. Germinating spores or resting zoospores may divide directly to form a new thallus or produce a germ tube and an irregular contracted filament which may branch at various angles. The organism, however, only forms a rudimentary substrate mycelium and lacks aerial hyphae. The pseudohyphal nature of the organization has been confirmed by electron microscopy (da Fonseca and Edwards, 1971).

Geodermatophili are catalase positive, hydrolyse starch but are not haemolytic. Colonies usually turn dark-grey, dark-brown or black in old cultures. The wall peptidoglycan contains *meso*-DAP as the diamino acid but whole-organism hydrolysates do not contain madurose. In preliminary polar lipid analyses phosphatidylglycerol, diphosphatidyglycerol, phosphatidylethanolamine, phosphatidylinositol and phosphatidylinositol mannosides were detected. The DNA base composition is 73–75 mol% G + C. The organism is found in soil.

The genus is monotypic and the type strain is *Geodermatophilus obscurus* ATCC 25078.

The genus *Geodermatophilus* was established by Luedemann (1968) for novel isolates from desert soils in the western United States. Strains have also been isolated from the Baltic Sea (Ahrens and Moll, 1970) and from soils collected at high altitudes on Mount Everest (Ishiguro and Wolfe, 1970). Surprisingly, *Geodermatophilus* has received little attention but was sharply separated from all other chemotype III taxa in the numerical phenetic survey of Goodfellow and Pirouz (1982).

8. Nocardioform Actinomycetes

This aggregate group includes organisms classified in the genera *Caseobacter, Corynebacterium, Mycobacterium, Nocardia, Rhodococcus* and the '*aurantiaca*' taxon. Strains in these taxa have many properties in common (Barksdale, 1970, 1981; Bradley and Bond, 1974; Barksdale and Kim, 1977; Goodfellow and Minnikin, 1977, 1981c, 1983; Goodfellow and Wayne, 1982) and there is some evidence that the aggregate taxon forms a recognizable phylogenetic branch (Fig. 1; Stackebrandt and Woese, 1981a, b; Stackebrandt *et al.*, 1982; Mordarski *et al.*, 1980a, b). In addition, corynebacteria, mycobacteria, nocardiae, rhodococci and aurantiaca strains have been recovered as discrete and recognizable phena in numerical phenetic surveys (Goodfellow *et al.*, 1982a, b).

Nocardioform actinomycetes are aerobic, Gram-positive, catalase-positive actinomycetes that show considerable morphological diversity. Caseobacters, corynebacteria, aurantiaca strains, many rhodococci and most mycobacteria are more or less amycelial whereas nocardiae, some rhodococci and a few mycobacteria typically form a branched mycelium which sooner or later fragments into bacillary and coccoid elements that may be markedly pleomorphic. Nocardiae and some mycobacteria produce aerial hyphae which range from being sparse and invisible to the naked eye to completely covering the substrate mycelium with a white down. The aerial hyphae of some nocardiae may differentiate into short chains of arthrospores. Nocardioform actinomycetes are, however, most easily recognized and best defined by chemical properties.

They contain *meso*-DAP as the diamino acid of the wall peptidoglycan, arabinose and galactose as major wall sugars, have mycolic acids, major amounts of straight-chain saturated and monounsaturated fatty acids, and diphosphatidylglycerol, phosphatidylinositol and phosphatidylinositol mannosides as major phospholipids (Table 2; Minnikin and Goodfellow, 1980). Glycolipids have not been systematically investigated but representative strains of *Corynebacterium*, *Mycobacterium*, *Nocardia* and *Rhodococcus* contain 6,6'-dimycolic esters of trehalose, the so-called cord factors. Representatives of these taxa have also been shown to be closely related using a variety of serological techniques. The most comprehensive serological studies have employed immunodiffusion techniques and common precipitinogens have been detected among corynebacteria, mycobacteria, nocardiae and rhodococci (Lind and Ridell, 1976, 1983).

The term nocardioform was introduced to describe a life-cycle considered to have evolutionary implications (Prauser, 1970, 1976a, 1978, 1981) but the label also provided a means of focussing attention on a relatively poorly studied group of actinomycetes (Goodfellow and Minnikin, 1977, 1978, 1981c). The application of modern taxonomic methods has revolutionized our understanding of the composition and relationships of nocardioform taxa and it is now clear that the latter fall into more than one phylogenetic group (Fig. 1). It is timely to restrict the term nocardioform actinomycetes to *Nocardia* and related taxa that have a wall chemotype IV and contain mycolic acids.

A. *Caseobacter* (Crombach, 1978)

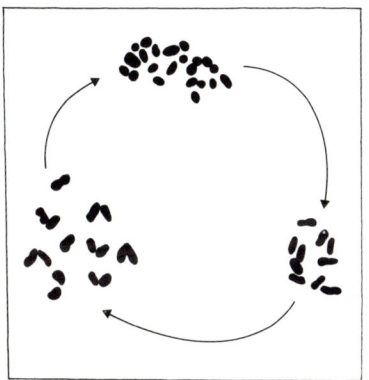

Aerobic, Gram-positive, non-motile, asporogenous actinomycetes which show a change of form during the growth cycle. Stationary phase cultures are composed of ovoid or coccoid cells which give rise to regular rods (0.8–1.2×2.0–4.0 μm) when transferred to fresh complex media. Cells occur singly, in pairs, and in clusters; palisade formation and V-forms may occur. Small grey-white, slightly pink or red, colonies are formed on yeast extract glucose agar. The organism can tolerate high concentrations of salt (up to 12%, w/v).

The wall peptidoglycan contains alanine, glutamic acid and *meso*-DAP; arabinose, galactose and mannose are the characteristic wall sugars. Caseobacters also have major amounts of straight-chain saturated and unsaturated fatty acids and tuberculostearic acid, dihydrogenated menaquinones with nine isoprene units, MK-$9(H_2)$, as the major isoprenologue and mycolic acids with 30 to 36 carbons. The fatty acid esters released on pyrolysis gas chromatography of mycolic esters consist of 14 to 18 carbon atoms. The G+C content of the DNA lies within the limit of 65–67 mol%. The organisms are an important component of the surface flora of both hard and soft cheeses.

The genus is monotypic, the type strain *Caseobacter polymorphus* NCDO 2097.

The genus *Caseobacter* was introduced by Crombach (1978) for certain 'grey-white' or 'non-orange' cheese coryneform bacteria first described by Mulder and Antheunisse (1963). The 'grey-white' isolates were subsequently found to be heterogeneous in wall composition (Keddie and Cure, 1977) and only those with a wall chemotype IV were included in the genus *Caseobacter*. Like corynebacteria, caseobacters contain short-chain mycolic acids (Collins *et al.*, 1982b) but can be distinguished from the latter on the basis of DNA base and fatty acid composition (Collins *et al.*, 1982c). They form a distinct DNA homology group and have relatively little DNA in common with *Arthrobacter*, *Brevibacterium*, *Corynebacterium* or *Rhodococcus equi* (Crombach, 1972, 1974; Suzuki *et al.*, 1981).

B. *Corynebacterium* (Lehmann and Neumann, 1896)

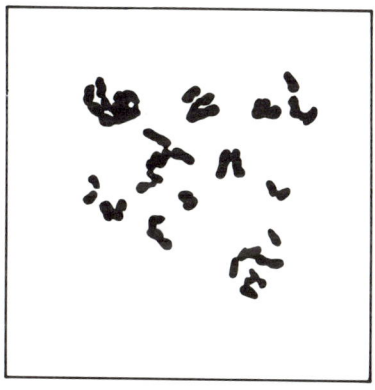

Gram-positive, non-acid-fast, asporogenous actinomycetes that form straight to slightly curved rods with uneven stained segments, and sometimes granules. Snapping division produces angular and palisade arrangements of cells which are frequently swollen at one or both ends. The organism is catalase positive, aerobic and facultatively anaerobic but grows under aerobic conditions. Carbohydrate metabolism is both oxidative and fermentative. Some pathogenic species produce exotoxins.

The wall peptidoglycan contains N-acetylated muramic acid, *meso*-DAP as the diamino acid, and is of the A1γ type. The polysaccharide moeity of the wall contains arabinose, galactose and usually mannose. Corynebacteria have major amounts of saturated and unsaturated fatty acids, dihydrogenated menaquinones with either eight or nine isoprene units, MK-$8(H_2)$, $9(H_2)$ as the major isoprenologue, diphosphatidylglycerol, phosphatidylinositol and phosphatidylinositol mannosides, and relatively low-molecular-weight mycolic acids with approximately 22 to 36 carbon atoms and up to two double bonds. The fatty acid esters released on pyrolysis gas chromatography of mycolic esters consist of 8 to 18 carbon atoms. The G+C content of the DNA is generally within the range 51–59 mol%. The organism is probably widely distributed in nature. Some strains are important pathogens of man and animals.

The type species is *Corynebacterium diphtheriae*, the type strain ATCC 27010.

The genus *Corynebacterium* has a long and involved taxonomic pedigree which has been the subject of detailed review (Barksdale, 1970, 1981; Goodfellow and Minnikin, 1981a). The taxon was introduced by Lehmann and Neumann (1896) to accommodate the diphtheria bacillus and closely related organisms but it was soon to encompass other bacteria, such as *Cnbc. pseudodiphtheriticum*, *Cnbc. pseudotuberculosis*, *Cnbc. pyogenes*, *Cnbc. renale* and *Cnbc. xerosis*, which were associated with, or caused, disease in animals. In subsequent years the genus became a repository not only for the animal parasitic species but also for a heterogeneous collection of morphologically similar plant pathogenic species and saprophytic species from a wide range of habitats.

Indeed, it was only with the introduction and application of modern taxonomic methods, especially wall and lipid analysis, that it proved possible to bring some order into the classification of *Corynebacterium* and related genera. It is now abundantly clear that the name *Corynebacterium* should be reserved for *Cnbc. diphtheriae*, a number of animal pathogenic and parasitic species, and several saprophytic species, and that the revised taxon has many properties in common with *Mycobacterium*, *Nocardia* and *Rhodococcus* (Barksdale, 1970, 1981; Schleifer and Kandler, 1972; Jones, 1978; Minnikin *et al.*, 1978a; Keddie and Jones, 1981; Goodfellow and Minnikin, 1981a, 1982; Goodfellow *et al.*, 1982b).

A large number of species labelled *Corynebacterium* are excluded from the redefined genus. They include the animal pathogenic species '*Cnbc.*' *haemolyticum* and '*Cnbc.*' *pyogenes* mentioned previously, '*Cnbc.*' *equi*, which has been reclassified in the genus *Rhodococcus* as *Rodc. equi* (Goodfellow and Alderson, 1977), and all of the plant pathogens, many of which have been assigned to the genera *Arthrobacter*, *Curtobacterium* and *Rhodococcus* (Keddie and Cure, 1977, 1978; Minnikin *et al.*, 1978a; Collins *et al.*, 1981b). Saprophytic taxa similarly excluded include '*Cnbc. autotrophicum*', '*Cnbc.* beticola' and '*Cnbc. nephridii*' which have the properties of Gram-negative bacteria (Collins and Jones, 1981a, 1982b), '*Cnbc. alkanum*', '*Cnbc. laevaniformis*' and '*Cnbc. manihot*' which contain lysine as the diamino acid in the wall peptidoglycan; '*Cnbc. barkeri*' which contains ornithine as the wall diamino acid, '*Cnbc. aquaticum*', '*Cnbc. mediolanum*' and '*Cnbc. okanaganae*' which possess 2,4-DAB as wall diamino acid and '*Cnbc.*' *hoagii*, '*Cnbc. hydrocarboclastus*' and '*Cnbc. rubrum*' which have properties consistent with their transfer to the genus *Rhodococcus* (Collins *et al.*, 1982b, c; Goodfellow *et al.*, 1982b). *Corynebacterium paurometabolum* should be transferred to the '*aurantiaca*' taxon (Collins and Jones, 1982c).

A number of saprophytic species currently classified in the genera *Arthrobacter*, *Brevibacterium* and *Microbacterium* conform to the redescribed genus *Corynebacterium*. Thus, the glutamic acid producing strains labelled '*Atbc. albidus*', *Brev. ammoniagenes*, *Brev. divaricatum*, '*Brev. flavum*', '*Brev. immariophilum*', *Brev. lactofermentum*', '*Brev. roseum*' and '*Miba. ammoniophilum*' should probably all be reduced to synonyms of *Cnbc. glutamicum* together with '*Cnbc. acetoacidophilum*' *Cnbc. callunae*, '*Cnbc. herculis*', *Cnbc. lilium* and '*Cnbc. melassecola*' (Abe *et al.*, 1967; Suzuki *et al.*, 1981; Collins *et al.*, 1982b). In addition, *Bacterionema matruchotii*, *Miba. flavum*, *Brev. liquefaciens* and *Brev. vitarumen* have been transferred to the genus *Corynebacterium* as *Cnbc. matruchotii* (Collins, 1982c; Goodfellow *et al.*, 1982b), *Cnbc. flavescens* (Barksdale *et al.*, 1979), *Cnbc. liquefaciens* and *Cnbc. vitarumen* (Lanéelle *et al.*, 1980), respectively.

Lipid markers may also be of value in the subgeneric classification of the genus *Corynebacterium*. Species can be divided into broad groups on the basis of menaquinone patterns (Collins and Jones, 1981a). Thus, most of the animal-associated strains (e.g. *Cnbc. diphtheriae, Cnbc. kutscheri, Cnbc. pseudotuberculosis* and *Cnbc. renale*) contain major amounts of dihydrogenated menaquinones with eight isoprene units whereas *Cnbc. bovis, Cnbc. glutamicum* and *Cnbc. matruchotii* have dihydrogenated menaquinones with nine isoprene units as the predominant isoprenologue. Variation also exists in the size of the mycolic acids (Collins *et al.*, 1982b). Two species, *Cnbc. bovis* and '*Cnbc. mycetoides*' are especially distinctive, the former contains exceptionally low-molecular-weight mycolic acids (22 to 32 carbon atoms), the mycolic esters of which yield fatty acid esters with 8 to 10 carbon atoms on pyrolysis gas chromatography, whereas '*Cnbc. mycetoides*' has major amounts of mycolates having a side chain containing odd numbers of carbons. Most of the remaining species, including *Cnbc. diphtheriae* and *Cnbc. glutamicum*, have mycolic acids with chain lengths between 26 and 36 carbon atoms and saturated side chains. *Corynebacterium bovis* and '*Cnbc. mycetoides*' can also be distinguished from other species of *Corynebacterium* on the basis of their non-hydroxylated fatty acid profiles (Collins *et al.*, 1982c). The profile of *Cnbc. bovis* is especially distinctive for, unlike those of all other corynebacteria, it contains tuberculostearic acid.

Little information exists on the numbers and distribution of corynebacteria in natural habitats. *Corynebacterium glutamicum* has been isolated from soil, animal and plant matter (Abe *et al.*, 1967), *Cnbc. matruchotii* forms part of the oral flora (Bowden and Hardie, 1978), and isolates with properties of corynebacteria have been reported from sea-water (Bousfield, 1978) and human skin (Pitcher and Noble, 1978). All saprophytic corynebacteria studied to date are nutritionally fastidious: *Cnbc. glutamicum* requires biotin and sometimes B vitamins, *Cnbc. flavescens* needs amino nitrogen, biotin and pantothenic acid (Keddie and Jones, 1981), and *Cnbc. matruchotii* grows well on brain–heart infusion agar in an atmosphere of 5% carbon dioxide and 95% nitrogen. *Corynebacterium diphtheriae*, the causative agent of diphtheria, is found in the upper respiratory tract and vagina, and can be cultivated on casein hydrolysate medium supplemented with glutamate, pantothenate and tryptophan (Barksdale, 1970).

C. *Mycobacterium* (Lehmann and Neumann, 1896)

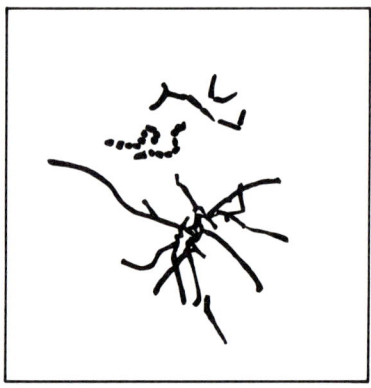

Aerobic, acid-alcohol-fast actinomycetes that usually form slightly curved or straight rods (0.2–0.6 × 1.0–10 μm) which occasionally branch. Extensively branched filaments may occur but they readily fragment into rods and coccoid elements. Aerial hyphae are normally absent. Strains are non-motile, do not form endospores, spores or capsules, produce acid from sugars oxidatively, are usually considered to be Gram-positive but are not readily stained by Gram's method.

The wall contains *meso*-DAP, arabinose and galactose and the peptidoglycan is of the A1γ type, the muramic acid residues are N-glycollated. Mycobacteria are rich in lipids. They contain major amounts of straight-chain saturated and unsaturated fatty acids and tuberculostearic acid, diphosphatidylglycerol, phosphatidylethanolamine, phosphatidylinositol and phosphatidylinositol mannosides, dihydrogenated menaquinones with nine isoprene units, MK-9(H_2), as the major isoprenologue and have a G+C ratio within the range 62–70 mol%. Mycobacterial mycolic acids are complex, have a high molecular weight (60 to 90 carbons), lack components with more than two points of unsaturation in the molecule and, on pyrolysis, release C_{22} to C_{26} straight-chain saturated acids. Most mycobacteria are saprophytes, some are occasional pathogens while others cause severe disease in man and animals.

The type species is *Mycobacterium tuberculosis*, the type strain ATCC 27294.

Early developments in mycobacterial systematics were hampered by the overreliance placed on a few morphological and staining properties and by the tendency to measure mycobacteria solely in terms of their relationships with *Mybc. tuberculosis* (Goodfellow and Minnikin, 1983). The dominance of *Mybc. tuberculosis* was reflected in the common use of the term 'atypical mycobacteria' for strains that could not be identified as *Mybc. tuberculosis* or *Mybc. bovis*. Indeed, it was not until the mid-fifties that a much needed systematic base was provided for the comparison of mycobacteria (Timpe and Runyon, 1954; Runyon, 1958, 1959). In Runyon's classification, mycobacteria, apart from those in the *Mybc. tuberculosis* complex and the non-cultivatable taxa, were divided into four overtly artificial groups based on growth rates and pigmentation properties. In recent years, Runyon's groups have been superceded by a more natural classification derived from the application of modern taxonomic techniques to mycobacterial systematics (Barksdale and Kim, 1977; Goodfellow and Minnikin, 1980, 1983; Goodfellow and Wayne, 1982). The collaborative surveys carried out under the *aegis* of the *International Working Group on Mycobacterial Taxonomy* (IWGMT) form an important landmark in the taxonomy of mycobacteria as over five hundred strains were screened for around one hundred thousand pieces of information (Wayne, 1978). These co-operative studies not only helped to provide comprehensive descriptions of both established and recently described species but also highlighted tests of value for the identification of fresh isolates (Kubica and Good, 1981; Wayne *et al.*, 1974, 1976, 1980, 1981).

Although the genus *Mycobacterium* is not formally divided into subgenera, it is almost essential to consider it in terms of two major categories. The so-called rapid-growers encompass strains of those species which, under optimal conditions of nutrition and temperature, yield, from dilute inocula, colonies on solid medium that are visible to the naked eye within 7 days. In contrast, the slow-growers require 7 days or more to yield visible colonies under comparable conditions. *Mycobacterium leprae* has not been grown *in vitro* but can be cultivated in mouse footpads (Shepard, 1960) or in the nine-banded armadillo (Kirchheimer and Storrs, 1971). Tsukamura (1967) proposed that the rapid growers be classified in a separate genus, *Mycomycobacterium*, but this proposal has not received much support. Indeed, measurements of DNA homology (Gross and Wayne, 1970; Bradley, 1973; Baess and Weis Bentzon, 1978), lipid analyses (Minnikin and Goodfellow, 1980; Minnikin *et al.*, 1980), immunodiffusion and immunoelectrophoretic studies (Stanford and Grange, 1974; Chaparas *et al.*, 1978a, b) and bacteriophage typing (Redmond *et al.*, 1979) have all given data that run counter to this proposal.

1. Slow-Growing Mycobacteria

The current taxonomy of the slow growers has developed from a nucleus of taxonomic analyses, especially those performed under the wing of the IWGMT. These analyses rested on large pools of data derived from a plethora of cultural, biochemical and physiological tests, many of which were designed or adapted for use with these 'difficult' organisms. Initially, the numerical phenetic studies were based on Runyon's Groups, three of which were composed of slow growers distinguished from one another on the grounds of pigment production and the influence of light thereon (Runyon, 1959), i.e. photochromogens (Group I), scotochromogens (Group II) and non-photochromogens (Group III). The international collaborators examined large sets of selected strains, analysed their pooled data numerically, and evaluated the numerical classifications obtained in the light of independent criteria derived from chemical, immunological, phage susceptibility and experimental pathogenicity studies (Wayne et al., 1971, 1978, 1981; Meissner et al., 1974). On the basis of these and similar studies the species containing slow-growing strains can be assigned to a number of aggregate taxa (Table 7).

The genus *Mycobacterium* was introduced by Lehmann and Neumann (1896) to accommodate the tubercle and leprosy bacilli, organisms that had previously been classified as '*Bacterium tuberculosis*' and '*Bacterium leprae*', respectively. The term 'tubercle bacilli' was subsequently applied to bacteria currently classified as *Mybc. tuberculosis*, *Mybc. avium* and *Mybc. bovis*. Classifications derived from the application of modern taxonomic methods unequivocally place *Mybc. avium* in a position quite distinct from the mammalian tubercle bacilli (Gross and Wayne, 1970; Wayne and Diaz, 1979). There is also strong evidence for believing that the species presently recognized in the *Mybc. tuberculosis* complex (Table 7) merely reflect bias in test selection. Thus, studies on the conservation of amino-acid sequences in bacterial catalase, lipid composition and immunoelectrophoretic analyses of shared antigens show that *Mybc. bovis* and *Mybc. tuberculosis* should be treated as subspecies of a single species (Wayne and Diaz, 1976, 1979; Minnikin and Goodfellow, 1980). Further, *Mybc. africanum* and *Mybc. microti* appear to represent intermediate forms between *Mybc. bovis* and *Mybc. tuberculosis* making the distinction between the latter even more questionable.

Members of the *Mybc. tuberculosis* complex can readily be distinguished from other species of *Mycobacterium* including those assigned to the *Mybc. avium* complex (Table 7). The distinction between *Mybc. avium* and *Mybc. intracellulare* is, however, a tenuous (Meissner et al., 1974; Wayne and Diaz, 1979), if not artificial, one based on agglutination serovars, with three serovars ascribed to *Mybc. avium* and over 20 additional ones assigned to *Mybc. intracellulare* (Schaefer, 1979) which is often associated with disease in man. Serovar 2 of *Mybc. avium* is usually associated with strains that have a higher virulence for birds than any of the other serovars but the virulence can readily be lost on subculture (Meissner et al., 1974). *Mycobacterium scrofulaceum* is often included with *Mybc. avium* and *Mybc. intracellulare* in the MAIS (*Mybc. avium–intracellulare–scrofulaceum*) complex because of taxonomic and disease relationships. *Mycobacterium xenopi* also has properties in common with *Mybc. avium* and *Mybc. intracellulare* but it exhibits a characteristic filamentous colony morphology, a distinct serovar and is unique amongst mycobacteria as it grows poorly at 37°C, preferring 42 to 45°C.

TABLE 7. Species and pathogenicity of slow growing mycobacteria*

Aggregate taxon	*Mybc.* species†	Patho-genicity	Aggregate taxon	*Mybc.* species†	Patho-genicity
Mybc. avium complex	*avium*	2	*Mybc. tuberculosis* complex	*africanum*	1
	intracellulare	2		*bovis*	1
	xenopi	3		*microti*	1
Mybc. gordonae complex	*asiaticum*	3		*tuberculosis*	1
	gordonae	5	Species with special growth requirements	*haemophilum*	2
	szulgae	2		*leprae*	1
Mybc. kansasii complex	*gastri*	5		*lepraemurium*	1
	kansasii	2		*paratuberculosis*	1
Mybc. scrofulaceum complex	*scrofulaceum*	2	Species with no special growth requirements	*farcinogenes*	1
	simiae	3		*malmoense*	2
Mybc. terrae complex	*nonchromogenicum*	5		*marinum*	2
	terrae	5		*ulcerans*	1
	triviale	5			

* 1, obligate pathogens; 2, usually as pathogens; 3, commonly as non-pathogens; 4, usually as non-pathogens; 5, have never been recorded as pathogens.
† Author citations can be found in Skerman *et al.* (1980) or Goodfellow and Wayne (1982).

Mycobacterium scrofulaceum and *Mybc. kansasii* are typically associated with clinical disease but contain strains that can be distinguished from one another and from other slow-growing mycobacteria. Thus, *Mybc. scrofulaceum* can be separated from members of the *Mybc. avium* complex using taxometric techniques (Wayne *et al.*, 1971, 1980, 1981; Meissner *et al.*, 1974) and by a number of immunologically based tests (Magnusson, 1967; Stanford and Grange, 1974; Wayne and Diaz, 1976, 1979). *Mycobacterium simiae*, originally isolated from monkeys (Karassova *et al.*, 1965), is a photochromogenic organism which is able to cause lesions in mice and pulmonary disease in man. The organism has little in common with the *Mybc. avium–intracellulare* complex or with the photochromogenic species *Mybc. kansasii* and *Mybc. marinum* (Baess and Magnusson, 1982). These workers also found *Mybc. simiae* serovar 1 and '*Mybc. habana*' to be identical. The integrity of *Mybc. kansasii* is supported by numerical phenetic analysis, antigenic composition, sensitin specificity and phage typing (Wayne *et al.*, 1978). Immunodiffusion studies indicate a close relationship between *Mybc. kansasii* and *Mybc. gastri* (Norlin *et al.*, 1969; Stanford and Grange, 1974) but strains in these species can be distinguished by numerical taxonomy, phage typing, sensitin and seroagglutination tests (Magnusson, 1971; Wayne *et al.*, 1978).

Non-pigmented slowly growing mycobacteria assigned to the *Mybc. terrae* complex (Table 7) can readily be distinguished from *Mybc. avium* and related strains by their very rapid hydrolysis of Tween 80 and strong catalase reaction. These organisms have not been implicated in human disease but are commonly encountered in sputum, presumably being derived from water and dust. *Mycobacterium marinum* grows well between 25 and 35°C but shows little or no growth at 37°C. It is found in aquatic habitats, causes disease in fish and nodular skin lesions in man (Schaefer, 1979), and forms a discrete numerical taxonomic cluster and a unique agglutinating serovar (Wayne *et al.*, 1978). *Mycobacterium ulcerans* also has a very restricted growth-range temperature and needs 30°C for primary isolation. It is also associated with skin lesions, but causes a more progressive malignant disease than *Mybc. marinum* and, unlike the latter, is found in tropical as opposed to temperate regions. A novel group of non-pigmented slow-growing mycobacteria have been isolated from a variety of natural habitats in Zaire (Portaels, 1980; Portaels *et al.*, 1982). These organisms produce a primary mycelium covered by abundant aerial hyphae but, unlike *Mybc. ulcerans*, they do not cause local swellings or ulceration when inoculated into footpads of mice. *Mycobacterium farcinogenes* also produces an extensive primary mycelium but strains of this taxon can be distinguished by their ability to form lesions of the lymphatic system or the parenchyma of African cattle, their positive malonamidase reaction and distinctive pathogenicity for guinea pigs (Chamoiseau, 1979). Mycolic acid and comparative immunodiffusion studies indicate a close relationship between *Mybc. farcinogenes* and the fast-growing taxa *Mybc. fortuitum* and *Mybc. senegalense* (Ridell *et al.*, 1979, 1982; Ridell, 1983). However, in a numerical phenetic survey representative strains of *Mybc. farcinogenes* formed a homogeneous phenon well separated from clusters corresponding to *Mybc. fortuitum* and *Mybc. senegalense* (Ridell and Goodfellow, 1983). DNA:DNA pairing data support the integrity of *Mybc. farcinogenes*, *Mybc. fortuitum* and *Mybc. senegalense* but also show that these species have more DNA in common with one another than with other species of the genus (Baess, 1982).

Mycobacterium gordonae encompasses a common group of saprophytic mycobacteria which are ubiquitous in tap water. Taxometric (Wayne *et al.*, 1971, 1981), immunodiffusion (Stanford and Grange, 1974) and lipid (Wayne *et al.*, 1974) analyses all support the status of this taxon. Much less is known about the two remaining species in the *Mybc. gordonae* complex. *Mycobacterium szulgai* is a cause of pulmonary disease in human beings (Medinger and Spagnolo, 1981) and can be recognized by its distinctive seroagglutination and unique TLC lipid pattern (Marks *et al.*, 1972), whereas *Mybc. asiaticum* is a recognizable but poorly described species which appears to be closely related to *Mybc. gordonae* (Wayne *et al.*, 1982).

Comparative systematic studies are also needed to determine the detailed relationships of *Mybc. haemophilum*, *Mybc. lepraemurium*, *Mybc. malmoense*, *Mybc. shimoidei* and *Mybc. paratuberculosis*. The latter can only be grown in media supplemented with mycobactins (Matthews *et al.*, 1978) and is the aetiological agent of Johne's disease, a chronic enteritis of ruminants. *Mycobacterium paratuberculosis* appears to be related to strains of the *Mybc. avium* complex, especially *Mybc. avium* (Thorel and Valette, 1976). *Mycobacterium haemophilum* was proposed for a strain isolated from granulomatous skin lesions of a patient with Hodgkin's disease and is unique amongst slow-growing mycobacteria in its dependence on haemin (Sompolinsky *et al.*, 1978). Strains identified as *Mybc. haemophilum* have been found to have a growth requirement for ferric ammonium citrate (Dawson and Jennis, 1980). The name *Mybc. shimoidei* has been revived by Tsukamura (1982) for novel isolates that caused a lung infection in man. *Mycobacterium lepraemurium* can be grown on 1% egg yolk medium and has been distinguished from *Mybc. avium*, *Mybc. intracellulare* and *Mybc. tuberculosis* in a numerical phenetic survey (Saito *et al.*, 1977a). However, the very poor growth of *Mybc. lepraemurium* and its slow metabolic activity make numerical taxonomy less than an ideal tool for establishing the relationships of this taxon.

The affinities of *Mybc. leprae* are even less clear than those of *Mybc. lepraemurium* though the ability to obtain relatively large amounts of bacilli from armadillos has facilitated studies on this organism. *Mycobacterium leprae* can be considered to be a *bona fide* member of the genus *Mycobacterium* and has a unique mycolic acid pattern (Draper *et al.*, 1982). The organism contains two mycolic acid types (Draper, 1976; Stanford *et al.*, 1977) which have been identified as an 'α-*mycolate*' with two *cis*-cyclopropane rings and a ketomycolate with an 83 carbon main component (Etémadi and Convit, 1974; Asselineau *et al.*, 1981; Draper *et al.*, 1982). Etémadi (1967) considered ketomycolates to be biosynthetic precursors of either methoxymycolates or ω-carboxymycolates and, in all other mycobacteria examined to date, ketomycolates co-occur with one of these two types (Minnikin and Goodfellow, 1980). Thus, the inability of *Mybc. leprae* to form other oxygenated mycolic acids from ketomycolates distinguishes leprosy bacilli from all other mycobacteria.

2. Fast-Growing Mycobacteria

Rapidly growing mycobacteria are common saprophytes in natural habitats but despite their ecological importance they have received less attention than their slow growing, more clinically significant, relatives. Indeed, the ability to cause disease in animals and humans appears to be restricted to *Mybc. chelonae*, *Mybc. fortuitum* and *Mybc. senegalense*. A host of taxometric studies have done much to unravel the complex systematics of fast-growing mycobacteria with important contributions being made by IWGMT collaborators (Kubica et al., 1972; Saito et al., 1977b). Numerical taxonomies have been evaluated by the application of immunologically based tests (Ridell et al., 1979; Lind and Ridell, 1982), bacteriocin and phage typing (Takeya and Tokiwa, 1972; Redmond et al., 1979), chemical (Minnikin and Goodfellow, 1980) and nucleic acid reassociation studies (Baess and Weis Bentzon, 1978; Mordarski et al., 1981b). On the basis of such studies, rapidly growing mycobacteria can be allotted to five aggregate taxa (Table 8) and most can be divided into non-photochromogens and scotochromogens.

Mycobacterium fortuitum is a well characterized species (Kubica et al., 1972; Tsukamura et al., 1979; Tsukamura, 1981; Vanden Berghe and Pattyn, 1979) which encompasses strains previously labelled '*Mybc. giae*', '*Mybc. minetti*' and '*Mybc. peregrinum*'. There is, however, abundant evidence that the taxon is heterogeneous (Jenkins et al., 1971; Kubica et al., 1972; Pattyn et al., 1974). Grange and Stanford (1974) recognized seven serotypes which showed some correlation with lipid, temperature and fermentation patterns. Serotype I was frequently associated with disease, whereas serotype II, although common in natural habitats, was not. *Mycobacterium chelonae* has many properties in common with *Mybc. fortuitum*, includes strains previously assigned to '*Mybc. abscessus*', '*Mybc. borstelense*' and '*Mybc. runyonii*', and can be divided into two subspecies, *Mybc. chelonae* subsp. *abscessus* and *Mybc. chelonae* subsp. *chelonae* (Kubica et al., 1972; Stanford et al., 1972). There is evidence of an increasing frequency of serious infection in humans associated with *Mybc. chelonae* and *Mybc. fortuitum* (Kubica and Good, 1981). The two species can be distinguished by mycolic acid and protein electrophoresis patterns (Haas et al., 1974; Minnikin et al., 1982), biochemical (Pattyn et al., 1974) and immunodiffusion tests (Kubica et al., 1972; Stanford et al., 1972; Pattyn et al., 1974). *Mycobacterium fortuitum* and *Mybc. senegalense* are clearly closely related on the basis of comparative immunodiffusion and mycolic acid analyses (Ridell, 1982; Ridell et al., 1982) but have been distinguished from one another, and from *Mybc. chelonae* in a numerical phenetic survey (Ridell and Goodfellow, 1982).

The three remaining non-photochromogenic species, *Mybc. agri*, *Mybc. chitae* and *Mybc. smegmatis*, form distinct clusters in numerical phenetic surveys (Kubica et al., 1972; Saito et al., 1977b; Tsukamura, 1981). The well-studied *Mybc. smegmatis* also forms a homogeneous group in immunodiffusion and immunological studies (Lind and Ridell, 1983), and exhibits species-specific sensitins (Magnusson, 1962) and phages (Kubica et al., 1972). Preliminary experiments show that *Mybc. smegmatis* has little DNA (less than 30%) in common with either *Mybc. farcinogenes* or *Mybc. phlei* (Baess and Weis Bentzon, 1978) but the large number of ribosomal precipitinogens shared by *Mybc. smegmatis* and *Mybc. phlei* suggests that these taxa are phylogenetically close (Ridell *et al.*, 1979). It is also interesting that *Mybc. farcinogenes* and *Mybc. smegmatis* give similar mycolic acid patterns on two-dimensional TLC of whole-organism methanolysates (Minnikin *et al.*, 1980; Ridell *et al.*, 1982).

TABLE 8. Species and pathogenicity of fast-growing mycobacteria

Aggregate taxon	*Mybc.* species	
Mybc. fortuitum complex	*chelonae*	4
	fortuitum	4
	senegalense	4
Mybc. parafortuitum complex	*aurum*	5
	'diernhoferi'	5
	neoaurum	5
	parafortuitum	5
	vaccae	5
Thermotolerant and related species	*flavescens*	5
	phlei	5
	thermoresistibile	4
Additional non-photochromogenic species	*agri*	5
	chitae	5
	smegmatis	5
Additional scotochromogenic species	*aichiense*	5
	chubuense	5
	duvalii	5
	gadium	5
	gilvum	5
	komossense	5
	obuense	5
	rhodesiae	5
	sphagni	5
	tokaiense	5

* See footnote to Table 7.

Mycobacterium phlei and *Mybc. thermoresistibile* are the only mycobacteria that can grow at 52°C, they have many additional properties in common but form distinct and homogeneous clusters in numerical phenetic analyses (Saito *et al.*, 1977b). The homogeneity of *Mybc. phlei* is also supported by sensitin testing (Magnusson, 1962), phage and bacteriocin typing (Baess and Weis Bentzon, 1969; Takeya and Tokiwa, 1972), and by immunodiffusion, immunoelectrophoretic and immunofluorescence studies (Norlin, 1965; Jones and Kubica, 1968). Strains of *Mybc. thermoresistibile* grow more slowly than those of *Mybc. phlei* but faster than slow-growing mycobacteria. *Mycobacterium flavescens* also has a growth rate intermediate between that of rapidly and slowly growing mycobacteria but it does not grow at 52°C. In numerical phenetic analyses, *Mybc. flavescens* can be separated from both rapid- (Kubica *et al.*, 1972; Saito *et al.*, 1977b) and slow-growing mycobacteria (Wayne *et al.*, 1971), though its metabolism and physiological activities are nearer those of the rapid growers (Wayne, 1967; Wayne *et al.*, 1981).

Strains assigned to the *Mybc. parafortuitum* complex (Table 8) have many properties in common and cannot easily be allotted to distinct species using conventional criteria. Saito *et al.* (1977b) found that *Mybc. vaccae* and other strains in the *Mybc. parafortuitum* complex, especially *Mybc. aurum*, formed a single serospecies while *Mybc. diernhoferi* comprised a second serospecies. Further, in contrast to *Mybc. aurum* and *Mybc. vaccae*, '*Mybc. diernhoferi*' and *Mybc. parafortuitum* can be distinguished on the basis of bacteriocin typing (Takeya and Tokiwa, 1972). Agglutination tests can, however, be used to distinguish *Mybc. aurum*, '*Mybc. diernhoferi*', *Mybc. parafortuitum* and *Mybc. vaccae* (Pattyn, 1970) while *Mybc. aurum*, *Mybc. neoaurum* and *Mybc. parafortuitum* give characteristic patterns of radioactive spots on TLC analysis of ethanol/diethyl ether extracts of cells incubated with [^{35}S]methionine (Tsukamura and Mizuno, 1978). Clearly, further comparative studies are needed to unscramble the confused taxonomy of the *Mybc. parafortuitum* complex.

Ten additional species of rapidly growing scotochromogenic mycobacteria are currently recognized, most on the basis of numerical phenetic analysis (Tsukamura *et al.*, 1981). Representatives of most of these species have not been widely distributed and their taxonomic status needs to be clarified. However, *Mybc. duvalii* and *Mybc. gilvum* have been shown to be distinct and homogeneous species in lipid (Tsukamura and Mizuno, 1978), serological (Stanford and Gunthorpe, 1971) and taxometric analyses (Tsukamura and Mizuno, 1977; Tsukamura *et al.*, 1981). A related group of bacteria also from clinical material, were described as *Mybc. gadium* by Casal and Calero (1974). *Mycobacterium komossense* and *Mybc. sphagni*, which accommodate strains isolated from sphagnum, appear to be good species on the basis of lipid, immunological and numerical phenetic studies (Kazda and Müller, 1979; Kazda, 1980).

D. *Nocardia* (Trevisan, 1889)

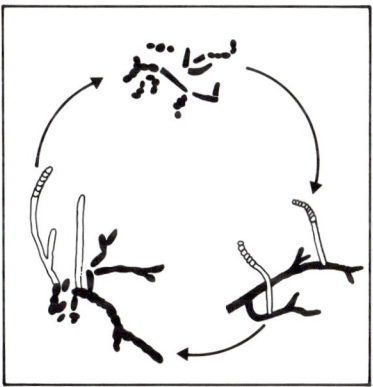

Aerobic, Gram-positive, acid to partially acid-fast fast actinomycetes which produce a primary mycelium that fragments into bacillary and coccoid elements. The latter germinate to form branched hyphae 0.5–1.0 μm in diameter. The organisms are non-motile, grow well on standard laboratory media at 30°C and usually produce an aerial mycelium that may differentiate into arthrospores. They are resistant to lysozyme, form acid from glucose oxidatively and use a wide range of organic compounds as sole sources of carbon for energy and growth.

The peptidoglycan backbone of nocardiae is composed of N-glycolylmuramic acid, N-acetylglucosamine, L-alanine, D-alanine, D-glutamic acid with *meso*-DAP as the diamino acid. The peptidoglycan is of the A1γ type. The polysaccharide fraction of the wall contains arabinose and galactose though glucose and mannose have been detected in some strains. Nocardiae characteristically contain tetrahydrogenated menaquinones with eight isoprene units, MK-$8(H_4)$, except *Nrda. amarae* where the predominant isoprenologue is dihydrogenated and has nine isoprene units (MK-$9(H_2)$), diphosphatidylglycerol, phosphatidylethanolamine, phosphatidylinositol and phosphatidylinositol mannosides, major amounts of straight-chain saturated and unsaturated fatty acids and tuberculostearic acids, and mycolic acids with 46 to 60 carbons and up to three double bonds. The fatty acid esters released on pyrolysis gas chromatography of mycolic esters contain 12 to 18 carbon atoms. The G+C content of the DNA ranges from 64 to 72 mol%. The genus is widely distributed and is abundant in soil. Some strains are pathogenic for man and animals.

The type species is *Nocardia asteroides*, the type strain ATCC 19247.

The long and confused taxonomic history of the genus *Nocardia* has been outlined by Lechevalier (1976). The use of modern taxonomic techniques underlined the heterogeneity of *Nocardia sensu* Waksman (1961), led to improved circumscription of the genus and provided a sound basis for future studies (Tsukamura, 1969, 1977; Goodfellow, 1971; Lechevalier and Lechevalier, 1974; Lechevalier, 1976; Minnikin and Goodfellow, 1976, 1980, 1981a; Bradley and Mordarski, 1976; Mordarski et al., 1977, 1978b, 1980a; Goodfellow and Minnikin, 1977, 1978, 1981b, c; Schaal and Reutersberg, 1978; Orchard and Goodfellow, 1980; Goodfellow et al., 1982a). The redefined genus accommodates the well established species *Nrda. amarae*, *Nrda. asteroides*, *Nrda. brasiliensis*, *Nrda. farcinica* (group Kyoto-1), *Nrda. otitidis-caviarum* (née *Nrda. caviae*) and the less well studied *Nrda. carnea*, *Nrda. transvalensis* and *Nrda. vaccinii*. It has recently been proposed that *Mips. brevicatena*, which has many properties in common with true nocardiae, be renamed *Nrda. brevicatena* comb. nov. (Goodfellow and Pirouz, 1982). In contrast, strains currently labelled '*Nrda. aerocolonigenes*', '*Nrda*'. *autotrophica* (née '*Nrda*'. *coeliaca*), '*Nrda*'. *mediterranea* and '*Nrda*'. *orientalis* have little in common with nocardiae (see page 129).

Nocardia brasiliensis, *Nrda. farcinica* (group Kyoto-1) and *Nrda. otitidis-caviarum* are all good taxospecies (Tsukamura, 1969, 1977; Goodfellow, 1971; Tsukamura et al., 1979). The same is true for *Nrda. amarae* but its assignment to the genus *Nocardia* is not unambiguous (Goodfellow et al., 1982a). *Nocardia amarae* strains fall on the periphery of the *Nocardia* cluster, contain dihydrogenated menaquinones with nine isoprene units, have mycolic acids which release C_{16} and C_{18} mono-unsaturated fatty acids on pyrolysis, and the type strain is not lysed by nocardiophages (Williams et al., 1980). In contrast, *Nrda. asteroides* has been found to be heterogeneous on the basis of genetical (Bradley and Mordarski, 1976; Franklin and McClung, 1976; Mordarski et al., 1977, 1978b), numerical phenetic (Goodfellow and Minnikin, 1978; Schaal and Reutersberg, 1978; Tsukamura et al., 1979; Orchard and Goodfellow, 1980; Ridell and Goodfellow, 1982), serological (Ridell, 1975; Magnusson, 1976; Pier and Fichtner, 1981), protein electrophoresis (Mauff et al., 1981), phage host range (Pulverer et al., 1975). pathogenicity (Uesaka et al., 1971) and physiological (Berd, 1973) studies but, as few strains are common to all of these investigations, it is difficult to determine whether or not, and to what extent, defined groups overlap. Schaal and Reutersberg (1978) recognized two well-defined taxa, *Nrda. asteroides* A and B, but it seems likely that the *Nrda. asteroides* complex encompasses several additional species (Orchard and Goodfellow, 1980).

Nocardiae are agents of nocardiosis and actinomycetoma in man but also cause a number of animal diseases (see Chapter 9). Many of the animal infections recognized relate to bovine mastitis infections (Koehne et al., 1981), but nocardiosis is not uncommon in dogs, fish, domestic and wild rodents (Goodfellow and Minnikin, 1981b; Orchard, 1981). *Nocardia asteroides* is the most important cause of nocardiosis which, in its pulmonary form, may lead to secondary and often fatal involvement with the brain, lungs and meninges alone or in combination with other organs. There is some evidence that the disease is transmissible in man (Houang et al., 1980).

Actinomycete mycetoma has a more limited distribution and the primary agent, *Nrda. brasiliensis* is endemic in certain subtropical regions such as the sugar-growing areas of Mexico. *Nocardia asteroides* and *Nrda. otitidis-caviarum* occasionally cause actinomycetoma. *Nocardia vaccinii* causes galls and bud proliferations on blueberry plants, a disease that has been reported on only one occasion (Demaree and Smith, 1952).

Nocardiae are widely distributed in soil where populations up to 7.3×10^4 dry weight have been reported from suspensions plated on to Diagnostic Sensitivity Test agar supplemented with tetracyclines (Orchard and Goodfellow, 1974; Orchard *et al.*, 1977; Orchard, 1981). They also form mutualistic associations with blood-sucking arthropods and occur in aquatic habitats where they have been implicated in the biodeterioration of natural rubber joints in water and sewage pipes (Hutchinson *et al.*, 1975; Orchard, 1981). Most of the strains from these habitats have been assigned to the *Nrda. asteroides* complex but their apparent predominance may merely reflect the isolation methods used. Large populations of *Nrda. amarae*, and smaller numbers of *Nrda. otitidis-caviarum*, have been found in foam on the surface of aeration tanks in activated-sludge sewage-treatment plants (Lechevalier and Lechevalier, 1974; Cross, 1981a).

E. *Rhodococcus* (Zopf, 1891; Tsukamura, 1974)

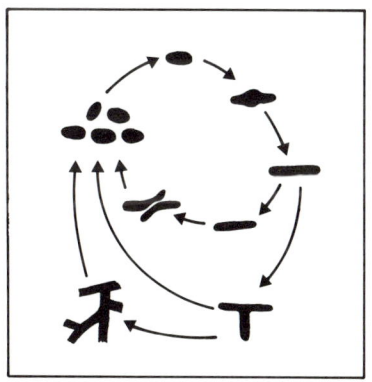

Aerobic, non-motile, Gram-positive actinomycetes that show considerable morphological diversity and can be acid-fast. In all strains the morphogenetic cycle may be assumed to start with the coccus or short rod stage with differences between organisms related to a succession of more or less complex morphological stages by which the completion of the growth cycle is achieved. Thus, cocci may merely germinate into short rods, form filaments with side projections, show elementary branching or, in the most advanced forms, produce extensively branched hyphae. In all cases, the next generation of cocci or short rods are produced by the fragmentation of the rods, filaments and hyphae. Some strains form feeble aerial hyphae, which may be branched, or aerial synnemata consisting of unbranched filaments that coalesce and project upwards. Most strains grow well on standard laboratory media at 30°C though some require thiamine. Colonies may be rough, smooth, mucoid or mycobacterial-like and are usually pigmented buff, orange or red, though colourless variants do occur. Rhodococci are sensitive to lysozyme, are arylsulphatase negative, produce acid from glucose oxidatively and are able to use a wide range of organic compounds as sole sources of carbon for energy and growth.

The peptidoglycan contains *meso*-DAP, *N*-glycollated muramic acid and is of the A1γ type. The predominant wall sugars are arabinose and galactose. Rhodococci contain dihydrogenated menaquinones with either eight or nine isoprene units, MK-$8(H_2)$ or MK-$9(H_2)$, as the major isoprenologue, diphosphatidylglycerol, phosphatidylethanolamine, phosphatidylinositol and phosphatidylinositol mannosides, major amounts of straight-chain saturated and unsaturated fatty acids and tuberculostearic acids, and mycolic acids with 32 to 66 carbons and up to four double bonds. The fatty acid esters released on pyrolysis gas chromatography of mycolic esters contain 12 to 18 carbon atoms. The G+C content of the DNA ranges from 63 to 72 mol%. The organisms are widely distributed but are notably abundant in soil and herbivorous dung. Some strains are pathogenic for man and animals.

The type species is *Rhodococcus rhodochrous*, the type strain ATCC 13808.

The long and chequered history of actinomycetes currently classified in the genus *Rhodococcus* has been considered in detail elsewhere (Cross and Goodfellow, 1973; Bradley and Bond, 1974; Bousfield and Goodfellow, 1976; Goodfellow and Alderson, 1977; Goodfellow and Minnikin, 1981c). The genus was resurrected to accommodate a heterogeneous group of bacteria previously classified as *Corynebacterium*, '*Gordona*', '*Jensenia*', '*Mybc*'. *rhodochrous* or assigned to the '*rhodochrous*' complex (Tsukamura, 1974; Goodfellow and Alderson, 1977), and now contains twelve species which collectively show considerable chemical and morphological variation (Table 9). Most of the recognized species were initially defined in numerical phenetic surveys (Tsukamura, 1974; Goodfellow and Alderson, 1977; Rowbotham and Cross, 1977; Goodfellow *et al.*, 1982b, c) but were subsequently shown to be homogeneous on both chemical and genetical criteria (Mordarski *et al.*, 1978a, 1980b, 1981a; Minnikin and Goodfellow, 1980; Suzuki *et al.*, 1981). *Rhodococcus* (née *Corynebacterium*) *equi* has recently been redescribed and shown to have properties consistent with its continued inclusion in the genus *Rhodococcus* (Goodfellow *et al.*, 1982c) while the type strain of the taxon clustered with other rhodococci in preliminary rRNA:DNA pairing experiments (Mordarski *et al.*, 1980a). There are also good grounds for transferring '*Cnbc*'. *fascians* to the genus *Rhodococcus* (Collins *et al.*, 1982a). *Rhodococcus* is phylogenetically distinct from both *Corynebacterium* and *Nocardia* (Stackebrandt and Woese, 1981b) and can be distinguished from the latter using chemical criteria (Minnikin *et al.*, 1978a; Minnikin and Goodfellow, 1980).

TABLE 9. Some morphological and chemical properties of *Rhodococcus* species

	Morphogenetic sequence[1]	$G+C$[2] (mol%)	Mycolic acids (total number of carbon atoms)[3]	Predominant menaquinone[4]
Rhodococcus				
bronchialis	rod–coccus	63–65	54–66	MK-9(H_2)
coprophilus	hypha–rod–coccus	67–70	38–48	MK-8(H_2)
equi	rod–coccus	70–72	30–38	MK-8(H_2)
erythropolis	elementary branching–rod–coccus	64–71	34–38	MK-8(H_2)
globerulus	elementary branching–rod–coccus	63–67	ND	MK-8(H_2)
luteus[5]	elementary branching–rod–coccus	64	ND	MK-8(H_2)
maris	rod–coccus	73	ND	MK-8(H_2)
rhodnii	elementary branching–rod–coccus	66	38–52	MK-8(H_2)
rhodochrous	elementary branching–rod–coccus	67–70	36–50	MK-8(H_2)
ruber	hypha–rod–coccus	69–73	40–50	MK-8(H_2)
rubropertinctus	rod–coccus	67–69	48–62	MK-9(H_2)
terrae	rod–coccus	64–69	52–64	MK-9(H_2)

ND, not determined. References: [1] Locci *et al.* (unpublished data); [2] Mordarski *et al.* (1980b, 1981a); Goodfellow *et al.* (1982a, b); [3] Collins *et al.* (1982b), Minnikin and Goodfellow (1976), Minnikin *et al.* (unpublished data); [4] Minnikin and Goodfellow (1980); [5] Nesterenko *et al.* (1982) and Collins *et al.* (unpublished data).

Rhodococci are widely distributed in nature and have been frequently isolated from soil, fresh water, and the gut contents of blood-sucking arthropods with which they may form a mutualistic association (Cross et al., 1976; Cross, 1981a; Goodfellow and Minnikin, 1981b). Thus, *Rodc. erythropolis*, *Rodc. luteus*, *Rodc. rhodochrous*, *Rodc. ruber* and *Rodc. terrae* have been isolated from soil, *Rodc. maris*, *Rodc. rubropertinctus* and *Rodc. rhodnii* from the intestinal tract of carp (*Cyprinus carpis*), cockroaches (*Rapa nui*) and the bed bug, *Rhodnius prolixus*, respectively and *Rodc. coprophilus* from herbivorous dung, whereas *Rodc. bronchialis* is associated with sputa of patients suffering from cavitary pulmonary tuberculosis and bronchiectasis. *Rhodococcus equi* is an important equine pathogen which occasionally infects other domestic animals, especially cattle and swine, and causes infection in human patients compromised by either immunosuppressive drug therapy and/or lymphoma (Goodfellow et al., 1982c). '*Corynebacterium*' *fascians* causes fasciation of sweet pea (Tilford, 1936). A number of techniques are recommended for the selective isolation of rhodococci (Barton and Hughes, 1981; Goodfellow and Minnikin, 1981b; Nesterenko et al., 1982).

Further comparative work is required to determine the taxonomic status of strains previously classified as '*Gordona aurantiaca*' (Tsukamura and Mizuno, 1971). Early numerical phenetic studies indicated the equivocal position of this species in the genus *Gordona* (Tsukamura, 1974, 1975) and the type strain fell outside the *Rhodococcus* cluster as defined by Goodfellow and Alderson (1977). Aurantiaca strains were subsequently found to contain characteristic mycolic acids, completely unsaturated menaquinones with nine isoprene units (MK-9) and formed a numerically defined taxon equivalent in rank to phena corresponding to the genera *Corynebacterium*, *Mycobacterium*, *Nocardia* and *Rhodococcus*. *Corynebacterium paurometabolum* and '*Mybc. album*' have properties in common with the 'aurantiaca' taxon (Goodfellow and Minnikin, 1980; Collins and Jones, 1982c).

9. Streptomycetes

The streptomycetes are common terrigenous Gram-positive sporoactinomycetes that are highly oxidative, form an extensive branching substrate and aerial mycelium and typically have a wall peptidoglycan containing LL-DAP as the diamino acid and glycine as the cross-linking amino acid but have no characteristic sugars (wall chemotype I). Fragmentation of the substrate mycelium is rare and spores are occasionally formed on the substrate hyphae. Species belonging to the most commonly isolated genus, *Streptomyces*, characteristically have aerial hyphae bearing long uniserial chains of spores (usually more than 50 spores) enclosed within a thin fibrous sheath but short chains are found in some species. The sheath consists of at least two layers, an inner mosaic of chitin rodlets and an outer lipid layer that renders the spore chain hydrophobic. The catenulate spores can often appear smooth but may have appendages (spines, hairs or warts) which can be seen as spore ornaments when the chains are viewed in silhouette under the transmission electron microscope or as shadowed mounts under the scanning electron microscope. The spore chains may be straight, wavy, hooked or even in the form of tight spirals. Spore walls can be unpigmented, when the sporulating hyphae appear white, or pigmented and then impart a grey, red, green, yellow or blue shade to the sporulating hyphae. Species growing on laboratory media can produce a range of pigments which colour the substrate mycelium and diffuse into the surrounding medium.

Streptomycetes have proved to be a rich source of antibiotics, vitamins, enzymes and enzyme inhibitors and consequently millions have been isolated from soils and sediments from all parts of the earth in a search for lucrative metabolites. Between 1940 and 1957 over 1,000 *Streptomyces* species were described (Pridham *et al.*, 1958) and by 1970 the total number had increased to about 3,000 though many had been inadequately described or cited in the patent literature (Trejo, 1970). Speciation was based upon aerial mycelium colour, spore chain morphology, spore ornamentation, the ability to produce soluble melanoid pigments in media containing tyrosine, and on the carbohydrate utilization spectrum and proteolytic activities. The possible permutation of these characters could theoretically give many thousands of species and so make identification of isolated strains very difficult.

In the early 1960s it became increasingly obvious that streptomycete taxonomy was encountering serious problems. In 1964 a collaborative study, the International *Streptomyces* Project (ISP), was organized to evaluate taxonomic characters and to obtain reliable descriptions of most of the available type strains after applying a small number of tests carried out under specified and standard conditions (Shirling and Gottlieb, 1966; Gottlieb and Shirling, 1967). Over 450 species were redescribed and type and neotype strains were deposited in four recognized culture collections (Shirling and Gottlieb, 1968a, b, 1969, 1972). Much of the information obtained was incorporated into the descriptions of 463 *Streptomyces* species contained in the most recent edition of *Bergey's Manual of Determinative Bacteriology* (Pridham and Tresner, 1974).

Meanwhile, numerical taxonomic studies (Silvestri *et al.*, 1962) had suggested that the genus was over-classified and that there might be as few as 25 centres of variation. The results of the most recent and more exhaustive numerical phenetic studies (Williams *et al.*, 1981, 1983a) have also suggested that there should be relatively few recognized cluster groups (or species) within the genus and that the heavy weighting given to morphological characters in past classifications was no longer justified.

During the last 20 years the streptomycetes have been further partitioned by proposals of new genera which were differentiated from *Streptomyces* by morphological criteria. *Actinosporangium* (Krasilnikov and Yuan, 1961) and *Actinopycnidium* (Krasilnikov, 1962) were reported to form sporangia and pycnidia respectively, but subsequent studies have shown the spore-containing structures to be artefacts produced by lysis of aerial hyphae resulting in dense spore droplets. We know of no recent evidence that would support the continued use or retention of these two generic names and indeed the evidence of numerical taxonomic studies (Williams *et al.*, 1981, 1983a) and phage sensitivity studies (Wellington and Williams, 1981) place the strains studied in the genus *Streptomyces*. *Actinopycnidium* and *Actinosporangium* should therefore be regarded as junior synonyms of *Streptomyces* and the named species relocated within that genus (Williams *et al.*, 1983a).

Chainia (Thirumalachar, 1955) was the generic name given to streptomycetes that form sclerotia either embedded in colonies growing on agar media or in shaken flasks. Such organisms appear to have been frequently isolated from arid soils in India but rarely reported in surveys conducted elsewhere. The sclerotia may be quite large (up to 100 μm in diameter) and consist of dense masses of hyphae containing oil droplets (Sharples and Williams, 1976) cemented together by an amorphous material containing L-2,3-diaminopropionic acid (Lechevalier *et al.*, 1973). The ability to form sclerotia is lost on continued cultivation on laboratory media. Species form chains of spores typical of the genus *Streptomyces*, they have a wall chemotype I, are lysed by *Streptomyces* phage (Wellington and Williams, 1981) and cluster with typical *Streptomyces* species in numerical phenetic analyses (Williams *et al.*, 1981, 1983a). The genus was regarded as a junior synonym of *Streptomyces* by Pridham and Tresner (1974) but was retained as a separate genus in the *Approved List of Bacterial Names* (Skerman *et al.*, 1980). Lechevalier *et al.* (1973) favoured the recognition of *Chainia* as a distinct genus 'until it has been demonstrated experimentally that ordinary streptomycetes can be induced to form truly plurilocular sclerotia'. We now advocate the converse argument that, until characters other than the transient ability to form sclerotia are produced to separate *Chainia* from *Streptomyces*, the genus must be regarded as a junior synonym of *Streptomyces* and its species located within that genus.

The genus *Microellobosporia* (Cross *et al.*, 1963) (see below) was originally proposed for species that formed club-shaped sporangia, containing a short linear row of non-motile spores, on both the aerial and substrate hyphae and was accordingly placed in the family *Actinoplanaceae*. Species were subsequently found to have a wall chemotype I, to be susceptible to *Streptomyces* phage (Prauser *et al.*, 1967; Wellington and Williams, 1981) and

Microellobosporia   *Elytrosporangium*

 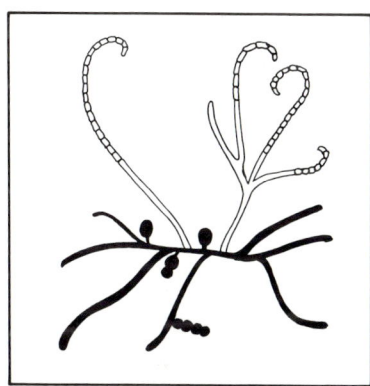

to cluster with *Streptomyces* species in numerical taxonomic analyses (Williams *et al.*, 1981, 1983a). What was originally regarded as a sporangial wall must now be equated with the sheath normally surrounding *Streptomyces* spore chains. The species originally included within the genus *Microellobosporia* characteristically had very short spore chains, larger spores than those normally found in *Streptomyces* species and consistently to form spores on the substrate mycelium.

Another genus, named *Elytrosporangium* by Falcao de Morais *et al.* (see above) (1966), also forms short chains of spores on the substrate mycelium but with typical *Streptomyces* spore chains on the aerial mycelium. Species in this genus have also been shown to have a wall chemotype I, to be sensitive to *Streptomyces* phage (Wellington and Williams, 1981) and to cluster with *Streptomyces* species in numerical phenetic analyses (Williams *et al.*, 1981, 1983a). Recent studies have shown that streptomycetes exhibiting a morphology typical of *Elytrosporangium* and *Microellobosporia* can be isolated from soil together with intermediates with longer chains of irregularly sized spores on the aerial mycelium or even single spores on the substrate mycelium (Cross and Al-Diwany, 1981). These morphological forms can be accommodated within *Streptomyces* providing that genus is enlarged to include wall chemotype I actinomycetes able to form spores on the substrate mycelium.

The genus *Kitasatoa* (Matsumae *et al.*, 1968) was also reported to form club-shaped sporangia on both substrate and aerial hyphae but the contained catenulate spores were motile when placed in water. Type strains in culture collections now form chains of non-motile spores typical of *Streptomyces*, they have a wall chemotype I, phospholipids similar to those of streptomycetes (Hasegawa *et al.*, 1979), are sensitive to *Streptomyces* phage (Wellington

Kitasatoa

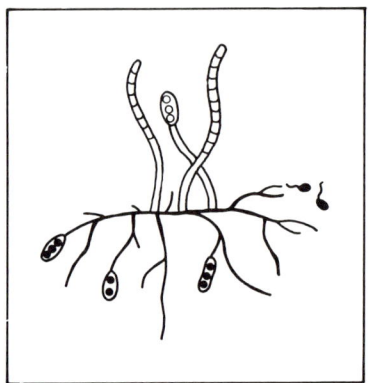

and Williams, 1981), share a relatively high DNA homology with streptomycetes (Kroppenstedt *et al.*, 1981) but were not recovered in the *Streptomyces* cluster-group in the numerical phenetic classification of Williams *et al.* (1981, 1983a). The status of the genus *Kitasatoa* remains uncertain and will probably remain so until new strains that consistently form motile spores are found and studied in detail. However, the chemical, genetical and numerical phenetic data provide sufficient grounds for reducing *Kits. diplospora*, *Kits. kauaiensis* and *Kits. nagasakiensis* to synonyms of *Kits. purpurea* (Williams *et al.*, 1983a).

Wall chemotype I streptomycete genera that are currently regarded as being distinct and recognizable are briefly circumscribed below.

A. *Intrasporangium* (Kalakoutskii *et al.*, 1967)

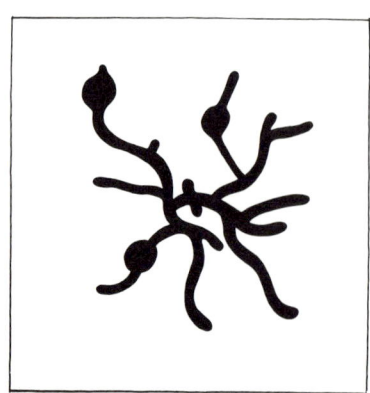

Aerobic actinomycetes with a fine branching substrate mycelium (0.4–1.2 μm diameter) that appears non-fragmented when the colony is examined *in situ*. Fragmentation can, however, be induced even in young hyphae by preparing a smear or placing a drop of water on a colony and viewing the mycelium beneath a coverslip. No aerial hyphae are formed on the smooth round glistening colonies (1.5 mm diameter) which are white or cream coloured. Characteristic vesicles (5–15 μm diameter) appear after 5 or 6 days incubation at 28°C and are most abundant after 12 days. These structures, termed sporangia by Kalakoutskii *et al.* (1967), are intercalary or sometimes form just below tips of hyphae. They were originally reported to contain one to more than 20 round or oval bodies (1–2 μm diameter) which exhibited Brownian movement within the sporangium. These bodies termed spores were observed to germinate by one or two germ tubes when placed on fresh medium. Subsequent ultrastructural studies on the type strain by Lechevalier and Lechevalier (1969) failed to confirm spore formation within the vesicles. These workers observed highly convoluted contents and bodies sometimes separated from the spore-bearing hyphae by septa, but no structures comparable to the sporangiospores of the *Actinoplanetes*, and they concluded that the vesicles were a characteristic cellular reaction to age or toxicity of the medium constituents.

Intrasporangium grows between 28 and 38°C, has a wall chemotype I (Prauser, 1967; Sukapure *et al.*, 1970) and its phospholipids are of the phospholipid IV type (Lechevalier *et al.*, 1977). The G+C content of the type strain has not been determined but that of a similar strain was reported to be 71 mol% (Sukapure *et al.*, 1970).

The type and only species is *Intrasporangium calvum*, the type strain ATCC 23552.

Comment: This unusual actinomycete was originally classified in the family *Actinoplanaceae* and later transferred to a family named *Streptosporangiaceae* by Krasilnikov (1970). Most workers now place the organism in the Streptomycetes because of its wall composition but numerical taxonomic studies (Williams *et al.*, 1981, 1983a) and phage sensitivity patterns (Wellington and Williams, 1981) have failed to reveal any relationships to the other wall chemotype I genera. *Intrasporangium calvum* was isolated by chance on a plate of meat extract peptone agar from the air in a school dining room. Other strains and species are required to fully expose the properties and status of this genus and to discover its natural habitat.

B. *Kineosporia* (Pagani and Parenti, 1978)

A genus so far represented by only a single strain. The colonies on agar are cream to orange in colour and lack aerial mycelium. Numerous small sporangia form at the tips of substrate mycelium hyphae and the layer that forms on the surface of colonies gives them a shiny or glossy appearance. The round to pyriform sporangia (1.0–2.0 μm diameter) are borne on short swollen sporophores and contain a single motile spore. Purified walls contain LL-DAP and glycine with only a trace of lysine, the whole-organism sugars are arabinose, galactose and xylose. Growth is good at 20 to 30°C but the species is unable to grow at 37°C and above. Data on mol% G+C and lipid composition are lacking. The species is non-pathogenic in mice.

The type strain is *Kineosporia aurantiaca*, ATCC 29727.

C. *Nocardioides* (Prauser, 1976b)

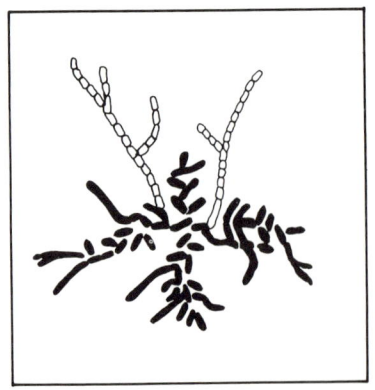

Aerobic, Gram-positive, non-acid-fast actinomycetes that form a primary mycelium (0.6–0.8 μm diameter) which fragments into irregular to rod- to coccus-like elements. The hyphae of the aerial mycelium (0.6–0.8 μm diameter) fragment into rod- to coccus-like elements with smooth surfaces.

The elements so formed give rise to new mycelia. The primary mycelium on oatmeal agar, yeast extract–malt extract agar and similar rich agar media is whitish to faint yellowish. Colonies are pasty with smooth to wrinkled and dull to bright surfaces. The aerial mycelium is thick, dense and chalky. Soluble pigments are not formed.

The organism contains LL-DAP and glycine in the wall peptidoglycan, is rich in *iso, anteiso*, straight-chain, unsaturated and tuberculostearic acids, contains tetrahydrogenated menaquinones with eight isoprene units, MK-$8(H_4)$, as the major isoprenologue and major amounts of diphosphatidylglycerol and phosphatidylglycerol with a number of incompletely characterized phospholipids. The G+C content of the DNA is within the range 66–67 mol%. The organism is found in soil.

The type species is *Nocardioides albus*, the type strain ATCC 27980.

Nocardioides is an unusual organism which was assigned to the family *Streptomycetaceae* mainly on the basis of wall composition (Prauser, 1976b). *Nocardioides* strains are attacked by phages of a taxon specific set, form a homogeneous group on the basis of lipid analyses (O'Donnell *et al.*, 1983) but their supra-generic affinities are uncertain. The organism has little DNA in common with *Streptomyces* (Tille *et al.*, 1978), is not lysed by phages virulent to streptomycetes and related wall chemotype I strains (Prauser and Falta, 1968, Wellington and Williams, 1981b), but the type strain was loosely associated with the *Streptomyces* cluster-group in a numerical phenetic study (Williams *et al.*, 1981, 1983a). It has recently been proposed that *Atbc. simplex* be transferred to the genus *Nocardioides* as *Nrdo. simplex* comb. nov. (O'Donnell *et al.*, 1983).

D. *Sporichthya* (Lechevalier *et al.*, 1968)

This unusual actinomycete was originally included in the family *Streptomycetaceae* because of its wall composition (wall chemotype I) but it exhibits no obvious relationships to the other actinomycete genera in this section and is included for convenience rather than to infer a taxonomic position.

Strains form an aerial mycelium but no substrate mycelium. Sparingly branched hyphae (0.5–1.2 μm diameter × 10–25 μm long) grow on the surface of solid media to which they are attached by holdfasts formed at the base of the hypha. These aerial hyphae divide to form a chain of rods and cocci which become motile in the presence of water. Young hyphae and elements are Gram-negative but older elements are Gram-positive. The walls have a typical Gram-positive composition and structure. Colonies in the form of a chalky-white crust are formed on nutritionally poor media. On richer media the colonies are glistening, dirty white mucoid masses composed of pleomorphic elements, some almost fish-shaped from which the generic name was derived.

The single species *Srch. polymorpha* was isolated from soil on water agar and grows between 22 and 44°C. Growth in static liquid media occurs as a white pellicle but virtually no growth occurs if the medium is agitated. Additional isolates are required for comparative studies before the affinities of *Sporichthya* can be established.

The type species is *Sporichthya polymorpha*, the type strain ATCC 23823.

E. *Streptomyces* (Waksman and Henrici, 1943)

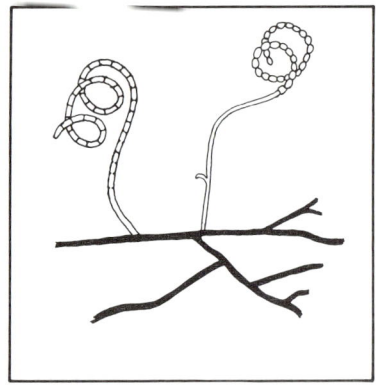

Aerobic, Gram-positive sporoactinomycetes with an aerial mycelium bearing chains of three to many spores and a substrate mycelium which can also bear short chains of spores. Species are highly oxidative and can include mesophiles, psychrotrophs and thermophiles, acidophiles, neutrophiles and alkalophiles, obligate halophiles and pathogens of man and animals. Some strains may form sclerotia which are particularly evident on agar media.

The wall peptidoglycan contains LL-DAP but it must be remembered that wall analyses have almost always been performed on hyphae grown in submerged culture. Unlike the hyphal wall, the spore wall is insensitive to lysozyme and may have a different composition as indicated by the presence of *meso*-DAP in the aerial mycelium of one species (Omura *et al.*, 1981). Streptomycetes contain major amounts of saturated, *iso*- and *anteiso*-fatty acids, possess either MK-$9(H_6)$ or MK-$9(H_8)$ as the predominant isoprenologue and typically contain diphosphatidylglycerol, phosphatidylethanolamine, phosphatidylglycerol and phosphatidylinositol dimannosides. The G+C of the DNA ranges from 69 to 78 mol%. The organism is common in soil and consequently can contaminate a wide variety of plant and animal products. The spores are washed into both fresh water and marine habitats and can reside and remain viable for long periods of time in sedimentary muds.

The type species is *Streptomyces albus*, the type strain ÀTCC 3004.

The long and complex history of the genus *Streptomyces* has already been touched upon and has been the subject of a number of reviews (Waksman, 1961, 1967; Cross and Goodfellow, 1973; Kutzner, 1981). In short, developments in streptomycete systematics were held up for many years by the absence of widely accepted and reliable characters for both classification and identification and by the requirements of 'Patent Law' which encouraged the description of a new species of *Streptomyces* for each strain found to produce a new antibiotic so workers adopted the more practical approach of proposing a new name and description rather than face the daunting task of 'identification' with previously described types. Sneath's (1970) view that numerical phenetic analysis was the only practical way of coping with the resultant overspeciation in the genus *Streptomyces* and that the examination of a few 'chosen' tests could not be expected to reveal natural phenetic groups has recently been endorsed by a comprehensive numerical phenetic survey on *Streptomyces* and related actinomycetes with a wall chemotype I (Williams *et al.*, 1983a).

In their analysis, Williams and his colleagues examined over 400 *Streptomyces* and marker cultures for 139 unit characters and analysed the data using conventional numerical taxonomic statistics. Most of the 394 *Streptomyces* type cultures fell into one large cluster-group in the analysis based on the simple matching coefficient (S_{SM}) and the unweighted average linkage algorithm. This, the *Streptomyces* cluster-group, also contained the marker strains of the genera *Actinopynidium, Actinosporangium, Chainia, Elytroporangium* and *Microellobosporia*. At the 77.5% S_{SM} similarity level, the strains in the *Streptomyces* cluster-group were recovered in 19 major and 40 minor clusters, with 18 strains forming single member clusters. The status of the latter as species was therefore confirmed. Most of the minor clusters, consisting of two to five strains, were also regarded as species. The major clusters contained between 6 and 71 strains, varied in their homogeneity and were provisionally regarded as species groups. The results of this study not only provide a basis for the reduction of the large number of *Streptomyces* species but also show that the earlier reliance on a limited number of subjectively chosen characters to define species groups or species resulted in artificial classification. The numerical phenetic study also provided data for the construction of a computer probability matrix for the classification of unknown isolates to the major *Streptomyces* cluster-groups (Williams *et al.*, 1983b).

F. *Streptoverticillium* (Baldacci, 1958)

Aerobic, Gram-positive sporoactinomycetes with a non-fragmenting substrate mycelium and a well developed aerial mycelium which bears short chains of spores. The aerial hyphae typically bear verticils at regular intervals, whorls of three or more short spore chains (monoverticillate sporophore), or whorls of three or more short branches which terminate in an umbel of short spore chains (umbellate monoverticillate sporophore) or bear primary or secondary whorls of short chains (biverticillate sporophore) (Baldacci *et al.*, 1966; Locci *et al.*, 1969; Locci and Petrolini, 1970). Spores with smooth or rough surfaces are usually borne in straight or flexuous spore chains. Spore chains show a characteristic twisting under the scanning electron microscope (Locci and Petrolini Baldan, 1971; Cross *et al.*, 1973).

Streptoverticillia contain LL-DAP in the wall peptidoglycan, major amounts of saturated, *iso-* and *anteiso-* fatty acids, MK-9(H_6) and MK-9(H_8), as the predominant isoprenologue and diphosphatidylglycerol, phosphatidylethanolamine, phosphatidylinositol and phosphatidylinositol mannosides. The G+C of the DNA ranges from 69 to 73 mol%. The organism is found in soil.

The type species is *Streptoverticillium baldaccii*, the type strain ATCC 23654.

Numerical phenetic data lend support for the continued recognition of the genus *Streptoverticillium* (Williams *et al.*, 1983a) but not for the current practice of classifying species into series based on colour of the aerial and substrate mycelium (Locci *et al.*, 1969; Baldacci and Locci, 1974). Comparative studies on additional *Streptoverticillium* and related strains are needed to determine speciation within the genus and its relationship with other wall chemotype I taxa. There are, however, good grounds for considering *Streptoverticillium* and *Streptomyces* to be closely related. Thus, strains from each genus have a wall chemotype I (M. P. Lechevalier and Lechevalier,

1970a), DNA rich in G+C (Pridham and Tresner, 1974a), share high DNA homology values (Kroppenstedt *et al.*, 1981), are lysed by the same phages (Prauser, 1976a; Wellington and Williams, 1981b) and contain similar lipids (Collins *et al.*, 1977, 1980a; Lechevalier *et al.*, 1977, 1981). Streptoverticillia do, however, show a much higher resistance to lysozyme (Kutzner *et al.*, 1978) and to neomycin than most *Streptomyces* species, except for the *Stmy. lavendulae* group. Despite a number of similarities to the *lavendulae* group of species, there appear to be strong practical arguments in favour of retaining the morphologically distinct genus *Streptoverticillium*.

10. Maduromycetes

The term Maduromycetes has been introduced for a collection of sporoactinomycetes that have a wall chemotype III and contain the sugar madurose (3-*O*-methyl-D-galactose). Although it seems more than likely that this aggregate group is artificial, it does provide at least a useful temporary resting place for a number of relatively poorly studied taxa. The group currently contains the genera *Actinomadura*, *Excellospora*, *Microbispora*, *Microtetraspora*, *Planobispora*, *Planomonospora*, *Spirillospora* and *Streptosporangium*, all of which can be readily distinguished from *Dermatophilus* and *Frankia*, which also contain madurose, on morphological grounds.

The removal of the genera *Planobispora*, *Planomonospora*, *Spirillospora* and *Streptosporangium* from the family *Actinoplanaceae* is indisputable as they differ significantly from other actinoplanetes in wall composition, morphology, DNA–DNA homology and RNA cistron similarity (Stackebrandt *et al.*, 1981). Representatives of these genera were considered to be genetically related by Farina and Bradley (1970) but Stackebrandt and his colleagues found *Spls. albida* to be genetically unrelated to any of the actinoplanetes in their study. These workers did, however, recover the three representative strains of *Planobispora*, *Planomonospora* and *Streptosporangium* in a single DNA homology group. Representative strains of *Actinomadura* and *Microtetraspora* showed little genetical similarity to the strains in this homology group or to one another.

The genera *Actinomadura*, *Excellospora*, *Microbispora* and *Microtetraspora* show no obvious relationships to the other members of the aggregate group other than the ability to form madurose (H. A. Lechevalier and Lechevalier, 1981). Indeed, the artificiality of this grouping has been further emphasized by the recent report of madurose in a wall chemotype I actinomycete with a streptomycete morphology (H. Weyland, 1982, Proceedings of the V International Symposium on Actinomycete Biology, Mexico). There is, however, some evidence from numerical phenetic studies (Goodfellow and Pirouz, 1982; Williams *et al.*, 1983a) that *Actinomadura* and members of the genera *Microbispora* and *Microtetraspora* share many properties in common.

A. *Actinomadura* (H. A. Lechevalier and Lechevalier, 1970)

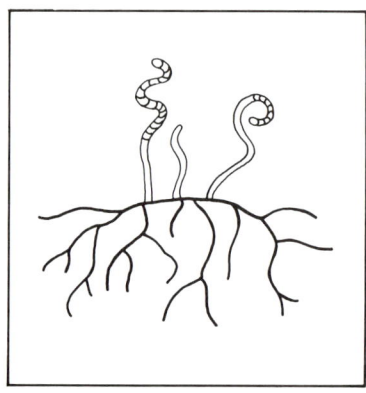

Gram-positive, non-acid-fast sporoactinomycetes with a branching substrate mycelium which only rarely becomes fragmented. An aerial mycelium is formed infrequently in some species (*Acmd. madurae* and *Acmd. pelletieri*) but is more common in most of the more recently described species which also exhibit chains of spores in the form of loops and spirals. Detailed studies on spore formation in *Acmd. verrucosospora* showed that the spore septum developed as a lateral ingrowth from one side of the sporogenous hypha (Soina et al., 1977) rather than the annular ingrowth commonly seen in other actinomycete genera. The transverse septa become irregularly thickened and appeared dumb-bell shaped in cross section in contrast to the more regular sporulation septa typical of other species such as *Acmd. fastidiosa* (Soina et al., 1975). A sheath enclosing the chain of spores may be ornamented with grape-like clusters of material to give a warty surface in *Acmd. verrucosospora* (Soina et al., 1977; Soina and Agre, 1979) or spines in the thermophilic species '*Acmd. flexuosa*' (Krasilnikov and Agre, 1964a).

Actinomadurae have a wall chemotype III but the peptidoglycan of some species has been found to contain small amounts of LL-DAP in addition to the *meso*-DAP isomer which might suggest an affinity with the Streptomycetes (Mordarska et al., 1972; Meyer and Sveshnikova, 1974). Whole-organism hydrolysates normally contain madurose but this sugar appears to be lacking in *Acmd. africana* and *Acmd. longispora* or present only as a trace in *Acmd. coeruleofusca* (Preobrazhenskaya and Sveshnikova, 1974). Such species lacking madurose might now be considered as possible members of the genus *Nocardiopsis* as suggested by Preobrazhenskaya et al.

(1977). *Actinomadura madurae* and *Acmd. pelletieri* contain predominantly straight-chain saturated and unsaturated fatty acids whereas *Acmd. helvata*, *Acmd. pusilla* and *Acmd. roseoviolacea* contain approximately equal amounts of straight and branched chain acids (Agre *et al.*, 1975). The menaquinones of *Actinomadura* species are partially hydrogenated with nine isoprene units (Collins *et al.*, 1977; Yamada *et al.*, 1977a, b; Athalye *et al.*, 1984) and representative strains have been found to contain DNA within the range 70–78 mol% $G+C$ (M. Mordarski *et al.*, unpublished data).

Actinomadura madurae and *Acmd. pelletieri* cause actinomycetic mycetoma of which Madura foot (maduramycosis) is a characteristic form. Infections caused by *Acmd. pelletieri* have only been found in Africa but *Acmd. madurae* is widespread, especially in the tropics and subtropics, although infection probably occurs through wounds with soil as the primary reservoir. *Acmd. pelletieri* has only been found in clinical material, whereas *Acmd. madurae* isolates from soil lack the red endopigment of clinical isolates and sporulate more readily (Lechevalier, 1981). Strains resembling these species have also been isolated from the water of streams (Lawson and Davey, 1972). Recently developed methods for the selective isolation of actinomadurae, which involve air drying and then dry heating soil at 100°C followed by the use of selective media, have shown them to be common in a variety of soils (Nonomura and Ohara, 1971a; Lavrova *et al.*, 1972; Preobrazhenskaya *et al.*, 1975, 1978; Chormonova, 1978; Athalye *et al.*, 1981).

The type species is *Actinomadura madurae*, the type strain NCTC 5654.

The genus *Actinomadura*, when originally proposed by H. A. Lechevalier and Lechevalier (1970) to house wall chemotype III actinomycetes previously included in the genus *Nocardia*, contained the three mesophilic species *Acmd. madurae*, *Acmd. pelletieri* and *Acmd. dassonvillei*. The latter species lacked madurose and was subsequently found to differ from the other species in spore morphology, fatty acid composition, pigmentation and in its polar lipid and menaquinone patterns. Meyer (1976) transferred this species to the new genus *Nocardiopsis* and the two genera have been recovered as distinct clusters in numerical phenetic studies (Alderson and Goodfellow, 1979; Goodfellow *et al.*, 1979; Goodfellow and Pirouz, 1982). Many of the recently described mesophilic *Actinomadura* species isolated during the search for new antibiotics (Nonomura and Ohara, 1971c; Preobrazhenskaya and Sveshnikova, 1974; Meyer, 1979; Huang, 1980) have not been studied in such great detail but preliminary numerical taxonomic studies on some species show that they have a high affinity with *Acmd. madurae* strains (Goodfellow and Pirouz, 1982). This same study, however, recovered *Acmd. madurae* and *Acmd. pelletieri* in separate clusters, which raised some doubt about the continued inclusion of these species in the same genus (Tsukamura, 1969; Goodfellow *et al.*, 1979).

B. *Excellospora* (Agre and Guzeva, 1975).

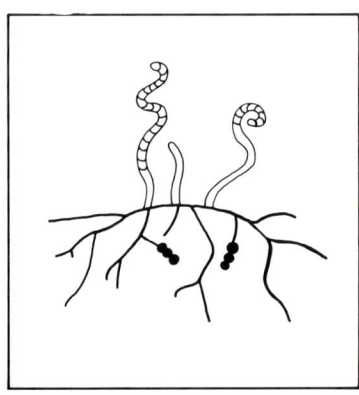

This genus was created to accommodate two thermophilic actinomycetes previously located in the genus *Micropolyspora*, '*Mips. viridinigra*' and '*Mips. rubrobrunea*' (Krasilnikov et al., 1968), and a newly isolated thermophilic strain named *Excellospora viridilutea*. The species formed short chains or single spores on both substrate and aerial hyphae (so distinguishing the genus from *Actinomadura*), had a wall chemotype III and whole-organism hydrolysates contained madurose (so separating the genus from *Micropolyspora*). Agre and Guzeva (1975) also placed emphasis on the predominance of C_{16}, C_{17} and C_{18} branched-chain fatty acids in *Excellospora* species for separating the genus from *Actinomadura* despite the heterogeneity of fatty acid types reported for species in this latter genus (Agre and Guzeva, 1975; Agre et al., 1975). Studies on the fine structure of *Excl. viridilutea* spores showed an irregular formation of sporulation septa resulting in a chain containing both mature and immature spores (Agre et al., 1977). The mature spores had thick walls, contained many vacuoles and were surrounded by a spiny sheath. Rapid autolysis of the aerial mycelium resulted in loss of viability. The three *Excellospora* species are distinguished primarily by colony pigmentation and would appear to be very similar if not synonymous (Williams and Wellington, 1981).

The type species is *Excellospora viridilutea*, the type strain INMI 187.

When Guzeva et al. (1972) studied the wall composition of '*Mips. rubrobrunea*' and '*Mips. viridinigra*' they included a strain of the thermophilic species '*Mips. flexuosa*' (syn '*Thermopolyspora flexuosa*', Krasilnikov and Agre, 1964) and found that all three species had a wall chemotype III and contained madurose. These authors also considered that the three species belonged to the same morphological group within *Micropolyspora*, referred to as the '*Mips. flexuosa*'-type, although the morphological characteristics of the sub-groups were not defined. '*Micropolyspora flexuosa*' was not included in the genus *Excellospora* by Agre and Guzeva (1975) and this remains inexplicable. It had been transferred earlier to the genus *Actinomadura* by Cross and Goodfellow (1973) because of its wall and sugar composition and the absence of spores on the substrate mycelium. The relationships of the thermophilic *Excellospora* species and '*Acmd. flexuosa*' to the mesophilic *Actinomadura* species have not been satisfactorily clarified. Thermophilic species were not included in the numerical taxonomic studies on *Actinomadura* by Goodfellow *et al.* (1979) and Goodfellow and Pirouz (1982).

C. *Microbispora* (Nonomura and Ohara, 1957)

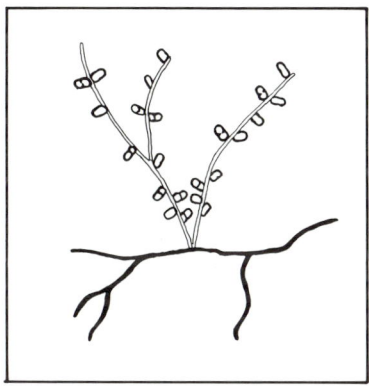

Gram-positive, non-acid-fast sporoactinomycetes growing as stable branched hyphae. A conspicuous aerial mycelium, usually pink in colour, bears the characteristic longitudinal pairs of spores which may be closely arranged along the aerial hyphae, giving the appearance of a catkin, or borne at longer intervals. They first appear as club-shaped initials which later become transformed into the paired spores visible under the light microscope. The sporophore of *Mibs. rosea* partially enclose the basal spore giving the appearance of a ball and socket joint (Williams, 1970). Cultures of *Mibs. aerata*, *Mibs. amethystogenes* and *Mibs. parva* produce bronze–violet irridescent crystals of iodinin (1,6-phenazinediol-5,10-dioxide) in certain solidified media, e.g. oatmeal agar, or release the compound as a red soluble pigment in submerged culture (Gerber and Lechevalier, 1964). The B-vitamins, particularly thiamine, are essential for the growth of some species. Mesophilic and facultatively thermophilic species have been described, the mesophiles can be aerobic to facultatively anaerobic. The normal habitat appears to be soil from which they can be isolated infrequently by dilution plating on media normally used to isolate streptomycetes. Heating air-dried soil at 120°C for 1 hour and the use of a selective agar medium enabled Nonomura and Ohara (1969a, 1971b) to isolate many species from Japanese soils. One species, *Mibs. rosea*, has been implicated in a case of pericarditis and pleuritis in man (Louria and Gordon, 1960).

Microbisporas have a wall chemotype III and whole-organism hydrolysates contain madurose. A G+C content of 73.7 mol% has been reported for *Mibs. rosea* (Lechevalier *et al.*, 1971) which also contains tetrahydrogenated menaquinones with nine isoprene units, MK-9[H$_4$], as the main component (Athalye *et al.*, 1983).

The type species is *Microbispora rosea*, the type strain ATCC 12950.

Microbispora species have been distinguished mainly by growth temperature requirements and relatively few physiological tests. We must agree with the comment of Williams and Wellington (1981) that the validity of separating some species by using only one or two characteristics is questionable. Indeed, the status of some strains currently included in the genus because of their morphology must now be questioned as a result of preliminary numerical phenetic studies (Goodfellow and Pirouz, 1982). Three strains, including the type species, were recovered in a well defined cluster, whereas *Mibs. echinospora* and *Mibs. thermodiastatica* showed little affinity to one another or to the *Mibs. rosea* cluster.

D. *Microtetraspora* (Thiemann *et al.*, 1968)

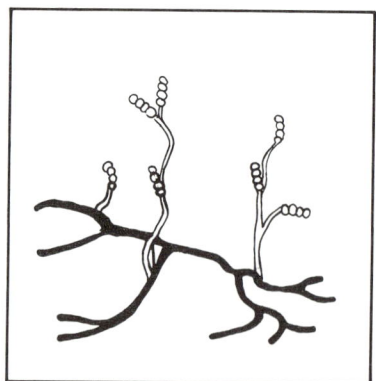

Colonies consisting of a non-fragmenting vegetative mycelium bear short, erect, sparsely branched aerial hyphae which carry the characteristic chains of four spores laterally on distinct sporophores. Chains of two or three spores can occasionally be seen and very rarely five spores but the majority of chains contain four spores. In older cultures the chains can fuse to give spore masses. Walls contain *meso*-DAP and no characteristic sugars but trace amounts of LL-DAP and glycine can often be present (Thiemann et al., 1968; Nonomura and Ohara, 1971a). The type species, *Mits. glauca*, together with *Mits. fusca* and *Mits. viridis*, have a phospholipid type PIV but the strain *Mits. viridis* var. intermedia (nonomura and Ohara, 1971a) has a phospholipid type PI (Lechevalier *et al.*, 1981). *Microtetrapora glauca* and *Mits. niveoalba* contain tetrahydrogenated menaquinones with nine isoprene units, MK-9(H_4), as the main component (Athalye *et al.*, 1983).

The type species is *Microtetraspora glauca*, the type strain ATCC 23057.

Five species have now been described *Mits. fusca*, *Mits. glauca* (Thiemann *et al.*, 1968), *Mits. viridis* (Nonomura and Ohara, 1971a), *Mits. niveoalba* (Nonomura and Ohara, 1971b) and *Mits. caesia* (Tomita *et al.*, 1980). They have been differentiated mainly on the basis of colour, carbon source utilization and biochemical reactions. However, *Mits. caesia* appears to be a problematical member of this genus because its inclusion would widen the original circumscription of the taxon. Its walls contain *meso*-DAP and galactose but lack madurose. Unlike other *Microtetraspora* species it has oval vesicles on the substrate mycelium which contain single spores or straight chains of two to several spores. Oval or banana-shaped spores with a single long polar flagellum are formed 'capriciously' when the substrate mycelium is suspended in water. This species shows some affinities to the genus *Streptoalloteichus* (Tomita *et al.*, 1978).

The genus would appear to be represented in low numbers in most soils but isolation requires the use of heating dry soil, to reduce the number of associated bacteria, and selective media (Nonomura and Ohara, 1971a). The latter authors also noticed the preferential growth of *Microtetraspora* colonies on soil particles placed on the low-nutrient isolation medium suggesting a requirement for growth factors supplied by the soil.

E. *Planobispora* (Thiemann and Beretta, 1968)

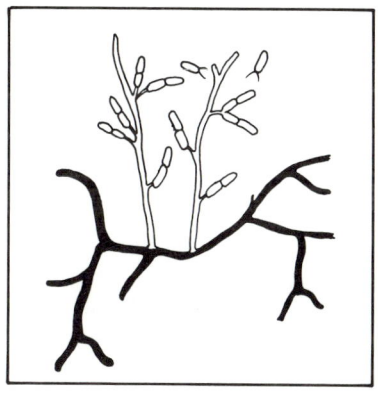

Gram-positive, non-acid-fast sporoactinomycetes which form a branched, non-fragmenting substrate mycelium that penetrates the agar medium and forms the colony which bears a powdery white (*Plob. longispora*) or rose coloured (*Plob. rosea*) aerial mycelium. Sporangia (1.0–1.2 × 6–8.0 μm) are formed singly or in groups along the aerial hyphae. They contain a longitudinal pair of fusiform straight or slightly curved spores (1.0–1.2 × 2.6–4.0 μm) and retain their integrity and shape after release of their spores (Thiemann, 1970). Detailed studies on spore formation have not been undertaken but it has been suggested that they are formed endogenously in a similar way to spore formation in *Dactylosporangium* and *Planomonospora* (Sharples *et al.*, 1974). The spores become motile after release and have peritrichous flagella. Walls contain *meso*-DAP and lack arabinose and galactose.

The type species is *Planobispora longispora*, the type strain ATCC 23867.

Strains would appear to be rare in soil and have only been isolated by Thiemann and Beretta (1968) using a hitherto undisclosed method.

F. *Planomonospora* (Thiemann *et al.*, 1976b)

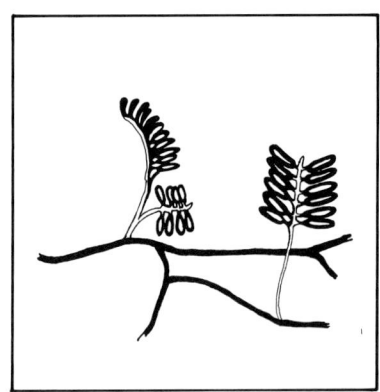

Gram positive, non-acid-fast sporoactinomycetes which form a substrate mycelium (0.6–0.8 μm) that grows into agar media and forms a compact colony which bears the obvious but sparsely branched aerial mycelium (0.5–1.0 μm). Elongated sporangia (1.5 μm wide × 3.5–4.5 μm long) are formed only on the aerial hyphae in a close double, parallel row (*Plom. parontospora*) or on both sides of the hyphae to give a characteristic palm leaf pattern (*Plom. venezuelae*; Thiemann, 1970). The sporangia contain a single fusiform spore which develops endogenously (Williams *et al.*, 1973; Sharples *et al.*, 1974) and is released through a definite apical pore. Spores (1.0 × 3.0–3.0 μm) become motile 30 to 40 minutes after release and have lophotrichous flagellation. Walls contain *meso*-DAP and madurose has been detected in whole-organism hydrolysates.

The type species is *Planomonospora parontospora*, the type strain ATCC 23863.

Strains have been isolated infrequently from temperate and tropical soils (Thiemann *et al.*, 1967b; Thiemann 1970; Schäfer, 1973). One strain was reported to produce the antibiotic sporangiomycin (Thiemann *et al.*, 1968) and another a protease inhibitor (Wingendar *et al.*, 1975).

G. *Spirillospora* (Couch, 1963)

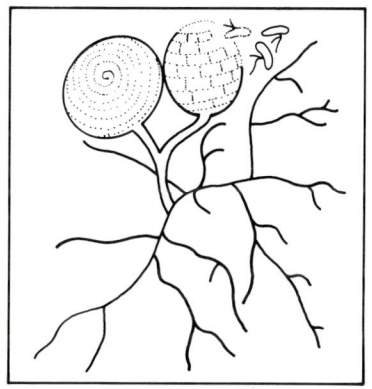

Mycelial actinomycetes with a fine (0.2–1.0 μm) branching mycelium penetrating an agar medium and forming pigmented colonies which carry a woolly white aerial mycelium. Spherical spore vesicles (sporangia), 5–24 μm in diameter, form on side branches of the aerial hyphae. The enclosing membrane is an extension of a sheath that surrounds the aerial hyphae and 'sporangiophores'. Within this sheath, branched hyphae coil and eventually become subdivided into compartments that remain arranged in parallel coils (spirals) until dehiscence (Lechevalier *et al.*, 1966b). Mature vesicles placed in water release motile rod-shaped spores (0.8–1.8 × 2.8–3.2 μm) with a sub-polar tuft of flagella (Higgins *et al.*, 1967; Schäfer, 1973). Occasionally, chains or coils of spores without an enclosing sheath can be observed on the aerial hyphae. The peptidoglycan contains *meso*-DAP (wall chemotype III, Lechevalier *et al.*, 1966a) with minor amounts of HO-DAP and galactose (Yamaguchi, 1965).

The type species is *Spirillospora albida*, the type strain ATCC 15331.

Strains have been isolated from soil by baiting with pollen (Couch, 1963) or mouse hair (Schäfer, 1973) and from leaf litter by subculturing washings on colloidal chitin agar (Willoughby, 1969a).

H. *Streptosporangium* (Couch, 1955)

The aerial mycelium bears single or clustered spore vesicles (sporangia) which are commonly 7–20 μm in diameter but may be up to 40 μm. Aplanospores form by the septation of a spiral, unbranched hypha within the spore vesicle (Lechevalier *et al.*, 1966b; Williams *et al.*, 1973; Sharples *et al.*, 1974). The sporangium is normally spherical. *Streptosporangium corrugatum* forms small globose spore vesicles (1.0–5.0 μm diameter) enclosing a coiled chain of about 20 spores together with club-shaped vesicles (0.75–1.0 μm wide × 3.5–8.0 μm long) enclosing a straight row of three to eight spores (Williams and Sharples, 1976). Three species, '*Stso. bovinum*' (Chaves Batista *et al.*, 1963), *Stso. roseum* (Couch, 1955) and *Stso. viridogriseum* (Okuda *et al.*, 1966) were also reported to form streptomycete-like chains of arthrospores. Most species grow best at temperatures between 25 and 30°C, exceptions are *Stso. nondiastaticum* (Nonomura and Ohara, 1969b) growing up to 42°C and *Stso. pseudovulgare* (Nonomura and Ohara, 1969b) which will grow at 55°C. The non-motile spores are released when the spore-vesicle is placed in water; the surrounding sheath is ruptured when the intersporal matrix swells to exert pressure. The spores are spherical, oval or short rods and usually smooth, though ridges were reported on the spores of *Spso. corrugatum* (Williams and Sharples, 1976).

Streptosporangia have a DNA base composition in the range of 69.5 to 71 mol% G+C (Jones and Bradley, 1964; Tsyganov *et al.*, 1966; Yamaguchi, 1967; Farina and Bradley, 1970). An exceptional strain, with an optimum growth temperature between 50 and 55°C and named *Stso. album* var. *thermophilum* (Manchini *et al.*, 1965), was found to have a lower G+C content of 49.0–53.7 mol% (Craveri *et al.*, 1965; Silvestri, 1970). Subsequent taxonomic studies on this strain by one of the authors (T.C.) proved it to be a *Thermoactinomyces* sp. Walls contain *meso*-DAP but lack galactose and arabinose (Becker *et al.*, 1965; Lechevalier *et al.*, 1966a) so placing them in the wall chemotype III classification of Lechevalier and Lechevalier (1970b). *Streptosporangium*, together with *Intrasporangium* and *Microbispora*, characteristically have a phospholipid pattern containing glucosamine (GluNU) together with phosphatidylinositol, phosphatidylethanolamine and diphosphatidylglycerol but no phosphatidylglycerol (phospholipid type PIV; Lechevalier *et al.*, 1977). In addition, streptosporangia contain phosphatidyl methyl ethanolamine which is absent from *Intrasporangium* and *Microbispora*.

The type species is *Streptosporangium roseum*, the type strain ATCC 12428.

Streptosporangia had been infrequently isolated from soil and dung (Couch, 1955) or leaf litter (Van Brummelen and Went, 1957; Protekhina, 1965), before the introduction of a specific isolation technique by Nonomura and Ohara (1969a) showed that they were a significant component of the actinomycete population in soils. Species have also been isolated from lake sediments (Johnston and Cross, 1976) and beach sand (Williams and Sharples 1976). '*Streptosporangium bovinum*' was isolated from infected bovine hooves (Chaves Batista *et al.*, 1963).

Streptosporangium indianesis (Gupta, 1965) does not form true spore vesicles, according to Schäfer (1969), and is probably a *Streptomyces* sp. with spore aggregates resulting from the autolysis of sporulating aerial hyphae. Motile spores have been reported for isolates labelled *Streptosporangium* type 1 isolated from stream water (Willoughby, 1969b), but the descriptions and photographs would suggest that these strains belong to the genus *Actinoplanes*. They can be clearly distinguished from the isolates labelled *Streptosporangium* type 2, isolated mostly from soil and lake muds, which had a typical and evident aerial mycelium.

Morphological features, differences in nitrate reduction, starch hydrolysis and pigmentation have been used to justify proposals for new species. We still await a definitive study of these and alternative characters for use in species differentiation and identification.

11. Thermomonosporas

An artificial grouping of mesophilic and thermophilic actinomycetes with a cell wall containing *meso*-DAP and no other characteristic sugars or amino acids (wall chemotype III). Few comparative studies have been done which would reveal relationships amongst the genera though *Thermomonospora* and *Nocardiopsis* differ from most other actinomycetes in containing menaquinones with unusually long, partially saturated isoprenyl side chains (MK-$10(H_4-H_8)$).

A. *Actinosynnema* (Hasegawa *et al.*, 1978)

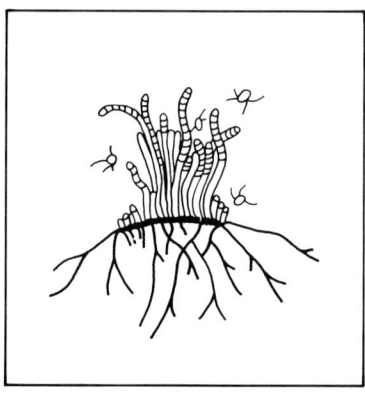

Aerobic, mesophilic actinomycetes with a finite substrate mycelium (0.5 μm diameter) penetrating nutrient agar media and forming yellow colonies with characteristic synnemata (up to 160–180 μm high) or dome-like bodies on their surfaces. Substrate hyphae growing in liquid culture may break up into motile elements with peritrichous flagella. A white to pale-yellow aerial mycelium, composed of long branching hyphae (0.5 to 1.0 μm diameter), originates as a tuft from the tips of the synnemata and occasionally from the surface of flat colonies. The aerial hyphae become subdivided into chains of smooth, rod-shaped spores which become motile with peritrichous flagella when flooded with nutrient broths.

The cell wall contains *meso*-DAP as the diamino acid with no other characteristic amino acids or sugars (wall chemotype III). Mycolic acids are absent and the phospholipid composition includes major amounts of phosphatidylinositol mannosides, phosphatidylinositol, phosphatidylethanolamine and diphosphatidylglycerol (Hasegawa *et al.*, 1979). The DNA base composition is 71 mol% G+C (Hasegawa *et al.*, 1982).

The type species is *Actinosynnema mirum*, the type strain IMRU 3971 (=ATCC 29888).

The type species was isolated from a blade of grass taken from river water and incubated on dilute yeast extract agar. Synnemata appeared on the surface of the blade and portions of the growth were streaked onto fresh medium. Other strains, isolated by the same method, were initially placed in the genus *Nocardia* (Higashide *et al.*, 1977; Tanida *et al.*, 1980a, b, c) but have recently been reclassified as a new species and subspecies in the genus *Actinosynnema* (Hasegawa *et al.*, 1982).

The 'spore-dome' actinomycetes described by Willoughby (1969a) show similarities to *Actinosynnema* in that they have dome-like projections on the surface of their colonies which release motile spores. A collection of similar spore-dome strains recently isolated by Makkar and Cross (1982) were all unable to form aerial mycelium and released motile spores directly from the colony surface. The majority of strains contained LL-DAP and glycine as wall components (wall chemotype I) but some contained major amounts of *meso*-DAP as well as the LL-DAP isomer. All spore-dome strains contained tetrahydrogenated menaquinones with nine isoprene units as the only isoprenologue (MK-$9(H_4)$) whereas *Acty. mirum* contained both MK-$9(H_4)$ and MK-$9(H_6)$. The relationship between *Actinosynnema* species and the so called 'spore-dome' actinomycetes requires further study.

B. *Nocardiopsis* (Meyer, 1976)

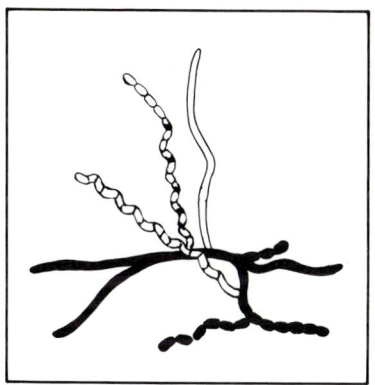

Aerobic, Gram-positive, non-acid-fast, non-motile sporoactinomycetes which form a branched substrate mycelium that fragments into rod and coccal elements. The organism produces abundant aerial hyphae which frequently form chains of smooth spores of variable length. During spore formation hyphae have a zig-zag appearance due to developing spores subtending angles of varying degree to their neighbouring spores. The spores are enclosed within a fibrillar sheath and have thickneneded polar walls.

The cell wall contains *meso*-DAP as the diamino acid with no other characteristic amino acids or sugars (wall chemotype III). Representative strains lack madurose and mycolic acids, contain major amounts of *iso*- and *anteiso*-branched chain fatty acids, have either tetra- or hexa-hydrogenated menaquinones with ten isoprene units, MK-$10(H_4, H_6)$, as the predominant isoprenologue, and possess large amounts of diphosphatidylglycerol, phosphatidylglycerol and phosphatidylcholine. The DNA base composition is within the range 70–76 mol% G+C (M. Mordarski *et al.*, unpublished data). The organism is common in soil, in stored products such as mouldy grain and hay, and has occasionally been implicated in ocular and pulmonary infections.

The type species is *Nocardiopsis dassonvillei,* the type strain ATCC 23218.
Nocardiopsis (Meyer, 1976) was proposed for *Acmd. dassonvillei,* which differs in many respects from *Actinomadura* species (Agre *et al.,* 1975; Williams *et al.,* 1976; Goodfellow *et al.,* 1979). *Nocardiopsis dassonvillei* appears to be closely related to *Streptomyces* with respect to biochemical, physiological and morphological properties (Gordon and Horan, 1968; Williams *et al.,* 1983a) but can be distinguished from the latter on the basis of wall (H. A. Lechevalier and Lechevalier, 1970) and lipid composition (Collins *et al.,* 1977; Lechevalier *et al.,* 1977) and by phage-host range studies (Wellington and Williams, 1981b). A second species *Nrdp. syringae,* was described by Gauze *et al.* (1977), but was not included on the *Approved Lists of Bacterial Names* (Skerman *et al.,* 1980). This organism appears to be similar to *Nrdp. dassonvillei,* being distinguished by its formation of lilac-coloured aerial mycelium. Preobrazhenskaya *et al.* (1978) noted a similarity between *Nocardiopsis* and some actinomadurae and recommended that *Acmd. africana, Acmd. coeruleofusca, Acmd. flava* and *Acmd. longispora* be transferred to the genus *Nocardiopsis.*

c. *Streptoalloteichus* (Tomita *et al.,* 1978)

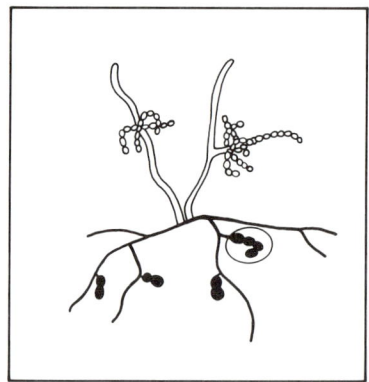

Colonies of this unusual organism bear an abundant cover of light yellowish beige or pale pinkish yellow aerial mycelium. Curved or L-shaped spore chains are short (10 to 20 spores) and branched, and usually appear in dense clusters, sometimes with free and longer spore chains protruding. Individual spores have a smooth surface and are oval to cylindrical in shape. Sclerotia are also formed among the aerial hyphae.

The substrate mycelium bears spherical to sub-spherical sporangia, usually with an uneven surface (1.5–4.5 × 2.7–7.0 μm) and containing one to four spores in a straight or V-shaped chain. Individual spores are rod-shaped or irregular, 0.9–1.5 × 1.2–4.0 μm in size, and are motile with polar flagella. Smaller peanut-shell shaped sporangia can also be seen.

The species grows well at 32–50°C with no growth at 12 or 56°C. Sodium chloride reduces growth when present in the medium at 5% (w/v), there is no growth with 7%. Purified walls contain *meso*-DAP with only a trace of glycine. Whole-organism sugars include large amounts of galactose and mannose with a smaller rhamnose component. No madurose was found.

The type species is *Streptoalloteichus hindustanus*, the type strain is ATCC 31217.

Streptoalloteichus hindustanus produces the nebramycin antibiotic factors originally found in culture filtrates of *Stmy. tenebrarius* (Higgens and Kastner, 1968). When *Stmy. tenebrarius* ATCC 17920 was examined again it was found to grow well at high temperatures, form spore chains in dense clusters, produce sclerotia and to have a carbohydrate utilization pattern similar to that of *Sall. hindustanus*. It also had a wall containing *meso*-DAP with little glycine and a whole-organism sugar pattern of galactose with small amounts of mannose and rhamnose. So *Stmy. tenebrarius* would appear to be a doubtful member of the genus *Streptomyces* but no sporangia could be found on the substrate hyphae and its relationship to *Streptoalloteichus* awaits further study.

D. *Thermomonospora* (Henssen, 1957)

Aerobic, Gram-positive, non-motile sporoactinomycetes typically forming a branched substrate mycelium and leathery colonies covered with aerial mycelium when grown on agar media. Single heat-sensitive spores are formed at the tips of short simple or branched sporophores borne laterally on the aerial hyphae or on both aerial and substrate hyphae (Cross and Goodfellow, 1973). Aerial mycelium production and sporulation of species in the white *Thermomonospora* group (*Temo. alba*, *Temo. curvata* and '*Temo. fusca*') is improved by growing strains on media at a high pH (>8.0; McCarthy, 1981). The spores may appear phase bright but are not endospores and, unlike the spores of *Thermoactinomyces* species, are very rapidly killed when aqueous suspensions are heated at 90°C ($D_{90°C}$ <2 minutes). Most commonly encountered *Thermomonospora* species are facultative thermophiles with a growth temperature range 30 to 55°C and an optimum growth temperature of 50°C. *Thermomonospora mesophila* (Nonomura and Ohara, 1971a) has a lower growth temperature range from 25 to 48°C and an optimum of 40°C.

The cell wall contains *meso*-DAP as the diamino acid without any other characteristic amino acids or sugars (wall chemotype III). *Thermomonospora albus*, *Temo. curvata* and '*Temo. fusca*' contain very complex mixtures of hydrogenated menaquinones with ten and eleven isoprene units in contrast to '*Temo. chromogena*' which has major amounts of tetrahydrogenated menaquinones with only nine isoprene units (Collins *et al.*, 1982f). There is no information available on the G+C content of the DNA for any member of the genus. Species can be isolated from soil and lake sediments but highest numbers occur in natural vegetable composts, prepared mushroom compost and overheated fodders (Lacey, 1973, 1978; McCarthy and Cross, 1981).

The type species is *Thermomonospora curvata*, the type strain ATCC 19995.

Henssen (1957) originally described three thermophilic species which were reported to form single spores on the aerial mycelium only. *Thermomonospora curvata*, the only species isolated in pure culture, was later formally designated the type species of the genus (Henssen and Schnepf, 1967). A recent numerical phenetic study (McCarthy, 1981) has confirmed the status of *Temo. curvata* and provided strong evidence for the formal recognition of another of Henssen's original species, '*Temo. fusca*', which had meanwhile been isolated in pure culture and described in detail (Crawford, 1975; Crawford and Gonda, 1977) but not formally recognized (Skerman *et al.*, 1980). A third cluster recovered in this numerical taxonomic study included the type strains of *Temo. alba* (Locci *et al.*, 1967; Cross and Goodfellow, 1973) and *Temo. mesouviformis* (Nonomura and Ohara, 1974) and the former name would have nomenclatural priority. The above three species, termed the 'white *Thermomonospora* group' because of their white aerial mycelium, showed little phenetic similarity to a large cluster of strains which included the type strains of '*Actinobifida chromogena*' (Krasilnikov and Agre, 1965), '*Temo. falcata*' (Henssen, 1970) and many similar actinomycetes isolated from mushroom compost (McCarthy and Cross, 1981). These '*chromogena*' strains with reddish brown colonies, a light-brown aerial mycelium and a different menaquinone composition (see above) had been provisionally included in the genus *Thermomonospora* (Cross, 1981b) because of their wall composition and morphology. This location is convenient and practical but may have to be reviewed in the light of future studies and comparisons.

The taxon previously named *Temo. viridis* (Küster and Locci, 1963) has now been transferred to the genus *Saccharomonospora* (Nonomura and Ohara, 1971a). '*Thermomonospora galeriensis* (also referred to as *Thermopolyspora galeriensis* and *Micromonospora galeriensis*; Szabo *et al.*, 1976; Válylyi-Nagy *et al.*, 1970) has a wall chemotype IV and was reported to produce single spores and a dark green pigment. It would also appear to belong to the genus *Saccharomonospora*.

Members of the white *Thermomonospora* group, especially '*Temo. fusca*', are cellulolytic and also degrade pectin, xylan and starch suggesting that they have a fundamental role in the breakdown of plant material in overheated natural substrates such as composts. They are currently being evaluated for cellulose bioconversion processes designed to produce single cell protein (Crawford *et al.*, 1973; Bellamy 1974, 1977; Humphrey *et al.*, 1977) or sugar syrups for fermentation to ethanol (Lee and Humphrey, 1979; Hägerdal *et al.*, 1979). Several of these publications have referred to a cellulolytic *Thermoactinomyces* strain for which the specific name '*Team. cellulosae*' was suggested (Vacca and Bellamy, 1976) but this organism has now been acknowledged to be a member of the genus *Thermomonospora* (Hägerdal *et al.*, 1980) and its correct identification must be established.

12. Micropolysporas

It has become increasingly evident that there are a number of sporoactinomycetes with a peptidoglycan containing *meso*-DAP and wall-associated arabino-galactan polymers but which do not contain mycolic acids. They cannot be accommodated with the nocardioform actinomycetes and, in order to direct future taxonomic studies to this relatively poorly studied assemblage, we propose to group them under the epithet 'Micropolysporas'. This aggregate group contains the currently recognized genera *Actinopolyspora*, *Pseudonocardia*, *Saccharomonospora* and *Saccharopolyspora* together with species that have been misclassified in the genera *Nocardia* and *Streptomyces* as a result of undue weight being given to morphological characters. The transfer of *Mips. brevicatena* to the genus *Nocardia* (Goodfellow and Pirouz, 1982) has left the remaining *Micropolyspora* species without a type species but the proposal to conserve the generic name *Micropolyspora* would provide a genus for some of these organisms and a convenient umbrella name for the group (McCarthy *et al.*, 1983).

It is not yet clear whether all of the genera in the aggregate group really merit generic status or if they collectively form a distinct evolutionary branch. They do, however, have a number of properties in common. They are all aerobic, Gram-positive, non-motile and catalase positive but they are morphologically somewhat heterogeneous. Thus, single or short chains of spores can be present either on the aerial mycelium or on both the aerial and the substrate mycelium. Fragmentation is generally much less pronounced than in nocardioform actinomycetes and may be partly due to localized areas of autolysis (Williams *et al.*, 1976). Further, in addition to having a wall chemotype IV, Micropolysporas contain major amounts of branched chain *iso* and *anteiso*-fatty acids, tetrahydrogenated menaquinones with nine isoprene units as the major isoprenologue and, with a single exception, DNA rich in G+C (Table 2). They show some variation in polar lipid composition but it is interesting that *Actinopolyspora*, *Micropolyspora* and *Saccharopolyspora* strains all contain phosphatidylcholine.

A. *Actinopolyspora* (Gochnauer *et al.*, 1975)

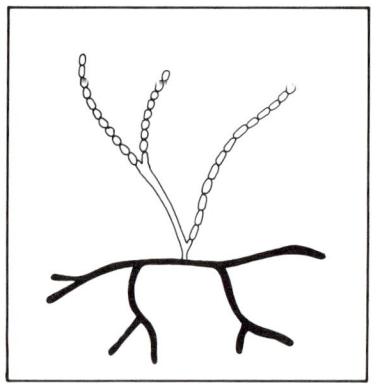

Gram-positive, acid-fast, sporoactinomycetes with a non-fragmenting substrate mycelium lacking spores and a white aerial mycelium that is subdivided into chains of smooth spores. The single species, *Acts. halophila*, was isolated as a contaminant in a culture medium containing 25% (w/v) sodium chloride and is unusual among actinomycetes in requiring a medium containing at least 10% (w/v) sodium chloride for growth. There is no growth in its absence, optimum growth occurs in concentrations of 15 to 20% and slow growth in sodium chloride concentrations as high as 30% (w/v). It is a mesophile with optimum growth at 37°C and a growth temperature range of 10 to 43°C.

The cell wall contains *meso*-DAP together with arabinose and galactose (wall chemotype IV) but mycolic acids are absent. Its lipid composition includes the phospholipids phosphatidylcholine, lysophosphatidylglycerol and phosphatidylglycerol, two glycolipids, tetrahydrogenated menaquinones with nine isoprene units ($MK-9(H_4)$), and the fatty acids are mostly branched chain iso-C_{15}–C_{17} and $anteiso$-C_{17}. The organism is sensitive to lysozyme and its DNA base composition was found to be 64.2 mol% G+C.

The type species is *Actinopolyspora halophila*, the type strain ATCC 27976.

This unusual halophilic actinomycete was given generic status largely on the basis of salt requirement, sporulation on the aerial mycelium only, acid fastness, growth temperature requirement and G+C content. Its current relocation in the Micropolyspora group indicates that there could be several related taxa which might now be included in a common genus. Salt requirement should be regarded as a species rather than a generic character.

B. *Micropolyspora nomen conservandum*

The genus *Micropolyspora* was originally erected to accommodate two actinomycete strains isolated from the sputa of patients who had undergone treatment for tuberculosis (Lechevalier *et al.*, 1961). They were given the specific name *Mips. brevicatena* and typically produced short chains of spores on both substrate and aerial mycelium. The name *Micropolyspora* later proved to be a source of nomenclatural confusion when it was realized that Shchepkina (1940) had applied a similar name, *Micropolispora*, to sporangia forming actinomycetes isolated from cotton bolls. The original cultures were not available for comparative study and their descriptions were not sufficiently detailed to permit assignment to a recognized or new actinomycete genus. Consequently H. A. Lechevalier (1968) requested that the genus name *Micropolyspora* be conserved and that *Micropolispora* be regarded as a *nomen dubium*.

Micropolyspora has remained, essentially, a form-genus and morphology has featured prominently if not exclusively in the description of new species.

Becker *et al.* (1965) demonstrated that *Micropolyspora* spp. had a wall chemotype IV and wall composition has been incorporated into the genus description (Cross and Goodfellow, 1973). This clearly improved the circumscription of the genus and as a result the species '*Mips. rubrobrunea*' and '*Mips. viridinigra*' (Krasilnikov *et al.*, 1968), which have a wall chemotype III, were transferred to the genus *Excellospora* (Agre and Guzeva, 1975).

Probably the most detailed studies on members of this genus have been directed towards *Mips. faeni* (Cross *et al.*, 1968), the main causative agent of farmer's lung and previously named '*Thermopolyspora polyspora*' (Corbaz *et al.*, 1963), and *Mips. rectivirgula* (Prauser and Momirova, 1970) previously named '*Thermopolyspora rectivirgula*' (Krasilnikov and Agre, 1964). They both produce chains of spores on the aerial and substrate mycelium, have a wall chemotype IV but lack mycolic acids (Mordarskaya *et al.*, 1973; Collins *et al.*, 1977; Kroppenstedt and Kutzner, 1978), and so differ from the type species, *Mips. brevicatena*, which contains mycolic acids (Collins *et al.*, 1977). *Micropolyspora faeni* and *Mips. rectivirgula* have recently been shown to be synonomous (Arden-Jones *et al.*, 1979) yet still exhibit the morphological characters proposed for the genus *Micropolyspora*. *Micropolyspora brevicatena*, on the other hand, has menaquinones (Collins *et al.*, 1977), phospholipids (Lechevalier *et al.*, 1977), fatty acids (Kroppenstedt and Kutzner, 1978) and mycolic acids typical of *Nocardia* and consequently has been transferred to the genus *Nocardia* as *Nrda. brevicatena* comb. nov. (Goodfellow and Pirouz, 1982). One implication of this reclassification is that the genus *Micropolyspora*, having lost its type species, would be nomenclaturally invalid (Lapage *et al.*, 1975). The other *Micropolyspora* spp. cited in the *Approved Lists of Bacterial Names* (Skerman *et al.*, 1980), namely *Mips. faeni* (syn. *Mips. rectivirgula*), *Mips. angiospora* (Zhukova *et al.*, 1968) and *Mips. internatus* (Agre *et al.*, 1974) cannot at present be accommodated in *Nocardia* or any other recognized genus of the *Actinomycetales*. The loss of the generic name, particularly because of its widespread use in the binomial *Mips. faeni*, would cause considerable confusion. We therefore propose to conserve the generic name *Micropolyspora* for sporoactinomycetes with chains of spores on the aerial and substrate mycelium and a wall envelope containing *meso*-DAP (wall chemotype IV) but lacking mycolic acids (McCarthy *et al.*, 1983). The lipid data so far available are given in Table 2. A more difficult problem concerns the proposal for a new type species. The most obvious candidate appears to be *Mips. faeni* because of the considerable information available on this species, the common usage of the binomial and to minimize confusion in the literature (McCarthy *et al.*, 1983).

C. *Pseudonocardia* (Henssen, 1957)

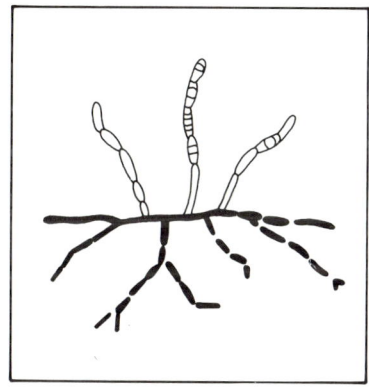

Aerobic, Gram-positive, non-acid-fast actinomycetes which form a non-fragmenting substrate mycelium (0.4 × 1.8 μm diameter) that has a characteristic zig-zag appearance in liquid and on solid media. The angularly displaced cells are a product of the acropetal budding of the growing hyphae, and they may or may not be separated by septa. The unbranched aerial hyphae (0.4–1.8 μm diameter) arise as lateral branches and become transformed into long cylindrical spores (2.5–5.0 μm long). In ageing cultures the spores may divide again to yield chains of spherical to short cylindrical spores (H. A. Lechevalier and Lechevalier, 1970). In such old cultures, the substrate hyphae also form terminal and intercalary spores. The spores may be smooth walled or spiny, are usually 0.5 to 1.0 μm wide × 1.5–3.0 μm long but both longer and broader spores have been observed.

Pseudonocardiae lack mycolic acids but have a cell wall which contains *meso*-DAP together with arabinose and galactose (wall chemotype IV). *Pseudonocardia thermophila* contains major amounts of branched chain *iso*- and *anteiso*-fatty acids, tetrahydrogenated menaquinones with nine isoprene units, MK-$9(H_4)$, as the major isoprenologue, phosphatidylcholine, phosphatidylethanolamine, phosphatidylinositol and phosphatidylmonoethanolamine, and its DNA base composition was found to be 79 mol% G+C (Lechevalier *et al.*, 1971). The organism has been isolated from manure and soil.

The type species is *Pseudonocardia thermophila*, the type strain ATCC 19285.

Henssen (1957) originally proposed the genus *Pseudonocardia* for a single species, *Psnc. thermophila*. The genus description was emended by Henssen and Schäfer (1971) when another species, *Psnc. spinosa*, was described. Strains of *Psnc. thermophila* form colourless or yellow–orange colonies on a variety of rich media under aerobic or anaerobic conditions at 40 to 50°C but grow poorly at 28°C. *Pseudonocardia spinosa* also forms yellow colonies covered by abundant white aerial hyphae but can be distinguished from the type species by its slow growth, mesophilic properties, absence of septa in substrate and aerial mycelium, by its less pronounced zig-zag growth and spiny spores.

Pseudonocardia is disinguished from related genera by certain unusual morphological properties but its detailed affinities to other wall chemotype IV taxa lacking mycolic acids will only be established when additional strains have been examined for a wide range of characters. To date, few strains have been isolated though high numbers of *Pseudonocardia*-like organisms have been reported from sugar cane bagasse (Lacey, 1973, 1978, 1981).

D. *Saccharomonospora* (Nonomura and Ohara, 1971c)

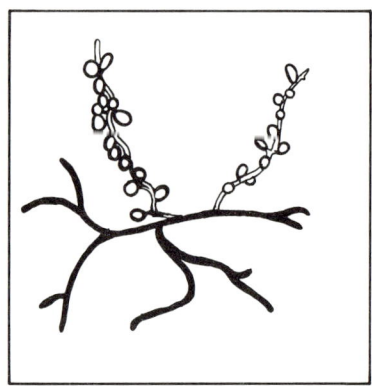

Aerobic, Gram-positive, non-fragmenting, non-motile sporoactinomycetes typically forming a branched substrate mycelium and leathery colonies covered with aerial mycelium when grown on agar media. Single heat-sensitive spores are densely packed along the aerial hyphae on unbranched sporophores of variable length (Cross and Lacey, 1970; Locci, 1971). Spores may be observed on hyphae at the surface of the agar and, in some strains, there have been reports of limited spore formation on the substrate hyphae (Corbaz *et al.*, 1963; Kalakoutskii *et al.*, 1968; Krasilnikov *et al.*, 1970; Prauser 1970; Agre *et al.*, 1974) and on the vegetative hyphae grown in submerged culture. Typical strains grown at 40 or 50°C initially form a white aerial mycelium which becomes a characteristic grey–green to dark green when sporulation occurs on continued incubation. Other strains only become pigmented when incubated at 40°C but at 30°C most strains remain white and may then appear similar to *Thermomonospora* species. Sporulation in most strains is also accompanied by the production of a dark green soluble pigment but one unusual strain (J. Lacey A969), which is similar to the type strain in all other respects (McCarthy, 1981), forms a conspicuous lilac to purple soluble pigment.

The cell wall contains *meso*-DAP as the diamino acid together with arabinose and galactose (wall chemotype IV). The species lacks mycolic acids but contains major amounts of *iso*- and *anteiso*-fatty acids, and tetrahydrogenated menaquinones with nine isoprene units as the major isoprenolog (Collins *et al.*, 1982). The G+C content of the DNA ranges from 74 to 75 mol%. Strains can be found in soil or lake sediments but are more frequently encountered in overheated vegetable composts and fodders.

The type species is *Saccharomonospora viridis*, the type strain ATCC 15386.

The genus currently contains one species, *Sacs. viridis* (syn. *Thermoactinomyces monosporus* Schütze, 1908; *Thermoactinomyces viridis* Schuurmans *et al.*, 1956; *Thermopolyspora glauca* Corbaz *et al.*, 1963; *Thermomonospora viridis* Küster & Locci, 1963). It is a commonly encountered facultatively thermophilic species and most isolates can be identified with little difficulty. A wall chemotype IV, the production of a greyish aerial mycelium and a dark green soluble pigment on certain media are characteristics of this species that are also exhibited by actinomycetes classified in the genus *Micropolyspora*. '*Micropolyspora caesia*' (Kalakoutskii, 1964) was found to be morphologically indistinguishable from *Sacs. viridis* (Lechevalier and Lechevalier, 1967; Williams *et al.*, 1976) but the former species was reported to form single and short chains of spores on the substrate mycelium in the original publication and the two species differed in their fatty acid composition (Kroppenstedt and Kutzner, 1978). A recent numerical phenetic study (McCarthy, 1981) recovered ten *Sacs. viridis* strains in a cluster defined at 81%S which included the type strain of *Sacs. viridis* and '*Mips. caesia*' but another numerical study (Goodfellow and Pirouz, 1982) separated the two taxa. '*Micropolyspora coerulea*' (Preobrazhenskaya *et al.*, 1973) and *Mips. internatus* (Agre *et al.*, 1974) also exhibit green pigmentation and mainly single spores. Again sporulation was observed on both aerial and substrate hyphae and, although spores were mainly single, the presence of short chains was reported.

Comment: The monospecific genus *Saccharomonospora*, forming single spores only on the aerial hyphae, is currently distinguished from *Micropolyspora* on morphological criteria: species in the latter genus form single and short chains of spores on both substrate and aerial hyphae. The distinction now appears very artificial and there is an urgent need for a detailed study of the relationships among the non-mycolic acid containing wall chemotype IV actinomycetes. Numerical phenetic studies have been of only limited value and alternative taxonomic methods must be applied to redress or validate the current emphasis on morphology.

E. *Saccharopolyspora* (Lacey and Goodfellow, 1975)

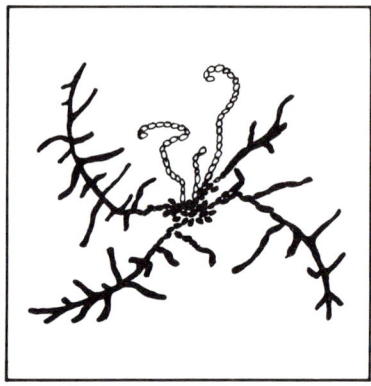

Aerobic, Gram-positive, non-acid-fast actinomycetes which form a well developed substrate mycelium (0.4–0.6 μm diameter) that fragments into rod-shaped elements (1.0 × 0.5 μm), more often in older parts of the colony and seldom near the growing margins. The white aerial mycelium (0.5–0.7 μm diameter) is characteristically segmented into bead-like chains of spores (0.7–1.3 × 0.5–0.7 μm) which are usually separated by lengths of 'empty' hyphae and retained in a distinctive hairy sheath. Colonies are thin, raised or convex, slightly wrinkled, mucoid or gelatinous in appearance and carry sparse aerial mycelium which is often produced in tufts. Optimum growth occurs at temperatures of 37 to 40°C and the growth temperature range extends from 25 to 50°C.

The cell wall contains *meso*-DAP together with arabinose and galactose. Saccharopolysporas lack mycolic acids, are sensitive to lysozyme, contain tetrahydrogenated menaquinones with nine isoprene units, MK-9(H_4), as the major isoprenolog and phosphatidylcholine, phosphalidylinositol and phosphatidylmonoethanolamine. The DNA base composition is 77 mol% G+C. The organism has only been isolated from sugar cane bagasse.

The type species is *Saccharopolyspora hirsuta*, the type strain ATCC 27875.

This genus, represented by the single species *Saps. hirsuta*, was shown to differ from typical members of the genera *Actinomadura*, *Mycobacterium*, *Nocardia* and *Rhodococcus*. However, its status may have to be reconsidered following comparisons with other members of the Micropolyspora aggregate group.

F. UNASSIGNED MICROPOLYSPORAS

It has become evident that several species, previously classified in the genera *Nocardia* and *Streptomyces*, cannot be accommodated in these or other actinomycete genera as they are currently defined (Table 10). Strains in these

TABLE 10. Unassigned micropolysporas which have a wall chemotype IV and lack mycolic acids

Species	Comments
'*Nrda. aerocolonigenes*' (Shinobu & Kawato, 1960) Pridham 1970	Only one of 14 strains formed aerial mycelium spores. Contain galactose but *no* arabinose according to Gordon *et al.* (1978). Traces of arabinose were detected by Pridham and Lyons (1969)
'*Stmy. africanus*' (Pjiper & Pullinger, 1927) Waksman & Henrici, 1948	Found to contain *meso*-DAP, galactose and a trace of arabinose by Pridham and Lyons (1969)
'*Stmy. albovinaceus*' (Kudrina) Pridham *et al.* 1958 Strain SC 3511 = ATCC 12951	Strain produces the antibiotic rifamycin B and differs from the type strain of "*Stmy. albovinaceus*" INA 273/53; ATCC 15823 (Pridham and Lyons, 1969)
'*Nrda.*' *autotrophica* (Takamiya & Tubaki, 1956), Hirsch, 1961	Amplified description of many strains given by Gordon *et al.* (1974)
'*Nrda. capreola*' (Stark *et al.*, 1963) Pridham 1970	Showed little affinity to *Nocardia* or *Streptomyces* in a numerical phenetic study (Williams *et al.*, 1983a)
'*Nrda. leishmanii*' Chalmers & Christopherson, 1916	Lipid composition studied by Yano *et al.* (1970)
'*Nrda.*' *mediterranea* (Margalith & Beretta 1960, 1961) Thiemann *et al.*, 1969	Chemical and numerical phenetic studies indicate a clear separation from *Nocardia* and *Streptomyces* (Alderson *et al.*, 1981)
Nrda. orientalis (Pittinger & Brigham, 1956) Pridham & Lyons, 1969	Seven out of 21 strains formed chains of spores on the aerial mycelium
'*Nrda. rugosa*' DiMarco & Spalla, 1957	Lipid composition studied by Bordet and Michel (1963)
'*Stmy. tolypophorus*' Shibata *et al.*, 1971	Found to be very similar to *Nrda. mediterranea* in a chemical and numerical phenetic study (Alderson *et al.*, 1981)

species have a cell wall with *meso*-DAP as the diamino acid and contain arabinose and galactose though the proportion of the two sugars may not be equal. Species that have been examined lack mycolic acids but show considerable variation in morphology, physiological and biochemical characters (Goodfellow and Minnikin, 1981b). Representatives of these taxa should be included in the detailed comparative studies needed to unravel the zones of variation encompassed in the Micropolyspora aggregate group. Such studies should include isolates from bagasse and fodder that have a wall chemotype IV and lack mycolic acids but which fall into two numerically defined clusters (Goodfellow *et al.*, 1979).

13. Thermoactinomyces

A. *Thermoactinomyces* (Tsiklinsky, 1899)

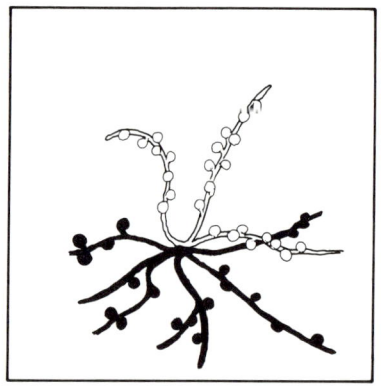

Gram-positive bacteria with extensive branching hyphae forming compact colonies differentiated into substrate and aerial mycelium. Single spores are borne laterally or terminally on both substrate and aerial hyphae. Spores are formed endogeneously and have the typical structure of endospores (Cross *et al.*, 1968a; Dorokhova *et al.*, 1968). They are phase bright, contain dipicolinic acid and exhibit heat resistance and longevity. The aerial mycelium is white in all species except *Team. dichotomicus* where a pale yellow mycelium covers the lemon yellow colonies. *Thermoactinomyces thalpophilus* forms dark-brown colonies and a soluble melanin pigment on media containing tyrosine (Cross, 1981b). The common species are thermophilic exhibiting optimum growth at 50°C with a growth temperature range from 30 to 60°C. An uncommon species, *Team. peptonophilus*, has a lower growth temperature range from 25 to 40°C.

The cell wall contains *meso*-DAP with no other characteristic amino acids or sugars (Becker *et al.*, 1965). Menaquinones are unsaturated with seven or nine isoprene units (MK-7 or MK-9; Collins *et al.*, 1982g). The DNA base composition of authentic species averages 53 mol% G+C (Fritzsche, 1967; Craveri and Manachini, 1966; Craveri *et al.*, 1966).

Thermoactinomyces species are common in natural high temperature habitats, such as leaf and compost heaps, and in over-heated stores of plant materials such as hay, bagasse and grain. The resistant spores are disseminated widely in soils and accumulate in fresh-water and marine sediments.

The type species is *Thermoactinomyces vulgaris*, the type strain KCC A-0162 (NCIB 11364, CBS 505.77).

The genus *Thermoactinomyces* was one of the earliest actinomycete genera to be named but it is now becoming increasingly evident that thermoactinomycetes should no longer be classified within the *Actinomycetales*. Their menaquinones, wall composition, dipicolinic acid containing endospores, low mol% G+C content and 16S RNA sequence data all suggest that the genus be reclassified within the family *Bacillaceae*. Their typical actinomycete-like morphology will, however, cause them to be studied and enumerated along the sporoactinomycetes and it is therefore useful to include them in this volume.

Thermoactinomyces vulgaris (syn. *Team. candidus*) and *Team. sacchari* have been implicated in the hypersensitivity diseases farmer's lung and bagassosis, respectively.

REFERENCES

Abe S., Takayama and Kinoshita, S. (1967). Taxonomical studies on glutamic-acid producing bacteria. *Journal of General and Applied Microbiology* **13**, 279–301.

Adam, A., Petit, J. F., Wietzerbin-Falszpan, J., Sinay, P., Thomas, D. W. and Lederer, E. (1969). L'acide *N*-glycolylmuramique, constituant des parois en *Mycobacterium smegmatis*: Identification par spectrometrie de marse. *FEBS Letters* **4**, 87–92.

Agre, N. S. and Guzeva, L. N. (1975). New actinomycete genus *Excellospora* gen. nov. *Mikrobiologiya* **44**, 518–523 (In Russian).

Agre, N. S., Guzeva, L. N. and Dorokhova, L. A. (1974). A new species of the genus *Micropolyspora-Micropolyspora internatus*. *Mikrobiologiya* **43**, 679–685.

Agre, N. S., Efimova, T. P. and Guzeva, L. N. (1975). Heterogeneity of the genus *Actinomadura* Lechevalier and Lechevalier. *Microbiologiya* **44**, 253–257. (In Russian).

Agre, N. S., Soina, S., Sokolov, A. A. and Guzeva, L. N. (1977). Sporogenesis in *Excellospora viridilutea* 187. *Izvestiia Akademii Nauk SSSR Seriya Biologischenskaya* 461–466.

Ahrens, R. and Moll, G. (1970). Ein neues knospendes Bakterium aus der Ostsee. *Archiv für Mikrobiologie* **70**, 243–265.

Alderson, G. and Goodfellow, M. (1979). Classification and identification of actinomycetes causing mycetoma. *Postepy Higieny I Medycyny Doswiadczalnej* **33**, 109–124.

Alderson, G., Goodfellow, M., Wellington, E. M. H., Williams, S. T., Minnikin, S. M. and Minnikin, D. E. (1981). Chemical and numerical taxonomy of *Nocardia mediterranea*. *Zentralblatt für Bakteriologie, Mikrobiologie und Hygiene. I. Abteilung, Supplement* **11**, 39–46.

An, C. S., Wills, J. W., Riggsby, W. S. and Mullin, B. C. (1983). Deoxyribonucleic acid base composition of twelve frankiae strains. *Canadian Journal of Botany* (in press).

Arden-Jones, M. P., McCarthy, A. J. and Cross, T. (1979). Taxonomic and serological studies on *Micropolyspora faeni* and *Micropolyspora* strains from soil bearing the specific epithet *rectivirgula*. *Journal of General Microbiology* **115**, 343–354.

Asselineau, C. and Asselineau, J. (1978a). Lipides spécifiques des mycobacteries. *Annales de Microbiologie (Institut Pasteur)* **129A**, 49–69.

Asselineau, C. and Asselineau, J. (1978b). Trehalose-containing glycolipids. *Progress in the Chemistry of Fats and Other Lipids* **16**, 59–99.

Asselineau, C., Clavel, S., Clément, F., Daffé, M., David, H., Lanéelle, M. A. and Promé, J. C. (1981). Constituants lipidiques de *Mycobacterium leprae* isolé de tatou infecté experimentalement. *Annales de Microbiologie (Institut Pasteur)* **132A**, 19–30.

Asselineau, J. (1966). *The Bacterial Lipids*. Hermann, Paris.

Athalye, M., Lacey, J. and Goodfellow, M. (1981). Selective isolation and enumeration of actinomycetes using rifampicin. *Journal of Applied Bacteriology* **51**, 289–297.

Athalye, M., Goodfellow, M. and Minnikin, D. E. (1984). Menaquinone composition in the classification of *Actinomadura* and related taxa. *Journal of General Microbiology* (in press).

Austwick, P. K. C. (1958). Cutaneous streptothricosis, mycotic dermatitis and strawberry footrot and the genus *Dermatophilus* Van Saceghem. *Veterinary Reviews and Annotations* **4**, 33–48.

Azuma, I., Thomas, D. W., Adam, A., Ghuysen, J. M., Bonaly, R., Petit, J. F. and Lederer, E. (1970). Occurrence of N-glycolylmuramic acid in bacterial cell walls. A preliminary survey. *Biochimica et Biophysica Acta* **208**, 444–451.

Baess, I. (1982). Deoxyribonucleic acid relatedness among species of rapidly growing mycobacteria. *Acta Pathologica et Microbiologica Scandinavica, Section B* **90**, 371–375.

Baess, I. and Magnusson, M. (1982). Classification of *Mycobacterium simiae* by means of comparative reciprocal intradermal sensitin testing on guinea-pigs and deoxyribonucleic acid hybridization. *Acta Pathologica et Microbiologica Scandinavica, Section B* **90**, 101–107.

Baess, I. and Weis Bentzon, M. (1969). Rapidly growing mycobacteria. Susceptibility to bacteriophages, reactions in the amidase test, production of acids from carbohydrates and growth at various temperatures. *Acta Pathologica et Microbiologica Scandinavica Section B* **75**, 331–347.

Baess, I. and Weis Bentzon, M. (1978). Deoxyribonucleic acid hybridization between different species of mycobacteria. *Acta Pathologica et Microbiologica Scandinavica, Section B* **86**, 71–76.

Baird-Parker, A. C. (1974). *Micrococcus*. In: *Bergey's Manual of Determinative Bacteriology, Eighth Edition* (R. E. Buchanan and N. E. Gibbons, eds.), pp. 478–483. The Williams & Wilkins Company, Baltimore.

Baker, D. and Torrey, J. G. (1979). The isolation and cultivation of actinomycetous root nodule endophytes. In: *Symbiotic Nitrogen Fixation in the Management of Temperate Forests* (J. C. Gordon, C. T. Wheeler and D. A. Perry, eds.), pp. 38–56. Forestry Research Laboratory, Corvallis, Oregon State University.

Baker, D., Pengelly, W. L. and Torrey, J. G. (1981). Immunochemical analysis of relationships among isolated frankiae (*Actinomycetales*). *International Journal of Systematic Bacteriology* **31**, 148–151.

Baldacci, E. (1958). Development in the classification of actinomycetes. *Giornale di Microbiologia* **6**, 10–27.

Baldacci, E. and Locci, R. (1961). Osservazioni e ricerche su *Micromonospora melanosporea* comb. nov. e descrizione di una nuovo sottospecie *M. melanosporea* subsp. *corymbica*. *Annali di Microbiologia ed Enzimologia* **11**, 1–12.

Baldacci, E. and Locci, R. (1966). A tentative arrangement of the genera in *Actinomycetales*. *Giornale di Microbiologia* **14**, 131–139.

Baldacci, E. and Locci, R. (1974). Genus II. *Streptoverticillium* Baldacci, 15, emend. mut. char. Baldacci, Farina and Locci 1966, 168. In *Bergey's Manual of Determinative Bacteriology, Eighth Edition* (R. E. Buchanan and N. E. Gibbons, eds.), pp. 829–842. Williams and Wilkins, Baltimore.

Baldacci, E., Farina, G. and Locci, R. (1966). Emendation of the genus *Streptoverticillium* (Baldacci 1958) and revision of some species. *Giornale di Microbiologia* **14**, 153–171.

Barksdale, L. (1970). *Corynebacterium diphtheriae* and its relatives. *Bacteriological Reviews* **34**, 378–422.

Barksdale, L. (1981). The genus *Corynebacterium*. In *The Prokaryotes: A Handbook of Habitats, Isolation and Identification of Bacteria, Volume* II, (M. P. Starr, H. Stolp, H. G. Trüper, A. Balows and H. G. Schlegel, eds.), pp. 1827–1837. Springer Verlag, Berlin.
Barksdale, L. and Kim, K. S. (1977). *Mycobacterium. Bacteriological Reviews* **41**, 217–372.
Barksdale, W. L., Li, K., Cummins, C. S. and Harris, H. (1957). The mutation of *Corynebacterium pyogenes* to *Corynebacterium haemolyticum. Journal of General Microbiology* **16**, 749–758.
Barksdale, L., Lanéelle, M. A., Pollice, M. C., Asselineau, J., Welby, M. and Norgard, M. V. (1979). Biological and chemical basis for the reclassification of *Microbacterium flavum* Orla-Jensen as *Corynebacterium flavescens* nom. nov. *International Journal of Systematic Bacteriology* **29**, 222–233.
Barton, M. D. and Hughes, K. L. (1981). Comparison of three techniques for isolation of *Rhodococcus (Corynebacterium) equi* from contaminated sources. *Journal of Clinical Microbiology* **13**, 219–221.
Becker, B., Lechevalier, M. P. and Lechevalier, H. A. (1965). Chemical composition of cell wall preparations from strains of various form-genera of aerobic actinomycetes. *Applied Microbiology* **13**, 236–243.
Becking, J. H. (1970). *Frankiaceae* fam. nov. (*Actinomycetales*) with one new combination and six new species of the genus *Frankia* Brunchorst 1886, 714. *International Journal of Systematic Bacteriology* **20**, 201–220.
Becking, J. H. (1974). *Frankiaceae* (Becking 1970). In *Bergey's Manual of Determinative Bacteriology* (R. E. Buchanan and N. E. Gibbons, eds.), pp. 701–706. Williams and Wilkins, Baltimore.
Becking, J. H. (1981). The genus *Frankia*. In: *The Prokaryotes: A Handbook of Habitats, Isolation and Identification of Bacteria, Volume* II (M. P. Starr, H. Stolp, H. G. Trüper, A. Balows and H. G. Schlegel, eds.), pp. 1990–2003. Springer Verlag, Berlin.
Bellack, S. and Jordan, H. V. (1972). Serological identification of rodent strains of *Actinomyces viscosus* and their relationship to actinomyces of human origin. *Archives of Oral Biology* **17**, 175–182.
Bellamy, W. D. (1974). Single cell proteins from cellulosic wastes. *Biotechnology and Bioengineering* **16**, 869–880.
Bellamy, W. D. (1977). Cellulose and lignocellulose digestion by thermophilic actinomyces for single-cell protein production. *Developments in Industrial Microbiology* **18**, 249–254.
Berd, D. (1973). Laboratory identification of clinically important aerobic actinomycetes. *Applied Microbiology* **25**, 665–681.
Beretta, G. (1973). *Actinoplanes italicus*, a new red pigmented species. *International Journal of Systematic Bacteriology* **23**, 37–42.
Bergey, D. H., Harrison, F. C., Breed, R. S., Hammar, B. W. and Huntoon, F. M. (1923). *Bergey's Manual of Determinative Bacteriology, First Edition*. Williams and Wilkins, Baltimore.
Berkeley, R. C. W. and Goodfellow, M. (1981). *The Aerobic Endospore-forming Bacteria: Classification and Identification*. Academic Press, London.
Bland, C. E. and Couch, J. N. (1981). The family *Actinoplanaceae*. In *The Prokaryotes: A Handbook of Habitats, Isolation and Identification of Bacteria, Volume* II (M. P. Starr, H. Stolp, H. G. Trüper, A. Balows and H. G. Schlegel, eds.), pp. 2004–2010. Springer Verlag, Berlin.
Bojalil, L. F., Cerbón, J. and Trujillo, A. (1962). Adansonian classification of mycobacteria. *Journal of General Microbiology* **28**, 333–346.
Bordet, C. and Michel, G. (1963). Étude des acides gras isolés de plusieurs espèces de *Nocardia. Biochimica et Biophysica Acta* **70**, 613–626.
Bordet, C., Karahjoli, M., Gateau, O. and Michel, G. (1972). Cell walls of nocardiae and related actinomycetes: identification of the genus *Nocardia* by cell wall analyses. *International Journal of Systematic Bacteriology* **22**, 251–259.
Bousfield, I. J. (1972). A taxonomic study of some coryneform bacteria. *Journal of General Microbiology* **71**, 441–455.
Bousfield, I. J. (1978). A taxonomy of coryneform bacteria from the marine environment. In: *Coryneform Bacteria* (I. J. Bousfield and A. G. Callely, eds.), pp. 217–233. Academic Press, London.

Bousfield, I. J. and Callely, A. G. (1978). *Coryneform Bacteria*. Academic Press, London.
Bousfield, I. J. and Goodfellow, M. (1976). The "*rhodochrous*" complex and its relationships with allied taxa. In: *The Biology of the Nocardiae* (M. Goodfellow, G. H. Brownell and J. A. Serrano, eds.), pp. 39–65. Academic Press, London.
Bowden, G. A. and Hardie, J. M. (1971). Anaerobic organisms from the human mouth. In: *Isolation of Anaerobes* (D. A. Shapton and R. G. Board, eds.), pp. 177–205. Society for Applied Bacteriology Technical Series No. 5. Academic Press, London.
Bowden, G. H. and Hardie, J. M. (1973). Commensal and pathogenic *Actinomyces* species in man. In: *Actinomycetales: Characteristics and Practical Importance* (G. Sykes and F. A. Skinner, eds.), pp. 277–299. Academic Press, London.
Bowden, G. H. and Hardie, J. M. (1978). Oral pleomorphic (coryneform) Gram-positive rods. In: *Coryneform Bacteria* (I. J. Bousfield and A. G. Callely, eds.), pp. 235–263. Academic Press, London.
Bowie, I. S., Grigor, M. R., Dunckley, G. G., Loutit, M. W. and Loutit, J. S. (1972). The DNA base composition and fatty acid constitution of some Gram-positive pleomorphic soil bacteria. *Soil Biology and Biochemistry* **4**, 397–412.
Braden, A. R. and Thayer, D. W. (1976). Serological study of *Cellulomonas*. *International Journal of Systematic Bacteriology* **26**, 123–126.
Bradley, S. G. (1973). Relationships among mycobacteria and nocardiae based upon deoxyribonucleic acid reassociation. *Journal of Bacteriology* **113**, 645–651.
Bradley, S. G. (1975). Significance of nucleic acid hybridization to systematics of actinomycetes. *Advances in Applied Microbiology* **19**, 59–70.
Bradley, S. G. and Bond, J. S. (1974). Taxonomic criteria for mycobacteria and nocardiae. *Advances in Applied Microbiology* **18**, 131–190.
Bradley, S. G. and Mordarski, M. (1976). Association of polydeoxyribonucleotides of deoxyribonucleic acids from nocardioform bacteria. In: *The Biology of the Nocardiae* (M. Goodfellow, G. H. Brownell and J. A. Serrano, eds.), pp. 310–336. Academic Press, London.
Breed, R. S. (1953). The *Brevibacteriaceae* fam. nov. of order *Eubacteriales*. *Riassunti delle Communicazione VI Congresso Internazionale di Microbiologia, Roma* **1**, 13–14.
Brock, D. W. and Georg, L. K. (1979). Determination and analysis of *Actinomyces israelii* serotypes by fluorescent-antibody procedures. *Journal of Bacteriology* **97**, 581–588.
Brock, D. W., Georg, L. K., Brown, J. M. and Hicklin, M. D. (1973). Actinomycosis caused by *Arachnia propionica*. *American Journal of Clinical Pathology* **59**, 66–77.
Brown, J. M., Georg, L. K. and Waters, L. C. (1969). Laboratory identification of *Rothia dentocariosa* and its occurrence in human clinical material. *Applied Microbiology* **17**, 150–156.
Brunchorst, J. (1886). Über einige Wurzelanschwellungen, besonders diejenigen von *Alnus* und den Elaeagnaceen. *Botanische Institut Tübingen* **2**, 151–177.
Buchanan, B. B. and Pine, L. (1962). Characterization of a propionic acid producing actinomycete, *Actinomyces propionicus*, sp. nov. *Journal of General Microbiology* **28**, 305–323.
Buchanan, R. E. (1917). Studies in the nomenclature and classification of the bacteria. II. The primary subdivisions of the *Schizomycetes*. *Journal of Bacteriology* **2**, 155–164.
Buchanan, R. E. (1918). Studies in the classification and nomenclature of the bacteria. The subgroups and genera of the *Actinomycetales*. *Journal of Bacteriology* **3**, 403–406.
Buchanan, R. E. and Gibbons, N. E. (eds.) (1974). *Bergey's Manual of Determinative Bacteriology, Eighth Edition*. The Williams and Wilkins Company, Baltimore.
Bullock, G. L. and Stuckey, H. M. (1975). Fluorescent antibody identification and detection of the *Corynebacterium* causing kidney disease in salmonids. *Journal of the Fisheries Research Board of Canada* **32**, 2224–2227.
Bullock, G. L., Stuckey, H. M. and Chen, P. K. (1974). Corynebacterial kidney disease of salmonids: growth and serological studies of the causative bacterium. *Applied Microbiology* **28**, 811–814.
Callaham, D., Del Tredici, P. and Torrey, J. G. (1978). Isolation and cultivation *in vitro* of the actinomycete causing root nodulation in *Comptonia*. *Science* **199**, 899–902.
Casal, M. and Calero, J. R. (1974). *Mycobacterium gadium* sp. nov. A new species of rapidly growing scotochromogenic mycobacteria. *Tubercle, London* **55**, 299–308.
Casida, L. E. (1965). Abundant microorganism in soil. *Applied Microbiology* **13**, 327–334.

Castellani, A., De Brito, M. M. X. and Pinto, M. R. (1959). An actinomycete isolated from an autochthonous case of mycetoma in Portugal. *Journal of Tropical Medicine and Hygiene* **62**, 27–36.
Chalmers, A. J. and Christopherson, J. B. (1916). A Sudanese actinomycosis. *Annals Tropical Medicine and Parasitology* **10**, 223–282.
Chamoiseau, G. (1979). Etiology of farcy in african bovines: Nomenclature of causal organisms *Mycobacterium farcinogenes* (Chamoiseau) and *Mycobacterium senegalense* (Chamoiseau) comb. nov. *International Journal of Systematic Bacteriology* **29**, 407–410.
Chaparas, S. D., Brown, T. M. and Hyman, I. S. (1978a). Antigenic relationships of various mycobacterial species with *M. tuberculosis*. *American Review of Respiratory Diseases* **117**, 1091–1097.
Chaparas, S. D., Brown, T. M. and Hyman, I. S. (1978b). Antigenic relationships among species of *Mycobacterium* studied by fused rocket immunoelectrophoresis. *International Journal of Systematic Bacteriology* **28**, 547–560.
Chaves Batista, A., Shome, S. K. and Americo de Lima, J. (1963). *Streptosporangium bovinum* sp. nov. from cattle hooves. *Dermatologia Tropica* **2**, 49–54.
Chormonova, N. T. (1978). Isolation of *Actinomadura* from soil samples on selective media with kanamycin and rifampicin. *Antibiotiki* **23**, 22–26.
Clark, J. B. (1972). Morphogenesis in the genus *Arthrobacter*. *CRC Critical Reviews in Microbiology* **1**, 521–544.
Clarke, F. E. (1951). The generic classification of certain cellulolytic bacteria. *Soil Science Society America Proceedings* **15**, 180–182.
Clarke, F. E. (1952). The generic classification of the soil corynebacteria. *International Bulletin of Bacteriological Nomenclature and Taxonomy* **2**, 45–56.
Clarke, F. E. (1953). Criteria suitable for species differentiation in *Cellulomonas* and a revision of the genus. *International Bulletin of Bacteriological Nomenclature and Taxonomy* **3**, 179–199.
Collins, M. D. (1982a). A note on the separation of natural mixtures of bacterial menaquinones using reverse-phase high-performance liquid chromatography. *Journal of Applied Bacteriology* **52**, 457–460.
Collins, M. D. (1982b). Lipid composition of *Agromyces ramosus* (Gledhill and Casida). *FEMS Microbiology Letters* **14**, 187–189.
Collins, M. D. (1982c). Reclassification of *Bacterionema matruchotii* (Mendel) in the genus *Corynebacterium*, as *Corynebacterium matruchotii* comb. nov. *Zentralblatt für Bakteriologie, Parasitenkunde, Infektionskrankheiten und Hygiene*. I. Abteilung Originale **C3**, 364–367.
Collins, M. D. and Jones, D. (1980). Lipids in the classification and identification of coryneform bacteria containing peptidoglycans based on 2,4-diaminobutyric acid (DAB). *Journal of Applied Bacteriology* **48**, 459–470.
Collins, M. D. and Jones, D. (1981a). Distribution of isoprenoid quinone structural types in bacteria and their taxonomic implications. *Microbiological Reviews* **45**, 316–354.
Collins, M. D. and Jones, D. (1981b). A note on the separation of natural mixtures of bacterial ubiquinones using reverse-phase partition thin-layer chromatography and high performance liquid chromatography. *Journal of Applied Bacteriology* **51**, 129–134.
Collins, M. D. and Jones, D. (1981c). Lipid composition of *Brevibacterium lyticum* (Takayama, Udagawa and Abe). *FEMS Microbiology Letters* **11**, 193–195.
Collins M. D. and Jones, D. (1981d). Lipid composition of the entomopathogen *Corynebacterium okanaganae*. *FEMS Microbiology Letters* **10**, 157–159.
Collins, M. D. and Jones, D. (1982a). Reclassification of *Corynebacterium pyogenes* (Glage) in the genus *Actinomyces*, as *Actinomyces pyogenes* comb. nov. *Journal of General Microbiology* **128**, 901–903.
Collins, M. D. and Jones, D. (1982b). Taxonomic studies on *Corynebacterium beticola* (Abdou). *Journal of Applied Bacteriology* **52**, 229–233.
Collins, M. D. and Jones, D. (1982c). Lipid composition of *Corynebacterium paurometabolum* (Steinhaus). *FEMS Microbiology Letters* **13**, 13–16.
Collins, M. D., Pirouz, T., Goodfellow, M. and Minnikin, D. E. (1977). Distribution of menaquinones in actinomycetes and corynebacteria. *Journal of General Microbiology* **100**, 221–230.

Collins, M. D., Goodfellow, M. and Minnikin, D. E. (1979). Isoprenoid quinones in the classification of coryneform and related bacteria. *Journal of General Microbiology* **110**, 127–136.

Collins, M. D., Goodfellow, M. and Minnikin, D. E. (1980a). Fatty acid, isoprenoid quinone and polar lipid composition in the classification of *Curtobacterium* and related taxa. *Journal of General Microbiology* **118**, 29–37.

Collins, M. D., Shah, H. N. and Minnikin, D. E. (1980b). A note on the separation of natural mixtures of bacterial menaquinones using reverse phase thin-layer chromatography. *Journal of Applied Bacteriology* **48**, 277–282.

Collins, M. D., Jones, D., Keddie, R. M. and Sneath, P. H. A. (1980c). Reclassification of *Chromobacterium iodinum* (Davis) in a redefined genus *Brevibacterium* (Breed) as *Brevibacterium iodinium* nom. rev. comb. nov. *Journal of General Microbiology* **120**, 1–10.

Collins, M. D., Ross, H. N. M., Tindall, B. J. and Grant, W. D. (1981a). Distribution of isoprenoid quinones in halophilic bacteria. *Journal of Applied Bacteriology* **50**, 559–565.

Collins, M. D., Jones, D. and Kroppenstedt, R. M. (1981b). Reclassification of *Corynebacterium ilicis* (Mandel, Guba & Litsky) in the genus *Arthrobacter*, as *Arthrobacter ilicis* comb. nov. *Zentralblatt für Bakteriologie Mikrobiologie und Hygiene. C. Originale Abteilung I*, **2**, 318–323.

Collins, M. D., Goodfellow, M. and Minnikin, D. E. (1982a). Polar lipid composition in the classification of *Arthrobacter* and *Microbacterium*. *FEMS Microbiology Letters* **15**, 299–302.

Collins, M. D., Goodfellow, M. and Minnikin, D. E. (1982b). A survey of the structure of mycolic acids in *Corynebacterium* and related taxa. *Journal of General Microbiology* **128**, 129–149.

Collins, M. D., Goodfellow, M. and Minnikin, D. E. (1982c). Fatty acid composition of some mycolic acid-containing coryneform bacteria. *Journal of General Microbiology* **128**, 2503–2509.

Collins, M. D., Jones, D. and Schofield, G. M. (1982d). Reclassification of *Corynebacterium haemolyticum* (Maclean, Liebow and Rosenberg) in the genus *Arcanobacterium* gen. nov., as *Arcanobacterium haemolyticum* nom. rev. comb. nov. *Journal of General Microbiology* **128**, 1279–1281.

Collins, M. D., Jones, D., Kroppenstedt, R. M. and Schleifer, K. H. (1982e). Chemical studies as a guide to the classification of *Corynebacterium pyogenes* and *Corynebacterium haemolyticum*. *Journal of General Microbiology* **128**, 335–341.

Collins, M. D., McCarthy, A. J. and Cross, T. (1982f). New highly saturated members of the vitamin K_2 series from *Thermomonospora*. *Zentralblatt für Bakteriologie, Parasitenkunde, Infektionskrankheiten und Hygiene. Originale* **C3**, 358–363.

Collins, M. D., Mackillop, G. C. and Cross, T. (1982g). Menaquinone composition of members of the genus *Thermoactinomyces*. *FEMS Microbiology Letters* **13**, 151–153.

Colmer, A. R. and McCoy, E. (1943). *Micromonospora* in relation to some Wisconsin lakes and lake populations. *Transactions of the Wisconsin Academy of Sciences, Arts and Letters* **35**, 187–220.

Colmer, A. R. and McCoy, E. (1950). Some morphological and cultural studies on lake strains of micromonosporae. *Transactions of the Wisconsin Academy of Sciences, Arts and Letters* **40**, 49–70.

Conn, H. J. and Dimmick, I. (1947). Soil bacteria similar in morphology to *Mycobacterium* and *Corynebacterium*. *Journal of Bacteriology* **54**, 291–303.

Corbaz, R., Gregory, P. H. and Lacey, M. E. (1963). Thermophilic and mesophilic actinomycetes in mouldy hay. *Journal of General Microbiology* **32**, 449–455.

Couch, J. N. (1949). A new group of organisms related to *Actinomyces*. *Journal of the Elisha Mitchell Scientific Society* **65**, 315–318.

Couch, J. N. (1950). *Actinoplanes*, a new genus of the *Actinomycetales*. *Journal of the Elisha Mitchell Scientific Society* **66**, 87–91.

Couch, J. N. (1955). A new genus and family of the *Actinomycetales*, with a revision of the genus *Actinoplanes*. *Journal of the Elisha Mitchell Scientific Society* **71**, 148–155.

Couch, J. N. (1963). Some new genera and species of the *Actinoplanaceae*. *Journal of the Elisha Mitchell Scientific Society* **79**, 53–70.

Couch, J. N. (1964). A proposal to replace the name *Ampullaria* Couch with *Ampullariella*. *Journal of the Elisha Mitchell Scientific Society* **80**, 29.

Couch, J. N. and Bland, C. E. (1974a). Genus I. *Actinoplanes*. In: *Bergey's Manual of Determinative Bacteriology, Eighth Edition* (R. E. Buchanan and N. E. Gibbons, eds.), pp. 708–710. Williams and Wilkins, Baltimore.

Couch, J. N. and Bland, C. E. (1974b). Genus IV. *Amorphosporangium* Couch 1963, 65. In: *Bergey's Manual of Determinative Bacteriology, Eighth Edition*, (R. E. Buchanan and N. E. Gibbons, eds.), pp. 715–716. Williams and Wilkins, Baltimore.

Coykendall, A. L. and Munzenmaier, A. J. (1979). Deoxyribonucleic acid hybridization among strains of *Actinomyces viscosus* and *Actinomyces naeslundii*. *International Journal of Systematic Bacteriology* **29**, 234–240.

Craveri, R. and Manachini, P. L. (1966). Base composition of DNA in *Streptomyces argenteolus* and *Thermoactinomyces vulgaris* cultivated at different temperatures. *Annali di Microbiologia ed Enzimologia* **16**, 1–3.

Craveri, R., Hill, L. R., Manachini, P. L. and Silvestri, L. S. (1965). Deoxyribonucleic acid base composition among thermophilic actinomycetes: the occurrence of two strains with low GC content. *Journal of General Microbiology* **41**, 335–339.

Craveri, R., Manachini, P. L. and Pacini, N. (1966). Deoxyribonucleic acid base composition of actinomycetes with different temperature requirements for growth. *Annali di Microbiologia ed Enzimologia* **16**, 115–117.

Crawford, D. L. (1975). Cultural, morphological and physiological characteristics of *Thermomonospora fusca* (strain 190 Th). *Canadian Journal of Microbiology* **21**, 1842–1848.

Crawford, D. L. and Gonda, M. A. (1977). The sporulation process in *Thermomonospora fusca* as revealed by scanning electron microscopy and transmission electron microscopy. *Canadian Journal of Microbiology* **23**, 1088–1095.

Crawford, D. L., McCoy, E., Harkin, J. M. and Jones, P. (1973). Production of microbial protein from waste cellulose by *Thermomonospora fusca*, a thermophilic actinomycete. *Biotechnology and Bioengineering* **25**, 833–843.

Crawford, J. M., Taubman, M. A. and Smith, D. J. (1978). The natural history of periodontal bone loss in germfree and gnotobiotic rats infected with periodontopathic microorganisms. *Journal of Periodontal Research* **13**, 316–325.

Crombach, W. H. J. (1972). DNA base composition of soil arthrobacters and other coryneforms from cheese and sea fish. *Antonie van Leeuwenhoek* **38**, 105–120.

Crombach, W. H. J. (1974). Relationships among coryneform bacteria from soil, cheese and sea fish. *Antonie van Leeuwenhoek* **40**, 347–359.

Crombach, W. H. J. (1978). *Caseobacter polymorphus* gen. nov., sp. nov., a coryneform bacterium from cheese. *International Journal of Systematic Bacteriology* **28**, 354–366.

Cross, T. (1968). Thermophilic actinomycetes. *Journal of Applied Bacteriology* **31**, 36–53.

Cross, T. (1981a). Aquatic actinomycetes: a critical survey of the occurrence, growth and role of actinomycetes in aquatic habitats. *Journal of Applied Bacteriology* **50**, 397–423.

Cross, T. (1981b). The monosporic actinomycetes. In: *The Prokaryotes: A Handbook of Habitats, Isolation and Identification of Bacteria, Volume II*, (M. P. Starr, H. Stolp, H. G. Trüper, A. Balows and H. G. Schlegel, eds.), pp. 2091–2101. Springer Verlag, Berlin.

Cross, T. and Al-Diwany, L. J. (1981). Streptomycetes with substrate mycelium spores: the genus *Elytrosporangium*. *Zentralblatt für Bakteriologie, Mikrobiologie und Hygiene. 1. Abteilung, Supplement* **11**, 59–65.

Cross, T. and Attwell, R. W. (1974). Recovery of viable thermoactinomycete endospores from deep mud cores. In: *Spore Research 1973* (A. N. Barker, G. W. Gould and J. Wolf, eds.), pp. 11–20. Academic Press, London.

Cross, T. and Goodfellow, M. (1973). Taxonomy and classification of the actinomycetes. In: *Actinomycetales: Characteristics and Practical Importance* (G. Sykes and F. A. Skinner, eds.), pp. 11–112. Academic Press, London.

Cross, T. and Lacey, J. (1970). Studies on the genus *Thermomonospora*. In *The Actinomycetales* (H. Prauser, ed.), pp. 211–219. Gustav Fischer Verlag, Jena.

Cross, T., Lechevalier, M. P. and Lechevalier, H. A. (1963). A new genus of the *Actinomycetales*: *Microellobosporia* gen. nov. *Journal of General Microbiology* **31**, 421–429.

Cross, T., Walker, P. D. and Gould, G. W. (1968a). Thermophilic actinomycetes producing resistant endospores. *Nature, London* **220**, 352–354.

Cross, T., MacIver, A. and Lacey, J. (1968b). The thermophilic actinomycetes in mouldy hay: *Micropolyspora faeni* sp. nov. *Journal of General Microbiology* **50**, 351–359.

Cross, T., Attwell, R. W. and Locci, R. (1973). Fine structure of the spore sheath in *Streptoverticillium* species. *Journal of General Microbiology* **75**, 421–424.

Cross, T., Rowbotham, T. J., Mishustin, E. N., Tepper, E. Z., Portaels, F. P., Schaal, K. P. and Bickenbach, H. B. (1976). The ecology of nocardioform actinomycetes. In: *The Biology of the Nocardiae* (M. Goodfellow, G. H. Brownell and J. A. Serrano, eds.), pp. 337–371. Academic Press, London.

Cummins, C. S. (1962). Chemical composition and antigenic structure of cell walls of *Corynebacterium, Mycobacterium, Nocardia, Actinomyces* and *Arthrobacter. Journal of General Microbiology* **28**, 35–50.

Cummins, C. S. and Harris, H. (1956). The chemical composition of the cell wall in some Gram-positive bacteria and its possible value as a taxonomic character. *Journal of General Microbiology* **14**, 583–600.

Cummins, C. S. and Harris, H. (1959). Taxonomic position of *Arthrobacter. Nature (London)* **184**, 831–832.

Cummins, C. S., Lelliott, R. A. and Rogosa, M. (1974). *Corynebacterium.* In *Bergey's Manual of Determinative Bacteriology, Eighth Edition* (R. E. Buchanan and N. E. Gibbons, eds.), pp. 602–617. Williams and Wilkins, Baltimore.

Da Fonseca, A. F. and Edwards, M. R. (1971). Ultrastructure of the actinomycete *Geodermatophilus obscurus. Proceedings of the 29th Meeting of the Electron Microscopy Society of America,* p. 248.

Davis, G. H. G. and Freer, J. H. (1960). Studies upon an oral-aerobic actinomycete. *Journal of General Microbiology* **23**, 163–178.

Dawson, D. J. and Jennis, F. (1980). Mycobacteria with a growth requirement for ferric ammonium citrate, identified as *Mycobacterium haemophilum. Journal of Chemical Microbiology* **11**, 190–192.

De Smedt, J. and De Ley, J. (1977). Intra- and intergeneric similarities of *Agrobacterium* ribosomal ribonucleic acid cistrons. *International Journal of Systematic Bacteriology* **27**, 222–240.

De Smedt, J., Bauwens, M., Tytgat, R. and De Ley, J. (1980). Intra- and intergeneric similarities of ribosomal ribonucleic acid cistrons of free-living, nitrogen-fixing bacteria. *International Journal of Systematic Bacteriology* **30**, 106–122.

Demaree, J. B. and Smith, N. R. (1952). *Nocardia vaccinii* n. sp. causing galls on blueberry plants. *Phytopathology* **42**, 249–252.

Dent, V. E., Hardie, J. M. and Bowden, G. H. (1976). A preliminary study of dental plaque on animal teeth. *Journal of Dental Research Special Issue* D. Abstract No. 85D, 127.

Di Marco, A. M. and Spalla, C. (1957). La produzione di cobalamine de fermentazione con nuova specie di *Nocardia: Nocardia rugosa. Giornale Microbiologie* **4**, 24–30.

Doi, R. H. and Igarashi, R. T. (1965). Conservation of ribosomal and messenger ribonucleic acid cistrons in *Bacillus* species. *Journal of Bacteriology* **90**, 384–390.

Döpfer, H., Stackebrandt, E. and Fiedler, F. (1982). Nucleic acid hybridization studies on *Microbacterium, Curtobacterium, Agromyces* and related taxa. *Journal of General Microbiology* **128**, 1697–1708.

Dorokhova, L. A., Agre, N. S., Kalakoutskii, L. V. and Krasilnikov, N. V. (1968). Fine structure of spores in a thermophilic actinomycete *Micromonospora vulgaris. Journal of General and Applied Microbiology* **14**, 295–303.

Draper, P. (1976). Cell walls of *Mycobacterium leprae. International Journal of Leprosy* **44**, 95–98.

Draper, P., Dobson, G., Minnikin, D. E. and Minnikin, S. M. (1982). The mycolic acids of *Mycobacterium leprae* harvested from experimentally infected nine-banded armadillos. *Annales de Microbiologie (Institut Pasteur)* **133B**, 39–47.

Dubnau, D., Smith, I., Morell, P. and Marmur, J. (1965). Gene conservation in *Bacillus* species. I. Conserved genetic and nucleic acid sequence homologies. *Proceedings of the National Academy of Sciences of the United States of America* **54**, 491–498.

Dye, D. W. and Kemp, W. J. (1977). A taxonomic study of plant pathogenic *Corynebacterium* species. *New Zealand Journal of Agricultural Research* **20**, 563–582.

Embley, T. M., Goodfellow, M. and Austin, B. (1982). A semi-defined growth medium for *Renibacterium salmoninarum. FEMS Microbiology Letters* **14**, 299–301.

Embley, T. M., Goodfellow, M., Minnikin, D. E. and Austin, B. (1983). Fatty acid, isoprenoid quinone and polar lipid composition in the classification of *Renibacterium salmoninarum. Journal of Applied Bacteriology* **55**, 31–37.
Ensign, J. C. (1978). Formation, properties and germination of actinomycete spores. *Annual Review of Microbiology* **32**, 185–219.
Erikson, D. (1935). Pathogenic aerobic organisms of the actinomyces group. *Medical Research Council (Great Britain), Special Report Series* **203**, 5–61.
Erikson, D. (1941). Studies on some lake mud strains of *Micromonospora. Journal of Bacteriology* **41**, 277–300.
Erikson, D. (1954). Factors promoting cell division in a 'soft' mycelial type of *Nocardia*: *Nocardia turbata* n. sp. *Journal of General Microbiology* **11**, 198–208.
Etémadi, A. H. (1967). Les acides mycoliques structure, biogenèse et inhérêt phylogénétique. *Exposés Annuels de Biochimie Medicale* **28**, 77–109.
Etémadi, A. H. and Convit, J. (1974). Mycolic acids from "noncultivable" mycobacteria. *Infection and Immunity* **10**, 235–239.
Falcao de Morais, J. O., Chaves Batista, A. and Massa, D. M. G. (1966). *Elytrosporangium*, a new genus of the *Actinomycetales. Mycopathologia et Mycologia Applicata* **30**, 161–171.
Farina, G. and Bradley, S. G. (1970). Reassociation of deoxyribonucleic acids from *Actinoplanes* and other actinomycetes. *Journal of Bacteriology* **102**, 30–35.
Feltham, R. K. A. (1981). A study of the taxonomy of the *Micrococcaceae. Zentralblatt für Bakteriologie, Mikrobiologie und Hygiene. 1. Abteilung, Supplement* **10**, 1–7.
Fiedler, F. and Kandler, O. (1973a). Die Amino-säuresequenz von 2,4-Diamino-buttersäure enthaltenden Mureinen bei verschiedenen coryneformen Bakterien und *Agromyces ramosus. Archiv für Mikrobiologie* **89**, 51–66.
Fiedler, F. and Kandler, O. (1973b). Die Mureintypen in der Gattung *Cellulomonas* Bergey *et al. Archiv für Mikrobiologie* **89**, 41–50.
Fiedler, F., Schleifer, K. H. and Kandler, O. (1973). Amino acid sequence of the threonine-containing murein of coryneform bacteria. *Journal of Bacteriology* **113**, 8–17.
Fiedler, F., Schäffler, M. J. and Stackebrandt, E. (1981). Biochemical and nucleic acid hybridisation studies on *Brevibacterium linens* and related strains. *Archives of Microbiology* **129**, 85–93.
Fillery, E. D., Bowden, G. H. and Hårdie, J. M. (1978). A comparison of strains of bacteria designated *Actinomyces viscosus* and *Actinomyces naeslundii. Caries Research* **12**, 299–312.
Fox, G. E., Magrum, L. J., Balch, W. E., Wolfe, R. S. and Woese, C. R. (1977a). Classification of methanogenic bacteria by 16S ribosomal RNA characterization. *Proceedings of the National Academy of Sciences, U.S.A.* **74**, 4537–4541.
Fox, G. E., Pechman, K. G. and Woese, C. R. (1977b). Comparative cataloguing of 16S ribosomal ribonucleic acid: Molecular approach to procaryotic systematics. *International Journal of Systematic Bacteriology* **27**, 44–57.
Franke, F. (1973). Untersuchungen zur Ätiologie der Gesaugeaktinomykose des Schweines. *Zentralblatt für Bakteriologie, Parasitenkunde, Infektionskrankheiten und Hygiene. Abteilung 1, Originale Reihe* **A223**, 111–124.
Franklin, A. A. Jr. and McClung, N. M. (1976). Heterogeneity among *Nocardia asteroides* strains. *Journal of General and Applied Microbiology* **22**, 151–159.
Fritzsche, H. (1967). Infra-red studies of deoxyribonucleic acids, their constituents and analogues. II. Deoxyribonucleic acids with different base composition. *Biopolymers* **5**, 863–870.
Fryer, J. L. and Sanders, J. E. (1981). Bacterial kidney disease of salmonid fish. *Annual Review of Microbiology* **35**, 273–298.
Gaertner, A. (1955). Über zwei ungewöhnliche Keratinophile Organismen aus Ackerboden. *Archiv für Mikrobiologie* **23**, 28–37.
Gauze, G. F., Sveshnikova, M. A., Ukholina, R. S., Komarova, G. N. and Bazhanova, V. S. (1977). Production of nocamycin, a new antibiotic, by *Nocardiopsis syringae* sp. nov. *Antibiotiki* **22**, 483–486. (In Russian).
Georg, L. K. and Brown, J. M. (1967). *Rothia*, gen. nov. An aerobic genus of the family *Actinomycetaceae. International Journal of Systematic Bacteriology* **17**, 79–88.
Georg, L. K. and Coleman, R. M. (1970). Comparative pathogenicity of various *Actinomyces* species. In *The Actinomycetales* (H. Prauser, ed.), pp. 35–45. Gustav Fischer Verlag, Jena.

Georg, L. K., Robertstad, G. N., Brinkmann, S. A. and Hicklin, M. D. (1965). A new pathogenic anaerobic *Actinomyces* species. *Journal of Infectious Diseases* **115**, 88–99.

Georg, L. K., Pine, L. and Gerencser, M. A. (1969). *Actinomyces viscosus* comb. nov., a catalase positive, facultative member of the genus *Actinomyces*. *International Journal of Systematic Bacteriology* **19**, 291–293.

Gerber, N. N. and Lechevalier, M. P. (1964). Phenazines and phenoxazinones from *Waksmania aerata* sp. nov. and *Pseudomonas iodina*. *Biochemistry* **3**, 598–602.

Gerber, N. N. and Lechevalier, M. P. (1976). Prodiginine (prodigiosin-like) pigments from *Streptomyces* and other aerobic actinomycetes. *Canadian Journal of Microbiology* **22**, 658–667.

Gerencser, M. A. (1979). The application of fluorescent antibody techniques to the identification of *Actinomyces* and *Arachnia*. In: *Methods in Microbiology, Volume 13* (T. Bergan and J. R. Norris, eds.), pp. 287–321. Academic Press, London.

Gerencser, M. A. and Slack, J. M. (1967). Isolation and characterization of *Actinomyces propionicus*. *Journal of Bacteriology* **94**, 109–115.

Gerencser, M. A. and Slack, J. M. (1969). Identification of human strains of *Actinomyces viscosus*. *Applied Microbiology* **18**, 80–87.

Gillis, M. and De Ley, J. (1980). Intra- and intergeneric similarities of the ribosomal ribonucleic acid cistrons of *Acetobacter* and *Gluconobacter*. *International Journal of Systematic Bacteriology* **30**, 7–27.

Gilmour, M. N. (1974). Genus *Bacterionema*. In: *Bergey's Manual of Determinative Bacteriology, Eighth Edition* (R. E. Buchanan and N. E. Gibbons, eds.), pp. 676–679. Williams and Wilkins, Baltimore.

Gledhill, W. E. and Casida, L. E. Jr. (1969a). Predominant catalase-negative soil bacteria. II. Occurrence and characterization of *Actinomyces humiferus*, sp. n. *Applied Microbiology* **18**, 114–121.

Gledhill, W. E. and Casida, L. E. Jr. (1969b). Predominant catalase negative soil bacteria. III. *Agromyces*, gen. n. microorganisms intermediary to *Actinomyces* and *Nocardia*. *Applied Microbiology* **18**, 340–349.

Gochnauer, M. B., Leppard, G. G., Komaratat, P., Kates, M., Novitsky, T. and Kushner, D. J. (1975). Isolation and characterization of *Actinopolyspora halophila*, gen. et sp. nov., an extremely halophilic actinomycete. *Canadian Journal of Microbiology* **21**, 1500–1511.

Goldfine, H. (1972). Comparative aspects of bacterial lipids. *Advances in Microbial Physiology* **8**, 1–58.

Goodfellow, M. (1971). Numerical taxonomy of some nocardioform bacteria. *Journal of General Microbiology* **69**, 33–80.

Goodfellow, M. and Alderson, G. (1977). The actinomycete-genus *Rhodococcus*: a home for the '*rhodochrous*' complex. *Journal of General Microbiology* **100**, 99–112.

Goodfellow, M. and Board, R. G. (eds.) (1980). *Microbiological Classification and Identification*. Academic Press, London.

Goodfellow, M. and Cross, T. (1974). Actinomycetes. In: *Biology of Plant Litter Decomposition* (C. H. Dickinson and G. J. F. Pugh, eds.), pp. 269–302. Academic Press, London.

Goodfellow, M. and Minnikin, D. E. (1977). Nocardioform bacteria. *Annual Review of Microbiology* **31**, 159–180.

Goodfellow, M. and Minnikin, D. E. (1978). Numerical and chemical methods in the classification of *Nocardia* and related taxa. *Zentralblatt für Bakteriologie, Parasitenkunde, Infektionskrankheiten und Hygiene. I. Abteilung, Supplement* **6**, 43–51.

Goodfellow, M. and Minnikin, D. E. (1980). Definition of the genus *Mycobacterium* vis-à-vis other taxa. In *1954 to 1979: Twenty Five Years of Mycobacterial Taxonomy* (G. P. Kubica and L. G. Wayne, eds.), pp. 115–130. U.S. Department of Health, Education and Welfare, Center for Disease Control, Atlanta.

Goodfellow, M. and Minnikin, D. E. (1981a). Introduction to coryneform bacteria. In: *The Prokaryotes: A Handbook of Habitats, Isolation and Identification of Bacteria, Volume II* (M. P. Starr, H. Stolp, H. G. Trüper, A. Balows and H. G. Schlegel, eds.), pp. 1811–1826. Springer Verlag, Berlin.

Goodfellow, M. and Minnikin, D. E. (1981b). The genera *Nocardia* and *Rhodococcus*. In: *The Prokaryotes: A Handbook of Habitats, Isolation and Identification of Bacteria, Volume II* (M. P. Starr, H. Stolp, H. G. Trüper, A. Balows and H. G. Schlegel, eds.), pp. 2016–2027. Springer Verlag, Berlin.

Goodfellow, M. and Minnikin, D. E. (1981c). Classification of nocardioform bacteria. *Zentralblatt für Bakteriologie, Mikrobiologie und Hygiene. I. Abteilung, Supplement II*, 7–16.
Goodfellow, M. and Minnikin, D. E. (1983). Circumscription of the genus. In: *The Mycobacteria: A Sourcebook, Volume I* (G. P. Kubica and L. G. Wayne, eds.), in press. Marcel Dekker, New York.
Goodfellow, M. and Pirouz, T. (1982). Numerical classification of sporoactinomycetes containing *meso*-diaminopimelic acid in the cell wall. *Journal of General Microbiology* **128**, 503–527.
Goodfellow, M. and Schaal, K. P. (1979). Identification methods for *Nocardia, Actinomadura* and *Rhodococcus.* In: *Identification Methods for Microbiologists* (D. W. Lovelock and F. A. Skinner, eds.), pp. 261–276. Academic Press, London.
Goodfellow, M. and Wayne, L. G. (1982). Taxonomy and nomenclature. In: *The Biology of the Mycobacteria, Volume 1. Physiology, Identification and Classification* (C. Ratledge and J. L. Stanford, eds.), pp. 472–521. Academic Press, London.
Goodfellow, M., Collins, M. D. and Minnikin, D. E. (1976). Thin-layer chromatographic analysis of mycolic acid and other long-chain components in whole organism methanolysates of coryneform and related taxa. *Journal of General Microbiology* **96**, 351–358.
Goodfellow, M., Orlean, P. A. B., Collins, M. D. and Alshamaony, L. (1978). Chemical and numerical taxonomy of strains received as *Gordona aurantiaca. Journal of General Microbiology* **109**, 57–68.
Goodfellow, M., Alderson, G. and Lacey, J. (1979). Numerical taxonomy of *Actinomadura* and related actinomycetes. *Journal of General Microbiology* **112**, 95–111.
Goodfellow, M., Minnikin, D. E., Todd, C., Alderson, G., Minnikin, S. M. and Collins, M. D. (1982a). Numerical and chemical classification of *Nocardia amarae. Journal of General Microbiology* **128**, 1283–1297.
Goodfellow, M., Weaver, C. R. and Minnikin, D. E. (1982b). Numerical classification of some rhodococci, corynebacteria and related organisms. *Journal of General Microbiology* **128**, 731–745.
Goodfellow, M., Beckham, A. R. and Barton, M. D. (1982c). Numerical classification of *Rhodococcus equi* and related actinomycetes. *Journal of Applied Bacteriology* **53**, 199–207.
Gordon, M. A. (1974). Genus *Dermatophilus.* In: *Bergey's Manual of Determinative Bacteriology, Eighth Edition* (R. E. Buchanan and N. E. Gibbons, eds.), pp. 723–724. Williams and Wilkins, Baltimore.
Gordon, M. A. (1976). Characteristics of *Dermatophilus congolensis,* its affinities with the actinomycetes and differentiation from *Geodermatophilus.* In: *Dermatophilus Infection in Animals and Man* (D. H. Lloyd and K. D. Sellars, eds.), pp. 187–201. Academic Press, London.
Gordon, M. A. and Perrin, U. (1971). Pathogenicity of *Dermatophilus* and *Geodermatophilus. Infection and Immunity* **4**, 29–33.
Gordon, R. E. and Horan, A. C. (1968). *Nocardia dassonvillei,* a macroscopic replica of *Streptomyces griseus. Journal of General Microbiology* **50**, 235–240.
Gordon, R. E., Barnett, D. A., Handerhan, J. E. and Pang, C. H. N. (1974). *Nocardia coeliaca, Nocardia autotrophica* and the nocardin strain. *International Journal of Systematic Bacteriology* **24**, 54–63.
Gordon, R. E., Mishra, S. K. and Barnett, D. A. (1978). Some bits and pieces of the genus *Nocardia: N. carnea, N. vaccinii, N. transvalensis, N. orientalis* and *N. aerocolonigenes. Journal of General Microbiology* **109**, 69–78.
Gottlieb, D. (1973). General consideration and implications of the *Actinomycetales.* In: *Actinomycetales: Characteristics and Practical Importance* (G. Sykes and F. A. Skinner, eds.), pp. 1–10. Academic Press, London.
Gottlieb, D. (1974). Actinomycetales. In: *Bergey's Manual of Determinative Bacteriology, Eighth Edition* (R. E. Buchanan and N. E. Gibbons, eds.), pp. 657–659. The Williams and Wilkins Company, Baltimore.
Gottlieb, D. and Shirling, E. B. (1967). Cooperative description of type cultures of *Streptomyces.* I. The International *Streptomyces* Project. *International Journal of Systematic Bacteriology* **17**, 315–322.
Grange, J. M. and Stanford, J. L. (1974). Re-evaluation of *Mycobacterium fortuitum* (synonym *Mycobacterium ranae*). *International Journal of Systematic Bacteriology* **24**, 320–329.

Grässer, R. (1957). *Vergleichende Untersuchungen an Actinomyceten von Mensch, Rind und Schwein.* Thesis, Leipzig.
Grein, A., Merli, S. and Spalla, C. (1980). New anthracycline glycosides from *Micromonospora*. I. Description of the producing strain. *Journal of Antibiotics* **33**, 1462–1467.
Griffiths, D. E. (1977). EEC regulations: some technical aspects of the intervention arrangements in milk products. *Dairy Industries International* **42**, 17–24.
Gross, D. C. and Vidaver, A. K. (1979). A selective medium for the isolation of *Corynebacterium nebraskense* from soil and plant parts. *Phytopathology* **69**, 82–87.
Gross, W. M. and Wayne, L. G. (1970). Nucleic acid homology in the genus *Mycobacterium*. *Journal of Bacteriology* **104**, 630–634.
Gupta, K. C. (1965). A new species of the genus *Streptosporangium* isolated from Indian soil. *Journal of Antibiotics Series A* **18**, 125–127.
Guzeva, L. N., Agre, N. S. and Sokolov, A. A. (1972). Taxonomy of actinomycetes forming catenate spores. *Mikrobiologiya* **41**, 1080–1085. (In Russian).
Guzeva, L. N., Efimova, T. P., Agre, N. S. and Krasilnikov, N. A. (1973). Fatty acids in the mycelia of actinomycetes that form catenate spores. *Mikrobiologiya* **42**, 26–31. (English translation, *Microbiology* **42**, 19–23).
Haas, H., Michel, J. and Sachs, T. (1974). Identification of *Mycobacterium fortuitum*, *Mycobacterium abscessus*, and *Mycobacterium borstelense* by polyacrylamide gel electrophoresis of their cell proteins. *International Journal of Systematic Bacteriology* **24**, 366–369.
Hagedorn, C. and Holt, J. G. (1975). A nutritional and taxonomic survey of *Arthrobacter* soil isolates. *Canadian Journal of Microbiology* **21**, 353–361.
Hägerdal, B., Harris, H. and Pye, E. K. (1979). Association of beta glucosidase with intact cells of *Thermoactinomyces*. *Biotechnology and Bioengineering* **21**, 345–356.
Hägerdal, B., Ferchak, J. D. and Pye, E. K. (1980). Saccharification of cellulose by the cellulolytic enzyme system of *Thermomonospora* sp. I. Stability of cellulolytic activities with respect to time, temperature and pH. *Biotechnology and Bioengineering* **22**, 1515–1526.
Hammond, B. F. (1970). Isolation and serological characterization of a cell wall antigen of *Rothia dentocariosa*. *Journal of Bacteriology* **103**, 634–640.
Han, Y. W. and Callihan, C. D. (1974): Cellulose fermentation: effect of substrate pretreatment on microbial growth. *Applied Microbiology* **27**, 159–165.
Hanton, W. K. (1968). *Amorphosporangium* (*Actinoplanaceae*). Report of motility and additional characters. *Journal of General Microbiology* **53**, 317–320.
Hanton, W. K. (1974). Genus VI *Pilimelia*. In: *Bergey's Manual of Determinative Bacteriology, Eighth Edition* (R. E. Buchanan and N. E. Gibbons, eds.), pp. 718–719. Williams and Wilkins, Baltimore.
Hardisson, C. and Suarez, J. E. (1979). Fine structure of spore formation and germination in *Micromonospora chalcea*. *Journal of General Microbiology* **110**, 233–237.
Harz, C. O. (1877). *Actinomyces bovis* ein neuer Schimmel in den Gweben des Rindes. *Deutsche Zeitschrift für Thiermedizin* **5**, 125–140.
Hasegawa, T., Lechevalier, M. P. and Lechevalier, H. A. (1978). New genus of the *Actinomycetales*: *Actinosynnema* gen. nov. *International Journal of Systematic Bacteriology* **28**, 304–310.
Hasegawa, T., Lechevalier, M. P. and Lechevalier, H. A. (1979). Phospholipid composition of motile actinomycetes. *Journal of General and Applied Microbiology* **25**, 209–213.
Hasegawa, T., Tanida, S., Hatano, K., Higashide, E. and Yoneda, M. (1982). Taxonomy of new motile actinomycetes. *International Journal of Systematic Bacteriology* **33**, 314–320.
Henssen, A. (1957). Beiträge zur Morphologie und Systematik der thermophilen Actinomyceten. *Archiv für Mikrobiologie* **26**, 373–414.
Henssen, A. (1970). Spore formation in thermophilic actinomycetes. In: *The Actinomycetales* (H. Prauser, ed.), pp. 205–210. Gustav Fischer Verlag, Jena.
Henssen, A. and Schäfer, D. (1971). Emended description of the genus *Pseudonocardia* Henssen and description of a new species *Pseudonocardia spinosa* Schäfer. *International Journal of Systematic Bacteriology* **21**, 29–34.
Henssen, A. and Schnepf, E. (1967). Zur Kenntnis thermophiler Actinomyceten. *Archiv für Mikrobiologie* **57**, 214–231.
Higashide, R., Asai, M., Ootsu, K., Tanida, S., Kosai, Y., Hasegawa, T., Kishi, T., Sugino, Y. and Yonedo, M. (1977). Ansamitocin, a group of novel maytansinoids antibiotics with antitumour properties from *Nocardia*. *Nature (London)* **270**, 721–722.

Higgens, C. E. and Kastner, R. E. (1968). Nebramycin, a new broad spectrum antibiotic complex. II. Description of *Streptomyces tenebrarius*. *Antimicrobial Agents and Chemotherapy* 1967, 324–331.

Higgins, M. L. (1967). Release of sporangiospores by a strain of *Actinoplanes*. *Journal of Bacteriology* **94**, 495–498.

Higgins, M. L., Lechevalier, M. P. and Lechevalier, H. A. (1967). Flagellated actinomycetes. *Journal of Bacteriology* **93**, 1446–1451.

Hirsch, P. (1961). Wasserstoffaktivierung und Chemoautotrophie bei Actinomyceten. *Archiv für Mikrobiologie* **39**, 360–373.

Holmberg, K. and Forsum, U. (1973). Identification of *Actinomyces*, *Arachnia*, *Bacterionema*, *Rothia* and *Propionibacterium* species by defined immunofluorescence. *Applied Microbiology* **25**, 834–843.

Holmberg, K. and Hallander, H. O. (1973). Numerical taxonomy and laboratory identification of *Bacterionema matruchotii*, *Rothia dentocariosa*, *Actinomyces naeslundii*, *Actinomyces viscosus*, and some related bacteria. *Journal of General Microbiology* **76**, 43–63.

Holmberg, K. and Nord, C. E. (1975). Numerical taxonomy and laboratory identification of *Actinomyces* and *Arachnia* and some related bacteria. *Journal of General Microbiology* **91**, 17–44.

Houang, E. T., Lovett, I. S., Thompson, F. D., Harrison, A. R., Joekes, A. M. and Goodfellow, M. (1980). *Nocardia asteroides* infection – a transmissible disease. *Journal of Hospital Infection* **1**, 31–40.

Howell, A., Jr. (1963). A filamentous microorganism isolated from periodontal plaque in hamsters. I. Isolation, morphology and general cultural characteristics. *Sabouraudia* **3**, 81–92.

Howell, A., Jr. and Jordan, H. V. (1963). A filamentous microorganism isolated from periodontal plaque in hamsters. II. Physiological and biochemical characteristics. *Sabouraudia* **3**, 93–105.

Huang, L. H. (1980). *Actinomadura macra*, sp. nov., the producer of antibiotics CP-47, 433 and CP-47, 434. *International Journal of Systematic Bacteriology* **30**, 565–568.

Humphrey, A. E., Moreira, A., Armiger, W. and Zabriskie, D. (1977). Production of single cell protein from cellulose wastes. *Biotechnology and Bioengineering, Symposium* **7**, 45–64.

Hungate, R. E. (1946). Studies on cellulose fermentation. II. An anaerobic cellulose decomposing actinomycete: *Micromonospora propionici* n. sp. *Journal of Bacteriology* **51**, 51–56.

Hunter, J. C., Eveleigh, D. E. and Casella, G. 1981. Actinomycetes of a salt marsh. *Zentralblatt für Bakteriologie, Mikrobiologie und Hygiene. 1. Abteilung, Supplement* **11**, 195–200.

Hutchinson, M., Ridgway, J. W. and Cross, T. (1975). Biodeterioration of rubber in contact with water, sewage and soil. In *Microbial Aspects of the Deterioration of Materials* (D. N. Lovelock and R. A. Gilbert, eds.), pp. 187–202. Academic Press, London.

Hütter, R. (1967). *Systematik der Streptomyceten*. S. Karger, Basel and New York.

Hyslop, N. St. G. (1980). Dermatophylosis (streptothricosis) in animals and man. *Comparative Immunology and Microbiology of Infectious Diseases* **2**, 389–404.

Ishiguro, E. E. and Wolfe, R. S. (1970). Control of morphogenesis in *Geodermatophilus*: Ultrastructural studies. *Journal of Bacteriology* **104**, 566–580.

Jayne-Williams, D. J. and Skerman, T. M. (1966). Comparative studies on coryneform bacteria from milk and dairy sources. *Journal of Applied Bacteriology* **29**, 72–92.

Jenkins, P. A., Marks, J. and Schaefer, W. B. (1971). Lipid chromatography and seroagglutination in the classification of rapidly growing mycobacteria. *American Review of Respiratory Disease* **103**, 179–187.

Jensen, H. L. (1953). The genus *Nocardia* (or *Proactinomyces*) and its separation from other *Actinomycetales*, with some reflections on the phylogeny of the actinomycetes. In: *Actinomycetales, Biology and Systematics* (E. Baldacci and P. Redaelli, eds.), pp. 69–88. Fondazione Emanuele Paterno, Rome.

Jensen, H. L. (1966). Some introductory remarks on the coryneform bacteria. *Journal of Applied Bacteriology* **29**, 13–16.

Johnston, D. W. and Cross, T. (1976). The occurrence and distribution of actinomycetes in lakes of the English Lake District. *Freshwater Biology* **6**, 457–463.

Johnson, J. L. and Cummins, C. S. (1972). Cell wall composition and deoxyribonucleic acid similarities among the anaerobic coryneforms, classical propionibacteria, and strains of *Arachnia propionica*. *Journal of Bacteriology* **109**, 1047–1066.

Jones, D. (1975). A numerical study of coryneform and related bacteria. *Journal of General Microbiology* **87**, 52–96.
Jones, D. (1978). An evaluation of the contribution of numerical taxonomy to the classification of the coryneform bacteria. In: *Coryneform Bacteria* (I. J. Bousfield and A. G. Cally, eds.), pp. 13–46. Academic Press, London.
Jones, D. and Sackin, M. J. (1980). Numerical methods in the classification and identification of bacteria with especial reference to the *Enterobacteriaceae*. In: *Microbiological Classification and Identification* (M. Goodfellow and R. G. Board, eds.), pp. 73–106. Academic Press, London.
Jones, D., Watkins, J. and Meyer, D. J. (1970). Cytochrome composition and effect of catalase on growth of *Agromyces ramosus*. *Nature (London)* **226**, 1249–1250.
Jones, L. A. and Bradley, S. G. (1964). Phenetic classification of actinomycetes. *Developments in Industrial Microbiology* **5**, 267–272.
Jones, W. D. Jr. and Kubica, G. P. (1968). Fluorescent antibody techniques with mycobacteria. III. Investigation of five serologically homogeneous groups of mycobacteria. *Zentralblatt für Bakteriologie, Parasitenkunde, Infektionskrankheiten und Hygiene. Abteilung I, Originale A* **207**, 58–62.
Jordan, H. V. and Hammond, B. F. (1972). Filamentous bacteria isolated from root surface caries. *Archives of Oral Biology* **17**, 1–12.
Jordan, H. V. and Sumney, D. L. (1973). Root surface caries: Review of the literature and significance of the problem. *Journal of Periodontology* **44**, 158–163.
Jordan, H. V., Fitzgerald, R. J. and Stanley, H. R. (1965). Plaque formation and periodontal pathology in gnotobiotic rats infected with an oral actinomycete. *American Journal of Pathology* **47**, 1157–1167.
Jordan, H. V., Keyes, P. H. and Bellack, S. (1972). Periodontal lesions in hamsters and gnotobiotic rats infected with *Actinomyces* of human origin. *Journal of Periodontal Research* **7**, 21–28.
Jordan, H. V., Bellack, S., Keyes, P. H. and Gerencser, M. A. (1974). Periodontal pathology and enamel caries in gnotobiotic rats infected with a unique serotype of *Actinomyces naeslundii*. International Association of Dental Research Abstract 73. *Journal of Dental Research* **53**, Special Issue, 73.
Kalakoutskii, L. V. (1964). A new species of the genus *Micropolyspora–Micropolyspora caesia* n. sp. *Mikrobiologiya* **33**, 858–862.
Kalakoutskii, L. V. and Kuznetsov, V. D. (1964). A new species of *Actinoplanes* Couch. *Mikrobiologiya* **33**, 553–560.
Kalakoutskii, L. V., Kirillova, I. P. and Krasilnikov, N. A. (1967). A new genus of the *Actinomycetales–Intrasporangium* gen. nov. *Journal of General Microbiology* **48**, 79–85.
Kalakoutskii, L. V., Agre, N. S. and Krasilnikov, N. A. (1968). Comparative study on some oligosporic actinomycetes. *Hindustan Antibiotics Bulletin* **10**, 254–268.
Kandler, O. (1970). Amino acid sequence of the murein and taxonomy of the genera *Lactobacillus, Bifidobacterium, Leuconostoc* and *Pediococcus*. *International Journal of Systematic Bacteriology* **20**, 491–507.
Kane, W. D. (1966). A new genus of Actinoplanaceae, *Pilimelia*, with a description of two species *Pilimelia terevasa* and *Pilimelia anulata*. *Journal of the Elisha Mitchell Scientific Society* **82**, 220–230.
Karassova, V. T., Weiszfeiler, J. and Krasznay, E. (1965). Occurrence of atypical mycobacteria in *Macacus rhesus*. *Acta Microbiologica Academiae Scientiarum Hungaricae* **12**, 275–282.
Karling, J. S. (1954). An unusual keratinophilic micro-organism. *Proceedings of the Indiana Academy of Science* **63**, 83–86.
Kaufmann, A., Fegan, J., Doleac, P., Gainer, C., Wittich, D. and Glann, A. (1976). Identification and characterization of a cellulolytic isolate. *Journal of General Microbiology* **94**, 405–408.
Kawamotoa, I., Oka, T. and Nara, I. (1981). Cell wall composition of *Micromonospora olivoasterospora*, *Micromonospora sagamiensis* and related organisms. *Journal of Bacteriology* **146**, 527–534.
Kazda, J. (1980). *Mycobacterium sphagni* sp. nov. *International Journal of Systematic Bacteriology* **30**, 77–81.
Kazda, J. and Müller, K. (1979). *Mycobacterium komossense* sp. nov. *International Journal of Systematic Bacteriology* **29**, 361–365.

Keddie, R. M. (1974a). Genus *Arthrobacter*. In: *Bergey's Manual of Determinative Bacteriology, Eighth Edition* (R. E. Buchanan and N. E. Gibbons, eds.), pp. 618–625. Williams and Wilkins, Baltimore.
Keddie, R. M. (1974b). Genus *Cellulomonas*. In: *Bergey's Manual of Determinative Bacteriology, Eighth Edition* (R. E. Buchanan and N. E. Gibbons, eds.), pp. 629–631. Williams and Wilkins, Baltimore.
Keddie, R. M. (1978). What do we mean by coryneform bacteria? In: *Coryneform Bacteria* (I. J. Bousfield and A. G. Cailely, eds.), pp. 1–12. Academic Press, London.
Keddie, R. M. and Bousfield, I. J. (1980). Cell wall composition in the classification and identification of coryneform bacteria. In: *Microbiological Classification and Identification* (M. Goodfellow and R. G. Board, eds.), pp. 167–188. Academic Press, London.
Keddie, R. M. and Cure, G. L. (1977). The cell wall composition and distribution of free mycolic acids in named strains of coryneform bacteria and in isolates from various natural sources. *Journal of Applied Bacteriology* **42**, 229–252.
Keddie, R. M. and Cure, G. L. (1978). Cell wall composition of coryneform bacteria. In: *Coryneform Bacteria* (I. J. Bousfield and A. G. Cailely, eds.), pp. 47–83. Academic Press, London.
Keddie, R. M. and Jones, D. (1981). Saprophytic, aerobic coryneform bacteria. In: *The Prokaryotes: A Handbook of Habitats, Isolation and Identification of Bacteria, Volume II*, (M. P. Starr, H. Stolp, H. G. Trüper, A. Balows and H. G. Schlegel, eds.), pp. 1838–1878. Springer Verlag, Berlin.
Keddie, R. M., Leask, B. G. S. and Grainger, J. M. (1966). A comparison of coryneform bacteria from soil and herbage: cell wall composition and nutrition. *Journal of Applied Bacteriology* **29**, 17–43.
Keyes, P. H. and Jordan, H. V. (1964). Periodontal lesions in the Syrian hamster–III. Findings related to an infectious and transmissible component. *Archives of Oral Biology* **9**, 377–400.
Kilian, M. (1978). Rapid identification of *Actinomycetaceae* and related bacteria. *Journal of Clinical Microbiology* **8**, 127–133.
Kirchheimer, W. F. and Storrs, E. E. (1971). Attempts to establish the armadillo (*Dasypus novemcinctus*) as a model for study of leprosy. I. Report of lepromatoid leprosy in an experimentally infected armadillo. *International Journal of Leprosy* **39**, 693–702.
Kloos, W. E., Tornabene, T. G. and Schleifer, K. H. (1974). Isolation and characterization of micrococci from human skin, including two new species: *Micrococcus lylae* and *Micrococcus kristinae*. *International Journal of Systematic Bacteriology* **24**, 79–101.
Koehne, G., Maddux, R. and Britt, J. (1981). Rapidly growing mycobacteria associated with bovine mastitis. *American Journal of Veterinary Research* **42**, 1238–1239.
Komura, I., Yamada, K., Otsuka, S. and Komagata, K. (1975). Taxonomic significance of phospholipids in coryneform and nocardioform bacteria. *Journal of General and Applied Microbiology* **21**, 251–261.
Krasilnikov, N. A. (1962). A new genus of ray fungus—*Actinopycnidium* n. gen. of family Actinoplanaceae. *Mikrobiologiya* **31**, 250–253. (English Translation 204–207).
Krasilnikov, N. A. (1970). *Ray Fungi: Higher Forms* (Trans). Nauka Publishers, Moscow.
Krasilnikov, N. A. and Agre, N. S. (1964). On two new species of *Thermopolyspora*. *Hindustan Antibiotics Bulletin* **6**, 97–107.
Krasilnikov, N. A. and Agre, N. S. (1965). The brown group of *Actinobifida chromogena* n. sp. *Mikrobiologiya* **34**, 284–291.
Krasilnikov, N. A. and Yuan, C. S. (1961). *Actinosporangium*, a new genus of the family Actinoplanaceae. *Izvestiya Akademii Nauk SSSR Seriya Biologischeskaya* **8**, 113–116.
Krasilnikov, N. A., Kalakoutskii, L. V. and Kirillova, N. V. (1961). A new genus of ray-fungi—*Promicromonospora*. *Bulletin of the Academy of Sciences U.S.S.R.* (*Series Biology*) **1**, 107–112.
Krasilnikov, N. A., Agre, N. S. and El-Reghistan, G. I. (1968). New thermophilic species of the genus *Micropolyspora*. *Mikrobiologiya* **37**, 1065–1075. (In Russian).
Krasilnikov, N. A., El-Registan, G. I., Ilyasova, V. B. and Agre, N. S. (1970). The infra-red spectra of the whole cells of ray fungi. In: *The Actinomycetales* (H. Prauser, ed.), pp. 293–298. Gustav Fischer Verlag, Jena.
Kroppenstedt, R. M. and Kutzner, H. J. (1976). Biochemical markers in the taxonomy of the Actinomycetales. *Experientia* **32**, 318–319.

Kroppenstedt, R. M. and Kutzner, H. J. (1978). Biochemical taxonomy of some problem actinomycetes. *Zentralblatt für Bakteriologie, Parasitenkunde, Infektionskrankheiten und Hygiene. I. Abteilung, Supplement* **6**, 125–133.

Kroppenstedt, R. M., Korn-Wendisch, F., Fowler, V. J. and Stackebrandt, E. (1981). Biochemical and molecular genetic evidence for a transfer of *Actinoplanes armeniacus* into the family *Streptomycetaceae*. *Zentralblatt für Bakteriologie, Parasitenkunde, Infektionskrankheiten und Hygiene. I. Abteilung, Originale* **C2**, 254–262.

Kubica, G. P. and Good, R. C. (1981). The genus *Mycobacterium* (except *M. leprae*). In: *The Prokaryotes: A Handbook of Habitats, Isolation and Identification of Bacteria, Volume II* (M. P. Starr, H. Stolp, H. G. Trüper, A. Balows and H. G. Schlegel, eds.), pp. 1962–1984. Springer Verlag, Berlin.

Kubica, G. P., Baess, I., Gordon, R. E., Jenkins, P. A., Kwapinski, J. B. G., McDurmont, C., Pattyn, S. R., Saito, H., Silcox, V., Stanford, J. L., Takeya, K. and Tsukamura, M. (1972). A cooperative numerical analysis of the rapidly growing mycobacteria. *Journal of General Microbiology* **73**, 55–70.

Küster, E. and Locci, R. (1963). Transfer of *Thermoactinomyces viridis* Schuurmans *et al.* 1956 to the genus *Thermomonospora* as *Thermomonospora viridis* (Schuurmans, Olson and San Clemente) comb. nov. *International Bulletin of Bacteriological Nomenclature and Taxonomy* **13**, 213–216.

Kutzner, H. J. (1981). The family *Streptomycetaceae*. In: *The Prokaryotes: A Handbook of Habitats, Isolation and Identification of Bacteria, Volume II* (M. P. Starr, H. Stolp, H. G. Trüper, A. Balows and H. G. Schlegel, eds.), pp. 2028–2090. Springer Verlag, Berlin.

Kutzner, H. J., Böttiger, V. and Heitzer, R. D. (1978). The use of physiological criteria in the taxonomy of *Streptomyces* and *Streptoverticillium*. *Zentralblatt für Bakteriologie, Parasitenkunde, Infektionskrankheiten und Hygiene. I. Abteilung, Supplement* **6**, 25–29.

Lacey, J. (1973). Actinomycetes in soils, composts and fodders. In: *Actinomycetales: Characteristics and Practical Importance* (G. Sykes and F. A. Skinner, eds.), pp. 231–251. Academic Press, London.

Lacey, J. (1978). Ecology of actinomycetes in fodders and related substrates. *Zentralblatt für Bakteriologie, Parasitenkunde, Infektionskrankheiten und Hygiene. I. Abteilung, Supplement* **6**, 161–170.

Lacey, J. (1981). Airborne actinomycete spores as respiratory allergens. *Zentralblatt für Bakteriologie, Mikrobiologie und Hygiene. I. Abteilung, Supplement* **11**, 243–250.

Lacey, J. and Goodfellow, M. (1975). A novel actinomycete from sugar-cane bagasse: *Saccharopolyspora hirsuta* gen. et sp. nov. *Journal of General Microbiology* **88**, 75–85.

Lanéelle, M.-A., Asselineau, J., Welby, M., Norgard, M. V., Imaeda, T., Pollice, M. C. and Barksdale, L. (1980). Biological and chemical bases for the reclassification of *Brevibacterium vitarumen* (Bechdel *et al.*) Breed (Approved Lists, 1980) as *Corynebacterium vitarumen* (Bechdel *et al.*) comb. nov. and *Brevibacterium liquefaciens* Okabayashi and Masu (Approved Lists, 1980) as *Corynebacterium liquefaciens* (Okabayashi & Masuo) comb. nov. *International Journal of Systematic Bacteriology* **30**, 539–546.

Lapage, S. P., Sneath, P. H. A., Lessel, E. F., Skerman, V. B. D., Seeliger, H. P. R. and Clark, W. A. (1975). *International Code of Nomenclature of Bacteria, 1976 Revision*. American Society for Microbiology, Washington D.C.

Lavrova, N. V., Preobrazhenskaya, T. P. and Sveshnikova, M. A. (1972). Isolation of actinomycetes on selective media with rubomycin. *Antibiotiki* **17**, 965–970.

Lawson, E. N. and Davey, L. M. (1972). A waterborne actinomycete resembling strains causing mycetoma. *Journal of Applied Bacteriology* **35**, 389–394.

Lechevalier, H. A. (1968). Status of the generic names *Micropolyspora* Lechevalier *et al.* 1961 and *Micropolispora* Shchepkina 1940 (*Actinomycetales*). *International Journal of Systematic Bacteriology* **18**, 203–206.

Lechevalier, H. A. and Holbert, P. E. (1965). Electron microscopic observation of the sporangial structure of a strain of *Actinoplanes*. *Journal of Bacteriology* **89**, 217–222.

Lechevalier, H. A. and Lechevalier, M. P. (1967). Biology of the actinomycetes. *Annual Review of Microbiology* **21**, 71–100.

Lechevalier, H. A. and Lechevalier, M. P. (1969). Ultramicroscopic structure of *Intrasporangium calvum* (*Actinomycetales*). *Journal of Bacteriology* **100**, 522–525.

Lechevalier, H. A. and Lechevalier, M. P. (1970). A critical evaluation of the genera of aerobic actinomycetes. In: *The Actinomycetales* (Prauser, H., ed.), pp. 393–405. Gustav Fischer Verlag, Jena.
Lechevalier, H. A. and Lechevalier, M. P. (1981). Introduction to the order *Actinomycetales*. In: *The Prokaryotes: A Handbook of Habitats, Isolation and Identification of Bacteria*, Volume II (M. P. Starr, H. Stolp, H. G. Trüper, A. Balows and H. G. Schlegel, eds.), pp. 1915–1922. Springer Verlag, Berlin.
Lechevalier, H. A., Solotorovsky, M. and McDurmont, C. I. (1961). A new genus of the *Actinomycetales*: *Micropolyspora* gen. nov. *Journal of General Microbiology* **26**, 11–18.
Lechevalier, H. A., Lechevalier, M. P. and Becker, B. (1966a). Comparison of the chemical composition of cell walls of nocardiae with that of other aerobic actinomycetes. *International Journal of Systematic Bacteriology* **16**, 151–160.
Lechevalier, H. A., Lechevalier, M. P. and Holbert, P. E. (1966b). Electron microscopic observation of the sporangial structure of strains of *Actinoplanaceae*. *Journal of Bacteriology* **92**, 1228–1235.
Lechevalier, H. A., Lechevalier, M. P. and Gerber, N. N. (1971). Chemical composition as a criterion in the classification of actinomycetes. *Advances in Applied Microbiology* **14**, 47–72.
Lechevalier, M. P. (1972). Description of a new species, *Oerskovia xanthineolytica* and emendation of *Oerskovia* Prauser *et al*. *International Journal of Systematic Bacteriology* **22**, 260–264.
Lechevalier, M. P. (1976). The taxonomy of the genus *Nocardia*: Some light at the end of the tunnel? In: *The Biology of the Nocardiae* (M. Goodfellow, G. H. Brownell and J. A. Serrano, eds.), pp. 1–38. Academic Press, London.
Lechevalier, M. P. (1977). Lipids in bacterial taxonomy – a taxonomists viewpoint. In: *CRC Critical Reviews in Microbiology*, pp. 109–210. CRC Press, Ohio.
Lechevalier, M. P. (1981). Ecological associations involving actinomycetes. *Zentralblatt für Bakteriologie, Mikrobiologie und Hygiene. 1. Abteilung, Supplement* **11**, 159–166.
Lechevalier, M. P. and Lechevalier, H. A. (1970a). Chemical composition as a criterion in the classification of aerobic actinomycetes. *International Journal of Systematic Bacteriology* **20**, 435–443.
Lechevalier, M. P. and Lechevalier, H. A. (1970b). Composition of whole-cell hydrolysates as a criterion in the classification of aerobic actinomycetes. In: *The Actinomycetales* (H. Prauser, ed.), pp. 311–316. Gustav Fischer Verlag, Jena.
Lechevalier, M. P. and Lechevalier, H. A. (1974). *Nocardia amarae* sp. nov. an actinomycete common in foaming activated sludge. *International Journal of Systematic Bacteriology* **24**, 278–288.
Lechevalier, M. P. and Lechevalier, H. A. (1975). Actinoplanete with cylindrical sporangia, *Actinoplanes rectilineatus* sp. nov. *International Journal of Systematic Bacteriology* **25**, 371–376.
Lechevalier, M. P. and Lechevalier, H. A. (1979). The taxonomic position of the actinomycetic endophytes. In: *Symbiotic Nitrogen Fixation in the Management of Temperate Forests* (J. C. Gordon, C. T. Wheeler and D. A. Perry, eds.), pp. 111–122. Forestry Research Laboratory, Corvallis, Oregon State University.
Lechevalier, M. P., Lechevalier, H. A. and Holbert, P. E. (1968). *Sporichthya*, un nouveau genre de *Streptomycetaceae*. *Annales de L'Institut Pasteur* **114**, 227–286.
Lechevalier, M. P., Lechevalier, H. A. and Heintz, C. E. (1973). Morphological and chemical nature of the sclerotia of *Chainia olivacea* Thirumalachar and Sukapure of the order *Actinomycetales*. *International Journal of Systematic Bacteriology* **23**, 157–170.
Lechevalier, M. P., De Bièvre, C. and Lechevalier, H. A. (1977). Chemotaxonomy of aerobic actinomycetes: phospholipid composition. *Biochemical Systematics and Ecology* **5**, 249–260.
Lechevalier, M. P., Stern, A. E. and Lechevalier, H. A. (1981). Phospholipids in the taxonomy of actinomycetes. *Zentralblatt für Bakteriologie, Mikrobiologie und Hygiene. I. Abteilung, Supplement* **11**, 111–116.
Lederer, E., Adam, A., Ciorbaru, R., Petit, J-F. and Wietzerbin, J. (1975). Cell walls of mycobacteria and related organisms; chemistry and immunostimulant properties. *Molecular and Cellular Biochemistry* **7**, 87–104.
Lee, S. E. and Humphrey, A. E. (1979). Use of continuous culture techniques for determining the growth kinetics of a cellulolytic *Thermoactinomyces* sp. *Biotechnology and Bioengineering* **21**, 1277–1288.

Lehmann, K. B. and Neumann, R. (1896). *Atlas und Grundis der Bakteriologie und Lehrbuch der speciellen Bakteriologischen Diagnostik*, First edition. J. F. Lehmann, München.
Lehmann, K. B. and Neumann, R. (1920). *Lehmann's Medizin. Handatlanten.* X. *Atlas und Grundriss der Bakteriologie und Lehrbuch der speziellen Bakteriologischen Diagnostik*, 6th edition. J. F. Lehmann, München.
Lehmann, K. B. and Neumann, R. (1927). *Bakteriologie insbesondere Bakteriologische.* II. *Allgemeine und spezielle Bakteriologie*, 7th edition. J. F. Lehmann, München.
Lelliott, R. A. (1966). The plant pathogenic coryneform bacteria. *Journal of Applied Bacteriology* **29**, 114–118.
Lesher, R. J., Gerencser, M. A. and Gerencser, V. F. (1974). Morphological biochemical and serological characterization of *Rothia dentocariosa*. *International Journal of Systematic Bacteriology* **24**, 154–159.
Lieske, R. (1921). *Morphologie und Biologie der Strahlenpilze*. Gebruder Borntraeger, Leipzig.
Lind, A. and Ridell, M. (1976). Serological relationships between *Nocardia*, *Mycobacterium*, *Corynebacterium* and the "*rhodochrous*" taxon. In: *The Biology of the Nocardiae* (M. Goodfellow, G. H. Brownell and J. A. Serrano, eds.), pp. 220–235. Academic Press, London.
Lind, A. and Ridell, M. (1983). Immunological classification: immunodiffusion and immunoelectrophoresis. In: *The Mycobacteria: A Source Book* (G. Kubica and L. G. Wayne, eds.), in press. Marcel Dekker, New York.
Lloyd, D. H. and Sellers, K. C. (eds.) (1976). *Dermatophilus Infection in Animals and Man.* Academic Press, London.
Locci, R. (1971). On the spore formation process in actinomycetes. IV. Examination by scanning electron microscopy of the genera *Thermoactinomyces*, *Actinobifida*, and *Thermomonospora*. *Revista di Patologia Vegetale* **7**, 63–80.
Locci, R. (1976). Developmental micromorphology of actinomycetes. In: *Actinomycetales. The Boundary Microorganisms* (T. Arai, ed.), pp. 249–297. Toppan Company, Tokyo.
Locci, R. (1978). Micromorphological development of *Actinomyces* and of related genera. *Zentralblatt für Bakteriologie, Parasitenkunde, Infektionskrankheiten und Hygiene. I. Abteilung, Supplement* **6**, 173–180.
Locci, R. (1981). Micromorphology and development of actinomycetes. *Zentralblatt für Bakteriologie, Mikrobiologie und Hygiene. 1. Abteilung, Supplement* **11**, 119–130.
Locci, R. and Petrolini, B. B. (1970). Morphology and development of *Streptoverticillium* species as examined by scanning electron microscopy. *Giornale di Microbiologia* **18**, 69–76.
Locci, R. and Petrolini-Baldan, B. (1971). On the spore formation process in actinomycetes. V. Scanning electron microscopy of some genera of *Actinoplanaceae*. *Rivista di Patologia Vegetale Pavia* **7** Supplemento Serie **IV**, 81–96.
Locci, R., Baldacci, E. and Petrolini, B. (1967). Contribution to the study of oligosporic actinomycetes. I. Description of a new species of *Actinobifida*: *Actinobifida alba* sp. nov. and revision of the genus. *Giornale di Microbiologia* **15**, 79–91.
Locci, R., Baldacci, E. and Petrolini, B. B. (1969). The genus *Streptoverticillium*. A taxonomic study. *Giornale di Microbiologia* **17**, 1–60.
Louria, D. B. and Gordon, R. E. (1960). Pericarditis and pleuritis caused by a recently discovered micro-organism, *Waksmania rosea*. *American Review of Respiratory Diseases* **81**, 83–88.
Luedemann, G. M. (1968). *Geodermatophilus*, a new genus of the *Dermatophilaceae* (*Actinomycetales*). *Journal of Bacteriology* **96**, 1848–1858.
Luedemann, G. M. (1974). The genus *Micromonospora*. In: *Bergey's Manual of Determinative Bacteriology, Eighth Edition* (R. E. Buchanan and N. E. Gibbons, eds.), pp. 846–855. Williams and Wilkins, Baltimore.
Luedemann, G. M. and Casmer, C. J. (1973). Electron microscope study of whole mounts and thin sections of *Micromonospora chalcea* ATCC 12452. *International Journal of Systematic Bacteriology* **23**, 243–255.
Ludwig, W., Seewaldt, E., Schleifer, K. H. and Stackebrandt, E. (1981). The phylogenetic status of *Kurthia zopfi*. *FEMS Microbiology Letters* **10**, 193–197.
Luscombe, B. M. and Gray, T. R. G. (1971). Effect of varying growth rate on the morphology of *Arthrobacter*. *Journal of General Microbiology* **69**, 433–434.
Luthy, P. (1974). *Corynebacterium okanaganae*, an entomopathogenic species of the *Corynebacteriaceae*. *Canadian Journal of Microbiology* **20**, 791–794.

McCarthy, A. J. (1981). Taxonomy and ecology of thermophilic actinomycetes. Doctoral Thesis, University of Bradford.

McCarthy, A. J. and Cross, T. (1981). A note on a selective medium for the thermophilic actinomycete *Thermomonospora chromogena*. *Journal Applied Bacteriology* **51**, 299–302.

McCarthy, A. J., Cross, T., Lacey, J. and Goodfellow, M. (1983). Conservation of the name *Micropolyspora* Lechevalier, Solotorovsky and McDurmont and designation of *Micropolyspora faeni* Cross, Maciver and Lacey as the type species of the genus. *International Journal of Systematic Bacteriology* **33**, 430–433.

McLean, R. A. and Sulzbacher, W. L. (1953). *Microbacterium thermosphactum*, spec. nov., a nonheat resistant bacterium from fresh pork sausage. *Journal of Bacteriology* **65**, 428–433.

Maclean, P. D., Liewbow, A. A. and Rosenberg, A. A. (1946). A haemolytic corynebacterium resembling *Cornebacterium ovis* and *Corynebacterium pyogenes* in man. *Journal of Infectious Diseases* **79**, 69–90.

Magnusson, M. (1962). Specificity of sensitins. III. Further studies in guinea pigs with sensitin of various species of *Mycobacterium* and *Nocardia*. *American Review of Respiratory Diseases* **86**, 395–404.

Magnusson, M. (1967). Identification of species of *Mycobacterium* on the basis of the specificity of the delayed type reaction in guinea pigs. *Zeitschrift für Tuberkulose und Erkrankungen der Thoraxorgane* **127**, 55–56.

Magnusson, M. (1971). A comparative study of *Mycobacterium gastri* and *Mycobacterium kansasii* by delayed type skin reactions in guinea pigs. *American Review of Respiratory Diseases* **104**, 377–384.

Magnusson, M. (1976). Sensitin tests as an aid in the taxonomy of *Nocardia* and its pathogenicity. In: *The Biology of the Nocardiae* (M. Goodfellow, G. H. Brownell and J. A. Serrano, eds.), pp. 236–265. Academic Press, London.

Makkar, N. S. and Cross, T. (1982). Actinoplanetes in soil and on plant litter from aquatic habitats. *Journal of Applied Bacteriology* **52**, 209–218.

Małuszyńska, G. M. and Janota-Bassalik, L. (1974). A cellulolytic rumen bacterium, *Micromonospora ruminantium* sp. nov. *Journal of General Microbiology* **82**, 57–65.

Manachini, P. L., Ferrari, A. and Craveri, R. (1965). Forme termofile di *Actinoplanaceae*. Isolamento e caratteristiche di *Streptosporangium album* var. *thermophilum*. *Annali di Microbiologia ed Enzimologia* **15**, 129–144.

Margalith, P. and Beretta, G. (1960). A new antibiotic producing *Streptomyces*: *Str. bellus* nov. sp. *Mycopathologia et Mycologia Applicata* **12**, 189–195.

Margalith, P. and Beretta, G. (1961). Rifomycin. XI. Taxonomic study on *Streptomyces mediterranei* nov. sp. *Mycopathologia et Mycologia Applicata* **13**, 321–330.

Marks, J., Jenkins, P. A. and Tsukamura, M. (1972). *Mycobacterium szulgai* – a new pathogen. *Tubercle, London* **53**, 210–214.

Matsumae, A. M. and Hata, T. (1968). In: Matsumae, A. M., Ohtani, M., Takeshima, H. and Hata, T., A new genus of *Actinomycetales*: *Kitasatoa* gen. nov. *Journal of Antibiotics (Tokyo)* **21**, 616–625.

Matthews, P. R. J., McDiarmid, A., Collins, P. and Brown, A. (1978). The dependence on some strains of *Mycobacterium avium* on mycobactin for initial and subsequent growth. *Journal of Medical Microbiology* **11**, 53–57.

Mauff, G., Herrmann, M. and Schaal, K. P. (1981). Electrophoretic protein patterns of nocardiae and their possible taxonomic relevance. *Zentralblatt für Bakteriologie, Mikrobiologie und Hygiene. 1. Abteilung, Supplement* **11**, 33–38.

Medinger, A. E. and Spagnolo, S. V. (1981). *Mycobacterium szulgae* pulmonary infection: the importance of knowing. *Southern Medical Journal* **74**, 85–86.

Meissner, G., Schröder, K. H., Amadio, G. E., Anz, W., Chaparas, S., Engel, H. W. B., Jenkins, P. A., Käppler, W., Kleeberg, H. H., Kubala, E., Kubin, M., Lauterbach, D., Lind, A., Magnusson, M., Mikova, Zd., Pattyn, S. R., Schaefer, W. B., Stanford, J. L., Tsukamura, M., Wayne, L. G., Willers, I. and Wolinsky, E. (1974). A cooperative numerical analysis of nonscoto – and nonphotochromogenic slowly growing mycobacteria. *Journal of General Microbiology* **83**, 207–235.

Melville, T. H. (1965). A study of the overall similarity of certain actinomycetes mainly of oral origin. *Journal of General Microbiology* **40**, 309–315.

Meyer, J. (1976). *Nocardiopsis*, a new genus of the order *Actinomycetales*. *International Journal of Systematic Bacteriology* **26**, 487–493.

Meyer, J. (1979). The new species of the genus *Actinomadura*. *Zeitschrift für Allgemeine Mikrobiologie* **19**, 37–44.

Meyer, J. and Sveshnikova, M. (1974). *Micromonospora rubra* Sveshnikova et al = *Actinomadura rubra* comb. nov. *Zeitschrift für Allgemeine Mikrobiologie* **14**, 167–170.

Minnikin, D. E. and Goodfellow, M. (1976). Lipid composition in the classification and identification of nocardiae and related taxa. In: *The Biology of the Nocardiae* (M. Goodfellow, G. H. Brownell and J. A. Serrano, eds.), pp. 160–219. Academic Press, London.

Minnikin, D. E. and Goodfellow, M. (1978). Polar lipids of nocardioform and related bacteria. *Zentralblatt für Bakteriologie, Parasitenkunde, Infektionskrankheiten und Hygiene. 1. Abteilung, Supplement* **6**, 75–83.

Minnikin, D. E. and Goodfellow, M. (1980). Lipid composition in the classification and identification of acid-fast bacteria. In: *Microbiological Classification and Identification* (M. Goodfellow and R. G. Board, eds.), pp. 189–256. Academic Press, London.

Minnikin, D. E. and Goodfellow, M. (1981a). Lipids in the classification of actinomycetes. *Zentralblatt für Bakteriologie, Parasitenkunde, Infektionskrankheiten und Hygiene. 1. Abteilung, Supplement* **11**, 99–109.

Minnikin, D. E. and Goodfellow, M. (1981b). Lipids in the classification of *Bacillus* and related taxa. In: *The Aerobic Endospore – forming Bacteria: Classification and Identification* (R. C. W. Berkeley and M. Goodfellow, eds.), pp. 59–90. Academic Press, London.

Minnikin, D. E., Alshamaony, L. and Goodfellow, M. (1975). Differentiation of *Mycobacterium*, *Nocardia* and related taxa by thin-layer chromatographic analysis of whole-cell methanolysates. *Journal of General Microbiology* **88**, 200–204.

Minnikin, D. E., Pirouz, T. and Goodfellow, M. (1977a). Polar lipid composition in the classification of some *Actinomadura* species. *International Journal of Systematic Bacteriology* **27**, 118–121.

Minnikin, D. E., Patel, P. V., Alshamaony, L. and Goodfellow, M. (1977b). Polar lipid composition in the classification of *Nocardia* and related bacteria. *International Journal of Systematic Bacteriology* **27**, 104–117.

Minnikin, D. E., Goodfellow, M. and Collins, M. D. (1978a). Lipid composition in the classification and identification of coryneform and related taxa. In: *Coryneform Bacteria* (I. J. Bousfield and A. G. Callely, eds.), pp. 85–160. Academic Press, London.

Minnikin, D. E., Collins, M. D. and Goodfellow, M. (1978b). Menaquinone patterns in the classification of nocardioform and related bacteria. *Zentralblatt für Bakteriologie, Parasitenkunde, Infektionskrankheiten und Hygiene. 1. Abteilung, Supplement* **6**, 85–90.

Minnikin, D. E., Collins, M. D. and Goodfellow, M. (1979). Fatty acid and polar lipid composition in the classification of *Cellulomonas*, *Oerskovia* and related taxa. *Journal of Applied Bacteriology* **47**, 87–95.

Minnikin, D. E., Hutchinson, I. G., Caldicott, A. B. and Goodfellow, M. (1980). Thin-layer chromatography of methanolysates of mycolic acid-containing bacteria. *Journal of Chromatography* **188**, 221–233.

Minnikin, D. E., Minnikin, S. M., Goodfellow, M. and Stanford, J. L. (1982). The mycolic acids of *Mycobacterium chelonei*. *Journal of General Microbiology* **128**, 817–822.

Mitsuoka, T., Morishita, Y., Terada, A. and Watanabe, K. (1974). *Actinomyces eriksonii* Georg, Robertstad, Brinkman und Hicklin 1965 identische mit *Bifidobacterium adolescentis* Reuter 1963. *Zentralblatt für Bakteriologie, Parasitenkunde, Infektionskrankheiten und Hygiene. I. Abteilung. Originale Reihe A* **226**, 257–263.

Moore, R. L. and McCarthy, B. (1967). Comparative study of ribosomal ribonucleic acid cistrons in enterobacteria and myxobacteria. *Journal of Bacteriology* **94**, 1066–1074.

Mordarska, H., Mordarski, M. and Goodfellow, M. (1972). Chemotaxonomic characters and classification of some nocardioform bacteria. *Journal of General Microbiology* **71**, 77–86.

Mordarskaya, G., Guzeva, L. N. and Agre, N. A. (1973). Lipids from the mycelia of thermophilic actinomycetes. *Mikrobiologiya* **42**, 165–166. (In Russian).

Mordarski, M., Schaal, K. P., Szyba, K., Pulverer, G. and Tkacz, A. (1977). Classification of *Nocardia asteroides* and allied taxa based upon deoxyribonucleic acid reassociation. *International Journal of Systematic Bacteriology* **27**, 66–70.

Mordarski, M., Goodfellow, M., Szyba, K., Pulverer, G. and Tkacz, A. (1978a). Deoxyribonucleic acid base composition and homology studies on *Rhodococcus* and allied taxa. *Zentralblatt für Bakteriologie, Parasitenkunde, Infektionskrankheiten und Hygiene. 1. Abteilung, Supplement* **6**, 99–106.

Mordarski, M., Schaal, K. P., Tkacz, A., Pulverer, G., Szyba, K. and Goodfellow, M. (1978b). Deoxyribonucleic acid base composition and homology studies on *Nocardia*. *Zentralblatt für Bakteriologie, Parasitenkunde, Infektionskrankheiten und Hygiene. 1. Abteilung, Supplement* **6**, 91–97.

Mordarski, M., Goodfellow, M., Tkacz, A., Pulverer, G. and Schaal, K. P. (1980a). Ribosomal ribonucleic similarities in the classification of *Rhodococcus* and related taxa. *Journal of General Microbiology* **118**, 313–319.

Mordarski, M., Goodfellow, M., Kaszen, I., Tkacz, A., Pulverer, G. and Schaal, K. P. (1980b). Deoxyribonucleic acid reassociation in the classification of the genus *Rhodococcus* Zopf 1891 (Approved Lists 1980). *International Journal of Systematic Bacteriology* **30**, 521–527.

Mordarski, M., Kaszen, I., Tkacz, A., Goodfellow, M., Alderson, G., Schaal, K. P. and Pulverer, G. (1981a). Deoxyribonucleic acid pairing in the classification of the genus *Rhodococcus*. *Zentralblatt für Bakteriologie, Mikrobiologie und Hygiene. 1. Abteilung, Supplement* **11**, 25–31.

Mordarski, M., Tkacz, A., Goodfellow, M., Schaal, K. P. and Pulverer, G. (1981b). Ribosomal ribonucleic acid similarities in the classification of actinomycetes. *Zentralblatt für Bakteriologie, Mikrobiologie und Hygiene. 1. Abteilung, Supplement* **11**, 79–85.

Morquer, R. and Comby, L. (1943). Affinités systématiques du genere *Micromonospora* Ørskov. I. La classification des *Actinomyces*. *Bulletin de la Société d'Histoire Naturelle, Toulouse* **78**, 23–28.

Mulder, E. G. and Antheunisse, J. (1963). Morphologie, physiologie et écologie des *Arthrobacter*. *Annales de l'Institut Pasteur* **105**, 46–74.

Mungelluzi, C. (1966). *Actinomonospora lusitanica* Castellani, De Brito & Pinto (1959). *Archives Italiennes Scienze Mediche Tropical Parasitologia* **47**, 33–38.

Nesterenko, O. A., Nogina, T. M., Kasumova, S. A., Kvasnikov, E. I. and Batrakov, S. G. (1982). *Rhodococcus luteus* nom. nov. and *Rhodococcus maris* nom. nov. *International Journal of Systematic Bacteriology* **32**, 1–14.

Nonomura, H. and Ohara, Y. (1957). Distribution of actinomycetes in soil. II. *Microbispora*, a new genus of *Streptomycetaceae*. *Journal of Fermentation Technology* **35**, 307–311.

Nonomura, H. and Ohara, Y. (1960). Distribution of the actinomycetes in soil. (V). The isolation and classification of the genus *Streptosporangium*. *Journal of Fermentation Technology* **38**, 405–409.

Nonomura, H. and Ohara, Y. (1969a). Distribution of actinomycetes in soil. (VI). A culture method effective for both preferential isolation and enumeration of *Microbispora* and *Streptosporangium* strains in soil (part I). *Journal of Fermentation Technology* **47**, 463–469.

Nonomura, H. and Ohara, Y. (1969b). Distribution of actinomycetes in soil. (VII). A culture method effective for both preferential isolation and enumeration of *Microbispora* and *Streptosporangium* strains in soil (part 2). Classification of the isolates. *Journal of Fermentation Technology* **47**, 701–709.

Nonomura, H. and Ohara, Y. (1971a). Distribution of actinomycetes in soil. VIII. Green-spore group of *Microtetraspora*, its preferential isolation and taxonomic characteristics. *Journal of Fermentation Technology* **49**, 1–7.

Nonomura, H. and Ohara, Y. (1971b). Distribution of actinomycetes in soil. IX. New species of the genus *Microbispora* and *Microtetraspora* and their isolation methods. *Journal of Fermentation Technology* **49**, 887–894.

Nonomura, H. and Ohara, Y. (1971c). Distribution of actinomycetes in soil. (X). New genus and species of monosporic actinomycetes. *Journal of Fermentation Technology* **49**, 895–903.

Nonomura, H. and Ohara, Y. (1971d). Distribution of actinomycetes in soil. XI. Some new species of the genus *Actinomadura* Lechevalier *et al. Journal of Fermentation Technology* **49**, 904–912.

Nonomura, H. and Ohara, Y. (1974). A new species of actinomycetes, *Thermomonospora mesouviformis* sp. nov. *Journal of Fermentation Technology* **52**, 10–13.
Nonomura, H. and Takagi, S. (1977). Distribution of actinoplanetes in soils of Japan. *Journal of Fermentation Technology* **55**, 423–428.
Nonomura, H., Hayakawa, M. and Iino, S. (1979). Classification of actinomycetes of the genus *Ampullariella* from soils of Japan. *Hakkogaku Kaishi* **57**, 79–85.
Norlin, M. (1965). Unclassified mycobacteria, a comparison between a serological and biochemical classification method. *Bulletin of the International Union against Tuberculosis* **36**, 25–32.
Norlin, M., Lind, A. and Ouchterlony, Ö. (1969). A serologically based taxonomic study of *Mycobacterium gastri*. *Zeitschrift für Immunitatsforschung, Allergie und Klinische Immunologie* **137**, 241–248.
O'Donnell, A. G., Goodfellow, M. and Minnikin, D. E. (1983). Lipids in the classification of *Nocardioides*: Reclassification of *Arthrobacter simplex* in the emended genus *Nocardioides* as *Nocardioides simplex* comb. nov. *Archives of Microbiology* **133**, 323–329.
Okuda, T., Furumai, T., Watanabe, E., Okugawa, T. and Kimura, S. (1966). Actinoplanaceae antibiotics. II. Studies of sporoviridin. 2. Taxonomic study of the sporoviridin producing microorganism: *Streptosporangium viridogriseum* nov. sp. *Journal of Antibiotics Series A* **19**, 121–127.
Omura, S., Iwai, Y., Takahashi, Y., Kojima, K. and Otoguro, K. (1981). Type of diaminopimelic acid different in aerial and vegetative mycelia of setamycin producing actinomycete KM-6054. *Journal of Antibiotics* **34**, 1633–1634.
Onisi, M. (1949). Study on the *Actinomyces* isolated from the deeper layers of carious dentine. *Shikazaku Zasehi* **6**, 273–282.
Onisi, M. and Nuckolls, J. (1958). Description of actinomycetes and other pleomorphic organisms recovered from pigmented carious lesions of the dentine of human teeth. *Oral Surgery Oral Medicine and Oral Pathology* **11**, 913–930.
Orchard, V. A. (1979). Effect of sewage sludge additions on *Nocardia* in soil. *Soil Biology and Biochemistry* **11**, 217–220.
Orchard, V. A. (1981). The ecology of *Nocardia* and related taxa. *Zentralblatt für Bakteriologie, Mikrobiologie und Hygiene. 1. Abteilung, Supplement* **11**, 167–180.
Orchard, V. A. and Goodfellow, M. (1974). The selective isolation of *Nocardia* from soil using antibiotics. *Journal of General Microbiology* **85**, 160–162.
Orchard, V. A. and Goodfellow, M. (1980). Numerical classification of some named strains of *Nocardia asteroides* and related isolates from soil. *Journal of General Microbiology* **118**, 295–312.
Orchard, V. A., Goodfellow, M. and Williams, S. T. (1977). Selective isolation and occurrence of nocardiae in soil. *Soil Biology and Biochemistry* **9**, 233–238.
Ordal, E. J. and Earp, B. J. (1956). Cultivation and transmission of the etiological agent of kidney disease in salmonid fishes. *Proceedings of the Society of Experimental Biology and Medicine* **92**, 85–88.
Orla-Jensen, S. (1919). *The Lactic Acid Bacteria*. Host and Son, Copenhagen.
Ørskov, J. (1923). *Investigations into the Morphology of the Ray Fungi*. Levin and Munksgaard, Copenhagen.
Ørskov, J. (1938). Untersuchungen über Strahlenpilze, reingezüchtet aus dänischen Erdproben. *Zentralblatt für Bakteriologie, Parasitenkunde, Infektionskrankheiten und Hygiene. Abteilung 2* **98**, 344–357.
Owen, R. J. and Hill, L. R. (1979). The estimation of base compositions, base pairing and genome sizes of bacterial deoxyribonucleic acids. In: *Identification Methods for Microbiologists* (F. A. Skinner, and D. W. Lovelock, eds.), pp. 277–296. Academic Press, London.
Owens, J. D. and Keddie, R. M. (1969). The nitrogen nutrition of soil and herbage coryneform bacteria. *Journal of Applied Bacteriology* **32**, 338–347.
Pagani, H. and Parenti, F. (1978). *Kineosporia*, a new genus of the Order *Actinomycetales*. *International Journal of Systematic Bacteriology* **28**, 401–406.
Palleroni, N. J. (1976). Chemotaxis in *Actinoplanes*. *Archives of Microbiology* **110**, 13–18.
Palleroni, N. J. (1979). New species of the genus *Actinoplanes*, *Actinoplanes ferrugineus*. *International Journal of Systematic Bacteriology* **29**, 61–65.

Palleroni, N. J. (1980). A chemotactic method for the isolation of *Actinoplanaceae*. *Archives of Microbiology* **128**, 53–58.
Pandhi, P. W. and Hammond, B. F. (1975). A glycolipid from *Rothia dentocariosa*. *Archives of Oral Biology* **20**, 399–401.
Parenti, F. and Coronelli, C. (1979). Members of the genus *Actinoplanes* and their antibiotics. *Annual Review of Microbiology* **33**, 389–411.
Parenti, F., Beretta, G., Berti, M. S. and Arioli, V. (1978). Teichomycins, new antibiotics from *Actinoplanes teichomyceticus*, nov. sp. I. Description of the producer strain, fermentation studies and biological properties. *Journal of Antibiotics* **31**, 276–283.
Pattyn, S. R. (1970). Agglutination with rapidly growing (Runyon's Group IV) mycobacteria. *Zentralblatt für Bakteriologie, Parasitenkunde, Infektionskrankheiten und Hygiene. Abteilung I. Originale A*, **215**, 99–105.
Pattyn, S. R., Magnusson, M., Stanford, J. L. and Grange, J. M. (1974). A study of *Mycobacterium fortuitum* (*ranae*). *Journal of Medical Microbiology* **7**, 67–76.
Pier, A. C. and Fichtner, R. E. (1981). Distribution of serotypes of *Nocardia asteroides* from animal, human, and environmental sources. *Journal of Clinical Microbiology* **13**, 548–553.
Pine, L. (1970). Classification and phylogenetic relationship of microaerophilic actinomycetes. *International Journal of Systematic Bacteriology* **20**, 445–474.
Pine, L. and Boone, C. J. (1967). Comparative wall analysis of morphological forms within the genus *Actinomyces*. *Journal of Bacteriology* **94**, 875–883.
Pine, L. and Georg, L. K. (1969). Reclassification of *Actinomyces propionicus*. *International Journal of Systematic Bacteriology* **19**, 267–272.
Pine, L. and Georg, L. K. (1974). Genus *Arachnia*. In: *Bergey's Manual of Determinative Bacteriology, Eighth Edition* (R. E. Buchanan and N. E. Gibbons, eds.), pp. 668–669. The Williams and Wilkins Company, Baltimore.
Pine, L. and Hardin, H. (1959). *Actinomyces israelii*: a cause of lacrimal canaliculitis in man. *Journal of Bacteriology* **78**, 164–170.
Pitcher, D. G. (1976). Arabinose with L-diaminopimelic acid in the cell wall of an aerobic coryneform organism isolated from human skin. *Journal of General Microbiology* **94**, 225–227.
Pitcher, D. G. and Noble, W. C. (1978). Aerobic diphtheroids of human skin. In *Coryneform Bacteria* (I. J. Bousfield and A. G. Callely, ed.), pp. 265–287. Academic Press, London.
Pittenger, R. C. and Brigham, R. B. (1956). *Streptomyces orientalis* n. sp., the source of vancomycin. *Antibiotics and Chemotherapy* **6**, 642–647.
Pjiper, A. and Pullinger, B. D. (1929). South African nocardioses. *Journal of Tropical Medicine and Hygiene* **30**, 153–156.
Pommer, E.-H. (1959). Über die Isolierung des Endophyten aus den Wurzelknöllchen von *Alnus glutinosa* Gaertn. und über erfolgreiche Reinfektionsversuche. *Bericht der Deutschen botanischen Gesellschaft* **72**, 138–150.
Pommier, M. T. and Michel, G. (1973). Phospholipid and acid composition of *Nocardia* and nocardoid bacteria as criteria of classification. *Biochemical Systematics* **1**, 3–12.
Portaels, F. (1980). Study of unclassified dapsone sensitive mycobacteria isolated from the environment in Zaire. *Annales de la Société belge Médicine tropicale* **60**, 381–386.
Portaels, F., Goodfellow, M., Minnikin, D. E., Minnikin, S. M. and Hutchinson, I. G. (1982). *Nocardia* – like mycobacteria from natural habitats in Zaire. *Annales de la Société belge de Médecine tropicale* **61**, 477–487.
Potekhina, L. L. (1965). *Streptosporangium rubrum* n. sp. – a new species of the *Streptosporangium* genus. *Mikrobiologiya* **34**, 292–299. (In Russian).
Prauser, H. (1967a). DAP-freie, gelbe Actinomyceten mit Tendenz sur Beweglichkeit. *Zeitschrift für Allegemeine Mikrobiologie* **7**, 81–83.
Prauser, H. (1967b). Contributions to the taxonomy of the *Actinomycetales*. *Publications of the Faculty of Sciences. J. E. Purkyně University, Brno*, **K40**, 196–199.
Prauser, H. (1970). Characters and genera arrangement in the *Actinomycetales*. In: *The Actinomycetales* (H. Prauser, ed.), pp. 407–418. Gustav Fischer, Jena.
Prauser, H. (1976a). Host-parasite relationships in nocardioform organisms. In: *The Biology of the Nocardiae* (M. Goodfellow, G. H. Brownell and J. A. Serrano, eds.), pp. 266–284. Academic Press, London.

Prauser, H. (1976b). *Nocardioides*, a new genus of the order *Actinomycetales*. *International Journal of Systematic Bacteriology* **26**, 58–65.

Prauser, H. (1978). Considerations on taxonomic relations among Gram-positive, branching bacteria. *Zentralblatt für Bakteriologie, Parasitenkunde, Infektionskrankheiten und Hygiene. 1. Abteilung, Supplement* **6**, 3–12.

Prauser, H. (1981). Nocardioform organisms: General characterization and taxonomic relationships. *Zentralblatt für Bakteriologie, Mikrobiologie und Hygiene. 1. Abteilung, Supplement* **11**, 17–24.

Prauser, H. and Bergholz, M. (1974). Taxonomy of actinomycetes and screening for antibiotic substances. *Postepy Higieny I Medycyny Doswiadczalnej* **28**, 441–457.

Prauser, H. and Falta, R. (1968). Phagensensibilität Zellwandzusammensetzung und Taxonomie von Actinomyceten. *Zeitschrift für allgemeine Mikrobiologie* **8**, 39–46.

Prauser, H. and Momirova, S. (1970). Phagensensibilität, Zellwandzusammensetzung und Taxonomie einiger thermophiler Actinomyceten. *Zeitschrift für Mikrobiologie* **10**, 219–222.

Prauser, H., Müller, L. and Falta, R. (1967). On the taxonomic position of the genus *Microellobosporia* Cross, Lechevalier and Lechevalier 1963. *International Journal of Systematic Bacteriology* **17**, 361–366.

Prauser, H., Lechevalier, M. P. and Lechevalier, H. A. (1970). Description of *Oerskovia* gen. n. to harbour Ørskov's motile *Nocardia*. *Applied Microbiology* **19**, 534.

Preobrazhenskaya, T. P. and Sveshnikova, M. A. (1974). New species of the genus *Actinomadura*. *Mikrobiologiya* **43**, 864–868.

Preobrazhenskaya, T. P., Ukholina, R. S., Nechaeva, N. P., Filicheva, V. A., Gavrilina, G. V., Kudinova, M. K., Borisova, V. N., Petukhova, N. M., Kovsharova, I. N., Proshlyakova, V. V. and Rossolimo, O. K. (1973). A new species of the genus *Micropolyspora* and its antibiotic properties. *Antibiotiki* **18**, 963–968.

Preobrazhenskaya, T. P., Lavrova, N. V., Ukholina, R. S. and Nechaeva, N. P. (1975). Isolation of new species of *Actinomadura* on selective media with streptomycin and bruneomycin. *Antibiotiki* **20**, 404–409.

Preobrazhenskaya, T. P., Sveshnikova, M. A. and Terekhova, L. P. (1977). Key for identification of the species of the genus *Actinomadura*. *Biology of Actinomycetes and Related Organisms* **12**, 30–38.

Preobrazhenskaya, T. P., Sveshnikova, M. A., Terekhova, L. P. and Chormonova, N. T. (1978). Selective isolation of soil actinomycetes. *Zentralblatt für Bakteriologie, Parasitenkunde, Infektionskrankheiten und Hygiene. I. Abteilung, Supplement* **6**, 119–123.

Pridham, T. G. (1970). New names and new combinations in the Order *Actinomycetales* Buchanan 1917. *United States Department of Agriculture Technical Bulletin* No. 1424.

Pridham, T. G. and Lyons, A. J. (1969). Progress in clarification of the taxonomic and nomenclatural status of some problem actinomycetes. *Developments in Industrial Microbiology* **10**, 183–221.

Pridham, T. C. and Tresner, H. D. (1974). Family VII. *Streptomycetaceae* Waksman and Henrici 1943, 339. In: *Bergey's Manual of Determinative Bacteriology, Eighth Edition* (R. E. Buchanan and N. E. Gibbons, eds.), pp. 747–829. Williams and Wilkins, Baltimore.

Pridham, T. G., Hesseltine, C. W. and Benedict, R. G. (1958). A guide for the classification of streptomycetes according to selected groups. *Applied Microbiology* **6**, 52–79.

Pridham, T. G., Lyons, A. J. and Seckinger, H. L. (1965). Comparison of some dried holotype and neotype specimens with their living counterparts. *International Bulletin of Bacteriological Nomenclature and Taxonomy* **15**, 191–237.

Priest, F. (1981). DNA homology in the genus *Bacillus*. In: *The Aerobic Endospore-forming Bacteria: Classification and Identification* (R. C. W. Berkeley and M. Goodfellow, eds.), pp. 33–57. Academic Press, London.

Pulverer, G. and Schaal, K. P. (1978). Pathogenicity and medical importance of aerobic and anaerobic actinomycetes. *Zentralblatt für Bakteriologie, Parasitenkunde, Infektionskrankheiten und Hygiene. 1. Abteilung, Supplement* **6**, 417–427.

Pulverer, G., Schütt-Gerowitt, H. and Schaal, K. P. (1975). Bacteriophages of *Nocardia asteroides*. *Medical Microbiology and Immunology* **161**, 113–122.

Rasmussen, E. G., Gibbons, R. J. and Socransky, S. S. (1966). A taxonomic study of fifty Gram positive anaerobic diphtheroids isolated from the oral cavity of man. *Archives of Oral Biology* **11**, 573–579.

Reddy, C. A. and Cornell, C. P. (1981). A proposal for the transfer of *Corynebacterium pyogenes* to the genus *Actinomyces* as *Actinomyces pyogenes* comb. nov. Abstract presented at the 62nd Conference of Research Workers in Diseases, Chicago, 1981.
Reddy, C. A., Cornell, C. P. and Fraga, A. M. (1982). Transfer of *Corynebacterium pyogenes* (Glage) Eberson to the genus *Actinomyces* as *Actinomyces pyogenes* (Glage) comb. nov. *International Journal of Systematic Bacteriology* **32**, 419–429.
Redmond, W. B., Bates, J. H. and Engel, H. W. B. (1979). Methods for bacteriophage typing of mycobacteria. In: *Methods in Microbiology, Volume 13* (T. Bergan and J. R. Norris, eds.), pp. 345–375. Academic Press, London.
Reller, L. B., Maddoux, G. L., Eckman, M. R. and Pappas, G. (1975). Bacterial endocarditis caused by *Oerskovia turbata*. *Annals of Internal Medicine* **83**, 664–666.
Richard, J. L., Thurston, J. R. and Pier, A. C. (1976). Comparison of antigens of *Dermatophilus congolensis* isolates and their use in serological tests in experimental and natural infections. In: *Dermatophilus Infection in Animals and Man* (D. H. Lloyd and K. C. Sellars, eds.), pp. 216–227. Academic Press, London.
Ridell, M. (1975). Taxonomic study of *Nocardia farcinica* using serological and physiological characters. *International Journal of Systematic Bacteriology* **25**, 124–132.
Ridell, M. (1983). Immunodiffusion analyses of *Mycobacterium farcinogenes*, *Mycobacterium senegalense* and some other mycobacteria. *Journal of General Microbiology* **129**, 613–619.
Ridell, M. and Goodfellow, M. (1983). Numerical classification of *Mycobacterium farcinogenes*, *Mycobacterium senegalense* and related taxa. *Journal of General Microbiology* **129**, 599–611.
Ridell, M., Baker, R., Lind, A. and Ouchterlony, Ö. (1979). Immunodiffusion studies of ribosomes in classification of mycobacteria and related taxa. *International Archives of Allergy and Applied Immunology* **59**, 162–172.
Ridell, M., Goodfellow, M., Minnikin, D. E., Minnikin, S. M. and Hutchinson, I. G. (1982). Classification of *Mycobacterium farcinogenes* and *Mycobacterium senegalense* by immunodiffusion and thin-layer chromatography of long chain components. *Journal of General Microbiology* **128**, 1299–1307.
Roberts, D. S. (1961). The life cycle of *Dermatophilus dermatonomus*, the causal agent of ovine mycotic dermatitis. *Australian Journal of Experimental Biology and Medical Science* **39**, 463–476.
Roberts, D. S. (1967). Chemotherapy of epidermal infection with *Dermatophilus congolensis*. *Journal of Comparative Pathology* **77**, 129–136.
Roberts, D. S. (1970). *Dermatophilus congolensis*, a zoopathogenic actinomycete with a motile infective stage. In: *The Actinomycetales* (H. Prauser, ed.), pp. 265–271. Gustav Fischer Verlag, Jena.
Roberts, D. S. (1981). The family *Dermatophilaceae*. In: *The Prokaryotes: A Handbook of Habitats, Isolation and Identification of Bacteria, Volume II* (M. P. Starr, H. Stolp, H. G. Trüper, A. Balows and H. G. Schlegel, eds.), pp. 2011–2015. Springer-Verlag, Berlin.
Rogosa, M. and Keddie, R. M. (1974a). *Brevibacterium*. In: *Bergey's Manual of Determinative Bacteriology, Eighth Edition* (R. E. Buchanan and N. E. Gibbons, eds.), pp. 625–628. Williams and Wilkins, Baltimore.
Rogosa, M. and Keddie, R. M. (1974b). *Microbacterium*. In: *Bergey's Manual of Determinative Bacteriology, Eighth Edition* (R. E. Buchanan and N. E. Gibbons, eds.), pp. 628–629. Williams and Wilkins, Baltimore.
Rogosa, M., Cummins, C. S., Lelliott, R. A. and Keddie, R. M. (1974). Coryneform group of bacteria. In: *Bergey's Manual of Determinative Bacteriology, Eighth Edition* (R. E. Buchanan and N. E. Gibbons, eds.), pp. 599–602. Williams and Wilkins, Baltimore.
Roth, G. D. (1957). Proteolytic organisms of the carious lesion. *Oral Surgery Oral Medicine and Oral Pathology* **10**, 1105–1117.
Roth, G. D. and Flanagan, V. (1969). The pathogenicity of *Rothia dentocariosa* inoculated into mice. *Journal of Dental Research* **48**, 957–958.
Roth, G. D. and Thurn, A. N. (1962). Continued study of oral *Nocardia*. *Journal of Dental Research* **41**, 1279–1292.
Rothwell, F. M. (1957). A further study of Karling's keratinophilic organism. *Mycologia* **49**, 68–72.

Rowbotham, T. J. and Cross, T. (1977). *Rhodococcus coprophilus* sp. nov.: an aerobic nocardioform actinomycete belonging to the "rhodochrous" complex. *Journal of General Microbiology* **100**, 123–138.
Runyon, E. H. (1958). Mycobacteria encountered in clinical laboratories. *Leprosy Briefs* **9**, 21.
Runyon, E. G. (1959). Anonymous mycobacteria in pulmonary disease. *The Medical Clinics of North America* **43**, 273–290.
Saito, H., Yamaoko, K. and Kiyotani, K. (1977a). In vitro properties of *M. lepraemurium* strain Keischicko. *International Journal of Systematic Bacteriology* **26**, 111–115.
Saito, H., Gordon, R. E., Juhlin, I., Käppler, W., Kwapinski, J. B. G., McDermont, C., Pattyn, S. R., Runyon, E. H., Stanford, J. L., Tarnok, I., Tasaka, H., Tsukamura, M. and Weiszfeiler, J. (1977b). Cooperative numerical analysis of rapidly growing mycobacteria. *International Journal of Systematic Bacteriology* **27**, 75–85.
Samsonoff, W. A., Detlesen, M. A., Fonseca, A. F. and Edwards, M. R. (1977). Deoxyribonucleic acid base composition of *Dermatophilus congolensis* and *Geodermatophilus obscurus*. *International Journal of Systematic Bacteriology* **27**, 22–25.
Sanders, J. E. and Fryer, J. L. (1980). *Renibacterium salmoninarum* gen. nov., sp. nov., the causative agent of bacterial kidney disease in salmonid fishes. *International Journal of Systematic Bacteriology* **30**, 496–502.
Scardovi, V. (1980). The genus *Bifidobacterium*. In: *The Prokaryotes: A Handbook of Habitats, Isolation and Identification of Bacteria, Volume* II (M. P. Starr, H. Stolp, H. G. Trüper, A. Balows and H. G. Schelegel, eds.), pp. 1951–1961. Springer Verlag, Berlin.
Scardovi, V., Trovatelli, L. D., Zani, G., Crociani, F. and Matteuzzi, D. (1971). Deoxyribonucleic acid homology relationships among species of the genus *Bifidobacterium*. *International Journal of Systematic Bacteriology* **21**, 276–294.
Schaal, K. P. and Pulverer, G. (1981). The genera *Actinomyces, Agromyces, Arachnia, Bacterionema* and *Rothia*. In: *The Prokaryotes: A Handbook of Habitats, Isolation and Identification of Bacteria, Volume* II (M. P. Starr, M. Stolp, H. G. Trüper, A. Balows and H. G. Schlegel, eds.), pp. 1923–1950. Springer Verlag, Berlin.
Schaal, K. P. and Reutersberg, H. (1978). Numerical taxonomy of *Nocardia asteroides*. *Zentralblatt für Bakteriologie, Parasitenkunde, Infektionskrankheiten und Hygiene. 1. Abteilung, Supplement* **6**, 53–62.
Schaal, K. P. and Schofield, G. M. (1981a). Current ideas on the taxonomic status of the Actinomycetaceae. *Zentralblatt für Bakteriologie, Mikrobiologie und Hygiene. 1. Abteilung, Supplement* **11**, 67–78.
Schaal, K. P. and Schofield, G. M. (1981b). Taxonomy of Actinomycetaceae. *Revue de Institut Pasteur de Lyon* **14**, 27–39.
Schaefer, W. B. (1979). Serological identification of atypical mycobacteria. In: *Methods in Microbiology, Volume* **13** (T. Bergan and J. R. Norris, eds.), pp. 323–343. Academic Press, London.
Schäfer, D. (1969). Eine neue *Streptosporangium* – art aus türkischer Steppenerde. *Archiv für Mikrobiologie* **66**, 365–373.
Schäfer, D. (1973). Beitrage zur Klassifizierung und Taxonomie der Actinoplanaceen. Dissertation: Phillipps Universitat Marburg.
Schleifer, K. H. (1970). Die Mureintypen in der Gattung *Microbacterium*. *Archiv für Mikrobiologie* **71**, 271–282.
Schleifer, K. H. and Kandler, O. (1967). Zur chemischen Zusammensetzung der Zellwand der Streptokokken. I. Die Aminosauresequenz des Mureins von *Str. thermophilus* und *Str. faecalis*. *Archiv für Mikrobiologie* **57**, 335–364.
Schleifer, K. H. and Kandler, O. (1970). Amino acid sequence of the murein of *Planococcus* and other *Micrococcaceae*. *Journal of Bacteriology* **103**, 387–392.
Schleifer, K. H. and Kandler, O. (1972). Peptidoglycan types of bacterial cell walls and their taxonomic implications. *Bacteriological Reviews* **36**, 407–477.
Schofield, G. and Schaal, K. P. (1982). A numerical taxonomic study of members of the Actinomycetaceae and related taxa. *Journal of General Microbiology* **127**, 237–259.
Schutze, H. (1908). Beiträge zur Kenntnis de thermophilen Aktinomyceten und ihrer Sporenbildung. *Archiv für Hygiene und Bakteriologie* **67**, 35–56.

Schuurmans, D. M., Olson, B. H. and San Clemente, C. L. (1956). Production and isolation of thermoviridin, an antibiotic produced by *Thermoactinomyces viridis* n. sp. *Applied Microbiology* **4**, 61–66.
Sebald, M. and Prévot, A. R. (1962). Étude d'une nouvelle espèce anaérobie stricte *Micromonospora acetoformici* n. sp. isolée de l'intestin posterieur de *Recticulitermes lucifugus var. saintonensis. Annales de l'Institut Pasteur* **102**, 199–214.
Seidl, P. H., Faller, A. H., Loider, R. and Schleifer, K. H. (1980). Peptidoglycan types and cytochrome patterns of strains of *Oerskovia turbata* and *O. xanthineolytica. Archives of Microbiology* **127**, 173–178.
Sharpe, M. E., Law, B. A. and Phillips, B. A. (1976). Coryneform bacteria producing methanethiol. *Journal of General Microbiology* **94**, 430–435.
Sharpe, M. E., Law, B. A., Phillips, B. A. and Pitcher, D. G. (1977). Methane thiol production by coryneform bacteria strains from dairy and human skin sources and *Brevibacterium linens. Journal of General Microbiology* **101**, 345–349.
Sharples, G. P. and Williams, S. T. (1976). Development and fine structure of sclerotia and spores of the actinomycete *Chainia olivacea. Microbios* **15**, 37–47.
Sharples, G. P., Williams, S. T. and Bradshaw, R. M. (1974). Spore formation in the Actinoplanaceae [*Actinomycetales*]. *Archiv für Mikrobiologie* **101**, 9–20.
Shaw, N. (1974). Lipid composition as a guide to the classification of bacteria. *Advances in Applied Microbiology* **17**, 63–108.
Shchepkina, T. V. (1940). Description of endoparasites of cotton fibres. *Bulletin of the Academy of Science USSR. Classe, Biological Sciences* **5**, 643–661 [in Russian].
Shepard, C. C. (1960). The experimental disease that follows the infection of human leprosy bacilli into foot pads of mice. *Journal of Experimental Medicine* **112**, 445–454.
Shibata, M., Hasegawa, T. and Higashide, E. (1971). Tolypomycin, a new antibiotic. I *Streptomyces tolypophorus* nov. sp. a new antibiotic tolypomycin producer. *Journal of Antibiotics* **24**, 810–816.
Shinobu, R. and Kawato, M. (1960). On *Streptomyces aerocolonigenes* nov. sp. forming secondary colonies on the aerial mycelia. *Botanical Magazine (Tokyo)* **73**, 212–216.
Shirling, E. B. and Gottlieb, D. (1966). Methods for characterization of *Streptomyces* species. *International Journal of Systematic Bacteriology* **16**, 313–340.
Shirling, E. B. and Gottlieb, D. (1968a). Co-operative description of type cultures of *Streptomyces*. II. Species descriptions from first study. *International Journal of Systematic Bacteriology* **18**, 69–189.
Shirling, E. B. and Gottlieb, D. (1968b). Co-operative description of type cultures of *Streptomyces*. III. Additional species descriptions from first and second studies. *International Journal of Systematic Bacteriology* **18**, 279–391.
Shirling, E. B. and Gottlieb, D. (1969). Co-operative description of type cultures of *Streptomyces*. IV. Species description from the second third and fourth studies. *International Journal of Systematic Bacteriology* **19**, 391–512.
Shirling, E. B. and Gottlieb, D. (1972). Co-operative description of type species of *Streptomyces*. V. Additional descriptions. *International Journal of Systematic Bacteriology* **22**, 265–394.
Shomura, T., Kojima, M., Yoshida, J., Ito, M., Amano, S., Totsugawa, K., Niwa, T., Inouye, S., Ito, T. and Niida, T. (1980). Studies on new aminoglycoside antibiotic, dactinicin. I. Producing organism and fermentation. *Journal of Antibiotics* **33**, 924–930.
Silvestri, L. G. (1970). The evolution of the thermophilic *Actinomycetales*: an apparent evolutionary paradox. In: *The Actinomycetales* (H. Prauser, ed.), pp. 239–243. Gustav Fischer Verlag, Jena.
Silvestri, L., Turri, M., Hill, L. R. and Gilardi, E. (1962). A quantitative approach to the systematics of actinomycetes based on overall similarity. In: *Microbial Classification* (G. C. Ainsworth and P. H. A. Sneath, eds.), pp. 333–360. Cambridge University Press, Cambridge.
Skerman, V. B. D., McGowan, V. and Sneath, P. H. A. (1980). Approved lists of bacterial names. *International Journal of Systematic Bacteriology* **30**, 225–420.
Skyring, G. W. and Quadling, C. (1969). Soil bacteria: principal component analysis of descriptions of named cultures. *Canadian Journal of Microbiology* **15**, 141–158.

Skyring, G. W. and Quadling, C. (1970). Soil bacteria: a principal component analysis and guanine-cytosine contents of some arthrobacter-coryneform soil isolates and of some named cultures. *Canadian Journal of Microbiology* **16**, 95–106.

Skyring, G. W., Quadling, C. and Rouatt, J. W. (1971). Soil bacteria: principal component analysis of physiological descriptions of some named cultures of *Agrobacterium*, *Arthrobacter* and *Rhizobium*. *Canadian Journal of Microbiology* **17**, 1299–1311.

Slack, J. M. (1974a). Genus *Actinomyces*. In: *Bergey's Manual of Determinative Bacteriology, Eighth Edition* (R. E. Buchanan and N. E. Gibbons, eds.), pp. 660–667. Williams and Wilkins, Baltimore.

Slack, J. M. (1974b). Family 1. *Actinomycetaceae*. In: *Bergey's Manual of Determinative Bacteriology, Eighth Edition* (R. E. Buchanan and N. E. Gibbons, eds.), pp. 658–659. Williams and Wilkins, Baltimore.

Slack, J. M. and Gerencser, M. A. (1975). *Actinomyces, Filamentous Bacteria. Biology and Pathogenicity*. Burgess Publishing Company, Minneapolis.

Slack, J. M., Landfried, S. and Gerencser, M. A. (1969). Morphological, biochemical and serological studies of 64 strains of *Actinomyces israelii*. *Journal of Bacteriology* **97**, 873–884.

Sneath, P. H. A. (1970). Application of numerical taxonomy to *Actinomycetales*: Problems and prospects. In: *The Actinomycetales* (H. Prauser, ed.), pp. 371–377. Gustav Fischer Verlag, Jena.

Sneath, P. H. A. (1976). An evaluation of numerical taxonomic techniques in the taxonomy of *Nocardia* and allied taxa. In: *The Biology of the Nocardiae*, (M. Goodfellow, G. H. Brownell and J. A. Serrano, eds.), pp. 74–101. Academic Press, London.

Sneath, P. H. A. (1978a). Classification of microorganisms. In: *Essays in Microbiology* (J. R. Norris and M. H. Richmond, eds.), ch. 9, pp. 1–31. John Wiley, London.

Sneath, P. H. A. (1978b). Identification of microorganisms. In: *Essays in Microbiology* (J. R. Norris and M. H. Richmond, eds.), ch. 10, pp. 1–32. John Wiley, London.

Sneath, P. H. A. and Johnson, R. (1972). The influence on numerical taxonomic similarities of errors in microbiological tests. *Journal of General Microbiology* **72**, 377–392.

Sneath, P. H. A. and Jones, D. (1976). *Brochothrix*, a new genus tentatively placed in the family *Lactobacillaceae*. *International Journal of Systematic Bacteriology* **26**, 102–104.

Sneath, P. H. A. and Sokal, R. R. (1973). *Numerical Taxonomy: The Principles and Practice of Numerical Classification*. W. H. Freeman, San Francisco.

Snyder, M. L., Slawson, M. S., Bullock, W. and Parker, R. B. (1967). Studies on oral filamentous bacteria. II. Serological relationships within the genera *Actinomyces*, *Nocardia*, *Bacterionema* and *Leptotrichia*. *Journal of Infectious Diseases* **117**, 341–345.

Socransky, S. S. (1970). Relationship of bacteria to the etiology of periodontal disease. *Journal of Dental Research* **49**, 203–222.

Socransky, S. S., Hubersak, C. and Propas, D. (1970). Induction of periodontal destruction in gnotobiotic rats by a human oral strain of *Actinomyces naeslundii*. *Archives of Oral Biology* **15**, 993–995.

Soina, V. S. and Agre, N. S. (1979). New elementary structures of spores in actinomycetes of the genera *Actinomadura* and *Streptomyces*. *Mikrobiologiya* **48**, 90–92. [In Russian].

Soina, V. S., Sokolov, A. A. and Agre, N. S. (1975). Ultrastructure of mycelium and spores of *Actinomadura fastidiosa* sp. nov. *Mikrobiologiya* **44**, 883–887. (In Russian).

Soina, V. S., Sokolov, A. A. and Agre, N. S. (1977). Election microscopic study of spore formation by *Actinomadura verrucosospora*. *Mikrobiologiya* **46**, 703–706. (In Russian).

Sompolinsky, D., Lagziel, A., Naveh, D. and Yonkilevitz, L. (1978). *Mycobacterium haemophilum* sp. nov., a new pathogen from humans. *International Journal of Systematic Bacteriology* **28**, 67–75.

Sottnek, F. O., Brown, J. M., Weaver, R. E. and Carroll, G. F. (1977). Recognition of *Oerskovia* species in the clinical laboratory: characterization of 35 isolates. *International Journal of Systematic Bacteriology* **27**, 263–270.

Stackebrandt, E. and Fiedler, F. (1979). DNA: DNA homology studies among strains of *Arthrobacter* and *Brevibacterium*. *Archives of Microbiology* **120**, 289–295.

Stackebrandt, E. and Kandler, O. (1974). Biochemischtaxonomische Untersuchungen an der Gattung *Cellulomonas*. *Zentralblatt für Bakteriologie, Parasitenkunde, Infektionskrankheiten und Hygiene. Abteilung I. Originale Reihe A* **228**, 128–135.

Stackebrandt, E. and Kandler, O. (1979). Taxonomy of the genus *Cellulomonas*, based on phenotypic characters and deoxyribonucleic acid – deoxyribonucleic acid homology, and proposal of seven neotype strains. *International Journal of Systematic Bacteriology* **29**, 273–282.

Stackebrandt, E. and Woese, C. R. (1979). A phylogenetic dissection of the family *Micrococcaceae*. *Current Microbiology* **2**, 317–322.

Stackebrandt, E. and Woese, C. R. (1981a). The evolution of prokaryotes. In: *Molecular and Cellular Aspects of Microbial Evolution* (M. J. Carlile, J. F. Collins and B. E. B. Moseley, eds.), pp. 1–31. Cambridge University Press, Cambridge.

Stackebrandt, E. and Woese, C. R. (1981b). Towards a phylogeny of the actinomycetes and related organisms. *Current Microbiology* **5**, 197–202.

Stackebrandt, E., Fiedler, F. and Kandler, O. (1978). Peptidoglycantyp und Zusammensetzung der Zellwandpolysaccharide von *Cellulomonas cartalyticum* und einigen coryneformen Organismen. *Archives of Microbiology* **117**, 115–118.

Stackebrandt, E., Lewis, B. J. and Woese, C. R. (1980a). The phylogenetic structure of the coryneform group of bacteria. *Zentralblatt für Bakteriologie, Parasitenkunde, Infektionskrankheiten und Hygiene. I. Abteilung, Originale Reihe* **C1**, 137–149.

Stackebrandt, E., Häringer, M. and Schleifer, K. H. (1980b). Molecular evidence for the transfer of *Oerskovia* species into the genus *Cellulomonas*. *Archives of Microbiology* **127**, 179–185.

Stackebrandt, E., Wunner-Füssl, B., Fowler, V. J. and Schleifer, K. H. (1981). Deoxyribonucleic acid homologies and ribosomal ribonucleic acid similarities among sporeforming members of the order *Actinomycetales*. *International Journal of Systematic Bacteriology* **31**, 420–431.

Stackebrandt, E., Ludwik, W., Seewaldt, E. and Schleifer, K. H. (1983). Phylogeny of sporeforming members of the order *Actinomycetales*. *International Journal of Systematic Bacteriology* **33**, 173–180.

Stanford, J. L. and Grange, J. M. (1974). The meaning and structure of species as applied to mycobacteria. *Tubercle, London* **55**, 143–152.

Stanford, J. L. and Gunthorpe, W. J. (1971). A study of some fast-growing scotochromogenic mycobacteria including species descriptions of *Mycobacterium gilvum* (new species) and *Mycobacterium duvalii* (new species). *British Journal of Experimental Pathology* **52**, 627–637.

Stanford, J. L., Gunthorpe, W. J., Pattyn, S. R. and Portaels, F. (1972). Studies on *Mycobacterium chelonei*. *Journal of Medical Microbiology* **5**, 171–182.

Stanford, J. L., Bird, R. G., Carswell, J. W., Draper, P., Lowe, C., McDougall, A. C., McIntyre, G., Pattyn, S. R. and Rees, R. J. W. (1977). A study of alleged leprosy bacillus strain H1-75. *International Journal of Leprosy* **45**, 101–106.

Stark, W. M., Higgens, C. E. and Wolfe, R. N. (1963). Capreomycin, a new antimicrobial agent produced by *Streptomyces capreolus* sp. n. *Antimicrobial Agents and Chemotherapy 1962*, 596–606.

Starr, M. P., Mandel, M. and Murata, N. (1975). The phytopathogenic coryneform bacteria in the light of DNA base composition and DNA: DNA segmental homology. *Journal of General and Applied Microbiology* **21**, 13–26.

Stodola, F. H., Lesuk, A. and Anderson, R. J. (1938). The chemistry of lipids of tubercle bacilli. LIV. The isolation and properties of mycolic acid. *Journal of Biological Chemistry* **126**, 505–513.

Suarez, J. E., Barbes, C. and Hardisson, C. (1980). Germination of spores of *Micromonospora chalcea*: physiological and biochemical changes. *Journal of General Microbiology* **121**, 159–167.

Sukapure, R. S., Lechevalier, M. P., Reber, H., Higgins, M. L., Lechevalier, H. A. and Prauser, H. (1970). Motile nocardoid *Actinomycetales*. *Applied Microbiology* **19**, 527–533.

Sundaram, T. K., Wright, I. P. and Wilkinson, A. E. (1980). Malate dehydrogenase from thermophilic and mesophilic bacteria. Molecular size, subunit structure, aminoacid composition, immunochemical homology and catalytic activity. *Biochemistry* **19**, 2017–2022.

Suzuki, K., Kaneko, T. and Komagata, K. (1981). Deoxyribonucleic acid homologies among coryneform bacteria. *International Journal of Systematic Bacteriology* **31**, 131–138.

Syed, S. A., Loesche, W. J., Paper, H. L., Jr. and Grenier, E. (1975). Predominant cultivable flora isolated from human root surface caries plaque. *Infection and Immunity* **11**, 727–731.

Szabo, I. M., Marton, M., Kulcsár, G. and Buti, I. (1976). Taxonomy of primycin producing actinomycetes. I. Description of the type strain of *Thermomonospora galeriensis*. *Acta Microbiologica Academiae Scientiarum Hungaricae* **23**, 371–376.

Szaniszlo, P. J. (1968). The nature of the intramycelial pigmentation of some *Actinoplanaceae*. *Journal of the Elisha Mitchell Scientific Society* **84**, 24–26.

Szaniszlo, P. J. and Gooder, H. (1967). Cell wall composition in relation to the taxonomy of some *Actinoplanaceae*. *Journal of Bacteriology* **94**, 2037–2047.

Takamiya, A. and Tubaki, K. (1956). A new form of streptomyces capable of growing autotrophically. *Archiv für Mikrobiologie* **25**, 58–64.

Takeya, K. and Tokiwa, H. (1972). Mycobacteriocin classification of rapidly growing mycobacteria. *International Journal of Systematic Bacteriology* **22**, 178–180.

Tanida, S., Haibara, K. and Asai, M. (1980a). Production of tomaymycin by an ansamitocin producing organism. *Journal Takeda Research Laboratories* **39**, 138–139.

Tanida, S., Hasegawa, T., Hatano, K., Higashide, E. and Yoneda, M. (1980b). Ansamitocins, maytansinoid antitumor antibiotics. Producing organism, fermentation and antimicrobial activities. *Journal of Antibiotics* **33**, 192–198.

Tanida, S., Hasegawa, T., Muroi, M. and Higashide, E. (1980c). Dnacins, new antibiotics. I. Producing organism, fermentation and antimicrobial activities. *Journal of Antibiotics* **33**, 1443–1448.

Tanner, R. S., Stackebrandt, E., Fox, G. E. and Woese, C. R. (1981). A phylogenetic analysis of *Acetobacterium woodii*, *Clostridium barkeri*, *Clostridium butyricum*, *Clostridium lituseburense*, *Eubacterium limosum*, and *Eubacterium tenue*. *Current Microbiology* **5**, 35–38.

Thayer, D. W., Yang, S. P., Key, A. B., Yang, H. H. and Barker, J. W. (1975). Production of cattle feed by the growth of bacteria on mesquite wood. *Developments in Industrial Microbiology* **16**, 465–474.

Thiemann, J. E. (1967). A new species of the genus *Amorphosporangium* isolated from Italian soil. *Mycopathologia Mycologia Applicata* **33**, 233–240.

Thiemann, J. E. (1970). Study of some new genera and species of the *Actinoplanaceae*. In: *The Actinomycetales* (H. Prauser, ed.), pp. 245–257. Gustav Fischer Verlag, Jena.

Thiemann, J. E. and Beretta, G. (1968). A new genus of the *Actinoplanaceae*: *Planobispora* gen. nov. *Archiv für Mikrobiologie* **62**, 157–166.

Thiemann, J. E., Pagani, H. and Beretta, G. (1967a). A new genus of the *Actinoplanaceae*: *Dactylosporangium* gen. nov. *Archiv für Mikrobiologie* **58**, 42–52.

Thiemann, J. E., Pagani, H. and Beretta, G. (1967b). A new genus of the *Actinoplanaceae*: *Planomonospora* gen. nov. *Giornale di Microbiologia* **15**, 27–38.

Thiemann, J. E., Pagani, H. and Beretta, G. (1968). A new genus of the *Actinomycetales*: *Microtetraspora* gen. nov. *Journal of General Microbiology* **50**, 295–303.

Thiemann, J. E., Coronelli, C., Pagani, H., Beretta, G., Tamoni, G. and Arioli, V. (1968). Antibiotic production by new form-genera of the *Actinomycetales*. I. Sporangiomycin, an antibacterial agent isolated from *Planomonospora parontospora* var *antibiotica* var. nov. *Journal of Antibiotics* **21**, 525–531.

Thiemann, J. E., Zucco, G. and Pelizza, G. (1969). A proposal for the transfer of *Streptomyces mediterranei* Margalith and Beretta 1960 to the genus *Nocardia* as *Nocardia mediterranea* [Margalith and Beretta] comb. nov. *Archiv für Mikrobiologie* **67**, 147–155.

Thirumalachar, M. J. (1955). *Chainia*, a new genus of the *Actinomycetales*. *Nature (London)* **176**, 934–935.

Thomas, S. B., Druce, R. G., Peters, G. J. and Griffiths, D. G. (1967). Incidence and significance of thermoduric bacteria in farm milk supplies: a reappraisal and review. *Journal of Applied Bacteriology* **30**, 265–298.

Thorel, M. F. and Valette, L. (1976). Etude de quelques souches de *M. paratuberculosis*: caracters biochimiques et activite allergique. *Annales Recherche Vétérinaires* **7**, 207–213.

Tilford, P. E. (1936). Fasciation of sweet peas caused by *Phytomonas fascians* n. sp. *Journal of Agricultural Research* **53**, 383–394.

Tille, D., Prauser, H., Szyba, K. and Mordarski, M. (1978). On the taxonomic position of *Nocardioides albus* Prauser by DNA: DNA-hybridization. *Zeitschrift für Allgemeine Mikrobiologie* **18**, 459–462.

Timpe, A. and Runyon, E. G. (1954). The relationship of 'atypical' acid-fast bacteria to human disease. *Journal of Laboratory and Clinical Medicine* **44**, 202–209.

Tjepkema, J. D., Ormerod, W. and Torrey, J. G. (1980). Vesicle formation and acetylene reduction activity in *Frankia* sp. CP I 1 cultured in defined nutrient media. *Nature (London)* **287**, 633–635.

Tomita, K., Uenoyama, Y., Numata, K-I., Sasahira, T., Hoshino, Y., Fujisawa, K-I., Tsukiura, H. and Kawaguchi, H. (1978). *Streptoalloteichus*, a new genus of the *Actinoplanaceae*. *Journal of Antibiotics* **31**, 497–510.

Tomita, K., Hoshino, Y., Sasahira, T., Hasegawa, K. and Akiyama, M. (1980). Taxonomy of the antibiotic BU 2313 producing organism, *Microtetraspora caesia* new species. *Journal of Antibiotics* **33**, 1491–1501.

Trejo, W. G. (1970). An evaluation of some concepts and criteria used in the speciation of streptomycetes. *Transactions of the New York Academy of Sciences* **32**, 989–997.

Trevisan, V. (1889). *I Generi e le Specie delle Bacteriacee*. Milano: Zanaboni and Gubuzzi.

Tribe, H. T. and Abu El-Souod, S. M. [1979]. Colonization of hair in soil-water cultures, with special reference to the genera *Pilimelia* and *Spirillospora* [*Actinomycetales*]. *Nova Hedwigia* **31**, 789–805.

Tsiklinsky, P. (1899). Sur les mucedinées thermophiles. *Annales de l' Institut Pasteur* **13**, 500–504.

Tsukamura, M. (1967). Identification of mycobacteria. *Tubercle, London* **48**, 311–338.

Tsukamura, M. (1969). Numerical taxonomy of the genus *Nocardia*. *Journal of General Microbiology* **56**, 265–287.

Tsukamura, M. (1971). Proposal of a new genus, *Gordona*, for slightly acid-fast organisms occurring in sputa of patients with pulmonary disease and in soil. *Journal of General Microbiology* **68**, 15–26.

Tsukamura, M. (1974). A further numerical taxonomic study of the *rhodochrous* group. *Japanese Journal of Microbiology* **18**, 37–44.

Tsukamura, M. (1975). Numerical analysis of the relationship between *Mycobacterium*, *rhodochrous* group and *Nocardia* by use of hypothetical median organisms. *International Journal of Systematic Bacteriology* **25**, 329–335.

Tsukamura, M. (1977). Extended numerical taxonomy study of *Nocardia*. *International Journal of Systematic Bacteriology* **27**, 311–323.

Tsukamura, M. (1981). Numerical analysis of rapidly growing, nonphotochromogenic mycobacteria, including *Mycobacterium agri* [Tsukamura 1972] Tsukamura sp. nov., nom. rev. *International Journal of Systematic Bacteriology* **31**, 247–258.

Tsukamura, M. (1982). *Mycobacterium shimoidei* sp. nov., nom., rev., a lung pathogen. *International Journal of Systematic Bacteriology* **32**, 67–69.

Tsukamura, M. and Mizuno, S. (1971). A new species *Gordona aurantiaca* occurring in sputa of patients with pulmonary disease. *Kekkaku* **46**, 93–98.

Tsukamura, M. and Mizuno, S. (1977). Numerical analysis of relationships among rapidly growing, scotochromogenic mycobacteria. *Journal of General Microbiology* **98**, 511–517.

Tsukamura, M. and Mizuno, S. (1978). A further study on the method of identification of mycobacteria by thin-layer chromatography after incubation with 35 S-methionine. *Kekkaku* **54**, 15–27.

Tsukamura, M., Mizuno, S., Tsukamura, S. and Tsukamura, J. (1979). Comprehensive numerical classification of 369 strains of *Mycobacterium*, *Rhodococcus* and *Nocardia*. *International Journal of Systematic Bacteriology* **29**, 110–129.

Tsukamura, M., Mizuno, S. and Tsukamura, S. (1981). Numerical analysis of rapidly growing, scotochromogenic mycobacteria, including *Mycobacterium obuense* sp. nov., nom., rev., *Mycobacterium rhodesiae* sp. nov., nom., rev., *Mycobacterium chubuense* sp. nov., nom., rev., and *Mycobacterium tokaiense* sp. nov., nom., rev. *International Journal of Systematic Bacteriology* **31**, 263–275.

Tsyganov, V. A., Namestnikova, V. P. and Krassykova, V. A. (1966). DNA composition in various genera of the *Actinomycetales*. *Mikrobiologiya* **35**, 92–95.

Tsyganov, V. A., Efimova, T. P., Zhukova, R. A. and Konev, Yu. E. (1970). New forms of actinomycetes and taxonomic significance of some of their metabolites. In: *The Actinomycetales* (H. Prauser, ed.), pp. 329–335. Gustav Fischer Verlag, Jena.

Uchida, K. and Aida, K. (1977). Acyl type of bacterial cell wall: its simple identification by colorimetric method. *Journal of General and Applied Microbiology* **23**, 249–260.

Uchida, K. and Aida, K. (1979). Taxonomic significance of cell wall acyl type in *Corynebacterium-Mycobacterium – Nocardia* group by a glycolate test. *Journal of General and Applied Microbiology* **25**, 169–183.

Uesaka, I., Oiwa, K., Yasuhira, K., Kobara, Y. and McClung, N. M. (1971). Studies on the pathogenicity of *Nocardia* isolates for mice. *Japanese Journal of Experimental Medicine* **41**, 443–457.

Van Brummelen, J. and Went, J. C. (1957). *Streptosporangium* isolated from forest litter in the Netherlands. *Antonie van Leeuwenhoek* **23**, 385–392.

Van Niel, C. B. (1946). The classification and natural relationships of bacteria. *Cold Spring Harbor Symposium on Quantitative Biology* **11**, 285–301.

Van Saceghem, R. (1915). Dermatose contagieuse [Impétigo contagieux]. *Bulletin de la Société de Pathologie Exotique* **8**, 354–359.

Vanden Berghe, D. A. and Pattyn, S. R. (1979). Comparison of proteins from *Mycobacterium fortuitum*, *Mycobacterium nonchromogenicum* and *Mycobacterium terrae* using flat bed electrophoresis. *Journal of General Microbiology* **111**, 283–291.

Vacca, J. G. and Bellamy W. D. (1976). Classification and morphological properties of a high temperature, celluloytic actinomyces. *Abstracts of the Annual Meeting of the American Society for Microbiology* p. 117.

Vacheron, M. J., Guinand, M., Michel, G. and Ghuysen, J. M. (1972). Structural investigations on cell walls of *Nocardia* sp. The wall lipid and peptidoglycan moieties of *Nocardia kirovani*. *European Journal of Biochemistry* **29**, 156–166.

Vályi-Nagy, T., Kulcsar, G., Szilagyi, I., Valu, G., Magyar, K., Kiss, G. H. and Horvath, I. (1970). Process for producing primycin. *U.S. Patent* **3**, 498,884.

Wagman, G. H., Toota, P. T., Patel, M., Marquez, J. A., Oden, F. M., Waitz, J. A. and Weinstein, M. J. (1975). A new polyene antifungal antibiotic produced by a species of *Actinoplanes*. *Antimicrobial Agents and Chemotherapy* **7**, 457–461.

Waksman, S. A. (1961). *The Actinomycetales, Volume II. Classification. Identification and Descriptions of Genera and Species*. Williams and Wilkins, Baltimore.

Waksman, S. A. (1967). *The Actinomycetes: A Summary of Current Knowledge*. The Ronald Press Company, New York.

Waksman, S. A. and Henrici, A. T. (1943). The nomenclature and classification of the actinomycetes. *Journal of Bacteriology* **46**, 337–341.

Wayne, L. G. (1967). Selection of characters for an Adansonian analysis of mycobacterial taxonomy. *Journal of Bacteriology* **93**, 1382–1391.

Wayne, L. G. (1978). Mycobacterial taxonomy: A search for discontinuities. *Annales de Microbiologie [Institut Pasteur]* **129A**, 13–27.

Wayne, L. G. and Diaz, G. A. (1976). Immunoprecipitation studies of mycobacterial catalase. *International Journal of Systematic Bacteriology* **26**, 38–44.

Wayne, L. G. and Diaz, G. A. (1979). Reciprocal immunologic distances of catalase derived from *Mycobacterium avium*, *M. tuberculosis* and closely related species. *International Journal of Systematic Bacteriology* **29**, 19–24.

Wayne, L. G., Dietz, T. M., Gernez-Rieux, C., Jenkins, P. A., Käppler, W., Kubica, G. P., Kwapinski, J. B. G., Meissner, G., Pattyn, S. R., Runyon, E. H., Schröder, K. H., Silcox, V. A., Tacquet, A., Tsukamura, M. and Wolinsky, E. (1971). A co-operative numerical analysis of scotochromogenic slowly growing mycobacteria. *Journal of General Microbiology* **66**, 255–271.

Wayne, L. G., Engback, H. C., Engel, H. W. B., Froman, S., Gross, W., Hawkins, J., Käppler, W., Karlson, A. G., Kleeberg, H. H., Krasnow, I., Kubica, G. P., McDurmont, C., Nel, E. E., Pattyn, S. R., Schröder, K. H., Showalter, S., Tarnok, I., Tsukamura, M., Vergmann, B. and Wolinsky, E. (1974). Highly reproducible techniques for use in systematic bacteriology in the genus *Mycobacterium*: Tests for pigment, urease, resistance to sodium chloride, hydrolysis of Tween 80 and β-galactosidase. *International Journal of Systematic Bacteriology* **24**, 412–419.

Wayne, L. G., Engel, H. W. B., Grassi, C., Gross, W., Hawkins, J., Jenkins, P. A., Käppler, W., Kleeberg, H. H., Krasnow, I., Nel, E. E., Pattyn, S. R., Richards, P. A., Showalter, S., Slosarek, M., Szabo, I., Tarnok, I., Tsukamura, M., Vergmann, B. and Wolinsky, E. (1976). Highly reproducible techniques for use in systematic bacteriology in the genus *Mycobacterium*: Tests for niacin and catalase and for resistance to isoniazid, thiophene 2-carboxylic hydrazide, hydroxylamine and *p*-nitrobenzoate. *International Journal of Systematic Bacteriology* **26**, 311–318.

Wayne, L. G., Andrade, L. B., Froman, S., Käppler, W., Kubala, E., Meissner, G. and Tsukamura, M. (1978). A co-operative numerical analysis of *Mycobacterium gastri*, *Mycobacterium kansasii* and *Mycobacterium marinum*. *Journal of General Microbiology* **109**, 319–327.

Wayne, L. G., Krichevsky, E. J., Love, L. L., Johnson, R. and Krichevsky, M. I. (1980). Taxonomic probability matrix for use with slowly growing mycobacteria. *International Journal of Systematic Bacteriology* **30**, 528–538.

Wayne, L. G., Good, R. C., Krichevsky, M. I., Beam, R. E., Blacklock, Z., Chaparas, S. D., Dawson, D., Froman, S., Gross, W., Hawkins, J., Jenkins, P. A., Juhlin, I., Käppler, W., Kleeberg, H. H., Krasnow, I., Lefford, M. J., Mankiewicz, E., McDurmont, C., Meissner, G., Morgan, P., Nel, E. E., Pattyn, S. R., Portaels, F., Richards, P. A., Rusch, S., Schröder, K. H., Silcox, V. A., Szabo, I., Tsukamura, M. and Vergmann, B. (1981). First report of the co-operative, open-ended study of slowly growing mycobacteria by the International Working Group on Mycobacterial Taxonomy. *International Journal of Systematic Bacteriology* **31**, 1–20.

Wellington, E. M. H. and Williams, S. T. (1981a). Transfer of *Actinoplanes armeniacus* Kalakoutskii and Kusnetsov to *Streptomyces armeniacus* [Kalakoutskii and Kusnetsov] comb. nov. *International Journal of Systmatic Bacteriology* **31**, 77–81.

Wellington, E. M. H. and Williams, S. T. (1981b). Host ranges of phages isolated to *Streptomyces* and other genera. *Zentralblatt für Bakteriologie, Mikrobiologie und Hygiene. I. Abteilung, Supplement* **11**, 93–98.

Weyland, H. (1969). Actinomycetes in the North Sea and Atlantic Ocean sediments. *Nature (London)* **223**, 858.

Weyland, H. (1981). Distribution of actinomycetes on the sea floor. *Zentralblatt für Bakteriologie, Mikrobiologie und Hygiene. I. Abteilung, Supplement* **11**, 185–193.

Williams, S. T. (1970). Further investigations of actinomycetes by scanning electron microscopy. *Journal of General Microbiology* **62**, 67–73.

Williams, S. T. and Sharples, G. P. (1976). *Streptosporangium corregatum* sp. nov., an actinomycete with some unusual morphological features. *International Journal of Systematic Bacteriology* **26**, 45–52.

Williams, S. T. and Wellington, E. M. H. (1980). Micromorphology and fine structure of actinomycetes. In: *Microbiological Classification and Identification* (M. Goodfellow and R. G. Board, eds.), pp. 139–165. Academic Press, London.

Williams, S. T. and Wellington, E. M. H. (1981). The genera *Actinomadura*, *Actinopolyspora*, *Excellospora*, *Microbispora*, *Micropolyspora*, *Microtetraspora*, *Nocardiopsis*, *Saccharopolyspora* and *Pseudonocardia*. In: *The Prokaryotes: A Handbook of Habitats, Isolation and Identification of Bacteria, Volume II* (M. P. Starr, H. Stolp, H. G. Trüper, A. Balows and G. H. Schlegel, eds.), pp. 2103–2117. Springer-Verlag, Berlin.

Williams, S. T., Sharples, G. P. and Bradshaw, R. M. (1973). The fine structure of the *Actinomycetales*. In: *Actinomycetales: Characteristics and Practical Importance* (G. Sykes and F. A. Skinner, eds.), pp. 113–130. Academic Press, London.

Williams, S. T., Sharples, G. P., Serrano, J. A., Serrano, A. A. and Lacey, J. (1976). The micromorphology and fine structure of nocardioform organisms. In: *The Biology of the Nocardiae* (M. Goodfellow, G. H. Brownell and J. A. Serrano, eds.), pp. 102–140. Academic Press, London.

Williams, S. T., Wellington, E. M. H. and Tipler, L. S. (1980). The taxonomic implications of the reactions of representative *Nocardia* strains to actinophage. *Journal of General Microbiology* **119**, 173–178.

Williams, S. T., Wellington, E. M. H., Goodfellow, M., Alderson, G., Sackin, M. and Sneath, P. H. A. (1981). The genus *Streptomyces* – a taxonomic enigma. *Zentralblatt für Bakteriologie, Mikrobiologie und Hygiene. 1. Abteilung, Supplement* **11**, 47–57.

Williams, S. T., Goodfellow, M., Alderson, G., Wellington, E. M. H., Sneath, P. H. A. and Sackin, M. J. (1983a). Numerical classification of *Streptomyces* and related genera. *Journal of General Microbiology* **129**, 1743–1813.

Williams, S. T., Goodfellow, M., Wellington, E. M. H., Vickers, J. C., Alderson, G., Sneath, P. H. A., Sackin, M. and Mortimer, A. M. (1983b). A probability matrix for identification of some streptomycetes. *Journal of General Microbiology* **129**, 1815–1830.

Willoughby, L. G. (1966). A conidial *Actinoplanes* isolate from Blelham Tarn. *Journal of General Microbiology* **44**, 69–72.

Willoughby, L. G. (1968). Aquatic *Actinomycetales* with particular reference to the *Actinoplanaceae*. *Veroff Institut Meeresforschung Bremerhaven* **3**, 19–26.

Willoughby, L. G. (1969a). A study on aquatic actinomycetes, their allochthonous leaf component. *Nova Hedwigia* **18**, 45–113.

Willoughby, L. G. (1969b). A study of the aquatic actinomycetes of Blelham Tarn. *Hydrobiologia* **34**, 465–483.

Willoughby, L. G., Baker, C. D. and Foster, S. E. (1968). Sporangium formation in the *Actinoplanaceae* induced by humic acids. *Experientia* **24**, 730–731.

Wingender, W., von Hugo, H., Frommer, W. and Schäfer, D. (1975). A protease inhibitor from *Planomonospora parontospora*. *Journal of Antibiotics* **28**, 611–612.

Woese, C. R. and Fox, G. E. (1977). Phylogenetic structure of the prokaryotic domain: the primary kingdoms. *Proceedings of the National Academy of Sciences, U.S.A.* **74**, 5088–5090.

Yamada, K. and Komagata, K. (1970a). Taxonomic studies on coryneform bacteria. II. Principal aminoacids in the cell wall and their taxonomic significance. *Journal of General and Applied Microbiology* **16**, 103–113.

Yamada, K. and Komagata, K. (1970b). Taxonomic studies on coryneform bacteria. III. DNA base composition of coryneform bacteria. *Journal of General and Applied Microbiology* **16**, 215–224.

Yamada, K. and Komagata, K. (1972a). Taxonomic studies on coryneform bacteria. IV. Morphological, cultural, biochemical and physiological characteristics. *Journal of General and Applied Microbiology* **18**, 399–416.

Yamada, K. and Komagata, K. (1972b). Taxonomic studies on coryneform bacteria. V. Classification of coryneform bacteria. *Journal of General and Applied Microbiology* **18**, 417–431.

Yamada, Y., Inouye, G., Tahara, Y. and Kondo, K. (1976). The menaquinone system in the classification of coryneform and nocardioform bacteria and related organisms. *Journal of General and Applied Microbiology* **22**, 203–214.

Yamada, T., Ishikawa, T., Tahara, Y. and Kondo, K. (1977a). The menaquinone system in the classification of the genus *Nocardia*. *Journal of General and Applied Microbiology* **23**, 207–216.

Yamada, T., Yamashita, M., Tahara, Y. and Kondo, K. (1977b). The menaquinone system in the classification of the genus *Actinomadura*. *Journal of General and Applied Microbiology* **23**, 331–335.

Yamaguchi, T. (1965). Comparison of the cell wall composition of morphologically distinct actinomycetes. *Journal of Bacteriology* **89**, 444–453.

Yamaguchi, T. (1967). Similarity in DNA of various morphologically distinct actinomycetes. *Journal of General and Applied Microbiology* **13**, 63–71.

Yano, I., Furukawa, Y. and Kusunose, M. (1970). α-Hydroxy fatty acid containing phospholipids of *Nocardia leishmanii*. *Biochimica et Biophysica Acta* **202**, 189–191.

Young, C. L. and Chapman, G. C. (1978). Ultrastructural aspects of the causative agent and renal histopathology of bacterial kidney disease in brook trout [*Salvelinus fontinalis*]. *Journal of the Fishery Research Board of Canada* **35**, 1234–1248.

Zhukova, R. A., Tsyganov, V. A. and Morozov, V. M. (1968). A new species of *Micropolyspora-Micropolyspora angiospora* sp. nov. *Mikrobiologiya* **37**, 724–728 [In Russian].

Zopf, W. (1891). Über Ausscheidung von Fettfarbstoffen [Lipochromen] seitens gewisser Spaltpilze. *Berichte der Deutschen botanisehen Gessellschaft* **9**, 22–28.

3
Morphology

R. LOCCI* and G. P. SHARPLES†

*Plant Pathology Department, University of Milan, Milan, Italy and
†Department of Biology, Liverpool Polytechnic, Liverpool, UK

1. Introduction 165
2. Mycelial organization of actinomycetes 166
   A. Substrate mycelium 166
   B. Aerial mycelium 169
   C. Special structures 170
3. Actinomycete cell structure 172
   A. Cytoplasmic inclusions 172
   B. Membrane structures 173
   C. Cell walls 176
   D. Cross-walls 178
   E. Staling phenomena 180
4. Actinomycete spores 180
   A. Patterns of spore development 180
   B. Spore delimitation 182
   C. Spore maturation 183
   D. Fibrous sheath 186
   E. Spore germination 188
5. Growth of actinomycete cells 188
   A. Polar growth of rods 188
   B. Apical extension of hyphae 189
   Acknowledgement 191
   References 191

1. Introduction

A morphological assessment of bacteria would appear at least a superfluity, if not a paradox, to students familiar with the apparent structural uniformity of, for example, *Staphylococcus aureus* or *Bacillus subtilis*. Yet morphological diversity is not exceptional in prokaryotes (Starr and Skerman, 1965; Dow *et al.*, 1976; Parish, 1979). Among the morphologically exotic forms, actinomycetes occupy a special position, exhibiting a structure which ranges from relatively 'simple' rods and cocci to a complex mycelial organization similar to that of some eukaryotes. It is beyond the scope of this chapter to catalogue all the information currently available on actinomycete morphology. Our intention, therefore, is to describe, as far as possible, the general features possessed by these organisms and to consider some aspects of their growth and ontogenesis.

2. Mycelial Organization of Actinomycetes

A. SUBSTRATE MYCELIUM

When an actinomycete propagule, such as a spore, is allowed to grow on a solid substrate, it usually gives rise to hyphae, which branch at intervals and spread radially. The resulting mycelium, consisting of hyphae that either penetrate the substrate or grow along its surface (Fig. 1a), has been variously described as 'primary', 'substrate' or 'vegetative' and represents one of the main characteristics of actinomycetes. The mycelial growth habit has been related to their ability to break down insoluble organic materials by extracellular enzymes (Chater and Merrick, 1979).

The branching of the hyphae is, most commonly, monopodial but dichotomous ramification may also occur, for example, in *Thermoactinomyces dichotomica* (Kalakoutskii and Agre, 1976). The emergence angle of side branches appears to vary to a certain extent in different actinomycetes and within an individual colony with the progressive increase of its radius (Kalakoutskii and Agre, 1976). A similar phenomenon is characteristic of some *Actinomyces* strains but, in most cases, it is caused by dislocation of rods, following early fragmentation (Locci, 1978a, 1981). Investigations on branching organization may reveal interesting patterns in actinomycetes; so far no systematic studies have been carried out.

In some microaerophilic forms, rhodococci and others, no true mycelium is produced (Fig. 1b). The original propagule, in these organisms, elongates by synthesizing new wall material at either or both poles and generates more or less extended filaments (Locci and Schaal, 1980). Filament formation, when arising in this way, may be considered as the first stage in actinomycete morphological differentiation. There is not a clear distinction, however, between bacteria proper and actinomycetes in this respect and various transitional situations can be recognized. The filamentous condition may be more or less prolonged and is followed by fragmentation into smaller units.

Further morphological development is represented by the appearance of elementary branching, a phenomenon that has been reported in amycelial bacteria (Germida and Casida, 1980; Locci and Schaal, 1980). Extreme care should be taken, however, in distinguishing between 'true' and 'false' branching. 'T', 'V', 'X' and 'Y'-shaped formations often derive from 'snapping' mechanisms and temporary adherence of rods after fission. Light microscopy is frequently unsuitable, because of its resolution limits, for differentiating spatially overlapping cells. Sideways displacement of rod-shaped cells, which remain in contact with adjacent elements, may result in a pseudomycelial organization. The true nature of this, although sometimes clarified by time-lapse microphotography, is most effectively studied with the scanning electron microscope (Locci, 1981).

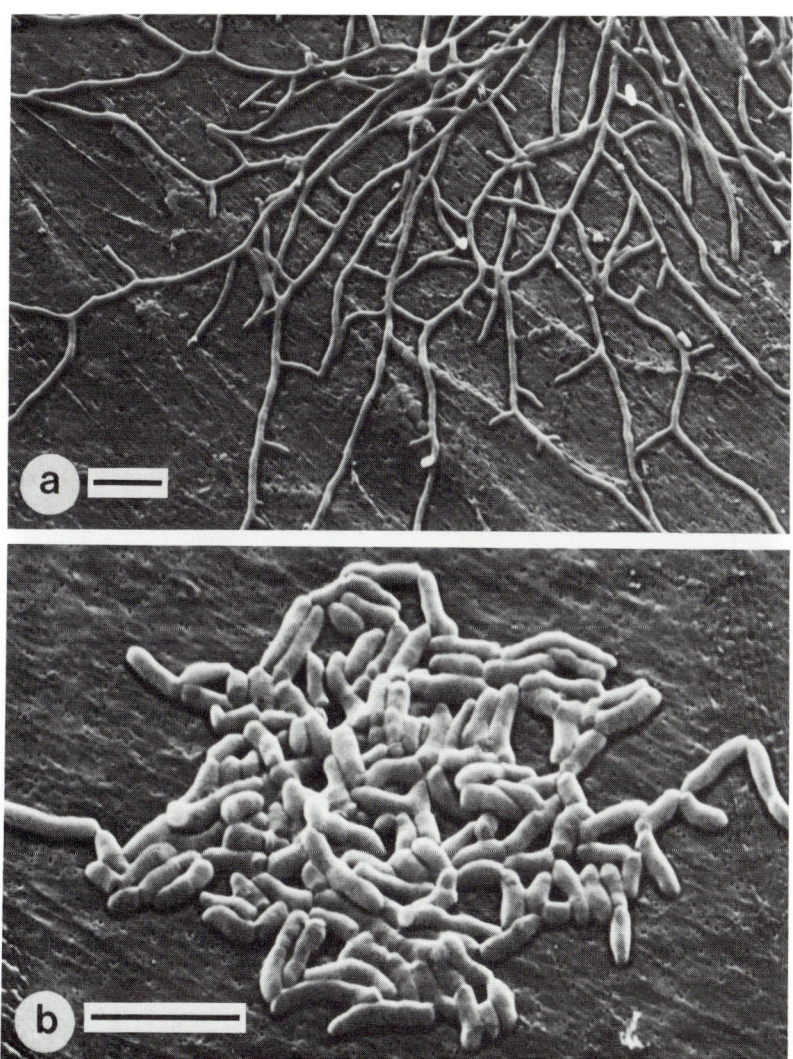

FIG. 1. (a) Branching hyphae of *Nocardia asteroides* and (b) rod-shaped cells of *Rhodococcus globerulus*. Bar marker represents 5.0 μm.

In morphologically more complex forms, such as species of the genus *Nocardia*, a true mycelium is formed, at least in the early stages of growth. This subsequently undergoes fragmentation into rods and cocci (Locci, 1976). Such fragmentation, which results from septation and disarticulation of hyphae, is a characteristic feature of so-called 'nocardioform' actinomycetes (Fig. 2a; Prauser, 1976, 1978). The timing of the fragmentation process is influenced by many environmental factors (Williams *et al.*, 1976) but these may well act through their effects on growth rates, which are known to be important in determining cell form (Williams and Wellington, 1980). The reduction of a branched organization into numerous autonomous elements appears to be related to the end of the growth cycle and, following exhaustion of nutrients, to the necessity of ensuring dispersion. As pointed out by Erikson (1953), such organisms, in which the mycelium readily fragments into rods and cocci, achieve a facility of dispersion similar to sporogenous actinomycetes. Thus, hyphal fragmentation occurs typically in taxa either devoid of spores or producing them only in limited amounts. The process is sometimes more complex, however, for the fragmented elements can immediately become new centres of growth, contributing to further development of the colony, i.e. the colony becomes typically polycentric (Locci, 1976). Arborescent-like arrangements of colonies, zig-zag and palisading effects represent the final products of this process. Fragmentation of hyphae, therefore, not only gives rise to more colony-forming units but also may result in an overall increase of cytoplasmic mass. In addition, fragmentation may give rise to motile elements. The genus *Oerskovia* forms abundant branched hyphae which eventually fragment into motile rod-shaped cells (Prauser *et al.*, 1970). Motile cocci in *Dermatophilus* and both motile rods and non-motile cocci in *Geodermatophilus* are formed from hyphae which are divided by longitudinal and transverse septa (Ishiguro and Wolfe, 1970).

The majority of spore-forming actinomycetes ('sporoactinomycetes'), including the genus *Streptomyces*, produce a non-fragmenting, branched substrate mycelium. Such a mycelium, which represents the most advanced condition for actinomycetes (Prauser, 1978), is typically fungal in organization, i.e. it is monocentric (Locci, 1976). It develops from the growth extension of a single propagule and "all portions of the growth remain in filamentous continuity with the original element" (Erikson, 1953). Transverse septa may be present, mainly in the older portions of the mycelium, but they do not normally allow disarticulation of hyphal elements (Williams *et al.*, 1973). In some cases, fragmentation of hyphae may occur. For instance, 6% of the *Streptomyces* species examined by Nikitina and Kalakoutskii (1971) had fragmented substrate hyphae when D-fructose was included in the culture medium, fragmentation being most marked in *Stmy. roseoflavus* var. *roseofungini* (Nikitina *et al.*, 1971). The formation of spores on the substrate mycelium occurs regularly in such genera as *Microellobosporia*,

Micropolyspora, Micromonospora and *Thermoactinomyces* (Cross and Goodfellow, 1973) and occasionally in the genus *Streptomyces* (Cross and Al-Diwany, 1981). Spore formation, however, is normally a feature of mycelial growth in the aerial environment (Kalakoutskii and Agre, 1976).

Hyphal growth and branching determine the appearance of the mycelium. The fact that some actinomycetes do not form a mycelium provides interesting opportunities for ontogenetic studies, particularly on the evolution from the 'undifferentiated' rod or coccus to the well-developed mycelial characteristics of the sporoactinomycetes. Three evolutionary levels, coryneforms, nocardioforms and sporoactinomycetes have been recognized by Prauser (1976, 1978). These levels, which are clearly illustrated by taxonomically distinct actinomycetes, may be observed in organisms of the same genus. Different species of the genus *Rhodococcus* can be arranged in a sequence based on the morphological complexity of their substrate mycelia. Species, such as *Rodc. bronchialis, Rodc. equi, Rodc. rubropertinctus, Rodc. sputi* and *Rodc. terrae*, show a growth cycle resembling that of the genus *Arthrobacter* (Clark, 1979), i.e. fluctuating from coccus to rod forms. Elementary branching characterizes *Rodc. erythropolis, Rodc. globerulus, Rodc. rhodnii* and *Rodc. rhodochrous*, while *Rodc. ruber* and *Rodc. coprophilus* form a true mycelium. Correspondingly, the completion of the developmental cycle requires progressively longer times, ranging from 24 hours (*Rodc. equi*) to several days (*Rodc. coprophilus*) (Locci *et al.*, 1982).

B. AERIAL MYCELIUM

After a certain amount of growth of the substrate mycelium, vertically developing filaments may be formed. In some cases, such aerial hyphae arise as side ramifications of single, isolated and scarcely branched substrate mycelia (Locci, 1976). Claims to the existence of special structures ('initial cells') controlling aerial filament formation (Klieneberger-Nobel, 1947) appear doubtful. With further development, a network of aerial hyphae may cover the colony surface, giving it a typical 'hairy' or 'powdery' appearance. The production of an aerial mycelium is influenced by a number of factors including the composition of the growth medium, the incubation temperature and the presence of specific stimulating compounds (Kalakoutskii and Agre, 1976; Williams *et al.*, 1976; Chater and Merrick, 1979; Pogell, 1979).

Reasons for making a distinction between substrate and aerial growth are more profound than their mere location. The two mycelia are different ontogenetically, morphologically, structurally and physiologically (Kalakoutskii and Agre, 1976; Ensign, 1978). In general, aerial growth appears to be less branched than the substrate mycelium and, in contrast to the latter, it is hydrophobic (Higgins and Silvey, 1966). Francisco and

Silvey (1971) claimed that the substrate mycelium of a *Streptomyces* species was facultatively aerobic, while aerial growth was obligately so. Aerial growth, in prokaryotes, is unusual and may represent an adaptation to the terrestrial environment. Thus, typical aquatic actinomycetes are usually devoid of such growth.

Growth in the aerial environment is not restricted to actinomycetes forming true substrate mycelia. In the genus *Sporichthya*, growth consists only of aerial filaments adhering by holdfasts to the surface of the growth medium (Williams, 1970). Single, mostly unbranched, vertical hyphae, frequently coalescing into 'synnemata', can often be observed in amycelial *Mycobacterium* species (Fig. 2b) and sometimes even in *Actinomyces bovis, A. naeslundii* and *A. viscosus* (Locci and Schofield, 1981).

In some nocardiae, aerial growth may be quite abundant, although substrate hyphal organization is rather elementary. If sporulation takes place, the entire filament is involved in the process, with no differentiation into sporophore and spore chain (Locci, 1976). On the other hand, aerial hyphae of streptomycetes form a mycelial 'mat' on which sporulating structures are differentiated. The complexity of this mat, however, can vary greatly. Sporulation may take place as soon as hyphae arise from the medium surface or only after the formation of a thick mycelial felt (Baldacci *et al.*, 1971a, b; Locci, 1976). In streptoverticillia, grown under favourable conditions, emerging hyphae build up a loosely knit web, consisting of long straight filaments which run almost parallel to the medium surface. Sporogenous hyphae occur as long threads, which form, at intervals, short verticillate branches. Each branch, in turn, bears terminal umbels of spore chains (Locci *et al.*, 1969; Locci and Baldan, 1970). A similar tendency to develop long aerial hyphae may be observed in some thermophilic actinomycetes (Locci, 1971, 1976). Hyphal arrangement in *Streptoverticillium* species confers a loose consistency to the aerial mat, which is 'cottony' in appearance, contrasting with the 'powdery' texture of most streptomycetes.

Aerial mycelium development usually ceases with the onset of sporulation. Lytic processes also take place in ageing colonies and may play a role in spore liberation of some actinomycetes (Locci, 1971, 1976). Some authors have stressed the parasitic nature of aerial growth on the substrate mycelium. According to Wildermuth (1970b), the latter undergoes extensive breakdown during aerial mycelium development. It has been suggested that the production of antibiotics by actinomycetes may prevent the invasion of lysing substrate mycelia by motile bacteria, thus promoting aerial growth in the absence of strong competition (Chater and Merrick, 1979).

C. SPECIAL STRUCTURES

Actinomycetes are known to form structures differing morphologically from the mycelial organizations so far discussed. They often resemble analogous fungal conformations and, although their functions have not been comparatively investigated, they are assumed to play a role similar to that of their eukaryotic counterparts.

Aerial hyphal aggregates are quite common amongst actinomycetes and can range from bundles in which individual filaments are still distinguishable to closely packed structures whose components are well cemented together

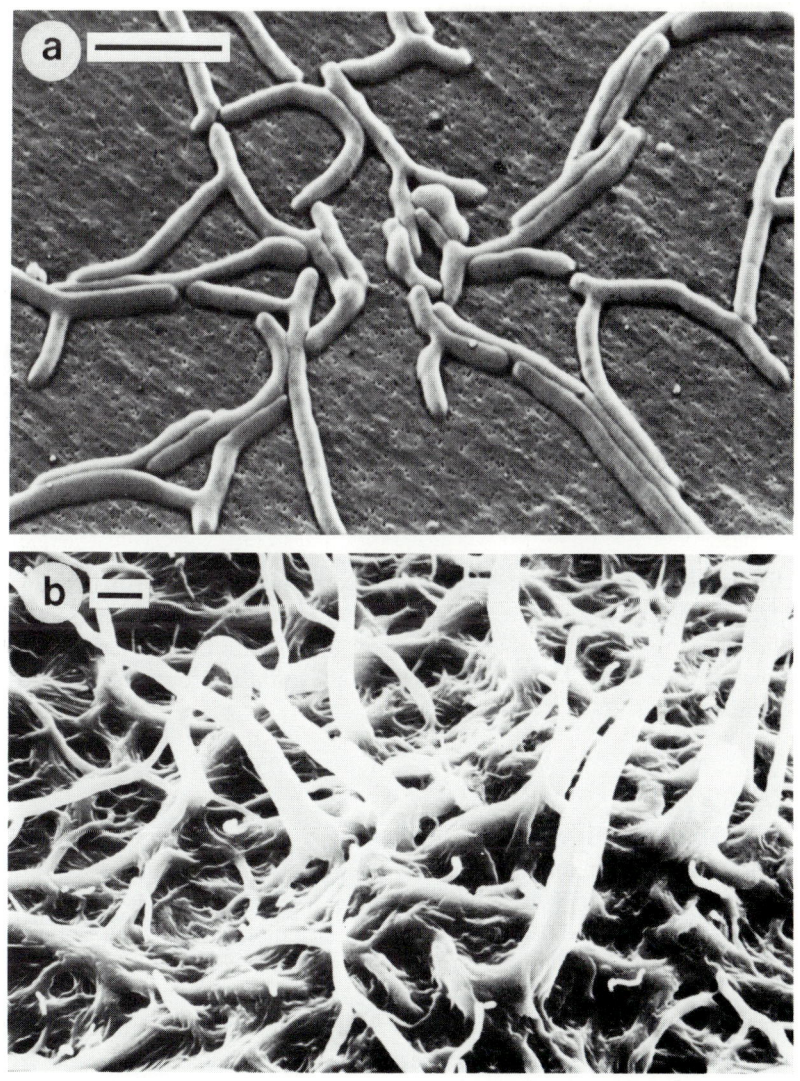

FIG. 2. (a) Fragmenting hyphae of *Rhodococcus rhodnii* and (b) synnemata of *Mycobacterium senegalense*. Bar marker represents 5.0 μm.

(Locci and Baldan, 1970; Locci, 1971, 1976, 1978a; Williams et al., 1973, 1976). They can form 'bridges' and fuse with neighbouring structures (Grein and Spalla, 1962). Commonly sub-conical in form, they have been variously described as 'aerial spikes', 'isarioid sporodochia', 'coremia' and 'synnemata'. Both 'coremium' and 'synnema' terminologies are mycological in origin and, for the reasons put forward by Langeron (1965), the latter should be preferred, since it is "expressive, not prejudicial from the systematic point of view and can designate both vegetative and sporiferous structures". The choice is reinforced by the recent discovery of actinomycetes possessing sporulating synnemata (Hasegawa et al., 1978).

Species of the genus *Chainia* regularly form sclerotia (Thirumalachar, 1955). These structures, like those formed by filamentous fungi, possess a pseudoparenchymatous tissue organization, consisting of a highly compacted mass of short cells, that are rounded or angular in shape (Lechevalier et al., 1973; Ganju and Iyengar, 1974; Sharples and Williams, 1976a). Hyphal aggregates of this type, although lacking an outer protective rind, presumably have some survival function. They contain substantial quantities of storage material, principally in the form of lipids, and exhibit the phenomenon of 'intrahyphal growth'. *Chainia* sclerotia are clearly distinguishable from the interwoven mass of undifferentiated hyphae that make up the so called 'granules' observed in other actinomycetes (Baldacci et al., 1966).

3. Actinomycete Cell Structure

Actinomycete cells have a typical prokaryotic organization, consisting of a fibrillar nuclear region and a granular cytoplasm with ribosomes (12 nm diameter) (Fig. 3a). The cytoplasm may also contain a variety of other inclusions depending on the organism, its age and the growth medium used (Williams et al., 1973).

A. CYTOPLASMIC INCLUSIONS

The most characteristic and frequent cytoplasmic inclusions that have been detected by both light and electron microscopy, particularly in nocardioform actinomycetes, are polyphosphate granules and lipid globules (Williams et al., 1973, 1976; Barksdale and Kim, 1977; Beaman et al., 1978). Polyphosphate granules stain with basic aniline dyes and, in thin sectioned material observed by transmission electron microscopy, appear as extremely electron-dense spherical bodies. Lipid globules, on the other hand, stain with Sudan Black B and, in thin sections, appear as clearly defined spherical electron-light areas (Fig. 3b). Their occurrence increases with the age of the culture (Dipersio and Deal, 1974). The precise composition of such vacuole-like areas in the majority of actinomycetes has not been established with certainty. In some nocardioform organisms, at least, they appear to represent sites of accumulation of neutral lipids, i.e. fatty acids, glycerides and waxes (Dipersio and Deal, 1974; Barksdale and Kim, 1977). In most bacteria, however, they are generally considered to contain polymers of β-hydroxybutyrate. A significant accumulation of poly-β-hydroxybutyrate (PHB) has been demonstrated in several species of the genus *Streptomyces* (Kannan and Rehacek, 1970). PHB 'granules', unlike lipid globules, are normally enclosed by a 'non-unit membrane', 2.0–4.0 nm thick, and have a stranded or particulate surface appearance in freeze-etched preparations (Dawes and Senior, 1973; Dunlop and Robards, 1973; Shively, 1974; Williams *et al.*, 1976).

Polysaccharide storage granules may also be present in the cytoplasm of some actinomycete cells. They are normally small electron-light round bodies, similar to lipid globules, but with a rather diffuse outline (Fig. 3c). Their empty appearance reflects the low electron-scattering power common to all polysaccharides treated with osmium tetroxide and glutaraldehyde fixatives. They may be electron-dense, however, in thin sections stained with lead citrate (Shively, 1974). Polysaccharide granules are possibly of widespread occurrence in actinomycetes but, because of their often somewhat indistinct appearance in thin sections, they are easily overlooked, unless specific cytochemical techniques are used. Such techniques have clearly demonstrated the presence of polyglucoside (glycogen) granules in *Nrda. asteroides* cells grown in nitrogen-limiting, glucose-supplemented medium (Dipersio and Deal, 1974). The accumulation of polysaccharide material has also been shown to occur during the early stages of sporogenesis and in 'old' disintegrating hyphae of *Stmy. viridochromogenes* but never in 'young' substrate and aerial hyphae (Manzanal *et al.*, 1981).

B. MEMBRANE STRUCTURES

The cytoplasm is surrounded by a typical 'unit membrane', 7.5–10.0 nm thick, composed of two electron-dense zones, each 2.5–3.0 nm thick, separated by a less dense zone. In some cases, the outer dense zone may be thicker and the cytoplasmic membrane thus has an asymmetrical appearance (Fig. 3d). An asymmetrical membrane profile is characteristic for cells of *Stmy. hygroscopicus* showing high metabolic activity (Gumpert, 1978). This asymmetry is considered to represent the 'normal' structure for cytoplasmic membranes of at least some Gram-positive bacteria (Silva *et al.*, 1976a, 1979). It has been shown, in eubacteria, that membrane structure may be influenced by the culture conditions (Santos-Mota *et al.*, 1972) and by the action of a number of membrane-damaging treatments (Silva *et al.*, 1976a, 1979). It may also be affected by the fixation procedures used (Silva, 1971;

Santos-Mota et al., 1972; Suganuma and Morioka, 1979). In chemically unfixed, freeze-etched preparations of actinomycete cells, the cytoplasmic membrane has a particulate surface structure (Wildermuth, 1971; Williams et al., 1972).

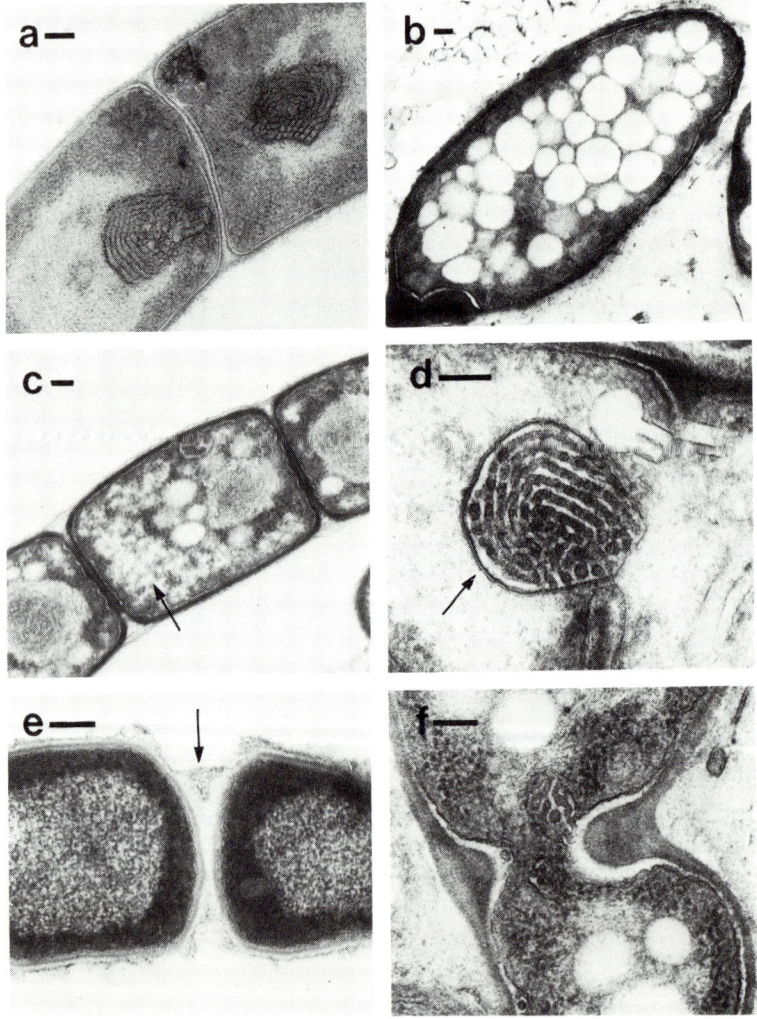

FIG. 3. (a) Prokaryotic cell structure of *Streptomyces aureofaciens*, (b) lipid globules in '*Streptomyces ostreogriseus*', (c) polysaccharide granules in '*Stmy. ostreogriseus*', (d) asymmetrical cytoplasmic membrane in *Microellobosporia flavea*, (e) accumulation of surface material at cell poles of *Nocardia otitidis-caviarum*, and (f) enlarged developing cross wall in '*Stmy. ostreogriseus*'. Bar marker represents 0.1 μm.

Mesosomes or mesosome-like structures, differing in shape, size, location and complexity, like those found in the eubacteria, have been observed in the cells of various representatives of the actinomycetes (Williams et al., 1973). These structures typically arise as 'sac-like' invaginations of the cytoplasmic membrane and contain membrane components, which become extruded upon plasmolysis or removal of the cell wall (Reusch and Burger, 1973; Ghosh, 1974; Greenawalt and Whiteside, 1975).

On the basis of their appearance in thin sections, mesosomes can be categorized as 'lamellar', 'vesicular' or 'tubular' but various combinations of these types may also occur. The significance of this polymorphism is not clear. It is evident that the form of the mesosome may be influenced by the fixation procedures used (Highton, 1969, 1970a, b; Burdett and Rogers, 1970). Such fixation effects seem to be variable, however, since specimens prepared by the same methods may possess structures widely different in complexity and form, as noted by Greenawalt and Whiteside (1975) and, more recently, by Beaman et al. (1978) for the mesosomes in the genus *Nocardia*. It is possible that the diversity observed could be caused by cell-to-cell variations, which will exist within one preparation due, for example, to localized differences in ionic environment. Rogers (1970) has proposed that the change of form, from lamellae to tubules to vesicles, represents the natural developmental sequence of bacterial mesosomes. It is also possible that the various forms of mesosome represent distinct organelles, with different origins and functions. A number of functions have been suggested but, despite extensive searching, a unique one for these structures has yet to be revealed. "The possibility", therefore, "that mesosomes are artefactual structures must now be considered seriously" (Salton and Owen, 1976).

Silva et al. (1976b) favoured the interpretation that mesosomes may well be artefacts produced by membrane-damaging fixation procedures and concluded that "a continuous cytoplasmic membrane, without infoldings (mesosomes), would be the real image of membrane organization in Gram-positive eubacteria".

Doubts about the existence of mesosomes in bacterial cells have been expressed by other workers. Nanninga (1971), using the technique of freeze-fracturing, observed only a few mesosomes in young unfixed cells of *Bacillus subtilis*, whereas numerous mesosomes were seen in osmium tetroxide fixed cells. It was concluded that mesosomes are either present in cells but require chemical fixation to stabilize their structure or are absent from young cells and chemical fixation influences membrane metabolism so that mesosomes are formed. It is clear that chemical fixation alters bacterial cell structure in such a way that mesosomes become more frequently detectable in freeze-etched/freeze-fractured replicas. This may be partly due to the effect of chemical fixatives on the location of the fracture plane, which, in turn, will have a marked effect on the ability to observe mesosomes (Higgins and Daneo-Moore, 1974; Ghosh and Nanninga, 1976).

The effect of fixation on the organization of mesosomes in freeze-fractures of *Streptococcus faecalis* was carefully analysed by Higgins *et al.* (1976). The kinetics of amino acid cross-linking by glutaraldehyde was shown to be well correlated with the frequency of mesosome observation but poorly correlated with the average area occupied by the mesosomes. This indicated that, although the frequency of freeze-fractures showing mesosomes may be related to the cross-linkage reaction, the area occupied by mesosomes is probably largely a function of structural rearrangements that occur after the glutaraldehyde fixative has been added. It was hypothesized that, upon fixation or the receipt of some physical insult (e.g. filtration, centrifugation, addition of large amounts of glycerol), mesosome precursors, found in undisturbed cells, undergo a change in state, which results in their visibility in freeze-fractures.

A variety of data, including those obtained by freeze-etching, strongly suggested to Beaman *et al.* (1978) that "mesosomes in *Nocardia* do exist in the natural state of the bacterial growth and cell differentiation".

Despite these studies, the fundamental question of whether the mesosome in eubacteria and in actinomycetes is altogether an artefactually generated structure still remains unresolved. The possibility exists, however, that two different populations of mesosome are present in bacterial cells, one affected and the other relatively unaffected by chemical fixation. It has been proposed that it is the large complex mesosomes that are the products of inefficient fixation, while the real mesosomes are small intrusions of the cytoplasmic membrane (Fooke-Achterrath *et al.*, 1974; Ghosh and Nanninga, 1976). It may be significant that lamellar membrane systems of a complex nature develop, in some actinomycetes, in association with unusual morphological growth forms (Beaman and Shankel, 1969; Cherny *et al.*, 1972a, b).

C. CELL WALLS

The cell walls of actinomycetes, like those of most Gram-positive bacteria, are relatively structureless in thin section. They may exhibit a variety of profiles (see Gumpert, 1978) but, most commonly, appear either homogeneously electron-dense or have a tribanded profile, composed of two electron-dense bands (zones) separated by a single less dense band. The significance of these differences in the appearance of the cell walls is not fully known. It is clear that the fixation and post-staining methods greatly influence the image of the cell wall (Nermut, 1967; Beveridge, 1978) but unfortunately the reasons for stain depositions in biological material are not sufficiently understood to allow deductions to be made.

The cell walls consist of peptidoglycan together with one or more associated polymers, such as anionic teichoic acids and neutral polysaccharides. Teichoic acids, which are present in most Gram-positive bacteria, account for up to 20% of the dry weight of streptomycete cell walls (Naumova et al., 1978). The tribanded appearance of cell walls has led to the view that the outer and inner dense bands might be enriched with anionic polymers, like the teichoic acids (Weibull, 1973). More recently, evidence has been presented which suggests that both the peptidoglycan and the non-peptidoglycan polymers are concentrated at the outer and inner surfaces of the cell walls (Tsien et al., 1978). It is interesting to note that the walls of nocardioform bacteria, which appear to lack teichoic acids (Barksdale and Kim, 1977), usually exhibit a homogeneous profile, without any banding, irrespective of the methods of fixation and staining (Aristarkhova, 1976a, b; Beadles et al., 1980). Such walls are relatively thin, however, and it is possible that the middle wall band is not easily resolvable.

Actinomycete cell walls are normally about 10–20 nm thick. Their thickness varies considerably, however, and often increases with the age of the culture. The thickness of the cell walls of *Nrda. asteroides*, for example, is greater in stationary phase cells than in those from the exponential phase (Beaman, 1975). Similarly, streptomycete cell walls become thicker during development (Cherny et al., 1972a), but then thinner, in the case of tetracycline and streptomycin producers, at the time of cell lysis and disintegration (Zaslavskaya et al., 1977, 1979). Wall thickening also occurs when protein synthesis is inhibited by the addition of antibiotics, such as chloramphenicol (Bewick and Williams, 1977).

Most actinomycetes, at some stage in their development, possess superficial surface layers of variable thickness and appearance on the outside of the 'true' cell wall. Diffuse capsule-like material is commonly observed in nocardioform actinomycetes, whereas sporoactinomycetes possess a surface layer normally described as the fibrous sheath. Such a distinction, based on differences apparent when specimens are viewed in thin sections, is somewhat arbitrary, since very little is known about the chemical composition of the surface material. In *Nrda. asteroides*, the capsule-like appearance of the cell surface becomes less pronounced with increasing culture age (Beaman, 1975). Nevertheless, a consistent feature of the surface material of *Nocardia* species is the manner in which it accumulates at the cell poles of fragmenting hyphae (Kawata and Inoue, 1965; Aristarkhova, 1976b; Fig. 3e). This is in marked contrast to the 'inert' behaviour of the fibrous sheath, which ruptures at the junction of adjacent spores in the genus *Streptomyces* (Williams and Sharples, 1970).

Much of the detailed information about the fine structure of actinomycete cell surfaces comes from studies using the techniques of carbon replication, negative staining and freeze-etching. Such studies have demonstrated clearly the ubiquitous presence of fibrillar surface elements, both in nocardioform actinomycetes (Williams et al., 1976) and in sporoactinomycetes (Kalakoutskii and Agre, 1976; Ensign, 1978).

D. CROSS-WALLS

Cross-walls form in the vegetative hyphae of actinomycetes to delimit hyphal fragments. They also occur in stable, non-fragmenting hyphae. A clear ultrastructural difference appears to exist between these cross-walls, the form of the cross-wall determining the subsequent behaviour of the hyphae (Williams et al., 1973).

Individual cross-walls, present in non-fragmenting vegetative hyphae, consist of a single wall layer. Such cross-walls, which have been classed as 'type 1' by Williams et al. (1973), are formed by the centripetal deposition of an annulus of wall material, in a manner typical of most Gram-positive eubacteria (Ellar et al., 1967; Higgins and Shockman, 1970; Fig. 4). In eubacteria, the young cross-wall, which is usually about twice the thickness of the cell wall, splits after completion and the daughter cells separate. In actinomycetes possessing type 1 cross-walls in their vegetative hyphae, the neighbouring compartments normally do not separate, although they may do so under certain adverse conditions (Carvajal, 1947; Hopwood, 1960; Lechevalier et al., 1961; Thirumalachar, 1970). Microplasmodesmata-like pores (4–10 nm diameter), when present, are presumed to allow some level of communication to be maintained between the compartments (Čáslavská et al., 1978; Gumpert, 1978; Strunk, 1978).

The major feature of the cross-walls, observed in fragmenting vegetative hyphae, is the presence of two separate cross-wall layers, both of which form the new polar walls of the neighbouring compartments. These cross-walls, classed as 'type 2' (Williams et al., 1973), develop from the centripetal extension of a two-layered ingrowth and are clearly designed from the outset to produce separate cells (Fig. 4).

It has been demonstrated that, in Gram-positive eubacteria, what happens between the initial elaboration of the cross-wall and the parting of the daughter cells, depends on the particular bacterium in question and the conditions of growth. In most cases, separation of the cross-wall begins peripherally, i.e. at the base of the cross-wall, and proceeds centripetally, following completion of the cross-wall. Observations on actinomycetes suggest that separation of type 1 cross-walls, when it occurs, follows this pattern (Williams et al., 1973). In *Bacillus cereus*, cross-wall splitting (separation) occurs after cross-wall growth (Highton and Hobbs, 1972). Chung (1973) noted, however, that splitting and bifurcation at the base of the cross-wall, in chloramphenicol-treated *Bacl. cereus*, may occur at any stage before or after cross-wall closure. Higgins and Shockman (1970) observed

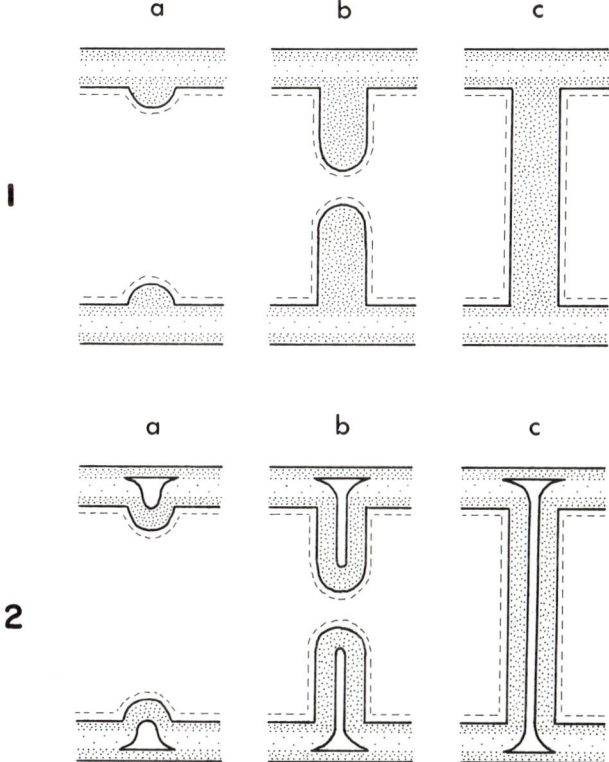

FIG. 4. Stages in the formation of types 1 and 2 cross-walls.

that, in *Streptococcus faecalis*, the separation of the cross-wall may occur more or less simultaneously with cross-wall formation. The double ingrowth of type 2 cross-walls in actinomycetes undergoes a similar simultaneous cross-wall growth and separation behaviour.

Centripetal cross-wall growth, in Gram-positive eubacteria, is considered to result from wall synthetic activity at or near to the leading edge of the developing cross-wall (Rogers et al., 1978). It is likely that similar mechanisms of wall growth occur in type 1 cross-walls of actinomycetes. During development of type 2 cross-walls, the hyphal wall appears to split tangentially into two layers and only the inner layer invaginates to produce the cross-wall (Williams et al., 1973). To explain these observations, two models of cross-wall growth are possible. The first assumes that, like type 1 cross-walls, the addition of new wall material takes place at the leading edge of the growing cross-wall and, at the same time, localized action of autolytic enzymes causes cross-wall splitting (separation). In the second model, the cross-wall is considered to be fed by an annulus of membrane at the base of the forming cross-wall. Wall synthesis, in this case, takes place by uniform intercalation, the action of autolytic enzymes, which make breaks in the wall, allowing extension of the cross-wall. Both mechanisms, of course, may occur and it is quite probable that secondary sites of wall synthesis can be initiated in either case.

E. STALING PHENOMENA

Growth of an actinomycete in batch culture eventually leads to the development of conditions that are unfavourable to its further growth. This 'cultural staling' may be manifested by a number of complex morphogenetic events, involving cytoplasmic 'vacuolation', hyphal swelling, disturbed wall synthesis, intrahyphal growth and autolysis (Wildermuth, 1970b; Cherny et al., 1972a, b; Gumpert, 1978; Figs. 3f, 5a–d). In cultures containing nutritionally-rich media, high levels of secondary metabolites are normally produced only at this stage, i.e. during the 'idiophase', when most of the cellular growth has occurred (Martin and Demain, 1980). It is not surprising, therefore, that several attempts have been made to link the above cytological features, in the genus Streptomyces, to the biosynthesis of a variety of secondary metabolites, including antibiotics (Ludvik et al., 1971; Kuryłowicz and Malinowski, 1972a, b; Collett and Jones, 1974; Zaslavskaya and Zhukov, 1976; Kuimova and Sokolov, 1977; Zaslavskaya et al., 1977, 1979; Čáslavská et al., 1978; Kuimova et al., 1978). This, as pointed out by Bewick et al. (1976), has proved difficult, however, and no clear relationship has yet been established.

4. Actinomycete Spores

Actinomycete spores have been categorized as being either endogenous or of 'hyphal origin' (Cross, 1970; Williams et al., 1973, 1976; Cross and Attwell, 1975). The major distinction between these categories relates to the mode of spore wall formation during sporogenesis. In addition, the presence or absence of a fibrous sheath overlying the wall of the sporogenous hyphae can influence the way in which the spores are borne and released.

A. PATTERNS OF SPORE DEVELOPMENT

Most actinomycete genera form spores which are of hyphal origin. They may be produced singly or in chains of 2 to more than 50, in some cases within 'spore vesicles' (Cross, 1970), and, apart from a few possible exceptions, develop as a result of septation and disarticulation (i.e. fragmentation)

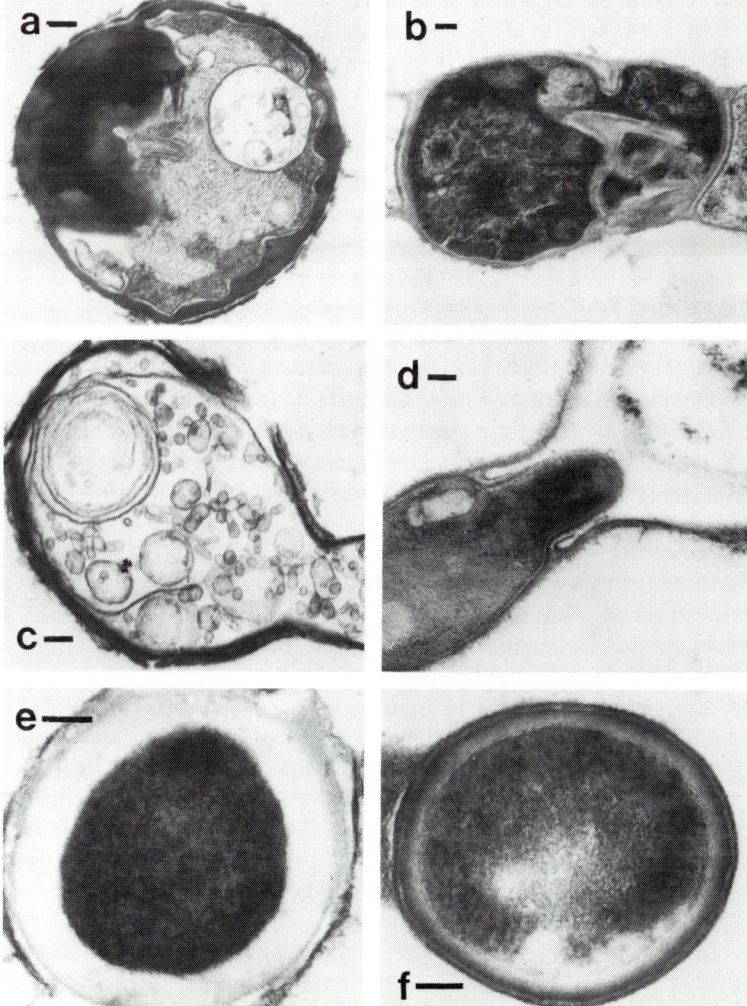

FIG. 5. (a) Irregular thickening of the cell wall in '*Streptomyces ostreogriseus*', (b) irregular cross-wall formation in *Streptomyces aureofaciens*, (c) autolysis in '*Stmy. ostreogriseus*' (d) intrahyphal growth in *Stmy. aureofaciens*, (e) spore of *Micropolyspora faeni* and (f) spore of '*Micromonospora melanosporea*'. Bar marker represents 0.1 μm.

of pre-existing hyphal elements. Using mycological terminology, this pattern of spore development would be described as 'thallic' and, more specifically, as 'thallic-arthric', when chains of spores are formed (Kendrick, 1971). The characteristic feature of these spores is that the spore wall is formed, at least in part, from all wall layers of the parent hypha.

In contrast, the wall of endogenous spores is formed from a *new* wall layer, produced within the parent hypha, i.e. the wall of the parent hypha does not contribute to the formation of the spore wall. Endogenous spores of the genus *Thermoactinomyces* are of the type designated as 'true endospores', similar to those of species of *Bacillus* and *Clostridium* (Cross et al., 1968; Dorokhova et al., 1970b; McVittie et al., 1972). In these genera, a spore septum forms to segregate a double-membrane enclosed forespore, containing cytoplasmic and nuclear materials. Subsequently, a cortex is deposited between the two membranes and a multilayered spore coat forms on the outside. Such a complex sequence of events does not occur in other endogenous spore-forming genera. The endogenously-produced 'zoospores' (Cross and Attwell, 1975) of the genera *Planomonospora* and *Dactylosporangium* are delimited simply by the formation of a spore wall between the cytoplasmic membrane and the parent hyphal wall (Sharples et al., 1974).

It is apparent, therefore, that the formation of the spore wall, in the actinomycetes, can occur in three ways.

(a) By the spore wall being formed, at least in part, from all wall layers of the parent hypha—*holothallic development*.
(b) By the spore wall being formed as a new layer or layers deposited between the cytoplasmic membrane and the wall of the parent hypha—*enterothallic development*.
(c) By the spore wall being formed as a new layer or layers synthesized around a delimited volume of protoplasm—'*true*' endospore development.

The terms 'holothallic' and 'enterothallic' have been proposed (Kendrick, 1971) for fungal conidium development in which, respectively, either "all wall layers of the condiogenous cell are involved in the formation of the conidium wall" or "the outer wall of the sporogenous cell is not involved in the formation of the spore wall". Clearly, the term 'holothallic' can be applied to actinomycete spores that have a hyphal origin. Moreover, 'enterothallic' appears to be the ideal descriptive term for endogenous spore formation in *Planomonospora* and *Dactylosporangium*.

B. SPORE DELIMITATION

All actinomycete spores, apart from 'true endospores', are delimited ultimately by cross-walls or so-called 'sporulation septa' (Wildermuth and Hopwood, 1970). In genera producing chains of spores, the sequence of formation of such septa may occur more or less simultaneously, e.g. in *Streptomyces* (Wildermuth and Hopwood, 1970), basipetally, e.g. in *Micropolyspora* (Dorokhova et al., 1969, 1970a), acropetally, e.g. in *Pseudonocardia*

(Henssen et al., 1981) or randomly, e.g. in *Nocardiopsis* (*Actinomadura*) (Williams et al., 1974).

Sporulation septa are more diverse in form than the cross-walls present in vegetative hyphae. They may be of 'type 1' or 'type 2' (Williams et al., 1973) or they may show variations on these basic types. The greatest diversity of septal form appears to exist in the genus *Streptomyces*. Hardisson and Manzanal (1976) have attempted to categorize the various septa and have proposed three major types (Fig. 6). In 'type I', the sporulation septum develops from the centripetal growth of two separate annuli or cross-walls that originate from the inner face of the hyphal wall. The two annuli remain as separate entities throughout the process. In 'type II', septum formation is initiated by deposition of a single wall layer on the inside of the hyphal wall. Two annuli, which develop at the leading edge of the developing cross-wall, subsequently join centrally to form the completed sporulation septum. In 'type III', the septum is formed from the centripetal growth of a single, thick cross-wall layer. On the basis of these septal types, Hardisson and Manzanal (1976) devised a tentative classification for some members of the genus *Streptomyces*.

The taxonomic and functional significance of variations in septal ultrastructure are unclear. It is important to consider that such variations may simply reflect a physiological peculiarity of the species concerned and it is premature, at this stage, to draw conclusions from limited ultrastructural data.

C. SPORE MATURATION

The completion of the sporulation septum marks the beginning of a series of morphological changes leading to the formation of a mature spore. The nature and extent of these changes, which commonly involve deposition of additional wall material, localized lysis and cleavage of septa, condensation of cytoplasmic contents and modification of spore shape, depend on the cellular organization of the mature spore compared to the parent hypha (Cross and Attwell, 1975; Kalakoutskii and Agre, 1976; Ensign, 1978).

With the exception of the thermoresistant endospores of the genus *Thermoactinomyces*, actinomycete spores, at maturity, are relatively unspecialized compartments of hyphae. They are bounded by walls, which are normally about 30–50 nm thick (cf. vegetative cell walls, 10–20 nm thick), but values as high as 70–100 nm have been quoted for the genus *Micropolyspora* (Dorokhova et al., 1969, 1970a) and as low as 10–14 nm for some members of the *Actinoplanaceae* (Sharples et al., 1974). The deposition of the additional wall material appears to take place after completion of the sporulation septum (Dorokhova et al., 1969; Wildermuth and Hopwood, 1970; Luedemann and Casmer, 1973; Manzanal and Hardisson, 1978; Henssen

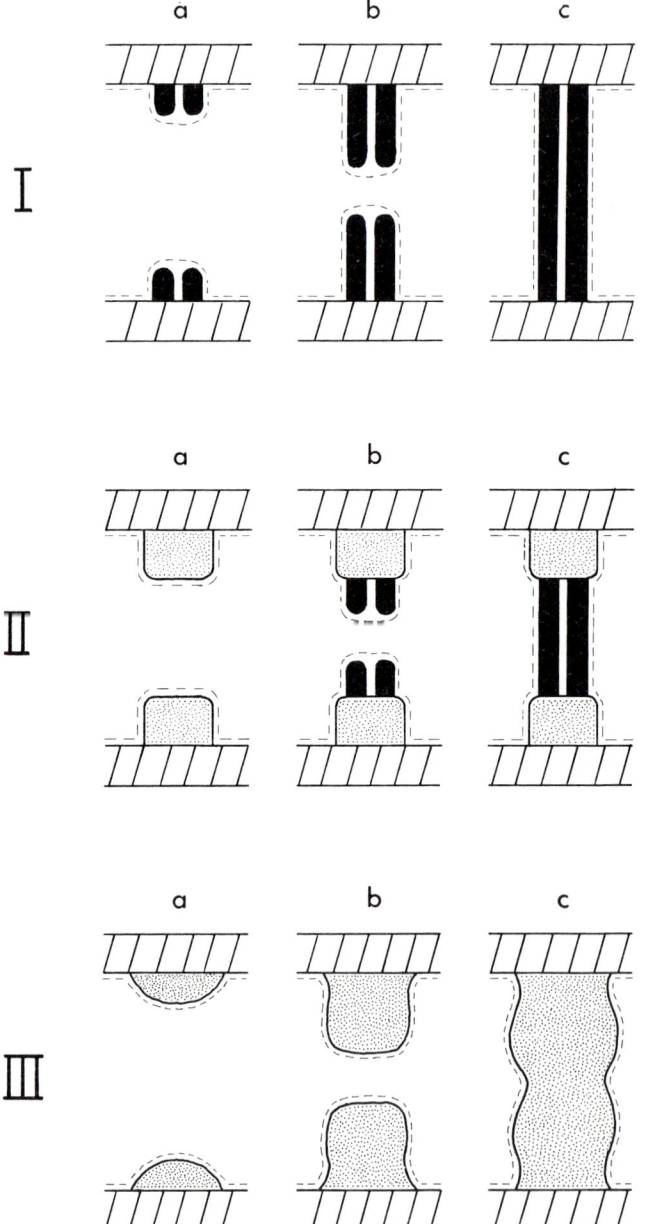

FIG. 6. Three proposed models for the formation of sporulation septa in different species of *Streptomyces* (after Hardisson and Manzanal, 1976). For details, see the text.

et al., 1981) but there are indications that wall thickening may be initiated prior to this (McVittie, 1974; Strunk, 1978). It has been suggested, moreover, that, in *Stmy. coelicolor*, spore wall synthesis may be a two-stage process, the first stage involving deposition of material of the same composition as the pre-sporulation hyphal wall, while further thickening is due to addition of material of a different composition (McVittie, 1974). It is difficult to distinguish any clear boundary between 'old' and 'new' wall in streptomycetes and the spore walls, although possessing zones of different electron density, consist of a single layer (Hardisson and Manzanal, 1976). A distinction between 'old' and 'new' wall is apparent, however, in the multilayered spore walls of the genus *Micromonospora* (Luedemann and Casmer, 1973; Stevens, 1975; Hardisson and Suárez, 1979). Multilayered walls have also been observed in spores of the genera *Planomonospora* and *Microbispora* (Sharples *et al.*, 1974; Sharples and Williams, 1976b). It should be borne in mind that stratification of walls "will be apparent in transmission electron microscopy only if there are differences in electron transparency of adjacent layers. Such differences could result directly from chemical differences, or be due to differential uptake of electron-dense material during fixation and staining procedures" (Gull and Trinci, 1971).

The increase in wall thickness, which occurs during spore maturation, is generally considered to be brought about by deposition of new wall material. It may, of course, result from expansion of wall material already present, as suggested for the thickening of the endospore cortex (Cross and Attwell, 1975). Such expansion, as pointed out by Cross and Attwell (1975), could facilitate 'rounding up', i.e the change from the cylindrical hyphal shape to the ovoid or spherical form of the spore, and may also account for the 'compressed' or condensed appearance of the contents of many mature spores (Figs. 5e, f). The change in spore shape and the alteration in appearance of the contents appear to parallel the increase in wall thickness in the genus *Streptomyces* (Hardisson and Manzanal, 1976).

Most actinomycete spores are ovoid or spherical in shape but, in the genus *Chainia*, they are roughly cuboid (Sharples and Williams, 1976a) and the 'enterothallic' spores of the genera *Planomonospora* and *Dactylosporangium* are claviform (Sharples *et al.*, 1974). In addition, spores of the genus *Micromonospora* have a somewhat 'warty' surface ornamentation which, in the absence of a fibrous sheath, is due to localized 'bulges' or thickenings of the wall (Luedemann and Casmer, 1973; Sharples and Williams, 1976b; Hardisson and Suárez, 1979). The 'warts', although present on the developing spore wall before completion of the sporulation septum, are more clearly visible after its completion (Hardisson and Suárez, 1979).

Conspicuous polar wall thickenings at the contact sites of adjacent spores have been observed in a number of genera producing spores of the 'thallicarthric' type. Such thickenings, which have also been described as interspleral 'pads' (Rancourt and Lechevalier, 1963, 1964; Williams et al., 1973; Sharples and Williams, 1976a) or 'nozzles' (Strunk, 1978), may develop either during or following sporulation septum formation and represent the final area of attachment of adjacent spores before their eventual separation. They were considered, in *Mips. rectivirgula*, to be temporary assisting structures, functional only during sporogenesis and not intrinsic to mature spores (Dorokhova et al., 1969, 1970a). The transient nature of the wall thickenings is similarly apparent in several species of the genus *Streptomyces* (Manzanal and Hardisson, 1978; Strunk, 1978; Henssen et al., 1981). Cytoplasmic continuity between adjacent spore compartments may be maintained, during the maturation process, by means of 'microplasmodesmata' within the 'nozzles' (Strunk, 1978).

D. FIBROUS SHEATH

The fibrous sheath is a relatively delicate, extracellular entity which covers the spores of most actinomycete genera during their formation. It appears to play no role in the spore formation process but can influence the way in which the delimited spores are borne and released (Williams et al., 1973; Sharples et al., 1974; Cross and Attwell, 1975; Ensign, 1978).

The characteristic surface structure of spores of different *Streptomyces* species is due to the presence of the fibrous sheath (Wildermuth, 1970a, 1972a, b; Dietz and Mathews, 1971; Williams et al., 1972). Spore surface structure, which may be categorized as smooth, spiny, warty or hairy (Tresner et al., 1961), has been used widely for the morphological characterization of species (Shirling and Gottlieb, 1976; Williams and Wellington, 1980). Intermediates between these categories do occur and further subdivisions have been proposed (Lyons and Pridham, 1971; Dietz and Mathews, 1972, 1977) but these have yet to be adopted generally in species descriptions. Current evidence indicates that spore surface structure is a very stable and useful taxonomic character (Williams and Wellington, 1980). Irregularities do occur in single strains but these "should be accepted without surprise" (Shirling and Gottlieb, 1976). A lack of surface ornamentation on a few spores among ornamented ones may be due to loss or disruption of the ornamented sheath material or may result from interference with the production and activities of a sheath transformation inducing agent(s) (Szabó, 1977; Szabó et al., 1977, 1979).

The sheath of a number of sporoactinomycetes, in addition to basic fibrillar elements, consists of hollow tubules, which, in some species, fragment into short 'rodlets' (Kalakoutskii and Agre, 1976; Ensign, 1978). Centrally-grooved structures, with a similar negatively-stained appearance to that of the tubules, have also been described in some streptomycetes

(Wildermuth *et al.*, 1971; Williams *et al.*, 1972; Gerasimov *et al.*, 1978). The size of these sheath elements exhibits great variation, fibrils normally having a diameter of about 4–10 nm and the rodlets or tubules a diameter of about 8–20 nm. There are indications that rodlets may, in some cases, develop from the fibrils. The rodlets of *Mips. angiospora* appear to be composed of a two-stranded helix, each strand having a diameter similar to that of a fibril (Takeo, 1976). Pairs of fibrils are also considered to constitute the structure of rodlets in *Stmy. coelicolor* (Smucker and Pfister, 1978).

The surface patterns created by the sheath elements show some characteristic features. The 'smooth' spores of *Stmy. coelicolor*, for example, have apparent interwoven pairs of parallel rodlets, described as the "rodlet mosaic" (Smucker and Pfister, 1978). Such a pattern, although also present on the sheath overlying the spores of *Stmy. griseus* and *Stmy. venezuelae*, is not evident in other smooth-spored *Streptomyces* species so far examined (Wildermuth, 1970a; Williams *et al.*, 1972). It is notable that the rodlet mosaic, which shows a marked similarity to the pattern present on the surfaces of certain fungal spores (Cole *et al.*, 1979; Dempsey and Beever, 1979), is considered by Smucker and Pfister (1978) to be situated, in *Stmy. coelicolor*, between the spore wall and an external granular matrix, the fibrous sheath being on the *outside* of the latter. This conflicts with earlier reports equating the rodlet mosaic with the fibrous sheath (Bradley and Ritzi, 1968; Wildermuth *et al.*, 1971).

The sheath elements appear to contribute to the formation of the surface ornaments of streptomycete spores, although the precise relationship is not clear. New material overlying the sheath may also be involved (Wildermuth, 1972a). Spines and hairs are composed of from one to several rod-like or tubular units, each of which is constructed, at least partly, from fibrillar elements (Wildermuth, 1972a, b; Williams *et al.*, 1972; Matselyukh, 1978). It remains to be seen whether warts are similarly constructed.

The chemical composition of the fibrous sheath has received relatively little attention. Characteristically, most streptomycete spores enclosed within a sheath are hydrophobic. This property has been considered to be due to the presence of surface lipids (Erikson, 1947; Kalakoutskii and Sokolov, 1961; Katz, 1963). Several lines of evidence indicate, however, that the principal component of the rodlet mosaic in *Stmy. coelicolor* is a polysaccharide, probably chitin (Smucker and Pfister, 1978). In contrast, it has been suggested that the rodlets of *Stmy. griseus*, which have a low content of nitrogen, amino sugars and carbohydrate, may be partly inorganic (Bradshaw and Williams, 1976) and, in *Stmy. roseoflavus* var. *roseofungini*, purified sheath material appears to contain a polyene antibiotic, fatty acids, calcium, magnesium and silicon (Cherny *et al.*, 1974; Pouzharitskaja *et al.*, 1974).

E. SPORE GERMINATION

Most actinomycete spores upon encountering favourable environmental conditions will normally germinate. Endospores of *Team. vulgaris*, however, are constitutively dormant and germination does not take place until they are 'activated' in some way (Cross and Attwell, 1975; Kalakoutskii and Agre, 1976; Ensign, 1978). An 'activation' treatment, consisting of a heat shock, is also required to initiate the germination of some streptomycete spores (Ensign, 1978).

On germination, spores swell, lose refractility and one or more germ tubes emerge. The spore contents, which are sometimes difficult to distinguish in thin sections, become clearly visible and storage granules, if present, disappear (Sharples and Williams, 1976b; Hardisson *et al.*, 1978; Hardisson and Suárez, 1979). The behaviour of the spore wall varies between different genera and three categories can be recognized (Sharples and Williams, 1976b).

(a) The germ-tube wall arises from a wall layer synthesized *de novo*, during germination, within the existing spore wall (e.g. *Microellobosporia, Thermoactinomyces*).
(b) The germ tube wall arises from an existing inner layer of the spore wall but this does not become visible until germination is initiated (e.g. *Micropolyspora, Streptomyces*).
(c) The germ-tube wall arises from an existing inner layer of the spore wall which is distinguishable in the dormant spore (e.g. *Microbispora, Micromonospora*).

These categories are similar to those recognized in the fungi, the main distinction being between germ-tube walls that arise from newly synthesized material and those arising from pre-existing material in dormant spores (Bartnicki–Garcia, 1968). Unfortunately, as is the case for fungi (Kahn, 1975), no clear taxonomic significance can be attached to these categories.

5. Growth of Actinomycete Cells

A. POLAR GROWTH OF RODS

The formation of rod-shaped cells is common amongst fragmenting actinomycetes. There may be analogies between these cells and the rods of other bacteria.

Cytoplasm is delimited by walls that must extend in accordance with an increase in cell volume. Scanning electron microscope (SEM) investigations on the rods of nocardioform (Locci, 1976) and microaerophilic (Locci, 1978a) actinomycetes have demonstrated that they elongate by apical growth.

Most of the evidence was gathered from serial observations of the 'budding' process in different organisms at various growth stages, using external markers (e.g. 'scars' or 'bands') present on the rods. Time-lapse microphotography confirmed the sequence of events. Polar growth occurs not only in newly-seeded propagules but also in rods produced by hyphal fragmentation *in situ*. Even cocci show polarity in that they tend to regrow from regions corresponding to the main axis of the filament, from which they originated by fragmentation (Locci, 1976, 1981).

Additional evidence of polar growth has been obtained by immunofluorescent labelling of facultatively anaerobic actinomycetes (Locci and Schaal, 1980). With this method, it was possible to follow elongation with time by examining rods after various incubation periods. Cell walls appear to be newly made at either or both poles, confirming the conclusions drawn from SEM observations. May (1963), with immunofluorescent labelling, has distinguished the different means of distribution of derived cell wall materials, during extension, as (a) uniform dispersion, (b) equatorial or interpolar extension and (c) bipolar extension leading to conservation at the centre of the cell ('interpolary conservation'). The latter, a characteristic of *Schizosaccharomyces pombe* (May, 1962), could also apply to actinomycete rods, where both unipolar and bipolar extension occurs.

Active synthesis at the apex of a cell should result in a debilitation of the wall structural strength in that region. This can be assessed by serial observations of the effects of unfavourable agents. In amycelial rhodococci exposed to lysozyme, developing rods do show initial wall damage at their tips (Fig. 7a). However, when non-elongating rods are exposed to lysozyme, lysis occurs over the whole wall (Fig. 7b), a situation similar to that observed in unbranched bacteria (Locci, 1980b, 1981; Schofield and Locci, 1981).

It is fruitless, at this stage, to enter into phylogenetic arguments. Apical growth, however, seems to occur even in the morphologically less differentiated members of the actinomycetes and, therefore, could represent a basic system for branching organisms. Indeed, some unbranched bacteria also grow in this manner. The concomitant possibility of bud formation (Hoffman and Frank, 1964) in organisms other than typical 'budding bacteria' (Hirsch, 1974) stresses once again the complexity of growth processes even in 'simple' forms and the need for further investigations into this poorly understood yet fundamental phenomenon.

B. APICAL EXTENSION OF HYPHAE

Fungal hyphae elongate by the commonly known process of apical growth. Essentially this is polarized, i.e. confined to the tip (the so-called extension zone), its shape approximating more closely to half ellipsoids of revolution than to hemispheres (Saunders and Trinci, 1979). By using external markers

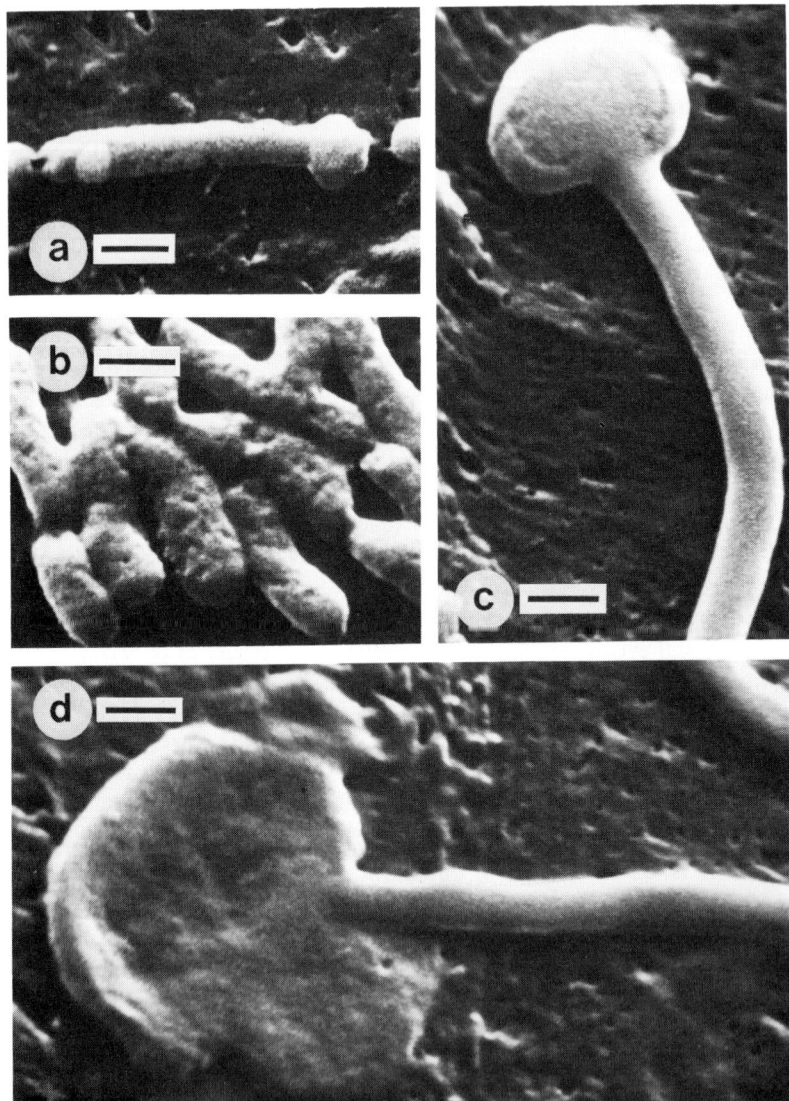

FIG. 7. Results of exposure of actinomycetes to lysozyme. (a) Developing filament of '*Gordona*' *aurantiaca*, (b) *Rhodococcus rhodochrous* rods and (c and d) hyphal apices of *Streptomyces viridochromogenes*. Bar marker represents 1.0 μm.

(e.g. side branch insertions), it has been possible to demonstrate that the hyphae of actinomycetes also develop by apical extension (Locci, 1981). In *Stmy. hygroscopicus*, hyphal growth is limited to the apical 20 μm, while side branches emerge in the 80–110 μm region (Schuhmann and Bergter,

1976). No branches are formed in the 'older' portions of the hyphae. A phenomenon similar to fungal 'apical dominance' seems to be operating (Kalakoutskii and Agre, 1976). Observations on '*Stmy. streptomycini*' have shown an inverse relationship between hyphal extension and side branch formation (Dmitrieva and Rodionova, 1971). Physiological differences, such as variations in isoelectric point (Hagedorn, 1955), enzyme activity (Giolitti, 1960) and temperature sensitivity (Boltyanskaya *et al.*, 1972), occur between apices and the rest of the hyphae.

The phenomenon of preferential incorporation of new wall material at the apex is associated with a different structural strength in this region. As with fungi, this would imply the existence of an apical 'unset' wall, elastic and capable of yielding and stretching in response to pressure differentials, and of a 'set' one, inelastic and incapable of further deformation (Da Riva Ricci and Kendrick, 1972). Thus an alternative approach to studying sites of hyphal extension is to follow the micromorphological effects of conditions interfering with growth. If there is structural differentiation along the hypha, exposure to adverse conditions should cause some differential behaviour. Because of the fragility of actinomycetes, membrane-transfer techniques have been used to study this effect (Locci, 1980a). In *Nrda. asteroides* exposed to gentamicin, the earliest symptoms consisted of alterations of the hyphal tip (Figs. 7c, d), followed by its degeneration. This reaction is typical of filamentous fungi exposed to a series of toxicants (Richmond, 1975). Similar results have been obtained by following the action of lysozyme on microcolonies of a number of actinomycetes, including mycelial rhodococci, showing that initially the most sensitive areas are the actively growing parts, i.e. hyphal apices (Locci, 1978b, 1980b; Schofield and Locci, 1981).

ACKNOWLEDGEMENT

The authors wish to thank Dr. S. T. Williams for much helpful discussion during the preparation of this chapter.

REFERENCES

Aristarkhova, V. I. (1976a). Electron microscope study of *Nocardia rubra* grown on a medium with phenol. *Mikrobiologiya* **45**, 284–287 (English translation).
Aristarkhova, V. I. (1976b). Fine structure of phenol-oxidising *Nocardia corallina*. *Mikrobiologiya* **45**, 436–440 (English translation).
Baldacci, E., Locci, R. and Rogers, J. L. (1966). Production of "granules" by *Actinomycetales*. *Giornale di Microbiologia* **14**, 173–184.
Baldacci, E., Locci, R. and Baldan, B. P. (1971a). On the spore formation process in actinomycetes. II. Sporulating structures of some *Streptomyces* species with smooth and spiny spore surface. *Rivista di Patologia Vegetale, Supplemento* **7**, 21–44.

Baldacci, E., Locci, R. and Baldan, B. P. (1971b). On the spore formation process in actinomycetes. III. Sporulation in *Streptomyces* species with hairy spore surface as detected by scanning electron microscopy. *Rivista di Patologia Vegetale, Supplemento* **7**, 45–61.

Barksdale, L. and Kim, K. S. (1977). *Mycobacterium*. *Bacteriological Reviews* **41**, 217–372.

Bartnicki-Garcia, S. (1968). Cell wall chemistry, morphogenesis and taxonomy of fungi. *Annual Review of Microbiology* **22**, 87–108.

Beadles, T. A., Land, G. A. and Knezek, D. J. (1980). An ultrastructural comparison of the cell envelopes of selected strains of *Nocardia asteroides* and *Nocardia brasiliensis*. *Mycopathologia* **70**, 25–32.

Beaman, B. L. (1975). Structural and biochemical alterations of *Nocardia asteroides* cell walls during its growth cycle. *Journal of Bacteriology* **123**, 1235–1253.

Beaman, B. L. and Shankel, D. M. (1969). Ultrastructure of *Nocardia* cell growth and development on defined and complex agar media. *Journal of Bacteriology* **99**, 876–884.

Beaman, B. L., Serrano, J. A. and Serrano, A. A. (1978). Comparative ultrastructure within the nocardiae. *Zentralblatt für Bakteriologie, Parasitenkunde, Infektionskrankheiten und Hygiene*. 1. *Abteilung, Supplement* **6**, 201–220.

Beveridge, T. J. (1978). The response of cell walls of *Bacillus subtilis* to metals and to electron microscopic stains. *Canadian Journal of Microbiology* **24**, 89–104.

Bewick, M. W. M. and Williams, S. T. (1977). Effects of addition of chloramphenicol on the growth and ultrastructure of *Streptomyces venezuelae*. *Microbios* **19**, 27–35.

Bewick, M. W. M., Williams, S. T. and Veltkamp, C. (1976). Growth and ultrastructure of *Streptomyces venezuelae* during chloramphenicol production. *Microbios* **16**, 191–199.

Boltyanskaya, E. V., Agre, N. S., Sokolov, A. A. and Kalakoutskii, L. V. (1972). Effects of changes in temperature and relative humidity on rings in colonies of *Thermoactinomyces vulgaris*. *Mikrobiologiya* **41**, 595–598 (English translation).

Bradley, S. G. and Ritzi, D. (1968). Composition and ultrastructure of *Streptomyces venezuelae*. *Journal of Bacteriology* **95**, 2358–2364.

Bradshaw R. M. and Williams, S. T. (1976). Chemical composition of the spore sheath of *Streptomyces griseus*. *Microbios* **15**, 57–65.

Burdett, I. D. J. and Rogers, H. J. (1970). Modification of the appearance of mesosomes in sections of *Bacillus licheniformis* according to the fixation procedures. *Journal of Ultrastructure Research* **30**, 354–367.

Carvajal, F. (1947). The production of spores in submerged cultures by some *Streptomyces*. *Mycologia* **39**, 426–440.

Čáslavská, J., Štastná, J., Ludvík, J. and Blumauerová, M. (1978). Ultrastructure of some species of *Streptomyces* during submerged cultivation. *Zentralblatt für Bakteriologie, Parasitenkunde, Infektionskrankheiten und Hygiene*. 1. *Abteilung, Supplement* **6**, 181–191.

Chater, K. F. and Merrick, M. J. (1979). Streptomycetes. In: *Developmental Biology of Prokaryotes* (J. H. Parish, ed.), pp. 93–114. Blackwell Scientific Publications, Oxford.

Chung, K. L. (1973). Influence of cell-wall thickness on cell division: Electron microscopic study with *Bacillus cereus*. *Canadian Journal of Microbiology* **19**, 217–221.

Cherny, N. E., Tikhonenko, A. S., Nikitina, E. T. and Kalakoutskii, L. V. (1972a). Ultrastructure of *Streptomyces roseoflavus* var. *roseofungini* and its stable nocardioform 'fructose' variant. *Cytobios* **5**, 7–24.

Cherny, N. E., Tikhonenko, A. S., Nikitina, E. T. and Kalakoutskii, L. V. (1972b). Intracytoplasmic membrane system in 'fructose' variant of *Streptomyces roseoflavus* var. *roseofungini* grown on meat-peptone agar. *Cytobios* **5**, 101–110.

Cherny, N. E., Tikhonenko, A. S., Kuroda, S. and Kalakoutskii, L. V. (1974). Purification and preliminary characterization of extracellular tubular material produced by 'fructose' mutant of *Streptomyces roseoflavus* var. *roseofungini*. *Microbios* **10**, 7–14.

Clark, J. B. (1979). Sphere-rod transitions in *Arthrobacter*. In: *Developmental Biology of Prokaryotes* (J. H. Parish, ed.), pp. 73–92. Blackwell Scientific Publications, Oxford.

Cole, G. T., Sekiya, T., Kasai, R., Yokoyama, T. and Nozawa, Y. (1979). Surface ultrastructure and chemical composition of the cell walls of conidial fungi. *Experimental Mycology* **3**, 132–156.

Collett, M. and Jones, G. H. (1974). Morphological changes accompanying actinomycin production in *Streptomyces antibioticus*. *Journal of Ultrastructure Research* **46**, 452–465.

Cross, T. (1970). The diversity of bacterial spores. *Journal of Applied Bacteriology* **33**, 95–102.
Cross, T. and Al-Diwany, L. J. (1981). Streptomycetes with substrate mycelium spores: the genus *Elytrosporangium. Zentralblatt für Bakteriologie, Mikrobiologie und Hygiene.* 1. Abteilung, Supplement **11**, 59–65.
Cross, T. and Attwell, R. W. (1975). Actinomycete spores. In: *Spores VI* (P. Gerhardt, R. N. Costilow and H. L. Sadoff, eds.), pp. 3–15. American Society of Microbiology, Washington.
Cross, T. and Goodfellow, M. (1973). Taxonomy and classification of the actinomycetes. In: *Actinomycetales: Characteristics and Practical Importance* (G. Sykes and F. A. Skinner, eds.), pp. 11–112. Academic Press, London.
Cross, T., Walker, P. D. and Gould, G. W. (1968). Thermophilic actinomycetes producing resistant endospores. *Nature (London)* **220**, 352–354.
Da Riva Ricci, D. and Kendrick, R. (1972). Computer modelling of hyphal tip growth in fungi. *Canadian Journal of Botany* **50**, 2455–2462.
Dawes, E. A. and Senior, P. J. (1973). The role and regulation of energy reserve polymers in microorganisms. *Advances in Microbial Physiology* **10**, 135–266.
Dempsey, G. P. and Beever, R. E. (1979). Electron microscopy of the rodlet layer of *Neurospora crassa* conidia. *Journal of Bacteriology* **140**, 1050–1062.
Dietz, A. and Mathews, J. (1971). Classification of *Streptomyces* spore surfaces into five groups. *Applied Microbiology* **21**, 527–533.
Dietz, A. and Mathews, J. (1972). Characterization of hairy-spored streptomycetes. *International Journal of Systematic Bacteriology* **22**, 173–177.
Dietz, A. and Mathews, J. (1977). Characterization of hairy-spored streptomycetes. II. Twelve additional cultures. *International Journal of Systematic Bacteriology* **27**, 282–287.
Dipersio, J. R. and Deal, S. J. (1974). Identification of intracellular polysaccharides in thin sections of *Nocardia asteroides. Journal of General Microbiology* **83**, 349–358.
Dmitrieva, S. V. and Rodionova, E. G. (1971). Cytological observations on developing culture of a streptomycin producer with special reference to its heterogeneity. *Antibiotiki* **16**, 529–534 (in Russian).
Dorokhova, L. A., Agre, M. S., Kalakoutskii, L. V. and Krasilnikov, N. A. (1969). Fine structure of sporulating hyphae and spores in a thermophilic actinomycete, *Micropolyspora rectivirgula. Journal de Microscopie (Paris)* **8**, 845–854.
Dorokhova, L. A., Agre, N. S., Kalakoutskii, L. V. and Krasilnikov, N. A. (1970a). A study of the morphology of two cultures belonging to the genus *Micropolyspora. Mikrobiologiya* **39**, 79–86 (English translation).
Dorokhova, L. A., Agre, N. S., Kalakoutskii, L. V. and Krasilnikov, N. A. (1970b). Electron microscopic study on spore formation in *Micromonospora vulgaris. Mikrobiologiya* **39**, 680–684 (English translation).
Dow, C. S., Westmacott, D. and Whittenbury, R. (1976). Ultrastructure of budding and prosthecate bacteria. In: *Microbial Ultrastructure* (R. Fuller and D. W. Lovelace, eds.), pp. 187–221 Academic Press, London.
Dunlop, W. F. and Robards, A. W. (1973). Ultrastructural study of poly-β-hydroxybutyrate granules from *Bacillus cereus. Journal of Bacteriology* **114**, 1271–1280.
Ellar, D. J., Lundgren, D. G. and Slepecky, R. A. (1967). Fine structure of *Bacillus megaterium* during synchronous growth. *Journal of Bacteriology* **94**, 1189–1205.
Ensign, J. C. (1978). Formation, properties and germination of actinomycete spores. *Annual Review of Microbiology* **32**, 185–219.
Erikson, D. (1947). Differentiation of the vegetative and sporogenous phases of the actinomycetes. 1. The lipid nature of the outer wall of the aerial mycelium. *Journal of General Microbiology* **1**, 39–44.
Erikson, D. (1953). Variation of mycelial pattern in sporogenous and asporogenous actinomycetes. In: *Actinomycetales: Morphology, Biology and Systematics* (E. Baldacci and P. Redaelli, eds.), pp. 102–121. Proceedings of VIth International Congress of Microbiology, Rome.
Fooke-Achterrath, M., Lickfield, K. G., Reusch, V. M., Aebi, U., Tschöpe, U. and Menge, B. (1974). Close to life preservation of *Staphylococcus aureus* mesosomes for transmission electron microscopy. *Journal of Ultrastructure Research* **49**, 270–285.
Francisco, D. E. and Silvey, J. K. G. (1971). The effect of carbon monoxide inhibition on the growth of an aquatic streptomycete. *Canadian Journal of Microbiology* **17**, 347–351.

Ganju, P. L. and Iyengar, M. R. S. (1974). Micromorphology of some sclerotial actinomycetes and development of their sclerotia. *Journal of General Microbiology* **82**, 35–48.
Gerasimov, V. N., Cherny, N. E., Kalakoutskii, L. V. and Tikhonenko, A. S. (1978). Fine structure of the outer sheath of aerial mycelium of *Actinomyces levoris*. *Mikrobiologiya* **47**, 64–67 (English translation).
Germida, J. J. and Casida, L. E. (1980). Myceloid growth of *Arthrobacter globiformis* and other *Arthrobacter* species. *Journal of Bacteriology* **144**, 1152–1158.
Ghosh, B. K. (1974). The mesosome: a clue to the evolution of the plasma membrane. *Sub-Cellular Biochemistry* **3**, 311–367.
Ghosh, B. K. and Nanninga, N. (1976). Polymorphism of the mesosome in *Bacillus lichenformis* (749/C and 749). *Journal of Ultrastructure Research* **56**, 107–120.
Giolitti, G. (1960). Some observations on the cytochemistry of *Streptomyces*. *Journal of General Microbiology* **23**, 83–86.
Greenawalt, J. W. and Whiteside, T. L. (1975). Mesosomes: membranous bacterial organelles. *Bacteriological Reviews* **39**, 405–463.
Grein, A. and Spalla, C. (1962). Studio sui coremi formati in culture di *Streptomyces peucetius*. *Giornale di Microbiologia* **10**, 175–184.
Gull, K. and Trinci, A. P. J. (1971). Fine structure of spore germination in *Botrytis cinerea*. *Journal of General Microbiology* **68**, 207–220.
Gumpert, J. (1978). Ultrastructure and modifiability of the cell envelope in *Streptomyces hygroscopicus*. *Zentralblatt für Bakteriologie, Mikrobiologie, Parasitenkunde, Infektionskrankheiten und Hygiene*. 1. Abteilung, Supplement **6**, 221–233.
Hagedorn, H. (1955). Untersuchungen über den isoelektrischen Punkt bei Aktinomyceten. *Protoplasma* **45**, 115–124.
Hardisson, C. and Manzanal, M. B. (1976). Ultrastructural studies of sporulation in *Streptomyces*. *Journal of Bacteriology* **127**, 1443–1454.
Hardisson, C. and Suárez, J. E. (1979). Fine structure of spore formation and germination in *Micromonospora chalcea*. *Journal of General Microbiology* **110**, 233–237.
Hardisson, C., Manzanal, M. B., Salas, J. A. and Suárez, J. E. (1970). Fine structure, physiology and biochemistry of arthrospore germination in *Streptomyces antibioticus*. *Journal of General Microbiology* **105**, 203–214.
Hasegawa, T., Lechevalier, M. P. and Lechevalier, H. A. (1978). New genus of *Actinomycetales*: *Actinosynnema* gen. nov. *International Journal of Systematic Bacteriology* **28**, 304–310.
Henssen, A., Weise, E., Vobis, G. and Renner, B. (1981). Ultrastructure of sporogenesis in actinomycetes forming spores in chains. *Zentralblatt für Backteriologie, Mikrobiologie und Hygiene*. 1. Abteilung, Supplement **11**, 137–146.
Higgins, M. L. and Daneo-Moore, L. (1974). Factors influencing the frequency of mesosomes observed in fixed and unfixed cells of *Streptococcus faecalis*. *Journal of Cell Biology* **61**, 288–300.
Higgins, M. L. and Shockman, G. D. (1970). Model for cell wall growth of *Streptococcus faecalis*. *Journal of Bacteriology* **101**, 643–648.
Higgins, M. L. and Silvey, I. K. (1966). Slide culture observations of two freshwater actinomycetes. *Transactions of the American Microscopical Society* **85**, 390–398.
Higgins, M. L., Tsien, H. C. and Daneo-Moore, L. (1976). Organization of mesosomes in fixed and unfixed cells. *Journal of Bacteriology* **127**, 1519–1523.
Highton, P. J. (1969). An electron microscopic study of cell growth and mesosomal structure of *Bacillus licheniformis*. *Journal of Ultrastructure Research* **26**, 130–147.
Highton, P. J. (1970a). An electron microscopic study of the structure of mesosomal membranes in *Bacillus licheniformis*. *Journal of Ultrastructure* **31**, 247–259.
Highton, P. J. (1970b). An electron microscopic study of mesosomes in *Bacillus subtilis*. *Journal of Ultrastructure Research* **31**, 260–271.
Highton, P. J. and Hobbs, D. G. (1972). Penicillin and cell wall synthesis: a study of *Bacillus cereus* by electron microscopy. *Journal of Bacteriology* **109**, 1181–1190.
Hirsch, P. (1974). Budding bacteria. *Annual Review of Microbiology* **28**, 392–444.
Hoffman, H. and Frank, M. E. (1964). 'Germination tube' growth in *Escherichia coli* microcultures. *Journal of Bacteriology* **88**, 1151–1154.

Hopwood, D. A. (1960). Phase-contrast observations on *Streptomyces coelicolor. Journal of General Microbiology* **22**, 295–302.
Ishiguro, E. E. and Wolfe, R. S. (1970). Control of morphogenesis in *Geodermatophilus*: ultrastructural studies. *Journal of Bacteriology* **104**, 566–580.
Khan, S. R. (1975). Wall structure and germination of spores in *Cunninghamella echinulata. Journal of General Microbiology* **90**, 115–124.
Kalakoutskii, L. V. and Agre, N. S. (1976). Comparative aspects of development and differentiation in actinomycetes. *Bacteriological Reviews* **40**, 469–524.
Kalakoutskii, L. V. and Sokolov, A. A. (1961). Heterogeneity of the cell walls of *Actinomyces violaceus* aerial mycelium. *Mikrobiologiya* **30**, 60–65 (English translation).
Kannan, L. V. and Rehacek, Z. (1970). Formation of poly-β-hydroxybutyrate by actinomycetes. *Indian Journal of Biochemistry* **7**, 126–129.
Katz, L. N. (1963). The chemical nature of the mycelial membrane and spores of *Actinomyces aureofaciens. Mikrobiologiya* **32**, 392–396 (English translation).
Kawata, T. and Inoue, T. (1965). Ultrastructure of *Nocardia asteroides* as revealed by electron microscopy. *Japanese Journal of Microbiology* **9**, 101–114.
Kendrick, W. B. (ed.) (1971). *Taxonomy of Fungi Imperfecti.* University of Toronto Press, Toronto.
Klieneberger–Nobel, E. (1947). The life cycle of sporing actinomycetes as revealed by a study of their structure and septation. *Journal of General Microbiology* **1**, 22–32.
Kuimova, T. F. and Sokolov, A. A. (1977). Ultrastructural changes in the mycelium of *Actinomyces hygroscopicus* var. *enhygrus*, a producer of proteolytic enzyme in the process of submerged fermentation. *Mikrobiologiya* **46**, 129–133 (English translation).
Kuimova, T. F., Soina, V. S., Sokolov, A. A. and Artamonova, O. I. (1978). Electron microscopic structure of mycelium of *Actinomyces chrysomallus*, the producer of the antibiotic chrysomallin, during deep fermentation. *Mikrobiologiya* **47**, 601–606 (English translation).
Kuryłowicz, W. and Malinowski, K. (1972a). Changes in the mycelial ultrastructure of *Streptomyces aureofaciens* during tetracycline biosynthesis. *Mikrobiologiya* **41**, 621–627 (English translation).
Kuryłowicz, W. and Malinowski, K. (1972b). Ultrastructure of the mycelium of *Streptomyces aureofaciens* in the course of the biosynthesis of tetracycline. *Postepy Higieny i Medycyny Doświadczalnej* **26**, 563–569.
Langeron, M. (1965). *Outline of Mycology.* I. Pitman and Sons Ltd., London.
Lechevalier, H. A., Solotorovsky, M. and McDurmot, C. (1961). A new genus of the Actinomycetales: *Micropolyspora* gen. nov. *Journal of General Microbiology* **26**, 11–18.
Lechevalier, M. P., Lechevalier, H. A. and Heintz, E. E. (1973). Morphological and chemical nature of the sclerotia of *Chainia olivacea* Thirumalachar and Sukapure of the order Actinomycetales. *International Journal of Systematic Bacteriology* **23**, 157–170.
Locci, R. (1971). On the spore formation process in actinomycetes. IV. Examination by scanning electron microscopy of the genera *Thermoactinomyces*, *Actinobifida* and *Thermomonospora. Rivista di Patologia Vegetale, Supplemento* **7**, 63–80.
Locci, R. (1976). Developmental micromorphology of actinomycetes. In: *Actinomycetes: The Boundary Micro-organisms* (T. Arai, ed.), pp. 249–297. University Park Press, Baltimore, London and Tokyo.
Locci, R. (1978a). Micromorphological development of *Actinomyces* and of related genera. *Zentralblatt für Bakteriologie, Parasitenkunde, Infektionskrankheiten und Hygiene.* **1.** *Abteilung, Supplement* **6**, 173–180.
Locci, R. (1978b). Micromorphological assessment of lysozyme activity against actinomycetes. *Annali di Microbiologia* **28**, 63–71.
Locci, R. (1980a). Response of developing branched bacteria to adverse environments. I. Membrane-transfer techniques for assessment and SEM visualization of drug activity against *Nocardia asteroides. Zentralblatt für Bakteriologie, Parasitenkunde, Infektionskrankheiten und Hygiene, Abteilung 1, Originale* **246**, 98–111.
Locci, R. (1980b). Response of developing branched bacteria to adverse environments. II. Micromorphological effects of lysozyme on some aerobic actinomycetes. *Zentralblatt für Bakteriologie, Parasitenkunde, Infektionskrankheiten und Hygiene, Abteilung 1, Originale* **247**, 374–382.

Locci, R. (1981). Micromorphology and development of actinomycetes. *Zentralblatt für Bakteriologie, Mikrobiologie und Hygiene. 1. Abteilung, Supplement* **11**, 119–130.
Locci, R. and Baldan, B. P. (1970). Morphology and development of *Streptoverticillium* species as examined by scanning electron microscopy. *Giornale di Microbiologia* **18**, 69–76.
Locci, R. and Schaal, K. P. (1980). Apical growth in facultative anaerobic actinomycetes as determined by immunofluorescent labelling. *Zentralblatt für Bakteriologie, Parasitenkunde, Infektionskrankheiten und Hygiene, Abteilung 1, Originale* **246**, 112–118.
Locci, R. and Schofield, G. M. (1981). Aerial growth in members of the genus *Actinomyces*. *Annali di Microbiologia* **31**, 109–113.
Locci, R., Baldacci, R. and Baldan, B. P. (1969). The genus *Streptoverticillium*. A taxonomic study. *Giornale di Microbiologia* **17**, 1–60.
Locci, R., Goodfellow, M. and Pulverer, G. (1982). Micromorphological, morphogenetic and chemical characteristics of rhodococci. *Abstracts of the 5th International Symposium on Actinomycetes Biology*, pp. 118–119.
Ludvik, J., Mikulik, K. and Vanek, Z. (1971). Fine structure of *Streptomyces aureofaciens* producing tetracycline. *Folia Microbiologica* **16**, 479–480.
Luedemann, G. M. and Casmer, C. J. (1973). Electron microscope study of whole mounts and thin sections of *Micromonospora chalcea* ATCC 12452. *International Journal of Systematic Bacteriology* **23**, 243–255.
Lyons, A. J. and Pridham, T. G. (1971). *Streptomyces torulosus* sp.n. an unusual knobby-spored taxon. *Applied Microbiology* **22**, 190–193.
McVittie, A. (1974). Ultrastructural studies on sporulation in wild-type and white colony mutants of *Streptomyces coelicolor*. *Journal of General Microbiology* **81**, 291–302.
McVittie, A., Wildermuth, H. and Hopwood, D. A. (1972). Fine structure and surface topography of endospores of *Thermoactinomyces vulgaris*. *Journal of General Microbiology* **71**, 367–381.
Manzanal, M. B. and Hardisson, C. (1978). Early stages of arthrospore maturation in *Streptomyces*. *Journal of Bacteriology* **133**, 293–297.
Manzanal, M. B., Braña, A. F. and Hardisson, C. (1981). Ultrastructural changes during the *Streptomyces* cell cycle. The occurrence of polysaccharide granules during growth and sporulation of *Streptomyces*. *Zentralblatt für Bakteriologie, Mikrobiologie und Hygiene. 1. Abteilung, Supplement* **11**, 147–152.
Martin, J. F. and Demain, A. L. (1980). Control of antibiotic biosynthesis. *Microbiological Reviews* **44**, 230–251.
Matselyukh, B. P. (1978). Ultrastructure of hairs on the spore surface of *Streptomyces olivaceus* VKX. *Zentralblatt für Bakteriologie, Parasitenkunde, Infektionskrankheiten und Hygiene. 1. Abteilung, Supplement* **6**, 235–260.
May, J. W. (1962). Sites of cell wall extension demonstrated by the use of fluorescent antibody. *Experimental Cell Research* **27**, 170–172.
May, J. W. (1963). The distribution of cell-wall label during growth and division of *Salmonella typhimurium*. *Experimental Cell Research* **31**, 217–220.
Nanninga, N. (1971). The mesosomes of *Bacillus subtilis* as affected by chemical and physical fixation. *Journal of Cell Biology* **48**, 219–224.
Naumova, I. B., Zaretskaya, M. S., Dmitrieva, M. F. and Streshinskaya, G. M. (1978). Structural features of teichoic acids of certain *Streptomyces* species. *Zentralblatt für Bakteriologie, Parasitenkunde, Infektionskrankheiten und Hygiene. 1. Abteilung, Supplement* **6**, 261–268.
Nermut, M. V. (1967). The ultrastructure of the cell wall of *Bacillus megaterium*. *Journal of General Microbiology* **49**, 503–512.
Nikitina, E. T. and Kalakoutskii, L. V. (1971). Induction of nocardioform growth in actinomycetes on media with D-fructose. *Zeitschrift für Allgemeine Mikrobiologie* **11**, 601–606.
Nikitina, E. T., Kasakova, G. G. and Kalakoutskii, L. V. (1971). Induction of unusual development in *Actinomyces roseoflavus* var. *roseofungini* on media with fructose. *Doklady Akademii Nauk SSSR* **196**, 448–450.
Parish, J. H. (ed.) (1979). *Developmental Biology of Prokaryotes*. Blackwell Scientific Publications, Oxford.
Pogell, B. M. (1979). Regulation of aerial mycelium formation in streptomycetes. In: *Genetics of Industrial Micro-organisms* (O. K. Sebek and A. I. Laskin, eds.), pp. 218–224. American Society of Microbiology, Washington.

Pouzharitskaja, L. M., Taptykova, S. D., Sarkisyan, S. T., Tulskii, S. V., Cherny, N. E. and Kalakoutskii, L. V. (1974). Composition of extracellular tubular structures produced by 'fructose' variant of *Actinomyces roseoflavus* var. *roseofungini*. *Antibiotiki* **19**, 963–966.
Prauser, H. (1976). New nocardioform organisms and their relationship. In: *Actinomycetes: The Boundary Micro-organisms* (T. Arai, ed.), pp. 193–207. University Park Press, Baltimore, London and Tokyo.
Prauser, H. (1978). Considerations on taxonomic relations among Gram-positive, branching bacteria. *Zentralblatt für Bakteriologie, Parasitenkunde, Infektionskrankheiten und Hygiene*, 1, *Abteilung, Supplement* **6**, 3–12.
Prauser, H., Lechevalier, M. P. and Lechevalier, H. (1970). Description of *Oerskovia* gen. n. to harbor Ørskov's motile *Nocardia*. *Applied Microbiology* **19**, 534.
Rancourt, M. W. and Lechevalier, H. A. (1963). Electron microscopic observations of the sporangial structure of an actinomycete, *Microellobospora flavea*. *Journal of General Microbiology* **31**, 495–499.
Rancourt, M. W. and Lechevalier, H. A. (1964). Electron microscopic study of the formation of spiny conidia in species of *Streptomyces*. *Canadian Journal of Microbiology* **10**, 311–316.
Reusch, V. M. and Burger, M. M. (1973). The bacterial mesosome. *Biochimica et Biophysica Acta* **300**, 79–104.
Richmond, D. V. (1975). Effects of toxicants on the morphology and fine structure of fungi. *Advances in Applied Microbiology* **19**, 289–319.
Rogers, H. J. (1970). Bacterial growth and the cell envelope. *Bacteriological Reviews* **34**, 194–214.
Rogers, H. J., Ward, J. B. and Burdett, I. D. J. (1978). Structure and growth of the walls of Gram-positive bacteria. In: *Relations between Structure and Function in the Prokaryotic Cell* (R. Y. Stanier, H. J. Rogers and J. B. Ward, eds.), pp. 139–176. Cambridge University Press, Cambridge.
Salton, M. R. J. and Owen P. (1976). Bacterial membrane structure. *Annual Review of Microbiology* **30**, 451–482.
Santos-Mota, J., Silva, M. T. and Guerra, F. C. (1972). Variations in the membranes of *Streptococcus faecalis* related to different culture conditions. *Archiv für Mikrobiologie* **83**, 293–302.
Saunders, P. T. and Trinci, A. P. J. (1979). Determination of tip shape in fungal hyphae. *Journal of General Microbiology* **110**, 469–473.
Schofield, G. M. and Locci, R. (1981). Micromorphological effects of lysozyme and of penicillin on microaerophilic actinomycetes. *Annali di Microbiologia* **31**, 61–65.
Schuhmann, E. and Bergter, F. (1976). Mikroskopische Untersuchungen zur Wachstuminetik von *Streptomyces hygroscopicus*. *Zeitschrift für Allgemeine Mikrobiologie* **16**, 201–215.
Sharples, G. P. and Williams, S. T. (1976a). Development and fine structure of sclerotia and spores of the actinomycete *Chainia olivacea*. *Microbios* **15**, 37–47.
Sharples, G. P. and Williams, S. T. (1976b). Fine structure of spore germination in actinomycetes. *Journal of General Microbiology* **96**, 323–332.
Sharples, G. P., Williams, S. T. and Bradshaw, R. M. (1974). Spore formation in the *Actinoplanaceae* (*Actinomycetales*). *Archiv für Mikrobiologie* **101**, 9–20.
Shirling, E. B. and Gottlieb, D. (1976). Retrospective evaluation of International *Streptomyces* Project taxonomic criteria. In: *Actinomycetes: The Boundary Micro-organisms* (T. Arai, ed.), pp. 9–41. University Park Press, Baltimore, London and Tokyo.
Shively, J. M. (1974). Inclusion bodies of prokaryotes. *Annual Review of Microbiology* **28**, 167–187.
Silva, M. T. (1971). Changes in the ultrastructure of the cytoplasmic and intracytoplasmic membranes of several Gram-positive bacteria by variations in OsO_4 fixation. *Journal de Microscopie (Paris)* **93**, 227–232.
Silva, M. T., Sousa, J. C. F. Macedo, M. A. E., Polónia, J. J. and Parente, A. M. (1976a). Effects of phenethyl alcohol on *Bacillus* and *Streptococcus*. *Journal of Bacteriology* **127**, 1359–1369.
Silva, M. T., Sousa, J. C. F., Polónia, J. J., Macedo, M. A. E. and Parente, A. M. (1976b). Bacterial mesosomes: real structures or artifacts? *Biochemica et Biophisica Acta* **443**, 92–105.
Silva, M. T., Sousa, J. C. F., Polónia, J. J. and Macedo, P. M. (1979). Effects of local anaesthetics on bacterial cells. *Journal of Bacteriology* **137**, 461–468.

Smucker, R. A. and Pfister, R. M. (1978). Characteristics of *Streptomyces coelicolor* A3(2) aerial spore rodlet mosaic. *Canadian Journal of Microbiology* **24**, 397–408.
Starr, M. P. and Skerman, V. B. D. (1965). Bacterial diversity: the natural history of selected morphologically unusual bacteria. *Annual Review of Microbiology* **19**, 407–454.
Stevens, R. T. (1975). Fine structure of sporogenesis and septum formation in *Micromonospora globosa* Kriss and *M. fusca* Jensen. *Canadian Journal of Microbiology* **21**, 1081–1088.
Strunk, C. (1978). Sporogenesis in *Streptomyces melanochromogenes*. *Archives of Microbiology* **118**, 309–316.
Suganuma, A. and Morioka, H. (1979). Morphological changes in membrane systems of staphylococci after different fixation procedures. *Journal of Electron Microscopy* **28**, 29–35.
Szabó, I. M. (1977). Are the morphogenetic changes in the sheath of sporulating hyphae of streptomycetes regulated by the tip-cell unit? *Actinomycetes* **12**, 21–27.
Szabó, I. M., Kondics, L., Marton, M. and Buti, I. (1977). Changes in the surface layer (sheath) of the cell wall of streptomycetes during sporulation. *Acta Microbiologica Academiae Scientiarum Hungaricae* **24**, 237–246.
Szabó, I. M., Pártay, G. and Szijártó, J. (1979). Types of the "irregular" ornamentation of the surface sheath of streptomycete spores and aerial hyphae. *Acta Microbiologica Academiae Scientiarum Hungaricae* **26**, 245–253.
Takeo, K. (1976). Existence of a surface configuration on the aerial spore and aerial mycelium of *Micropolyspora*. *Journal of General Microbiology* **95**, 17–26.
Thirumalachar, M. J. (1955). *Chainia*, a new genus of the *Actinomycetales*. *Nature (London)* **176**, 934–935.
Thirumalachar, M. J. (1970). Evaluation of some characters used in the taxonomy of *Actinomycetales*. In: *The Actinomycetales* (H. Prauser, ed.), pp. 425–429. Gustav Fischer Verlag, Jena.
Tresner, H. D., Davies, M. C. and Backus, E. J. (1961). Electron microscopy of *Streptomyces* spore morphology and its role in species differentiation. *Journal of Bacteriology* **81**, 70–80.
Tsien, H. C., Shockman, G. D. and Higgins, M. L. (1978). Structural arrangement of polymers within the wall of *Streptococcus faecalis*. *Journal of Bacteriology* **133**, 372–386.
Weibull, C. (1973). Electron microscope studies on aldehyde-fixed, unstained microbial cells. *Journal of Ultrastructure Research* **43**, 150–159.
Wildermuth, H. (1970a). Surface structure of streptomycete spores as revealed by negative staining and freeze-etching. *Journal of Bacteriology* **101**, 318–322.
Wildermuth, H. (1970b). Development and organisation of the aerial mycelium in *Streptomyces coelicolor*. *Journal of General Microbiology* **60**, 43–50.
Wildermuth, H. (1971). The fine structure of mesosomes and plasma membrane in *Streptomyces coelicolor*. *Journal of General Microbiology* **68**, 53–63.
Wildermuth, H. (1972a). The surface structure of spores and aerial hyphae in *Streptomyces viridochromogenes*. *Archiv für Mikrobiologie* **81**, 309–320.
Wildermuth, H. (1972b). Morphological surface characteristics of *Streptomyces glaucescens* and *S. acrimycini*, two streptomycetes with hairy spores. *Archiv für Mikrobiologie* **81**, 321–332.
Wildermuth, H. and Hopwood, D. A. (1970). Septation during sporulation in *Streptomyces coelicolor*. *Journal of General Microbiology* **60**, 51–59.
Wildermuth, H., Wehrli, E. and Horne, R. W. (1971). The surface structure of spores and aerial mycelium in *Streptomyces coelicolor*. *Journal of Ultrastructure Research* **35**, 168–180.
Williams, S. T. (1970). Further investigations of actinomycetes by scanning electron microscopy. *Journal of General Microbiology* **62**, 67–73.
Williams, S. T. and Sharples, G. P. (1970). A comparative study of spore formation in two *Streptomyces* species. *Microbios* **2**, 17–26.
Williams, S. T. and Wellington, E. M. H. (1980). Micromorphology and fine structure of actinomycetes. In: *Microbiological Classification and Identification* (M. Goodfellow and R. G. Board, eds.), pp. 139–165. Academic Press, London.
Williams, S. T., Bradshaw, R. M., Costerton, J. W. and Forge, A. (1972). Fine structure of the spore sheath of some *Streptomyces* species. *Journal of General Microbiology* **72**, 249–258.
Williams, S. T., Sharples, G. P. and Bradshaw, R. M. (1973). The fine structure of the *Actinomycetales*. in: *Actinomycetales: Characteristics and Practical Importance* (G. Sykes and F. A. Skinner, eds.), pp. 113–136. Academic Press, London.

Williams, S. T., Sharples, G. P. and Bradshaw, R. M. (1974). Spore formation in *Actinomadura dassonvillei* (Brocq-Rousseu) Lechevalier and Lechevalier. *Journal of General Microbiology* **84**, 415–419.

Williams, S. T., Sharples, G. P., Serrano, J. A., Serrano, A. A. and Lacey, J. (1976). The micromorphology and fine structure of nocardioform organisms. In: *The Biology of the Nocardiae* (M. Goodfellow, G. H. Brownell and J. A. Serrano, eds.), pp. 102–140. Academic Press, London.

Zaslavskaya, P. L. and Zhukov, V. G. (1976). Morphological and functional changes in the mycelium of active and inactive variants of *Actinomyces parvullus* during growth and biosynthesis of actinomycin D. *Mikrobiologiya* **45**, 885–890 (English translation).

Zaslavskaya, P. L., Makarevich, V. G. and Slugina, M. D. (1977). Morphological study of the development of *Actinomyces aureofaciens* under conditions of controlled and uncontrolled fermentation. *Mikrobiologiya* **46**, 233–238 (English translation).

Zaskavskaya, P. L., Zhukov, V. G., Kornitskaya, E. Y., Tovarova, I. I. and Khokhlov, A. S. (1979). Influences of A-factor on the ultrastructure of the A-factor deficient mutant of *Streptomyces griseus*. *Microbios* **25**, 145–153.

4
Genetics of the Nocardioform Bacteria

G. H. BROWNELL* and KATHERINE DENNISTON†

*Department of Cell and Molecular Biology, Medical College of Georgia, Augusta, Georgia 30912, USA, and †Division of Molecular Virology and Immunology, Georgetown University, Rockville, Maryland 20852, USA.

1. Introduction	201
2. Nocardia	203
A. Nocardia asteroides	203
B. 'Nocardia' mediterranea	204
C. Other nocardial species	207
3. Rhodococcus	210
A. Mating factors and mechanisms	211
B. Linkage models	212
C. Lysogeny and prophage inheritance	215
D. Protoplast formation and transfection	216
E. Phage ϕEC	216
4. Mycobacterium	218
A. Genetic transformation	218
B. Transduction	219
C. Conjugation	220
D. Extrachromosomal elements	222
E. Lysogeny and phage conversion	223
5. Conclusions	223
References	224

1. Introduction

Nocardioform actinomycetes are a group of Gram-positive, non-spore forming organisms that delay cell division and, as a consequence, produce elongated or filamentous structures that ultimately undergo fragmentation into cocci or rod-shaped forms. Although all members of the group show some degree of acid-fastness, depending on age and staining technique, it is basically the reproductive characteristics that define these organisms. The term nocardioform was first coined by Prauser (1967) and was readily accepted because it allowed reference to a group of organisms that cannot be assigned to a single genus. Classification of these organisms remains difficult but has been improved by the application of modern taxonomic techniques especially numerical taxonomy (Chapter 2). Nocardioform bacteria can usually be distinguished from sporoactinomycetes even though some of the latter may show some fragmentation of the vegetative mycelium. From a genetical standpoint the grouping of actinomycetes based on their nocardioform characteristics may be more significant than that of their genus descriptions since genetic recombination is known to occur between genera (Vezina, 1970; Adams and Brownell, 1976; Brownell, 1978a) and because assignment of organisms to a genus is subject to change as biochemical and genetic information accumulates.

The genetic potential of nocardioform actinomycetes has still to be adequately investigated. The lack of fundamental knowledge regarding the genetics of these bacteria may be attributed, in part, to the difficulties found in manipulating them for genetic studies. When compared to the genetically well-characterized Eubacteriales, nocardioform actinomycetes are more difficult to study because of their slow growth, tendency to clump and ability to form coenocytic structures. Further, many investigators have made little effort to follow through their primary observations. Examples of phenomena that have not been characterized beyond the initial observations abound in the literature and, as a consequence, the basic knowledge needed to stimulate interest in these organisms has not accumulated.

It is premature to evaluate the impact of the relatively new techniques of gene cloning and recombination by protoplast fusion (Hopwood *et al.*, 1977; Baltz, 1978) on genetic exploration of the nocardioform actinomycetes for no reports of such studies have been made at the time of this review. However, it seems likely that an interest in these biochemically diverse organisms will soon develop, particularly in light of reports that *Esch. coli* cannot express certain traits cloned from actinomycetes (Horinouchi *et al.*, 1980). This observation should encourage the development of cloning systems in actinomycetes.

Only a few species of *Nocardia*, *Rhodococcus* and *Mycobacterium* have been sufficiently studied to warrant review. There is some information regarding genetic interaction within the type species, *Nocardia asteroides*, as well as within species currently labelled '*Nrda*'. *mediterranea* and '*Nrda*'. *opaca*. The genus *Rhodococcus* currently contains 12 species and accommodates strains which were formerly assigned to a multiplicity of taxa (Chapter 2; Bousfield and Goodfellow, 1976). Our scant knowledge of the genus *Mycobacterium* is particularly surprising as this taxon contains important human pathogens. With the exception of a few drug resistance studies, our understanding of genetic interactions within this genus comes exclusively from studies on *Mybc. smegmatis*. There is little or no genetic information on the important pathogens, *Mybc. intracellulare*, *Mybc. kansasii*, *Mybc. leprae*, *Mybc. scrofulaceum*, or on the *Mybc. tuberculosis* complex.

2. Nocardia

Genetic recombination has been demonstrated in *Nrda. asteroides*, '*Nrda*'. *mediterranea*, '*Nrda. opaca*' and '*Nrda*'. *restricta*. A linkage model is available only for '*Nrda*'. *mediterranea*. '*Nocardia*' *mediterranea* was originally classified as *Streptomyces mediterranei* (Margalith and Beretta, 1960), was reclassified as *Nocardia* (Thiemann et al., 1969) but can readily be distinguished from both *Nocardia* and *Streptomyces* as it forms a taxon in its own right (Chapter 2; Alderson et al., 1981). Nevertheless, the genetic behaviour of '*Nrda*'. *mediterranea* and the arrangement of genes on the chromosome are testimony to a relatively close relationship to *Streptomyces* species (Schupp et al., 1975). Information regarding the genetics of '*Nrda*'. *restricta* is limited though it is known that mutant substrains of this species will recombine with one of the *Rodc. erythropolis* mating types (Vezina, 1970). Indeed, strains labelled '*Nrda*'. *restricta* were classified in the genus *Rhodococcus* by Goodfellow and Alderson (1977). Similarly, '*Nrda. opaca*' strains have been found to have many properties in common with rhodococci (Gordon and Mihm, 1957; Alshamaony et al., 1976). Genetic analysis of *Nrda. asteroides* has just begun, but it is clear that this taxon accommodates a very heterogeneous group of actinomycetes that can be classified into subgroups on the basis of biochemical, genetical, physiological and serological properties (Chapter 2).

There have been a few reports of lysogenic *Nocardia* and temperate nocardiophage (Prauser, 1976; Andrzejewski et al., 1978) but there is no evidence for transduction in the genus. However, since many species seem to carry prophage, transduction would seem to be a likely mode of genetic exchange in this genus. To date, a lack of genetically marked strains has precluded a satisfactory analysis of this phenomenon. Similarly, there have been no reports to indicate whether genetic exchange by DNA-mediated transformation occurs in the genus *Nocardia*. Efforts have been made to develop a phage system for classification but a practical system for phage typing has yet to emerge (Prauser, 1976).

A. *Nocardia asteroides*

The discovery of genetic recombination between strains designated *Nrda. asteroides* was made by Kasweck and his colleagues (Kasweck et al., 1981; Kasweck and Little, 1982). These workers generated a variety of auxotrophic mutants with both ultraviolet irradiation and *N*-methyl-*N'*-nitronitrosoguanidine and tested for genetic compatibility by mixed growth on trypticase-peptone agar containing 1%(w/v) glucose. The study utilized strains from diverse origins as well as from different subgroups (Table 1) defined in a numerical phenetic survey (Schaal and Reutersberg, 1978). Two major subgroups, designated A and B, were defined at the 65% similarity level using the simple matching coefficient. Kasweck *et al.* (1981) selected strains from subgroups A and B and did test crosses with a variety of auxotrophic mutants of each strain. All the mutants showed interstrain fertility (Table 2).

Thus, it appears that interstrain compatibility like that observed in the genus *Mycobacterium* (Mizuguchi *et al.*, 1976) exists in the case of *Nrda. asteroides*. The fertility pattern between the mutant strains is shown in Fig. 1; strains yielding fertile crosses are connected by solid lines.

Crosses between *Nrda. asteroides* N66 (*his-3, leu-1*) and N135 (*met-3, phe-3*), both from group B but from different origins, were analysed for the recombinant recovery frequencies of the nine possible recombinant class types. None of the possible twelve linear or three circular permutations for the four genes fit the observed recombinant recovery frequencies (Kasweck *et al.*, 1981). Evolutionary divergence of regions of the parental genomes might explain segregation data that does not conform to anticipated events. Thus, linkage models based on recombination frequencies between compatible *Nrda. asteroides* strains are likely to show abnormalities similar to those found in the *Rhodococcus* system. A tentative linkage model showing the least ambiguous gene array is shown in Fig. 2. In this model the *his-2, trp-2* and *bio-2* genes may be at either the 23 or 79 position on the map (Kasweck *et al.*, 1981).

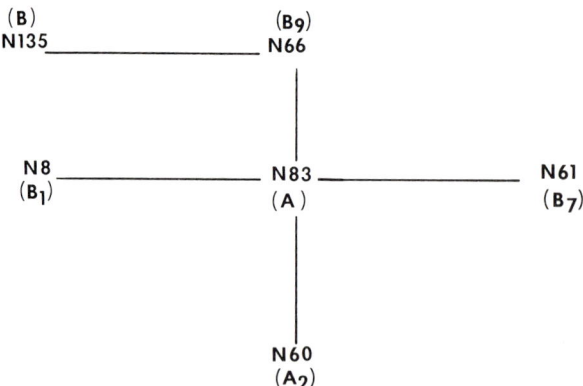

FIG. 1. Compatible mating combinations of strains designated *Nocardia asteroides*. The lines connect strains shown to produce recombinant progeny. The letter in parentheses indicates the proposed subgroup of the strains.

B. '*Nocardia*' *mediterranea*

A body of information exists on the genetic nature of the rifamycin producing strain '*Nrda*'. *mediterranea* ATCC 13685 thanks to the efforts of Schupp

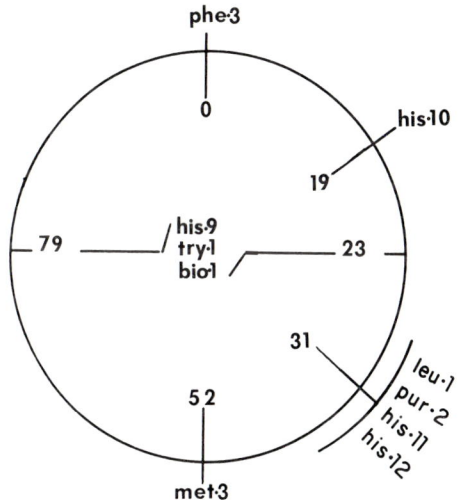

FIG. 2. A preliminary linkage model constructed from matings of *Nocardia asteroides* subgroup B strains. The genes depicted in the centre of the circle have two alternative map locations.

TABLE 1. Designation and origin of compatible *Nocardia asteroides*

Strain[a] designation	Classification	Type[b]	Source and other designations
N8 (KK1)	*Nocardia asteroides*	B_1	Hygiene Institute, Cologne, strain Rostock
N60 (KK2)	*Nrda. asteroides*	A_2	M. Tsukamura, *Nrda. asteroides* 23046 = M-129
N61 (KK3)	*Nrda. asteroides*	B_7	M. Tsukamura, *Nrda. farcinica* 23061 = M-133
N66 (KK4)	*Nrda. asteroides*	B_9	ATCC 3318, *Nrda. farcinica*
N83 (KK5)	*Nrda. asteroides*	A	J. Andrzejewski, *Nrda. asteroides* IMRU W3599
N135 (KK6)	*Nrda. asteroides*	B	Hygiene Institute, Cologne, B791

[a] *Nrda. asteroides* strains N8–N135 were redesignated KK1 through KK6 for genetic studies (Kasweck and Little, 1982).
[b] Subgroups proposed by Schaal and Reutersberg (1978).

and his colleagues (Schupp *et al.*, 1975; Schupp and Nuesch, 1979). '*Nocardia*' *mediterranea* is a self-fertile organism which yields recombinants at high frequencies (1×10^{-4} to 1×10^{-5}). The techniques used for linkage analysis in *Stmy. coelicolor* (Hopwood, 1967) were found to be suitable for

TABLE 2. Recombinant recovery frequency of crosses of compatible *Nocardia asteroides* mutant substrains

Substrain		Recombinant recovery frequency	loci	reversion frequencies
N8-2 (*his-1*)	×N83-4 (*ala-3*)	3.5×10^{-7}	*his-1*	1.8×10^{-8}
	×N83-5 (*his-15*)	1.6×10^{-5}	*ala-3*	7.1×10^{-7}
N60-1 (*his-4*)	×W83-3 (*leu-10*)	6.8×10^{-7}	*his-15*	9.5×10^{-8}
	×W83-6 (*leu-11*)	1.3×10^{-6}	*his-4*	3.3×10^{-8}
N61-2 (*ala-2*)	×N83-6 (*leu-3*)	3.9×10^{-7}	*ala-2*	1.0×10^{-6}
	×N83-7 (*leu-4*)	1.5×10^{-6}	*leu-10*	7.2×10^{-8}
N66-13 (*ile-1*)	×N135-2 (*phe-1*)	9.9×10^{-5}	*leu-11*	4.4×10^{-7}
N66-16 (*leu-1*)	×N135-1 (*met-1*)	1.9×10^{-4}	*ile-1*	5.9×10^{-6}
	×N135-2 (*phe-1*)	1.9×10^{-4}	*phe-1*	9.5×10^{-9}
	×N135-6 (*met-3*)	3.6×10^{-4}	*leu-1*	1.3×10^{-9}
N66-43 (*his-3*)	×N135-2 (*phe-1*)	4.6×10^{-6}	*met-1*	5.9×10^{-8}
	×N135-4 (*met-1*)	5.4×10^{-4}	*met-3*	3.0×10^{-8}
	×N135-6 (*met-3*)	2.6×10^{-5}	*his-3*	2.5×10^{-7}

construction of linkage models of '*Nrda*'. *mediterranea* (Fig. 3). A comparison of the linkage models of '*Nrda*'. *mediterranea* and *Stmy. coelicolor* A3 showed a remarkable similarity in the relative locations of auxotrophic traits (Schupp *et al.*, 1975). An analysis of unselected traits indicated that the chromosome of '*Nrda*'. *mediterranea* contains regions with negative interference, a behaviour consistent with the formation of incomplete zygotes from matings of isogenic strains.

Both parents can serve as donors of genetic material and, because of the nature of the recombinant types produced, it appears that a rather large portion of the chromosome may be transferred to a recipient strain. No fertility factors of either the plasmid-related type in *Stmy. coelicolor* or of chromosomal origin like *Rodc. erythropolis* have been found in '*Nrda*'. *mediterranea*.

As with other nocardioform actinomycetes, many of the colony-forming units (c.f.u.) from crosses involving '*Nrda*'. *mediterranea* are multinucleate and possibly some are coenocytic fragments. However, there are apparently no ambiguities in the segregation data which might be expected to occur as a result of additional recombination events associated with multiple mating rounds.

Preliminary studies on the genes responsible for rifamycin biosynthesis by '*Nrda*'. *mediterranea* have shown that the enzyme catalysing the final step in rifamycin B synthesis is encoded by a chromosomal locus (Schupp and Nuesch, 1979). Six mutants in rifamycin synthesis (*ans-1* to *ans-6*) were isolated and mutant loci mapped between *str-2* and *pro-1* on the '*Nrda*'. *mediterranea* linkage map (Fig. 3). In the absence of a gene complementation system, the investigators were unable to determine whether or not the six markers were allelic. They were, however, able to rule out plasmid involvement in this step in the synthesis of rifamycin.

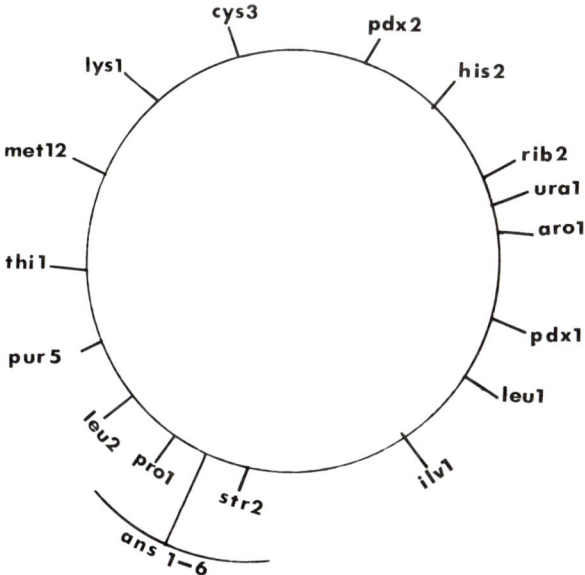

FIG. 3. A linkage model produced from mating of mutant strains of '*Nocardia*' *mediterranea* ATCC 13685.

C. OTHER NOCARDIAL SPECIES

Genetic recombination between mutant strains of '*Nrda. opaca*' CBS 332.61 and '*Nrda*'. *restricta* B-226 has been reported, although there are few facts regarding the phenomenon. While matings between multiple auxotrophic strains of '*Nrda. opaca*' were found to be fertile and to yield recombinants at frequencies ranging from 1.2×10^{-5} to 2.5×10^{-7} (Krallmann-Wenzel and Tárnok, 1978: Krallmann-Wenzel, 1981) no linkage map has been constructed for this species.

As in the case of '*Nrda*'. *mediterranea* there is no evidence for fertility factors and it is not known whether these self-compatible strains will recombine with other related nocardioform actinomycetes.

A transferable plasmid has been described in '*Nrda. opaca*' strain 1b (Reh, 1981; Reh and Schlegel, 1981). The plasmid bears traits allowing chemolithoautotrophic growth (aut^+) which includes genes for hydrogenase, phosphoribulokinase and ribulosebisphosphate carboxylase production.

The evidence for the existence of this extrachromosomal element lies in its spontaneous, irreversible loss at high frequency (0.3%) and in the fact that it can be transferred to other strains independent of any chromosomal traits. In matings between aut^+ and aut^- strains, the aut^- parental type strains are converted into the aut^+ genotype at a frequency of about 1×10^{-3} with no apparent reassortment of any chromosomal genes. Transfer of the plasmid is believed to be by a conjugal mechanism. Optimum transfer occurs during growth on enriched culture medium and between older cells.

The 'Nrda. opaca' 1b strain was found to be able to transfer the aut^+ trait to other strains of opaca ('Nrda. opaca' ATCC 17039) as well as to Rodc. erythropolis (Reh, 1981; Reh and Schlegel, 1981). However, spontaneous loss of the plasmid encoded traits was very high (20–100%) in all the recipient strains except those derived from the 'Nrda. opaca 1b'. These recipient strains also prove to be very poor donor strains. For example, when Rodc. erythropolis (aut^+) was tested for its ability to transfer the plasmid, it was found to be unable to do so with one exception. Rhodococcus erythropolis MR1621 did transfer the aut^+ trait to 'Cnbc. hydrocarboclastus' MR171 at low frequency (Table 3).

'Nocardia opaca' MR11 appears to be a very effective donor strain with the exception of the 'Nrda. opaca' ATCC 17039 recipient. The difference in the frequency of spontaneous loss of the plasmid in the transconjugants (19.8% for MR1621 and 29.4% for MR151 compared to 0.3% for the MR11 strain) is no doubt a contributory factor to transfer efficiency, however, it does not account for the difference in MR11 and MR151 to service as donors to MR152 strains. Thus, it may be that there is a compatibility combination for donor–recipient strains similar to fertility in matings involving recombination between chromosomal traits.

It was disappointing that attempts to demonstrate a difference in plasmid profiles between aut^+ and aut^- strains failed (Reh, 1981). Three plasmids, 11, 22 and 90 megadaltons, were found in both aut^- and aut^+ variants of 'Nrda. opaca' 1b as well as a single 90 megadalton plasmid in both the Rodc. erythropolis Ce3 aut^- wild-type and the aut^+ transconjugants.

Genetic studies were initiated on 'Nrda'. restricta by Vezina (1970) primarily because this organism produces 1-dehydrogenase which causes the aromatization of various 19-substituted steroids such as equilin (Chapter 7). Vezina demonstrated that mutants of 'Nrda'. restricta were self-infertile but would recombine with those generated from the Rodc. erythropolis mat-cE mating type strains. The 'Nrda'. restricta mutants failed to recombine with those originating from the Rodc. erythropolis mat-cE strains which are also self-incompatible. These observations suggest that 'Nrda'. restricta may possess a mat-cE like fertility factor(s). Analysis of unselected markers from matings of 'Nrda'. restricta and Rodc. erythropolis has generated the linear model shown in Fig. 4 (C. Vezina, personal communication).

TABLE 3. The interstrain transfer of autotrophic growth (aut$^+$) from 'Nocardia opaca' 1b and other transconjugants[a]

Species	Recipient strain	Donor strains[b]			
		Nrda. opaca MR11 (0.3)[c]	Nrda. opaca MR1863	Rodc. erythropolis MR1621 (19.8)[c]	Nrda. opaca MR151 (29.4)[c]
'Nrda. opaca' 1b	MR1450	+	–	–	+
'Nrda. opaca' 1b	MR1463	++	+	–	–
Rodc. erythropolis Ce3	MR1235	++	–	–	–
Rodc. erythropolis cE2	MR132	++	–	–	–
'Cnbc. hydrocarboclastus'	MR171	++	–	+	–
'Nrda. opaca' ATCC 17039	MR152	–	+	–	++

[a] Adapted from Reh and Schlegel (1981).
[b] The donor strains were obtained by agar matings with 'Nrda. opaca' MR11 and selected by their ability to grow on MM agar in a hydrogen/oxygen/carbon dioxide atmosphere. +–, no transfer; +, 1–20 colonies; ++, confluent growth.
[c] Percentage of spontaneous loss of the aut$^+$ genotype in donor strains.

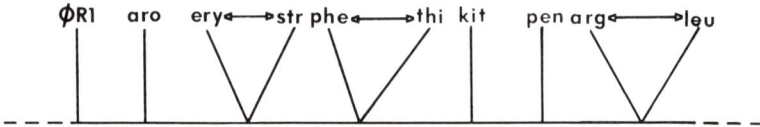

FIG. 4. A preliminary linkage model constructed from mating of '*Nocardia*' *restricta* and *Rhodococcus erythropolis mat-Ce* mating type strains. The arrows connect closely linked loci that may have reversed linear orders (*aro*, aromatization, *kit*, kitasamycin resistant, $\phi R1$, nocardiophage R1).

3. Rhodococcus

From a genetic aspect, one of the most extensively studied nocardioform actinomycetes is *Rodc. erythropolis*. Recombination in this organism was first reported by Adams and Bradley in 1963. *Rhodococcus erythropolis* is one of the ten species of *Rhodococcus* cited on the Approved Lists of Bacterial Names (Skerman et al., 1980). The *Rodc. erythropolis* mating system is an interstrain recombination event between organisms once considered as distinct species. Strains belonging to Group A ('*Jensenia canicruria*', ATCC 11048; '*Nrda. canicruria*', ATCC 17896) will recombine with organisms belonging to Group B (*Nrda. erythropolis*, ATCC 4277 and 17895; '*Nrda. opaca*' 765A and '*Nrda*'. *restricta* B-311; Brownell, 1978a). Several of these strains, together with the type strains of *Nrda. calcarea* (Metcalf and Brown, 1957) and *Nrda. globerula* (Gray, 1928; Waksman and Henrici, 1948), have been reduced to synonyms of *Rodc. erythropolis* (Goodfellow and Alderson, 1977; Goodfellow et al., 1982). However, mutants derived from cultures of the latter two strains do not reassort chromosomal genes with either group A or group B and they are also self-compatible (G. H. Brownell, unpublished results). They can, however, serve as recipients of an actinophage, ϕEC, found in lysogenic group A strains that is probably transferred by a conjugal mechanism (Brownell, 1978a, b). Thus, it is important that the original designations of these strains are not lost.

A short-coming in the *Rhodococcus* mating system is that the parental strains are of heterologous origin and evolutionary divergencies are reflected in the genetic data. This makes the system much less appealing than either the Eubacteriales or *Streptomyces* for studying fundamental genetic principles. In addition, there are still unanswered questions regarding the basic nature of this mating system including the mechanism(s) of recombination and the role of the mating factor(s). The absence of this type of information further discourages investigators from utilizing the *Rhodococcus* system as a genetic tool.

On the other hand, the *Rhodococcus* system can be genetically manipulated and is currently the best choice from the nocardioform actinomycetes for a gene cloning system. Temperate phages are available to serve as cloning vectors (Brownell et al., 1980) as well as mutant host strains that could satisfy most of the containment criteria. Such a cloning system could lend itself to studies of leprosy, nocardiosis and tuberculosis, by providing a vehicle for elucidating mechanisms of resistance, biochemical dependences on the host and possibly for producing antigens which could be used for immunization and epidemiology.

A. MATING FACTORS AND MECHANISMS

There are at least two gene products that are known to be required for mating and subsequent recombinant recovery between compatible *Rhodococcus* strains. These genes, *mat-E* and *mat-C* (E representing *erythropolis* and C *canicruria* since recombination was originally observed between strains designated *Nrda. erythropolis* and '*Nrda. canicruria*'), have been mapped on the *Rhodococcus* chromosome (Fig. 5). A fertile combination is one in which one parental type possesses the *mat-C* gene (group A strains) and the other the *mat-E* gene (group B strains). Matings can result in progeny that cross with both group A and B strains (*mat-CE* recombinants) or with neither (*mat-ce* recombinants). The production of these recombinants led investigators to speculate that *mat-c* and *mat-e* were alleles to the *mat-C* and *mat-E* genes respectively and, if these genes were inherited, the strains would lose the phenotype necessary for mating (Brownell and Kelly, 1969). On these criteria, the *mat-CE* recombinants should be self-fertile and circumvent the problems of strain heterology. Genetic analysis, however, has shown the *mat-CE* recombinants revert to either *mat-C* or *mat-E* mating behaviour when subjected to mutagenesis (G. H. Brownell, unpublished results). Since it has been shown that the mating factors are not of plasmid origin but rather a chromosomal borne trait there is no reason to believe that they should be more susceptible to mutation than other cell traits; consequently, the *mat-CE* progeny are probably derived from relatively stable heterogenotes (Brownell and Walsh, 1970). Recombinants that maintain a heterozygous state have been described by Adams and Mayberry (1978). Their 'pseudoprototrophs' result from complementation of several auxotrophic traits that map in the same general location on the chromosome as the *mat-E* mating gene.

In the absence of a true *mat-CE* recombinant, the existence of the *mat-c* and *mat-e* alleles is in doubt. The non-fertile recombinant may result from the exclusion of the *mat-C* and *mat-E* genes and there may be no alternate or mutant form of the mating genes.

Nothing is known about the phenotype(s) of the controlling genes other than that one cell must produce the *mat-E* gene product and the other *mat-C* gene product for zygote formation. It is known that both parental cells must be metabolically active and the inhibition of either parent precludes recombinant formation (Adams, 1964). It is possible that one gene controls the formation of pili-like structures, similar to those observed in *Mycobacterium* (Mizuguchi et al., 1978) while the other provides its surface receptor. Filaments or fibre-like structures have been observed on the surface of these organisms (J. A. Serrano, personal communication). It is also possible that the mating mechanism is merely a hyphal fusion at growing points that require some specific surface interactions coded by the *mat-C* and *mat-E* genes.

Recombinant recovery between compatible *Rhodococcus* strains also requires prolonged cell contact (18 to 26 hours) prior to recombinant selection (Adams, 1964). Brownell (1974) was able to reduce this preincubation time to only 4 hours by initiating crosses with older cell culture. Thus, one of the mating gene products may be synthesized exclusively in older stationary phase cells. A similar observation was made by Reh and Schlegel (1981) for the transfer of a plasmid bearing the aut^+ trait. It was during the older 'biphasic' growth that the conjugal transfer appeared to occur. A parental input ratio of 1:1 at a density of one 'cell' per 100 μm appears to be optimal for recombinant recovery (Kasweck, 1978). Broth-mated strains generally produce very low recombinant recoveries (Brownell and Runner, 1975).

Although there is good evidence for post-zygotic gene exclusion (Adams, 1968; Brownell, 1975a, 1978a; Adams and Mayberry, 1978) there are no data to support a pre-zygotic exclusion of the type that accompanies gene polarity transfer seen in donor–recipient matings. On the contrary, mating frequently results in the production of some 'recombinant' colonies that segregate out both parental and recombinant cells during unselected growth (Brownell and Walsh, 1970). Thus, the best guess for the mating mechanism is still one that leads to a cell in which parental nucleotides initially occupy a common cytoplasm. In the absence of selective pressure these cells may segregate out parental phenotypes, recombine to form haploid recombinants or form 'pseudoprototrophs' due to gene complementation (Brownell and Walsh, 1970; Adams and Mayberry, 1978; Brownell, 1978a).

B. LINKAGE MODELS

Approximately 65 genetic traits have been employed in the development of the *Rodc. erythropolis* linkage map (Fig. 5). The broken lines separate genes that map distal to one another. This phenomenon is a reflection of strain heterology and is caused by aberrant gene segregation within certain regions of the chromosome (Folkens and Adams, 1970; Adams, 1974; Brownell, 1975b). It is for this reason that the linkage model is separated into two regions, A and B. Traits that fall in region A (*purB2* to *hisA*) will reassort with respect to their physical distance so long as the genes used

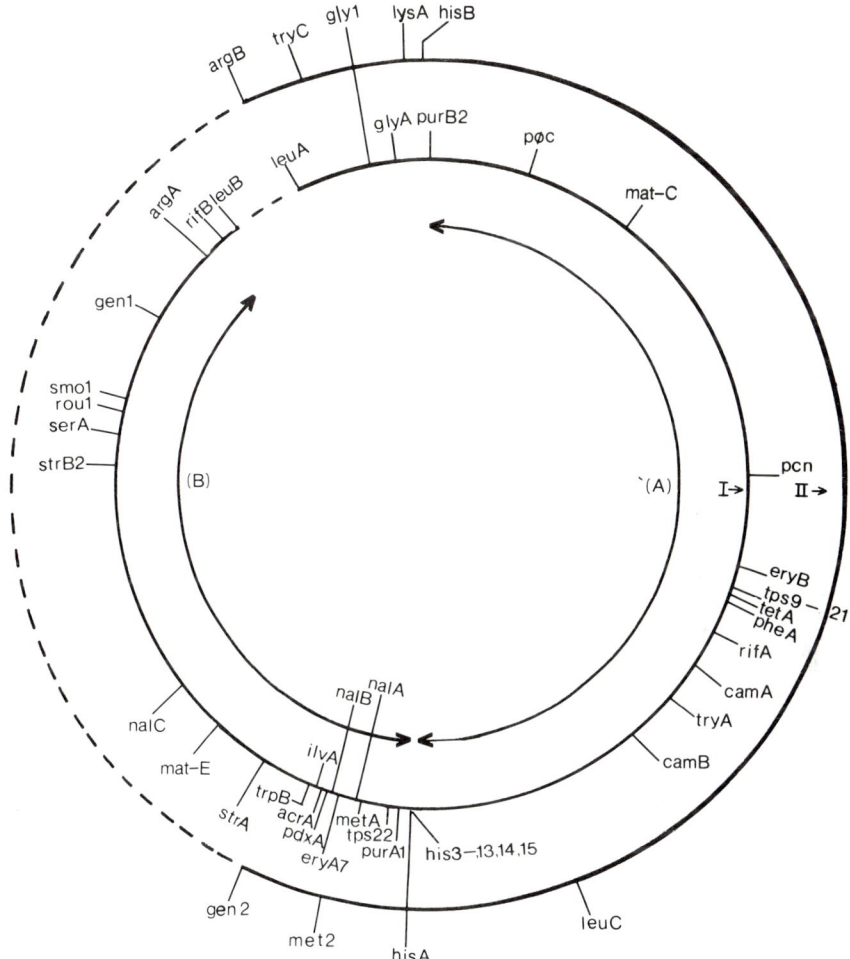

FIG. 5. A linkage map of *Rhodococcus erythropolis*. Linkage model I was constructed from matings between *mat-Ce* and *mat-cE* mating type strains (*Rodc. erythropolis* by '*Nrda. canicruria*'). Linkage model II was produced from matings between *mat-Me* and *mat-cE* parental strains (*Rodc. erythropolis* by *Rodc. rhodochrous*). The regions of the chromosome demarcated by arrows A and B represent the portions of the chromosome where genes are not effected by strain heterology so long as selected traits are within the same region. The broken lines separate genes whose linkage is not known. For details see Adams (1974) or Brownell (1975b, 1977).

for recombinant selection are also located within this region. The same behaviour is true for genes located in region B, selection of traits that lie between *hisA* and *argA* produce unambiguous segregation patterns for unselected genes within region B (Brownell, 1978a). When genes in both regions A and B are employed in recombinant selection, the matings either fail to yield recombinants or do so at low frequencies. When recombinants are obtained by selection of genes within one region, however, the genes in the alternative region are frequently involved in unselected cross-overs and produce aberrant recombinant class types. Thus, the *purB2* and *argA* genes map as though they were distal when in fact they are more closely linked to one another than to, say, the *hisA* gene (Fig. 5). Considerable effort has been spent in trying to visualize a zygote which could produce this kind of segregation. One approach has been to study the unselected markers in one region after selection for markers in another region. Such studies led to reports on non-reciprocal inheritance and partial exclusion of chromosomal genes amongst recombinants (Adams, 1968; Brownell, 1975a). The aberrant segregation of genes outside the selected regions has initiated a controversy over post- versus pre-zygotic gene exclusion. Brownell and his colleagues favoured a post-zygotic exclusion due to their ability to recover heterogenotes (Brownell and Walsh, 1970). The latter may be analogous to the heter-clones detected in *Streptomyces* (Hopwood et al., 1973) since there is unstable complementation of genes over a portion of the chromosome. Brownell and Walsh (1970) observed that certain portions of the 'recombinants' would continue to segregate out new class types during unselected growth. The traits involved were always in the unselected region and all of those in the mutant strains could be involved in the complementation. They concluded therefore that all the genes were present in the zygote. On the basis of DNA homology studies, Clark and Brownell (1972) postulated that a type of synaptic misalignment could occur between the chromosomes and that a recombination event could exclude certain genes from the progeny. The DNA within a region is similar enough for homologous recombination, however, because of the presence of unique nucleotide sequences between the two regions, synaptic alignment cannot occur (Clark and Brownell, 1972; Brownell, 1978a).

Adams and Mayberry (1978) presented a good argument for pre-zygotic exclusion. They obtained recombinants with genes that originated from a *mat-Ec* parental type and mapped in the B region, i.e. *met-A*, *pdx-A* (see Fig. 5). Their hybrids were fertile with the *mat-Ec* mating type strains so they could be backcrossed and that portion of region B extending from *his-A* to *met-A* or *pdx-A* would be homologous in the two strains (Adams and Mayberry, 1978). They were unable to find any increase in recombination between genes in that portion of region B but the matings did produce an extraordinarily high number of heterogenotes. Heterogenotes are usually found in less than 5% of the selected population (Brownell and Walsh, 1970; Folkens and Adams, 1970) but the backcrosses performed by Adams and Mayberry yielded 99% heterogenotes which were termed 'pseudoprototrophs'. The workers observed that only the wild-type alleles present in the *mat-Ec* parental type determined the production of the pseudoprototrophs and not those in the recombinant parental strain. In addition, neither the his^+ nor *arg*1 alleles could be recovered in segregants from pseudoprototrophs and thus must have been excluded from the production of the latter (Adams and Mayberry, 1978). Unfortunately, matings of their recombinants with the *mat-Ce* parental strains were not possible because they were not fertile.

Despite the aberrant segregation of genes outside a selected region, genetic analysis within the regions is straight forward. Since inter-strain fertility is obviously a natural phenomenon in these nocardioform actinomycetes, our understanding of the events seems justified.

C. LYSOGENY AND PROPHAGE INHERITANCE

Lysogenic strains of *Rodc. erythropolis* have been recovered after infection by actinophage ϕC and ϕEC (Brownell *et al.*, 1967; Crockett and Brownell, 1972). Recombinant analysis of matings between non-lysogenic and lysogenic strains allowed the mapping of the prophage integration loci (*pϕe* and *pϕc*, see Fig. 5) on the *Rodc. erythropolis* chromosome (Brownell, 1976). Although linkage between chromosomal-borne traits and the prophage allowed for an unambiguous mapping of an integrated phage genome, inheritance of just the prophage by non-lysogenic parental strains was found to be 16 to 19% whereas recombination between chromosomal traits was only about 1.0×10^{-5} per c.f.u. per ml (Brownell, 1973; Brownell and Adams, 1976). Control experiments showed that these non-recombinant lysogenic recipients were not generated by lysogenization with free phage since the acquisition required direct cell contact. In addition, the prophage was inherited only by cultures representing a fertile combination of mating factors, i.e. *mat-C* and *mat-E* strains (Brownell, 1975c, 1978b; Brownell and Adams, 1976). Thus, it was postulated that a non-integrated state of the prophage must also exist. The results of Brownell and his colleagues also indicate that there is competition between the transfer of the integrated and non-integrated prophage since most recombinants had either an integrated prophage or were non-lysogenic. Lysogens that bear integrated and/or non-integrated prophage could be generated by a phage that required integration for replication and produced a few copies before complete repression of replication or by readsorption of phage by lysogenic cells produced by induced hosts.

Further evidence for the existence of a non-integrated form of the prophage came from the observation that the prophage could be transferred to other strains in which recombination between chromosomal genes could not be demonstrated. A few strains of *Rodc. erythropolis* ('*Nrda*'. *globerula*, ATCC 9356, 15903, 19370; '*Nrda*'. *calcarea* ATCC 1936) were found to have acquired the prophage after mixed growth with *Rodc. erythropolis* lysogens (Brownell, 1978a, b). Attempts to demonstrate recombination between auxotrophic mutants of these organisms and *Rodc. erythropolis* have been unsuccessful. A mechanism for transfer of the prophage has not been postulated nor have induced phage been adequately studied for their transducing potential.

D. PROTOPLAST FORMATION AND TRANSFECTION

Several species of *Rhodococcus* have been shown to produce transfectable protoplasts (Brownell, 1981). The procedure required preincubation of cells in glycine (3–5%, w/v) followed by treatment with lysozyme (5 mg ml^{-1}). Protoplasts are generated after 15 to 20 minutes incubation in a support medium similar to that employed by Okanishi *et al.* (1974) but D-mannitol is used to provide the proper osmolarity.

The transfectability of the *Rhodococcus* protoplasts with actinophage ϕEC DNA using a procedure essentially the same as that employed for transfecting *Streptomyces* (Bibb *et al.*, 1978). The fraction of transfected protoplasts was found to be about 1×10^{-4} and the highest transfection frequency observed was around 1×10^{-5} infective centres per μg of DNA (Brownell *et al.*, 1982).

To date there have been no reports of protoplast fusion on *Rhodococcus*. This is somewhat surprising since the application of this technique could answer many of the questions still to be resolved regarding mating factors and gene linkage. The technique has been successfully employed in *Streptomyces* studies (Baltz, 1978; Hopwood and Wright, 1979; Baltz and Matsushima, 1980).

E. PHAGE ϕEC

Nocardioform phages have not been well characterized at the molecular level though studies have begun on phage ϕEC so that it can be used as a cloning vector in its *Rodc. erythropolis* host. ϕEC is a temperate phage isolated from soil enriched with *Rodc. erythropolis* (Brownell *et al.*, 1967). The virion is composed of a linear double-stranded DNA molecule with an average length of about 47.0 kilobase pairs (kbp). The head (52 ± nm in diameter) and tail (10 nm wide × 192 ± nm long) have a typical actinophage morphology. The ϕEC phage has a relatively narrow host range infecting only a few species of *Rhodococcus* (Brownell *et al.*, 1967) and, like many of the mycobacteriophages, is readily inactivated by many lipid solvents, except for ether and ethanol (Tokunaga *et al.*, 1970; Brownell and Crockett, 1971). Its receptor or attachment site on the host can be readily solubilized with ethanol which in turn can inactivate φEC phage suspensions (Brownell and Crockett, 1971).

The φEC DNA has been analysed by digestions with 32 restriction endonucleases (Table 4) and five enzymes were found to be suitable for restriction endonucleases mapping: BamHI, HpaI, PvuII, XbaI and XhoI (Brownell et al., 1980, 1981). A restriction map revealed at least three potential cloning sites in the φEC genome: BamHI with two sites, and XbaI and PvuII which cut the molecule three times (Fig. 6).

φEC deletion mutants have been isolated in order to map the non-essential regions of the phage genome, to remove unwanted target sites and reduce the genome in size so that it can accept foreign DNA fragments. A chelating agent (20 mM EDTA) was found to lower φEC phage titres by about 10^{-6} p.f.u. ml^{-1} and the EDTA-resistant phage proved to be deletion mutants (Brownell et al., 1981); three of the mutants are depicted in Fig. 6. The φdel-6a mutant has deletions that span both BamHI sites as well as the left terminal XhoI target sequence. The φdel-5 mutant demonstrates that another region of nonessential genes lies near the middle of the φEC genome. The φdel-6 mutant with a single BamHI site in a non-essential region should serve to test the cloning potential of the φEC vector. The φEC attachment or integration locus has been tentatively located near the centre of the φEC genome. This was determined from a Southern blotting experiment (Southern, 1975) wherein DNA from a Rodc. erythropolis lysogen was digested with XhoI, transferred to membrane filters and probed with radioactive φEC DNA. Radioautography revealed that the internal 11 kbp XhoI fragment was not present in the integrated phage DNA but two new smaller bands were present (G. H. Brownell, unpublished results).

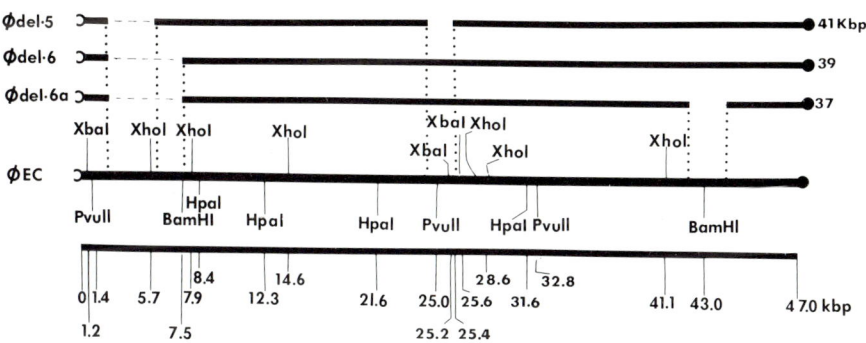

FIG. 6. A restriction endonuclease map of actinophage φEC. The lower line is a physical scale in kilobase pairs (kbp). The next band specifies endonuclease restriction sites. The three upper bands are deletion mutants, their broken lines indicating the extent of the deletion. The symbols at the ends of the lines represent cohesive termini.

TABLE 4. Restriction endonucleases digestion profile of actinophage φEC DNA

Enzyme	Number of fragments produced
AluI	+50
AvaI	+50
AvaII	+50
BamHI	3
BglI	+50
BglII	0
BstEII	+25
BstNI	+50
EcoRI	0
HaeII	+50
HaeIII	+50
HhaI	+50
HgiAI	+25
HincII	+50
HinfI	+50
HindIII	0
HpaI	5
HhpI	+50
KpnI	10
MboI	+50
MboII	+50
MspI	+50
PstI	0
PvuI	ca.21
PvuII	4
SacI	ca.25
SalI	0
Sau3A	+50
SmaI	0
TaqI	+50
XbaI	4
XhoI	7

4. Mycobacterium

A. GENETIC TRANSFORMATION

Transfer of genetic information in mycobacteria by DNA-mediated transformation has met with mixed success. In 1954 Katanuma and Nakasato reported that streptomycin resistance could be transferred to streptomycin sensitive strains by transformation. Subsequently, others (Tsukamura et al., 1960; Juhasz et al., 1971) reported successful transformation of both drug resistance and xylose-utilization markers into competent mycobacteria.

Other workers, however, have been unable to show transfer of genetic markers by transformation (Bloch et al., 1959; Bradley, 1970; Tárnok and Bönicke, 1970). At present, there is inconclusive evidence on which to define the conditions that will render mycobacterial cells competent for transformation or on the state of the DNA that should be used for transformation. Thus, it is possible that those experiments that were successful represent transfer of a plasmid or prophage carrying genetic markers rather than direct transfer of genomic DNA.

Although the uptake of bacterial DNA is ambiguous, successful transfection with phage DNA has been demonstrated in *Mybc. smegmatis* (Tokunaga and Sellers, 1964; Tokunaga and Nakamura, 1968), *Mybc. tuberculosis* (Tokunaga and Nakamura, 1967) and in *Mybc. phlei* (Kitahara and Sellers, 1975).

B. TRANSDUCTION

There are few reports concerning the exchange of genetic information by phage-mediated transduction. Unfortunately, the phages involved in these events are, in general, only poorly characterized. The first report of the transduction of genetic information was by Juhasz (1960) who found that streptomycin resistance could be transferred from *Mybc. bovis* BCG to *Mybc. phlei* by the filtered culture media of the former. Although it was demonstrated that this transfer of genetic information was deoxyribonuclease resistant, it was not unequivocally demonstrated that the exchange of genetic information was mediated by bacteriophage. Subsequently, Gelbart and Juhasz (1970) reported the transduction of xylose utilization from one strain of *Mybc. phlei* (xyl^+) to another (xyl^-). They also suggested that the genetic transfer was DNA-mediated (i.e. transformation) but the data are ambiguous and this claim must be treated with scepticism until stronger evidence is obtained. In further studies, these workers were unable to demonstrate transduction of other traits, such as resistance to streptomycin, ethambutol, rifampicin and cycloserine, using the same phage–host system (Gelbart and Juhasz, 1973). It is possible that phage Bo2 integrates into the genome of *Mybc. phlei* within or very near the xylose-utilization locus thereby yielding the efficient transfer of that marker and the inefficient transduction of others, not unlike λ.

In 1972, Jones and David demonstrated the transduction of streptomycin resistance from a resistance to a drug-sensitive strain of *Mybc. smegmatis*. These workers showed that streptomycin resistance was encoded by a plasmid but they did not attempt to purify the plasmid by physical means. In a later report (Jones et al., 1974), the transduction of the presumptive R-plasmid from *Mybc. smegmatis* to *Mybc. tuberculosis* H37Rv was demonstrated but once again no attempt was made to characterize the plasmid physically. It is not possible to say whether the streptomycin resistance transferred by Juhasz was chromosome- or plasmid-encoded.

At about the same time that Gelbart and Juhasz reported the transduction of xylose utilization, Sundar Raj and Ramakrishnan (1970) reported a generalized transducing phage, I3. This phage was used successfully to transduce a number of auxotrophic, drug-resistant and drug-sensitive markers in *Mybc. smegmatis* and was subsequently characterized in some detail. Kozloff *et al.* (1972) showed that phage I3, unlike other mycobacteriophages, possessed a contractile tail. The genetic material was shown to be double-stranded DNA with a guanine (G) plus cytosine (C) composition of 65 mol% (Gadagkar and Gopinathan, 1976). The interaction of this temperate phage with the host cell during lysogeny has also been genetically characterized to some extent. Gopinathan *et al.* (1978) showed that three distinct phage genes were involved in the establishment and maintenance of lysogeny. Finally, a *Mybc. smegmatis* mutant, presumably in adenylate cyclase although cyclic AMP levels were not directly measured, was isolated. This mutant host could not be lysogenized (Gopinathan *et al.*, 1978).

C. CONJUGATION

Knowledge of genetic recombination *via* conjugation in the genus *Mycobacterium* is due primarily to the efforts of Mizuguchi, Tokunaga and Suga. Initial experiments were made on two strains of *Mybc. smegmatis* which were designated Jucho and Lacticola (Mizuguchi and Tokunaga, 1971) but subsequently 19 compatible substrains of *Mybc. smegmatis* were identified and classified into five different mating types, A to E (Mizuguchi *et al.*, 1976). Eight substrains belonging to mating type A (Jucho) were fertile with substrains of types C (Rabinowitsch), D (PM-5-7) and E (Nishi). Mating type B (ATCC 607) recombined with mating types D and E, whereas mating type C strains were fertile with substrain types A, D and E (Table 5). The PM-5, -7 and Nishi strains were fertile with all of the organisms belonging to the other mating groups (Mizuguchi *et al.*, 1978). As with *Rhodococcus*, all strains belonging to the same mating type were infertile.

On the basis of the dominance of certain genetic traits from parental strains in the recombinant genome, a donor–recipient relationship has been postulated for the mating groups (see Fig. 7). A polarized transfer of genetic material was also suspected when ultraviolet-sensitive mutants of PM strains failed to yield recombinants with Rabinowitsch, Nishi or ATCC 607 mutant substrains (Mizuguchi, 1974). The same PM mutants, however, were fertile with the Jucho strains which suggested to the investigators that the transfer was bidirectional in this mating combination.

Linkage maps of *Mybc. smegmatis* are not available, except for a clustering of ribosomal genes which are linked to loci for arginine biosynthesis (Fig.

8). Linkage could not be demonstrated between any other loci analysed including *his, leu, met, str* and *arg*. The investigators have postulated that a mating type-specific restriction endonuclease may be responsible for generating what appears to be an unlinked transfer of donor DNA. However, evidence for a host-specific modification-restriction system on DNA from bacteriophage HC and PL initially propagated in ATCC 607 was inconclusive (Mizuguchi *et al.*, 1978).

Genetic analysis of drug resistance has been studied in some detail in *Mybc. smegmatis* (Yamada *et al.*, 1972, 1974, 1976; Mizuguchi *et al.*, 1974; Suga and Mizuguchi, 1974). The products of viomycin and capreomycin resistance genes were found to be components of the 50S (*vic*A) and 30S (*vic*B) ribosomal subunits. Genes that confer resistance to neomycin-kanamycin (*nek*), erythromycin (*ery*) and streptomycin (*str*) were found to alter the ribosomes and, with the exception of the *str* locus, were found to be closely linked on the *Mybc. smegmatis* chromosome (Fig. 8).

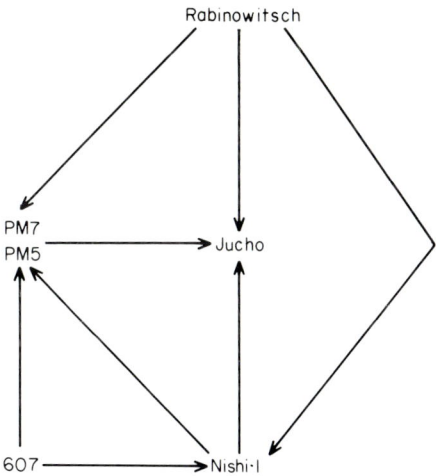

FIG. 7. Compatible mating combinations and postulated donor–recipient relations (arrows) of *Mycobacterium smegmatis*.

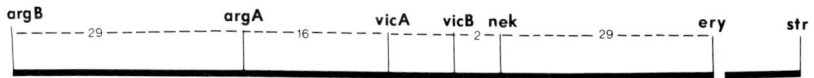

FIG. 8. A linkage map of *Mycobacterium smegmatis* constructed from crosses between strains Rabinowitsch and PM5.

TABLE 5. Mating type compatibility among *Mycobacterium smegmatis* substrains

Mating type and strains included	A	B	C	D	E
A. Jucho Dencho Li Jakeuchi Smegma (Y) Shiba A 288 Sekiguchi	−	−	+	+	+
B. ATCC 607	−	−	−	+	+
C. Rabinowitsch Lactocola Smegma P SN2 ATCC 14468 Butyricum Shiba A 496	+	−	−	+	+
D. PM5 PM7	+	+	+	−	+
E. Nishi 1	+	+	+[a]	+	+

[a] There is one mutant type C which is infertile in crosses with Nishi 1. This strain seems to have a mutation in the gene responsible for mating with the type E strain.

D. EXTRACHROMOSOMAL ELEMENTS

Jones and his colleagues (Jones and David, 1972; Jones et al., 1974) suggested that an unstable resistance of *Mybc. smegmatis* ATCC 607 to streptomycin may be determined by an extrachromosomal element. Their evidence is based on the transducibility of the element at high frequency by mycobacteriophage D29 and its sensitivity to ultraviolet light and acriflavins. These investigators have not, however, produced any physical evidence for the presence of plasmid DNA.

Plasmid DNA has been isolated from several strains of the *Mybc. avium–intracellulare* complex (Crawford and Bates, 1979). Three plasmids, pLR1, pLR2 and pLR3, were found to have molecular weights of 11.2, 18.3 and 107×10^6, respectively. Restriction-modification activity has been associated with the plasmid-bearing strains (J. T. Crawford, personal communication). Phage JFZ has a plating efficiency of 10^{-4} to 10^{-5} on plasmid-bearing hosts compared to *Mybc. smegmatis* ATCC 607.

E. LYSOGENY AND PHAGE CONVERSION

Host–phage relationships in the genus *Mycobacterium* have been extensively reviewed by Grange (1975) and Grange and Redmond (1978). It is not known how important a role phage conversion plays in genetic interactions of these organisms, but naturally occurring lysogens appear to be fairly common. The establishment of lysogeny has been reported to alter growth rates, enzyme activities, nutritional requirements, antigenic structure and colony morphology (Jones and White, 1968; Juhasz *et al.*, 1969; Jones and Beam, 1969; Jones and David, 1970; Mankiewicz and Tamari, 1972). In one report, infection by mycobacteriophage MC4 resulted in alterations in wall side-chain fatty acid, pigmentation, colony morphology and production of acid from some carbohydrates (Hawley *et al.*, 1976).

Lysogeny in the genus *Mycobacterium* was once considered to be a relatively rare event because of the numerous reports of failure in attempts to detect lysogens and because of the low number of lysogenic strains observed among strains tested. It now appears that in many of these earlier studies phage may not have been detected in culture lysates because of the lack of appropriate indicator strains. This view was first proposed by Bönicke (1969) after spot-testing culture filtrates from 10 lysogens on 84 strains and observing only one sensitive indicator strain for one of the phage. Isolating a cured parental cell from one of the lysogenic populations served as an indicator strain for all 10 phages. Thus, many of the temperate mycobacteriophage seem to have very narrow host ranges and the absence of suitable indicator strains may preclude their detection. Further evidence was provided by Grange and Bird (1978) who used electron microscopy to confirm the presence or absence of phage in culture filtrates of *Mybc. chelonei*, *Mybc. fortuitum* and *Mybc. kansasii*. They observed morphologically complete phage in 21 of 274 strains examined but found sensitive indicator strains for only eight of the phage.

5. Conclusions

The literature dealing with the genetics of nocardioform actinomycetes has been reviewed and a summary of the important points made. Unfortunately, much of the information is of a preliminary nature as only a few detailed studies have been carried out. This lack of a sustained research effort has resulted in only fragmentary information which is difficult to organize into a coherent review. One can only conclude that many fundamental questions still remain unanswered and little progress has been made since the literature was reviewed several years ago. This is a disappointing conclusion, for it is not clear why nocardioform actinomycetes remain relatively unexplored, particularly at the molecular level as the group includes strains of medical and commercial importance (Chapters 6 and 8). Many observations have been made. What is needed now is their examination with currently available molecular techniques.

REFERENCES

Adams, J. N. (1964). Recombination between *Nocardia erythropolis* and *Nocardia canicruria*. *Journal of Bacteriology* **88**, 865–867.

Adams, J. N. (1968). Partial exclusion of the *Nocardia erythropolis* chromosome in nocardial recombinants. *Journal of Bacteriology* **96**, 1750–1759.

Adams, J. N. (1974). Linkage map and list of markers for *Nocardia erythropolis*. In: *Handbook of Microbiology*, (A. I. Laskin and H. A. Lechevalier, eds.), pp. 661–664. CRC Press, Cleveland.

Adams, J. N. and Bradley, S. G. (1963). Recombination events in the bacterial genus *Nocardia*. *Science* **140**, 1392–1394.

Adams, J. N. and Brownell, G. H. (1976). Genetic studies in *Nocardia erythropolis*. In: *The Biology of the Nocardiae* (M. Goodfellow, G. H. Brownell and J. A. Serrano, eds.), pp. 285–309. Academic Press, London.

Adams, J. N. and Mayberry, K. J. C. (1978). Recovery of unusual heterogenotes from a recombinant of *Nocardia erythropolis*. In: *Genetics of the Actinomycetales*, (E. Freerksen, I. Tárnok and J. H. Thumin, eds.), pp. 149–161. Gustav Fischer Verlag, Stuttgart.

Alderson, G., Goodfellow, M., Wellington, E. M. H., Williams, S. T., Minnikin, S. M. and Minnikin, D. E. (1981). Chemical and numerical taxonomy of *Nocardia mediterranea*. *Zentralblatt für Bakteriologie, Mikrobiologie und Hygiene*. 1. Abteilung, Supplement **11**, 39–46.

Alshamaony, L., Goodfellow, M. and Minnikin, D. E. (1976). Free mycolic acids as criteria in the classification of *Nocardia* and the '*rhodococcus*' complex. *Journal of General Microbiology* **92**, 188–199.

Andrzejewski, J., Müller, G., Röhrscheidt, E. and Pietkiewicz, D. (1978). Isolation, characterization and classification of a *Nocardia asteroides* bacteriophage. *Zentralblatt für Bakteriologie, Parasitenkunde, Infektionskrankheiten und Hygiene*. 1. Abteilung, Supplement **6**, 319–326.

Baltz, R. H. (1978). Genetic recombination in *Streptomyces fradiae* by protoplast fusion and cell regeneration. *Journal of General Microbiology* **107**, 92–102.

Baltz, R. H. and Matsushima, P. (1980). Application of protoplast fusion, site directed mutagenesis and gene complication to antibiotic yield improvement in *Streptomyces*. *Actinomycetes* **15**, 18–34.

Bibb, M. J., Ward, J. M. and Hopwood, D. A. (1978). Transformation of plasmid DNA into *Streptomyces* at high frequency. *Nature, London* **274**, 398–400.

Bloch, H., Walter, A. and Yamamura, Y. (1959). Failure of deoxyribonucleic acid from mycobacteria to induce bacterial transformation. *American Review of Respiratory Diseases* **80**, 911.

Bönicke, R. (1969). Lysogeny among mycobacteria. *Folia Microbiologica* **14**, 297–301.

Bousfield, I. J. and Goodfellow, M. (1976). The '*rhodochrous*' complex and its relationships with allied taxa. In: *The Biology of the Nocardiae* (M. Goodfellow, G. H. Brownell and J. A. Serrano, eds.), pp. 39–65 Academic Press, London.

Bradley, S. G. (1970). New horizons in the genetic research on *Mycobacterium*, *Nocardia* and *Actinomyces*. In: *Host–virus Relationships in Mycobacterium, Nocardia and Actinomyces* (S. E. Juhasz and G. Plummer), pp. 3–14. Charles G. Thomas, Springfield, VA.

Brownell, G. H. (1973). Plasmid-like inheritance of the ϕEC prophage from matings of lysogenic nocardiae. *Genetics* **74**, 32.

Brownell, G. H. (1974). The effects of parental cell age on the recovery of nocardial recombinants. *Proceedings of the 1st International Congress on the Biology of the Nocardiae*. pp. 74–75. McGowen Printing Co., Augusta, U.S.A.

Brownell, G. H. (1975a). Non-reciprocal inheritance of gentamicin resistance in *Nocardia* recombinants. *Recent Developments in Nocardial Biology* **10**, 42–45.

Brownell, G. H. (1975b). Linkage maps of *Nocardia erythropolis* mating type strains. *Recent Developments in Nocardial Biology* **10**, 84–95.

Brownell, G. H. (1975c). The transfer of the ϕEC-plasmid between mat-Ce lysogens and other nocardioform organisms. *Recent Developments in Nocardial Biology* **10**, 15–19.
Brownell, G. H. (1976). Chromosomal location of the ϕC-prophage in *Nocardia erythropolis*. *Journal of General Microbiology* **94**, 217–221.
Brownell, G. H. (1977). Genetic recombination in *Nocardia erythropolis*: A new mating type strain. *Developments in Industrial Microbiology* **17**, 409–417.
Brownell, G. H. (1978a). Genetic interactions in the genus *Nocardia*. In: *Genetics of the Actinomycetales* (E. Freerksen, I. Tárnok and J. H. Thumin, eds.), pp. 137–148. Gustav Fischer Verlag, Stuttgart.
Brownell, G. H. (1978b). Plasmid transfer between *Nocardia erythropolis* and other nocardioform organisms. *Zentralblatt für Bakteriologie, Parasitenkunde, Infektionskrankheiten und Hygiene.* 1. Abteilung, Supplement **6**, 313–317.
Brownell, G. H. (1981). A protoplasting and transfecting procedure for *Nocardia (Rhodococcus)* species. *Actinomycetes* **16**, 54–56.
Brownell, G. H. and Adams, J. N. (1976). Inheritance of nocardiophage ϕEC plasmids in matings of nocardial lysogens. *Journal of Bacteriology* **126**, 1104–1107.
Brownell, G. H. and Crockett, J. K. (1971). Inactivation of nocardiophage ϕC and ϕEC by extracts of phage attachable cells. *Journal of Virology* **8**, 894–899.
Brownell, G. H. and Kelly, K. L. (1969). Inheritance of mating factors in nocardial recombinants. *Journal of Bacteriology* **99**, 25–36.
Brownell, G. H. and Runner, R. R. (1975). Recovery of nocardial recombinants from liquid media. *Canadian Journal of Microbiology* **21**, 1032–1040.
Brownell, G. H. and Walsh, R. S. III (1970). Heterogenomic recombinants from compatible nocardiae. *Journal of Bacteriology* **104**, 79–86.
Brownell, G. H., Adams, J. N. and Bradley, S. G. (1967). Growth and characterization of nocardiophages for *Nocardia canicruria* and *Nocardia erythropolis* mating types. *Journal of General Microbiology* **47**, 247–256.
Brownell, G. H., Enquist, L. W. and Denniston-Thompson, K. (1980). An analysis of the genome of actinophage ϕEC. *Gene* **12**, 311–314.
Brownell, G. H., Saba, J. A., Enquist, L. W. and Denniston-Thompson, K. (1981). The development of a gene cloning system for the actinomycetes. In: *Genetics and Physiology of Actinomycetes* (S. G. Bradley, ed.), pp. 264–283. Technical Information Service, National Springfield, VA.
Brownell, G. H., Saba, J. A., Denniston, K. and Enquist, L. W. (1982). The development of a *Rhodococcus*-Actinophage gene cloning system. *Developments in Industrial Microbiology* **23**, (in press).
Clarke, J. E. and Brownell, G. H. (1972). Genophore homologies among compatible nocardiae. *Journal of Bacteriology* **109**, 720–729.
Crawford, J. T. and Bates, J. H. (1979). Isolation of plasmids from mycobacteria. *Infection and Immunity* **24**, 979–981.
Crockett, J. K. and Brownell, G. H. (1972). Isolation and characterization of a lysogenic strain of *Nocardia erythropolis*. *Journal of Virology* **10**, 737–745.
Folkens, A. T. and Adams, J. M. (1970). Instability of nocardial recombinants. *Bacteriological Proceedings* p. 41.
Gadagkar, R. R. and Gopinathan, K. P. (1976). Physicochemical characterization of mycobacteriophage 13. *Proceedings of the 45th Annual Meeting of the Society of Biological Chemists*, India 1976.
Gelbart, S. M. and Juhasz, S. E. (1970). Genetic transfer in *Mycobacterium phlei*. *Journal of General Microbiology* **64**, 253–254.
Gelbart, S. M. and Juhasz, S. E. (1973). Transduction in *Mycobacterium phlei*. *Antoine van Leeuwenhoek* **39**, 1–12.
Goodfellow, M. and Alderson, G. (1977). The actinomycete genus *Rhodococcus*: home of the 'rhodochrous' complex. *Journal of General Microbiology* **100**, 99–122.
Goodfellow, M., Beckham, A. R. and Barton, M. D. (1982). Numerical classification of *Rhodococcus equi* and related actinomycetes. *Journal of Applied Bacteriology* **53**, 199–207.
Gopinathan, K. P., Saroja, D., Gadagkar, R. R. and Ramakrishnan, T. (1978). Control of lysogeny in mycobacteria. In: *Genetics of the Actinomycetales* (E. Freerksen, I. Tárnok and J. H. Thumin, eds.), pp. 237–249. Gustav Fischer Verlag, Stuttgart.

Gordon, R. E. and Mihm, J. M. (1957). A comparative study of some strains received as nocardiae. *Journal of Bacteriology* **73**, 15–27.

Grange, J. M. (1975). The genetics of mycobacteria and mycobacteriophages, a review. *Tubercle* **56**, 227–238.

Grange, J. M. and Bird, R. G. (1978). Lysogeny in the genus *Mycobacterium*. In: *Genetics of the Actinomycetales* (E. Freerksen, I. Tárnok and J. H. Thumin, eds.), pp. 243–249. Gustav Fischer Verlag, Stuttgart.

Grange, J. M. and Redmond, W. B. (1978). Host-phage relationships in the genus *Mycobacterium* and their clinical significance. *Tubercle* **59**, 203–225.

Gray, P. H. H. (1928). The formation of indigotin from indol by soil bacteria. *Proceedings of the Royal Society* **B102**, 263–280.

Hawley, R. I., Imaeda, T. and Mann, N. (1976). Isolation and characterization of Nocardia-like variants of *Mycobacterium smegmatis*. *Canadian Journal of Microbiology* **22**, 1480–1491.

Hopwood, D. A. (1967). Genetic analysis and genome structure in *Streptomyces coelicolor*. *Bacteriological Reviews* **31**, 373–403.

Hopwood, D. A. and Wright, H. M. (1979). Factors affecting recombination frequency in protoplast fusions of *Streptomyces coelicolor*. *Journal of General Microbiology* **111**, 137–143.

Hopwood, D. A., Chater, F. K., Dowding, J. E. and Vivian, A. (1973). Advances in *Streptomyces coelicolor* genetics. *Bacteriological Reviews* **51**, 371–405.

Hopwood, D. A., Wright, H. M., Bibb, M. J. and Cohen, S. N. (1977). Genetic recombination through protoplast fusion in *Streptomyces*. *Nature, London* **268**, 171–174.

Horinouchi, S., Uozumi, T. and Beppu, T. (1980). Cloning of *Streptomyces* DNA into *Escherichia coli*: Absence of heterospecific gene expression of *Streptomyces* genes in *E. coli*. *Agricultural Biology and Chemistry* **44**, 367–381.

Jones, W. D. and Beam, R. E. (1969). Lysogeny in the mycobacteria. II. Alterations of bacterial antigens mediated by mycobacteriophage. *Canadian Journal of Microbiology* **15**, 1112–1114.

Jones, W. D. and David, H. L. (1970). Biosynthesis of a lipase by *Mycobacterium smegmatis* ATCC 607 infected with mycobacteriophage D29. *American Review of Respiratory Diseases* **102**, 818–820.

Jones, W. D. and David, H. L. (1972). Preliminary observations on the occurrence of a streptomycin R-factor in *Mycobacterium smegmatis* ATCC 607. *Tubercle* **53**, 35–42.

Jones, W. D. and White, A. (1968). Lysogeny in mycobacteria. I. Conversion of colony morphology, nitrate reductase and Tween 80 hydrolysis of *Mycobacterium* sp. ATCC 607 associated with lysogeny. *Canadian Journal of Microbiology* **14**, 551–555.

Jones, W. D., Beam, R. E. and David, H. L. (1974). Transduction of a streptomycin R-factor from *Mycobacterium smegmatis* to *Mycobacterium tuberculosis* H37RV. *Tubercle* **55**, 72–80.

Juhasz, S. E. (1960). Intraspecific hybridization among mycobacteria. *Nature, London* **185**, 265.

Juhasz, S. E., Gelbart, S. M. and Harize, M. (1969). Phage-induced alteration of enzymic activity in lysogenic *Mycobacterium smegmatis* strains. *Journal of General Microbiology* **56**, 251–255.

Juhasz, S. E., Gelbart, S. M. and DeSalle, L. (1971). Genetic transfer in *Mycobacterium phlei*. *Bacteriological Proceedings*, p. 35.

Kasweck, K. L. (1978). Population interaction affecting pseudo-prototrophic recombinant production in *Nocardia erythropolis*. In: *Genetics of the Actinomycetales* (E. Freerksen, I. Tárnok and J. H. Thumin, eds.), pp. 163–185. Gustav Fischer Verlag, Stuttgart.

Kasweck, K. L. and Little, M. L. (1982). Genetic recombination in *Nocardia asteroides*. *Journal of Bacteriology* **149**, 403–406.

Kasweck, K., Little, M. and Alvarado, F. (1981). Genetic recombination in *Nocardia asteroides*. *Actinomycetes* **15**, 101–110.

Katanuma, N. and Nakasato, H. (1954). A study of the mechanism of the development of streptomycin resistant organisms by addition of deoxyribonucleic acid prepared from resistant bacilli. *Kekkaku* **29**, 19.

Kitahara, K. and Sellers, M. I. (1975). Transfection of *Mycobacterium phlei*. *Annali Sclavo* **17**, 605–606.

Kozloff, L. M., Sundar, Raj C. V., Nagaraja Rao, R., Chapman, V. A. and Delong, S. (1972). Structure of a transducing mycobacteriophage. *Journal of Virology* **9**, 390–393.

Krallmann-Wenzel, K. (1981). Recombination in self-compatible strains of *Nocardia opaca*. *Actinomycetes* **16**, 11–16.

Krallmann-Wenzel, K. and Tárnok, I. (1978). Recombination studies on auxotrophic mutants of *Nocardia opaca*. In: *Genetics of the Actinomycetales* (E. Freerksen, I. Tárnok and J. H. Thumin, eds.), pp. 187–191. Gustav Fischer Verlag, Stuttgart.

Mankiewicz, E. and Tamari, M. G. (1972). Lysogeny and deoxyribonuclease production in mycobacteria. *American Review of Respiratory Diseases* **106**, 609–610.

Margalith, P. and Beretta, G. (1960). Rifomycin. XI. Taxonomic study on *Streptomyces mediterranei* nov. sp. *Mycopathologia et Mycologia Applicata* **13**, 312–330.

Metcalf, G. and Brown, M. (1957). Nitrogen fixation by new species of *Nocardia*. *Journal of General Microbiology* **17**, 567–572.

Mizuguchi, Y. (1974). Effect of ultraviolet sensitive mutants on gene inheritance in mycobacterial matings. *Journal of Bacteriology* **117**, 913–916.

Mizuguchi, Y. and Tokunaga, T. (1971). Recombination between *Mycobacterium smegmatis* strains Jucho and Lacticola. *Japanese Journal of Microbiology* **15**, 359–366.

Mizuguchi, Y., Suga, K., Masuda, K. and Yamada, T. (1974). Genetic and biochemical studies on drug-resistant mutants in *Mycobacterium smegmatis*. *Japanese Journal of Microbiology* **18**, 457–462.

Mizuguchi, Y., Suga, K. and Tokunaga, T. (1976). Multiple mating types of *Mycobacterium smegmatis*. *Japanese Journal of Microbiology* **20**, 435–443.

Mizuguchi, Y., Suga, K. and Tokunaga, T. (1978). Genetic recombination in mycobacteria. In: *Genetics of the Actinomycetales* (E. Freerksen, Tárnok, I. and J. H. Thumin, eds.), pp. 111–120. Gustav Fischer Verlag, Stuttgart.

Okanishi, M., Suzuki, K. and Umezawa, H. (1974). Formation and reversion of *Streptomyces* protoplasts: Cultural conditions and morphological study. *Journal of General Microbiology* **80**, 389–400.

Prauser, H. (1976). Contributions to the taxonomy of the Actinomycetales. *Publications of the Faculty of Sciences, J. E. Purkyně University, Brno K.* **40**, pp. 196–199.

Reh, M. (1981). Chemolithoautotrophy as an autonomous and transferable property of *Nocardia opaca* 1b. *Zentralblatt für Bakteriologie, Mikrobiologie und Hygiene. 1. Abteilung, Supplement* **11**, 577–583.

Reh, M. and Schlegel, H. G. (1981). Hydrogen autotrophy as a transferable genetic character of *Nocardia opaca* 1b. *Journal of General Microbiology* **126**, 327–336.

Schaal, K. P. and Reutersberg, H. (1978). Numerical taxonomy of *Nocardia asteroides*. *Zentralblatt für Bakteriologie, Parasitenkunde, Infektionskrankheiten und Hygiene. 1. Abteilung, Supplement* **6**, 53–62.

Schupp, R. and Nuesch, J. (1979). Chromosomal mutations in the final step of rifamycin biosynthesis. *FEMS Microbiology Letters* **6**, 23–27.

Schupp, R., Hütter, R. and Hopwood, D. A. (1975). Genetic recombination in *Nocardia mediterranei*. *Journal of Bacteriology* **121**, 128–136.

Skerman, V. B. D., McGowan, V. and Sneath, P. H. A. (1980). Approved Lists of Bacterial Names. *International Journal of Systematic Bacteriology* **30**, 225–420.

Southern, E. M. (1975). Detection of specific sequences among DNA fragments separated by gel electrophoresis. *Journal of Molecular Biology* **98**, 180–182.

Suga, K. and Mizuguchi, Y. (1974). Mapping of antibiotic resistance markers in *Mycobacterium smegmatis*. *Japanese Journal of Microbiology* **18**, 139–147.

Sundar, Raj, C. V. and Ramakrishnan, T. (1970). Transduction in *Mycobacterium smegmatis*. *Nature, London* **228**, 280–281.

Tárnok, I. and Bönicke, R. (1970). Problems of genetic transformation of mycobacteria by deoxyribonucleic acid. *Bulletin of the International Union Against Tuberculosis* **43**, 210–213.

Thiemann, J. E., Zucco, G. and Pelizza, G. (1969). A proposal for the transfer of *Streptomyces mediterranei*, Margalith and Beretta 1960 to the genus *Nocardia* as *Nocardia mediterranei* (Margalith and Beretta) *comb. nov. Archiv für Mikrobiologie* **67**, 147–155.

Tokunaga, T. and Nakamura, R. M. (1967). Infection of *Mycobacterium tuberculosis* with deoxyribonucleic acid extracted from mycobacteriophage B1. *Journal of Virology* **1**, 448–449.

Tokunaga, T. and Nakamura, R. M. (1968). Infection of competent *Mycobacterium smegmatis* with deoxyribonucleic acid extracted from bacteriophage B1. *Journal of Virology* **2**, 110–117.

Tokunaga, T. and Sellers, M. I. (1964). Infection of *Mycobacterium smegmatis* with D29 phage DNA. *Journal of Experimental Medicine* **119**, 139–149.

Tokunaga, T., Kataoka, T. and Suga, K. (1970). Phage inactivation by an ethanol-ether extract of *Mycobacterium smegmatis*. *American Review of Respiratory Diseases* **101**, 309–313.

Tsukamura, M., Hasimoto, H. and Noda, Y. (1960). Transformation of isoniazid and streptomycin resistance in *Mycobacterium avium* by deoxyribonucleate derived from isoniazid- and streptomycin-double resistance cultures. *American Review of Respiratory Diseases* **81**, 403–406.

Vezina, C. (1970). The genetics of *Nocardia restrictus*. *Abstract 1st International Symposium on the Genetics of Industrial Microorganisms* p. 67.

Waksman, S. A. and Henrici, A. T. (1948). Family II. *Actinomycetaceae* Buchanan. In: *Bergey's Manual of Determinative Bacteriology*, 6th Edition (R. S. Breed, E. G. D. Murray and A. P. Hitchens, eds.). Williams and Wilkins, Baltimore.

Yamada, T., Masuda, K., Shoki, K. and Hori, M. (1972). Analysis of ribosomes from viomycin-sensitive and resistant strains of *Mycobacterium smegmatis*. *Journal of Bacteriology* **112**, 1–6.

Yamada, T., Masuda, K., Shoji, K. and Hori, M. (1974). Pleiotropic antibiotic resistance mutations associated with ribosomes and ribosomal subunits in *Mycobacterium smegmatis*. *Antimicrobial Agents and Chemotherapy* **6**, 46–53.

Yamada, T., Mizuguchi, Y. and Suga, K. (1976). Localization of co-resistance to streptomycin, kanamycin, capreomycin and tuberactinomycin in core particles derived from viomycin-resistant *Mycobacterium smegmatis*. *Journal of Antibiotics* **29**, 1124–1126.

5
Streptomyces Genetics

K. F. CHATER and D. A. HOPWOOD
John Innes Institute, Norwich, UK

1. Introduction	230
2. The genetic material of *Streptomyces*	231
A. Chromosomal DNA *in vivo* and *in vitro*	231
B. Genome complexity	231
C. Restriction enzyme analysis of chromosomal DNA	231
D. Plasmid and phage DNA	232
3. Natural gene exchange	232
A. Transduction	233
B. Conjugation	233
4. Protoplast fusion	244
5. Liposome-mediated transformation by chromosomal DNA	245
6. Genetic analysis	246
A. Chromosomal linkage mapping by conjugation	246
B. Complementation and dominance tests	249
C. Chromosomal mapping by protoplast fusion	250
D. The detection of extrachromosomal inheritance	251
E. The detection of transposable genetic elements	253
7. The application of genetic analysis to interesting aspects of *Streptomyces* biology	256
A. The regulation of gene expression in primary metabolism	256
B. Differentiation	258
C. Antibiotic production	259
D. Antibiotic resistance	259
E. Restriction-modification (RM) systems	264
F. Exoenzymes	266
8. Gene cloning in *Streptomyces*	267
A. The introduction of plasmids and phage DNA into *Streptomyces* protoplasts	268
B. Plasmid vector molecules	269
C. Phage vector molecules	273
D. Uses of bifunctional replicons and the question of heterologous gene expression	276
9. General conclusions and future prospects	277
References	278

THE BIOLOGY OF THE ACTINOMYCETES
ISBN 0-12-289670-X

Copyright © 1983 by Academic Press, London
All rights of reproduction in any form reserved

1. Introduction

In a volume of this kind it is not necessary to justify the study of streptomycetes, but it may be appropriate to explain why their genetics are of interest. Broadly, there are three reasons: first, as a central part of a complete description of the group; second, as a tool to analyse, and thus to describe more completely, their other aspects (e.g. their structure, physiology and ecology); and third, as the means by which to optimize their exploitation for human welfare. The historical phases of laboratory studies of *Streptomyces* genetics illustrate this progression. From the first reports of recombination (Sermonti and Spada-Sermonti, 1955, 1956; Hopwood, 1957, 1959) until the early 1970s almost all research was concerned with describing the genetic architecture and behaviour of the chromosome (e.g. Hopwood, 1967a; Sermonti, 1969; Friend and Hopwood, 1971; Alačević, 1976; Baumann *et al.*, 1974), with the emphasis later shifting towards sexual biology and the roles of plasmids (Hopwood *et al.*, 1973; Hopwood and Wright, 1976b; Bibb *et al.*, 1977; Friend *et al.* 1978). Meanwhile, the 1970s have also seen the use of the genetic system to analyse primary metabolism (e.g. carbon catabolite repression; Hodgson, 1980, 1982), antibiotic biosynthesis (e.g. Wright and Hopwood, 1976a,b; Kirby and Hopwood, 1977; Rudd and Hopwood, 1979, 1980; Rhodes *et al.*, 1981), differentiation (Chater and Merrick, 1979) and *Streptomyces* phages (Lomovskaya *et al.*, 1980). Most recently of all, it has been possible to capitalize on the accumulated knowledge of the genetic system by imposing artificial methods of gene exchange on the organisms, such as protoplast fusion (Baltz, 1978; Hopwood and Wright, 1978) and recombinant DNA systems (Bibb *et al.*, 1980a, b; Suarez and Chater, 1980b; Thompson *et al.*, 1980: reviewed by Chater *et al.*, 1982b and Hopwood and Chater, 1982). The widespread surge of industrial interest in these genetic techniques testifies to the likelihood that tangible human benefits will soon be felt from the application of such sophisticated genetics to organisms destined for fermenters (Queener and Baltz, 1979; Hopwood and Chater, 1980).

In this chapter we aim to reflect all these aspects of *Streptomyces* genetics, while reducing technical details (such as linkage analysis) to a minimum. Later sections, especially that on gene cloning, deal with what are at present the most actively studied areas. It will be noticed that one organism, *Stmy. coelicolor* A3(2), dominates the chapter. This results from the fact that, among the first streptomycetes selected for recombination studies, this strain gave early results that were satisfactorily interpretable in terms of a linkage map (Hopwood, 1959). The tendency has therefore been to consider *Stmy. coelicolor* the organism of choice for any fundamental study of *Streptomyces* biology that might benefit from the application of genetics, obviously a desirable strategy if the limited effort applied to *Streptomyces* genetics was not to be dissipated over many different organisms. Although the findings from such studies generally appear to be widely applicable to other streptomycetes (Hopwood and Merrick, 1977), there are (hopefully exceptional) cases where even with the help of the *Stmy. coelicolor* model, genetic analysis has been all but impossible (e.g. *Stmy. albus* G: Chater and Wilde, 1980). For the study of such organisms the newer, artificial techniques of genetic manipulation are crucial.

2. The Genetic Material of *Streptomyces*

A. CHROMOSOMAL DNA *IN VIVO* AND *IN VITRO*

Optical and electron microscopical studies of *Stmy. coelicolor* A(2) have shown that, as in other bacteria, the DNA is present in the cells in a condensed form, often in multiple copies per hyphal cell but usually only as a single copy in spores (Hopwood and Glauert, 1960a, b). It is possible to isolate *Stmy. coelicolor* DNA in condensed, fast-sedimenting units (Westpheling, 1980) which, like the nucleoids isolated in similar conditions from *Escherichia coli* and *Bacillus subtilis* (Worcel and Burgi, 1974; Guillen *et al.*, 1978), presumably individually consist of a continuous, circular DNA molecule organized into a number of supercoiled domains (the 'folded chromosome'), together with protein and other molecules, some of which may be required to stabilize the structure. However, most conditions for cell lysis result in the breakdown of this organized body and the release of the DNA from its supercoiled state. It then becomes very sensitive to shearing and can only be purified in the form of linear fragments. Such DNA solutions have been used in studies of a number of physical attributes of *Streptomyces* DNA.

B. GENOME COMPLEXITY

By renaturation analysis (Benigni *et al.*, 1975) an estimate of 10.5×10^3 kb was obtained for the amount of unique DNA in the *Streptomyces* genome (i.e. about three times more than for *Esch. coli* or *Bacl. subtilis*). In addition, Antonov *et al.* (1977) found that about 5% of the genome consisted of repetitive sequences (at about four copies per haploid genome) and 2% consisted of 'foldback' DNA; the significance of these various repeated sequences has not been investigated, nor is it known how characteristic they are of streptomycetes in general.

C. RESTRICTION ENZYME ANALYSIS OF CHROMOSOMAL DNA

Streptomyces DNA is remarkable for its high content (about 73%) of guanine and cytosine (G+C). It is therefore not surprising to find that (by and large) restriction enzymes recognizing sites low in G+C (e.g. EcoRI, HindIII, HpaI, KpnI) cut the DNA into relatively few and predominantly large fragments, whereas those recognizing sites high in G+C (e.g. BamHI, PstI, SalGI, SstII) give many, and predominantly small, fragments (unpublished results of several laboratories). With certain combinations of enzymes and DNA we have seen some relatively intense bands after agarose gel electrophoresis, suggestive of repeated DNA sequences, which may conceivably be related to the rapidly re-annealing fraction observed in renaturation analysis. In some variant strains, specific sets of DNA sequences may be greatly amplified (Robinson et al., 1981; Schrempf, 1982; Ono et al., 1982) but the implications of these observations are not yet clear.

D. PLASMID AND PHAGE DNA

Streptomyces plasmids are known to exist in a wide range of sizes (less than 4 kb to at least 170 kb) and copy numbers per chromosome (from about unity to several hundreds), and to be involved in the control of a wide range of phenotypic characters, including fertility, antibiotic production and resistance, and differentiation, as we shall see at various points in this chapter. DNA of SCP2 and SLP1 has the same buoyant density in CsCl gradients as host DNA and thus probably the same G+C content (Bibb et al., 1977, 1981) (other *Streptomyces* plasmids have not been studied in this way). Too few plasmids have yet been analysed with restriction enzymes to allow any predictions to be made about target site frequencies in relation to those for chromosomal DNA. It is, however, very clear that phage DNA departs significantly in this regard from chromosomal DNA. Those virulent and temperate phage DNA species described by Chater (1980), which covered a range of genome sizes from 40 to 100 kb, varied in G+C content from 55% to 73%, and some restriction enzyme target sites (particularly those for *Bam*HI and *Pst*I) were much rarer than in chromosomal or plasmid DNA. We shall consider the physical analysis of some selected plasmid and phage genomes in a later section dealing with the development of DNA cloning systems in *Streptomyces*.

3. Natural Gene Exchange

As far as we are aware, there has been no well documented report of chromosomal recombination through a natural transformation system in any streptomycete, but a generalized transduction system is now well-established in *Stmy. venezuelae* (see below). The great majority of the numerous reports of genetic recombination in members of the genus doubtless reflect the operation of conjugation systems, the strongest evidence, in the absence of any morphological information on the mechanism of mating, being the invariable finding that recombinants regularly inherit groups of distantly linked markers from each parent with no tendency for preferential inheritance of very short chromosomal segments. This result is incompatible with the properties of any known transduction system in eubacteria and is very difficult if not impossible to reconcile with a transformation mechanism. (As a matter of interest, the converse result–the finding that nearly all recombinants in a mixed culture inherited single markers from each parent–pointed to a transformation system in *Thermoactinomyces vulgaris* and this was later established by the use of purified DNA (Hopwood and Wright, 1972)). In those cases where plasmid sex factors have been implicated in gene exchange–*Stmy. coelicolor* A3(2), '*Stmy. lividans*' and *Stmy. rimosus*– evidence for a conjugation system is particularly strong. However, we need to keep constantly in mind that use of the term 'conjugation' to describe phenomena in streptomycetes does not imply a close mechanistic similarity to the system of plasmid-determined gene exchange first encountered and extensively studied, but still not fully understood, in F-containing derivatives of *Esch. coli* K-12, and later in other Gram-negative bacteria (Rowbury, 1977). Indeed, some considerations discussed below make this rather unlikely.

A. TRANSDUCTION

The only well documented report of transduction in a streptomycete is that of Stuttard (1979) on *Stmy. venezuelae*, involving the narrow host-range phage ϕSVI. Several phenotypically unrelated and genetically distantly linked markers were successfully transduced, so that the system is an example of generalized transduction. Yields of prototrophic transductants were significantly increased by ultraviolet irradiation of the phage preparations before addition to auxotrophic recipients. Evidence suggestive of localized mutagenesis and co-transduction has recently been reported (Stuttard, 1981). There is no doubt that the availability of a convenient transduction system would be most valuable for fine genetic analysis in *Stmy. coelicolor* and other streptomycetes and it is to be hoped that this promising result in *Stmy. venezuelae* will soon be extended both in that organism and in others.

B. CONJUGATION

(a) *The Identification and Properties of Sex Plasmids in* Streptomyces coelicolor *A3(2)*

First reported in 1957 (Hopwood, 1957) and used to construct the beginnings of a linkage map 2 years later (Hopwood, 1959), the recombination system of this isolate was subsequently found to depend on the activity of at least two sex plasmids, SCP1 (Vivian, 1971) and SCP2 (Bibb *et al.*, 1977). Recently, two further plasmids, SLP1 and SLP4, have been identified by the transfer of DNA, either naturally or artificially, from SCP1$^-$SCP2$^-$ *Stmy. coelicolor* A3(2) to '*Stmy. lividans*' 66 (Hopwood *et al.*, 1979; Bibb *et al.*, 1980, 1981). The SLP1 series of autonomous plasmids (SLP1.1, SLP1.2, etc.) thus revealed in '*Stmy. lividans*' appear to originate by 'looping out' of part of the chromosome of *Stmy. coelicolor*. The organization of SLP4 in *Stmy. coelicolor* is unstudied. Since no strains of *Stmy. coelicolor* A3(2) lacking SLP1 or SLP4 DNA are available, nothing is known of their possible role in its fertility system (although we do know that SLP1 but not SLP4 is a fertility plasmid in '*Stmy. lividans*'); the information on the properties of SCP1 and SCP2 which follows has been obtained with strains of *Stmy. coelicolor* all presumably harbouring SLP1 and SLP4 DNA.

The 'wild-type' *Stmy. coelicolor* A3(2) isolate carries SCP1 and SCP2 each in the autonomous state (SCP1$^+$SCP2$^+$). Since each plasmid is lost spontaneously from about 0.03% to about 1% of the spores arising from a plasmid-containing culture (Vivian and Hopwood, 1970; Bibb *et al.*, 1977; Bibb and Hopwood, 1981) and at higher frequency (2 to 25%) when protoplasts are prepared and allowed to regenerate (Hopwood, 1981a), strains lacking one or both plasmids can be isolated reproducibly. In the case of SCP1, the most convenient phenotypic character for monitoring plasmid loss is the lack of the ability to produce the SCP1-coded antibiotic methylenomycin A (see below). Another useful criterion is the change in recombination frequency when SCP1 is lost; this can be observed on individual colonies by the plate-crossing technique (Hopwood *et al.*, 1969) and is best seen in tests against an NF strain (which has SCP1 integrated into the chromosome: see below); SCP1$^-$ cultures give considerably higher levels of recombinants with NF strains than do SCP1$^+$ strains (Fig. 1) and so can be easily recognized (Hopwood *et al.*, 1969). On the other hand, the recombination frequencies in crosses of SCP2$^+$ and SCP2$^-$ strains are too low for satisfactory testing of colonies in plate-crosses and so SCP2$^-$ strains are usually isolated, not from SCP2$^+$ cultures, but from strains carrying the SCP2* variant form of the plasmid, which promotes an enhanced level of recombination. Loss of this level (Bibb *et al.*, 1977), correlated with loss of the lethal zygosis phenotype (see below) shown by SCP2* strains against a pre-existing SCP2$^-$ culture (Bibb and Hopwood, 1981), identifies rare individual colonies of SCP2$^-$ variants arising from an SCP2* culture.

Both SCP1 and SCP2 (or SCP2*) are transmissible plasmids, neither being dependent on the other for transfer to a strain lacking the corresponding plasmid. Indeed, in any mixed culture ('cross') of a plasmid$^-$ strain with a plasmid$^+$ strain (containing SCP1 itself in any of its known forms, see below, or SCP2 or SCP2*), virtually the whole of the plasmid$^-$ genotype

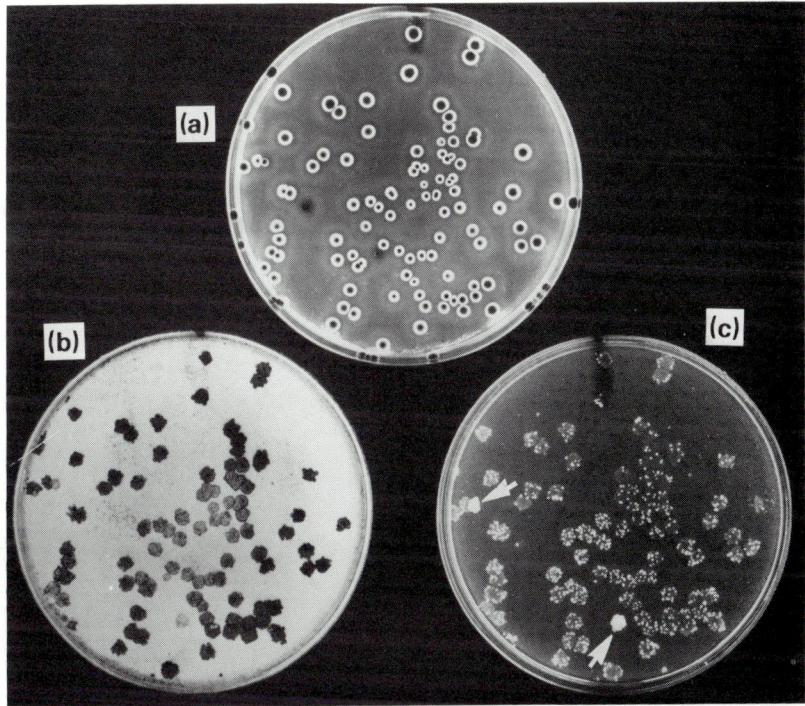

FIG. 1. Recognition of SCP1⁻ derivatives of an SCP1⁺ culture of *Streptomyces coelicolor* A3(2). The plate of colonies (a) was replica plated to a lawn of an NF strain on non-selective medium (b) and this in turn to a plate of selective medium (c). Most of the colonies (SCP1⁺) give a small number of recombinant progeny but two SCP1⁻ derivatives give rise to confluent patches of colonies (arrows). (From Hopwood et al., 1973.)

comes to contain the plasmid during the course of the cross (a period of several days of mixed growth followed by sporulation). When crosses are made between chromosomally marked strains carrying different variants of SCP1 (SCP1 and SCP1-prime: Kirby, 1976) or SCP2 (SCP2 and SCP2*: Bibb and Hopwood, 1981) the plasmid variants fail to re-assort randomly with respect to the two chromosomal genotypes. Instead, each plasmid tends to remain associated with its parental genotype, a phenomenon called 'entry disadvantage' by Kirby (1976). It may reflect a combination of phenomena comparable with the surface exclusion and incompatibility shown by some eubacterial plasmids, but the mycelial growth habit and differentiation into spores shown by streptomycetes, and the current impossibility of studying the transfer process itself over a short time period, make the situation very hard to analyse.

Another phenomenon that may also parallel a property of some eubacterial plasmids, derepression of transfer functions in newly infected cells (Stocker et al., 1963), has been shown to operate for SCP2 by the analysis of triparental matings between SCP2*, SCP2$^+$ and SCP2$^-$ strains (Bibb and Hopwood, 1981); the results suggest that SCP2 and SCP2* may become temporarily derepressed for transfer functions on entering SCP2$^-$ mycelium. Furthermore, the best available explanation for the SCP2* type of variant is that it is a partially depressed form of SCP2.

The frequencies of recombination of chromosomal markers observed in crosses of strains with various combinations of the autonomous SCP1 and either SCP2 or SCP2* plasmids are somewhat variable, depending on factors such as the 'balance' between the two parental types and their background genotypes, but characteristic average values are nevertheless seen. Such frequencies, taken from Bibb and Hopwood (1981), are shown in Table 1.

TABLE 1. Approximate recombination frequencies in crosses involving SCP1, SCP2 and SCP2* (from Bibb and Hopwood, 1981)

	SCP1$^-$ SCP2$^-$	SCP1$^-$ SCP2$^+$	SCP1$^-$ SCP2*	SCP1$^+$ SCP2$^-$	SCP1$^+$ SCP2$^+$	SCP1$^+$ SCP2*
SCP1$^-$ SCP2$^-$	10^{-8}	—	—	—	—	—
SCP1$^-$ SCP2$^+$	10^{-6}	10^{-7}	—	—	—	—
SCP1$^-$ SCP2*	10^{-3}	10^{-5}	10^{-3}	—	—	—
SCP1$^+$ SCP2$^-$	10^{-6}	10^{-5}	10^{-2}–10^{-3}	10^{-5}	—	—
SCP1$^+$ SCP2$^+$	10^{-5}	10^{-6}	10^{-3}	10^{-5}	10^{-5}	—
SCP1$^+$ SCP2*	10^{-2}–10^{-3}	10^{-4}	10^{-2}–10^{-3}	10^{-2}–10^{-3}	10^{-3}	10^{-3}

Even in the absence of both sex plasmids a low level of recombination occurs (about 10^{-8}) which is not an artefact caused by reverse mutation or some kind of complementation without recombination since new combinations of parental auxotrophic markers are regularly observed; whether this recombination involves the activity of SLP4 or the integrated SLP1 sequences referred to above, or some other unidentified plasmid, or is independent of any plasmid, is not known. Either SCP1 or SCP2 alone, in one or both parents, enhances the recombination frequency by up to three orders of magnitude. In the case of SCP1, the presence of the plasmid in both parents produces a slightly higher frequency of recombinants than when only one parent contains it, whereas $SCP2^+ \times SCP2^+$ crosses are at least ten times *less* fertile than $SCP2^+ \times SCP2^-$ crosses. Neither plasmid appears either to help or to hinder recombination promoted by the other. Crosses in which one of the parents contains SCP2* are considerably more fertile than in its absence and, as in the case of SCP1, but in contrast to that of SCP2, its presence in both parents does not lower this level of recombination.

Little is known about how these autonomous plasmids promote the transfer of chromosomal markers in crosses. Certainly SCP1 can interact physically with the chromosome, since examples of stable integration of the plasmid into the chromosome have been characterized genetically. The first example gave rise to the so-called NF strains (Hopwood *et al.*, 1969, 1973) and several others have been described (Hopwood and Wright, 1976b). The converse situation, the integration of chromosomal regions into the plasmid to give SCP1-primes, also occurs (Hopwood and Wright, 1976a, b). In crosses involving any of these variants, very high frequencies of recombination occur for particular markers – those on one or both sides of the integration points of SCP1 into the chromosome, or the markers carried on an SCP1-prime plasmid, are donated with high frequency to $SCP1^-$ strains. Thus a polarity in marker transfer from the $SCP1^+$ to the $SCP1^-$ parent in a cross could be expected, reflecting the presence of occasional donor variants within the $SCP1^+$ culture, and such a polarity has been reported (Hopwood *et al.*, 1973); however, this may well not be the whole story, just as some, but not all, of the recombination in $F^+ \times F^-$ matings in *Esch. coli* K-12 is due to the presence of Hfr variants within the F^+ population (Evanchik *et al.*, 1969).

In contrast, evidence for physical interaction of SCP2 or SCP2* with the chromosome is lacking. Not only is it hard to discern any meaningful polarity of marker transfer in $SCP2^+ \times SCP2^-$ or $SCP2^* \times SCP2^-$ crosses (Bibb and Hopwood, 1981) but a search for cases of stable integration of SCP2 into the chromosome, or examples of SCP2-primes, has proved fruitless (J. A. Ewing, personal communication).

Because of the mycelial growth habit of the organisms, kinetic analysis of the process of marker transfer in situations that might, in principle, be capable of analysis (such as matings of an NF donor strain with an $SCP1^-$ recipient) is difficult. In such matings, a gradient of increasing marker transfer with time was reported in experiments in which attempts were made to interrupt the mating process by grinding and re-plating a mixed culture, or by transfering the culture from a non-selective to a selective medium (Sermonti and Puglia, 1976). The apparent kinetics of marker transfer deduced from these experiments indicated a slow rate of chromosomal transfer. Thus there are indications that more extended heterozygous regions occur in later-formed compared with earlier-formed zygotes, but the significance of this is not clear and it should not be seen as a close parallel of events in *Esch. coli* Hfr × F⁻ crosses.

The examples of integrated SCP1 donors so far described present some interesting features. In NF, in which SCP1 behaves as if integrated in the 9 o'clock region of the circular chromosome (Hopwood *et al.*, 1969), a bidirectional gradient of inheritance of donor markers by recombinants is observed in matings with SCP1⁻ recipients: most recombinants inherit NF markers from regions close to, *and on either side of*, the 9 o'clock region, with decreasing frequencies for markers progressively further from 9 o'clock, *in both directions*. Essentially all progeny inherit the NF character itself. On the simplest hypothesis, the zygotes in NF × SCP1⁻ matings contain the complete SCP1 chromosome and a fragment of the NF chromosome, of variable length but always including the 9 o'clock region. Haploid recombinants arise by crossing-over on either side of 9 o'clock and, for some unknown reason, only those recombinants inheriting this region from the NF parent survive. (This does not appear to be due to a simple selective advantage for the methylenomycin-producing (and methylenomycin resistant) NF class over the sensitive non-producing SCP1⁻ class since methylenomycin non-producing mutants of an NF strain showed the same phenomenon: Kirby and Hopwood, 1977.) Interestingly, protoplast fusions (see below) between NF and SCP1⁻ strains give rise to equal inheritance of donor and recipient markers for all regions of the linkage map, as expected – presumably (almost) complete diploids occur instead of the merozygotes produced by mating – but the obligate inheritance of NF amongst the progeny is still found (D. A. Hopwood and H. M. Wright, unpublished results).

Other SCP1-integrated donors differ from NF in giving a unidirectional gradient of marker transfer to the progeny (Vivian and Hopwood, 1973; Hopwood and Wright, 1976b); obligate inheritance of the donor character is again observed. The existence of such *unidirectional* donors suggests that the bidirectional donation property of NF is not an inevitable consequence of SCP1 behaviour on integration and indeed there are some reasons to believe that the formation of bidirectional donors may have involved a second event which occurred after a primary recombinational interaction of SCP1 with the chromosome. NF itself arose in an early cross between marked derivatives of the wild-type and its origin could not be investigated (Vivian and Hopwood, 1970) but bidirectional donors very similar to NF arose from cultures containing an 'unstable unidirectional donor' (Vivian and Hopwood, 1973) – probably an SCP1-prime carrying an unmarked chromosomal region and capable of promoting chromosome transfer by virtue of the homology between exo- and endogenote (Hopwood and Wright, 1976b). Another example of the production of a bidirectional donor in two steps was reported by Hopwood and Wright (1976a), in which a characterized SCP1-prime integrated into the chromosome by crossing-over at the appropriate region of homology.

It should be noted that all studies so far carried out on the SCP1 donors were done with cultures now known to contain SCP2 (and indeed most existing NF strains appear to carry SCP2*; M. J. Bibb, unpublished results). It is therefore possible that some of the characteristics of crosses of such donors with SCP1⁻ strains might be different if SCP2⁻ strains were studied. The characteristics described for NF×NF matings (Vivian and Hopwood, 1970) might well reflect the activity of SCP2* in the cultures since the finding of apparently random inheritance of markers from all regions of the linkage map is at first sight not what would have been expected from strains with SCP1 integrated at the 9 o'clock position. This is therefore an interesting situation for future analysis. (The suggestion by Sermonti and Puglia (1976) that, in NF×SCP1⁻ matings, the earliest non-parental products are SCP1⁺ strains in which the integrated SCP1 would have detached itself from the NF chromosome could conceivably have reflected the presence of SCP2* in the NF culture, so that these early-produced progeny, with a rather low level of recombination, possibly carried SCP2*, and not SCP1 as was claimed.)

(b) *Plasmids and Lethal Zygosis*

The known plasmids of *Stmy. coelicolor* and of other streptomycetes exhibit a property that has been called lethal zygosis (Bibb *et al.*, 1977), by analogy with a phenomenon associated with the F plasmid of *Esch. coli* K-12 in which F⁻ cells tend to be killed by multiple mating reactions with Hfr strains (Skurray and Reeves, 1973). It was first described for *Streptomyces* when patches of an SCP2* variant were replica-plated to a lawn of an SCP2⁻ culture; the resulting SCP2* patches were surrounded by narrow zones in which growth (particularly sporulation) of the SCP2⁻ culture was retarded (Fig. 2). A diffusible inhibitor to account for this effect was virtually

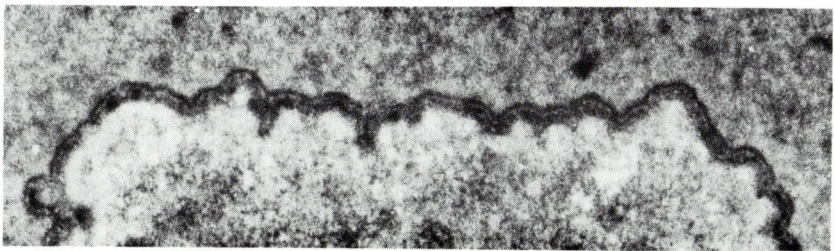

FIG. 2. Zones of lethal zygosis surrounding a patch of an SCP2* culture of *Streptomyces coelicolor* A3(2) replicated to a background of an SCP2⁻ strain (magnification ×10).

ruled out by the observation that the width of the inhibition zone was constant, even around the complex contour of an SCP2* patch (Bibb et al., 1977). Moreover, single spores of an SCP2* strain developing within a confluent lawn of an SCP2⁻ culture gave rise to small circular zones (Fig. 3a) – called 'pocks' – in which a minute region of growth of the SCP2* strain was surrounded by inhibition zones exactly similar to those seen around large replicated patches of growth (Bibb et al., 1978), which would not have been expected if the phenomenon were due to a diffusible agent.

Recognition of the lethal zygosis (ltz) reaction of SCP2* on SCP2⁻ (and later, on certain media, but to a lesser extent, of SCP2⁺ on SCP2⁻: Bibb and Hopwood, 1981) led to a re-interpretation of the similar pocks which had been seen on transfer of SCP1 into 'Stmy. lividans' (Hopwood and Wright, 1973) and which had been assumed at first to be due to the production of methylenomycin (the antibiotic was not chemically identified at the time) by SCP1⁺ colonies inhibiting the methylenomycin-sensitive SCP1⁻ background. Later it was found that methylenomycin non-producing mutants gave rise to the same pock reaction, ruling out this explanation. That transfer of SCP1 into the SCP1⁻ background was needed for the expression of the inhibitory response was shown by the lack of pock formation when mutants of SCP1 defective in transfer were studied (Kirby, 1976).

In the case of SCP2*, individual pocks in an SCP2⁻ lawn were analysed for chromosomal genotype and plasmid status in the various concentric zones (Fig. 3b). It was found (Bibb and Hopwood, 1981) that markers of the parental SCP2* strain (in the parental genotype and in recombinants) were confined to the small central area of growth, all the remaining zones containing only the SCP2⁻ genotype. However, the SCP2* plasmid was present in all the zones, even up to the edge of the confluent growth at the edge of the pock. This result indicates that the plasmid is capable of invading the recipient mycelium over a distance of a millimetre or more, well beyond any spread of the parental SCP2* mycelium or its chromosomal markers.

Whatever its basis, lethal zygosis is a most useful phenomenon in the experimental genetics of *Streptomyces*. It led to the recognition of the SLP1 family of plasmids (Bibb et al., 1980b) (we then thought that they originated from 'Stmy. lividans' 66 rather than from Stmy. coelicolor A3(2) as we now know – see above), and to the development of a transformation system for introducing plasmid DNA into *Streptomyces* protoplasts (Bibb et al., 1978). It also allowed detection of the transfer of SCP1 and SCP2* into 'Stmy. lividans' 66 and Stmy. parvulus ATCC 12434 (Hopwood and Wright, 1973; Bibb and Hopwood, 1981). Moreover, the ltz reaction is shown by a number of newly discovered plasmids (e.g. Figs. 3c and 3d) identified by physical studies in other members of the 'Stmy. violaceoruber' species group (to which Stmy. coelicolor A3(2) and 'Stmy. lividans' 66 both belong), upon

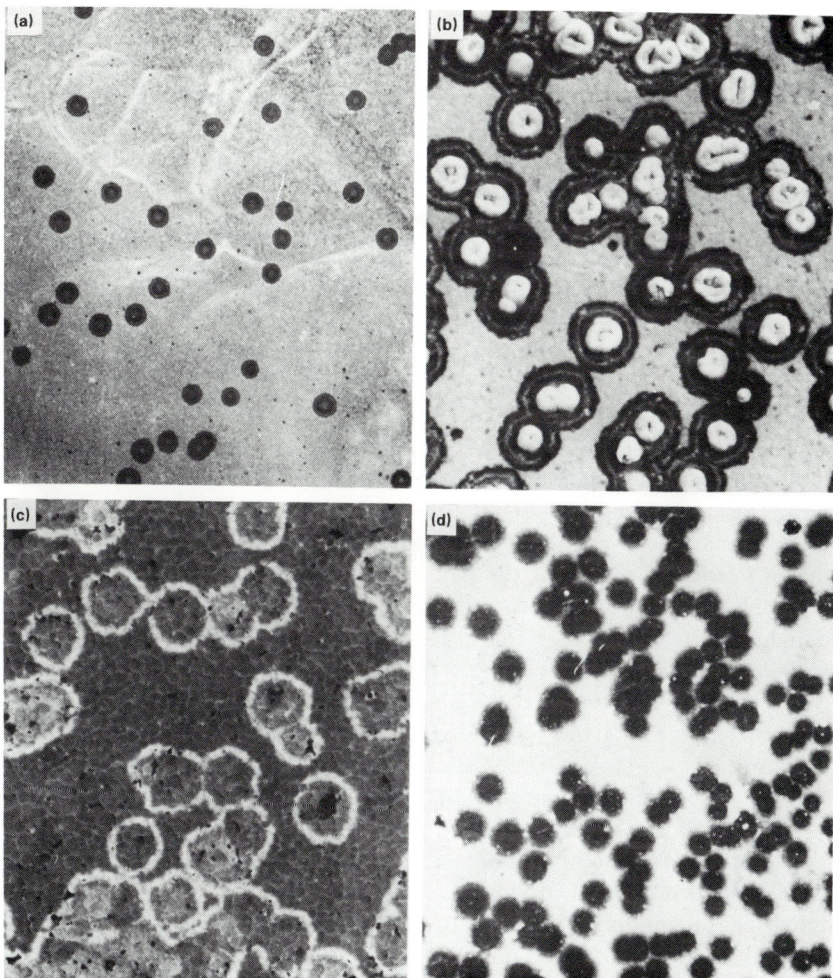

FIG. 3. 'Pocks' produced by various plasmids when individual plasmid-containing spores grow in a background of plasmid-free individuals (Magnification ×4). (a) SCP2* in *Streptomyces coelicolor* A3(2). (b) SCP2* in *Stmy. coelicolor* A3(2), with SCP1 present in the cultures; note the very distinct zones of growth and inhibition in each pock. (c) pIJ101 in '*Stmy. lividans*' 66. (d) pIJ101 in *Stmy. parvulus* ATCC 12434).

plasmid transfer by conjugation or transformation to *Stmy. coelicolor* A3(2) and '*Stmy. lividans*' 66 (D. A. Hopwood and H. M. Wright, unpublished results). There is therefore reason to think that the presence of virtually any conjugative *Streptomyces* plasmid could be revealed by such a reaction on a suitable medium which allows the right balance in growth rates of the background plasmid⁻ lawn and the plasmid-containing mycelium.

(c) *The Mechanism of Conjugation in* Streptomyces

Although we do not know the mechanism whereby plasmids, with or without accompanying chromosomal markers, are transferred between the hyphae of streptomycetes, we are certainly in a better position to attack this problem than we were a few years ago, with the recognition of a number of conjugative plasmids and the physical isolation of several of them.

SCP1 is hard to isolate (Hopwood *et al.*, 1979; Westpheling, 1980), probably partly because it is large (at least 150 kb), and it has not been satisfactorily characterized physically. On the other hand, SCP2 (and SCP2*, which does not differ from it structurally in any recognizable way) is rather easily isolated as ccc DNA molecules of 31 kb, existing in the mycelium in low copy number: estimates of about one (Bibb *et al.*, 1977) or four to five (Schrempf and Goebel, 1977) copies per chromosome have been obtained. When SCP2* was cleaved with restriction enzymes and religated, reduced replicons were obtained, about 15 kb in size, which showed all the known properties associated with SCP2* (transfer, chromosomal mobilization and lethal zygosis) but were highly unstable (Bibb *et al.*, 1980a). (It was suggested that a particular region of the parent SCP2* molecule, outside the 15 kb segment, was responsible for accurate distribution of plasmid copies to daughter cells as has been found for an *Esch. coli* plasmid: Meacock and Cohen, 1980.) Evidently the reduced SCP2* replicon, large enough to carry only about 15–20 average-sized genes, was normally transmissible, even if unstable, suggesting the possibility of a rather simple mode of conjugal transfer in *Streptomyces*. The same conclusion arises from the findings that SLP1.6, only about 10 kb in size (Bibb *et al.*, 1980b), is transferable in '*Stmy. lividans*', and that a plasmid called pIJ101, one of the series of new '*Stmy. violaceoruber*' plasmids referred to above, is only 9 kb in size and transferable in several species (Hopwood *et al.*, 1981; Kieser *et al.*, 1982). An ambiguity, however, surrounds conclusions of self transmissibility of these plasmids; they might perhaps depend on specific transfer functions provided by other plasmids present, undetected, in the same strain, or perhaps provided by the chromosome. In the latter case, the problem becomes partly semantic and hinges on what is useful to describe as self-transmissibility; any plasmid, even F which codes for specific gene products that assemble the sex pilus needed for transfer, requires the products of chromosomal genes (the rest of the cell) in order to transfer itself to another cell, and indeed there are host mutations which abolish conjugation (Hanekes and Hoekstra, 1976). Probably the best approach is to seek to identify, by mutation or deletion, particular regions of *Streptomyces* plasmids that are essential for transmissibility, and then to try to determine their roles. This approach has begun for pIJ101 and SCP2 in our laboratory.

pIJ101 is the largest member of a family of four plasmids present in the strain in which it was discovered ('*Stmy. lividans*' ISP 5434); the others (pIJ102, pIJ103 and pIJ104) are smaller (3.9–4.7 kb); they, and pIJ101, all share a common region (the 3.9 kb of the smallest member of the series), and indeed all the DNA of each plasmid appears to be included in larger members of the series. Only pIJ101 behaves as an efficiently transferable plasmid (Kieser et al., 1982). Therefore, the segment of about 4.3 kb which distinguishes pIJ101 from the largest of the other plasmids presumably carries transfer functions. This was confirmed by the cloning of foreign DNA segments carrying drug resistance markers into various sites in pIJ101; cloning into certain sites in the unique segment of pIJ101, but not into sites in the region common to the whole family of plasmids, abolishes efficient transferability. These studies have established that the transfer function(s) lie within a segment of only 2 kb (Kieser et al., 1982). Thus, some streptomycete plasmids, at least, carry genes needed for their own efficient transfer rather than being passively transferred from mycelium to mycelium by some non-specific, possibly plasmid-independent, hyphal fusion mechanism.

Perhaps one of the most interesting aspects of *Streptomyces* plasmids is the possibility that they could promote their own transfer, not only between unconnected hyphae, but also within a continuous segment of mycelium. One could imagine a selective advantage for such a property of plasmids dependent on mycelial hosts; conceivably the spread of SCP2* within a pock on a background of SCP2$^-$ mycelium, described earlier, is a manifestation of such a function. Indeed, there are *in vitro* derived 'small pock' derivatives of pIJ101 which appear to lack this 'spreading' function while retaining interstrain transmissibility (Kieser et al., 1982). Thus, although certain properties of *Streptomyces* plasmids may well reflect comparable phenomena already encountered in Gram-negative bacteria, some, at least, will probably turn out to be interestingly different.

(d) Conjugation in Other Streptomycetes

A considerable list of examples of conjugal gene exchange was published by Hopwood and Merrick (1977) and the reader is referred to that review for the available information, often very meagre, on these systems. Apart from *Stmy. coelicolor* A3(2), only in *Stmy. rimosus* has a sex plasmid been implicated in gene exchange in a genetically extensively studied organism. The linkage map of more than one strain of *Stmy. rimosus* has been built up by means of the conjugation system over the years (Friend and Hopwood, 1971; Alačević et al., 1973) but only recently has a sex plasmid responsible for gene transfer been recognized. The plasmid, SRP1, was detected by the chance origin of SRP1$^-$ strains within a pedigree of cultures (Friend et al., 1978). As in the case of SCP2, SRP1$^+$ × SRP1$^-$ crosses are more fertile than SRP1$^+$ × SRP1$^+$ crosses (10^{-3} to 10^{-4} compared with 10^{-5} to 10^{-7}); moreover, a low level of true recombination was detected in the absence of SRP1 (about 10^{-8}), and new evidence has shown that this is partly due to the presence of a second sex plasmid, SRP2 (E. J. Friend, personal communication). SRP1 and SRP2 have not been characterized physically.

In 'Stmy. reticuli', a conjugative plasmid was found to be necessary for antibiotic production and resistance, sporulation and melanin production and this plasmid, of about 75 kb, has been isolated, but not yet analysed in detail from the point of view of chromosomal gene exchange (Schrempf and Goebel, 1979).

4. Protoplast Fusion

Streptomyces genetics entered a new phase in 1977 with the discovery that very frequent chromosomal recombination occurred when protoplasts of two genetically marked derivatives of the same strain were mixed, treated briefly with polyethylene glycol (PEG) and allowed to regenerate into a mycelial, sporulating culture (Hopwood *et al.*, 1977). The technique appeared to be generally applicable to streptomycetes, since all five strains tested (*Stmy. coelicolor*, '*Stmy. lividans*,' *Stmy. parvulus*, *Stmy. griseus* and *Stmy. acrimycini*) yielded frequent recombinants and, in *Stmy. coelicolor* A3(2), recombination did not require the presence of either of the known sex plasmids, SCP1 or SCP2. This report was soon followed by others, notably those of Baltz (1978) on *Stmy. fradiae* and *Stmy. griseofuscus* and of Ochi *et al.* (1979) on *Stmy. parvulus* and *Stmy. antibioticus*. These authors made a direct selection on the regeneration medium for prototrophic colonies produced by fusion of protoplasts of two auxotrophic strains. Although this allowed direct estimation of the frequency of colonies containing prototrophic recombinants, the amount of information obtainable was limited by the selection. We have adopted instead the approach of regenerating the fused protoplasts under non-selective conditions, thereby allowing the recovery of all possible genotypes of progeny: four in a two-factor cross, 64 in a six-factor cross, etc. In *Stmy. coelicolor*, recombinants constitute 10–20% of the total spores on the regenerated cultures following protoplast fusion. It was found that the total recombination frequency was about the same as the frequency of colonies containing any recombinants; hence the estimates of the 'apparent recombination frequency' (2–25%) made by Ochi *et al.* (1979) are in the same range as the estimates of true recombination frequency in our work.

From the detailed analysis of a six-factor cross (Hopwood and Wright, 1978), we deduced that the primary fusion products contain essentially complete genomes of the two parents, in contrast to the merozygotes produced in a conjugal cross. At some time between the onset of regeneration and the production of spores on the regenerated colonies, several rounds of recombination can apparently occur, since many individual colonies contain a series of recombinant genotypes; but the process is complete.by sporulation (and probably long before), since all the spores behave as pure haploids. The analysis of genotypes to be found in single regenerating colonies showed that groups of parental alleles were always missing and it was suggested that genome fragmentation might account for this; it is probably more likely that it reflects, instead, a built-in 'sampling error' such that the hyphae that arise from individual fusion bodies happen not to include all the genotypes potentially available within the fusion body (Hopwood, 1981c). By the analysis of fusions between four differently marked parents it was deduced that at least four protoplasts can participate in individual fusion and recombination events, but the average number involved is quite low (only two in these experiments, when we include the protoplasts which fail to fuse at all).

The frequency of recombinants amongst the regenerating fused protoplast population was increased even further if the parental protoplasts were irradiated with ultra violet light immediately before fusion, since unfused protoplasts would fail to survive, while fusion allowed the repair by recombination of lethal hits at different sites in the chromosomes of different members of the population, thus providing an enrichment for recombinants (Hopwood and Wright, 1979). Baltz (1980) confirmed the same phenomenon by the enhanced survival of a single irradiated protoplast preparation when treated with PEG compared with an untreated control. Recently we have found that very heavily irradiated protoplasts can give recombinants when fused with unirradiated protoplasts, the difference in ultraviolet killing kinetics for recombinants in such a fusion and for protoplasts untreated with PEG being quite striking (Hopwood and Wright, 1981).

Interspecific fusions, although they have been reported to yield recombinants (Godfrey et al., 1978), have been very little investigated and this topic is worthy of considerable further study. A number of factors, including homology or inhomology of the genomes of the two strains, the operation of restriction-modification systems and interference by the secondary metabolites of one strain on the metabolism of the other might be expected to influence the outcome of interspecific fusions, which are typically very infertile (Hopwood, 1981a, c).

5. Liposome-mediated Transformation by Chromosomal DNA

Since fusion between protoplasts is so efficient at high PEG concentrations, it is not surprising that the same conditions allow efficient fusion of lipid vesicles (liposomes) with protoplasts, with the consequent introduction of the aqueous contents of the liposomes into the cytoplasm. Like many other biologically active molecules, DNA may be introduced in this way: thus

Makins and Holt (1981) found that when *Stmy. coelicolor* DNA was entrapped in liposomes made with extracted *Streptomyces* lipids, fusion with homologous protoplasts carrying complementary genetic markers resulted in transformation. Remarkably, as many as 10% of regenerated protoplasts were transformed for any of a number of genetic markers tested. The same high frequency was observed whether auxotrophs were transformed to prototrophy or prototrophs to auxotrophy. No linkage was detected between markers loosely linked in conjugational genetic analysis. This procedure should prove valuable for fine structure analysis and strain construction; and further studies will reveal whether it will be useful for localized mutagenesis or for screening DNA fragments for biological activity.

6. Genetic Analysis

A. CHROMOSOMAL LINKAGE MAPPING BY CONJUGATION

The first steps in the construction of a *Streptomyces* linkage map, in *Stmy. coelicolor* A3(2), depended on the analysis of the recombinant progeny arising in four-factor crosses by selecting colonies on four different media, each selecting for a different pair of markers, one from each parent, and leaving two markers (and therefore four genotypes) unselected (Hopwood, 1959, 1972). This procedure allows the recovery of nine out of 14 genotypes of recombinant progeny, including two complementary pairs. When the members of these pairs are found to have approximately equal frequencies, the *average* contribution of genetic material to the zygotes by the two parents may be assumed to be equal (although each zygote is incomplete, containing a whole chromosome from one parent and a fragment from the other: Hopwood, 1967a). Thus the frequency of any recombinant genotype can be taken as an estimate of the frequency of the complementary pair to which it belongs, and the total analysis provides enough information for the estimation of *relative* linkage distances between all pairs of markers on a circular linkage map. If necessary, appropriate recombinant genotypes from the first cross can be used to make two more crosses, with the same markers in new coupling arrangements, to confirm the map (Hopwood, 1959). This type of analysis served to establish the outlines of a linkage map in several other species: *Stmy. rimosus* (Friend and Hopwood, 1971), *Stmy. glaucescens* (Baumann et al., 1974), '*Nrda.*' *mediterranea* (Schupp et al., 1975) and *Stmy. acrimycini* (Wright and Hopwood, 1977). Once a few markers are located, others can be added to the map by making a simple selection for a marker from each parent, on opposite sides of the linkage map, and considering the segregations of the new marker(s) and of the non-selected markers of known map location (Hopwood, 1972).

The above procedures provide a means of rather coarse mapping. More detailed mapping can be done by making a selection for closer markers (Hopwood and Chater, 1974) but, although some groups of closely linked mutations have been analysed (e.g. Harold and Hopwood, 1970; Chater, 1972), a really efficient fine mapping procedure is lacking: hence the desirability of developing transduction as a generally available mapping tool. Liposome-mediated transformation may also prove valuable in this context.

In the special situation of 'ultra-fertile' crosses between NF and SCP1⁻ strains of *Stmy. coelicolor* A3(2) (Hopwood *et al.*, 1969), the total progeny can be recovered on a non-selective medium and characterized for the segregation of all markers. This provides a good coarse mapping procedure for markers on the left half of the linkage map (an example was given by Hopwood and Chater, 1974): donor markers in the right half are inherited too rarely by recombinants for efficient mapping (but a series of donors with SCP1 integrated into the chromosome at different points could, in principle, be used to cover the whole linkage map).

Heteroclones, which are recovered on selective media along with haploid recombinants when selection is made for closely-linked complementary markers (Hopwood and Sermonti, 1962; Hopwood *et al.*, 1963), arise from partially diploid recombinants containing genomes that behave formally as having terminally redundant regions (Hopwood, 1967a) and may actually contain tandem duplications (Anderson and Roth, 1977). These heterozygous colonies produce predominantly haploid spores, representing a series of recombinant genotypes in respect of the heterozygous region in a particular colony, whose frequencies can be used for mapping purposes (Hopwood and Sermonti, 1962). Alternatively, map information can be obtained by scoring a series of heteroclones, selected to be heterozygous at a particular pair of closely linked loci, for heterozygosity of a set of non-selected markers; this analysis served to establish the relative distances between the main landmarks on the *Stmy. coelicolor* map (Hopwood, 1966, 1967a).

The current linkage map of *Stmy. coelicolor* A3(2), derived from a combination of all these mapping procedures, is shown in Fig. 4. The map shows two unusual features. One is that two long arcs, on opposite sides of the map, contain very few markers. Since the mapping units are recombination frequencies (together with the probability of the ends of the heterozygous regions in heteroclones falling in particular map intervals, a parameter which may be related to recombination: Hopwood, 1966), the long 'silent' quadrants may represent regions of more frequent recombination per unit length, rather than long regions of DNA containing no genes recognizable by mutations of the kind so far studied, but the possibility of DNA with special functions occupying these regions cannot be eliminated. The second feature is the symmetry of the map: genes involved in the control of different steps in the same biosynthetic pathway tend to be located diametrically

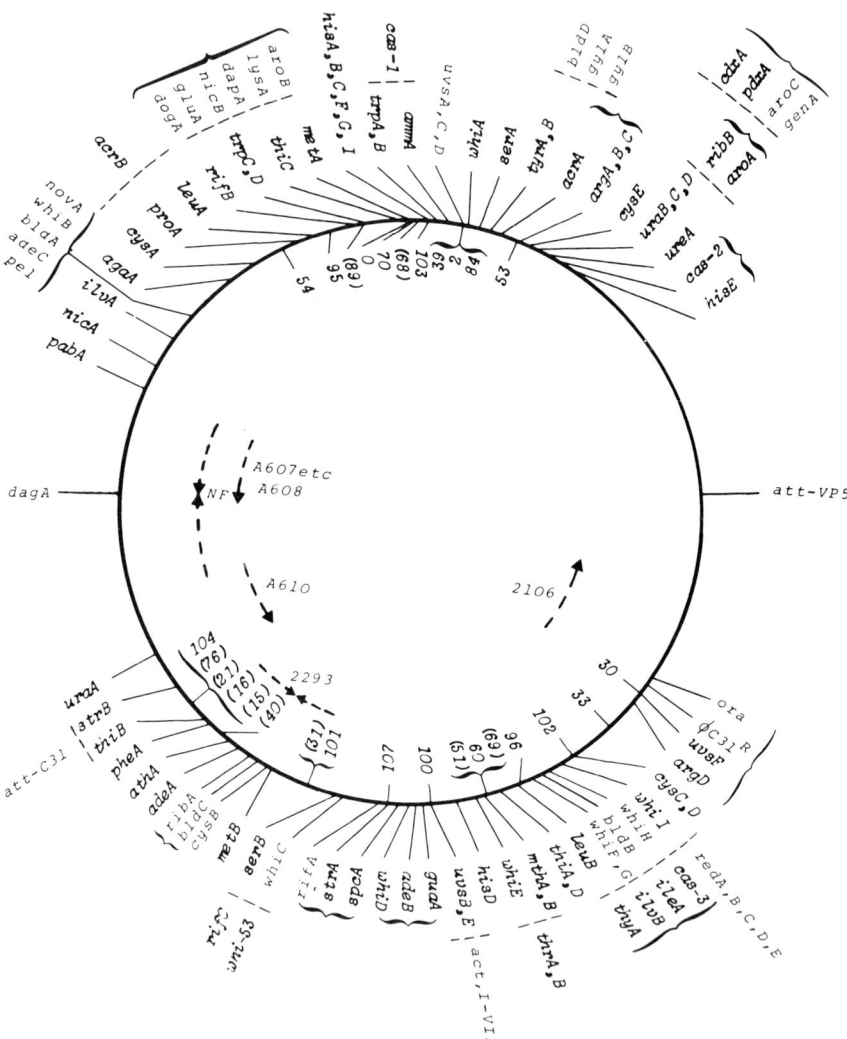

FIG. 4. Linkage map of *Streptomyces coelicolor* A3(2). Gene symbols are those in Hopwood *et al.* (1973) with the following additions: *act*, actinorhodin synthesis (Rudd and Hopwood, 1979); *dag*, diffusible agarase (Hodgson and Chater, 1981); *dog*, 2-deoxyglucose resistance (D. A. Hodgson and E. T. Seno, personal communication); *gen*, gentamicin resistance (Hopwood and Wright, 1976); *gyl*, glycerol utilization (E. T. Seno, personal communication); *nov*, novobiocin-resistance (D. A. Hopwood and H. M. Wright, unpublished results); *ora*, orange pigment production; *red*, red pigmented antibiotic synthesis (Rudd and Hopwood, 1980). Arrows indicate donation of markers by bidirectional or unidirectional donors.

opposite each other, a situation that may reflect an evolutionary tandem duplication of the whole genome, followed by loss of most of the redundant DNA (Hopwood, 1967b). There are also some strong indications that the linkage maps of several different streptomycetes bear a considerable similarity to each other (Friend and Hopwood, 1971), just as do those of enteric bacteria such as *Esch. coli* and *Salmonella typhimurium*. This similarity should not be overemphasized, especially since in very few cases is it known that genes recognized by auxotrophic mutations of a particular class, and apparently occupying comparable positions on the linkage map, are actually functionally identical, but there seems little doubt that the various linkage maps are related to some degree.

B. COMPLEMENTATION AND DOMINANCE TESTS

In spite of their uses, the analysis of heteroclones is relatively laborious for mapping purposes in relation to the information obtained; the calculations can be computerized (Alačević, 1976) but collecting the data remains time-consuming. Moreover, because of selective effects occurring during the growth of the heteroclone colonies, the analysis of haploid recombinant frequencies within individual heteroclones can be quantitatively misleading. However, heteroclones have another application; they provided the first possibility for dominance and complementation testing in *Streptomyces*. This use is probably confined to nutritional (Engel, 1973; Carere *et al.*, 1973) and certain recessive sensitive mutations (Harold and Hopwood, 1970), complementation between which can be used to select heteroclones. It is doubtful whether mutations concerned with characters such as antibiotic synthesis or differentiation, which are amongst the more interesting systems to analyse, can be used to select heteroclones; moreover, in the case of mutations affecting events within the aerial mycelium and spores, heterozygosity is unlikely to last long enough for efficient complementation. Naturally occurring plasmid-primes, such as those reported for SCP1 (see above), offer a better system, but it has not proved possible to isolate these efficiently for any given chromosomal region. Probably the construction of partial diploids by gene cloning (see below) is the best hope for analysing the functional relationships of genes.

Heterokaryons have been described as another approach to complementation testing. It is not clear how true heterokaryosis can occur in a bacterium like a streptomycete, which lacks a nuclear membrane and any known mechanisms for ensuring even a reasonably stable retention of genetically unlike chromosomes in the cell without recombining with each other. However, colonies answering to the operational definition of heterokaryons in filamentous fungi – unstable prototrophs arising by contact between two auxotrophic strains and giving rise to the two parental genotypes but not recombinants – were reported early in the study of *Streptomyces* genetics (Bradley and Lederberg, 1956). In these cases, they apparently constituted the only prototrophic colonies arising from matings but, in several other species (e.g. *Stmy. rimosus*: Friend and Hopwood, 1971; *Stmy. glaucescens*: Baumann *et al.*, 1974; *Stmy. albus* G: Chater and Wilde, 1980), they occurred in various proportions along with recombinants. In *Stmy. coelicolor* A3(2) they are found only rarely under the usual conditions of plating spores from a mixed culture of two auxotrophs on a selective medium (Hopwood and Sermonti, 1962), but are generated in large numbers when the two strains are mixed on cellophane lying on a non-selective medium and then transferred to a selective medium after a few hours of growth (Sermonti, 1969). In these conditions their formation is apparently unaffected by the complete absence of both SCP1 and SCP2 from the cross (D. A. Hopwood and H. M. Wright, unpublished results). They have been used to study complementation between auxotrophic mutations (Carere *et al.*, 1973) and to isolate newly induced auxotrophic mutations arising by co-mutation (Carere and Randazzo, 1976).

In examining published genetic studies on *Streptomyces* heterokaryons, we can find no well-documented *genetic* results (as opposed to the micromanipulation evidence of Bradley and Lederberg, 1956) that rigorously exclude the possibility that the so-called heterokaryons are really a manifestation of nutritional cross-feeding between the two parental strains growing in close proximity in conditions of nutrient limitation. (The transmission of melanin production in apparent heterokaryons of '*Stmy. scabies*' (Gregory and Huang, 1964) might have resulted from normal conjugal transfer between such closely growing hyphae.) Future attempts to analyse the nature of heterokaryons might profitably use well-defined temperature-sensitive mutations for macromolecular synthesis as the basis for selection, thus excluding nutritional cross-feeding as an explanation of any 'heterokaryons' that should arise.

C. CHROMOSOMAL MAPPING BY PROTOPLAST FUSION

Crossing-over in a given chromosomal segment is considerably more frequent in a protoplast fusion than in a natural mating: this probably reflects the facts that more of the genome is diploid after fusion than after mating (in which on average only about one fifth of the donor genome is transferred to each zygote), and that several rounds of recombination occur after protoplast fusion (Hopwood and Wright, 1978). This might be expected to make the detection of linkage over long distances more ambiguous in a fusion than in a mating. However, since the products of multiple crossing-over (four and six crossovers) in, say, a six-factor fusion cross are less frequent than those of double crossing-over (Hopwood and Wright, 1978), it must be possible to choose an arrangement of markers that minimizes the total number of crossovers and to generate a linkage map, and indeed retrospective analysis of the data of Hopwood and Wright (1978) shows that the correct map order of the markers can be deduced. Baltz (1980) gave an example of mapping unknown genes by protoplast fusion in *Stmy. fradiae*. Thus protoplast fusion should provide a useful routine mapping tool, especially in those strains in which conjugal recombination is rare or does not yield equal frequencies of complementary recombinant genotypes. Fusion is certainly a very efficient way of generating complex genotypes, by 'adding together' separately-induced auxotrophic mutations, which might then be used for mapping purposes by conventional mating (Hopwood *et al.*, 1977). Heteroclones have been reported after protoplast fusion but their detailed analysis has not been described (Baltz, 1980).

D. THE DETECTION OF EXTRA-CHROMOSOMAL INHERITANCE

There are several criteria that can be used to establish that a particular phenotypic property displayed by a microorganism is controlled by a plasmid (Hopwood, 1978). These include: a failure to locate a relevant genetic determinant on the chromosomal linkage map in a cross between strains differing in the character in question; conjugal transfer of the character, from a strain possessing it to one lacking it, at a much higher frequency than that of chromosomal recombination (infectious transfer); an absolute correlation between the presence of the character and that of extrachromosomal DNA; and a high frequency of loss ('instability') of the character, particularly after treatment with 'curing' agents known to cause the loss of plasmids during the reproduction of bacteria.

The first criterion of non-linkage of a genetic determinant to the chromosomal linkage map is subject to certain caveats. The linkage map should be either rather well-marked, or it should be relatively short (in crossover units). In streptomycetes the latter situation prevails. Thus the rarity of crossover events makes it virtually certain that any stable new chromosomal marker will show unambiguous linkage to a known marker in any single cross. Lack of linkage is manifested most simply by the finding that unacceptably high proportions of multiple crossovers would be needed to explain an observed set of progeny, no matter in which map interval the marker is located (e.g. Akagawa *et al.*, 1975; Freeman *et al.*, 1977). Although this kind of evidence has usually been regarded as indicating extra-chromosomal inheritance, it is also the behaviour that one might expect in either of two hypothetical cases in which the genetic determinant is physically a part of the chromosome: namely, induction of a gene to change from some non-expressing to an expressing state when both forms of the gene are present in the same cytoplasm (e.g. in a conjugational zygote); and induction of transposition when a transposable genetic element (see below) is introduced into a recipient lacking such a transposon (as has been reported, albeit at a low level, with two transposons of enteric bacteria: Beck *et al.*, 1980; Biek and Roth, 1980). In crosses where chromosomal recombinants are rare, but heterokaryons common, lack of linkage is shown when a character reassorts with the two parental chromosomal genotypes emerging from a heterokaryon, or fails to segregate from a heterokaryon when both parental genotypes do so (Gregory and Huang, 1964). (The interpretation of such results is not affected by the question, discussed earlier, of the exact nature of the 'heterokaryons'.) A comparable situation might be expected to apply in protoplast fusions, and there is one report of plasmid determination of a character (antibiotic production) when a high proportion of progeny from a fusion between antibiotic producing and non-producing strains had the chromosomal genotype of the non-producing parent but had acquired the ability to produce antibiotic (Ochi and Katz, 1978). However, chromosomal recombination in fusions may often be so frequent as to make it possible that apparently parental progeny are actually recombinants involving genetically unmarked regions of the chromosome.

Infectious conjugal transfer as unambiguous evidence of the presence of a plasmid (e.g. Vivian, 1971) may be expected to be particularly widely applicable in streptomycetes if, as seems likely at present, the great majority of their plasmids, even small ones, are transmissible (see above). However, there are special circumstances – see the discussion of unstable chloramphenicol resistance referred to below – where it may not be easy to be certain whether or not a truly infectious transfer of a character has actually taken place.

The unambiguous correlation of a phenotype with the presence of plasmid DNA has been particularly troublesome in streptomycetes since there are well established cases in which a genetically well defined plasmid long defied physical isolation or recognition, for example SCP1 in *Stmy. coelicolor* A3(2) (Westpheling, 1980). Such cases of 'difficult' plasmids may well be very common in streptomycetes. However, a related criterion – the ability to convert a strain lacking a phenotype into one manifesting it by transformation of protoplasts with small quantities of plasmid DNA, even amounts too small for physical characterization by the normal procedures – may help redress the balance since the transformation of *Streptomyces* protoplasts (at least by plasmids of up to 30 kb) is particularly efficient (Bibb *et al.* 1978).

The 'instability' of a phenotypic character is a particularly problematical criterion of plasmid involvement (Hopwood, 1978). Recent work has served to emphasize the unreliability of 'curing' by such agents as acridines or ethidium bromide as evidence of plasmid loss, since such agents are now known to cause frequent re-arrangements of DNA, either extrachromosomal or chromosomal (Schrempf and Goebel, 1979; Kieser *et al.*, 1981), rather than simply the total loss of plasmids. However, a potentially very useful procedure has recently been suggested, with the observation that either of the known sex plasmids of *Stmy. coelicolor* A3(2), SCP1 and SCP2, can be absent from quite a high frequency (2–25%) of individual colonies when protoplasts are prepared and allowed to regenerate (Hopwood, 1981a). Therefore the finding that a new phenotype reproducibly appears after such a mild 'curing' procedure as protoplasting has the potential of being quite a reliable indicator of plasmid involvement.

(a) The Detection of Transposable Genetic Elements

In eubacteria, the genetic determinants of many of the phenotypic properties (especially drug resistance, toxin production and catabolic pathways) commonly associated with plasmids are often part of a small genetic unit (a transposon) able to insert into new sites on the same or a different replicon present in a common cytoplasm (Starlinger and Saedler, 1976; Kleckner, 1977). This remarkable discovery has led to a radical reappraisal of ideas about genome evolution (Cohen, 1976). Moreover, the detailed properties of transposons have led to the development of exciting new procedures for genetic analysis and manipulation *in vivo* (Kleckner *et al.*, 1977). In any organism of real genetic interest (such as *Stmy. coelicolor* A3(2)) it has therefore become important to investigate endogenous transposons. In this section we review at some length the evidence purporting to show the existence of endogenous transposons in streptomycetes, and in doing so discuss the criteria that should be regarded as acceptable. We shall first focus on the phenomenon of chloramphenicol resistance in *Stmy. coelicolor* A3(2), which has received the most attention in the literature.

Streptomyces coelicolor A3(2) is naturally resistant to about 20 μg of chloramphenicol per ml through an undetermined, chloramphenicol acetyltransferase-independent, mechanism. Several years ago three laboratories independently discovered that sensitive mutants (Cml^S), resistant to only 1 μg ml^{-1}, occurred spontaneously at a very high frequency (about 0.1 to 2%) (Danilenko *et al.*, 1977; Sermonti *et al.*, 1977; Freeman *et al.*, 1977), and reverted to the original phenotype (Cml^R) at a lower, but detectable frequency (about 10^{-5}–10^{-3}) to give colonies that again segregated Cml^S derivatives at a high frequency (Freeman *et al.*, 1977). These first studies agreed in finding that Cml^R was not obviously genetically linked to the chromosome, these mapping attempts being complicated by the rarity of Cml^S progeny (either parental or recombinant) among the spores harvested from mixed cultures of Cml^R and Cml^S strains. An extrachromosomal location for the determinant was therefore suggested. However, a subsequent paper by Sermonti *et al.* (1978) suggested that (a) *cml* might be linked to the *argBAC* region, partially to account for the frequent occurrence of *argB* mutations among Cml^S mutants, (b) *cml* might be specified by a transposon, because of an apparent 'tendency of the active (*cml*) element to be transposed to the homologous region in zygotes', and (c) insertion sequence (IS)-like elements (Starlinger and Saedler, 1976) might be involved in the reversible Cml^R–Cml^S events. In a more recent paper Sermonti *et al.* (1980) have gone beyond these speculative models to state that *cml* is 'a jumping gene'. In the (necessarily detailed) analysis that follows we show that this statement is based on a mistaken interpretation of data presented in that paper, and thus that there is still no good evidence that *cml* is part of a transposon in *Stmy. coelicolor* A3(2).

The relevant data concern the segregation of Cml^R/Cml^S in crosses between strains carrying standard chromosomal markers. The crux of the argument used (most simply illustrated by reference to the data for cross 316 × 506 in Table 2 of Sermonti *et al.*, 1980) was that, although any one chromosomal map location for *cml* required that abnormally large numbers of recombinants should have arisen through multiple recombination events, and was therefore rejected (as had been done earlier on similar data by Freeman *et al.* (1977)), nonetheless within particular sub-classes of the recombinant population a particular map location for *cml* did not implicate multiple crossing-over in any of the recombinants. This was taken to give a real *cml* map location. Examination of a different sub-class of recombinants from the same cross also gave a *cml* location requiring no multiple crossover classes among that group of recombinants; but the location was different from that previously obtained. This, it was claimed, implied that taking part in the cross were two parental sub-populations *with different locations for cml*, hence proving transposability for the gene. Unfortunately there is a serious fallacy in this analysis. Careful examination of the proposed map locations for Cml^R in this cross reveals that no standard marker is present in one of the arcs between the initially selected marker and the marker 'arbitrarily' used as a second selection point in this unconventional mode of analysis; into this arc the *cml* marker was placed. However, in these circumstances, no multiple crossover recombinants involving *cml* could possibly be detected so that it is not surprising that they were completely absent from the progeny. All that the analysis is able to reveal is that *cml* does not map in the alternative arc, in which other markers are available to test for the frequency of multiple crossover classes. Similar criticisms apply to other data in the same paper. Thus, although the conclusion of this paper is not necessarily incorrect, the major argument on which it is founded is based on a misconception.

A number of other phenomena have been reported which may conceivably involve transposons in *Streptomyces*, for example, the frequent occurrence of mutants lacking aerial mycelium in some strains, often accompanied by mutation to arginine requirement and/or loss of secondary metabolites

(Redshaw et al., 1979; Nakano and Ogawara, 1980); but genetic and physical analysis of these examples has generally proved difficult and inconclusive. What is clear from certain systems is that genetic instabilities are often accompanied by DNA rearrangements, which might involve the activity of transposon- or IS-like elements. Thus, high frequency mutations in *Stmy. glaucescens* to streptomycin-sensitivity, which have a chromosomal location, are accompanied by DNA rearrangements detectable by restriction enzyme analysis (Kieser et al., 1980); and in '*Stmy. reticuli*' ethidium bromide-induced mutations leading to loss of a variety of characters (melanin production, aerial mycelium formation, leucomycin production) are accompanied by deletions and rearrangements within a 75 kb plasmid (Schrempf and Goebel, 1979).

Thus there are many phenomena in *Streptomyces* which might involve transposon-like elements, but no rigorous demonstration that these elements are present. We believe that too many speculative and unproductive models involving *Esch. coli* analogies, and too few critical experiments, are being brought to bear on these phenomena. The advent of endogenous *Streptomyces* DNA cloning systems should permit a more systematic approach to be made. It is noticeable that no publications exist documenting direct attempts to demonstrate transposons, for example by looking for transfer of drug resistance by plasmids or temperate phages passaged through a drug-resistant organism. From the point of view of future experimental exploitation of transposons as genetic tools, as well as the physical analysis needed to characterize them, this kind of approach seems to be the most appropriate. Attempts in our laboratory, currently confined to the use of deletion mutants of temperate phages as potential vectors for transposons, have so far proved negative (S. G. Foster, unpublished results). An attempt has been described by Danilenko et al. (1979) to clone kanamycin resistance from *Stmy. rimosus* into *Esch. coli* and then to look genetically and physically in *Esch. coli* for evidence of transposition: but the apparently successful result they obtained must be viewed with considerable caution because the transposon thus identified was not tested for homology with DNA of *Stmy. rimosus* and may therefore have originated by accidental contamination at some stage of the experiment (a serious danger in work involving powerful selection for these highly mobile genetic elements). This is particularly important because the 'new' transposon very closely resembles a previously characterized transposon (Tn 903; Nomura et al., 1978) of *Esch. coli*.

At least three transposons of enteric bacteria show a strong preference for insertion into regions of DNA rich in adenosine (A) and thymidine (T) (Miller et al., 1980). The very low A+T content of *Streptomyces* DNA raises the possibility that similar transposons in *Streptomyces* would have a very limited choice of target sites, or that their transposition mechanisms might have novel features. However, the Gram-negative transposon Tn5 has recently been shown to be able to transpose to various sites in *Streptomyces* phage DNA present in *Esch. coli* as part of a bifunctional replicon (S. G. Foster, personal communication).

7. The Application of Genetic Analysis to Interesting Aspects of *Streptomyces* Biology

A. THE REGULATION OF GENE EXPRESSION IN PRIMARY METABOLISM

Although many of the mutations used to construct the basic linkage map of *Stmy. coelicolor* A3(2) block processes in primary metabolism, especially the biosynthesis of amino acids, purines, pyrimidines and vitamins (Hopwood *et al.*, 1973), no sustained attempt has been made to use these mutants in more penetrating biochemical/genetical studies, perhaps because it has been felt that such unusual features of *Streptomyces* as differentiation and antibiotic synthesis were of more immediate interest. The dawning of the DNA cloning era in *Streptomyces* has made it clear that primary metabolic processes should not have been ignored for so long: if we want to understand differentiation and antibiotic production at the molecular level, knowledge of gene expression during normal vegetative growth will be crucial. In our laboratory, this consideration has led to investigations of apparently constitutive promoters for genes with enzymically and genetically 'convenient' products – particularly those encoding antibiotic inactivating enzymes – and to the study of carbon catabolite repression. Although these studies are at a relatively preliminary stage, we shall briefly summarize current progress.

(a) Neomycin Phosphotransferase

Many streptomycetes are specifically resistant to the antibiotics they produce, and this often involves the production of enzymes that modify the secondary metabolite. The genes controlling such functions are particularly suitable for DNA cloning and genes for neomycin phosphotransferase, neomycin acetyltransferase, and viomycin phosphotransferase have been cloned into plasmid vectors inhabiting *Stmy. lividans* 66, a strain that usually lacks these functions (Thompson *et al.*, 1980, 1982a). One of the resistance genes, encoding neomycin phosphotransferase, is very efficiently expressed in the strain containing the recombinant plasmid (the enzyme is apparently the most abundant protein species in high-speed supernatants of cell-free extracts; C. J. Thompson and J. M. Ward, unpublished work). This observation, combined with extensive *in vitro* characterization of the gene, makes the promoter of the gene a potential target for physical characterization of a 'basic' *Streptomyces* promoter expressed during vegetative growth and its interaction with RNA polymerase. This should be valuable in interpreting the structures of promoters subject to more complex regulation.

(b) Carbon Catabolite Repression

There is circumstantial evidence to implicate carbon catabolite repression in the control of differentiation (see below), in the production of some antibiotics (Martin and Demain, 1980), and in production of the commercially useful enzyme glucose isomerase (Sanchez and Quinto, 1975). Cyclic AMP mediates carbon catabolite repression in *Esch. coli*, where there have been extensive molecular and genetical studies of the mode of action of cyclic AMP at the promoter level; but cyclic AMP may not be involved in this process in some Gram-positive bacteria (Ullman, 1974). Thus the results of studies on carbon catabolite repression in *Streptomyces* cannot fail to be interesting both industrially and for basic research. The physiological studies by Hodgson clearly established that glucose repression in *Stmy. coelicolor* A3(2) operates at the transcription level on genes concerned with the uptake of several carbon sources – glycerol, arabinose, fructose and galactose – and that this repression can be manifested by the non-utilizable glucose analogue 2-deoxyglucose (DOG). Mutants (Dog^R) able to utilize any of these four carbon sources in the presence of DOG were easy to isolate, and proved to differ from the parent strain in their resistance to DOG on at least 13 carbon sources (Hodgson, 1980, 1982). Thus, utilization of all these carbon sources involves steps regulated by a common catabolite repression system known in some cases to operate at the transcription level. The majority class (85%) of Dog^R mutants are unable to utilize glucose as sole carbon source (Hodgson, 1982), and glucose kinase activity was low (less than 7% of the wild-type control) in cell-free extracts of three representative glucose-utilizing or glucose-non-utilizing Dog^R mutants (Seno and Chater, 1983). Thus glucose kinase may play a significant role in mediating carbon catabolite repression. E. T. Seno (unpublished work) has recently been able to isolate a group of mutants pleiotropically unable to utilize a wide range of carbon sources even when glucose is absent. Conceivably, some of these mutants may also be altered in carbon catabolite repression. By molecular cloning of the genes identified by these two classes of pleiotropic mutants we hope to investigate their function in detail. Such studies will also require the availability of cloned genes subject to carbon catabolite repression. A good candidate for the latter purpose is provided by genes for glycerol metabolism, at least two of which are closely linked to each other and to the *argA* gene; and a cloned DNA fragment which 'complements' mutations in both genes has recently been obtained (E. T. Seno, unpublished work).

B. DIFFERENTIATION

The *Streptomyces* colony is among the most complex structures formed by bacteria. Although there are considerable difficulties in the physiological and biochemical analysis of the processes involved in this differentiation process, its genetic analysis can realistically be attempted in *Stmy. coelicolor* A3(2). Our approach to this subject (reviewed by Chater and Merrick, 1979) has started from the isolation of mutants defective either in the production of aerial mycelium (*bld* mutants) or in its metamorphosis into mature spore chains (*whi* mutants). (These are recognizable by their retention, after prolonged incubation, of the appearance of 'juvenile' colonies.) From genetic mapping, four *bld* (Merrick, 1976) and eight *whi* loci (Chater, 1972) have been identified at different positions on the chromosomal linkage map, many being represented by several mutations. Mutations mapping at the same locus invariably share a common morphological phenotype which, at least for *bldB* and *bldC* mutants and most of the *whi* mutants, closely resembles a stage seen during normal development. Constructed double *whi* mutants always showed the morphology of one of the singly mutant parent strains. This analysis suggested a functional sequence for the *whi* gene products (Chater, 1975).

Two classes of *bld* mutants (*bldA* and *D*) exhibited an unusual phenotype which is not usually seen with the wild-type. Instead of forming normal aerial mycelium at 2–4 days, colonies develop a 'prostrate' surface layer in which the hyphae exhibit extensive fragmentation and aberrant lateral wall and cross-wall formation. The fragmentation is blocked in *bldA* mutants into which *whiA, B* or *H* mutations blocking sporulation septation are inserted, indicating that the fragmentation involves sporulation functions. However, a *whiG* mutation, which also blocks sporulation septation, did not block the aberrant fragmentation of *bldA* mutants, indicating that the *whiG* function can be bypassed in some circumstances and that it has no direct obligatory role in sporulation septation (Chater and Merrick, 1979). The *bldA* mutant phenotype is interesting in another context: it is expressed only in the presence of certain carbon sources (glucose, gluconate and cellobiose among those tested; Merrick, 1976; Chater and Merrick, 1976; Hodgson, 1980), all of which are apparently effective mediators of carbon catabolite repression, whereas carbon sources whose utilization is subject to carbon catabolite repression (such as mannitol, arabinose, fructose, galactose and glycerol) phenotypically suppress the *bldA* phenotype and allow the development of apparently normal aerial mycelium and spores. This may indicate a connection between carbon catabolite repression and certain differentiation functions, which should become clearer as the genetic analysis of carbon catabolite repression of primary metabolism proceeds (see above).

Further conventional genetic analysis of *bld* and *whi* mutants has been impeded by the absence of suitable systems for functional tests (i.e. of dominance and complementation). However, the possibility of using the available mutants to identify differentiation genes introduced *in vitro* into suitable DNA cloning vectors should soon remedy this deficiency, in addition to allowing the study of these genes *in vitro*. Furthermore, the use of relatively new techniques like gene fusion (Franklin, 1978) to analyse the expression of differentiation functions will also become feasible with the availability of cloned differentiation genes.

C. ANTIBIOTIC PRODUCTION

This important aspect of *Streptomyces* genetics has been the subject of several recent reviews (Hopwood and Merrick, 1977; Hopwood, 1978, 1979b; Okanishi, 1979) and is summarized in Table 2. The main interest has been focussed on the location of the genes (plasmid-borne or chromosomal) and their organization (clustered or scattered). Genetic evidence for plasmid linkage of structural genes for methylenomycin A production by *Stmy. coelicolor* A3(2) and for chromosomal linkage of the structural genes for synthesis of actinorhodin and a red prodigiosin-like antibiotic in the same strain, is well established. Moreover, in each of the latter cases, all mutations specifically leading to non-production were closely linked, suggesting the involvement of groups of genes possibly controlled by one or only a few promoters. In these cases, many of the mutants were relatively stringently defined as being defective in structural genes because they could take part in co-synthesis reactions with other mutants. In many streptomycetes there also exists another class of mutant, in which the mutational defect appears to be regulatory, and which occurs at relatively high frequency, particularly after treatment with intercalating dyes. In at least one case (chloramphenicol production in *Stmy. venezuelae*) there is convincing genetic evidence that such mutations are non-chromosomal, while the few structural gene mutations analysed did have chromosomal locations. In many other cases there is no such convincing evidence, and the possibility has constantly to be borne in mind that the dye treatment used to induce the mutations may cause internal rearrangements of DNA within replicons at frequencies higher than those at which it causes plasmid loss. (This phenomenon has been amply demonstrated by Schrempf and Goebel, 1979.)

The nature of the regulation imposed by plasmid borne genes on the phenotypic expression of chromosomal structural genes for antibiotic synthesis is still undetermined.

D. ANTIBIOTIC RESISTANCE

Streptomycetes are a rich source of antibiotic resistance genes, particularly those giving resistance to endogenously produced antibiotics; indeed, it has

TABLE 2. Genetic control of antibiotic production

Genetic control	Strength of evidence	Antibiotic	Producing strain	Notes	References
Structural genes chromosomal (no plasmid involvement detected)	Unambiguous	Actinorhodin	*Stmy. coelicolor* A3(2)	At least 7 clustered genes	Rudd and Hopwood (1979)
	Unambiguous	Red antibiotic	*Stmy. coelicolor* A3(2)	At least 5 clustered genes	Rudd and Hopwood (1980)
	Unambiguous	Oxytetracycline	*Stmy. rimosus*	2 gene clusters (9 genes)	Rhodes *et al.* (1981)
	Unambiguous	Rifamycin	'*Nrda*'. *mediterrenea*	At least 3 clustered genes	Ghisalbra *et al.* (1982)
	Weak	Zorbomycin zorbonomycin	*Stmy. bikiniensis* var. *zorbonensis*	3 genes, probably not more than 1 structural	Coats and Roeser (1971)
Structural genes plasmid-borne (no direct chromosomal involvement detected)	Unambiguous	Methyleno-enomycin A	*Stmy. coelicolor* A3(2)	At least 4 or 5 structural genes on plasmid SCP1	Kirby and Hopwood (1977), Hornemann and Hopwood (1981)
	Indicative	Tylosin	*Stmy. fradiae*	High frequency transfer of resistance and production to mutants	Baltz *et al.* (1981)

Structural genes chromosomal, regulatory gene(s) plasmid-borne	Unambiguous	Chloramphenicol	Stmy. venezuelae	At least 2 chromosomal structural genes. Non-chromosomal regulatory mutants lose 'flower shaped' DNA	Akagawa et al. (1975, 1979), Okanishi and Umezawa (1978)
	Indicative	Turimycin (leucomycin)	Stmy. hygroscopicus	Infectious transfer of production into some but not all non-producers	Noack et al. (1978)
	Indicative	Holomycin	Stmy. clavuligerus	One chromosomal locus; infectious transfer of production into regulatory mutants	Kirby (1978)
	Weak	Aureothricin	'Stmy. kasugaensis'	Possible correlation of plasmid DNA loss with some non-producers	Okanishi (1979)

Continued on next page

TABLE 2 (cont.)

Genetic control	Strength of evidence	Antibiotic	Producing strain	Notes	References
Unspecified production gene(s) plasmid-borne	Strong	Leucomycin	'Stmy. reticuli'	Plasmid DNA loss or rearrangement with loss of production, repaired after infectious transfer	Schrempf and Goebel (1979)
	Indicative	Actinomycin D	Stmy. parvulus and Stmy. antibioticus	Evidence of high frequency transfer in protoplast fusions	Ochi and Katz (1978)
	Indicative	Antimycin	M.506	Non-producers lack CCC DNA	El-Kersh and Vezina (1978)
	Indicative	Istamycin	'Stmy. tenjimariensis'	Non-producers lack a CCC DNA species	Okami (1979)
	Indicative	Lankacidins	Streptomyces sp.	Non-producers lack a linear plasmid DNA	Hayakawa et al. (1979)
	Indicative	Kasugamycin	'Stmy. kasugaensis'	Non-producers lack a 'flower shaped' DNA	Okanishi and Umezawa (1978)
	Indicative	Spiramycin	Stmy. ambofaciens	Non-producers lack CCC DNA	Ōmura et al. (1979)

In addition to the cases cited, non-producers were isolated at high frequency after 'curing agent' treatments in the following situations: oxytetracycline (one strain only of Stmy. rimosus: Boronin and Sadovnikova, 1972); chloramphenicol (Stmy. venezuelae 3022a: Michelson and Vining, 1978); kanamycin (Stmy. kanamyceticus: Chang et al., 1978; Hotta et al., 1979); neomycin (Stmy. fradiae: Yagisawa et al., 1978); paromomycin (Stmy. rimosus f. paromomycinus: White and Davies, 1978); streptomycin (Stmy. bikiniensis: Shaw and Piwowarski, 1977); ribostamycin ('Stmy. ribosidificus': Nojiri et al., 1980).

been postulated that this may be the origin of the plasmid-borne resistance determinants widespread among eubacteria (Benveniste and Davies, 1973). Only a few genetic studies of resistances of this kind (i.e. to an organism's own antibiotic) have been reported, and there is a single unambiguous demonstration of plasmid-linkage of the gene(s) concerned. This involves resistance to methylenomycin A in *Stmy. coelicolor* A3(2), in which the antibiotic production genes are linked to the same plasmid (SCP1) (Vivian, 1971; Kirby *et al*., 1975; Wright and Hopwood, 1976a; Kirby and Hopwood, 1977). The DNA encoding methylenomycin A resistance has been cloned into *Streptomyces* plasmid vectors by Bibb *et al*. (1980a), and shown to hybridize to high molecular weight plasmid DNA from SCP1$^+$ *Stmy. coelicolor* A3(2) (Westpheling, 1980) and to a plasmid of about 170 kb (pSV1) from an independently isolated methylenomycin A-producing strain (Aguilar and Hopwood, 1982).

Unfortunately, the mechanism of methylenomycin resistance is unknown. This impediment is not a problem with resistance to several other antibiotics, where particular enzymic functions have been clearly implicated. Thus, a measure of resistance to neomycin or viomycin in the producer strains is conferred by enzymes that phosphorylate or acetylate the antibiotics (Davies and Smith, 1978; Skinner and Cundliffe, 1981), and thiostrepton resistance in '*Stmy. azureus*' (the producer of thiostrepton) and erythromycin resistance in *Stmy. erythreus* (the producer of erythromycin) are due to enzymes that modify the ribosomal target sites involved in binding the antibiotics (Cundliffe, 1978; Skinner and Cundliffe, 1982). Thompson *et al*. (1980, 1982a, b, c) have cloned these genes into *Streptomyces* plasmid vectors (thereby demonstrating rigorously their involvement in conferring drug resistance) and a start has been made with their use as probes to detect the location of the genes in various fractions of the DNA of the organisms from which the genes were originally isolated. This approach should provide firm evidence of plasmid location (where this situation prevails). It has already been shown that a physically isolable plasmid of *Stmy. fradiae* previously thought (on circumstantial evidence) to carry neomycin resistance genes has no homology with the isolated genes for neomycin phosphotransferase and acetyltransferase (Komatsu *et al*., 1981).

Antibiotic inactivating enzymes in *Streptomyces* are not confined to the producers of the relevant antibiotics. Thus, although five strains out of 21 examined by Shaw and Hopwood (1976) possessed chloramphenicol acetyltransferase (CAT), the chloramphenicol producing strain *Stmy. venezuelae* lacked such an enzyme. (On the assumption that production of active antibiotic is of selective advantage to the producing organism this was to be expected since known CAT-producers, including *Stmy. acrimycini*, rapidly inactivate chloramphenicol in the surrounding medium, whereas this does not apply to aminoglycoside antibiotics in contact with organisms possessing modifying enzymes for this class of antibiotics.) In one of the five CAT⁺ strains (*Stmy. acrimycini*) the loss of CAT by a chromosomal mutation (Wright and Hopwood, 1977) was shown to be one of two mutational routes by which a 10-fold reduction in chloramphenicol resistance could arise (the nature of the defect resulting from another, more frequent, kind of mutation was not determined).

Generally, where streptomycetes are resistant to antibiotics which they do not themselves produce (Freeman *et al.*, 1977; Freeman and Hopwood, 1978) the resistance mechanisms involved have not been determined. However (as indicated for chloramphenicol resistance in *Stmy. coelicolor* A3(2), in the section on possible transposable elements) their genetic determination is of great interest because of its frequent instability: resistance is often lost and regained at relatively high frequencies. An interesting feature of such 'mutation' events is that resistances to more than one antibiotic are lost simultaneously at frequencies considerably higher than the product of the frequencies of the individual events (Danilenko *et al.*, 1977). This may indicate that some particular physiological state has a general stimulatory effect on such events.

F. RESTRICTION-MODIFICATION (RM) SYSTEMS

Because of their value in DNA cloning and DNA analysis, type II (Arber, 1974) restriction enzymes of varying site specificity have been looked for and found in an enormous range of prokaryotic organisms (Roberts, 1980). *Streptomyces* spp have proved to be a good source of such enzymes – at least 28 have been recorded (Lomovskaya *et al.*, 1980), with at least eight different target sites having been identified; at least five are useful for DNA cloning work (Roberts, 1980). However, little is known about the biology and genetics of such enzymes in any bacteria, and even less about the modification enzymes that usually accompany them *in vivo* and which protect a cell's own DNA against the action of the restriction enzymes. The analysis of these aspects of restriction (R) and modification (M) enzymes in *Streptomyces* may be particularly appropriate because of the wide range of such enzymes available in a group of organisms that have common sensitivity to many phages (which can potentially be used *in vivo* to recognize RM systems) and in which there are realistic prospects of genetic analysis. Such studies might also lead to the development of strains making high levels of the enzymes, which would not only make them more easily available to DNA researchers but also provide material for enzymological studies (usually restriction enzymes are found as minute quantities of enzymically highly active protein in cell-free extracts). Moreover, since restriction is likely to be a significant barrier to inter-species gene transfer, either natural or artificial, an understanding of *Streptomyces* RM systems should have significant practical benefits.

The interactions of *Streptomyces* RM systems with phages have recently been reviewed (Chater, 1978; Lomovskaya *et al.*, 1979, 1980), so we shall confine ourselves here mainly to other aspects.

(a) *RM Systems Detected* in vivo

Lomovskaya *et al.* (1980) tabulated 12 strains in which RM systems could be detected by phage tests, and more have come to light since then (C. S. Stuttard, personal communication; K. S. Cox and R. H. Baltz, personal communication; K. F. Chater, unpublished results). It begins to look as though lack of RM systems is the exception rather than the rule among streptomycetes, but that their detection by phage tests may not always be straightforward. This is partly due to the possession by some phages of systems that overcome host-specific restriction (Chater and Carter, 1978, 1979); partly because some hexanucleotide target sites are rare in phage DNA (Chater, 1977, 1980); and partly because relatively non-specific host-induced DNA modification may protect DNA against certain restriction enzymes (K. F. Chater, unpublished results). An apparent RM system of particular significance for gene cloning has been detected in *Stmy. coelicolor* A3(2): plasmid SCP1 can be reintroduced by conjugation into *Stmy. coelicolor* A3(2) only at a very low frequency from '*Stmy. lividans*' 66 or *Stmy. parvulus* ATCC 12434, even though transfer in the other direction is relatively efficient (Hopwood and Wright, 1973 and unpublished results). This raises the possibility that this barrier may impede both the cloning of DNA from other sources into *Stmy. coelicolor*, and the reintroduction of *Streptomyces–Esch. coli* bifunctional replicons (see below) from *Esch. coli* into *Stmy. coelicolor* A3(2). Unfortunately, no phages restricted by *Stmy. coelicolor* A3(2) have been discovered, which has made it difficult to isolate R^- mutants or to analyse the system genetically, nor has any restriction enzyme been detected in cell-free extracts (K. F. Chater, unpublished results).

(b) *Genetic Studies*

Genetic analysis has been carried out only on one RM system – the *Sal*GI system specified by *Stmy. albus* G. Chater and Wilde (1976, 1980) isolated a series of RM mutants (R^-M^+, R^-M^-, R^+M^{tps}) and showed that representative mutations of all three types were closely linked genetically. No evidence was found of extrachromosomal inheritance by mapping studies, but it is interesting to note that R^-M^- mutations could be isolated spontaneously at relatively high frequency (10^{-2} to 10^{-5}) as the major class of survivors when an R^+M^{tps} mutant was plated at the non-permissive temperature. This may indicate that the *Sal*GI RM genes are present on a special genetic element (such as a plasmid or prophage); or that they are subject to some instability, perhaps like that seen for various drug resistances (see above); or that they are easily deleted, conceivably through the action of the SalGI restriction enzyme itself (the *Esch. coli* restriction enzyme EcoRI can bring about recombination at EcoRI target sites *in vivo*: Chang and Cohen, 1977).

No clear evidence has yet emerged concerning the effects of endogenous RM systems on inter-strain gene transfer in *Streptomyces*, and it remains a distinct possibility that difficulties experienced in obtaining inter-specific recombinants either by conjugation or following protoplast fusion (D. A. Hopwood and H. M. Wright, unpublished results) may at least partly reflect the action of restriction enzymes. Such effects may become more amenable to analysis where RM genes have been cloned into genetically well-understood hosts.

(c) Taking Advantage of Streptomyces Modification Systems

The development of endogenous *Streptomyces* DNA cloning systems (see below) raises the possibility of taking advantage of *in vivo* modification systems. A particular case in point is afforded by the SalGI system. Cloned DNA propagated in a SalGI R^-M^+ mutant would be modified at all its SalGI target sites (GTCGAC) which would include a fraction of the HincII sites (GTPyPuAC) and TaqI sites (TCGA). Such *in vivo*-modified sites have been shown to be resistant to cleavage by the respective enzymes (K. F. Chater, unpublished results). These effects could prove useful in restriction mapping of DNA and in DNA cloning.

F. EXOENZYMES

Streptomycetes are rich producers of exoenzymes of which many (proteases, amylases, lipases, nucleases, etc.) are related rather obviously to their ecology (i.e. their utilization of macromolecules in the soil); others may be involved in resistance to antibiotics (e.g. β-lactamases); and some have no obvious importance to the organism (e.g. tyrosinase). Very few of these enzymes have been studied genetically, which is unfortunate both because of their ecological importance and because of their potential for exploitation for academic and industrial purposes.

(a) Tyrosinase

The tyrosinase involved in melanin formation by *Stmy. glaucescens* has been extensively studied by R. Hütter and his co-workers in Zürich. Mutants lacking tyrosinase (Mel$^-$) are frequent among the survivors of various treatments (e.g. curing agents, storage at $-20°C$), which might have been taken to indicate plasmid involvement; but careful genetic mapping (Hütter et al., 1981) has shown that at least one class of mutation (*melA*) has a chromosomal location (close to *ura-2* and *ura-3*), while a second class (*melB*) probably also maps to the chromosome (apparently near to *his-2*). With a third class (*melC*), it has been difficult to interpret the results of crosses. No conclusive genetic mapping data have been obtained with either of two other classes of mutants affecting tyrosinase–namely, those unable to excrete the enzyme or those in which its regulation is abnormal (Hütter et al., 1981). (The former mutants appeared to be specific to tyrosinase in that extracellular laminarinase and chitinase were normal.)

(b) Agarase

Streptomyces coelicolor A3(2) is unusual in being able to use agar as sole carbon source (Stanier, 1942). This is facilitated by the production of an extracellular diffusible agarase activity (the Dag^+ phenotype) which has recently been studied genetically (Hodgson and Chater, 1981). Two independent Dag^- mutations detected in the John Innes Institute Culture Collection have both been mapped to the middle of the '9 o'clock silent region' of the map. Interestingly, both the mutations are 100% linked to the integrated 'NF' form of plasmid SCP1, and one of them (*dagA*1) has apparently resulted from some kind of insertional inactivation of *dagA* by SCP1. The unusual location of *dagA*, its apparent insertional inactivation by plasmid integration, and the occurrence of spontaneous unselected mutations to Dag^-, all suggest the possibility that an unusual genetic element is concerned.

From the limited information available at present, it seems that many extracellular enzymes in streptomycetes show interesting genetic determination (e.g. instability). This, and the possibility of exploiting extracellular enzymes as a tool for the export of useful proteins into the culture medium through gene fusion techniques, should ensure that more information on extracellular enzymes will be gathered in the near future.

8. Gene Cloning in *Streptomyces*

No biologist can be unaware of the impact of the gene cloning techniques developed, mainly in *Esch. coli*, in the last decade. The techniques have led not only to an explosion of information about the structure, regulation and evolution of genes from organisms ranging from viruses to man, but also to the concept of industrial genetic engineering, as highlighted by the construction of *Esch. coli* strains able to make various important eukaryotic gene products (e.g. human insulin, interferon, growth hormone). Although the cloning of *Streptomyces* DNA into *Esch. coli* would certainly be useful for analysing the molecular biology of streptomycetes, and might even lead to the synthesis by *Esch. coli* of some *Streptomyces* products of industrial importance, our prejudice (as die-hard *Streptomyces* geneticists) has been that endogenous systems of gene cloning for *Streptomyces* would confer special benefits in addition (and often complementary) to those consequent on cloning *Streptomyces* DNA into *Esch. coli*. The large number of mutants already available for important *Streptomyces* properties could then be exploited in identifying cloned genes, and the cloned DNA could be put to work in dominance, complementation and fine structure mapping tests. *Streptomyces* promoters would be recognized by their natural RNA polymerase, allowing penetrating studies of transcriptional regulation to be done. Moreover, many *Streptomyces* products (most obviously antibiotics) may be lethal if made in a host that does not have the appropriate resistance mechanisms. From an industrial standpoint, gene cloning within *Streptomyces* offers the most direct route to assorting parts of antibiotics into new combinations, thus potentiating the development of new products; it should also allow the introduction of genes allowing growth on cheap growth media, and the application of increased gene dosage and/or gene expression to yield increases for existing products. (For a discussion of these possibilities, see Hopwood and Chater, 1980; Hopwood, 1981b.)

Effective gene cloning systems have now been developed for *Streptomyces*, using both phage (Suarez and Chater, 1980b; Chater *et al.*, 1981c), and plasmid vectors (Bibb *et al.*, 1980a; Thompson *et al.*, 1980; Hopwood *et al.*, 1981). In this section we will briefly review what has been and can be achieved with existing vectors and finally discuss possible future developments. For information about the basic principles of cloning, for example the use of restriction enzymes, a useful general review is by Old and Primrose (1981). More detailed accounts of the techniques and applications of DNA cloning in *Streptomyces* are given by Chater *et al.* (1982b) and Hopwood and Chater (1982).

A. THE INTRODUCTION OF PLASMID AND PHAGE DNA INTO *STREPTOMYCES* PROTOPLASTS

Although there had been earlier reports of DNA uptake by 'competent' cells and by protoplasts, no system efficient or reproducible enough for use in gene cloning, or allowing for the identification of individual transformation or transfection events, was available until the discovery that polyethylene glycol (PEG) greatly increased the amount of plasmid DNA taken into protoplasts (Bibb *et al.*, 1978). The modified procedures now used for transformation and transfection in cloning experiments in our laboratory have given improved results over this original procedure. In the basic procedure, a loose pellet of protoplasts is resuspended rapidly in a small volume of DNA solution containing suitable ions and an osmotic stabilizing agent, usually sucrose. PEG 1000 to a final concentration of 20% is added and, within a minute, the mixture is ready for plating on a regeneration medium. Plasmid transformants may be detected directly on the regeneration plates or by replating after regeneration, either by the lethal zygosis reaction (they give rise to pocks; see above) or by selection (suitably delayed to allow for phenotypic expression) for a plasmid-borne character such as drug-resistance (now that such markers have been introduced into plasmids). Transfection events, in which phage DNA has been introduced, may be detected as plaques on the regeneration plates (Chater, 1979; Suarez and Chater, 1980a; Isogai et al., 1980; Krügel et al., 1980) (this is helped by the use of soft agar overlays containing spores of a phage-sensitive indicator strain). Stable SCP2* derivatives of reduced size and containing drug resistance markers for selection have been constructed (D. J. Lydiate, personal communication) and are now being successfully used for shotgun cloning of large DNA fragments (F. Malpartida and H. Ikeda, personal communication).

The frequencies obtained for plasmid transformation by this means are usually very high (about 10–80% of regenerating protoplasts: Bibb et al., 1978, 1980), whereas the highest transfection frequency obtained until recently was about 1% of regenerating protoplasts (for ϕC31 DNA with 'Stmy. lividans' protoplasts – other phage–host combinations were much less efficient: Suarez and Chater, 1980a, 1981). This transformation frequency was sufficient to enable Bibb et al. (1980) to detect particular cloned genes in shotgun cloning experiments with plasmid vectors. However, the systematic variation of parameters involved in transformation (Thompson et al., 1982a) has shown that the original procedure could be made somewhat more efficient and reproducible, a study that has facilitated cloning work in our laboratory. A more dramatic improvement on existing frequencies was needed to make shotgun cloning with phage vectors a manageable undertaking: and this has recently been provided by the discovery that the addition of suspensions containing small, positively charged, liposomes (prepared from commercial lipid preparations in the absence of DNA) to transfection mixtures, together with the use of high (about 50%) PEG concentrations, gives at least 100 times more transfection events than the original transfection procedure (Rodicio and Chater, 1982). This modification also gave significant stimulation of plasmid transformation (though transformation by chromosomal DNA was not detectably stimulated).

B. PLASMID VECTOR MOLECULES

Mention has already been made of the three plasmid vector molecules currently in use: SCP2*, SLP1.2 and pIJ101. Their restriction maps are given in Fig. 5 and restriction enzyme site frequencies are listed in Table

3. In this section we have summarized the main features of each plasmid relevant to its use in cloning.

(a) SCP2*

This self-transmissible plasmid has an apparently narrow host range (so far known to include *Stmy. coelicolor* A3(2), '*Stmy. lividans*' 66 and *Stmy. parvulus* ATCC 12434). It has a low copy number (estimates vary from one to five; Bibb *et al.*, 1977; Schrempf and Goebel, 1977). Although its *stable* maintenance depends on its retention of genetic information contained in the second largest *Pst*I fragment (Fig. 5a), all the functions essential for its

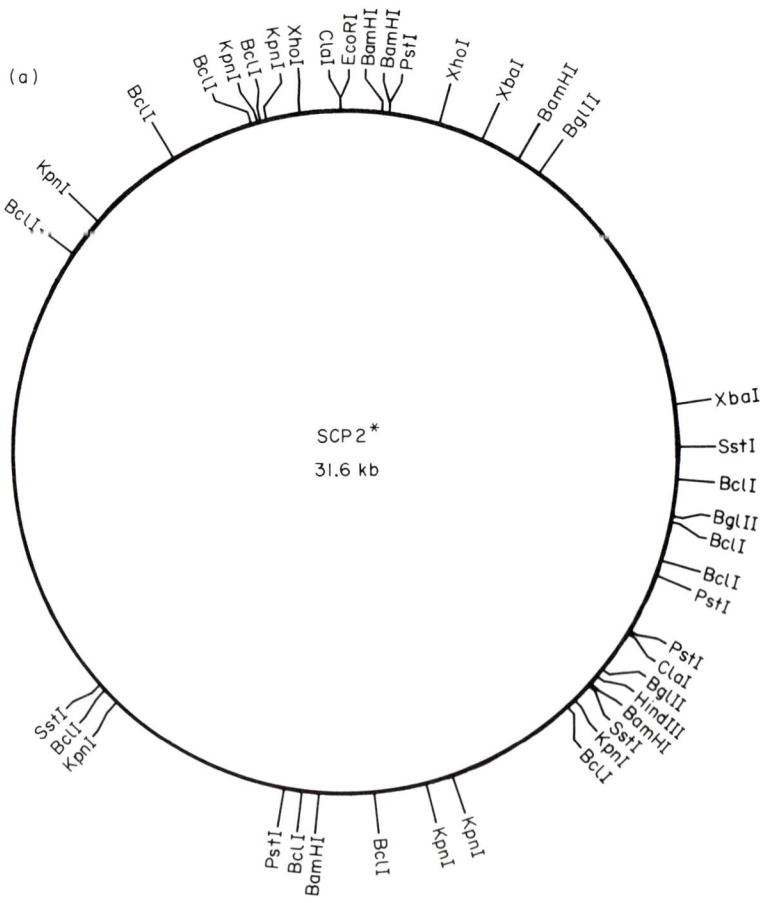

FIG. 5. Restriction enzyme cleavage maps of plasmids. (a) SCP2*, from Bibb *et al.* (1977) and D. J. Lydiate, personal communication. (b) SLP1.2, from Bibb *et al.* (1980a, b) and Thompson *et al.* (1982c). (c) pIJ101, from Kieser *et al.* (1982).

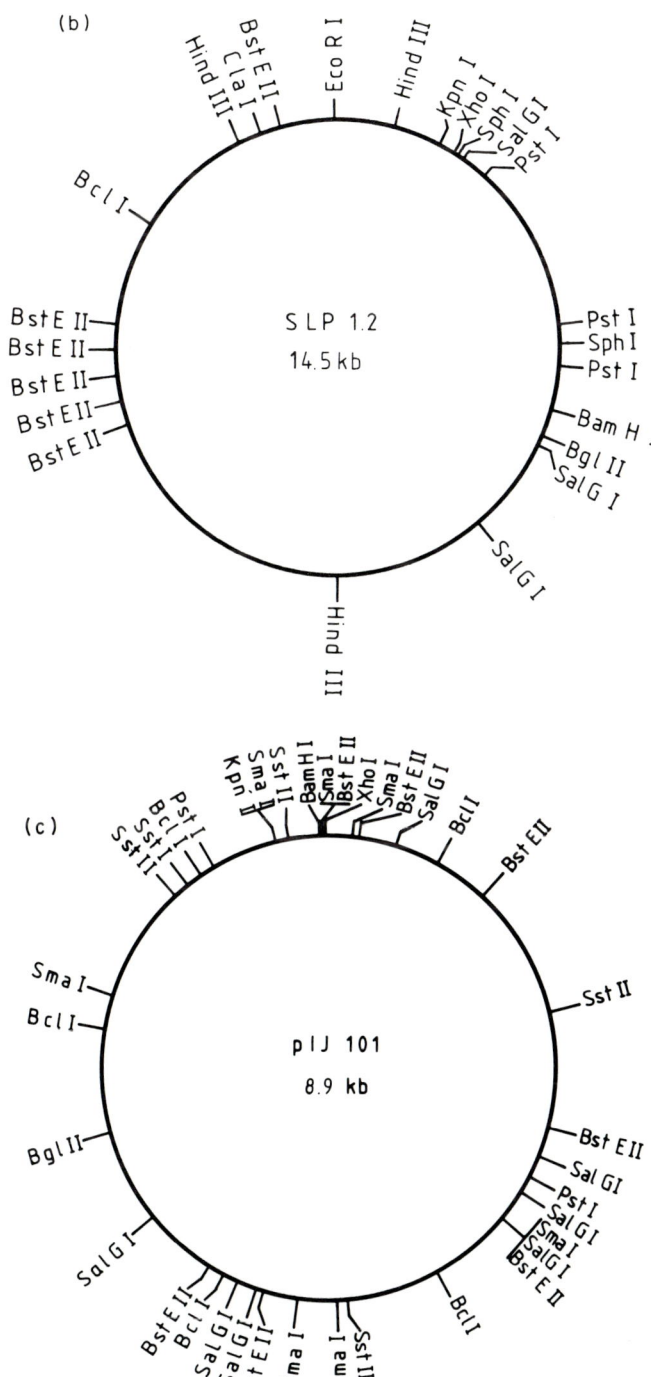

TABLE 3. Restriction enzyme target sites in plasmid and phage vectors

Restriction enzyme	Number of target sites in[a]			
	SCP2	SLP1.2	pIJ101	φC31
BamHI	5	1	1	0
BclI	11	1	5	c. 14
BglII	3	1	1	0
BstEII	>14	6	7	>15 (FspAI)
ClaI	2	1	0	5
EcoRI	1	1	0	6
HindIII	1	3	0	c. 13
HpaI	ND	ND	0	3
KpnI	7	1	1	5
PstI	4	3	2	0
SalGI	⩾20	3	8	>15
SmaI	c. 18	⩾6	7	>15
SphI	8	1	ND	8
SstI	3	0	1	0
SstII	>16	⩾20	4	?
XbaI	2	0	0	1
XhoI	2	1	1	0

[a] ND, Not done.
Based on information from Bibb et al. (1977), Schrempf and Goebel (1977) and D. J. Lydiate (personal communication) for SCP2*; Bibb et al. (1981) and Thompson et al. (1982c) for SLP1.2; Kieser et al. (1982) for pIJ101; and Chater et al. (1981b and unpublished results) and J. E. Harris (personal communication) for φC31 DNA.

replication are contained within the largest PstI fragment (Bibb et al., 1980). It is easily introduced into Stmy. coelicolor A3(2), in which host its biological properties have been analysed (Bibb et al., 1977; Bibb and Hopwood, 1981) (see Section 3). Stable SCP2* derivatives of reduced size and containing drug resistance markers for selection have recently been constructed (D. J. Lydiate, personal communication) and are now being successfully used for shotgun cloning of large DNA fragments (F. Malpartida and H. Ikeda, personal communication).

(b) SLP1.2

This self-transmissible plasmid has a copy number estimated at four or five (Bibb et al., 1980) and, like SCP2*, has a restricted host range ('Stmy. lividans' 66 and 'Stmy. reticuli' (H. Schrempf, personal communication) have so far been found to be hosts for it). Its failure to enter Stmy. coelicolor A3(2) at a high frequency is presumably due to the stable presence in that strain of the integrated form of the plasmid from which autonomous SLP1.2 was originally derived. This disadvantage is offset by the very convenient distribution of restriction enzyme target sites in SLP1.2 (Fig. 5b), which has made it an excellent vector for the cloning of antibiotic resistance genes from other streptomycetes (Bibb *et al.*, 1980; Thompson *et al.*, 1980, 1982a). Derivatives containing two drug resistance determinants, of which one can be used for clone recognition by insertional inactivation, the other for vector selection, are described by Thompson *et al.* (1982c): one of these (pIJ41) was used in the shotgun cloning of a tyrosinase gene from *Stmy. antibioticus* (E. Katz, personal communication) and of a gene for PABA synthetase from *Stmy. griseus* (Gil and Hopwood, 1983).

(c) pIJ101

This interesting small self-transmissible plasmid originated in a strain ('*Stmy. lividans*' ISP 5434) taxonomically related to *Stmy. coelicolor* A3(2). Unlike SLP1.2 or SCP2, pIJ101 is present in very large copy numbers (40 to 300 per chromosome) (Hopwood *et al.*, 1981; Kieser *et al.*, 1982). It is therefore easy to isolate in large quantities, which has led to its extensive characterization and which makes it suitable for rapid screening of individual clones (Kieser *et al.*, 1982). It has a relatively wide host range (13, including *Stmy. coelicolor* A3(2), out of 18 different streptomycetes tested were able to receive and maintain pIJ101 or its derivatives), and several convenient restriction enzyme sites are available for DNA insertions (Fig. 5c). This plasmid appears to be so favourable as a cloning vector that it is relevant to spell out its disadvantages (actual or possible). First, it was originally difficult to obtain transformants containing only one pIJ101-derived plasmid, but this problem was overcome by using a transfer-defective derivative (of which a number have been isolated: Kieser *et al.*, 1982). Second, some cloned genes might be lethal when present in multiple copies in the cell. Third, genetic analysis (especially complementation and dominance testing) is in principle best done with a low copy number (preferably only one) of the cloned gene. Nevertheless, pIJ101 derivatives carrying antibiotic resistance determinants for vector selection have been used successfully for the primary shotgun cloning of *Stmy. coelicolor* glycerol utilization genes (E. T. Seno, personal communication) and an *O*-methyltransferase gene involved in undecylprodigiosin biosynthesis in *Stmy. coelicolor* (Feitelson and Hopwood, 1983).

C. PHAGE VECTOR MOLECULES

The molecular biology and genetics of *Streptomyces* phages has recently been reviewed (Lomovskaya *et al.*, 1980) and will not be discussed here, except to point out that both temperate and virulent phages are easily obtained for many streptomycetes. Either temperate or virulent phages could be used as vectors, but temperate phages have the obvious advantage of being able to lysogenize, and may therefore be stably maintained in a host cell. Thus attempts to develop phage cloning vectors have mainly centred on temperate phages. Most temperate phages so far examined resemble coliphage λ in possessing DNA with cohesive ends, which is likely to be packaged into phage particles during lytic infections by a process in which multimeric genomes are cut by an endonuclease recognizing a specific base sequence present once per genome length. It follows that the insertion of extra DNA into such genomes (as would be done during cloning) would result in an increase in the amount of DNA packaged in each phage particle. Obviously there would be a limit on the amount of overpackaging tolerated, and this would limit the amount of foreign DNA that could be inserted. For this reason it is desirable to delete inessential regions from the phage genome and thus to make more room for inserting foreign DNA. This is most commonly achieved by selecting mutants resistant to chelating agents (i.e. as in phage λ; Parkinson and Husky, 1971), a procedure that has given deletion mutants with at least five different *Streptomyces* phages (Pal 6, Rosner *et al.*, 1980; VP5, R4 and ϕC31, Chater, 1980; and SH10, Klaus *et al.*, 1981).

Although other phages (notably R4 and SH10) have great potential as wide host range *Streptomyces* DNA cloning vectors, the most highly evolved series of vectors has so far been obtained from ϕC31. Figure 6 is a restriction map of ϕC31 DNA (a list of restriction enzyme target site frequencies is given in Table 3). A continuous length of about 20% (about 8 kb) of the DNA has been defined as inessential for plaque formation by the analysis of deletion mutants, but no single deletion spans the whole of this region. This is because the phage is unable to package less than 37 kb, i.e. 90% of the wild-type genome length (Chater *et al.*, 1981a). The multiplication of the phage DNA as a smaller replicon might, in principle, be achieved by making, *in vitro*, a ϕC31 molecule containing an *Esch. coli* plasmid, and then allowing the hybrid molecule to replicate as a plasmid in *Esch. coli*, in which mode the size of the molecule should not affect its ability to replicate. Suarez and Chater (1980b) described an experiment in which the small multi-copy *Esch. coli* plasmid pBR322 (4.362 kb) (Bolivar *et al.*, 1977) was inserted into one of the six *Eco*RI sites of a ϕC31 deletion mutant lacking about 3.3 kb of its DNA. The hybrid molecule was recognized by transfection into '*Stmy. lividans*' protoplasts, followed by hybridization of DNA present in the plaque to ^{32}P-labelled pBR322 (the procedure of Benton and Davis, 1977). The hybrid could easily be transformed into *Esch. coli*, where it replicated as a plasmid which could, in turn, be used to transfect '*Stmy. lividans*' protoplasts without any evidence of DNA alteration. The hybrid phage, being overpackaged by about 3% compared with the wild-type

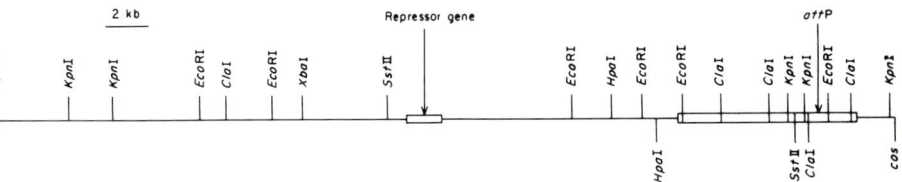

FIG. 6. Restriction enzyme cleavage map of φC31 DNA (Chater et al., 1981a, b, c and unpublished observations). Boxes indicate regions of the DNA encoding functions inessential for plaque formation, as defined by various deletion mutants. The whole φC31 DNA molecule contains 41.27 kilobase pairs (estimated by summing the sizes of EcoRI fragments).

φC31, was used in the isolation of a new series of deletion mutants. One group of deletions had lost parts of pBR322 and the other had lost segments of the main dispensable region, usually including the att site needed for insertion of the prophage in the formation of a lysogen (Chater et al., 1981a, 1982a). One of the latter mutants (φC31 Δ W17::pBR322) has been useful in the development of cloning vectors and has the following relevant properties: (a) it can replicate as a fairly wide host range *Streptomyces* phage or as an *Esch. coli* plasmid, (b) it has single BamHI and PstI recognition sites located in regions of the hybrid molecule inessential for replication in either *Esch. coli* or *Streptomyces*, (c) it can accommodate up to 4 kb of DNA inserted into either of these sites (30% of plaques contained inserted DNA in a 'shot-gun' cloning experiment of this kind: Chater et al., 1981c) and still be able to multiply in either mode, (d) it can accommodate up to 7 kb of DNA as a replacement for the segment of pBR322 DNA between the BamHI and PstI sites (at least 60% of plaques contained inserted DNA in a 'shotgun' cloning experiment of this kind: Chater et al., 1981c), but then loses the ability to replicate in *Esch. coli*, and (e) although the deletion of the phage's att site means that it cannot establish normal lysogeny, this does not prevent the detection of clones, since a superinfecting att-deleted phage can enter the chromosome by normal recombination if homologous DNA, such as a φC31 prophage, already resides there. The latter feature was exploited in the subcloning of a thiostrepton resistance gene into φC31 Δ W17::pBR322 (Chater et al., 1982a). Further *in vivo* and *in vitro* manipulations of one of the phages so obtained gave vectors such as φC31 KC400 (Harris et al., 1983), which can incorporate PstI fragments of foreign DNA of up to 6 kb while retaining an intact viomycin phosphotransferase gene. Using cloned DNA as a region of homology φC31 KC400 can integrate into the host from which the cloned DNA originated, giving viomycin resistant lysogens. Such prophage integration events are mutagenic in some circumstances. This property can be exploited, as in the mutational cloning of genes for methylenomycin biosynthesis and in gene disruption analysis of cloned DNA involved in glycerol metabolism (K. F. Chater and C. J. Bruton, unpublished results).

D. USES OF BIFUNCTIONAL REPLICONS AND THE QUESTION OF HETEROLOGOUS GENE EXPRESSION

The ϕC31–pBR322 chimaera has recently been augmented by several *Streptomyces* plasmid–*Esch. coli* plasmid bifunctional replicons. One of these (pIJ28) consists of SLP1.2 into which separate insertions of a neomycin phosphotransferase gene (into the *Bam*HI site) and pBR322 (into the *Bcl*I site) have been made (Thompson *et al.*, 1982c). Another example, pIJ361, made by Kieser *et al.* (1982), was made by fusing a pIJ101 derivative and a pBR322 derivative. Schottel *et al.* (1981) have made other hybrid plasmids comprising derivatives of SLP1.2, and the *Esch. coli* plasmids pACYC177 and pACYC184. The resulting molecules can be transformed between '*Stmy. lividans*' 66 and *Esch. coli* in either direction. Such vectors provide an obvious opportunity of taking advantage of a range of powerful genetic techniques available at present only in *Esch. coli*, such as transposon mutagenesis and *lac* fusions, to modify cloned *Streptomyces* DNA, following which the modified DNA can, in principle, be reintroduced into *Streptomyces*. Rigorously testing the expression of *Streptomyces* DNA in *Esch. coli* and vice-versa is simplified; thus it appears that pBR322-specified tetracycline-resistance (Chater *et al.*, 1982a), pACYC184-specified chloramphenicol resistance (Shottle *et al.*, 1981), and kanamycin resistance specified by Tn5 (S. G. Foster and J. A. Gil, personal communication) or pACYC177 (Shottle *et al.*, 1981) can all be expressed in at least some *Streptomyces* strains. As for expression of *Streptomyces* genes in *Esch. coli*, *Stmy. fradiae* neomycin phosphotransferase expression is at best very low in *Esch. coli*; but *Stmy. vinaceus* viomycin phosphotransferase can be expressed in *Esch. coli* (probably dependent on a promoter site provided by the vector), providing a particularly useful selective marker for work with bifunctional replicons (C. J. Thompson and J. M. Ward, unpublished results; Kieser *et al.*, 1982). In addition to these major advantages, there are several relatively minor but, nonetheless, interesting possibilities for taking advantage of bifunctionality, some of which may be of as much interest to *Esch. coli* genetic engineers as to those working on *Streptomyces*. We have already mentioned the possibility of maintaining unpackageable ϕC31 DNA molecules as *Esch. coli* plasmids; the converse, in which, say, phage λ DNA was cloned into a *Streptomyces* plasmid might also provide benefits in the use of λ DNA as a cloning vector. DNA sequences that are lethal when cloned into one organism, for example many T4 genes in *Esch. coli* (A. J.

Mileham, personal communication), might very well be cloned successfully into the other. As discussed earlier, DNA modification systems peculiar to one or the other host may be of use in removing particular restriction enzyme target sites from the hybrid molecule.

9. General Conclusions and Future Prospects

There has been a significant shift in emphasis in *Streptomyces* genetics in the last 10 years. Thus, although a 1973 review by Hopwood *et al.* was mainly concerned with traditional chromosomal genetics, it was possible in 1979 reviews (Chater, 1979; Hopwood, 1979a) to be mainly concerned with protoplast fusion, 'nonchromosomal' genetics and the imminent prospect of endogenous gene cloning; and the present chapter reviews the successful achievement of the latter prospect for some genes. We have constantly referred to cloning as the next stage of analysis in most of the important research areas, and indeed the existing cloning systems have already been used for the cloning of many different kinds of genes. If the detailed study of such clones is to be undertaken, it is likely that further new technical developments will become important. If we wish to analyse the promoters of, for example, differentiation genes whose products are unknown, ways will have to be found to fuse them to genes such as *lacZ* which have easily assayable products. Experiments using a chloramphenicol acetyltransferase gene from *Esch. coli* for such purposes have recently been reported (Bibb and Cohen, 1982). Conversely, if we wish to isolate the unknown protein product of an interesting gene, fusion to an easily controlled efficient promoter may also be desirable. The *in vitro* analysis of cloned DNA will require background work to be done on the endogenous transcription and translation systems. In the midst of all these exciting molecular approaches, it may almost seem that conventional chromosomal genetics is a thing of the past. We wish to emphasize that chromosomal genetics will continue to be important in *Stmy. coelicolor* A3(2) and that its absence from a strain interesting for other reasons will greatly limit what can be done even with the help of cloning: for example, in the analysis of complex biochemical processes, involving genes that are not closely linked; in the construction of strains containing in combination mutations previously available only in separate strains; and in the unravelling of the interactions between plasmid and chromosome in antibiotic synthesis and differentiation. In all these contexts, protoplast fusion may be of great importance in facilitating the isolation of recombinants; but natural genetic exchange is often just as effective, and less time-consuming. It follows that by far the most productive approach to *Streptomyces* genetics will be one that integrates the 'old' and the 'new' genetics.

REFERENCES

Aguilar, A. and Hopwood, D. A. (1982). Determination of methylenomycin A synthesis by the pSV1 plasmid from *Streptomyces violaceus-ruber* SANK 95570. *Journal of General Microbiology* **128**, 1893–1901.
Akagawa, H., Okanishi, M. and Umezawa, H. (1975). A plasmid involved in chloramphenicol production in *Streptomyces venezuelae*: evidence from genetic mapping. *Journal of General Microbiology* **90**, 336–346.
Akagawa, H., Okanishi, M. and Umezawa, H. (1979). Genetic and biochemical studies of chloramphenicol non-producing mutants of *Streptomyces venezuelae* carrying plasmid. *Journal of Antibiotics* **32**, 610–620.
Alačević, M. (1976). Some recent advances in *Streptomyces rimosus* genetics. In: *2nd International Symposium on the Genetics of Industrial Microorganisms* (Macdonald, K. D., ed.), pp. 513–519. Academic Press, London.
Alačević, M., Strašek-Vešligaj, M. and Sermonti, G. (1973). The circular linkage map of *Streptomyces rimosus*. *Journal of General Microbiology* **77**, 173–185.
Anderson, R. P. and Roth, J. R. (1977). Tandem genetic duplications in phage and bacteria. *Annual Review of Microbiology* **31**, 473–505.
Antonov, P. P., Ivanov, I. G. and Markov, G. G. (1977). Heterogeneity of *Streptomyces* DNA. *FEBS Letters* **79**, 151–154.
Arber, W. (1974). DNA modification and restriction. *Progress in Nucleic Acid Research and Molecular Biology* **14**, 1–37.
Baltz, R. H. (1978). Genetic recombination in *Streptomyces fradiae* by protoplast fusion and cell regeneration. *Journal of General Microbiology* **107**, 93–102.
Baltz, R. H. (1980). Genetic recombination by protoplast fusion in *Streptomyces*. *Developments in Industrial Microbiology* **21**, 43–54.
Baltz, R. H., Seno, E. T., Stonesifer, J., Matsushima, P. and Wild, G. M. (1981). Genetics and biochemistry of tylosin production by *Streptomyces fradiae*. *Microbiology 1981*, 371–375.
Baumann, R., Hütter, R. and Hopwood, D. A. (1974). Genetic analysis in a melanin-producing streptomycete, *Streptomyces glaucescens*. *Journal of General Microbiology* **81**, 463–474.
Beck, C. F., Moyed, H. and Ingraham, J. L. (1980). The tetracycline-resistance transposon Tn10 inhibits translocation of Tn10. *Molecular and General Genetics* **179**, 453–455.
Benigni, R., Antonov, P. P. and Carere, A. (1975). Estimate of the genome size by renaturation studies in *Streptomyces*. *Applied Microbiology* **30**, 324–326.
Benton, W. D. and Davis, R. W. (1977). Screening λ gt recombinant clones by hybridization to single plaques *in situ*. *Science* **196**, 180–182.
Benveniste, R. and Davies, J. (1973). Aminoglycoside antibiotic-inactivating enzymes in actinomycetes similar to those present in clinical isolates of antibiotic-resistant bacteria. *Proceedings of the National Academy of Sciences of the United States of America* **70**, 2276–2280.
Bibb, M. J. (1978). Genetic and physical studies of a *Streptomyces coelicolor* plasmid. PhD Thesis: University of East Anglia, Norwich.
Bibb, M. J. and Cohen, S. N. (1982). Gene expression in *Streptomyces*: construction and application of promoter-probe plasmid vectors in *Streptomyces lividans*. *Molecular and General Genetics* **182**, 265–277
Bibb, M. J. and Hopwood, D. A. (1981). Genetic studies of the fertility plasmid SCP2 and its SCP2* variants in *Streptomyces coelicolor* A3(2). *Journal of General Microbiology* **126**, 427–442.
Bibb, M. J., Freeman, R. F. and Hopwood, D. A. (1977). Physical and genetic characterization of a second sex factor, SCP2, for *Streptomyces coelicolor* A3(2). *Molecular and General Genetics* **154**, 155–166.
Bibb, M. J., Ward, J. M. and Hopwood, D. A. (1978). Transformation of plasmid DNA into streptomycetes at high frequency. *Nature, London* **274**, 398–400.
Bibb, M. J., Schottel, J. L. and Cohen, S. N. (1980a). A DNA cloning system for interspecies gene transfer in antibiotic-producing *Streptomyces*. *Nature, London* **284**, 526–531.
Bibb, M. J., Ward, J. M. and Hopwood, D. A. (1980b). The development of a cloning system for *Streptomyces*. *Developments in Industrial Microbiology* **21**, 56–64.

Bibb, M. J., Ward, J. M., Kieser, T., Cohen, S. N. and Hopwood, D. A. (1981). Excision of chromosomal DNA sequences from *Streptomyces coelicolor* forms a novel family of plasmids detectable in *Streptomyces lividans*. *Molecular and General Genetics* **184**, 230–240.
Biek, D. and Roth, J. (1980). Regulation of Tn5 transposition in *Salmonella typhimurium*. *Proceedings of the National Academy of Sciences of the United States of America* **77**, 6047–6051.
Bolivar, F., Rodriguez, R., Greene, P. J., Betlach, M., Heyneker, H. L., Boyer, H. W., Crosa, J. and Falkow, S. (1977). Construction and characterization of new cloning vehicles. A multipurpose cloning system. *Gene* **2**, 95–113.
Boronin, A. M. and Sadovnikova, L. G. (1972). Elimination by acridine dyes of oxytetracycline resistance in *Actinomyces rimosus*. *Genetika* **8**, 174–176.
Bradley, S. G. and Lederberg, J. (1956). Heterokaryosis in *Streptomyces*. *Journal of Bacteriology* **72**, 219–225.
Carere, A. and Randazzo, R. (1976). Co-mutation in *Streptomyces*. In: *Second International Symposium on the Genetics of Industrial Microorganisms* (K. D. Macdonald, ed.), pp. 573–581. Academic Press, London.
Carere, A., Russi, S., Bignami, M. and Sermonti, G. (1973). An operon for histidine biosynthesis in *Streptomyces coelicolor*. I. Genetic evidence. *Molecular and General Genetics* **123**, 219–224.
Chang, L. T., Behr, D. A. and Elander, R. P. (1978). Effects of plasmid curing agents on cultural characteristics and kanamycin formation in a production strain of *Streptomyces kanamyceticus*. *Abstracts of 3rd International Symposium on the Genetics of Industrial Microorganisms*, p. 36. Madison, Wisconsin.
Chang, S. and Cohen, S. N. (1977). *In vivo* site-specific genetic recombination promoted by the *Eco*RI restriction endonuclease. *Proceedings of the National Academy of Sciences of the United States of America* **74**, 4811–4815.
Chater, K. F. (1972). A morphological and genetic mapping study of white colony mutants of *Streptomyces coelicolor*. *Journal of General Microbiology* **72**, 9–28.
Chater, K. F. (1975). Construction and phenotype of double sporulation deficient mutants in *Streptomyces coelicolor* A3(2). *Journal of General Microbiology* **87**, 312–325.
Chater, K. F. (1977). A site-specific endodeoxyribonuclease from *Streptomyces albus* CMI 52766 sharing site-specificity with *Providencia stuartii* endonuclease *Pst*I. *Nucleic Acids Research* **4**, 1989–1998.
Chater, K. F. (1978). Restriction in *Streptomyces*. *Zentralblatt für Bakteriologie, Parasitenkunde, Infektionskrankheiten und Hygiene. 1. Abteilung, Supplement* **6**, 303–311.
Chater, K. F. (1979). Some recent developments in *Streptomyces* genetics. In *Genetics of Industrial Microorganisms* (O. K. Sebek and A. I. Laskin, eds.), pp. 123–133. American Society for Microbiology, Washington D.C.
Chater, K. F. (1980). Actinophage DNA. *Developments in Industrial Microbiology* **21**, 65–74.
Chater, K. F. and Carter, A. T. (1978). Restriction of a bacteriophage in *Streptomyces albus* P (CMI52766) by endonuclease *Sal*PI. *Journal of General Microbiology* **109**, 181–185.
Chater, K. F. and Carter, A. T. (1979). A new wide host range temperate bacteriophage, R4, of *Streptomyces* and its interaction with some restriction-modification systems. *Journal of General Microbiology* **115**, 431–442.
Chater, K. F. and Merrick, M. J. (1976). Approaches to the study of differentiation in *Streptomyces coelicolor* A3(2). In: *Second International Symposium of Industrial Microorganisms* (K. D. Macdonald, ed.), pp. 583–593. Academic Press, London.
Chater, K. F. and Merrick, M. J. (1979). *Streptomyces*. In: *Developmental Biology of Prokaryotes* (J. G. Parish, ed.), pp. 93–114. Blackwell Scientific Publications, Oxford.
Chater, K. F. and Wilde, L. C. (1976). Restriction of a bacteriophage of *Streptomyces albus* G involving endonuclease *Sal*I. *Journal of Bacteriology* **128**, 644–650.
Chater, K. F. and Wilde, L. C. (1980). *Streptomyces albus* G mutants defective in the *Sal*GI restriction-modification system. *Journal of General Microbiology* **116**, 323–334.
Chater, K. F., Bruton, C. J., Springer, W. and Suarez, J. E. (1981a). Dispensable sequences and packaging constraints of DNA from the *Streptomyces* temperate phage ϕC31. *Gene* **15**, 249–256.
Chater, K. F., Bruton, C. J. and Suarez, J. E. (1981b). Restriction mapping of the DNA of the *Streptomyces* temperate phage ϕC31 and its derivatives. *Gene* **14**, 183–194.

Chater, K. F., Bruton, C. J., Suarez, J. E. and Springer, W. (1981c). *Streptomyces* phages and their applications in DNA cloning. In: *Microbiology 1981* (D. Schlessinger, ed.), pp. 380–383. American Society for Microbiology, Washington.

Chater, K. F., Bruton, C. J., King, A. A. and Suarez, J. E. (1982a). The expression of *Streptomyces* and *Escherichia coli* drug resistance determinants cloned into the *Streptomyces* phage ϕC31. *Gene* **19**, 21–32.

Chater, K. F., Hopwood, D. A., Kieser, T. and Thompson, C. J. (1982b). Gene cloning in *Streptomyces*. *Current Topics in Microbiology and Immunology* **96**, 69–95.

Coats, J. H. and Roeser, J. (1971). Genetic recombination in *Streptomyces bikiniensis* var. *zorbonensis*. *Journal of Bacteriology* **105**, 880–885.

Cohen, S. N. (1976). Transposable genetic elements and plasmid evolution. *Nature, London* **263**, 731–738.

Cundliffe, E. (1978). Mechanisms of resistance to thiostrepton in the producing organism *Streptomyces azureus*. *Nature, London* **272**, 792–795.

Danilenko, V. N., Puzynina, G. G. and Lomovskaya, N. D. (1977). Multiple drug resistance in actinomycetes. *Genetika* **10**, 1831–1842.

Danilenko, V. N., Yankoviskii, N. K., Kaluzhskii, V. E., Moshentzeva, V. N., Sladkova, I. A., Kozlov, U. I., Redorenko, V. A., Rebentish, B. A., Lomovskaya, N. D. and Debabov, V. G. (1979). Study of the structure and functioning of an actinomycete kanamycin transposon transferred *in vitro* into *Escherichia coli* K12. *Doklady Academii Nauk USSR* **244**, 1235–1238.

Davies, J. and Smith, D. I. (1978). Plasmid-determined resistance to antimicrobial agents. *Annual Review of Microbiology* **34**, 469–518.

El-Kersh, T. and Vezina, C. (1978). Plasmid determination of antimycin A production in auxotrophs of *Streptomyces* sp. M-506. *Abstracts of 3rd International Symposium on the Genetics of Industrial Microorganisms*, p. 5. Madison, Wisconsin.

Engel, P. P. (1973). Genetic control of tryptophan biosynthesis in *Streptomyces coelicolor*. In: *Genetics of Industrial Microorganisms* (Z. Vanek, Z. Hoštálek and J. Cudlin, eds.), pp. 125–147. Academia, Prague.

Evanchik, Z., Stacey, K. A. and Hayes, W. (1969). Ultraviolet induction of chromosome transfer by autonomous sex factors in *Escherichia coli*. *Journal of General Microbiology* **56**, 1–14.

Feitelson, J. S. and Hopwood, D. A. (1983). Cloning of a *Streptomyces* gene for a O-methyltransferase involved in antibiotic biosynthesis. *Molecular and General Genetics* **190**, 394–398.

Franklin, N. C. (1978). Genetic fusions for operon analysis. *Annual Review of Genetics* **12**, 193–221.

Freeman, R. F. and Hopwood, D. A. (1978). Unstable naturally occurring resistance to antibiotics in *Streptomyces*. *Journal of General Microbiology* **106**, 377–381.

Freeman, R. F., Bibb, M. J. and Hopwood, D. A. (1977). Chloramphenicol acetyltransferase-independent chloramphenicol resistance in *Streptomyces coelicolor*. A3(2). *Journal of General Microbiology* **98**, 453–465.

Friend, E. J. and Hopwood, D. A. (1971). The linkage map of *Streptomyces rimosus*. *Journal of General Microbiology* **68**, 187–197.

Friend, E. J., Warren, M. and Hopwood, D. A. (1978). Genetic evidence for a plasmid controlling fertility in an industrial strain of *Streptomyces rimosus*. *Journal of General Microbiology* **106**, 201–206.

Ghisalba, O., Auden, J. A. L., Schupp, T. and Nuësch, J. (1982). The fermentation and biosynthesis of rifamycins. In: *Antibiotic Production by Fermentation Technology* (E. J. Vandamme, ed.) (in press).

Gil, J. A. and Hopwood, D. A. (1983). Cloning and expression of a p-aminobenzoic acid synthetase gene of the candicidin-producing *Streptomyces griseus*. *Gene*, in press.

Godfrey, O., Ford, L. and Huber, M. L. B. (1978). Interspecies matings of *Streptomyces fradiae* with *Streptomyces bikiniensis* mediated by conventional and protoplast fusion techniques. *Canadian Journal of Microbiology* **24**, 994–997.

Gregory, K. F. and Huang, J. C. C. (1964). Tyrosinase inheritance in *Streptomyces scabies*. I. Genetic recombination. *Journal of Bacteriology* **87**, 1281–1286.

Guillen, N., Le Hegarat, F., Fleury, A.-M. and Hirschbein, L. (1978). Folded chromosomes of vegetative *Bacillus subtilis*: composition and properties. *Nucleic Acids Research* **5**, 475–489.

Hanekes, L. M. and Hoekstra, W. P. M. (1976). Characterisation of an *Escherichia coli* K-12 F⁻ Con⁻ mutant. *Journal of Bacteriology* **125**, 593–600.

Harold, R. J. and Hopwood, D. A. (1970). Ultraviolet-sensitive mutants of *Streptomyces coelicolor*. II. Genetics. *Mutation Research* **10**, 439–448.

Harris, J. E., Chater, K. F., Bruton, C. J. and Piret, J. M. (1983). The restriction mapping of *c* gene deletions in *Streptomyces* bacteriophage ϕC31 and their use in cloning vector development. *Gene* **22**, 167–174.

Hayakawa, T., Tanaka, T., Sakaguchi, K., Otake, B. and Yonehara, H. (1979). A linear plasmid-like DNA in *Streptomyces* sp. producing lankacidin group antibiotics. *Journal of General and Applied Microbiology* **25**, 255–260.

Hodgson, D. A. (1980). Carbohydrate utilisation in *Streptomyces coelicolor* A3(2). PhD Thesis: University of East Anglia, Norwich.

Hodgson, D. A. (1982). Glucose repression of carbon source uptake in *Streptomyces coelicolor* A3(2) and its perturbation in mutants resistant to 2-deoxyglucose. *Journal of General Microbiology* **128**, 2417–2430.

Hodgson, D. A. and Chater, K. F. (1981). A chromosomal locus controlling extracellular agarase production by *Streptomyces coelicolor* A3(2), and its inactivation by chromosomal integration of plasmid SCP1. *Journal of General Microbiology* **124**, 339–348.

Hopwood, D. A. (1957). Genetic recombination in *Streptomyces coelicolor*. *Journal of General Microbiology* **16**, ii–iii.

Hopwood, D. A. (1959). Linkage and the mechanism of recombination in *Streptomyces coelicolor*. *Annals of the New York Academy of Sciences U.S.A.* **81**, 887–898.

Hopwood, D. A. (1966). Lack of constant genome ends in *Streptomyces coelicolor*. *Genetics* **54**, 1177–1184.

Hopwood, D. A. (1967a). Genetic analysis and genome structure in *Streptomyces coelicolor*. *Bacteriological Reviews* **31**, 373–403.

Hopwood, D. A. (1967b). A possible circular symmetry of the linkage map of *Streptomyces coelicolor*. *Journal of Cellular Physiology* **70**, Supplement **1**, 7–10.

Hopwood, D. A. (1972). Genetic analysis in microorganisms. In: *Methods in Microbiology*, Volume 7B (J. R. Norris and D. W. Ribbons, eds.), pp. 29–158. Academic Press, London.

Hopwood, D. A. (1978). Extrachromosomally determined antibiotic production. *Annual Review of Microbiology* **32**, 373–392.

Hopwood, D. A. (1979a). The many faces of recombination. In: *Genetics of Industrial Microorganisms* (O. K. Sebek and A. I. Laskin, eds.), pp. 1–9. American Society for Microbiology, Washington D.C.

Hopwood, D. A. (1979b). Genetics of antibiotic production by actinomycetes. *Journal of Natural Products* **42**, 596–602.

Hopwood, D. A. (1981a). Uses of protoplasts in the genetic manipulation of streptomycetes. *Zentralblatt für Bakteriologie, Parasitenkunde, Mikrobiologie und Hygiene*. I. Abteilung, Supplement **11**, 523–531.

Hopwood, D. A. (1981b). Future possibilities for the discovery of new antibiotics by genetic engineering. In: *Beta-lactam Antibiotics*. (M. R. J. Salton and G. D. Schockman, eds.), pp. 585–598. Academic Press, New York.

Hopwood, D. A. (1981c). Genetic studies with bacterial protoplasts. *Annual Review of Microbiology* **35**, 237–272.

Hopwood, D. A. and Chater, K. F. (1974). *Streptomyces coelicolor*. In: *Handbook of Genetics*, Volume 1 (R. C. King, ed.), pp. 237–255. Plenum Press, New York and London.

Hopwood, D. A. and Chater, K. F. (1980). Fresh approaches to antibiotic production. *Philosophical Transactions of the Royal Society, London B*, **290**, 313–328.

Hopwood, D. A. and Chater, K. F. (1982). DNA cloning in *Streptomyces*: systems and strategies. In: *Genetic Engineering*, Volume 4 (J. K. Setlow and A. Hollaender, eds.), pp. 119–145. Plenum Press, New York and London.

Hopwood, D. A. and Glauert, A. M. (1960a). Observations on the chromatinic bodies of *Streptomyces coelicolor*. *Journal of Biophysical and Biochemical Cytology* **8**, 257–265.

Hopwood, D. A. and Glauert, A. M. (1960b). The fine structure of *Streptomyces coelicolor*. II. The nuclear material. *Journal of Biophysical and Biochemical Cytology* **8**, 267–278.

Hopwood, D. A. and Merrick, M. J. (1977). Genetics of antibiotic production. *Bacteriological Reviews* **41**, 595–635.

Hopwood, D. A. and Sermonti, G. (1962). The genetics of *Streptomyces coelicolor*. *Advances in Genetics* **11**, 273–342.
Hopwood, D. A. and Wright, H. M. (1972). Transformation in *Thermoactinomyces vulgaris*. *Journal of General Microbiology* **71**, 383–398.
Hopwood, D. A. and Wright, H. M. (1973). Transfer of a plasmid between *Streptomyces* species. *Journal of General Microbiology* **77**, 187–195.
Hopwood, D. A. and Wright, H. M. (1976a). Genetic studies on SCP1-prime strains of *Streptomyces coelicolor* A3(2). *Journal of General Microbiology* **95**, 107–120.
Hopwood, D. A. and Wright, H. M. (1976b). Interaction of the plasmid SCP1 with the chromosome of *Streptomyces coelicolor* A3(2). In: *Second International Symposium on the Genetics of Industrial Microorganisms* (K. D. Macdonald, ed.), pp. 607–619. Academic Press, London.
Hopwood, D. A. and Wright, H. M. (1978). Bacterial protoplast fusion: recombination in fused protoplasts of *Streptomyces coelicolor*. *Molecular and General Genetics* **162**, 307–317.
Hopwood, D. A. and Wright, H. M. (1979). Factors affecting recombinant frequency in protoplast fusions of *Streptomyces coelicolor*. *Journal of General Microbiology* **111**, 137–143.
Hopwood, D. A. and Wright, H. M. (1981). Protoplast fusion in *Streptomyces*: fusions involving ultraviolet-irradiated protoplasts. *Journal of General Microbiology* **126**, 21–27.
Hopwood, D. A., Sermonti, G. and Spada-Sermonti, I. (1963). Heterozygous clones in *Streptomyces coelicolor*. *Journal of General Microbiology* **30**, 249–260.
Hopwood, D. A., Harold, R. J., Vivian, A. and Ferguson, H. M. (1969). A new kind of fertility variant in *Streptomyces coelicolor*. *Genetics* **62**, 461–477.
Hopwood, D. A., Chater, K. F., Dowding, J. E. and Vivian, A. (1973). Recent advances in *Streptomyces coelicolor* genetics. *Bacteriological Reviews* **37**, 371–405.
Hopwood, D. A., Wright, H M., Bibb, M. J. and Cohen, S. N. (1977). Genetic recombination through protoplast fusion in *Streptomyces*. *Nature, London* **268**, 171–173.
Hopwood, D. A., Bibb, M. J., Ward, J. M. and Westpheling, J. (1979). Plasmids in *Streptomyces coelicolor* and related species. In: *Plasmids of Medical, Environmental and Commercial Importance* (K. N. Timmis and A. Pühler, eds.), pp. 245–258. Elsevier/North Holland Biomedical Press, Amsterdam.
Hopwood, D. A., Thompson, C. J., Kieser, T., Ward, J. M. and Wright, H. M. (1981). Progress in the development of plasmid cloning vectors for *Streptomyces*. In: *Microbiology 1981* (D. Schlessinger, ed.), pp. 376–379. American Society for Microbiology, Washington.
Hornemann, U. and Hopwood, D. A. (1981). Biosynthesis of methylenomycin A: a plasmid-determined antibiotic. In: *Antibiotics, Biosynthesis, Volume IV* (J. W. Corcoran, ed.), pp. 123–131. Springer Verlag, Berlin, Heidelberg.
Hotta, K., Okami, Y. and Umezawa, H. (1977). Elimination of the ability of a kanamycin-producing strain to biosynthesize deoxystreptamine moiety by acriflavine. *Journal of Antibiotics* **30**, 1146–1149.
Hütter, R., Kieser, T., Crameri, R. and Hintermann, G. (1981). Chromosomal instability in *Streptomyces glaucescens*. *Zentralblatt für Bakteriologie, Mikrobiologie und Hygiene*. 1. Abteilung, Supplement **11**, 551–559.
Isogai, T., Takahashi, H. and Saito, H. (1980). High frequency protoplast transfection of *Streptomyces parvulus* 2297 with actinophage R4 DNA. *Agricultural Biology and Chemistry* **44**, 2425–2428.
Kieser, T., Hintermann, G., Crameri, R. and Hütter, R. (1981). Restriction analysis of *Streptomyces* DNA. *Zentralblatt für Bakteriologie, Mikrobiologie und Hygiene*. 1. Abteilung, Supplement **11**, 561–562.
Kieser, T., Hopwood, D. A., Wright, H. M. and Thompson, C. J. (1982). pIJ101, a multi-copy broad host-range *Streptomyces* plasmid: functional analysis and development of DNA cloning vectors. *Molecular and General Genetics* **185**, 223–238.
Kirby, R. (1976). Genetic studies on *Streptomyces coelicolor* plasmid one. PhD Thesis: University of East Anglia, Norwich.
Kirby, R. (1978). An unstable genetic element affecting the production of the antibiotic holomycin by *Streptomyces clavuligerus*. *FEMS Microbiology Letters* **3**, 283–286.
Kirby, R. and Hopwood, D. A. (1977). Genetic determination of methylenomycin synthesis by the SCP1 plasmid of *Streptomyces coelicolor* A3(2). *Journal of General Microbiology* **98**, 229–252.

Kirby, R., Wright, L. F. and Hopwood, D. A. (1975). Plasmid-determined antibiotic synthesis and resistance in *Streptomyces coelicolor*. *Nature, London* **254**, 265–267.

Klaus, S., Krügel, H., Süss, F., Neigenfind, M., Zimmerman, I. and Taubeneck, U. (1981). Properties of the temperate actinophage SH10. *Journal of General Microbiology* **123**, 269–279.

Kleckner, N. (1977). Translocatable elements in prokaryotes. *Cell* **11**, 11–23.

Kleckner, N., Roth, J. and Botstein, D. (1977). Genetic engineering *in vivo* using translocatable elements. *Journal of Molecular Biology* **116**, 125–159.

Komatsu, K., Leboul, J., Harford, S. and Davies, J. (1981). Studies of plasmids in neomycin-producing *Streptomyces fradiae*. In: *Microbiology 1981* (D. Schlessinger, ed.), pp. 384–387. American Society for Microbiology, Washington, D.C.

Krügel, H., Fiedler, G. and Noack, D. (1980). Transfection of protoplasts from *S. lividans* 66 with actinophage SH10 DNA. *Molecular and General Genetics* **177**, 297–300.

Lomovskaya, N. D., Voeykova, R. A., Sladkova, I. A., Chinenova, T. A., Mkrtumian, N. M. and Slavinskaya, E. V. (1979). Genetic relationships between actinomycetes and actinophages. In: *Genetics of Industrial Microorganisms* (O. K. Sebek and A. I. Laskin, eds.), pp. 141–146. American Society for Microbiology, Washington D.C.

Lomovskaya, N. D., Chater, K. F. and Mkrtumian, N. M. (1980). Genetics and molecular biology of *Streptomyces* bacteriophages. *Microbiological Reviews* **44**, 206–229.

Makins, J. F. and Holt, G. (1981). Liposome-mediated transformation of streptomycetes by chromosomal DNA. *Nature, London* **293**, 671–673.

Martin, J. F. and Demain, A. L. (1980). Control of antibiotic synthesis. *Microbiological Reviews* **44**, 230–251.

Meacock, P. A. and Cohen, S. N. (1980). Partitioning of bacterial plasmids during cell division – a cis-acting locus that accomplishes stable plasmid inheritance. *Cell* **20**, 529–542.

Merrick, M. J. (1976). A morphological and genetic mapping study of bald colony mutants of *Streptomyces coelicolor*. *Journal of General Microbiology* **96**, 299–315.

Michelson, A. M. and Vining, L. C. (1978). Loss of chloramphenicol production in strains of *Streptomyces* species 3022a treated with acriflavine and ethidium bromide. *Canadian Journal of Microbiology* **24**, 662–669.

Miller, J. H., Calos, M. P., Halas, D., Hofer, M., Buchel, D. E. and Müller-Hill, B. (1980). Genetic analysis of transposons in the *lac* region of *Escherichia coli*. *Journal of General Microbiology* **114**, 1–18.

Nakano, M. M. and Ogawara, H. (1980). Multiple effects induced by unstable mutation in *Streptomyces lavendulae*. *Journal of Antibiotics* **33**, 420–425.

Noack, D., Roth, M. and Zippel, M. (1978). Extrachromosomal control of growth and antibiotic production in *Streptomyces hygroscopicus* JA6599. In: *Genetics of the Actinomycetales* (E. Freerksen, I. Tárnok and J. H. Thumin, eds.), pp. 15–17. Gustav Fischer Verlag, Stuttgart.

Nojiri, C., Watabe, H., Katsumata, K., Yamada, Y., Murakami, T. and Kumata, Y. (1980). Isolation and characterisation of plasmids from parent and variant strains of *Streptomyces ribosidificus*. *Journal of Antibiotics* **33**, 118–121.

Nomura, M., Yamagishi, H. and Oka, A. (1978). Isolation and characterization of transducing coliphage fd carrying kanamycin resistance gene. *Gene* **3**, 39–51.

Ochi, K. and Katz, E. (1978). The possible involvement of a plasmid(s) in actinomycin synthesis by *Streptomyces parvulus* and *Streptomyces antibioticus*. *Journal of Antibiotics* **31**, 1143–1148.

Ochi, K., Hitchcock, M. J. M. and Katz, E. (1979). High-frequency fusion of *Streptomyces parvulus* or *Streptomyces antibioticus* protoplasts induced by polyethylene glycol. *Journal of Bacteriology* **139**, 984–992.

Okami, Y. (1979). Antibiotics from marine microorganisms with reference to plasmid involvement. *Journal of Natural Products* **42**, 583–595.

Okanishi, M. (1979). Plasmids and antibiotic synthesis in *Streptomyces*. In: *Genetics of Industrial Microorganisms* O. K. Sebek and A. I. Laskin, eds.), pp. 134–140. American Society for Microbiology, Washington, D.C.

Okanishi, M. and Umezawa, H. (1978). Plasmids involved in antibiotic production in streptomycetes. In: *Genetics of the Actinomycetales* (E. Freerksen, I. Tárnok and J. H. Thumin, eds.), pp. 19–38. Gustav Fischer Verlag, Stuttgart.

Old, R. W. and Primrose, S. B. (1981). *Principles of Gene Manipulation*. Blackwell Scientific Publications, Oxford.
Ōmura, S., Ikeda, H. and Kitao, C. (1979). The detection of a plasmid in *Streptomyces ambofaciens* KA-1028 and its possible involvement in spiramycin production. *Journal of Antibiotics* **32**, 1058–1060.
Ono, H., Hintermann, G., Crameri, R., Wallis, G. and Hütter, R. (1982). Reiterated DNA sequences in a mutant strain of *Streptomyces glaucescens* and cloning of the sequence in *Escherichia coli*. *Molecular and General Genetics* **186**, 106–110.
Parkinson, J. S. and Husky, R. J. (1971). Deletion mutants of bacteriophage lambda. I. Isolation and initial characterisation. *Journal of Molecular Biology* **56**, 369–384.
Queener, S. W. and Baltz, R. H. (1979). Genetics of industrial microorganisms. *Annual Reports on Fermentation Processes* **3**, 5–45.
Redshaw, P. A., McCann, P. A., Pentella, M. A. and Pogell, B. M. (1979). Simultaneous loss of multiple differentiation functions in aerial mycelium-negative isolates of streptomycetes. *Journal of Bacteriology* **137**, 891–899.
Rhodes, P. M., Winskill, N., Friend, E. J. and Warren, M. (1981). Biochemical and genetic characterisation of *Streptomyces rimosus* mutants impaired in oxytetracycline biosynthesis. *Journal of General Microbiology* **124**, 329–338.
Roberts, R. J. (1980). Restriction and modification enzymes and their recognition sequences. *Gene* **8**, 329–343.
Robinson, M., Lewis, E. and Napier, E. (1981). Occurrence of reiterated DNA sequences in strains of *Streptomyces* produced by an interspecific protoplast fusion. *Molecular and General Genetics* **182**, 336–340.
Rodicio, M. R. and Chater, K. F. (1982). Small DNA-free liposomes stimulate transfection of *Streptomyces* protoplasts. *Journal of Bacteriology* **151**, 1078–1085.
Rosner, A., Gutstein, R. and Aviv, H. (1980). Isolation of viable deletion mutants of *Streptomyces* actinophage (Pa16) and their molecular characterization. *Molecular and General Genetics* **178**, 337–341.
Rowbury, R. J. (1977). Bacterial plasmids with particular reference to their replication and transfer properties. *Progress in Biophysical and Molecular Biology* **31**, 271–317.
Rudd, B. A. M. and Hopwood, D. A. (1979). Genetics of actinorhodin biosynthesis by *Streptomyces coelicolor* A3(2). *Journal of General Microbiology* **114**, 35–43.
Rudd, B. A. M. and Hopwood, D. A. (1980). A pigmented mycelial antibiotic in *Streptomyces coelicolor*: control by a chromosomal gene cluster. *Journal of General Microbiology* **119**, 333–340.
Sanchez, S. and Quinto, C. M. (1975). D-Glucose isomerase: constitutive and catabolite repression-resistant mutants of *Streptomyces phaeochromogenes*. *Applied Microbiology* **30**, 750–754.
Schottel, J., Bibb, M. J. and Cohen, S. N. (1981). Cloning and expression in *Streptomyces lividans* of antibiotic resistance genes derived from *Escherichia coli*. *Journal of Bacteriology* **146**, 360–368.
Schrempf, H. (1982). Plasmid loss and changes within the chromosomal DNA of *Streptomyces reticuli*. *Journal of Bacteriology* **151**, 701–707.
Schrempf, H. and Goebel, W. (1977). Characterization of a plasmid from *Streptomyces coelicolor* A3(2). *Journal of Bacteriology* **131**, 251–258.
Schrempf, H. and Goebel, W. (1979). Functions of plasmid genes in *Streptomyces reticuli*. In: *Plasmids of Medical, Environmental and Commercial Importance* (K. N. Timmis and A. Pühler, eds.), pp. 259–268. Elsevier/North Holland Biomedical Press, Amsterdam.
Schupp, T., Hütter, R. and Hopwood, D. A. (1975). Genetic recombination in *Nocardia mediterranei*. *Journal of Bacteriology* **121**, 128–136.
Seno, E. T. and Chater, K. F. (1983). Glycerol catabolic enzymes and their regulation in wild-type and mutant strains of *Streptomyces coelicolor* A3(2). *Journal of General Microbiology* **129**, 1403–1413.
Sermonti, G. (1969). *Genetics of Antibiotic-Producing Microorganisms*. Wiley-Interscience, London.
Sermonti, G. and Puglia, A. M. (1976). Progressive fertilisation in *Streptomyces coelicolor*. In: *Second International Symposium on the Genetics of Industrial Microorganisms* (K. D. Macdonald, ed.), pp. 565–572. Academic Press, London.

Sermonti, G. and Spada-Sermonti, I. (1955). Genetic recombination in *Streptomyces*. *Nature (London)* **176**, 121.
Sermonti, G. and Spada-Sermonti, I. (1956). Gene recombination in *Streptomyces coelicolor*. *Journal of General Microbiology* **15**, 609–616.
Sermonti, G., Petris, A., Micheli, M. and Lanfaloni, L. (1977). A factor involved in chloramphenicol resistance in *Streptomyces coelicolor* A3(2): its transfer in the absence of the fertility factor. *Journal of General Microbiology* **100**, 347–353.
Sermonti, G., Petris, A., Micheli, M. R. and Lanfaloni, L. (1977). A factor involved in chloramphenicol resistance in *Streptomyces coelicolor* A3(2): its transfer in the absence of the fertility factor. *Journal of General Microbiology* **100**, 347–353.
Sermonti, G., Lanfaloni, L. and Micheli, M. R. (1980). A jumping gene in *Streptomyces coelicolor* A3(2). *Molecular and General Genetics* **177**, 453–458.
Shaw, P. D. and Piwowarski, J. (1977). Effects of ethidium bromide and acriflavine on streptomycin production by *Streptomyces bikiniensis*. *Journal of Antibiotics, Japan* **30**, 404–408.
Shaw, W. V. and Hopwood, D. A. (1976). Chloramphenicol acetylation in *Streptomyces*. *Journal of General Microbiology* **94**, 159–166.
Skinner, R. H. and Cundliffe, E. (1981). Resistance to the antibiotics viomycin and capreomycin in *Streptomyces* which produce them. *Journal of General Microbiology* **120**, 95–104.
Skinner, R. H. and Cundliffe, E. (1982). Dimethylation of adenine and the resistance of *Streptomyces erythreus* to its product erythromycin. *Journal of General Microbiology* **128**, 2411–2416.
Skurray, R. A. and Reeves, P. (1973). Characterisation of lethal zygosis associated with conjugation in *Escherichia coli* K-12. *Journal of Bacteriology* **113**, 58–70.
Stanier, R. Y. (1942). Agar utilising strain of the *Actinomyces coelicolor* species group. *Journal of Bacteriology* **44**, 555–570.
Starlinger, P. and Saedler, H. (1976). IS-elements in microorganisms. *Current Topics in Microbiology and Immunology* **75**, 111–152.
Stocker, B. A. D., Smith, S. M. and Ozeki, H. (1963). High infectivity of *Salmonella typhimurium* newly infected by the ColI factor. *Journal of General Microbiology* **30**, 201–221.
Stuttard, C. S. (1979). Transduction of auxotrophic markers in a chloramphenicol-producing strain of *Streptomyces*. *Journal of General Microbiology* **110**, 479–482.
Stuttard, C. S. (1981). Localised mutagenesis by transduction in *Streptomyces* Spec. 3022a. *Zentralblatt für Bakteriologie, Infektionskrankheiten und Hygiene. I. Abteilung. Supplement* **11**, 533–537.
Suarez, J. E. and Chater, K. F. (1980a). Polyethylene glycol-assisted transfection of *Streptomyces* protoplasts. *Journal of Bacteriology* **142**, 8–14.
Suarez, J. E. and Chater, K. F. (1980b). DNA cloning in *Streptomyces*: a bifunctional replicon comprising pBR322 inserted into a *Streptomyces* phage. *Nature, London* **286**, 527–529.
Suarez, J. E. and Chater, K. F. (1981). Development of a DNA cloning system in *Streptomyces* using phage vectors. *Ciencia Biologica* **6**, 99–110.
Thompson, C. J., Ward, J. M. and Hopwood, D. A. (1980). DNA cloning in *Streptomyces*: resistance genes from antibiotic producing species. *Nature, London* **286**, 525–527.
Thompson, C. J., Ward, J. M. and Hopwood, D. A. (1982a). Cloning of antibiotic resistance and nutritional genes in *Streptomyces*. *Journal of Bacteriology* **151**, 668–677
Thompson, C. J., Skinner, R. H., Thompson, I., Ward, J. M., Hopwood, D. A. and Cundliffe, E. (1982b). Biochemical characterisation of resistance determinants cloned from antibiotic-producing *Streptomyces*. *Journal of Bacteriology* **151**, 678–685.
Thompson, C. J., Kieser, T., Ward, J. M. and Hopwood, D. A. (1982c). Physical analysis of antibiotic-resistance genes from *Streptomyces* and their use in vector construction. *Gene* **20**, 51–62.
Ullman, A. (1974). Are cyclic AMP effects related to real physiological phenomena? *Biochemical and Biophysical Research Communications* **57**, 348–352.
Vivian, A. (1971). Genetic control of fertility in *Streptomyces coelicolor* A3(2): plasmid involvement in the interconversion of UF and IF strains. *Journal of General Microbiology* **69**, 353–364.
Vivian, A. and Hopwood, D. A. (1970). Genetic control of fertility in *Streptomyces coelicolor* A3(2): the IF fertility type. *Journal of General Microbiology* **64**, 101–117.

Vivian, A. and Hopwood, D. A. (1973). Genetic control of fertility in *Streptomyces coelicolor* A3(2): new kinds of donor strains. *Journal of General Microbiology* **76**, 147–162.

Westpheling, J. (1980). Physical studies of *Streptomyces* plasmids. PhD Thesis. University of East Anglia, Norwich.

White, T. J. and Davies, J. (1978). Possible involvement of plasmids in the biosynthesis of paromomycin. *Abstracts of 3rd International Symposium on the Genetics of Industrial Microorganisms*, p. 39. Madison, Wisconsin.

Worcel, A. and Burgi, E. (1974). Properties of a membrane attached form of the folded chromosome of *Escherichia coli*. *Journal of Molecular Biology* **82**, 91–105.

Wright, L. F. and Hopwood, D. A. (1976a). Identification of the antibiotic determined by the SCP1 plasmid of *Streptomyces coelicolor* A3(2). *Journal of General Microbiology* **95**, 96–106.

Wright, L. F. and Hopwood, D. A. (1967b). Actinorhodin is a chromosomally-determined antibiotic in *Streptomyces coelicolor* A3(2). *Journal of General Microbiology* **96**, 289–297.

Wright, H. M. and Hopwood, D. A. (1977). A chromosomal gene for chloramphenicol acetyltransferase in *Streptomyces acrimycini*. *Journal of General Microbiology* **102**, 417–421.

Yagisawa, M., Huang, T.-S. R. and Davies, J. (1978). The possible role of plasmids in neomycin biosynthesis and modification. *Abstracts of the 3rd International Symposium on the Genetics of Industrial Microorganisms*, p. 37. Madison, Wisconsin.

6
Transformation of Xenobiotics

WANDA PECZYŃSKA-CZOCH and M. MORDARSKI

Institute of Immunology and Experimental Therapy, Polish Academy of Sciences, ul. Czerska 12, Wrocław 53–114, Poland

1. General characteristics of microbial transformations of xenobiotics	287
2. Hydrocarbons	289
A. Aliphatic hydrocarbons	290
B. Aromatic hydrocarbons	291
C. Microbial cooxidation	291
3. Miscellaneous compounds of synthetic origin	296
4. Pesticides	300
5. Steroids	302
A. General characteristics of microbial transformations of steroids	303
B. Degradation of the steroid nucleus	307
C. Degradation of the steroid side chain	307
6. Terpenes	310
7. Cannabinoids	312
8. Alkaloids	314
A. Vinca alkaloids	315
B. Yohimbine	319
C. Other alkaloids	319
9. Antibiotics	322
A. Tetracyclines	324
B. 16-Membered macrolides	324
C. Anthracyclines	326
D. Rifamycins	328
10. General conclusions	331
References	332

1. General Characteristics of Microbial Transformations of Xenobiotics

The idea of taking advantage of microbial enzymic reactions in organic chemistry started when Peterson and Murray (1952) performed the stereospecific hydroxylation of steroids with *Rhizopus arrhizus*. The discovery of the phenomenon of microbial conversion of complicated organic compounds provided the impetus to employ microorganisms in the production of novel agents widely used in medical therapy. Gradually, microbial methods of preparing new derivatives of known substrates were applied to modify organic compounds of diverse structure such as hydrocarbons, terpenes, steroids, alkaloids, antibiotics and amino acids.

The microbial transformation of xenobiotics can be defined as the structural modification of components foreign to an organism's metabolism which occur in its chemical environment. Such modification encompasses any change in the substrate moiety under study whether by addition, modification or degradation. When the basic structure of the resultant product remains unaltered, for example, the biotransformation of steroids yields compounds with the cyclopentanoperhydrophenanthrene nucleus retained, the microbial transformation can be applied to the formation of the desired derivatives on a preparative scale.

If the enzymic system present in the microbial cell is able to degrade the xenobiotic molecule completely, the final products will be carbon dioxide and water. The transformation of a xenobiotic molecule associated with complete degradation is considered to be a biodegradation process.

The most characteristic reactions to which xenobiotics are subjected (Kieslich, 1976) are as follows: oxidative reactions; reductive reactions; hydrolytic reactions; dehydration and condensation reactions; degradation reactions; formation of new carbon–carbon or heteroatoms bonds and isomerization and rearrangements. The variety of enzymes involved in such reactions makes them attractive as specific catalysts in organic chemistry, biochemistry and biotechnology.

The high selectivity of enzymic reactions leads to the creation of new chiral centres in the products formed *via* the stereospecific reduction of carbonyl groups, hydroxylation or the introduction of other substituents. As a result, the compounds obtained are pure enantiomers as opposed to the racemic mixtures created in analogous chemical reactions. An additional advantage of microbial or enzymic reactions is the mild conditions of pH, temperature, the solvents employed and the specificity of the transformations performed (Jones *et al.*, 1976; Rosazza. 1980). Recently, great interest has centered on the immobilization of intact microbial cells and crude or purified enzymes on various supports and their industrial application (Abbott, 1977; Chibata and Tosa, 1977; Bernath *et al.*, 1977).

Another useful property of microorganisms is their ability to decompose organic wastes of industrial origin. Thus, pollutants such as oil spills, biocides and carcinogenic hydrocarbons may be subjected to microbial attack yielding products that are less toxic to the environment, as the result of biotransformation and subsequent biodegradation.

No attempt has been made in this chapter to give a comprehensive survey of microbial reactions as these can be found in the literature. Instead, an attempt has been made to fashion a framework of some representative conversions performed by actinomycetes, organisms that are now believed to have considerably more potential in microbial transformations than was previously recognized. Members of the genera *Nocardia* and *Streptomyces* are considered in some detail because of their ability to perform highly selective chemical modifications of complicated compounds of natural and synthetic origin in some cases without further degradation of the transformed substrate. In this respect they seem to have an advantage over the pseudomonads and other bacteria in the accumulation of the desired products of microbial reactions.

The identity of actinomycetes associated with specific microbial conversions has often to be treated with caution, particularly in the case of organisms labelled *Nocardia* or *Streptomyces*. The application of modern taxonomic methods have helped to clarify the composition and internal structure of the genus *Streptomyces* (Williams *et al.*, 1981) while organisms previously assigned to the genus *Nocardia* on the basis of a few subjectively chosen characters are now classified in genera such as *Actinomadura*, *Oerskovia* and *Rhodococcus* (Chapter 2; Minnikin and Goodfellow, 1980; Goodfellow and Minnikin, 1981). Although it seems likely that many of the microbial transformations associated with nocardiae are, strictly speaking, due to rhodococci this is not always easy to prove given the poor descriptions of many test strains. In this review, therefore, the label *Nocardia* is used but it should be interpreted broadly to include rhodococci and other closely related actinomycetes.

2. Hydrocarbons

The primary enzymic attack on the hydrocarbon molecule prior to biotransformation or biodegradation is the introduction of a hydroxy group in the xenobiotic moiety. Hydroxylation as a process catalysed by oxidases and oxygenases, enzymes belonging to the oxidoreductase class, consists of the introduction of an oxygen atom into the molecule of the substrate resulting in the activation of the non-active carbon atom. A characteristic of such reactions is that the oxygen atom introduced is derived from molecular oxygen. Such modified compounds are susceptible to further enzymic attack resulting in some cases in the total biodegradation of the molecule. The transformation does, however, often stop on the introduction of the hydroxy group in certain positions of the molecule.

Hydroxylation reactions as an initial stage of xenobiotic transformations are widely distributed and are of especial interest in the actinomycetes. They can be used for obtaining commercially valuable compounds and play an important role in the degradation of xenobiotic compounds in nature. Actinomycetes indigenous in soil and water are probably the first line of attack on hydrocarbon molecules as these compounds are not broken down by the majority of microorganisms. The compounds attacked in this way may then be accessible for further conversion performed by other microorganisms. In general, actinomycetes have the ability to hydroxylate aliphatic chains of hydrocarbons in the terminal and subterminal positions and this is frequently followed by shortening of the transformed chains. Actinomycetes are also able to cleave the aromatic rings of compounds of both natural and synthetic origin. The current review presents a selection from the literature chosen to give a general idea of the major pathways of hydrocarbon transformation and degradation of *Nocardia* strains. More comprehensive reviews are available (Van der Linden and Thijsse, 1965; Raymond and Jamison, 1971; Kieslich, 1976; Tárnok, 1976; Golovlev *et al.*, 1978).

A. ALIPHATIC HYDROCARBONS

The ability of *Nocardia* strains to metabolize hydrocarbons has been recognized for many years and provides a basis for differentiating these organisms into two groups: those able and those unable to utilize paraffins. Under optimal conditions some nocardiae are capable of very rapid growth on n-paraffinic hydrocarbons. The longer chain hydrocarbons, i.e. hexadecane, are oxidized more rapidly than the shorter chain ones (Webley and DeKock, 1952). Representatives of several species of *Nocardia* have been examined for their ability to metabolize a wide variety of hydrocarbons. Nocardial enzymic reactions have been categorized into three groups: oxidation, which results in biodegradation; utilization of the metabolites so formed in the biosynthesis of cell constituents; and cooxidation reactions resulting in products where the initial structure of the hydrocarbon is intact. There is evidence that most of the oxidized hydrocarbons enter the usual metabolic pathways which result in cell biosynthesis. The initial ω-oxidation of the aliphatic chain is followed by β-oxidation yielding fatty acids (Webley *et al.*, 1955). With certain exceptions a parallel with fatty acid metabolism was noted for branched alkane utilization. A study of the effect of substitution on alkane metabolism by nocardiae was carried out by McKenna (1966) who found that any substitution such as methyl-, ethyl-, or phenyl-, decreased the availability of alkanes for growth.

The intermediates accumulated in the degradation of alkanes, besides carboxylic acids, are limited to monoalkenes. Wagner *et al.* (1967) isolated 1-hexadecene from a hexadecane grown *Nocardia* strain. Abbott and Casida (1968) obtained a mixture of internal monoalkenes from the conversion of hexadecane and octadecane by resting cells of '*Nrda.*' *salmonicolor* PSU-N-18. Accumulation of these alkenes was accomplished by growing the cells on glucose followed by incubation with the alkane.

B. AROMATIC HYDROCARBONS

There are only a few reports on the degradation of aromatic hydrocarbons by *Nocardia* strains. In general, the biodegradation of unsubstituted aromatic hydrocarbons proceeds with hydroxylation of the aromatic nucleus and subsequent oxidation results in the cleavage of the aromatic system. The products are subject to further degradation. The degradation of benzene has been observed for only a very limited number of nocardiae. According to Van der Linden and Thijsse (1965) the degradation of benzene proceeds *via* oxidation to catechol resulting in the formation of muconic acid which can be metabolized further *via* 3-oxoadipic acid to succinate and acetyl-CoA (Fig. 1).

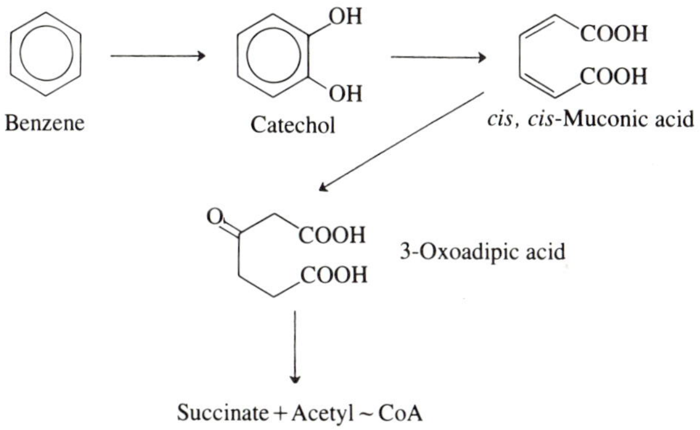

FIG. 1. The pathway in microbial benzene degradation.

The accumulation of salicylic acid during growth of nocardiae on naphthalene in a simple salts medium has been reported. According to Kieslich (1976) the utilization of naphthalene by nocardiae follows the pathway of naphthalene degradation established for soil pseudomonads. The salicylic acid so formed is converted into catechol and subjected to further degradation entering the metabolic pathway typical for benzene degradation (Fig. 2).

C. MICROBIAL COOXIDATION

According to Foster (1962) the term 'cooxidation' describes a fermentation system in which microorganisms oxidize 'nongrowth' substrate in the presence of a compound that is utilized as a source of carbon and energy for cell synthesis. The phenomenon of cooxidation in *Nocardia* strains was first observed by Davis and Raymond (1961) who noted that when nocardiae were cultivated on mineral oil in the presence of naphthenes as non-growth

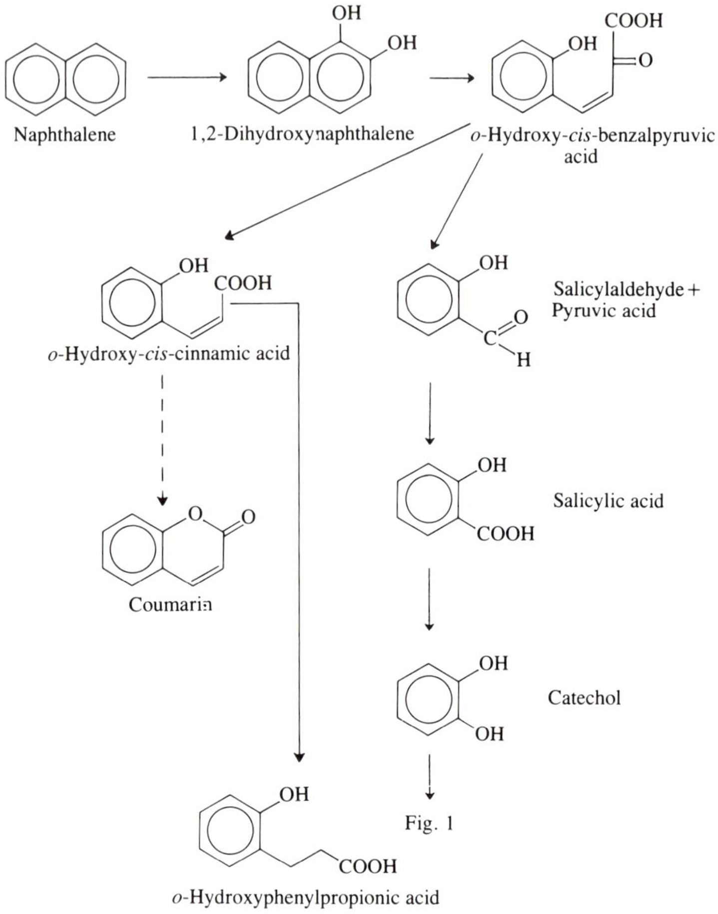

FIG. 2. The pathways in microbial naphthalene degradation.

substrates, naphthenic acid accumulated as a product of cooxidation. They also noted that, if the cyclic hydrocarbons subjected to cooxidation had a short alkyl substituent and normal paraffins, such as n-decane or n-octodecane, were provided for growth, then good yields were obtained for a number of products (Fig. 3).

The aromatic hydrocarbon dihydroxylation performed by nocardiae under cooxidative conditions was elaborated by Raymond et al. (1967). The cultures used in their studies were cultivated on n-paraffinic hydrocarbons as nutrient substrates with the addition of aromatic hydrocarbons which could not support the growth of the microorganisms. Strains of 'Nrda.' corallina, 'Nrda.' minima and 'Nrda.' salmonicolor, in addition to their

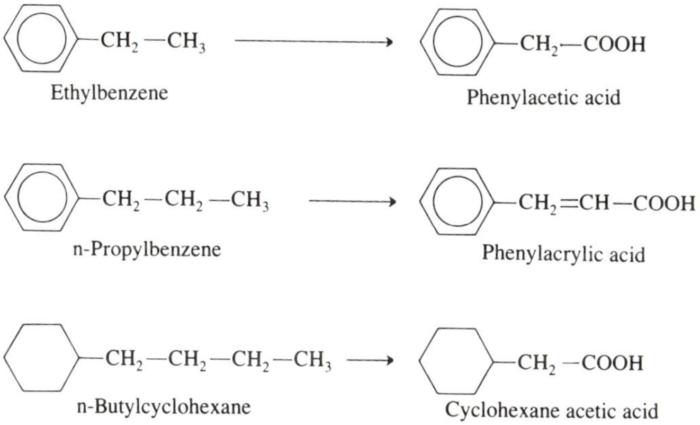

FIG. 3. Transformation of cyclic hydrocarbons by *Nocardia* under cooxidative conditions.

dihydroxylating abilities, were shown to oxidize methyl substituents on both mono- and di-cyclic aromatics. The representative products derived from the 'non-growth' aromatic hydrocarbon *p*-xylene during growth of cultures on hexadecanes are shown in Fig. 4. No evidence of further degradation of these hydroxylation products has been found.

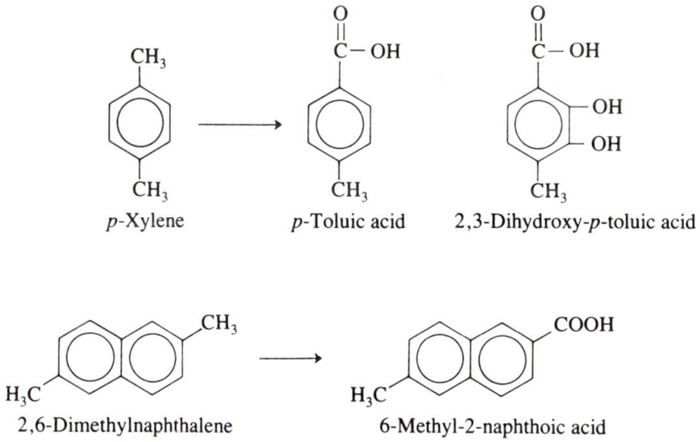

FIG. 4. Transformation of aromatic hydrocarbons by *Nocardia* sp. under cooxidative conditions.

Besides the several nocardiae which, under cooxidative conditions, can carry out *o*-hydroxylation of the aromatic ring, Jamison *et al.* (1969) described a strain of '*Nrda.*' *corallina* that possessed two pathways for the oxidation of *p*-xylene. When '*Nrda.*' *corallina* V-49 was grown on n-hexadecane with p-xylene as the co-oxidizable substrate, products of dihydroxylation and further degradation of the aromatic nucleus were detected. One of the observed pathways led to 2,3-dihydroxy-p-toluic acid formation whereas the other resulted in ortho ring cleavage of 3,6-dimethylpyrocatechol (Fig. 5).

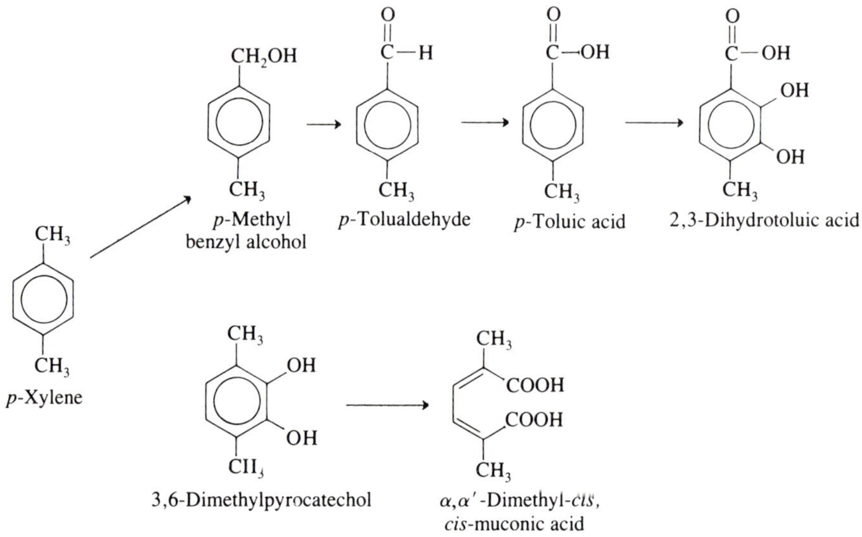

FIG. 5. The possible pathways in the conversion of p-xylene.

Further study of p-xylene oxidation (Golovlev et al., 1978) by nocardiae growing at the expense of n-hexadecane showed that the products obtained varied according to the culture used. Under cooxidative conditions nocardiae were able to hydroxylate one or both methyl substituents of p-xylene and/or to introduce an hydroxy group into the aromatic ring. Again, the products derived from the splitting of the aromatic nucleus were not observed (Fig. 6).

The application of microbial transformations to the production of commercially important industrial chemicals from petroleum based raw material is currently receiving considerable attention. Raymond and Jamison (1971) reported the production of phenylacetic acid from ethyl and butylbenzene by representative *Nocardia* species growing on aliphatic hydrocarbons. On the other hand, Cripps and his colleagues (1978) reported the degradation of acetophenone and 1-phenylethanol in two different pathways by *Nocardia* strain T5. The acetophenone was degraded to catechol while utilizing these hydrocarbons as the sole source of carbon and energy. The catechol was then utilized *via* ortho ring fission, while the 1-phenylethanol degradation proceeded with retention of the side chain to yield hydroxyethyl catechol.

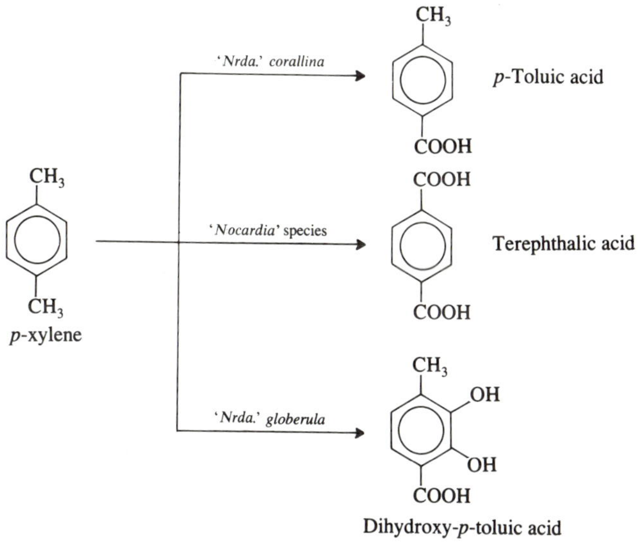

FIG. 6. Transformation of p-xylene by different 'Nocardia' strains.

The ring cleavage proceeded via meta fission, a type of conversion rather unusual for nocardiae.

An interesting example of the conversion of ethylbenzene into 1-phenylethanol under cooxidative conditions was reported by Cox and Goldsmith (1979) using a strain labelled 'Nrda. tartaricans'. This subterminal oxidative reaction with ethylbenzene has not been previously reported for nocardiae. The conversion of ethylbenzene into 1-phenylethanol does not occur when 'Nrda. tartaricans' ATCC 31190 is grown on glucose as the source of carbon and energy. If hexadecane is added as the only source of carbon the microorganism appeared to be able to form the aromatic alcohol and ketone. The mechanism of 1-phenylethanol and acetophenone production from ethylbenzene proceeded via the subterminal hydroxylation process and further oxidation of the resultant secondary alcohol to ketone (Fig. 7).

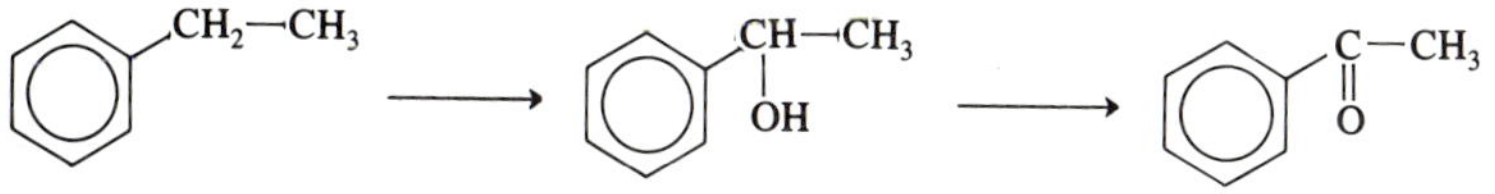

FIG. 7. Conversion of ethylbenzene into acetophenone by 'Nocardia tartaricans'.

The course of transformation of hydrocarbons depends also on the nature of the co-substrate. Thus, p-ksylene was oxidized to terephthalic acid by the Nocardia strain B-293 when growing at the expense of n-alkanes. The same organism growing on the medium with glucose produced mainly *p*-toluic acid; when butyrate was used as co-substrate the main product of transformation was *p*-hydroxymethylbenzoic acid (Golovlev *et al.*, 1978).

The most characteristic features of nocardial hydrocarbon oxidation can be summarized as follows: aliphatic hydrocarbons are degraded predominantly *via* the β-oxidation route; aromatic hydrocarbons are subjected to microbial degradation by hydroxylation of the aromatic nucleus and further ortho- or meta- cleavage of aromatic rings; in cyclic hydrocarbons substituted with aliphatic chain hydrocarbons under cooxidative conditions, oxidation of the paraffinic chain usually occurs; the course of transformation is very much influenced by the type and position of substituents; biotransformation which takes place only under cooxidative conditions depends on the nature of the co-substrate and the type of reaction performed is also strain dependent.

It is apparent that many of the aromatic hydrocarbons which are not able to support growth can be partially oxidized by nocardiae. Given the fact that *Nocardia* appears to be a soil actinomycete, it is particularly significant that cultures can not only accumulate products, but have, as an additional advantage, the possibility of decomposing xenobiotics in nature.

3. Miscellaneous Compounds of Synthetic Origin

The peculiar ability of nocardiae to accumulate products of transformation without their further utilization has been applied to obtain novel derivatives of organic compounds on a preparative scale.

The oxidative attack of the alkyl group tends to affect its terminal and subterminal positions. ω-Hydroxylation of lower alkyl groups is frequently accompanied by parallel β-oxidation resulting in the shortening of the alkyl chain and the subsequent oxidation of the primary alcohol to an acid structure. The oxidative attack which affects the subterminal position involves the formation of ketones as subsequent products. The microbial hydroxylation of various alkyl-substituted compounds has often been successfully employed in organic chemistry to selectively prepare hydroxylated products.

Siewert *et al.* (1973) performed the hydroxylation of various alkyl substituents of the pyrimidine moiety of 4-(2-pyrimidynylaminosulphonyl)-phenylacetamides with various *Streptomyces* strains. The hydroxy group was introduced in the primary, secondary and tertiary positions.

Some of the hydroxylated compounds are pharmacologically active metabolites of compounds with an antidiabetic action. In the case of branched and unbranched alkyl substituents, streptomycetes can be used to effect hydroxylation in the subterminal position (Fig. 8). The ability to hydroxylate organic compounds in the terminal and subterminal positions has been employed in the conversion of cinerone(2-2'-*cis*-butenyl(3-methyl-2-cyclo-

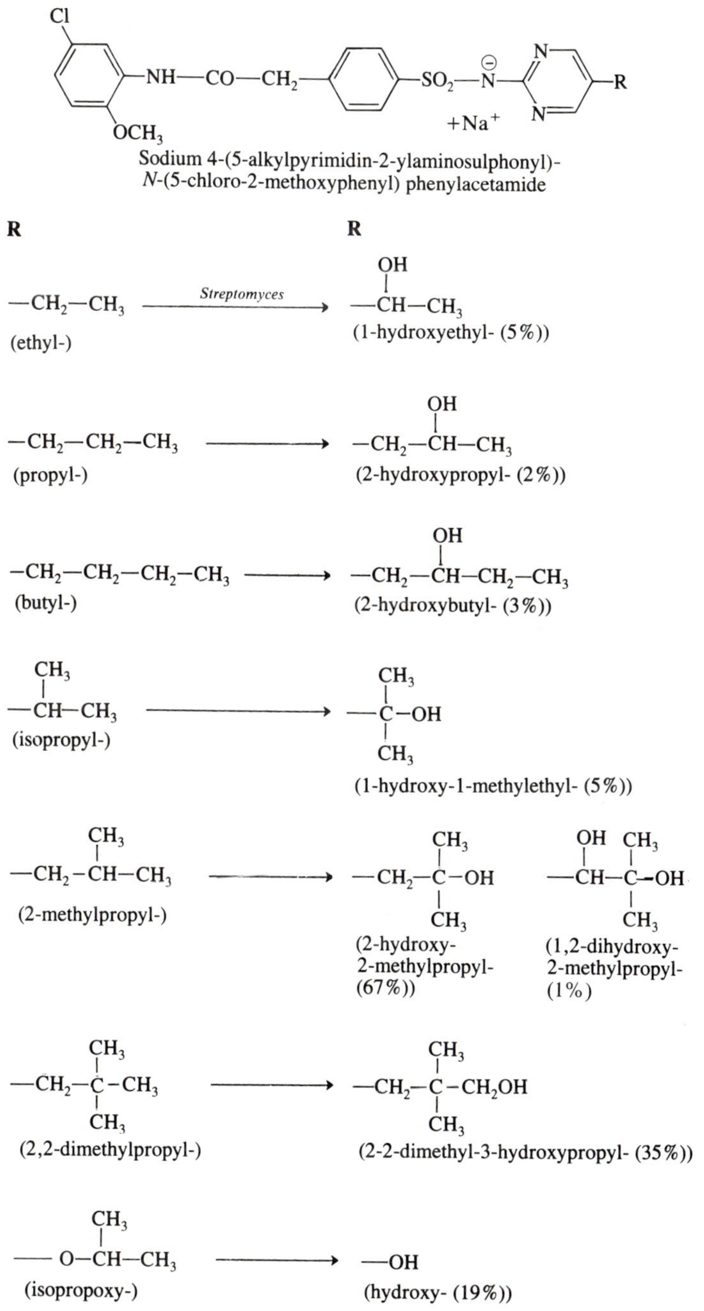

Sodium 4-(5-alkylpyrimidin-2-ylaminosulphonyl)-
N-(5-chloro-2-methoxyphenyl) phenylacetamide

R

—CH₂—CH₃ →(*Streptomyces*) —CH(OH)—CH₃
(ethyl-) (1-hydroxyethyl- (5%))

—CH₂—CH₂—CH₃ → —CH₂—CH(OH)—CH₃
(propyl-) (2-hydroxypropyl- (2%))

—CH₂—CH₂—CH₂—CH₃ → —CH₂—CH(OH)—CH₂—CH₃
(butyl-) (2-hydroxybutyl- (3%))

—CH(CH₃)—CH₃ → —C(CH₃)(CH₃)—OH
(isopropyl-) (1-hydroxy-1-methylethyl- (5%))

—CH₂—CH(CH₃)—CH₃ → —CH₂—C(CH₃)(CH₃)—OH   —CH(OH)—C(CH₃)(CH₃)—OH
(2-methylpropyl-) (2-hydroxy-2-methylpropyl- (67%)) (1,2-dihydroxy-2-methylpropyl- (1%))

—CH₂—C(CH₃)(CH₃)—CH₃ → —CH₂—C(CH₃)(CH₃)—CH₂OH
(2,2-dimethylpropyl-) (2-2-dimethyl-3-hydroxypropyl- (35%))

—O—CH(CH₃)—CH₃ → —OH
(isopropoxy-) (hydroxy- (19%))

FIG. 8. Microbial hydroxylation of 4-(2-pyrimidynyl-aminosulphonyl)phenyl-acetamides (Siewert *et al.*, 1973).

penten-1-one); **I**), to cinerolone (2-2'-*cis*-butenyl(3-methyl-4-hydroxy-2-cyclopenten-1-one); **II**), by a number of streptomycetes, other bacteria, and fungi (Fig. 9). *Streptomyces aureofaciens* strains yielded 42% of cinerolone.

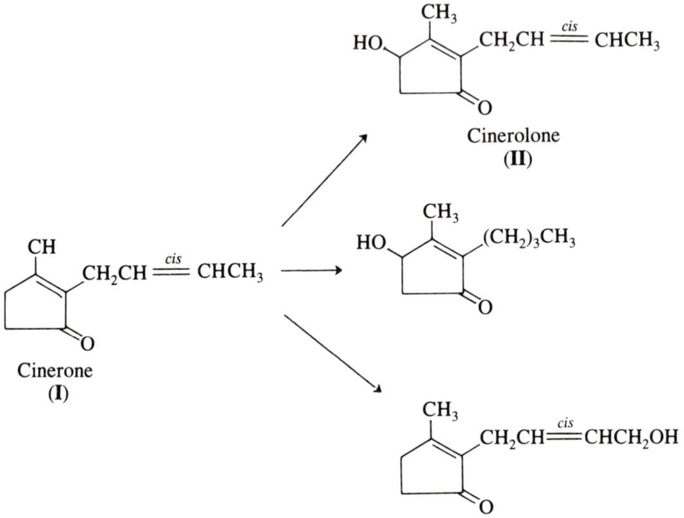

FIG. 9. Microbial transformation of cinerone to cinerolone by *Streptomyces aureofaciens*.

Various 2-substituted 3-methyl-4-hydroxycyclopentenones are chemically valuable intermediates which form the naturally occurring insecticides, pyrethrins and their analogues, when esterified with chrysantemic acid. Since 2-substituted 3-methyl-cyclopentenones are easier to synthesize than the corresponding 4-hydroxy compounds, advantage has been taken of the microbial potential to introduce specifically the hydroxy group in the desired position (Tabenkin *et al.*, 1969).

The ability of nocardiae to utilize several nitroaromatic compounds is also well known. The work of Cartwright and Cain (1959) and Cain (1966a, b) have elucidated the pathways of utilization of *p*- and *o*-nitrobenzoic acids. The first step in the conversion of nitrocompounds is the reduction of nitrogroups to amino-groups and the removal of the nitrogen from the aromatic compound (Fig. 10). The aromatic hydroxylated compounds obtained are subjected to further degradation before entering known metabolic cycles. It is also worth noting that some streptomycetes possess the ability to oxidize aromatic amino- groups to nitro-groups (Kawai *et al.*, 1965).

The degradation of halogenated hydrocarbons proceeds in an analogous way. Tranter and Cain (1967) and Cain *et al.* (1968) established the metabolic pathways of 2-fluoro-nitrobenzoate degradation. Products analogous to those from nitrobenzoate were formed and it was found that the fluoro-compounds were more easily oxidized than chloro-, bromo-, and iodo-derivatives (Fig. 11).

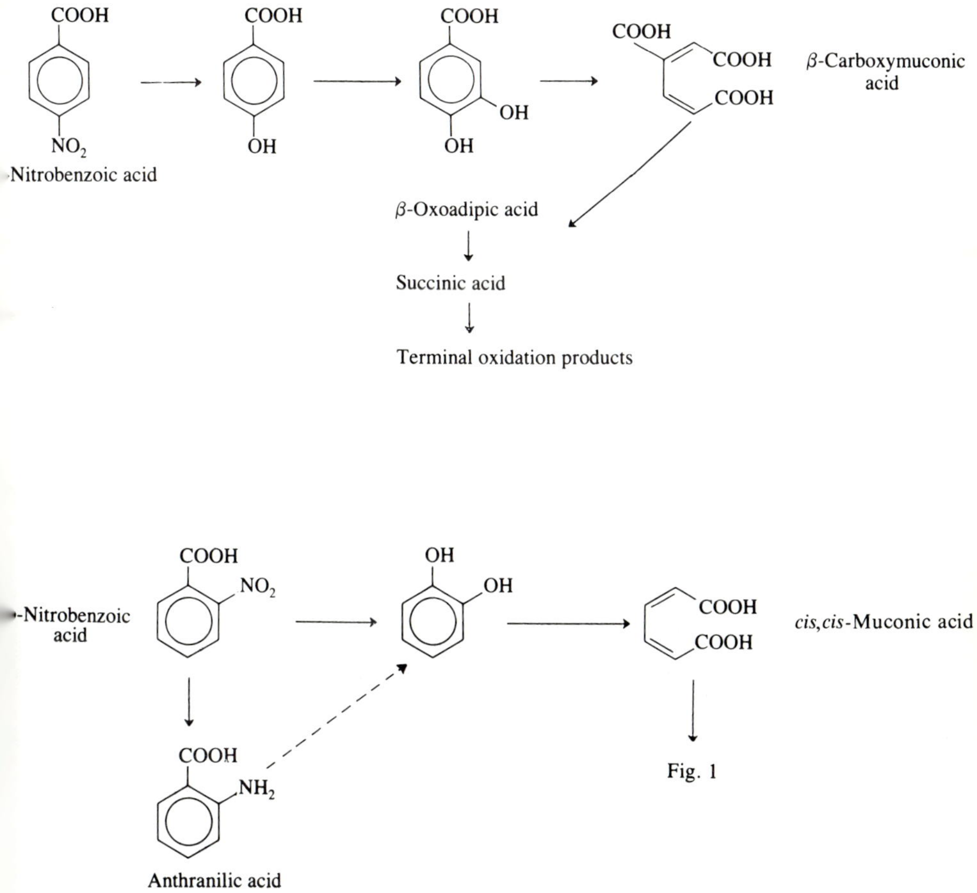

FIG. 10. Pathways in the biodegradation of *p*- and *o*-nitrobenzoic acid.

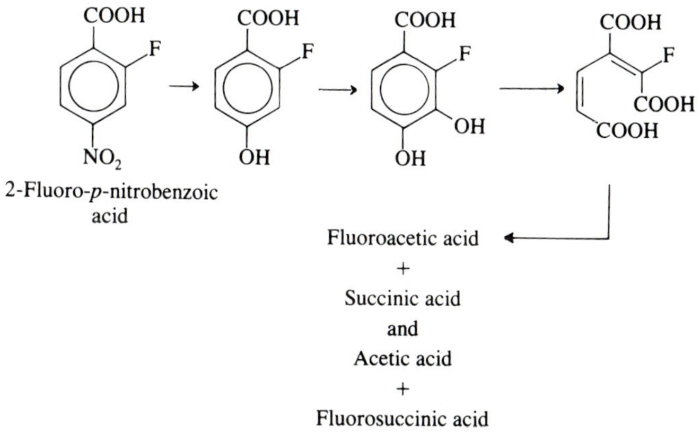

FIG. 11. Pathways in the biodegradation of 2-fluoro-*p*-nitrobenzoic acid.

4. Pesticides

The ever increasing pollution of the environment by various kinds of xenobiotics, including pesticides, is causing dramatic and unexpected changes in the biosphere. Microorganisms able to degrade such organic wastes play an important, but still not fully appreciated, role in the natural turnover of these contaminants. Unfortunately, little is known of the mechanism of decomposition of xenobiotics. The biodegradation of foreign organic compounds is usually associated with the activity of pseudomonads, fungi and arthrobacters. In this respect the actinomycetes have received less attention than these other groups of microorganisms.

Chemicals in the soil can be degraded by spontaneous decomposition, the action of light or metabolic transformation by living organisms. Populations of microorganisms present in soil may adapt to the presence of pesticides by changes in species diversity or by adaptation of enzyme systems, so that a pesticide may be more rapidly metabolized by an already adapted microbial population in a soil previously treated with the same or with a related pesticide.

Studies on the effect of pesticides on soil microorganisms have proved extremely difficult given the heterogeneity of the environment and its microbial content, the lack of knowledge on many of the microbial processes operative in soil and our inability to isolate more than a small fraction of the viable soil microbial flora. As a result, laboratory studies with pure and mixed microbial cultures have often been used to determine or predict the ecological effects of pesticides. Such simplified systems are, however, most unlikely to reproduce accurately the complex soil ecosystem. Indeed, it is quite possible that the extrapolation of laboratory studies to the soil situation may be very misleading when compared with the true effects of pesticides on soil microorganisms *in situ*. It is also possible that, in natural habitats, microorganisms may act synergistically to effect the degradation of pesticides.

In general, microorganisms possess the enzymic capacity to degrade a large proportion of naturally occurring chemicals although many xenobiotics remain resistant to microbial attack. The recalcitrance of pesticides in soil may be due to a number of factors such as the lack of suitable enzymic potential for degrading the pesticide, inability to penetrate the microbial cell, the steric configuration of the pesticide molecule preventing or hindering the enzymic attack, and the concentration or physical nature of the pesticide in the environment. From studies in nutrient media, it has become evident that many microorganisms cannot use pesticides as a sole source of carbon and energy but will metabolize them in the presence of alternative substrates. This phenomenon has been termed cometabolism, an extension of the original cooxidative process described by Raymond and

Jamison (1971). The occurrence of this process with both pesticidal and nonpesticidal chemicals presents evidence for its existence in natural habitats. It is, therefore, possible that the stability of some pesticides in the environment may, in part, be due to the lack of suitable co-metabolites.

There are, however, only a few reports of pesticide biodegradation by actinomycetes. There is no doubt that these bacteria play an important role in the decomposition of many kinds of chemicals, probably due to their ability to perform terminal and subterminal hydroxylation of aliphatic chains and the possibility of hydroxylation and often further degradation of substituted aromatics. A comprehensive review of pesticide biodegradation by microorganism has been presented by Hill and Wright (1978).

Actinomycetes are able to degrade certain pesticides and a few examples are given below. Thus, the herbicide dalapon, 2,2-dichloropropionic acid, was degraded by selected *Nocardia* strains isolated from soil (Hirsch and Alexander, 1960). These strains were able to degrade 24% of the dalapon in 8 days and, when yeast extract was added to the medium, up to 42% of the pesticide was broken down. The insecticide heptachlor (1,4,5,6,7,8,8-heptachloro-3a,4,7,7a-tetrahydro-4,7-endomethanoindene), a compound extremely recalcitrant to microbial degradation and toxic for animals, insects and humans (Martin, 1974), was transformed to its epoxide by a *Nocardia* strain (Miles *et al.* 1969) (Fig. 12).

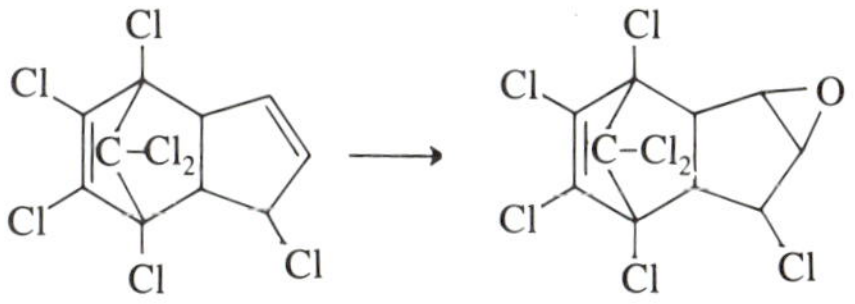

FIG. 12. Microbial transformation of heptachlor.

For several years, a considerable amount of attention has been paid to the microbial conversion of DDT (1,1,1-trichloro-2,2-di-4-chlorophenyl-ethane). Chacko *et al.* (1966) reported that this extremely recalcitrant insecticide was partially dechlorinated by some actinomycetes, but not by fungi, yielding DDD (1,1-dichloro-2,2-di-4-chlorophenyl ethane; Fig. 13). The most effective dechlorinating actinomycetes were classified as '*Nrda.*' *erythropolis*, *Stmy. aureofaciens*, '*Stmy. cinnamoneous*' and *Stmy. virido-chromogenes*. The same strains were effective in the degradation of the pesticide PCNB (pentachloronitrobenzene) which was primarily reduced to the corresponding chloroaniline (Chacko *et al.*, 1966) and then subjected to further degradation (Fig. 14).

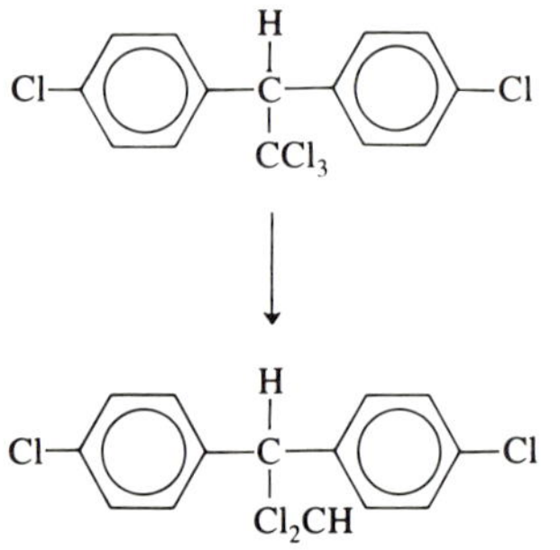

FIG. 13. Microbial transformation of DDT.

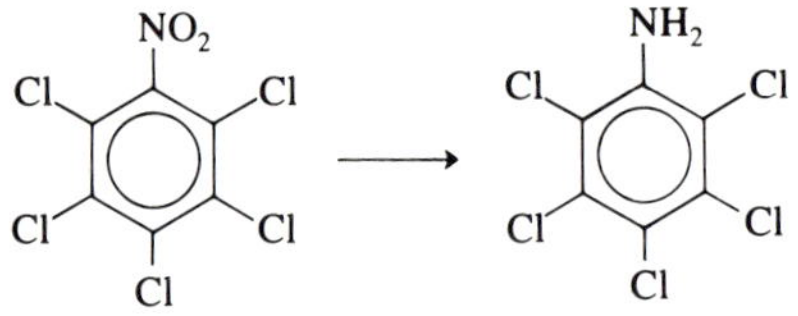

FIG. 14. Microbial transformation of pentachloronitrobenzene.

5. Steroids

Several reviews on the microbial transformation of steroids have appeared since the first report by Mammoli and Vercellone (1937). Over the years tremendous progress has been made in employing microorganisms to produce new, attractive steroid compounds (Capek et al., 1966; Charney and Herzog, 1967; Iizuka and Naito, 1967; Achrem and Titov, 1970; Laskin and Lechevalier, 1974). The microorganisms able to perform the bioconversion of steroids belong to such diverse groups as bacteria, fungi and actinomycetes. Members of the latter group are able to perform many different types of transformation of steroid molecules.

The basic structures of some steroid compounds are shown in Fig. 15. Most of the bioconversions performed by actinomycetes are oxidative or reductive processes, with the retention of the basic cyclopentanoperhydrophenanthrene system, such as hydroxylation, epoxidation, oxidation of alcohols, reduction of ketones, introduction of carbon–carbon double bonds or their reduction, aromatization, and degradative processes such as cleavage of the steroid nucleus and side chain degradation. Although transformation of steroids has been observed in a relatively small number of actinomycetes, some interesting and valuable examples of conversion of many kinds of steroid derivatives exist, a few of which are presented below.

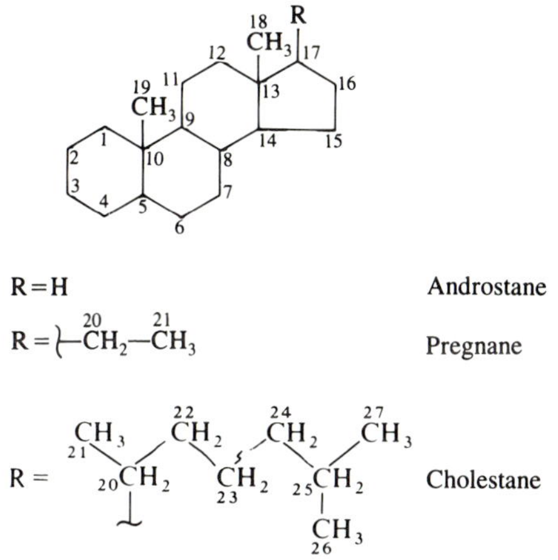

R = H   Androstane

R = $-CH_2-CH_3$ Pregnane

R = (cholestane side chain)   Cholestane

FIG. 15. The basic structures of some steroid compounds.

A. GENERAL CHARACTERISTICS OF MICROBIAL TRANSFORMATION OF STEROIDS

(a) Hydroxylation

Steroid hydroxylases exist in most microorganisms and are able to carry out hydroxylation in almost any position of the steroid nucleus though some carbon atoms of the steroid moiety do seem to be discriminated against. Thus, hydroxylations at positions 2-α, 3-β, 4-β and 5-α are rare, although the majority of microorganisms investigated are able to introduce an oxygen function usually at C-11 of the steroid molecule. The actinomycetes, however, favour hydroxylation at positions 9-α, 11-α and 16-α. The ability of actinomycetes to perform 16-α hydroxylation of the steroid nucleus has been employed to obtain many valuable derivatives in the androstan and pregan series. Compounds difficult to prepare by chemical synthesis have often been of great pharmacological importance as anti-inflammatory and anti-allergic agents, geriatric drugs or as oral contraceptives.

An interesting method for the microbial preparation of triamcinolone, an antiinflammatory drug, has been described by Lee et al. (1969). In this method, a mixed culture of '*Atbc. simplex* and '*Stmy. roseochromogenes*' was used to insert a double bond and hydroxy group in the desired position of the substrate, 9α-fluorohydrocortisone (I). The '*Atbc.*' simplex strain effected the dehydrogenation of the starting material leading to the formation of the intermediate 1-dehydro-9α-fluoro-hydrocortisone (II). In the second step of the conversion, the 16α-hydroxylation of the by-product was performed by the *Stmy. roseochromogenes* strain yielding 1-dehydro-9α-fluoro-16α-hydroxy-hydrocortisone, i.e. triamcinolone (III) (Fig. 16).

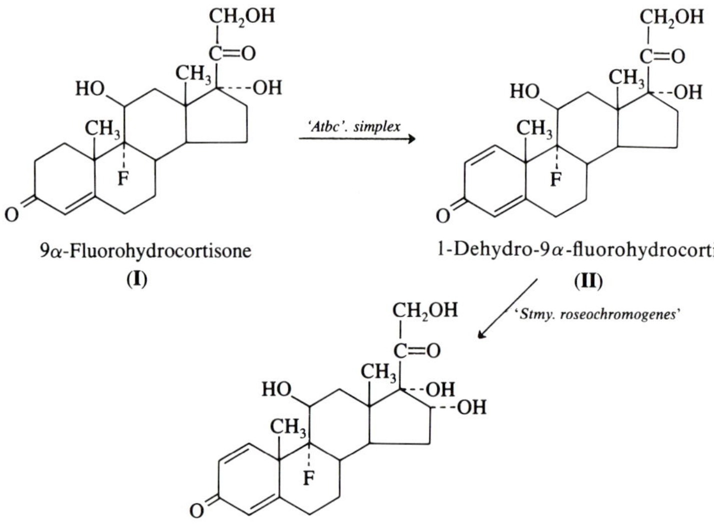

9α-Fluorohydrocortisone
(I)

1-Dehydro-9α-fluorohydrocortisone
(II)

1-Dehydro-9α-fluoro-16α-hydroxy-hydrocortisone (triamcinolone)
(III)

FIG. 16. Microbial conversion of 9α-fluorohydrocortisone to triamcinolone.

The dehydrogenation of the substrate by resting cells of '*Atbc*'. *simplex* was completed after 48 hours and 16-α-hydroxylation brought about the formation of the final product, triamcinolone, a yield of 89% being obtained after 72 hours of transformation. In pure culture, '*Atbc.*' *simplex* showed not only Δ^1 dehydrogenase action, but also the presence of a 20-ketoreductase. Fortunately, in the mixed culture the 20-ketoreductase was completely repressed. Attempts to replace the culture of '*Atbc.*' *simplex* with a '*Nrda.*' *restrictus* strain, known to be a good source of dehydrogenase, were unsuccessful. During incubation of the substrate in a mixed culture with '*Stmy.*' *roseochromogenes*, the Δ^1 dehydrogenase was repressed; the final product of the reaction was only the 16-α hydroxylated steroid produced as a result of '*Stmy.*' *roseochromogenes* action. Both Δ^1 dehydrogenase and 16α-hydroxylase were, however, fully active in pure cultures of the two strains.

(b) *Epoxidation*

The epoxidation reactions of unsaturated steroid molecules are closely related to the hydroxylation reactions. There are only rare examples in the literature of microbial epoxidation reactions performed by *Nocardia* strains. Sih (1962b) reported the epoxidation of androst-9(11)-ene-3,17-dione and related compounds by *Nocardia* strain ATCC 13934. The substrate was microbiologically converted into androst-9α-11α-oxido-3,17-dione (Fig. 17).

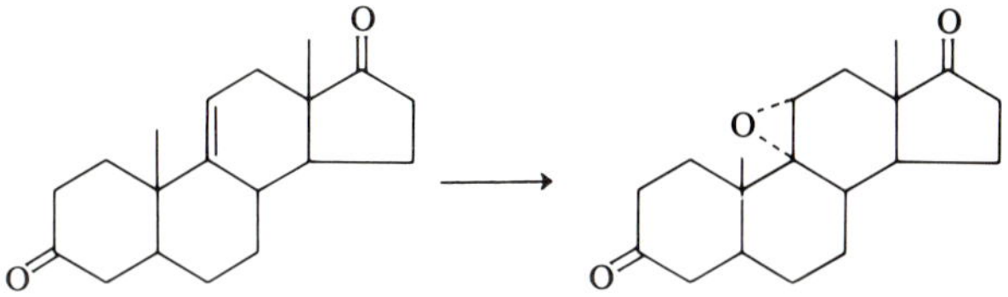

FIG. 17. Microbial conversion of an unsaturated steriod molecule into its epoxide.

(c) Ring Dehydrogenation

The conversion of 3-oxygenated pregnanes and allopregnanes to the corresponding 1,4-pregnadienes using '*Nrda. blackwellii*' was reported by Stoudt et al. (1958). This conversion is particularly significant since quite a number of chemical steps are replaced in any synthetic approach to 1-dehydrocortisone and 1-dehydrocortizol using a pregnane or allopregnane intermediate. The conversion is quite versatile in that it can be applied equally well to 3-keto or to 3-hydroxy derivatives (Fig. 18).

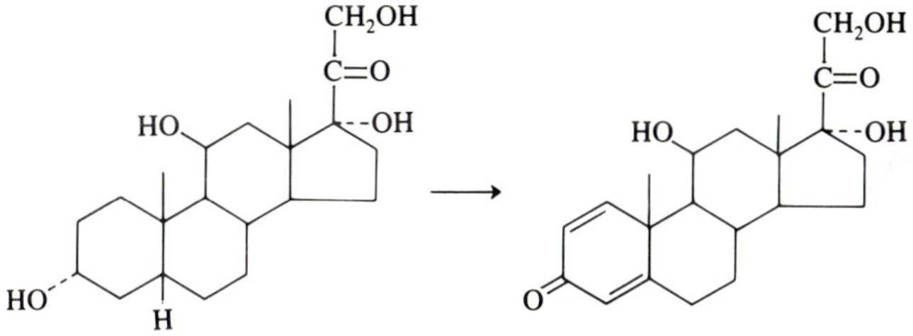

FIG. 18. The formation of prednisolone by '*Nocardia blackwellii*'.

(d) Aromatization of the Steroid Nucleus

The ability of *Nocardia* strains to aromatize ring A of 19-nor or 19-hydroxy steroids has been reported by several investigators including Achrem and Titov (1970). Sih (1962a) performed the transformation of 19-hydroxy-androst-4-en-3,17-dione (**I**), and then, of 19-hydroxy-cholest-4-en-3-one (**II**) (Sih and Wang, 1965) with selected *Nocardia* strains obtaining estrone (**III**) (Fig. 19).

In the search for steroid substrates of synthetic origin for microbial preparation of estrogens, several compounds containing a methyl group at the 19-position have been screened. It was shown that microbial C-19 hydroxylation is rare. However, Kluepfel and Vézina (1970) found that several actinomycetes were able to aromatize androsta-1,4,7-triene-3,17-dione (**I**) into equiline (**II**) and equilenine (**III**). This transformation reaction probably proceeded through 19-hydroxylation followed by a reverse aldol reaction (Fig. 20).

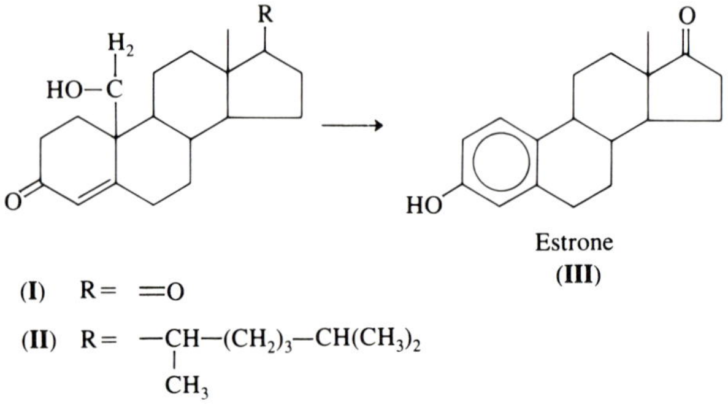

(I) R= =O
(II) R= —CH—(CH$_2$)$_3$—CH(CH$_3$)$_2$
         |
         CH$_3$

FIG. 19. Transformation of 19-hydroxy steroids into estrone by '*Nocardia*' strains.

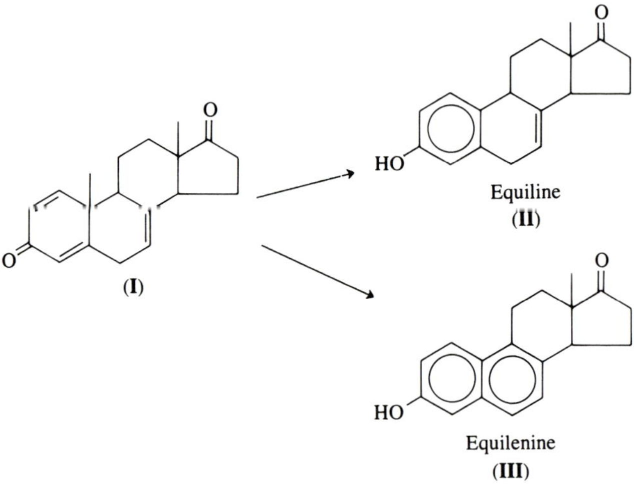

FIG. 20. Aromatization of the steroid nucleus.

(e) *Reduction of Ketones and Oxidation of Alcohols*

For several years a problem of considerable interest has been the enzymic reduction of compounds containing a carbonyl group to chiral alcohols and selective oxidation of alcohols to ketones. Knowledge of the redox reactions of hydroxy and carbonyl groups is very useful in explaining problems concerned with the stereochemistry and the biosynthesis of compounds of natural origin and provides methods for obtaining optically pure enantiomers by microbial or enzymic resolution of racemates or by directed asymmetric synthesis. The application of oxidoreductases isolated from actinomycetes is well known. Many microbial transformations of this type are performed by *Nocardia* and *Streptomyces* strains.

Agnelli and co-workers (1963) performed the selective reduction of 21-methyl-11,17-dihydroxypregna-1,4,-diene-3,20,21-trione (I) to its 20-hydroxy derivatives (II). The reaction was carried out with a *Stmy. erythraeus* strain with high stereospecificity; only one of the three keto groups present in the steroid molecule was reduced (Fig. 21).

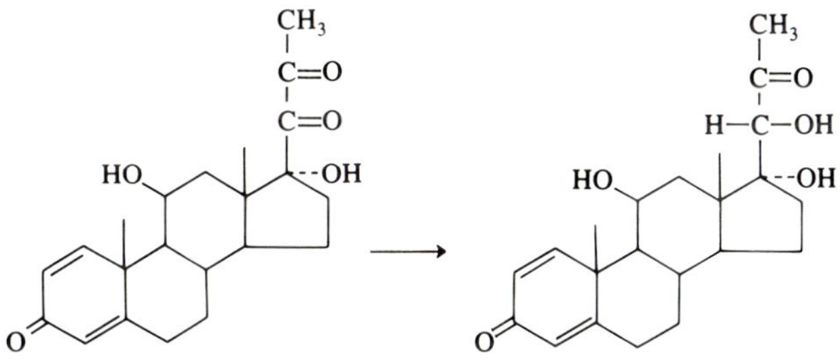

FIG. 21. Selective reduction of a ketosteroid performed by *Streptomyces erythraeus*.

B. DEGRADATION OF THE STEROID NUCLEUS

In systematic studies on the ability of microorganisms to decompose cholesterol, complete degradation of this compound was affected by actinomycetes and many other bacteria. It has also been shown that *Nocardia* and *Rhodococcus* strains are able to grow on testosterone as a sole source of carbon for energy and growth (Goodfellow, 1971).

Studies on the microbial utilization of steroids helped to elucidate the general pathways of steroid nucleus degradation among members of the genera *Arthrobacter, Mycobacterium, Nocardia* and *Pseudomonas* (Martin, 1977). Steroids containing the 3β-hydroxy-Δ^5 configuration are oxidized first to the corresponding 3-keto-Δ^4 compounds. The primary oxidation of the 3β-hydroxy function is followed by the isomerization of the Δ^5 double bond. Depending on the organism studied, the further metabolism of 3-keto-Δ^4 compounds involves 9α-hydroxylation followed by C-1 and then C-2 dehydrogenation or *vice versa*. The resulting metabolite undergoes simultaneous aromatization with the cleavage of ring B *via* a non-enzymic reverse aldol-type reaction to produce the 9,10-secophenolic derivative (Fig. 22).

C. DEGRADATION OF THE STEROID SIDE CHAIN

The ever increasing demand for pharmacologically active steroids has encouraged many laboratories to search for new sources of steroid structure compounds. In principle, the production of steroids involves two methods, total chemical synthesis and partial chemical synthesis from such naturally occurring steroids as diosgenin, stigmasterol and cholesterol.

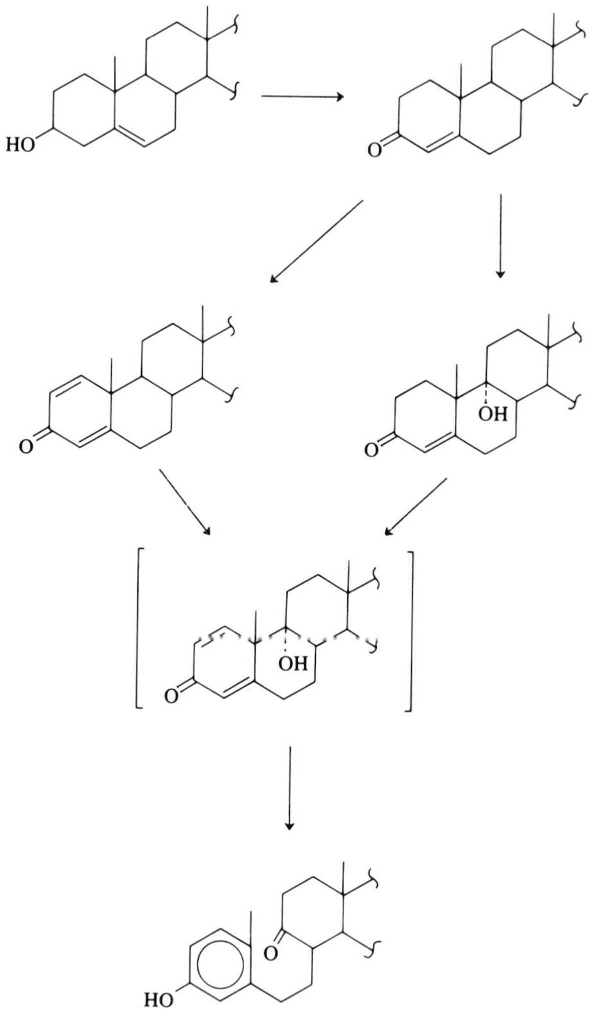

FIG. 22. Degradation of the steroid nucleus.

As a result of the shortage of diosgenin, prepared from the *Dioscorea* plant, new chemical and especially microbiological methods for the modification of cholesterol and β-sitosterol have been developed recently. The ability of nocardiae to perform hydroxylation of hydrocarbons have been employed for the activation and further degradation of the side chain of C_{27-29} steroids without further degradation of the steroid nucleus. In a series of brilliant experiments, Sih *et al.* (1967a, b, 1968a, b) elucidated the pathway whereby the C-17 steroid side chain was degraded during the microbial conversion of cholesterol into C-17 ketosteroids. Sih's work with members of the genera *Arthrobacter, Cornyebacterium, Mycobacterium* and *Nocardia* established that the shortening of the side chain occurred in a similar way to the β-oxidation of fatty acids. Following C-27 hydroxylation and presumably also the oxidation of the resulting alcohol to the C-27 carboxylic acid, the propionic acid and acetic acid produced were removed, resulting in the formation of the C-17 keto function (Figs 23 and 24).

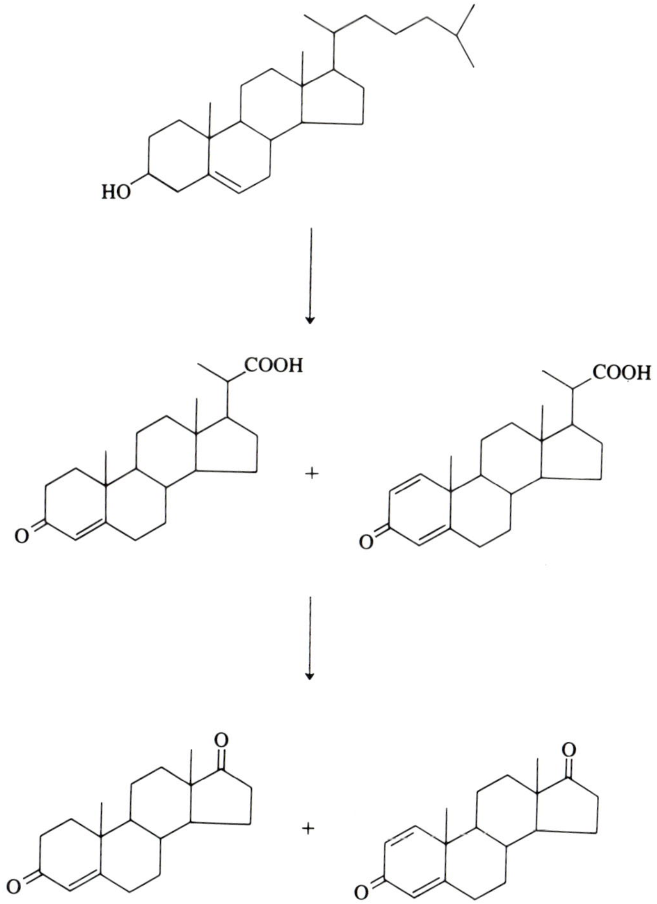

FIG. 23. Degradation of the side chain of cholesterol.

It is worth emphasizing that the enzymic reactions involved in the side chain degradation are frequently associated with B-ring cleavage. If the structure of the side chain is modified so that the enzymes normally involved in the degradation are unable to catalyse the fission, the ring system will

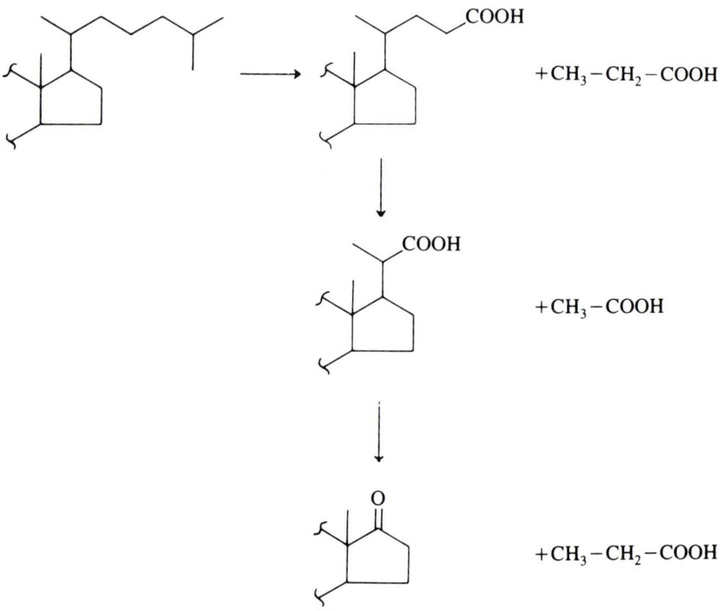

FIG. 24. Degradation of the side chain of cholesterol.

be attacked resulting in the accumulation of metabolites with partially oxidized ring structures. Conversely, in substrates with a modified ring structure blocking C-1(2)-dehydrogenation or 9α-hydroxylation, the side chain will be degraded resulting in the formation of 17-ketosteroids (Martin, 1977).

6. Terpenes

Terpenoid compounds are of widespread occurrence in nature and serve essential functions in the microbial, plant and animal kingdoms. The microbial conversion of terpenes primarily involves the breakdown and degradations of these natural products. The susceptibility of terpenes to microbial and chemical degradation is responsible for the limited application of microbial reactions to their transformation. Most of the work on this problem has been limited to the biodegradation of terpenes. Little attention has been paid to the microbial transformation of terpenes by actinomycetes (Kieslich, 1976; Sebek and Kieslich, 1977).

Davis and Raymond (1961) performed the hydroxylation of *p*-cymene (**I**; *p*-isopropyltoluene) in the presence of *n*-hexadecane as co-substrate, obtaining *p*-cumic acid (**II**; *p*-isopropylbenzoic acid) (Fig. 25).

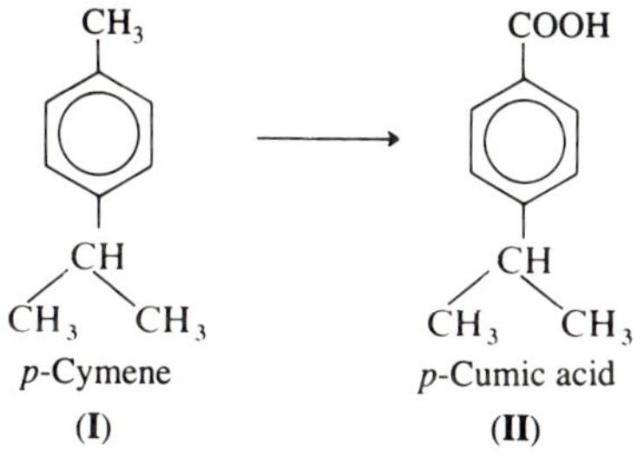

FIG. 25. Transformation of *p*-cymene to *p*-cumic acid by *Nocardia* strains.

It is interesting that the isopropyl side chain of *p*-cymene is more susceptible to chemical oxidation than the methyl substituent in the para-position. The enzymic system of *Nocardia* strain 107-332 has the ability to oxidize selectively the isopropyl group. Thus, biotransformations may be achieved by selected strains yielding the desired product. The introduction of hydroxy groups into α-kessyl alcohol (I), a constituent of several different species of Japanese valerian, is extremely difficult by chemical procedures. Hikino et al. (1968) achieved a microbial hydroxylation with the fungi *Cunninghamella blakesleeana*, *Corticium sassaki*, *Corticium centrifugum* and with the actinomycete *Stmy. aureofaciens*. The main product in each case was kessyl glycol (II), minor products of hydroxylation at C-4 (III), and the product of oxidation at C-8 (IV) (Fig. 26).

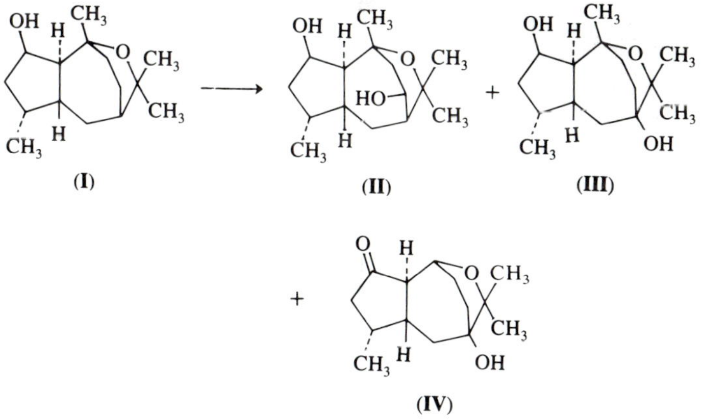

FIG. 26. Microbial transformation of α-kessyl alcohol by *Streptomyces aureofaciens*.

The amount of kessyl glycol produced varied depending on the strain; *Stmy. aureofaciens* gave the highest yield producing up to 87% kessyl glycol. Attempts to produce artemesin by microbial means were made by Hikino et al. (1970); α-santonin (I) was incubated with *Stmy. aureofaciens* yielding major amounts of 1,2-dihydrosantonin (II) instead of the desired product of hydroxylation (Fig. 27).

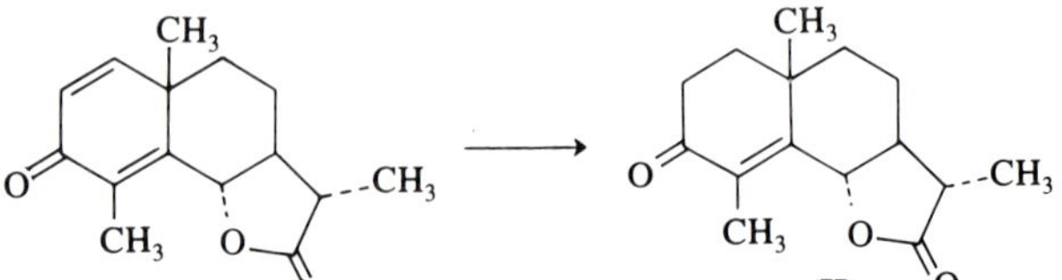

FIG. 27. Microbial transformation of α-santonin by *Streptomyces aureofaciens*.

7. Cannabinoids

Among the naturally occurring cannabinoids Δ^9-tetrahydrocannabinol is believed to be the pharmacologically active constituent of marihuana, *Cannabis sativa*. A variety of pharmacological activities often resulting in tachycardia and euphoria has been attributed to this type of compound. In contrast, Δ^9-tetrahydrocannabinols are believed to possess blood-pressure lowering, anti-anxiety and antidepressant properties.

In attempts to minimize the undesirable, while enhancing the desired clinical effects, new cannabinoids have been prepared by chemical synthesis and microbial conversion. This work has led to the production of a new group of clinically useful agents. Fukuda *et al.* (1977) performed the microbial conversion of sythetically prepared Δ6a,10a-tetrahydrocannabinol (I). In the course of the screening programme, 97 strains of bacteria, 175 actinomycetes and 86 fungi were examined, about 18% of these organisms being able to modify cannabinoids. Among the strains tested *Streptomyces* A41596, '*Mybc.*' *rhodochrous* ATCC19068 and *Bacillus cereus* NRRL B-8172 were selected to perform the transformation of $\Delta^{6a,10a}$-tetrahydrocannabinol giving yields of about 7% of the various products (Fig. 28). Transformation of $\Delta^{6a,10a}$-tetrahydrocannabinol (I) led to a side-chain hydroxylated derivative (V) using *Bacl. cereus*, a C-7 oxygenated compound (II) using a *Streptomyces* sp., and a pair of diastereomeric tertiary hydroxy compounds (IVa), (IVb) and cannabinol (III) with '*Mybc.*' *rhodochrous*.

Recently, a most promising microbial transformation was performed at Elli Lilly Laboratories (Archer *et al.*, 1979), with a synthetic cannabinoid, nabilone. Nabilone is believed to be an anti-emetic agent in cancer chemotherapy, an ocular pressure-reducing agent in glaucoma patients and an anti-anxiety agent. This promising compound has recently been the subject of extensive clinical evaluation. The microbial conversion of nabilone can, hopefully, lead to the production of novel derivatives with improved or more varied pharmacological properties than the parent compound. In the

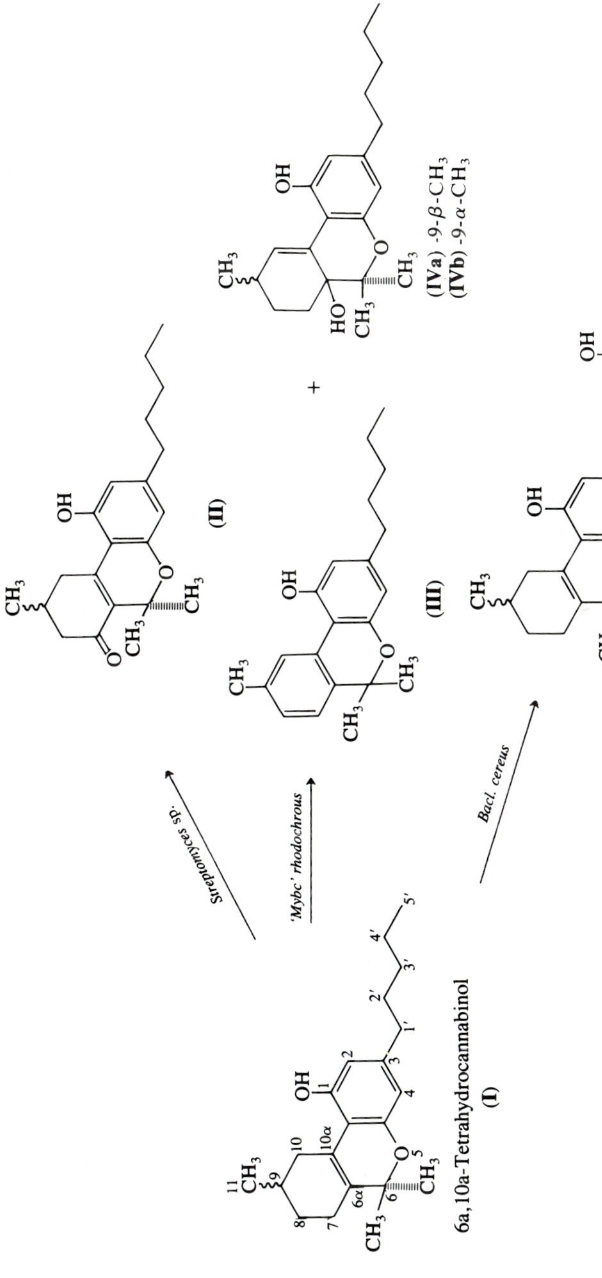

FIG. 28. Biotransformation of $\Delta^{6a,10a}$-tetrahydrocannabinol.

course of screening hundreds of actinomycetes, bacteria and fungi were tested for their ability to modify nabilone; 50% of the actinomycetes produced transformation products, the corresponding values for fungi and bacteria were 30 and 20% respectively.

The conversion of nabilone (I) by '*Nrda*' *salmonicolor* ATCC 19149 yielded products of the oxidation of the 1,1-dimethylheptyl side-chain, the carboxylic acids having varying chain lengths. These acids belong to two groups of compounds; one containing a carbonyl group at C-9 and terminal carboxylic acid side-chains of 5, 7 and 9 carbons (II) and the other having, presumably, a 9S hydroxyl group with terminal carboxylic acid side-chains of 5, 7 and 9 carbon atoms (III) (Fig. 29).

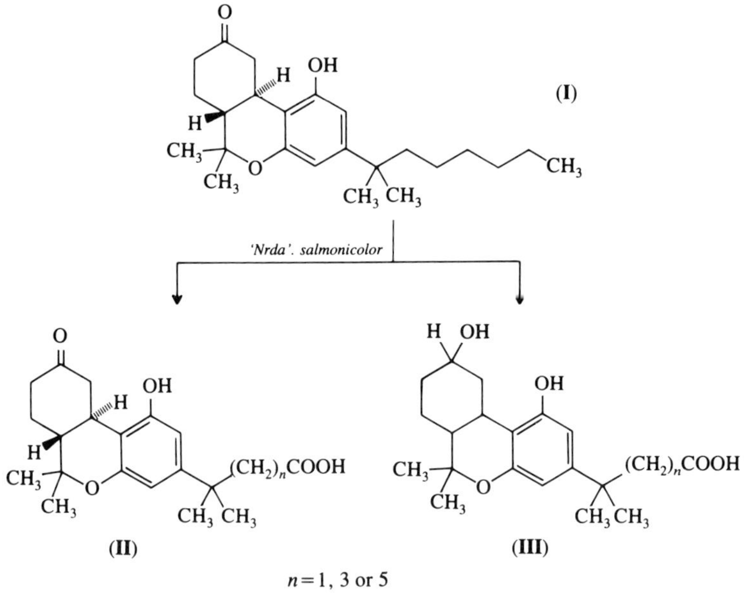

FIG. 29. Microbial transformation of nabilone (I) by '*Nocardia*' *salmonicolor*.

The formation of acid derivatives requires a shortening of the 1,1-dimethyl heptyl side-chain of nabilone. Such compounds were probably obtained as products of β-oxidation mechanisms, similar to the microbial oxidation of alkanes (Van der Linden and Thijsse, 1965).

8. Alkaloids

Compounds with an alkaloid structure are of great interest and physiological importance. Many of them have been found not only to have various pharmacological activities, but to be active against tumours and other forms of cancer. They represent many structural types of compounds and as such form a heterogeneous group whose activity–structure relationships cannot be generalized. It seems that the variety of structural types is the most noteworthy feature of these compounds. Alkaloids that possess a basic nitrogen, which can be reversibly protonated near physiological pH values, may be transported across membranes as free bases and be trapped in cells, and thus might result in increased contact with and effect on intracellular sites of action. The relatively large percentage of nitrogen-containing compounds which are effective in all areas in medicine seems to support this supposition.

Recently a lot of attention has been paid to the microbial transformation of alkaloids. In comparison with alicyclic compounds, alkaloids with heterocyclic structural elements are generally more difficult to transform (Kieslich, 1976). A heterocyclic N-atom is frequently subjected to undesirable degradation yielding products with an altered alkaloid structure. On the other hand, ring linkages via such heteroatoms may yield sterically unfavourable substrate structures that are resistant to enzymic attack or are only attacked non-specifically. Although alkaloids belong to a class of compounds with heterogeneous structures that are often difficult to transform, microbial transformation can result in unpredictable structural modification, which can provide a variety of novel compounds that may have desirable pharmacological activities.

A. VINCA ALKALOIDS

The members of the vinca alkaloid group presently being investigated as antitumour agents are derived from the periwinkle plant, *Catharanthus roseus* G. Don, which has been widely used in folk medicine for treating gingivitis, diabetes and in arresting bleeding. Since Beer (1955) and Johnson *et al.* (1959) reported their antineoplastic activity, vinca alkaloids have become the best antitumour agents isolated from plants, after the anthracycline antibiotics. Of the large number of compounds isolated from the *Catharanthus* plant, a few possess antitumour activity but only two, vincristine and vinblastine, which are dimeric in structure, have proved to be of clinical value, the others are more toxic and are unpredictable in effect. The vinca alkaloids occur in a very low concentration in *Catharanthus* plants, for example, 1 kg of dried leaves yields only 3 mg of vincristine. The most abundant of the *Catharanthus* alkaloids is the biologically inactive compound, vindoline (I)(Fig. 30). Its structure is found to be a part of the dimeric, antitumoric vinblastine (I) and vincristine (II). The term 'dimeric' has been applied to the structures of these compounds although in each case the two portions of the molecules are not identical. They are composed

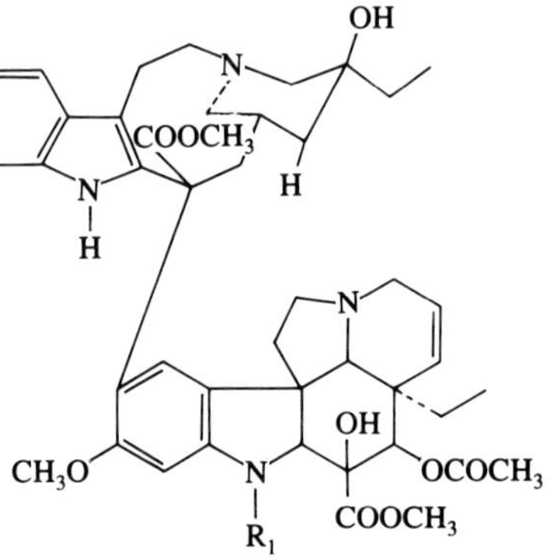

(I) $R_1=CH_3$ Vinblastine
(II) $R_1=CHO$ Vincristine

FIG. 30. Representatives of the vinca alkaloid group.

of a vindoline and catharanthine moiety. It has been established that certain features of the chemical structures of the alkaloids cannot be modified without loss of biological activity. The total inactivation of vinca alkaloids is achieved by acetylation of the free hydroxyl groups. All of the naturally occurring active alkaloids possess an acetyl group at C-17 of the vindoline moiety; removal of this group results in loss of activity. It has been shown that the naturally occurring group at C-17 need not be acetyl for activity. The replacement of this group with a wide variety of acyl and aminoacyl functions yield compounds of differing activity (Johnson et al., 1966). Change of the basic indole nitrogen in the catharanthine moiety by amide formation leads to total loss of activity, while the reduction of double bonds at C-14 and C-15 yields a less potent derivative.

Since vindoline is more readily available than vincristine and vinblastine, it was used as the base from which to prepare new derivatives. Although vinca alkaloids have rather resisted simple chemical synthesis, due to their structural complexity, microbial transformations have been employed to prepare potentially important vinca alkaloid derivatives.

In the Elli Lilly Laboratories a microbial screening programme has been in progress since the early 1960s and several hundred actinomycetes used to try and convert vinca alkaloids, including vindoline, into active compounds. Reported microbial reactions include deacetylation, N-demethylation, O-demethylation and the formation of a group of unusual ether derivatives. The best known vindoline metabolites include: deacetylvindoline (III) (Neuss et al., 1973, 1974a, b); deacetyldihydrovindoline ether (VII); dihydrovindoline ether (V); 3-acetonyldihydrovindoline ether (VI); a ring concentration product known as 16 dehydroxy-14,15-epoxy-14-oxo-3-norvindoline (VIII); N-demethylvindoline (IV); O-demethylvindoline (II) and a dimer consisting of two dihydrovindoline ether moieties linked through their 3 and 14 positions (see Fig. 31; Fig. 33; IX).

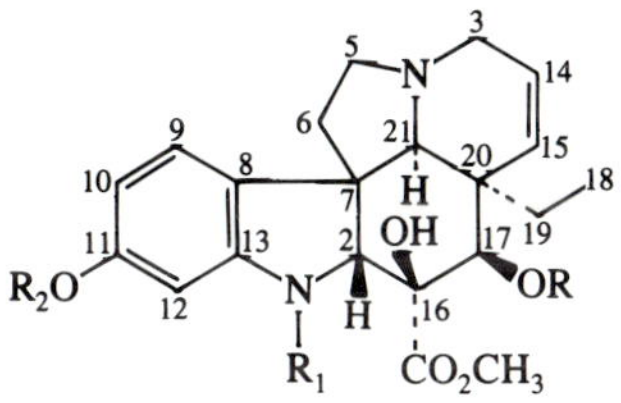

(I) $R_1 = R_2 = CH_3$, $R = COCH_3$
(II) $R_1 = CH_3$, $R_2 = H$, $R = COCH_3$
(III) $R_1 = CH_3$, $R_2 = CH_3$, $R = H$
(IV) $R_1 = H$, $R_2 = CH_3$, $R = COCH_3$

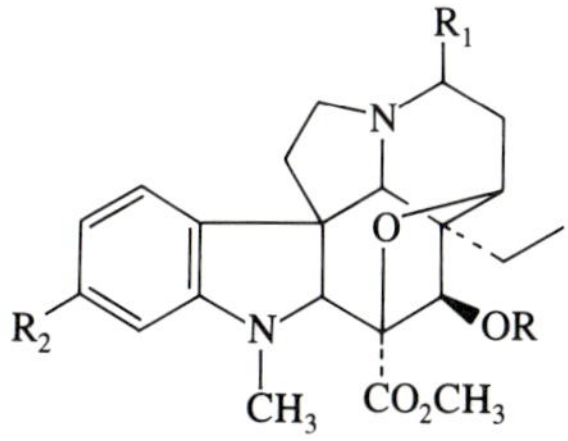

(V) $R_1 = H$, $R_2 = OCH_3$, $R = COCH_3$
(VI) $R_1 = CH_2COCH_3$, $R_2 = OCH_3$, $R = COCH_3$
(VII) $R_1 = R = H$, $R_2 = OCH_3$

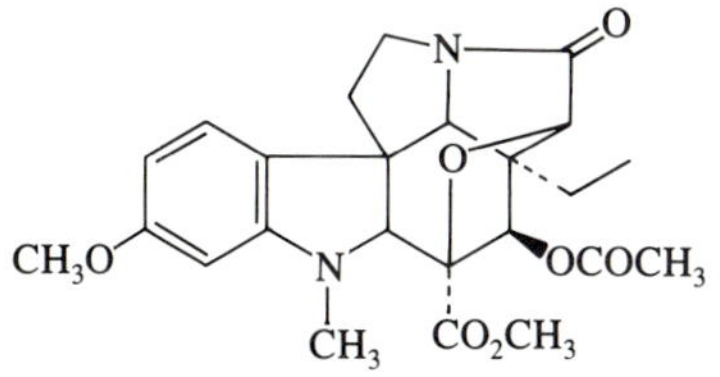

(VIII)

FIG. 31. Products of microbial conversion of vindoline. For details, see the text.

Mallet et al. (1964) screened over 400 microorganisms for their ability to transform vindoline; about 100 actinomycetes and bacteria attacked the compound, the best result being obtained with Stmy. cinnamonesis which transformed O-acetylvindoline (I) to vindoline (II), deacetylvindoline (III) and deacetyldihydrovindoline ether (IV) (Fig. 32).

An unusual dimer formation has been observed by Nabih et al. (1978). The transformation of vindoline conducted with Stmy. griseus UI 1158 yielded a dihydrovindoline ether and a novel dimeric vindoline metabolite which consisted of two dihydrovindoline moieties joined by a carbon–carbon bond (Fig. 33). Since the biologically active vincristine and vinblastine are dimeric in nature and include the vindoline moiety, the transformation of the latter into a dimer is an interesting finding promising further structural activity relationships in the vinca alkaloid field.

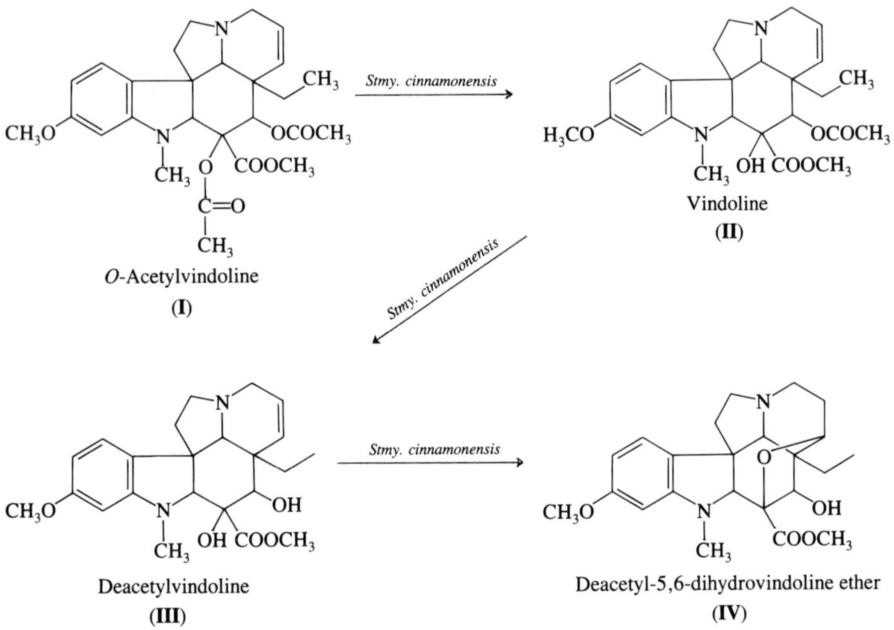

FIG. 32. Transformation of vindoline by *Streptomyces cinnamonensis*.

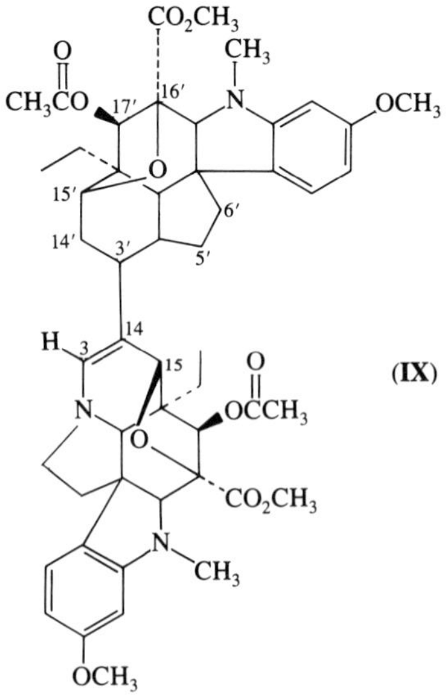

FIG. 33. Product of microbial transformation of vindoline by *Streptomyces griseus* strain UI 1158.

B. YOHIMBINE

Yohimbine, the indole type of alkaloid belonging to Rauwolfia alkaloids group, is isolated from the bark of *Corynanthe yohimbe*. The interest in the pharmacological activity of Rauwfolia alkaloids encouraged attempts to modify some of these compounds by microorganisms. The results obtained in an alkaloid screening programme (Kieslich, 1976) showed that the yohimbine-type alkaloids can be hydroxylated in the 10, 11 and 18 positions. The hydroxylations performed by *Streptomyces* strains proceeded with substrate specificity (Hartman *et al.*, 1964). *Streptomyces rimosus* transformed yohimbine and corynanthine to 10-hydroxy derivatives (Fig. 34).

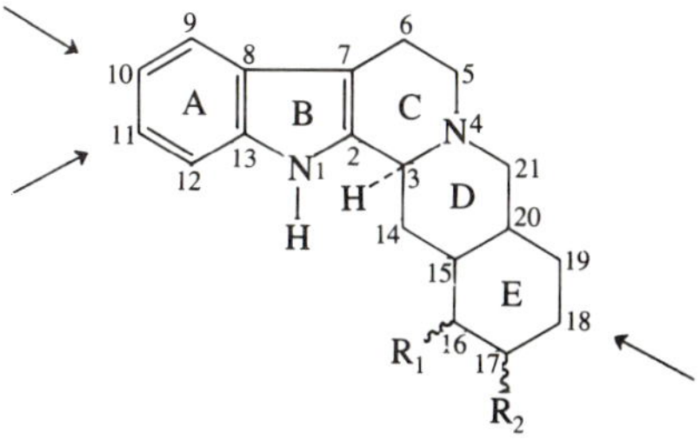

Yohimbine $R_1 = -COOCH_3\alpha, R_2 = -OH\alpha$
α-Yohimbine $R_1 = -COOCH_3\beta, R_2 = -OH\alpha$
β-Yohimbine $R_1 = -COOCH_3\alpha, R_2 = -OH\beta$
Corinanthine $R_1 = -COOCH_3\beta, R_2 = -OH\beta$

FIG. 34. Microbial transformation of yohimbine by *Streptomyces* strains.

Streptomyces platensis effects the hydroxylation of yohimbine, α-yohimbine and corynanthine also in the 10 position. In contrast, *Stmy. fulvissimus* hydroxylates only α-yohimbine yielding the 11-hydroxy derivative while the hydroxylation of β-yohimbine was not observed in *Streptomyces* strains.

C. OTHER ALKALOIDS

Microbial *O*-dealkylation reactions are generally considered to be mediated by a monooxygenase enzyme process requiring pyridine nucleotides and oxygen. Examples of *O*-dealkylation of some aryl ethers performed by microorganisms are substances such as nicotine, reserpine, thebaine and griseofulvin (Kieslich, 1976) and such transformations have been well described in the literature. It has been recognized that many *Streptomyces* strains are able to carry out selective *O*-demethylation.

Rosazza *et al.* (1975) in the couse of microbial transformations of 10,11-dimethoxyaporphine (I) found, among others, two strains of *Streptomyces* which were able to convert the substrate into a mixture of apocodeine (**II**) and isoapocodeine (**III**) (Fig. 35).

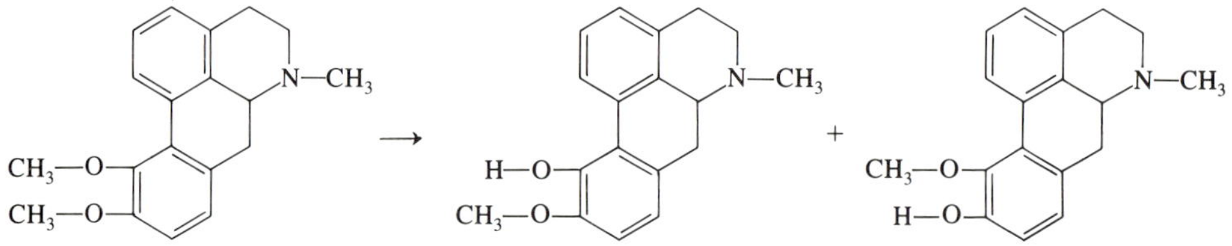

10,11-Dimethoxyaporphine (**I**) → Apocodeine (**II**) (24%) + Isoapocodeine (**III**) (20%)

FIG. 35. Microbial O-demethylation of 10, 11-dimethoxyaporphine.

In the search for a 'microbial model of mammalian metabolism', Rosazza *et al.* (1977) studied the O-demethylation of papaverine (**I**) with selected microorganisms. Strains belonging to the fungal taxa *Aspergillus, Cunninghamella*, and to the genus *Streptomyces*, actively metabolized the substrate in a manner similar to that of mammals. *Streptomyces griseus* UI 1158 and *Stmy. platensis* ATCC 13865 selectively demethylated papaverine; among other products of transformation, 4' O-desmethylpapaverine (**II**), a major mammalian metabolite, was found (Fig. 36).

Using *Streptomyces* strains, Davis and Rosazza (1976) performed microbial transformation of tetrandrine and Nabih *et al.* (1977) effected the conversion of thalicarpine; both of the compounds possessing antitumour activity. d-Tetrandrine (**I**) isolated from the plant *Stephania tetrandra* S. Moore was selectively demethylated yielding N'-nor-d-tetrandrine (**II**), cycleanorine, the naturally occurring compound of the plant *Cyclea pellata* (Fig. 37).

The microbial conversion with '*Stmy. punipalus*' NRRL 3529 of another antitumour compound, thalicarpine (**I**), led to (+)-hernandinol (**III**) formation in a two-step reaction. The first step was the oxidative cleavage of the isoquinoline ring and hernandaline (**II**) formation then reduction of the aldehyde to the alcohol, hernandinol; a compound present in the plant *Hernanda ovigieria* (Fig. 38). There is probably some similarity between microbial and plant transformation of thalicarpine.

The activity of colchicine and its derivatives resemble that of the vinca alkaloids. This class of agent share many similarities in their biochemical and biological action, mainly in their ability to produce metaphase arrest, although they possess disparate chemical structures.

Colchicine is a tropolone derivative containing three rings. Several derivatives of colchicine have been prepared of which the most useful clinically is decetyl N-methylcolchicine. Colchicine has four largely equivalent

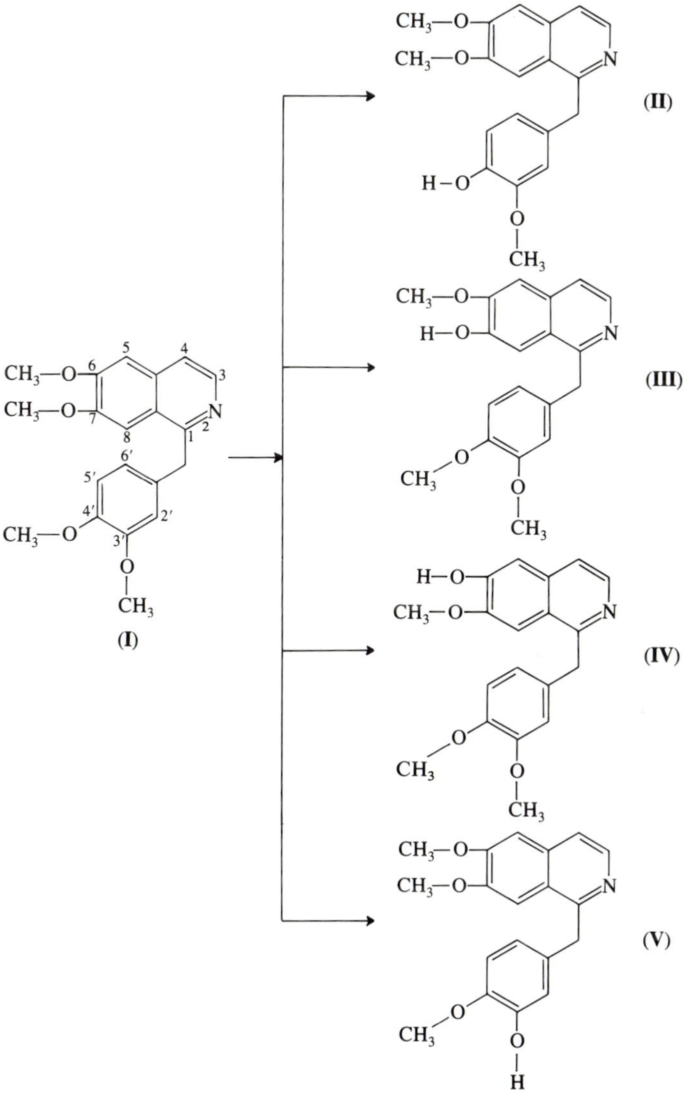

FIG. 36. Microbial O-demethylation of papaverine.

methoxy-groups and therefore selective fission of any particular ether bond cannot be achieved chemically. In order to prepare new derivatives of colchicine (I) a microbial transformation was carried out with a strain of *Stmy. griseus*, monodemethylcolchicine being formed (II) (Roussel, 1963). It is interesting that demethylation occurs on the 7-membered ring. The methoxy group at the 10 position is attacked selectively to produce the

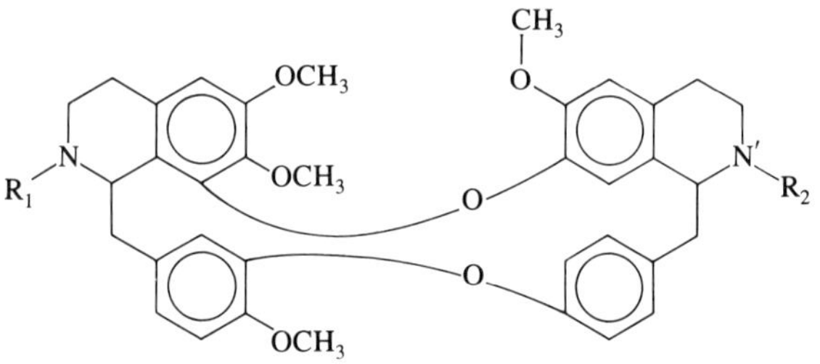

R₁ = R₂ = −CH₃   d-Tetrandrine (I)
R₁ = CH₃, R₂ = H   N'-nor-d-Tetrandrine (cycleanorine) (II)

FIG. 37. Microbial N-demethylation of d-tetrandrine.

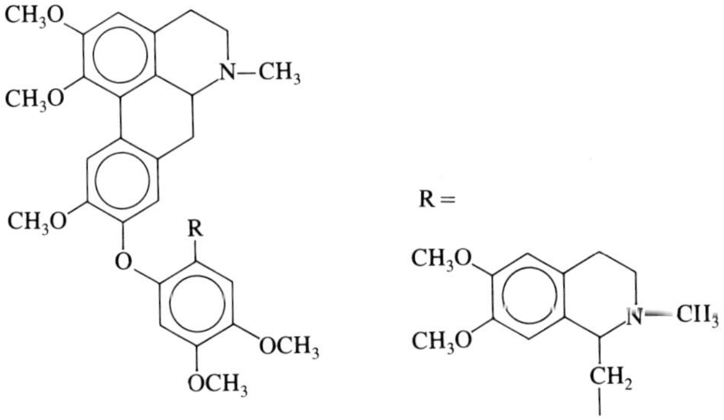

− thalicarpine (I)

R = −CHO − hernandaline (II)

R = −CH₂OH − hernandalinol (III)

FIG. 38. Microbial transformation of thalicarpine.

phenolic hydroxy group (Fig. 39). The analogous compound thiocolchicine (III) is transformed in a similar way. The substitution on the 7-membered ring is necessary to retain biological activity. The selective demethylation of the 7-membered ring can provide valuable substrate to produce novel colchicine derivatives (IV).

9. Antibiotics

In the search for novel pharmaceutical agents with antibacterial, antifungal and antitumour activity, significant progress has been made in the field of antibiotic biosynthesis, the structural elucidation of these compounds and their chemical microbial deactivation and degradation (Korzybski *et*

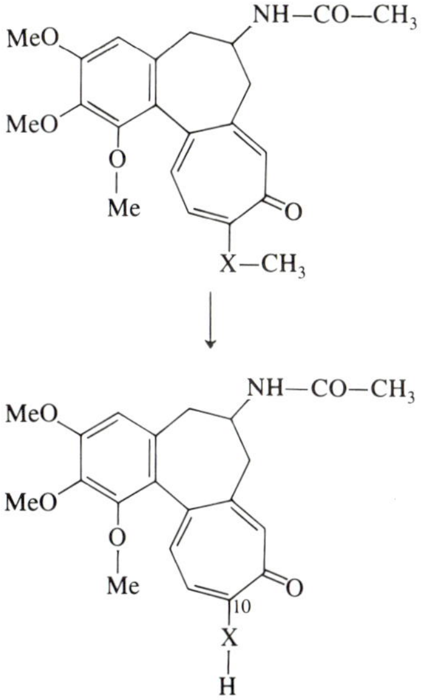

X = O Colchicine (I)
X = S Thiocolchicine (III)

X = O A-mono-demethylcolchicine (II)
X = S A-mono-demethylthiocolchicine (IV)

FIG. 39. Microbial demethylation of colchicine and thiocolchicine.

al., 1978). In recent years, exciting advances have been made in preparing new antibiotics by controlled and mutational biosynthesis and by the microbial transformation of known antibiotics to novel derivatives (Sebek and Perlman, 1971; Sebek, 1974; Kieslich, 1976; Shibata and Ueyda, 1978). The latter approach can be defined as the microbial transformation of xenobiotics and hopefully may lead to the discovery of compounds with reduced toxicity, enhanced desirable biological activity and improved clinical effects. Some of the advances made over the years are described below, though some of the examples given may not fulfil all of the criteria for xenobiotic transformations.

It has been recognized that microbial transformation (Rosazza and Smith, 1979) can mimic many of the kinds of biotransformation of antibiotics and other xenobiotics that are observed in mammals. In general, selected groups of microorganisms may serve to produce metabolites similar to those formed in the mammalian metabolic system. Thus, such metabolites, which are difficult to obtain as products of mammalian biconversion, would be easily available using routine fermentation scale-up techniques for complete structural elucidation and biological evaluation.

In most cases it is difficult to deal with all aspects of microbial transformation of antibiotics from a general point of view given the immense complexity of the problem. The following section is therefore restricted to a consideration of some typical examples of antibiotic conversion by actinomycetes.

A. TETRACYCLINES

Controlled biosynthesis of antibiotics (Shibata and Uyeda 1978) occurs when certain precursors added to the growing culture of the antibiotic-producing microorganism, are subjected to microbial conversion. In this way, novel antibiotics with basic structures related to the applied precursors can be synthesized. This method can be controlled by selection of the proper strain, or its mutants, and a suitable precursor of microbial or synthetic origin. The studies of McCormick et al. (1966) and McCormick (1965) with tetracyclines have elucidated certain problems of controlled biosynthesis of some representatives of the tetracycline family of antibiotics.

Tetracyclines belong to the group of antibiotics that have a perhydronaphthacene skeleton in common. Although a very large number of tetracycline derivatives have been studied over the past 20 years, an exact definition of the basic structural requirement for antibiotic activity has yet to be determined. Consideration of all of the known tetracycline variants suggests that the minimum structural requirement might be represented by structures 1–4 in Fig. 40.

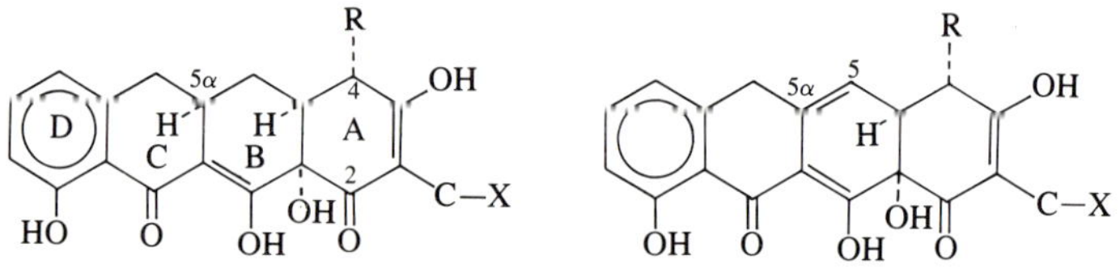

(1) R=H
(2) R=−N(CH$_3$)$_2$

(3) R=H
(4) R=−N(CH$_3$)$_2$

FIG. 40. Basic structural requirements for tetracycline activity.

The tetracycline moiety possesses two chromophoric groups, the A- and BCD-rings separated by a 12a-hydroxyl function. Modification of the chromophoric groups destroys the antibacterial activity while the stereochemistry of the 12a-hydroxyl group is essential for the biological activity of tetracyclines; 12a-epianalogues lack biological activity (Blackwood and English, 1977).

In order to try and explain some of the biosynthetic routes in tetracycline formation McCormick and Sjolander (1969) converted pretetramides into tetracyclines. Pretetramides, compounds believed to be intermediates in tetracycline biosynthesis, were microbially hydroxylated at the C-6 and 12a positions and the chromophore system essential for tetracycline activity was formed. Depending on the strain of *Stmy. aureofaciens*, tetracycline, 7-chlorotetracycline and 7-hydroxytetracycline were produced. Apparently, the corresponding demethyltetracyclines were obtained as products of the transformation of 6-demethyl-pretetramides (Fig. 41).

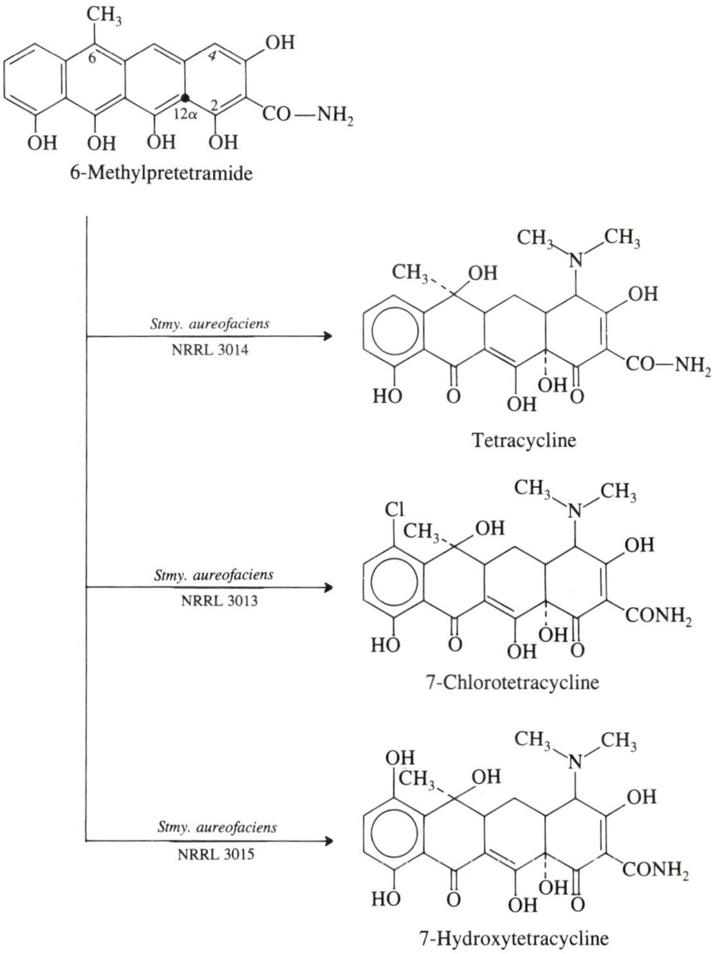

FIG. 41. Microbial conversion of 6-methylpretetramide into tetracycline derivatives.

The common reaction of all of the tested compounds was: hydroxylation at C-6 and C-12a, and the introduction of a dimethylamino group in the C-4 position.

B. 16-MEMBERED MACROLIDES

The acylation reaction of the hydroxyl groups of antibiotics is a property widely distributed among microorganisms, and is one that results in the inactivation of the antibiotic. Microbial acylation of chloramphenicol, kanamycin and spiramycin I to spiramycin II by *Stmy. ambofaciens* or acylation of rifamycin S to rifamycin B by '*Stmy.*' *mediterranei* is well known (Sebek and Perlman, 1971; Sebek, 1974). Beside the products of inactivation, selective acylation of the antibiotic moiety can lead to the production of novel derivatives and facilitates the establishment of the structure-activity relationships of antibiotics.

It is well known that amongst the 16-membered macrolides of the leucomycin and maridomycin groups there is a strong relationship between the biological activity, the position and the kind of acyl substitution that occurs (Omura and Nakagawa, 1975). The comparison of the biological properties of individuals of the leucomycin group has shown that there is higher activity where antibiotics have a non-acylated hydroxy group at C-3 of the lactone ring. It has also been shown that the acyl group at C-4 of mycarose plays an important role. The biological activity of antibiotics increases with the length of the acyl substituent in the following order: acetyl-, propionyl-, butyryl- and isovaleryl-.

The specific cleavage of the acyl substituent at the C-4 position by microbial or enzymic means causes significant lowering of antibacterial activity whereas the introduction of the acyl substituent at the C-4 position restores activity. It is possible that there is a relationship between increased size of the acyl substituent and increases in the lipophilicity of the compound which allows better penetration of the antibiotic into the microbial cell.

Omura *et al.* (1976) found that leucomycin A_1 was converted into leucomycin A_3, the 3-0-acetyl derivative of leucomycin A_1, by washed mycelia of '*Stmy. kitasatoensis*' 66-14-3 (Fig. 42).

In the course of screening, *Actp. missouriensis* IMRU 824 (AY B-886) was found to be most effective in transforming leucomycin A_5 (Singh and Rakhit, 1979). It was found that the leucomycin-transforming activity of the cells was greatly enhanced if the cells were introduced with demycarosyl 3-deacetyl leucomycin. Using induced cells, leucomycin A_5 was quantitatively transformed into one main product, leucomycin V (Fig. 43).

Selective deacylation provides the substrates for chemical acylation at the C-4 position. Novel derivatives of leucomycins were obtained by the introduction at C-4 of a benzoil substituent. The biological activity of the products with such modifications was significantly enhanced (Omura and Nakagawa, 1975).

C. ANTHRACYCLINES

The anthracycline antibiotics are almost universal antitumour agents and several including adriamycin and daunomycin are useful clinically against neoplastic diseases. Attempts have therefore been made to prepare new derivatives of these compounds by means of microbial transformations. It appears that microbial reduction of the side chain carbonyl group, *N*-

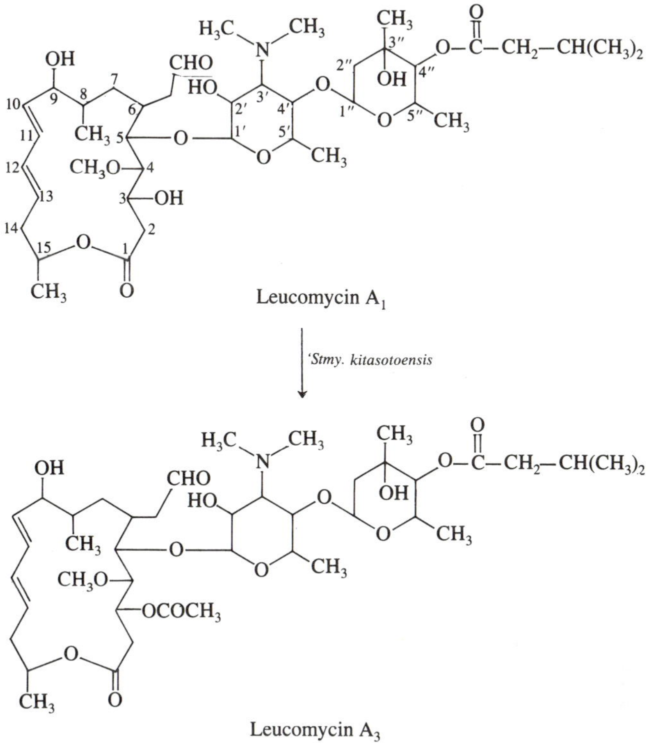

FIG. 42. Microbial transformation of leucomycin A_1 into leucomycin A_3.

acylation and glycoside cleavage by hydrolysis or reduction are the major reactions obtained.

The transformation of daunomycin (**I**) by selected *Streptomyces* strains led to the reduction of the carbonyl group at C-13 yielding daunomycinol (**II**) (Florent *et al.*, 1975; Aszalos *et al.*, 1977), or to the reductive cleavage of the glycoside bond yielding an aminosugar and an aglycone (Fig. 44). The 13-hydroxy derivative of daunomycin was described as a major mammalian metabolite and possessed lower antitumour activity than the parent compound. Presumably, the more polar alcohol is less capable of penetrating cancer cell membranes (Bachur *et al.*, 1976).

Daunomycin, transformed by crude enzyme preparations of *Stmy. steffiburgensis*' (Marshall *et al.*, 1976) in the presence of pyridine nucleotides, was converted into an aglycone and daunosamine (**IV**). The aglycone, 7-deoxydaunomycinone (**III**), obtained as a 66% yield, was further reduced to 7-deoxydaunomycinol (**V**) with a yield of 31%. Recently, an analogous reductive cleavage of the glycoside bond as a result of microbial

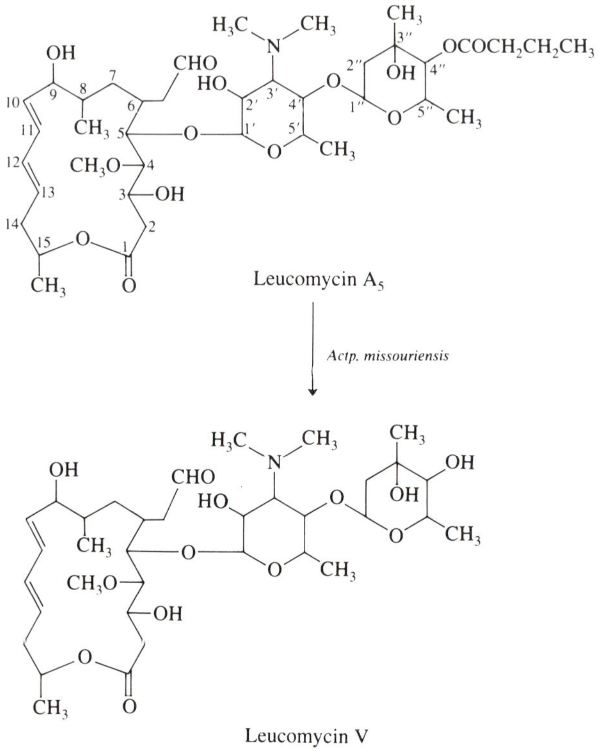

FIG. 43. Microbial transformation of leucomycin A_5 into leucomycin V.

transformation of nogalamycin and its derivatives has been observed (Rueckert et al., 1979).

Most of the microbial reactions reported have been observed in mammals and, unfortunately, most have been found to be inapplicable in the preparation of new and useful anthracyclines.

D. RIFAMYCINS

The rifamycins appear to be a group of antibiotics produced by a strain known variously as '*Stmy.*' *mediterranei* or '*Nrda.*' *mediterranea* (Alderson et al., 1981). This large family of antibiotics includes representatives which are closely related. Small modifications of the antibiotic moiety can cause dramatic changes in the biological activity of rifamycin (Lancini and Zanichelli, 1977). The minimal requirements for rifamycin activity appear to be connected with the presence of two hydroxy groups in positions 21 and 23 on the ansa chain and of two polar groups, either free or hydroxyl

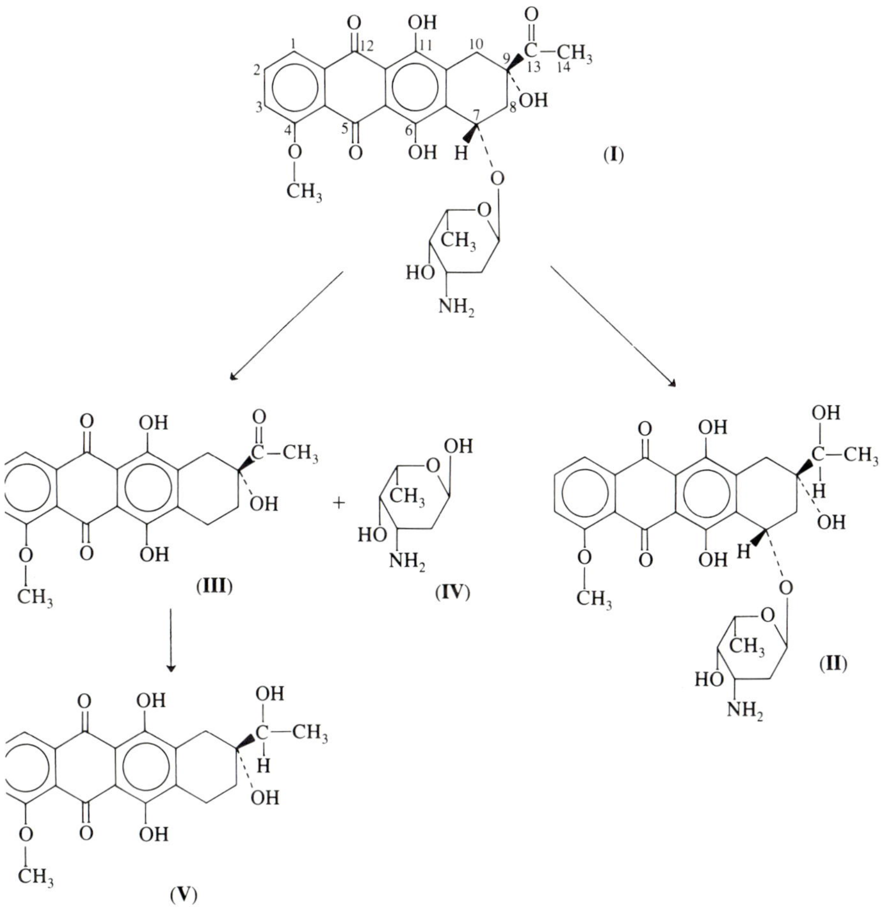

FIG. 44. Microbial transformation of daunomycin.

or carbonyl, at positions 1 and 8 of the naphthoquinone ring, together with a conformation of the ansa chain which results in certain specific geometrical relationships among these four functional groups (Fig. 45). According to Lancini and Zanichelli (1977) rifamycin SV (I) and its quinone form, rifamycin S (II), are precursors of rifamycin B (III) and probably of rifamycin O (IV). In the late stages of fermentation with the *Stmy. mediterranei* strain, rifamycin SV is transformed into rifamycin B and subsequently to rifamycin Y. Beside rifamycin B and Y, small amounts of rifamycin L (V), i.e. glycol esters of rifamycin SV are produced. Splitting the glycolic acid moiety from the molecule of rifamycin B increases the biological activity of the rifamycin S or SV produced.

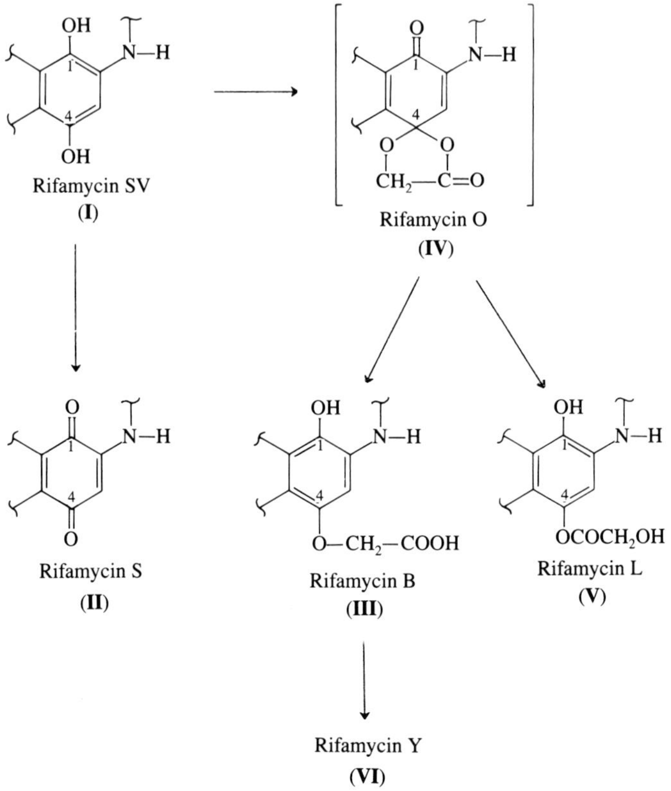

FIG. 45. The rifamycins (Korzybski et al., 1978)

It is possible that hydroxylation is the final step in the biosynthesis of active antibiotics or leads to their inactivation; the biogenesis of rifamycins illustrates this problem. Rifamycins are produced as an antibiotic complex composed of at at least five antibiotic substances. When the actinomycete is cultured in the presence of diethylbarbituric acid, only two compounds are obtained: rifamycin B and, in a smaller amount, rifamycin Y (Lancini et al., 1967). The difference in chemical structure of these compounds is in the degree of oxidation of the aliphatic ring. Rifamycin B is biologically active due to transformation of the O or S compound but rifamycin Y does not possess any antibiotic activity. When radiolabelled [^{14}C]rifamycin B was added to the growing culture of 'Stmy.' mediterranei, radiolabelled rifamycin Y was found to be a product of transformation (Fig. 46). There was no evidence of the reverse reaction. Transformation of rifamycin B consists of the introduction of a hydroxy group in the C-20 position with the simultaneous oxidation of the hydroxy group at C-21.

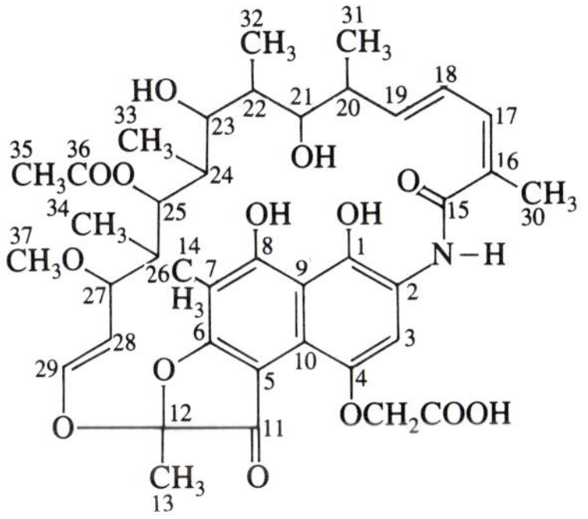

Rifamycin B
(III)

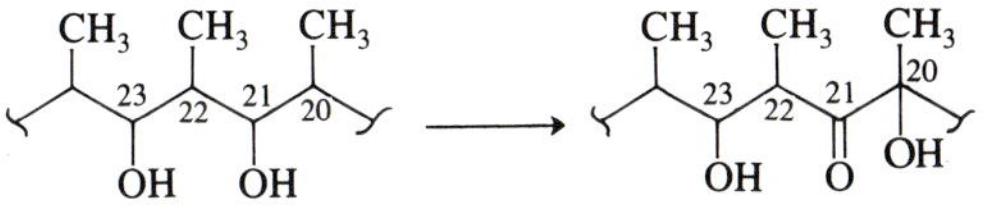

Rifamycin B (III) → Rifamycin Y (VI)

FIG. 46. Transformation rifamycin B into rifamycin Y.

10. General Conclusions

The ability of actinomycetes to perform a variety of microbial conversions of organic compounds is an important factor in the complicated processes of biodegradation of pollutants in soil and water and in the biodegradation of pesticides, oil spills, chemical waste decomposition and the like. The almost universal distribution of actinomycetes in natural habitats suggests the universality of their action on xenobiotics including recalcitrant compounds. This problem, still not fully understood, has recently been the subject of extensive study, resulting in the use of microorganisms for the turnover of industrial and agricultural contaminants.

The application of actinomycetes in bioorganic chemistry to perform microbial transformations of organic compounds has become a field of great interest and of commercial value. The ability to effect subtle modifications of complicated organic compounds has been employed in the production of several interesting products, some of which have been subjected to further chemical synthesis to provide valuable pharmaceutical agents. Some of the latter have become interesting as compounds with enhanced biological activity and can serve as model compounds to clarify the pathways of xenobiotic transformations in mammalian systems. The transformations performed by actinomycetes usually do not completely degrade the substrates. In most cases, the modification is limited to hydroxylation, mutual transformation of carbonyl and hydroxy groups, selective O- and N-demethylation, hydrolysis of ester and amide bonds and isomerization, although some actinomycetes are able to utilize hydrocarbons and related compounds as a source of carbon and energy.

Because of the great importance of antibiotics, a lot of effort has been directed towards their biosynthesis during the last 30 years. At present more and more attention is being paid to biotransformation and biodegradation processes performed by actinomycetes in the hope of using them to produce novel derivatives of known chemical compounds or to restore the natural state of the environment by removing contaminants whose numbers grow due to increasing industrialization. Actinomycetes are also a well known source of enzymes. Thus the production of glucose isomerase from different streptomycetes has been used for several years in the food industry and 6-aminopenicillic acid has been obtained from enzymic hydrolysis of penicillin by an acylase isolated from *Streptomyces* strains in the pharmaceutical industry. The isolation of cholesterol oxidase from *Nocardia* strains on a commercial scale is a good source of a valuable enzyme applied in medical diagnosis in the production of enzymic electrodes, and in the application of enzymic thermistore.

REFERENCES

Abbott, B. J. (1977). Immobilized cells. In: *Annual Reports of Fermentation Processes, Volume 1* (D. Perlman, ed.), pp. 205–234. Academic Press, New York.

Abbott, B. J. and Casida, L. E. Jr. (1968). Oxidation of alkanes to internal monoalkenes by a *Nocardia. Journal of Bacteriology* **96**, 925–930.

Achrem, A. A. and Titov, J. A. (1970). *Steroidy i Mikroorganizmy*. Nauka Press, Moscow.

Agnello, E. J., Figor, S. K., Hughes, G. M., Ordway, H. W., Pinson, R., Bloom, B. M. and Laubach, G. D. (1963). A stereospecific synthesis of C-21-methylated corticosteroids. *Journal of Organic Chemistry* **28**, 1531–1539.

Alderson, G., Goodfellow, M., Wellington, E. M. H., Williams, S. T., Minnikin, S. M. and Minnikin, D. E. (1981). Chemical and numerical taxonomy of *Nocardia mediterranea*. *Zentralblatt für Bakteriologie, Parasitenkunde, Infektionskrankheiten und Hygiene. 1. Abteilung, Supplement* **11**, 39–46.

Archer, R. A., Fukuda, D. S., Kossoy, A. D. and Abbott, B. J. (1979). Microbiological transformations of nabilone, a synthetic cannabinoid. *Applied and Environmental Microbiology* **37**, 965–971.

Aszalos, A. A., Bachur, N. B., Hamilton, B. K., Langlykke, A. F., Roller, P. P., Sheikh, M. Y., Statphin, M. S., Thomas, M. C., Warenheim, D. A. and Wright, L. H. (1977). Microbial reduction of the side-chain carbonyl of daunorubicin and N-acetyldaunorubicin. *Journal of Antibiotics* **30**, 50–58.

Bachur, N. R., Steele, M., Meriweather, P. W. and Hildebranch, R. (1976). Cellular pharmacodynamics of several anthracycline antibiotics. *Journal of Medicinal Chemistry* **19**, 651–654.

Beer, C. T. (1955). The leucopenic action of extracts of *Vinca rosea*. *Annual Report of the British Empire Cancer Campaign for Cancer Research* **33**, 487–488.

Bernath, F. R., Venkatasubramanian, K. and Vieth, W. R. (1977). Immobilized enzymes. *Annual Reports on Fermentatition Processes* **1**, 235–266.

Blackwood, R. K. and English, A. R. (1977). Structure-activity relationships in the tetracycline series. In: *Structure-Activity Relationships Among the Semisynthetic Antibiotics* (D. Perlman, ed.), pp. 397–426. Academic Press, New York.

Cain, R. B. (1966a). Introduction of an anthranilate oxidation system during the metabolism of ortho-nitrobenzoate by certain bacteria. *Journal of General Microbiology* **42**, 197–217.
Cain, R. B. (1966b). Utilization of anthranilic and nitrobenzoic acid by *Nocardia opaca* and *Flavobacterium*. *Journal of General Microbiology* **42**, 219–235.
Cain, R. B., Tranter, E. K. and Darrach, J. A. (1968). The utilization of some halogenated aromatic acids by *Nocardia*. *Biochemical Journal* **103**, 211–227.
Capek, A., Hanc, O. and Tadra, M. (1966). *Microbial Transformations of Steroids*. Academia, Praha.
Cartwright, N. J. and Cain, R. B. (1959). Bacterial degradation of the nitrobenzoic acids. *Biochemical Journal* **71**, 248–261.
Chacko, C. I., Lockwood, J. I. and Zabik, M. (1966). Chlorinated hydrocarbon pesticides, degradation by microbes. *Science* **154**, 893–895.
Charney, W. and Herzog, H. L. (1967). *Microbial Transformation of Steroids*. Academic Press, New York.
Chibata, I. and Tosa, T. (1977). Transformation of organic compounds by immobilized microbial cells. *Advances in Applied Microbiology* **22**, 1–25.
Cox, D. P. and Goldsmith, C. D. (1979). Microbial conversion of ethylbenzene to 1-phenyl-ethanol and acetophenone by *Nocardia tartaricans* ATCC 31190. *Applied and Environmental Microbiology* **38**, 514–520.
Cripps, R. E., Trudgill, P. W. and Whateley, J. R. (1978). The metabolism of 1-phenyl-ethanol and acetophenone by *Nocardia* T5 and an *Arthrobacter* species. *European Journal of Biochemistry* **86**, 175–186.
Davis, J. B. and Raymond, R. L. (1961). Oxidation of alkyl substituted cyclic hydrocarbons by *Nocardia* during growth on n-alkanes. *Applied Microbiology* **9**, 383–388.
Davis, P. J. and Rosazza, J. P. (1976). Microbial transformation of natural antitumour agents. 2. Studies with d-tetrandrine and laudanosine. *Journal of Organic Chemistry* **41**, 2548–2551.
Florent, J., Lunel, J. and Renaut, J. (1975). *Neue Vorfahren zur Herstellung des Antibiotikums 20*, 789. Ger Offen 2,456,139 May 28, 1975, *Chemical Abstracts* **83**: 112355q.
Foster, J. W. (1962). In: '*Oxigenases*' (O. Hayaishi, ed.), pp. 241–271. Academic Press, New York.
Fukuda, D., Archer, R. A. and Abbott, B. J. (1977). Microbial transformation of $\Delta^{6a,10a}$ tetrahydrocannabinol. *Applied and Environmental Microbiology* **33**, 1134–1140.
Golovlev, L., Golovleva, L. A., Eroshina, N. V. and Skryabin, G. K. (1978). Microbiological transformation of xenobiotics by *Nocardia*. *Zentralblatt für Bakteriologie, Parasitenkunde, Infektionskrankheiten und Hygiene. 1. Abteilung, Supplement* **6**, 269–283.
Goodfellow, M. (1971). Numerical taxonomy of some nocardioform bacteria. *Journal of General Microbiology* **69**, 33–80.
Goodfellow, M. and Minnikin, D. E. (1981). The genera *Nocardia* and *Rhodococcus*. In: *The Prokaryotes: A Handbook of Habitats, Isolation and Identification of Bacteria* (M. P. Starr, H. G. Stolp, A. Trüper, A. Balows and H. G. Schlegel, eds.), pp. 2016–2927. Springer Verlag, Berlin.
Hartman, R. E., Kraus, E. F., Andres, W. W. and Patterson, E. L. (1964). Microbial hydroxylation of indole alkaloids. *Applied Microbiology* **72**, 138–140.
Hikino, H., Tokuoka, Y., Hikino, H. and Takemoto, T. (1968). Microbial transformation of α-kessyl alcohol to kessyl glycol and kessane-2,7-diol. *Tetrahedron* **24**, 3147–3152.
Hikino, H., Tokuoka, Y. and Takemoto, T. (1970). Biochemical synthesis. VIII. Microbial transformation of α-santonin to 1,2-dihydro-α-santonin. *Chemical and Pharmaceutical Bulletin (Tokyo)* **18**, 2127–2128.
Hill, I. R. and Wright, S. J. L. (1978). *Pesticide Microbiology: Microbiological Aspects of Pesticide Behaviour in the Environment*. Academic Press, London.
Hirsch, P. and Alexander, M. (1960). Microbial decomposition of halogenated propionic and acetic acid. *Canadian Journal of Microbiology* **6**, 241–249.
Iizuka, H. and Naito, A. (1967). *Microbial Transformation of Steroids and Alkaloids*. University Park Press, State College of Pennsylvania, U.S.A.
Jamison, V. M., Raymond, R. L. and Hudson, J. O. (1969). Microbial hydrocarbons co-oxidation. III. Isolation and characterization of an α, α′-dimethyl-*cis,cis*-muconic acid-producing strain of *Nocardia corallina*. *Applied Microbiology* **17**, 853–856.

Johnson, I. S., Wright, H. F. and Svoboda, G. H. (1959). Experimental basis for clinical evaluation of antitumour principles derived from *Vinca rosea* Linn. *Journal of Laboratory and Clinical Medicine* **54**, 830.

Johnson, I. S., Hargrove, W. W., Harris, P. N., Wright, H. F. and Boder, G. B. (1966). Preclinical studies with vinglycinate, one of a series of chemically derived analogs of vinblastine. *Cancer Research* **26**, 2431–2436.

Jones, J. B., Sih, C. J. and Perlman, D. (1976). Application of biochemical systems. *Organic Chemistry*, Volume X. J. Wiley, Interscience Publishers, New York.

Kawai, S., Kobayashi, K., Oshima, T. and Egami, F. (1965). Studies on oxidation of p-aminobenzoate to p-nitrobenzoate by *Streptomyces thioluteus*. *Archives of Biochemistry and Biophysics* **112**, 537–543.

Kieslich, K. (1976). *Microbial Transformations of Non-steroid Cyclic Compounds*. John Wiley Interscience Publishers, New York.

Kleupfel, D. and Vézina, C. (1970). Microbial aromatization of androsta-1,4,7-triene-3,17-dione. *Applied Microbiology* **20**, 515–516.

Korzybski, T., Kowszyk-Gindifer, Z. and Kuryłowicz, W. (1978). *Antibiotics, Origin Nature and Properties*. American Society for Microbiology, Washington D.C.

Lancini, G. and Zanichelli, W. (1977). Structure–activity relationships in rifamycins. In: *Structure–Activity Relationships among the Semisynthetic Antibiotics* (D. Perlmann, ed.), pp. 531–600. Academic Press, New York.

Lancini, G. C., Thiemann, J. E., Sartori, G. and Sensi, P. (1967). Biogenesis of rifamycins. The conversion of rifamycin B into rifamycin Y. *Experimentia* **23**, 899–900.

Laskin, A. I. and Lechevalier, H. (ed.) (1974) *CRC Handbook of Microbiology*, Volume IV, CRC Press, Ohio. Cleveland,

Lee, B. K., Ryu, D. Y., Thoma, R. W. and Brown, W. E. (1969). Introduction and repression of steroid hydroxylases and dehydrogenases in mixed culture fermentations. *Journal of General Microbiology* **55**, 145–153.

McCormick, J. R. D. (1965). Biosynthesis of the tetracyclines. In: *Biogenesis of Antibiotic Substances* (Z. Vanněk and Z. Hošťálek, eds.), pp. 73–91. Czechoslovak Academy of Sciences, Praha.

McCormick, J. R. D. and Sjolander, N. O. (1969). Biological transformation of α-6-deoxytetracyclines to tetracyclines. *Chemical Abstracts* **70**, 113 857y.

McCormick, J. R. D., Newell, O. and Johnson, S. J. (1966). Biological transformation of 1,3,10,11,12-pentahydroxynaphthacene-2-carboxamides to tetracycline antibiotics. *Chemical Abstracts* **64**, 19522a.

McKenna, E. J. (1966). PhD Thesis, University of Iowa, Ames, Iowa, U.S.A.

Mallet, G. E., Fukuda, D. S. and Gorman, M. (1964). Microbial conversion of *Catharanthus* alkaloids. *Lloydia* **27**, 334–339.

Mammoli, L. and Vercellone, A. (1937). Biochemische Unwandlung von Δ^4 Androstendion in Δ^4-Testosteron. Ein Beitrag zur Genase des Keimdrüsenhormons. *Vorläufige Mitteilungen, Berlin* **70**, 470, 2079.

Marshall, V. P., Reisender, E. A. and Wiley, P. F. (1976). Bacterial metabolism of daunomycin. *Journal of Antibiotics* **29**, 966–968.

Martin, A. (1974). Microbial transformation of environmental pollutants. *Advances in Applied Microbiology* **18**, 1–64.

Martin, C. K. A. (1977). Microbial cleavage of sterol side-chains. *Advances in Applied Microbiology* **22**, 29–58.

Miles, J. R. W., Tu, C. M. and Harris, C. R. (1969). Metabolism of heptachlor and its degradation products by soil microorganisms. *Journal of Economic Entomology* **62**, 1334–1338,

Minnikin, D. E. and Goodfellow, M. (1980). Lipid composition in the classification and identification of acid-fast bacteria. In: *Microbiological Classification and Identification* (M. Goodfellow and R. E. Board, eds.), pp. 189–256. Academic Press, London.

Nabith, T., Davis, P. J., Caputo, J. P. and Rosazza, J. P. (1977). Microbial transformation of natural antitumour agents. 3. Conversion of thalicarpine to (+)-hernandalinol by *Streptomyces punipalus*. *Journal of Medicinal Chemistry* **20**, 914–917.

Nabih, T., Liesa Youel and Rosazza, J. P. (1978). Microbial transformation of natural antitumour agents. 5. Structure of a novel vindoline-dimer produced by *Streptomyces griseus*. *Journal of the Chemical Society, Perkin Transactions* I, 751–761.

Neuss, N., Fukuda, D. S., Brannon, D. R. and Huckstep, L. L. (1973). Vinca alkaloids. XXXII. Microbiological conversion of vinoline, a major alkaloid from *Vinca rosea L. Helvetica Chimica Acta* **56**, 2418–2426.

Neuss, N., Fukuda, D. S., Brannon, D. R. and Huckstep, L. L. (1974a). Vinca alkaloids. XXXIV. Preparation of des-N-methylvindoline by microbial conversion of vindoline, a major alkaloid of *Vinca rosea.* Linn. *Helvetica Chimica Acta* **57**, 1891–1893.

Neuss, N., Mallett, G. E., Brannon, D. R., Mabe. J. A., Hopton, H. R. and Huckstep, L. L. (1974b). Vinca alkaloids. XXXIII. Microbiological transformation of vincaleukoblastine (VLD, vinblastine) an antitumour alkaloid from *Vinca rosea.* Linn. *Helvetica Chimica Acta* **57**, 1886–1891.

Omura, S. and Nakagawa, A. (1975). Chemical and biological studies on 16-membered macrolide antibiotics. *Journal of Antibiotics* **28**, 401–433.

Omura, S., Miyazawa, J., Takeshima, H., Kitao, C., Atsumi, K. and Aizawa, H. (1976). Biconversion of leucomycins and its regulation by butyrate in a producing strain. *Journal of Antibiotics* **29**, 1131–1133.

Peterson, D. H. and Murray, H. C. (1952). Microbiological oxygenation of steroids at carbon 11. *Journal of American Chemical Society* **74**, 1871–1872.

Raymond, R. L. and Jamison, V. W. (1971). Biochemical activities of *Nocardia. Advances in Applied Microbiology* **14**, 93–120.

Raymond, R. L., Jamison, V. S. and Hudson, J. O. (1967). Microbial hydrocarbon cooxidation. I. Oxidation of mono- and diacyclic hydrocarbons by soil isolates of the genus *Nocardia. Applied Microbiology* **15**, 857–865.

Rosazza, J. P. (1980). Microbial transformation as an approach to analogue development. In: *Anticancer Agents Based On Natural Product Models* (J. M. Cassady and J. D. Douros, eds), pp. 437–463. Academic Press, New York.

Rosazza, J. P. and Smith, R. V. (1979). Microbial models for drug metabolism. *Advances in Applied Microbiology* **25**, 169–203.

Rosazza, J. P., Stocklinski, A. W., Gustafson, M. A. and Adrian, J. (1975). Microbial models of mammalian metabolism. O-dealkylation of 10,11-dimethoxyaporfine. *Journal of Medicinal Chemistry* **18**, 791–794.

Rosazza, J. P., Kammer, M., Youel, L., Smith, R. V., Erhardt, P. W., Troung, D. H. and Leslie, S. W. (1977). Microbial models of mammalian metabolism. *O*-Demethylation of papaverine. *Xenobiotica* **7**, 133–143.

Roussel (1963). UCLAF Brit., 923, 421, April 1963. *Chemical Abstracts* **59**, 13320a.

Rueckert, P. W., Wiley, P. F., McGowern, J. P. and Marshall, V. P. (1979). Mammalian and microbial cell-free conversion of anthracycline antibiotics and analogs. *Journal of Antibiotics* **32**, 141–147.

Sebek, O. K. (1974). Microbial conversion of antibiotics. *Lloydia* **37**, 115–133.

Sebek, O. K. and Kieslich, K. (1977). Microbial transformation of organic compounds. In: *Annual Reports on Fermentation Processes. Volume 1* (D. Perlman, ed.), pp. 267–297. Academic Press, New York.

Sebek, O. K. and Perlman, D. (1971). Microbial transformations of antibiotics. *Advances in Applied Microbiology* **14**, 123–150.

Shibata, M. and Uyeda, M. (1978). Microbial transformations of antibiotics. In *Annual Reports on Fermentation Processes, Volume 2* (D. Perlman, ed.), pp. 267–303. Academic Press, New York.

Siewert, G., Kieslich, K., Hoyer, G. A. and Rosenberg, D. (1973). Hydroxylierung von 5-alkyl-2-(benzolsulfonylamino) pyrimidinen und strukturwerwandten Antidiabetika. *Chemische Berichte* **106**, 1290–1302.

Sih, C. J. (1962a). Enzymatic mechanism of steroid ring A aromatization. *Biochemical and Biophysical Research Communications* **7**, 87–89.

Sih, C. J. (1962b). Microbiological transformation of steroids. *Journal of Bacteriology* **84**, 382.

Sih, C. J. and Wang, K. C. (1965). A new route to estrone from sterols. *Journal of the American Chemical Society* **87**, 1387–1388.

Sih, C. J., Tai, H. H. and Tsong, Y. Y. (1967a). The mechanism of microbial conversion of cholesterol into 17-keto steroids. *Journal of the American Chemical Society* **89**, 1957–1958.

Sih, C. J., Wang, K. C. and Tai, H. H. (1967b). C_{22} acid intermediates in the microbiological cleavage of the cholesterol side-chain. *Journal of the American Chemical Society* **89**, 1956.

Sih, C. J., Wang, K. C. and Tai, H. H. (1968a). Mechanism of steroid oxidation by microorganisms. XIII. C_{22} acid intermediates in the degradation of the cholesterol side-chain. *Biochemistry* **7**, 796–807.

Sih, C. J., Tai, H. H., Tsong, Y. Y., Lee, S. S. and Coombe, R. G. (1968b). Mechanism of steroid oxidation by microorganisms. XIV. Pathway of cholesterol side-chain degradation. *Biochemistry* **7**, 808–818.

Singh, K. and Rakhit, S. (1979). Microbial transformation of leucomycin A5. *Journal of Antibiotics* **32**, 78–80.

Stoudt, T. H., McAller, W. J., Kozłowski, M. A. and Marlatt, V. (1958). The microbial dehydrogenation of some pregnanes and allopregnanes to 1,4-pregnadienes. *Archives of Biochemistry and Biophysics* **74**, 280.

Tabenkin, B., LeMahieu, R. A., Berger, J. and Kierstead, R. W. (1969). Microbiological hydroxylation of cinerone to cinerolone. *Applied Microbiology* **17**, 714–717.

Tárnok, I. (1976). Metabolism in *Nocardia* and related bacteria. In: *The Biology of the Nocardiae* (M. Goodfellow, G. H. Brownell and J. A. Serrano, eds.), pp. 451–500. Academic Press, London.

Tranter, E. K. and Cain, R. B. (1967). The degradation of fluoroaromatic compounds to fluorocitrate and fluoroacetate by bacteria. *Biochemical Journal* **103**, 22–23.

Van Der Linden, A. C. and Thijsse, G. J. E. (1965). The mechanisms of microbial oxidations of petroleum hydrocarbons. *Advances in Enzymology* **27**, 469–546.

Wagner, F., Zahn, W. and Buhring, U. (1967). 1-Hexadecene, an intermediate in the microbial oxidation of n-hexadecane *in vivo* and *in vitro*. *Angewandte Chemie* **6**, 359–360.

Webley, D. M. and DeKock, P. C. (1952). The metabolism of some saturated aliphatic hydrocarbons, alcohols and fatty acids by *Proactinomyces opacus* Jensen (*Nocardia opaca* Waksman and Henrici). *Biochemical Journal* **51**, 371–375.

Webley, D. M., Duff, R. B. and Farmer, V. C. (1955). β-Oxidation of fatty acids by *Nocardia opaca*. *Journal of General Microbiology* **13**, 361–369.

Williams, S. T., Wellington, E. M. H., Goodfellow, M., Alderson, G., Sackin, M. and Sneath, P. H. A. (1981). The genus *Streptomyces* – a taxonomic enigma. *Zentralblatt für Bakteriologie, Parasitenkunde, Infektionskrankheiten und Hygiene. 1. Abteilung, Supplement* **11**, 47–58.

7
Actinomycete Envelope Lipid and Peptidoglycan Composition

D. E. MINNIKIN and A. G. O'DONNELL

Department of Organic Chemistry, The University, Newcastle upon Tyne, UK

1. Introduction 337
2. Plasma membrane lipids 338
   A. Amphipathic polar lipids and constituent fatty acids . 338
   B. Lipid-soluble quinones and pigments 347
3. Peptidoglycans 350
4. Wall membrane lipids 355
5. Conclusions 377
6. Acknowledgements 381
   References 382

1. Introduction

Bacterial envelopes have, as their innermost layer, a permeability barrier enclosing the bacterial cytoplasm. Many Gram-positive bacteria have been shown to have a rigid matrix of peptidoglycan and associated polymers surrounding the cytoplasmic membrane (for a review see Rogers *et al.*, 1980). Certain actinomycetes appear to have this simple type of envelope organization but some are characterized by the presence of an additional covalently bound external lipid matrix often associated with a variety of free lipids of unusual structure (Goren and Brennan, 1979; Minnikin and Goodfellow, 1980, 1981; Minnikin, 1982). Bacterial envelopes contain, of course, many types of functional proteins but attention will be restricted here to considerations of the composition, interrelation and possible role of lipid and cell wall components.

Plasma membranes are based on a bilayer system composed of amphipathic polar lipids in intimate association with specific membrane proteins. Amphipathic polar lipids consist of hydrophilic head groups usually linked to two hydrophobic aliphatic chains. Phospholipids are the most common polar lipids but glycolipids and acylated ornithine or lysine amides also fall into this category. The hydrophobic interior of plasma membranes is considered to provide a location for menaquinones which are the principal type of respiratory lipoquinones found in actinomycetes. Lipid-soluble pigments are also associated with plasma membranes.

The plasma membranes of actinomycetes are supported externally by a cage of polymeric peptidoglycan consisting of strands of amino-sugar-based polysaccharides linked together with peptide units. Peptidoglycans of actinomycetes are probably linked, in many instances, to other polysaccharides since characteristic sugars are released on hydrolysis of isolated cell walls. In the actinomycete envelopes, which have an external lipid layer, a peptidoglycan-linked arabinogalactan is esterified by mycolic acids which are high molecular weight 2-alkyl-branched 3-hydroxy long-chain acids. This mycolic acid-based external wax layer may provide, particularly in the mycobacteria, an anchorage for some very characteristic complex lipids (Minnikin, 1982).

The boundary between actinomycetes and the so-called coryneform bacteria has become increasingly blurred in recent years and the selection of organisms for consideration will follow broadly the ideas developed in Chapters 1 and 2. The distribution and chemotaxonomic value of the lipids of actinomycetes and related organisms have been the subject of extensive recent reviews (Minnikin and Goodfellow, 1976, 1980, 1981; Lechevalier, 1977; Minnikin et al., 1978) and the structures of peptidoglycans have also been surveyed (Schleifer and Kandler, 1972).

2. Plasma Membrane Lipids

A. AMPHIPATHIC POLAR LIPIDS AND CONSTITUENT FATTY ACIDS

A functional plasma membrane requires the presence of a suitable selection of the various structural types of amphipathic polar lipids. Such lipids interact together to form a bilayer system with the constituent fatty acid chains interlocked together and the polar head groups exposed to aqueous environments. Lipid bilayer systems, composed of amphipathic lipids, are inherently stable but in natural membranes specific functional proteins are incorporated.

The physical properties of the hydrophobic core of plasma membranes are achieved to a large extent by having a suitable mix of both relatively fluid and relatively solid fatty acids esterified to the polar head group. Examples of the most common actinomycete polar lipid fatty acids and their melting points are given in Table 1. Straight-chain and *iso*-acids are high melting but the remaining acids are all relatively low melting.

Several different general types of fatty acid mixtures are encountered in actinomycetes and related bacteria and some of these are summarized in Table 2. At one extreme straight-chain fatty acids occur mixed with fluid

TABLE 1. Structure and properties of representative actinomycete fatty acids

Fatty acid type	Common example and structure	Melting point[a] (°C)
Straight-chain	Hexadecanoic $CH_3 \cdot (CH_2)_{14} \cdot COOH$	61–62
Cis-unsaturated	Oleic $CH_3 \cdot (CH_2)_7 \cdot \underset{cis}{CH=CH} \cdot (CH_2)_7 \cdot COOH$	13–14
Tuberculostearic	10-Methyloctadecanoic $CH_3 \cdot (CH_2)_7 \cdot \underset{\underset{CH_3}{\vert}}{CH} \cdot (CH_2)_8 \cdot COOH$	10.3–11.7
Cis-cyclopropane	11,12-Methyleneoctadecanoic $CH_3 \cdot (CH_2)_5 \cdot \underset{cis}{CH - CH} \cdot (CH_2)_9 \cdot COOH$ with CH_2 bridge	28–29
Iso	13-Methyltetradecanoic $CH_3 \cdot \underset{\underset{CH_3}{\vert}}{CH} \cdot (CH_2)_{11} \cdot COOH$	52.5–53
Anteiso	12-Methyltetradecanoic $CH_3 \cdot CH_2 \cdot \underset{\underset{CH_3}{\vert}}{CH} \cdot (CH_2)_{10} \cdot COOH$	24

[a] Data from Asselineau (1966) and commercial catalogues.

monounsaturated acids and acids biosynthetically derived from unsaturated acids such as tuberculostearic and cyclopropane acids (Table 1). The simplest example is found in representatives of *Corynebacterium* and certain strains of *Actinomyces* in which unsaturated acids are the only fluid components (Table 2). Other strains of *Actinomyces* contain additional cyclopropane acids. Many mycolic acid-containing bacteria and certain actinomadurae contain tuberculostearic acid. At the other extreme, actinomycete fatty acid patterns have *iso* fatty acids as their main relatively solid base, although smaller amounts of straight-chain acids are usually present. The fluid components in patterns of this type consist solely of *anteiso* acids, no unsaturated acids or their derivatives being present. Examples (Table 2) of such fatty acid profiles, containing only straight-chain, *iso* and *anteiso* acids, are shown by representatives of *Streptomyces* and related organisms, *Actinopolyspora, Cellulomonas, Oerskovia, Micropolyspora faeni* and *Thermomonospora*. Between these relatively simple fatty acid mixtures, certain bacteria contain combinations of both types and examples are shown in Table 2.

TABLE 2. Typical mixtures of long-chain fatty acids found in actinomycetes[a]

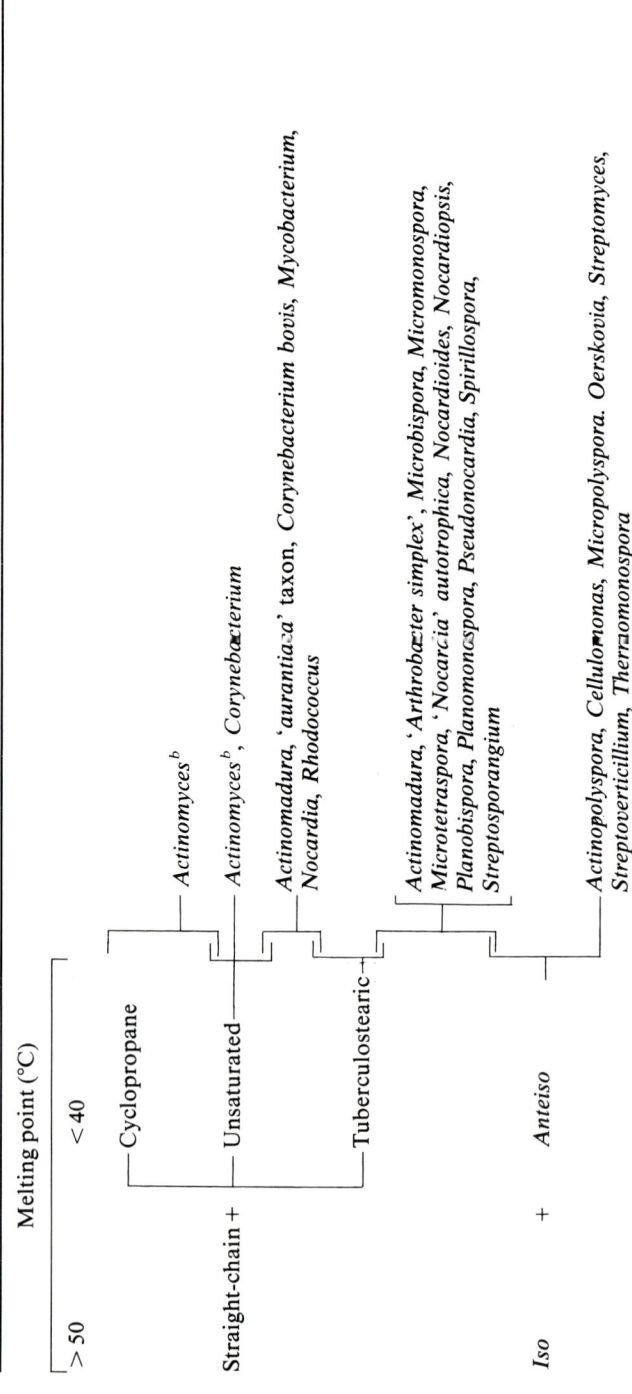

[a] Data from Lechevalier (1977), Lechevalier et al. (1977), Kroppenstedt and Kutzner (1978), Minnikin and Goodfellow (1980, 1981), Minnikin et al. (1978).
[b] Precise distribution of cyclopropane fatty acids among species of *Actinomyces* requires further study.

Phospholipids are the most common actinomycete polar lipids and their structures are summarized in Table 3. The polar amphipathic glycolipids of actinomycetes have not been studied systematically but the structures given in Tables 4 and 5 have been reported. Acylated long-chain lysine and ornithine amides (Table 6) are a further class of phosphorus-free polar lipids.

TABLE 3. Actinomycete phospholipids

$$\begin{array}{l} \text{CH}_2\cdot\text{OOC}\cdot\text{R}' \\ | \\ \text{R}\cdot\text{COO}\cdot\text{CH} \\ | \quad\quad\quad\text{O} \\ | \quad\quad\quad \| \\ \text{CH}_2\cdot\text{O}\cdot\text{P}\cdot\text{OY} \\ \quad\quad\quad | \\ \quad\quad\quad\text{OH} \end{array}$$

R, R' = long-chain alkyl

Overall charge	Polar head group substituent (Y)	Name and abbreviation
1+	Glycerol	Phosphatidylglycerol (PG)
2+	Phosphatidylglycerol	Diphosphatidylglycerol (DPG)
1+	Butane-2,3-diol	Phosphatidylbutanediol (PB)
1+	Inositol	Phosphatidylinositol (PI)
1+	Acylated mannosylinositols	Phosphatidylinositol mannosides (PIM)
0	Ethanolamine	Phosphatidylethanolamine (PE)
0	Choline	Phosphatidylcholine (PC)
0	Methylethanolamine	Phosphatidylmethylethanolamine (PME)

All actinomycetes contain acidic polar lipids and in certain cases (Table 7) these are the sole components. In other cases, acidic lipids occur together with a variety of neutral or amphoteric polar lipids (Table 7). As outlined above for fatty acids, the various structural types of polar lipids show certain interrelationships. It has been suggested, originally by Wilkinson (1968), that acidic glycolipids in certain pseudomonads may replace acidic phospholipids. In support of this view, Minnikin et al. (1974a) showed that in phosphate-limited chemostat cultures of *Pseudomonas diminuta* the usual acidic phospholipids were absent but high proportions of acidic glycolipids were recorded. Studies on the effect of growth environment on the polar lipids of various bacilli demonstrated an apparent interchangeability of diglycosyl diacylglycerol and phosphatidylethanolamine (PE) (Minnikin et al., 1971, 1972) and this latter neutral amphoteric lipid may be replaced by zwitterionic ornithine amide lipids (Wilkinson, 1972; Minnikin and Abdolrahimzadeh, 1974) in certain pseudomonads. A similar

TABLE 4. Neutral amphipathic glycolipids in actinomycetes

Source	Type	Structure	Reference
Mycobacterium fortuitum[a]	Trehalose diesters (**A**-symmetric, **B**-asymmetric)	**A** $R' = R'' = CH_3 \cdot (CH_2)_{14} \cdot CO$, $R''' = H$ **B** $R' = R''' = CH_3 \cdot (CH_2)_{16} \cdot CO$, $R'' = H$	Vilkas and Rojas (1964) Vilkas et al. (1968)
'*Nocardia polychromogenes*'	Diglucosyl diacylglycerol	Glu-(1→?)-Glu-(1→1)-diacylglycerol	Khuller and Brennan (1972)
Rothia dentocariosa	Dimannosyl diacylglycerol	Man-(1→3)-Man-(1→1)-diacylglycerol	Pandhi and Hammond (1975)
Actinomyces viscosus	Monogalactosyl diacylglycerol	Gal-(1→1)-diacylglycerol	Yribarren et al. (1974) Pandhi and Hammond (1978)

[a] A diacylated trehalose has been characterized from *Micromonospora* (Tabaud et al., 1971).

TABLE 5. Acidic amphipathic glycolipids from actinomycetes

Source	Type	Structure	Reference
Streptomyces LA 7017	Acylated glucosylgalacturonosyl diacylglycerol		Batrakov and Bergelson (1978)
	Glucosylglucuronosyl diacylglycerol		
Streptomyces sp.	Glucuronosyl diacylglycerol		

R = mainly *anteiso*-C_{15} and C_{17} and *iso*-C_{16} acid for all three above glycolipids

continued on following page

TABLE 5 (cont.)

Source	Type	Structure	Reference
'*Mycobacterium paraffinicum*'	Triacylated trehalose 6-succinate	Structure showing trehalose with CH$_2$·OCO·(CH$_2$)$_2$·COOH, OR', OR'', OH, HOH$_2$C groups; R' = CH$_3$·(CH$_2$)$_6$·CC, R'' = CH$_3$·(CH$_2$)$_8$·CO	Batrakov et al. (1977)
Nocardia otitidis-caviarum (neé *caviae*)	Diacylated glyceric acid glucoside	Structure showing glucoside linked to glyceric acid with COOH, CH$_2$OH, OR, OH groups; R = CH$_3$·(CH$_2$)$_{12,14,16}$·CO	Pommier and Michel (1981)

TABLE 6. *Actinomycete* amphipathic amino acid amides

Source	Type	Structure	Reference
Mycobacterium bovis BCG	Basic acylated ornithine amide	$\overset{\mid}{(CH_2)_3 \cdot NH_2}$ $R' \cdot CO \cdot NH \cdot CH \cdot COO \cdot CH_2 CH_2 \cdot OOC \cdot R''$ $R' = $ 3-OH acid, $R'' = $ tuberculostearate (Table 1)	Promé *et al.* (1969)
Streptomyces sioyaensis and '*Stmy. toyocaensis*'	Basic acylated ornithine and lysine amides	$\overset{\mid}{(CH_2)_{3,4} \cdot NH_2} \quad \overset{\mid}{CH_3}$ $R' \cdot CO \cdot NH \cdot CH \cdot COO \cdot CH_2 CH_2 CH \cdot OOC \cdot R''$ $R' = $ 3-OH acid, $R = $ alkyl	Kawanami (1971)
Streptomyces globisporus, '*Stmy. aureoverticillus*', and *Stmy. olivaceus*	Neutral acylated ornithine amide	$\overset{\mid}{(CH_2)_3 \cdot NH_2}$ $R \cdot CH \cdot CH_2 \cdot CO \cdot NH \cdot CH \cdot COOH$ $\overset{\mid}{R \cdot COO}$ $R = $ alkyl	Batrakov and Bergelson (1978)
Streptomyces 660–15	Neutral acylated ornithine amide	$\overset{\mid}{(CH_2)_3 \cdot NH_3}$ $R' \cdot CO \cdot NH \cdot CO \cdot O \cdot CH \cdot COOH$ $\overset{\mid}{R''}$ $R' = $ 3-OH acid, $R'' = $ alkyl	

TABLE 7. Typical mixtures of polar lipids found in actinomycetes

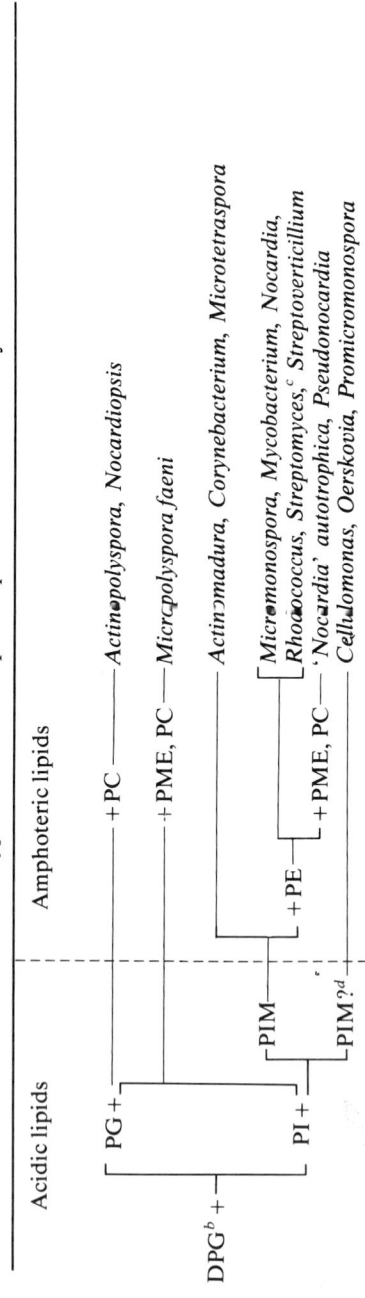

[a] For source of data see Minnikin and Goodfellow (1981).
[b] See Table 3 for abbreviations.
[c] Acidic glycolipids (Table 5) and amino acid amides (Table 6) have been characterized from a few streptomycetes.
[d] Glycophospholipids chromatographically similar to PIMs (Minnikin et al., 1979) but possibly corresponding to uncharacterized glucosamine-containing lipids (Lechevalier et al., 1977).

interchangeability of PE and ornithine amide lipids in streptomycetes has been discussed by Batrakov and Bergelson (1978).

B. LIPID-SOLUBLE QUINONES AND PIGMENTS

The plasma membrane lipid bilayer system provided by the interaction of amphipathic polar lipids and membrane proteins produces a stable matrix for the incorporation of other functional lipids. The best-studied of these in actinomycetes are the menaquinones and pigments, the latter being of both carotenoid and prodiginine types. Mention should be made here of the possible presence of hopanoid triterpenes, whose presence has been indicated in a streptomycete strain (Rohmer et al., 1979).

The menaquinones found in actinomycetes (Table 8) vary both in the length and degree of hydrogenation of their side chains and their distribution has been reviewed (Minnikin et al., 1978; Minnikin and Goodfellow, 1980, 1981; Collins and Jones, 1981a). Menaquinones are located in plasma membranes and are principally involved in electron transport and respiration (for a recent review see Taber, 1980). No detailed explanation has been provided for the wide range isoprenoid side-chains found in actinomycete menaquinones but it is likely that they are designed to have suitable physical properties for interaction with the fatty acid components of the host membrane. The highly lipophilic character of isoprenoid quinones suggests that their location is within the hydrophobic core of the plasma membrane. Since there appears to be little correlation between the lengths of isoprenoid (Table 8) and fatty acid (Table 2) chains, it is possible that menaquinones are located with their chains perpendicular to, rather than parallel with, the amphipathic polar lipid fatty acid chains. The functional naphthoquinone unit might then be located in suitable proximity to other components of the respiratory chain, the role of the isoprenoid side chain being as a hydrophobic anchor to maintain such a location. Mention should also be made here of 'mavioquinone', a minor non-isoprenoid lipoquinone found in *Mycobacterium avium* (see Minnikin, 1982).

The distribution of lipid-soluble pigments in actinomycetes has not been studied so systematically but the main types are carotenoids and prodiginines. A detailed discussion of the wide range of carotenoid structures is inappropriate here; the distribution of microbial carotenoids has been reviewed (Liaaen-Jensen and Andrewes, 1972) and data relevant to actinomycetes can be found in certain more recent reviews on lipids (Minnikin and Goodfellow, 1976; Minnikin et al., 1978; Minnikin, 1982). Carotenoids fall into two main structural types in actinomycetes, apolar varieties and xanthophylls which have oxygen functions such as sugar substituents. In the case of xanthophylls from '*Nocardia kirovani*', it was shown that acyl substituents were present on the sugar moieties (Vacheron

TABLE 8. Menaquinones of actinomycetes

Structure		Abbreviation: MK-n or MK-n (H$_x$) if $x/2$ double bonds are hydrogenated

Major isoprenologue(s)	Examples of taxa
MK-8(H$_2$)	*Corynebacterium diphtheriae*[a] *Rhodococcus coprophilus*[a] *Rodc. equi*,[a] *Rodc. erythropolis*,[a] *Rodc. rhodnii*,[a] *Rodc. rhodochrous*,[a] *Rodc. ruber*[a]
MK-8(H$_4$)	'*Nocardia*' *autotrophica*,[b] *Nrda. asteroides*,[a] *Nrda. brasiliensis*, *Nrda. otitidis-caviarum*[a]
MK-9	'*aurantiaca*' taxon,[a] *Cnbc. paurometabolum*[c]
MK-9 (H$_2$)	*Cnbc. bovis*,[a] *Cnbc. glutamicum*,[a] *Mycobacterium* spp.,[a,d] *Nrda. amarae*,[c] *Rodc. bronchialis*,[a] *Rodc. rubropertinctus* (syn. *corallinus*), *Rodc. terrae*[a]
MK-9 (H$_4$)	*Actinopolyspora halophila*,[f] *Cellulomonas* spp.,[g] *Oerskovia* spp.,[g] *Pseudonocardia thermophila*[h]
MK-9 (H$_4$, H$_6$, H$_8$)	*Actinomadura madurae*,[h] *Actm. pelletieri*,[h] *Micropolyspora faeni*,[h] *Streptomyces* spp.[h,i]
MK-10 (H$_4$, H$_6$, H$_8$)	*Actinomyces israelii*,[g] *Actn. viscosus*,[g] *Nocardiopsis dassonvillei*[g]

[a] Minnikin *et al.* (1978), Minnikin and Goodfellow (1980, 1981).
[b] Minnikin and Goodfellow (1980, 1981). These menaquinones chromatograph differently on reverse-phase media from those of mycolic acid containing strains of *Nocardia* and therefore probably have distinct structures (D. E. Minnikin, M. Goodfellow, G. Alderson and M. D. Collins, unpublished data; R. M. Kroppenstedt, personal communication).
[c] Collins (1978), Collins and Jones (1981b).
[d] For *Mybc. phlei*, the next but one isoprene unit from the naphthoquinone nucleus is hydrogenated (see Minnikin, 1982).
[e] Goodfellow *et al.* (1982).
[f] Collins *et al.* (1981).
[g] Minnikin *et al.* (1979).
[h] Minnikin and Goodfellow (1981), Batrakov and Bergelson (1978).
[i] Evidence for the location of the hydrogenated double bonds in the menaquinone of *Stmy. olivaceus* has been reviewed by Batrakov and Bergelson (1978).

et al., 1970). Examples of carotenoids found in actinomycetes are provided in Table 9.

The primary role of carotenoids appear to be as photoprotective agents (Krinsky, 1976) and, for *Micrococcus luteus*, it has been shown that

TABLE 9. Examples of types of carotenoid structures[a]

Name	Structure	Source
Lycopene		*Cnbc. diphtheriae*, *Mybc. kansasii*
β-Carotene		*Mycobacterium* spp., *Cnbc. fascians*
Isorenieratene (leprotene)		*Mybc. phlei*, *Mybc. kansasii*
Phlei-xanthophyll		*Mybc. phlei*, '*Nrda. kirovani*'

[a] For source of data see Minnikin *et al.* (1978), Minnikin (1982).

carotenoids protect menaquinone function (Turner and Prebble, 1980). There is evidence that carotenoid glycosides span bilayer membranes (Thirkell and Hunter, 1969; Yamamoto and Bangham, 1978) and it has been proposed, without substantive evidence, that such a structural role is of significant importance (Rohmer et al., 1979). In a detailed discussion of the carotenoids of mycobacteria (Minnikin, 1982), it has been argued that the photoprotective role is paramount though the correct structural location may be of obvious importance in the expression of such a function.

Prodiginine pigments were first characterized from *Serratia marcescens* but very characteristic types have also been found in actinomadurae and streptomycetes (Table 10) (Gerber, 1975; Gerber and Lechevalier, 1976). Antibiotic and antimalarial properties have been noted (Gerber, 1975; Gerber and Lechevalier, 1976) for these pigments. It is conceivable that prodiginines have photoprotective properties similar to those postulated for carotenoids but their characteristic tri-pyrrole nucleus (Table 10) suggests that they might well have a role in metal ion binding. The cyclic prodiginines, for example, are reminiscent of porphyrins. The antibiotic properties of prodiginines may, therefore, be related to possible action as ionophores, causing leakage of essential metal ions, and such ionophoric behaviour may be significant in the uptake of such metal ions.

3. Peptidoglycans

The peptidoglycans of actinomycetes appear to be of the most common type based on a linear glycan composed of alternative $1 \to 4$-linked glucosamine and muramic acid units (Fig. 1). These glycan chains are cross-linked by peptide sub-units which, through variation in their amino acid composition and mode of cross-linkage, account for the majority of differences in peptidoglycan structures. Despite the general uniformity of the glycan strands, atypical structures have been reported (Brumfitt et al., 1958; Araki et al., 1971; Fleck et al., 1971; Hayashi et al., 1973; Uchida and Aida, 1977). In the actinomycetes, the substitution of N-acetylmuramic acid by N-glycolylmuramic acid (CH_3CO- replaced by $HOCH_2CO-$) has been found (Table 11). Such studies suggest that the presence of N-glycolylmuramic acid is of potential value in classification and identification (see Chapter 2).

The alternating L- and D-amino acid sequence of the peptidoglycan tetrapeptide adds structural strength (D,L heteropolymers are stronger than either L or D homopolymers) and allows the amino acid side chains to align on one side of the peptide thereby enabling intrapeptidoglycan hydrogen bonding to occur (Wheat, 1980). The presence of D-amino acids and lack of aromatic amino acids means that peptidoglycan is not susceptible to

TABLE 10. Prodiginine pigments from actinomadurae and streptomycetes[a]

Structure		Name	Source
	$R = -(CH_2)_8 \cdot CH_3$	Nonylprodiginine	Actm. madurae, Actm. pelletieri
	$R = -(CH_2)_{10} \cdot CH_3$	Undecylprodiginine	Actm. pelletieri, Streptomyces sp. CWW6, Stmy. longisporuber, Streptoverticillium sp. 26-1
	$R = -(CH_2)_9-$	Cyclononylprodiginine	Actm. madurae
	$R = -CH \cdot (CH_2)_9-$ $\quad\;\; \mid$ $\quad\;\; CH_3$	Methyl cyclodecyl-prodiginine	Actm. pelletieri, Streptoverticillium sp. 26-1
	$R = -CH \cdot (CH_2)_8-$ $\quad\;\; \mid$ $\quad\;\; CH_2 \cdot CH_3$	Ethyl cyclononyl-prodiginine	
		Metacycloprodigiosin	Streptomyces sp. 13-4, Stmy. longisporuber
		Butylcycloheptyl-prodigiosin	Streptomyces sp. Y-42, Streptoverticillium sp. 26-1

[a] Data from Gerber (1975), Gerber and Lechevalier (1976), Wasserman et al. (1976a, b).

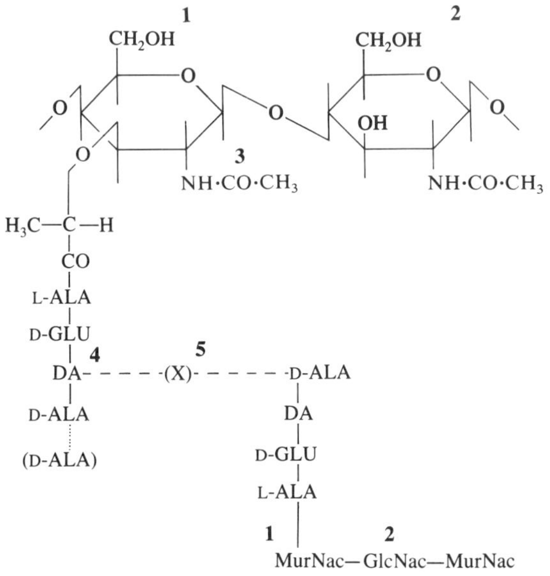

```
         1                    2
       CH₂OH                CH₂OH
      ┌───O                ┌───O
   O─┤    ├─O─┐         ┌─┤    ├─O
    ─┤    ├─  ├─O─┐  OH ─┤    ├─
      └───┘    └──┘       └───┘
            3
         NH·CO·CH₃            NH·CO·CH₃
  H₃C─C─H
      │
      CO
      │
    L-ALA
      │
    D-GLU
      │ 4         5
    DA ─ ─ ─ (X) ─ ─ ─ ─ D-ALA
      │                    │
    D-ALA                  DA
      ⋮                    │
   (D-ALA)               D-GLU
                          │
                        L-ALA
                  1       │  2
                   MurNac—GlcNac—MurNac
```

FIG. 1. Schematic representation of the primary structure of a typical peptidoglycan. Notes: 1, N-acetylmuramic acid; 2, N-acetylglucosamine; 3, replaceable by N-glycolyl; 4, diamino acid; 5, interpeptide bridge, replaceable by direct linkage.

L-proteases, such as trypsin and chymotrypsin, which are often used to remove contaminating protein from wall preparations. The peptide substituent is bound through its N-terminus to the carboxyl group of muramic acid (Fig. 1). Usually L-alanine is bound to muramic acid followed by D-glutamic acid, which in turn is linked to an L-diamino acid at position 3. Position 4 is D-alanine. The free amino group of the L-diamino acid either forms a peptide bond to the carboxyl group of D-alanine on an adjacent peptide subunit or is substituted through an interpeptide bridge. These interpeptide bridges cross-link the peptide subunits and usually extend between amino acids at positions 3 and 4. In an a minority of cases, they extend from the free carboxyl of D-glutamic acid to D-alanine. Most variations of the peptide moiety occur in the interpeptide bridge and in the method of cross-linkage (Schleifer and Kandler, 1972; Rogers et al., 1980).

Variations in peptidoglycan structure, first extensively studied by Cummins and Harris (1956), have been shown to be important in bacterial systematics (Schleifer and Kandler, 1972; Keddie and Bousfield, 1980; Seidl et al., 1980). In their original paper, Cummins and Harris (1956) noted that their test strains contained either lysine or diaminopimelic acid (DAP) in the wall and, like Work and Dewey (1953) who had previously investigated the distribution of DAP in bacteria, they thought the presence or absence of the diamino acid an important taxonomic character. Additional studies

TABLE 11. Distribution of acyl type of muramic acid among actinomycetes

Acyl type	Taxon
N-Glycolylmuramic acid	*Mycobacterium*,[a-f] *Nocardia*,[a,b,g,h] *Rhodococcus*,[b,i] *Actinoplanes*,[j] *Amorphosporangium*,[j] *Ampullariella*,[j] *Dactylosporangium*,[j] *Micromonospora*,[j] '*Corynebacterium*',[a,b,k]
N-Acetylmuramic acid	*Actinomadura*,[a,b] *Oerskovia*,[a,b] *Streptomyces*,[a,b,f] *Corynebacterium*[a,b]

[a] Uchida and Aida (1977).
[b] Uchida and Aida (1979).
[c] Petit et al. (1969).
[d] Petit et al. (1970).
[e] Adam et al. (1969).
[f] Azuma et al. (1970).
[g] Guinand et al. (1970).
[h] Bordet et al. (1972).
[i] Glycolylation suppressed by glycerol in medium. Organisms studied were labelled as strains of '*Gordona*' but are probably members of *Rhodococcus* (Goodfellow and Alderson, 1977).
[j] Kawamoto et al. (1981).
[k] These organisms are probably members of *Rhodococcus* (Goodfellow and Alderson, 1977).

demonstrated that DAP-containing organisms could be further differentiated according to whether they possessed the *meso* or LL isomer (Hoare and Work, 1956, 1957; Cummins and Harris, 1958). The discovery that certain taxa contained not DAP but other diamino acids, such as ornithine, lysine, diaminobutyric acid or hydroxydiaminopimelic acid (Perkins and Cummins, 1964; Perkins, 1965), in their walls has established the diamino acid as the major taxonomic character provided by qualitative wall analyses. This situation has, to a certain extent, dictated the development of wall analysis and many procedures have been designed primarily to determine the diamino acid composition (Becker et al., 1964; Johnson and Cummins, 1971; Berd, 1972; Staneck and Roberts, 1974). In addition, other investigations have increased the taxonomic potential of wall analysis by elucidating the composition of the peptide subunit. In general, such analyses have relied on either paper chromatography (Boone and Pine, 1968) or thin-layer chromatography (Richter, 1977; Harper and Davis, 1979) and as such have provided only qualitative data.

Quantitative wall profiles have usually been obtained using an amino acid analyser (Schleifer and Kandler, 1972; Beaman, 1975; Seidl et al., 1980; Kawamoto et al., 1981) although the analysis of amino acids by gas chromatography (Moss et al., 1971; Hŭsek and Macek, 1975; Ogata et al., 1975; O'Donnell et al., 1982) seems a promising and simplified approach to obtaining quantitative data. Further taxonomic information can be obtained by investigating the amino acid sequence and mode of cross-linkage. There have only been two attempts to classify peptidoglycans in this way, the first by Ghuysen (1968) and the second by Schleifer and Kandler (1972). Both systems regard the mode of cross-linkage as being most important, but the classification proposed by Schleifer and Kandler is the more comprehensive and can be divided into three parts. In the first of these, either the capital letter A or B represents the major class of cross-linking (Table 12), the second, a number, refers to the type of bridge or lack of it and the third, a Greek letter, identifies the amino acid found at position 3 in the primary chain. A summary of the classification of

TABLE 12. Peptidoglycan classification[a]

Position of cross-link	Peptide bridge	Amino acid at position 3
Peptidoglycan A Cross-linkage between positions 3 and 4 of two peptide subunits	1. None	α L-lysine β L-ornithine γ meso-diaminopimelic acid
	2. Polymerized subunits	α L-lysine
	3. Monocarboxylic L-amino acid or glycine or oligopeptides thereof	α L-lysine β L-ornithine γ LL-diaminopimelic acid
	4. Contains a dicarboxylic amino acid	α L-lysine β L-ornithine γ meso-diaminopimelic acid δ L-diaminobutyric acid
Peptidoglycan B Cross-linkage between positions 2 and 4 of two peptide subunits	1. Contains an L-amino acid	α L-lysine β L-homoserine γ L-glutamic acid δ L-alanine
	2. Contains a D-amino acid	α L-ornithine L-homoserine L-diaminobutyric acid

[a] According to Schleifer and Kandler (1972); modified by Rogers et al. (1980).

Schleifer and Kandler (1972) is set out in Table 12. The detailed determination of peptidoglycan types undoubtedly provides valuable taxonomic data but, because of the specialized techniques involved, such procedures have not been applied to large numbers of strains.

For the actinomycetes the introduction of systematic wall analysis has proven to be particularly important and many genera now contain details of their wall structure as part of their genus descriptions (see Chapter 2). The introduction of wall chemotypes (Lechevalier et al., 1966; Lechevalier and Lechevalier, 1970; Lechevalier, 1976) provided a method of classifying actinomycetes into a number of groups using qualitative wall data (Table 13). Partly on the basis of such data many strains formerly assigned to the genus *Nocardia* have been transferred to *Actinomadura*, *Oerskovia* and *Rothia* (Minnikin and Goodfellow, 1980). Simple wall composition analyses do not, however, distinguish between *Corynebacterium*, *Mycobacterium*, *Nocardia*, *Rhodococcus* and the '*aurantiaca*' taxon or between them and sporoactinomycete taxa such as *Actinopolyspora*, *Micropolyspora*, *Pseudonocardia*, *Saccharomonospora* and *Saccharopolyspora* (Minnikin and Goodfellow, 1980). Strains belonging to these taxa all have a wall chemotype IV, that is they contain major amounts of *meso*-DAP, arabinose and galactose. As noted in previous sections lipid analyses can be used to distinguish between various acid-fast and other bacteria with a wall chemotype IV (Minnikin and Goodfellow, 1980). These findings, together with the lack of correlation between morphology and wall chemotype (Williams and Wellington, 1980), suggest that, rather than grouping together taxa of similar wall type, this approach be used to exclude from a taxon organisms with a different wall chemotype, an important point recognized by Lechevalier et al. (1971).

4. Wall Membrane Lipids

The cell envelopes of mycobacteria, nocardiae, rhodococci, corynebacteria, 'aurantiaca' strains and related organisms contain substantial amounts of characteristic covalently bound long-chain fatty acids, the so-called mycolic acids (Minnikin and Goodfellow, 1980). These acids are considered to be esterified to an arabinogalactan linked to peptidoglycan (Misaki et al., 1974; Lederer et al., 1975). A variety of very complex free lipids may be associated with this wall lipid organelle, particularly in mycobacteria (Minnikin, 1982). Such a lipid-rich layer in the walls of mycolic acid-containing bacteria may be considered as a second membrane permeability barrier in addition to the plasma membrane.

Mycolic acids vary considerably in structure, ranging from relatively simple mixtures of saturated and unsaturated acids found in

TABLE 13. Relationship between wall chemotype[a] and peptidoglycan type[b] in actinomycetes

Wall type[a]	Major wall amino acids	Other distinguishing characteristics	Peptidoglycan type[b]	Genera[c]
I	gly, LL-DAP[d]	—	A3γ	*Streptomyces, Intrasporangium*
II	gly, *meso*-DAP or OH-DAP	—	A1γ	*Actinoplanes, Micromonospora*
III	*meso*-DAP	Madurose[e]	A1γ	*Actinomadura madurae, Dermatophilus, Microbispora*
IV	*meso*-DAP	Arabinose, galactose	A1γ	*Corynebacterium, Mycobacterium, Nocardia*
V	lys, orn, glu	—	—	*Actinomyces israelii*[f]
		—	—	*Actm. bovis*
VI	lys, asp — ser, thr	—	A4α	*Oerskovia xanthineolytica*[g]
	lys, glu, thr	Galactose	A4α	*Oerskovia turbata*[g]
	asp	—	A4β	*Cellulomonas flavigena*[g]
	orn — glu	—	A4β	*Cellulomonas cellasea,*[g] *Cell. gelida*[g]
VII	DAB, gly	—	B2α	*Agromyces*

[a] Based on Lechevalier and Lechevalier (1970).
[b] Abbreviations and partial data from Schleifer and Kandler (1972).
[c] Only representative genera. For more detailed list see Chapter 2.
[d] Abbreviations: asp, aspartic acid; DAB, diaminobutyric acid; DAP, diaminopimelic acid; glu, glutamic acid; gly, glycine; lys, lysine; orn, ornithine; ser, serine; thr, threonine.
[e] Only found in whole cells.
[f] Originally considered to be a type B peptidoglycan (Table 12) (Schleifer and Kandler, 1972) but recent results suggest an A-type (K. H. Schleifer, personal communication).
[g] Data from Seidl *et al.* (1980) and Stackebrandt *et al.* (1980).

Corynebacterium to very complex mixtures characteristic of mycobacteria (Minnikin and Goodfellow, 1980). The overall structures and distribution of representative non-mycobacterial mycolic acids are summarized in Table 14. Mycobacterial mycolic acids lacking oxygen functions, in addition to the hydroxy acid unit, are usually termed α-mycolic acids and in Table 15 a selection of these are given. These α-mycolates often occur with a selection of acids containing oxygen functions such as keto, methoxy and wax-ester functions (Minnikin, 1982; Minnikin and Goodfellow, 1980). The distribution of these types of mycolates can be readily assessed by thin-layer chromatography of whole-organism acid methanolysates, wax-ester mycolates being detected as ω-carboxymycolates and 2-eicosanol and related alcohols (Minnikin *et al.*, 1980). In this latter study, characteristic new types of oxygenated mycolic acids were discovered in methanolysates of *Mybc. fortuitum* and *Mybc. smegmatis*. Structural studies on these novel mycolates revealed that they were derived by acid methanolysis of alkali-stable mycolates containing a *trans*-epoxide function (Minnikin *et al.*, 1982); in independent studies, an epoxymycolate has also been characterized from *Mybc. fortuitum* (Daffé *et al.*, 1981). The structures of representative oxygenated mycobacterial mycolates are provided in Table 16; further details may be found in a recent review (Minnikin, 1982). Studies of a wide range of mycobacteria have shown that four principal patterns of mycolic acids are found (see Chapter 2) and these are summarized in Table 17. A distinct pattern, consisting only of α- and ketomycolates, is characteristic of *Mybc. leprae* (Young, 1980, 1982; Draper *et al.*, 1982) but the presence of a methoxymycolate has also been suggested (Asselineau *et al.*, 1981).

Considerations of the range of structures of mycolic acids in natural mixtures and some physical studies on trehalose mycolates (Durand *et al.*, 1979) resulted in the recent proposal (Minnikin, 1982) that these acids interact, in mycobacteria, to produce a functional external lipid permeability barrier. The general arrangement of the proposed mycolic acid-based matrix is shown in Fig. 2 which also shows the incorporation of certain complex free lipids whose structures will be given later. As indicated in Fig. 2 the way that mycolic acids interact with each other may produce a relatively closely packed inner permeability barrier. The outer regions of the mycolic acid-based lipid matrix are more open but the relatively regular spacing of the points of unsaturation and oxygen functions in mycobacterial mycolates (Tables 15 and 16) indicates that two further distinct zones may exist. The so-called 'parallel binding region' (Fig. 2) could act as an anchorage for the fatty acid chains of complex free lipids and some of these surface lipids might interact further with the terminal mycolate chains in the 'hydrophobic interaction region'. The very characteristic free lipids associated with mycobacterial wall membranes have very unusual structures and examples will be provided in the following sections.

TABLE 14. Non-mycobacterial mycolic acids[a]

General structure[b]

$$CH_3 \cdot (CH_2)_l \cdot [CH\!=\!\!CH \cdot (CH_2)_m]_n \cdot \underset{\underset{\textstyle cis}{}}{C}H \cdot \underset{}{C}H \cdot (CH_2)_x \cdot [CH\!=\!\!CH \cdot (CH_2)_y]_z \cdot CH_3$$
$$\overset{OH}{}\overset{COOH}{}$$
$$cis \qquad\qquad\qquad cis$$

l, m, x, y all > 1; $0 \leq n \leq 4$; $z = 0$ or 1

Total number of carbons	Number of double bonds	Major esters released on pyrolysis	Taxon
22–32	0–2	8:0, 10:0	*Corynebacterium bovis*
28–36	0–2	12:0–18:0[c]	*Corynebacterium spp.*[d] *Caseobacter polymorphus*
34–52	0–3	12:0–16:0	*Rhodococcus coprophilus, Rodc. equi, Rodc. erythropolis, Rodc. rhodnii, Rodc. rhodochrous, Rodc. ruber*
46–54	0–3	16:0, 16:1, 18:0, 18:1	*Nocardia amarae*
46–60	0–3	12:0–18:0	*Nrda. asteroides, Nrda. brasiliensis, Nrda. otitidis-caviarum*
48–64	1–4	16:0, 18:0	*Rodc. bronchialis, Rodc. rubropertinctus* (syn *corallinus*) *Rodc. terrae*
68–74	1–5	20:0, 20:1, 22:0, 22:1	'*aurantiaca*' taxon

[a] Data from Minnikin *et al.* (1978), Minnikin and Goodfellow (1980), Collins *et al.* (1982), Goodfellow *et al.* (1982).
[b] Determined by pyrolysis gas chromatography or mass spectrometry of the methyl ester. $R' \cdot CH(OH) \cdot CH(R'') \cdot COOCH_3 \rightarrow R' \cdot CHO + R'' \cdot COOCH_3$.
[c] Monounsaturated esters also present in variable amounts (Collins *et al.*, 1982).
[d] Including a number of taxa bearing other labels (Collins *et al.*, 1982).

Dimycolates of trehalose, known also as 'cord-factors', have been the subject of intensive study since they have some toxicity and antitumour properties in conjunction with other mycobacterial cell wall components (Asselineau and Asselineau, 1978a; Goren and Brennan, 1979). Trehalose mycolates are also implicated in mycolic acid biosynthesis perhaps playing a dramatic role in transferring mycolic acids to and from the basal arabinogalactan in order to maintain a suitably balanced mycolic acid

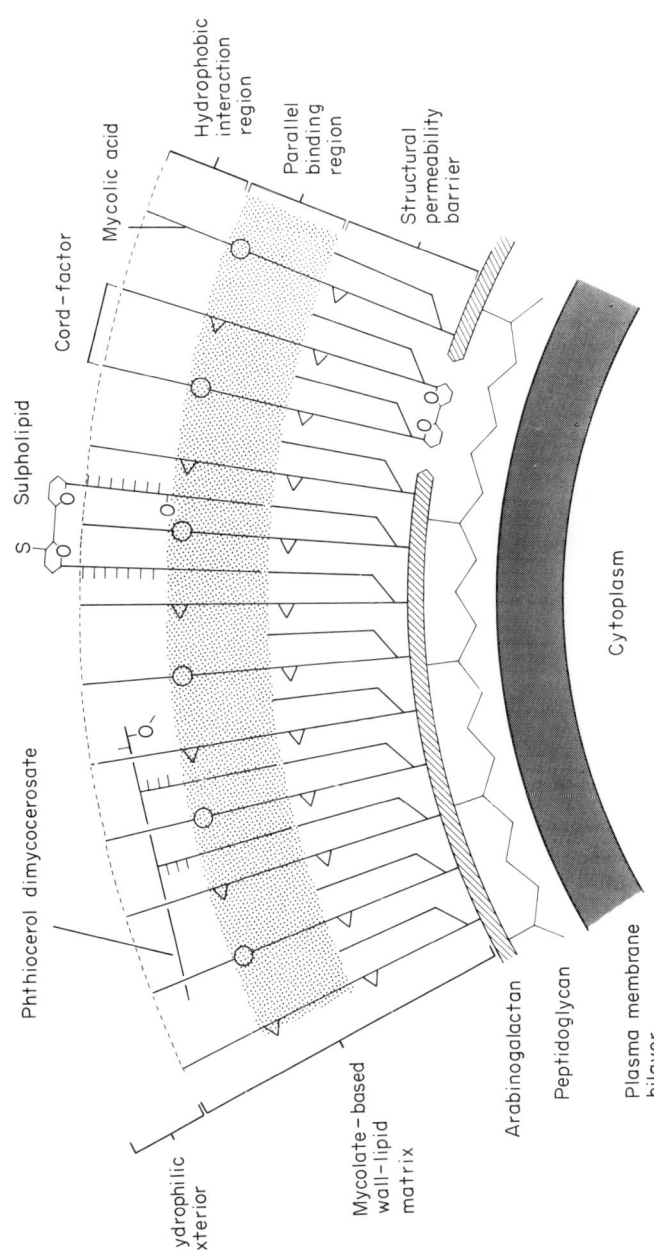

FIG. 2. Mycobacterial external membrane model, based on mycolic acids covalently bound to peptidoglycan-linked arabinogalactan. For a detailed discussion of the arguments leading to this chemical model see Minnikin (1982). Unsaturations, whether *cis* or *trans* double bonds or cyclopropane rings, are represented by triangles and oxygen functions are indicated by circles. (Reproduced with permission from *The Biology of the Mycobacteria* (1982). (C. Ratledge and J. L. Stanford, eds.), Academic Press, London.)

TABLE 15. Major components of representative mycobacterial α-mycolic acids[a]

Structure	Source
(a) $CH_3 \cdot (CH_2)_{17} \cdot CH=CH \cdot (CH_2)_{14} \cdot \underset{\underset{OH}{\|}}{CH} \cdot \underset{\underset{cis}{}}{CH}=CH \cdot (CH_2)_{17} \cdot \underset{\underset{COOH}{\|}}{CH} \cdot CH \cdot (CH_2)_{21} \cdot CH_3$	
(b) $CH_3 \cdot (CH_2)_{17} \cdot CH=CH \cdot (CH_2)_{13} \cdot \underset{\underset{cis}{}}{CH} \underset{\underset{CH_3}{\|}}{-} CH \cdot CH=CH \cdot (CH_2)_{17} \cdot \underset{\underset{OH}{\|}}{CH} \cdot \underset{\underset{trans}{}}{CH} \cdot CH \cdot (CH_2)_{21} \cdot CH_3$	Mybc. smegmatis
(c) $CH_3 \cdot (CH_2)_{17} \cdot CH \overset{CH_2}{\underset{}{\diagup\diagdown}} CH \cdot (CH_2)_{13} \cdot \underset{\underset{cis}{}}{CH} \underset{\underset{CH_3}{\|}}{-} CH \cdot CH=CH \cdot (CH_2)_{17} \cdot \underset{\underset{trans}{}}{CH} \cdot \underset{\underset{OH}{\|}}{CH} \cdot CH \cdot (CH_2)_{21} \cdot CH_3$	
(d) $CH_3 \cdot (CH_2)_{17} \cdot CH=CH \cdot (CH_2)_{17} \cdot \underset{\underset{cis}{}}{\underset{\underset{OH}{\|}}{CH}} \cdot CH \cdot (CH_2)_{21} \cdot CH_3$	

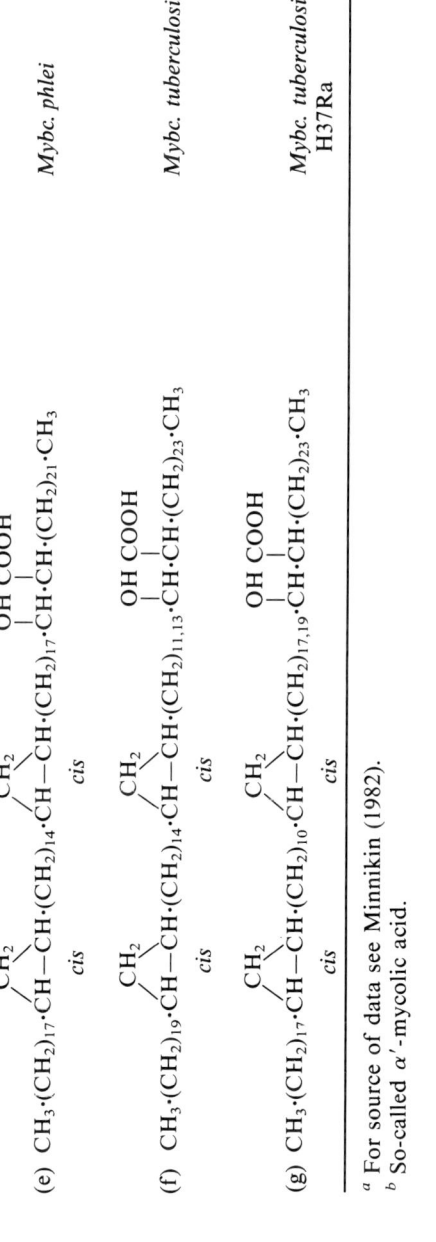

(e) $CH_3 \cdot (CH_2)_{17} \cdot \overset{\overset{\displaystyle CH_2}{\frown}}{CH-CH} \cdot (CH_2)_{14} \cdot CH \cdot (CH_2)_{17} \cdot \overset{\overset{\displaystyle OH \ COOH}{| \ \ |}}{CH-CH} \cdot (CH_2)_{21} \cdot CH_3$ Mybc. phlei
                                cis   cis

(f) $CH_3 \cdot (CH_2)_{19} \cdot \overset{\overset{\displaystyle CH_2}{\frown}}{CH-CH} \cdot (CH_2)_{14} \cdot CH \cdot (CH_2)_{11,13} \cdot \overset{\overset{\displaystyle OH \ COOH}{| \ \ |}}{CH-CH} \cdot (CH_2)_{23} \cdot CH_3$ Mybc. tuberculosis
                                  cis   cis

(g) $CH_3 \cdot (CH_2)_{17} \cdot \overset{\overset{\displaystyle CH_2}{\frown}}{CH-CH} \cdot (CH_2)_{10} \cdot CH \cdot (CH_2)_{17,19} \cdot \overset{\overset{\displaystyle OH \ COOH}{| \ \ |}}{CH-CH} \cdot (CH_2)_{23} \cdot CH_3$ Mybc. tuberculosis H37Ra
                                  cis   cis

[a] For source of data see Minnikin (1982).
[b] So-called α'-mycolic acid.

TABLE 16. Major components of representative mycobacterial oxygenated mycolic acids[a]

Type	Structure	Source
Keto	(a) $CH_3 \cdot (CH_2)_{17} \cdot \underset{\underset{CH_3O}{|}}{CH} \cdot \underset{\underset{O}{\|}}{C} \cdot (CH_2)_{16} \cdot \overset{\overset{CH_2}{\diagup \diagdown}}{CH — CH} \cdot (CH_2)_{19} \cdot \underset{\underset{OH}{\|}}{CH} \cdot \underset{\underset{COOH}{\|}}{CH} \cdot (CH_2)_{23} \cdot CH_3$ *cis* (b) $CH_3 \cdot (CH_2)_{17} \cdot \underset{\underset{CH_3O}{\|}}{CH} \cdot \underset{\underset{O}{\|}}{C} \cdot (CH_2)_{16} \cdot \overset{CH_3}{CH} — \overset{CH_2}{CH} \cdot (CH_2)_{18} \cdot \underset{\underset{OH}{\|}}{CH} \cdot \underset{\underset{COOH}{\|}}{CH} \cdot (CH_2)_{23} \cdot CH_3$ *trans*	*Mybc. tuberculosis*
Methoxy	(c) $CH_3 \cdot (CH_2)_{17} \cdot \underset{\underset{CH_3OCH_3}{\|}}{CH} \cdot CH \cdot (CH_2)_{16} \cdot \overset{\overset{CH_2}{\diagup \diagdown}}{CH — CH} \cdot (CH_2)_{17} \cdot \underset{\underset{OH}{\|}}{CH} \cdot \underset{\underset{COOH}{\|}}{CH} \cdot (CH_2)_{23} \cdot CH_3$ *cis* (d) $CH_3 \cdot (CH_2)_{17} \cdot \underset{\underset{CH_3OCH_3}{\|}}{CH} \cdot CH \cdot (CH_2)_{16} \cdot \overset{CH_3}{CH} — \overset{CH_2}{CH} \cdot (CH_2)_{16} \cdot \underset{\underset{OH}{\|}}{CH} \cdot \underset{\underset{COOH}{\|}}{CH} \cdot (CH_2)_{23} \cdot CH_3$ *trans*	

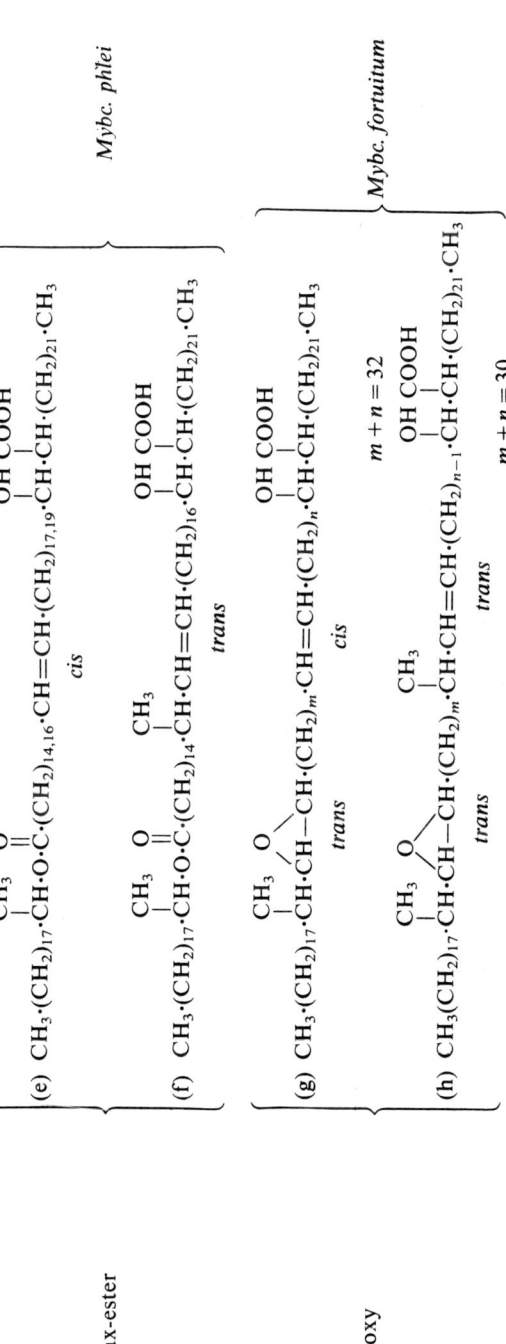

[a] For source of data see Minnikin (1982); Minnikin et al. (1982).

TABLE 17. Patterns of mycobacterial mycolic acids[a]

α-Mycolate (~80 carbons)
- +Ketomycolate
  - +Methoxymycolate
    - *Mybc. leprae*
    - *Mybc. tuberculosis, Mybc. bovis, Mybc. kansasii, Mybc. microti, Mybc. gordonae*
  - +Wax ester mycolate
    - *Mybc. avium, Mybc. intracellulare, Mybc. paratuberculosis, Mybc. phlei*
- +Epoxymycolate
  - *Mybc. farcinogenes, Mybc. fortuitum, Mybc. senegalense, Mybc. smegmatis*
- +α'-Mycolate (~62 carbons)
  - *Mybc. chelonei*

[a] For source of data see Minnikin (1982), Minnikin *et al* (1982).

composition (Minnikin, 1982). As indicated in Fig. 2 cord-factors may also assist in maintaining the integrity of the outer wax layers by plugging any gaps between the bound mycolic acid chains (Minnikin, 1982).

A selection of mycobacterial dimycolates of trehalose are shown in Table 18; monomycolates of trehalose have also been characterized from *Mybc. tuberculosis* (Kato and Maeda, 1974) and *Mybc. phlei* (Promé et al., 1976). Dimycolates of trehalose have been isolated from other mycolic acid-containing taxa, as reviewed by Asselineau and Asselineau (1978a) and Minnikin and Goodfellow (1980). In more recent studies, dimycolates of trehalose were partially characterized from nocardiae, rhodococci (Silva et al., 1979) and corynebacteria (Ioneda and Silva, 1979). In a particularly interesting study (Pommier and Michel, 1979), it was shown that trehalose mono esters contained mono and diunsaturated 52 to 58 carbon mycolates but diesters of trehalose had only diunsaturated 54 to 58 carbon mycolates. Trehalose mycolates are also excreted into culture media in substantial amounts in certain corynebacteria and rhodococci and their surfactant properties may be of value in the uptake of hydrocarbon substrates (Suzuki et al., 1969; Rapp et al., 1979; Cooper and Zajic, 1980).

Characteristic glycolipids, known as 'mycosides' are found in certain mycobacteria and are considered to be located on the cell surface since there is a clear correlation between their presence and colony morphology (Goren and Brennan, 1979; Minnikin, 1982). The so-called C-mycosides (Table 19) are acetylated glycopeptidolipids and, by means of a mild alkali deacetylation procedure, it has been shown that two separate families of relatively polar and apolar lipids occur together (Goren and Brennan, 1979; Brennan, 1981). The polar glycopeptidolipids are serologically active and type specific and correspond to the Schaefer antigens used in a seroagglutination procedure for the identification of representatives of *Mybc. avium*, *Mybc. intracellulare* and *Mybc. scrofulaceum*. Using specifically devised thin-layer chromatographic systems (Brennan et al., 1978, 1982) it was possible to obtain characteristic patterns of deacetylated polar C-mycosides for individual serotypes of this group of bacteria. A limited number of structural studies have been carried out on C-mycosides (Asselineau and Asselineau, 1978b; Goren and Brennan, 1979; Brennan, 1981; Minnikin, 1982) but all these lipids appear to be based on an acylated tripeptide-alaninol core with attached sugar substituents; the essential structures of the glycopeptidolipids of *Mybc. intracellulare* serotype 9 are shown, as an example, in Table 19. A simplified form of C-mycoside lacking a sugar substituent on *allo*-threonine was characterized from *Mycobacterium* sp. 378, an organism associated with bovine farcy (Lanéelle et al., 1971).

The other class of mycobacterial mycosides are lipids based on a phenolphthiocerol lipid core and the four examples that have been isolated are shown in Table 20. Minor variants of the mycosides from *Mybc. kansasii*

TABLE 18. Mycobacterial trehalose dimycolates (cord-factors)

Structure: [trehalose dimycolate structure with CH₂O·Mycolate and O·Mycolate groups on two sugar rings]

Taxon	Cord factor[a]	Mycolic acid composition[b]	
Mybc. phlei	CS_A CS_B CS_C	di – α α + wax ester di – wax ester	Promé et al. (1976)
Mybc. bovis AN-5, Mybc. tuberculosis (Aoyama B and Peurois)	I II III IV V VI	di – α α + methoxy α + keto di – methoxy methoxy + keto di – keto	
Mybc. bovis BCG	I III VI	di – α α + keto di – keto	Strain et al. (1977)
Mybc. avium, Mybc. phlei	I III VI	di – α α + γ[c] di – γ[c]	
Mybc. smegmatis	—	di – α	Mompon et al. (1978)

[a] Components of total cord factors separated by thin-layer chromatography of trimethylsilyl ether derivatives.
[b] For examples of mycolic acid structures see Tables 13 and 14.
[c] Not determined whether these 'γ'-mycolic acids were of keto or wax ester type.

and Mybc. marinum, with mycolic acids esterified to sugars, have been characterized by Gastambide-Odier (1973a, b). A so-called attenuation indicator lipid, detected only in attenuated strains of Mybc. tuberculosis, is also shown in Table 20. Structurally-related waxes, the phthiocerol dimycocerosates, have been characterized from Mybc. tuberculosis (Table 21) and detected in Mybc. bovis (Randall and Smith, 1964; Koul and Gastambide-Odier, 1977), Mybc. marinum (Navalkar et al., 1965) and Mybc. leprae (Young, 1982). A distinct family of αβ-unsaturated 2,4,6-trimethyl acids, the mycolipenic acids, have been isolated only from Mybc. tuberculosis but their cellular location is unknown (Minnikin, 1982).

Multimethyl branched long-chain acids, the phthioceranic acid family, are the major acyl substituents of a group of sulpholipids from *Mybc. tuberculosis* based on trehalose sulphate (Table 22). It has been claimed recently that related sulpholipids occur widely in mycobacteria and nocardiae (Prabhudesai *et al.*, 1981). Complex mixtures of polyunsaturated phleic acid (Fig. 3) esters of trehalose have been characterized from *Mybc. phlei* and they have also been detected in *Mybc. smegmatis* (Asselineau *et al.*, 1969, 1972).

$$CH_3 \cdot (CH_2)_m \cdot (CH=CH \cdot CH_2CH_2)_n \cdot COOH$$

cis

$m = 14, n = 4-6$ or $m = 12, n = 5$ or 6

FIG. 3. Structures of phleic acids from *Mycobacterium phlei*.

In addition to the glycopeptidolipids found in certain mycobacteria (Table 19), sugar-free peptidolipids have been isolated from a few mycobacteria and other mycolic acid-containing actinomycetes (Table 23). '*Mycobacterium rubrum*', which is probably better placed with the rhodococci (Collins *et al.*, 1982), is a hydrocarbon-oxidizing organism whose surface-active peptidolipid (Table 23) is possibly involved in hydrocarbon uptake (Krasil'nikov and Koronelli, 1971).

The unusual structures of the free lipids from mycobacterial walls require special explanations regarding their interaction with other envelope components. It has been proposed that the long-chain components interact in a specific way with a bound monolayer of mycolic acids (Fig. 2) (Minnikin, 1982). In particular, it has been suggested that the multimethyl branched fatty acid constituents of lipids, such as the phthiocerol dimycocerosates (Table 21), phenol-phthiocerol mycosides (Table 20) and trehalose sulphates (Table 22), may be designed to be inserted to various depths into a mycolic acid matrix (Fig. 2) (Minnikin, 1982). The straight-chain portion of these mycocerosic and phthioceranic acids (Tables 20-22) may be associated with the ordered central portions of the mycobacterial mycolic acid main chains with the methyl branches interacting with the more disordered terminal mycolate chains. The phleic acid constituents of the trehalose phleates (Fig. 3) are also constructed to a similar pattern having long straight terminal chains with a more random portion composed, in this case, of unsaturated repeating units (Minnikin, 1982).

The balance between the lengths of the straight-chain portions and the methyl branched or unsaturated regions of the mycocerosic and phthioceranic acids or phleic acids, respectively, may well be important in governing the location of the parent lipids. The phthioceranic acids in the trehalose sulphates (Table 22) have relatively long (10 to 20 carbons)

TABLE 19. Mycobacterial alaninol tripeptide based glycopeptidolipids (C-mycosides)[a]

Structure: Fatty acyl·Phe·*allo*—Thr·Ala·alaninol—sugar$_1$
                                               |
                                            O—sugar$_2$

Taxon	Fatty acyl	Sugar$_1$	Sugar$_2$
Mycobacterium sp. 1217	$C_{25}H_{51}$·CH(OH)·CH_2·CO	2,3,4-Me-Rha[b]	Ac_2-6-d Tal
Mybc. scrofulaceum	$CH_3(C_nH_{2n-2})$·CH(OCH$_3$)·CO ($n = 28, 29, 30, 31$)	3,4-Me-Rha	Ac_2-6-d Tal
Mybc. smegmatis (neé *butyricum*)	$C_{23}H_{47}$·CH=CH·CH(OCH$_3$)·CH_2·CO	2,3,4-Me-Rha	Ac_2-6-d Tal
Mycobacterium sp. 378 (*Mybc. senegalense*)	$C_{25}H_{51}$·CH(OCH$_3$)·CH_2·CO	3-Me-Rha or 3,4-Me-Rha	None

Mybc. intracellulare serotype 9	Not determined	3-Me-Rha 3,4-Me-Rha	6-d Tal 3-Me-6-d Tal 6-d Tal	Apolar
			6-d Tal-(2$\overset{\alpha}{\leftarrow}$1)- Rha-(3$\overset{\alpha}{\leftarrow}$1)- 2,3-Me-Fuc(4$\overset{\alpha}{\leftarrow}$1) 2,3-Me-Fuc	Polar

[a] For source of data see Goren and Brennan (1979), Brennan (1981), Minnikin (1982).
[b] Abbreviations: 3-Me-Rha, 3-O-methylrhamnose; 3,4-Me-Rha, 3,4-di-O-methylrhamnose; 2,3,4-Me-Rha, 2,3,4-tri-O-methylrhamnose; Ac$_2$-6-d Tal, diacetyl-6-deoxytalose; 2,3-Me-Fuc, 2,3-di-O-methylfucose.

TABLE 20. Mycobacterial mycosides and waxes based on phenolphthiocerol

Structure:

$$\text{RO}-\underset{}{\bigcirc}-(CH_2)_n \cdot \underset{OR'}{CH} \cdot CH_2 \cdot \underset{OR'}{CH} \cdot (CH_2)_4 \cdot \underset{CH_3}{CH} \cdot \underset{OCH_3}{CH} \cdot CH_2 \cdot CH_3$$

Taxon	Mycoside	n	R'	R
Mybc. kansasii [a,b]	A	16, 17, 18	12:0–18:0, tuberculostearate, mycocerosate	2,4-di-O-methylrhamnose, 2-O-methylrhamnose, 2-O-methylfucose
Mybc. bovis [a]	B	14–18	16:0 mycocerosates	2-O-methyl-D-rhamnose
Mybc. marinum [a]	G	18–22	mycocerosates	α-3-O-methyl-L-rhamnose
Mybc. leprae [b]	'L'	Not determined	Not determined	3,6-di-O-methylglucose, 2,3-di-O-methylrhamnose, 3-O-methylrhamnose
Mybc. tuberculosis [a]	Attenuation indicator	16, 18	16:0, 18:0, mycocerosate	CH_3

[a] For detailed references see Minnikin (1982).
[b] Hunter and Brennan (1981).

TABLE 21. Phthiocerol dimycocerosate waxes of *Mycobacterium tuberculosis*[a]

Phthiocerol family	Mycocerosates	
	$\underset{}{\text{CH}_3\cdot(\text{CH}_2)_{20,22}\cdot\overset{\text{O}}{\overset{\|}{\text{C}}}\cdot\text{O}\cdot\text{CH}\cdot\text{CH}_2\cdot\overset{\text{O}}{\overset{\|}{\text{C}}}\cdot\text{O}\cdot\text{CH}\cdot(\text{CH}_2)_4\cdot\overset{\overset{\text{CH}_3}{\|}}{\text{CH}}}-\begin{bmatrix}\overset{\text{OCH}_3}{\|}\\-\text{CH}\cdot\text{CH}_2\cdot\text{CH}_3\\\overset{\text{OCH}_3}{\|}\\-\text{CH}\cdot\text{CH}_3\\\overset{\text{O}}{\|}\\=\text{C}\cdot\text{CH}_2\cdot\text{CH}_3\end{bmatrix}$	Phthiocerol A Phthiocerol B Phthiodiolone

Mycocerosic (mycoceranic) acids $\text{CH}_3\cdot(\text{CH}_2)_{21}\cdot\left[\overset{\overset{\text{CH}_3}{\|}}{\text{CH}_2\cdot\text{CH}}\cdot\right]_{3,4}\cdot\text{COOH}$

[a] For original references see Minnikin (1982).

TABLE 22. Acylated trehalose sulphates of *Mycobacterium tuberculosis* H37Rv[a]

Structure

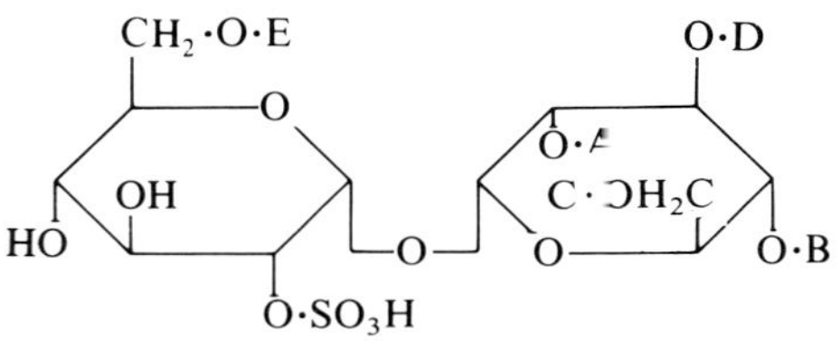

A–E = H, 16:0, 18:0, phthioceranate or hydroxyphthioceranate

Phthioceranic acids

$$CH_3\text{-}(CH_2)_{14}\text{-}CH_2\text{-}(CH\text{-}CH_2)_n\text{-}CH\text{-}COOH$$
with CH₃ and CH₃ branches (L)
$n = 4-9$

Hydroxyphthioceranic acids

$$CH_3\text{-}(CH_2)_{14}\text{-}CH_2\text{-}(CH\text{-}CH_2)_n\text{-}CH\text{-}COOH$$
with OH, CH₃ and CH₃ branches (L)
$n = 2$ or $4-9$

Order of elution[b]	Component	Trehalose substitution	16:● or 18:0	Relative acyl substitution Phthioceranate	Hydroxyphthioceranate
1	SL-II'	A,B,C,E	1	0	3
2	SL-II	A,B,D,E	1	0	3
3	SL-I	A,C,D,E	1	1	2
4	SL-I'	A,C,D,E	1	2	1
5	SL-III	A,C,D	1	0	2

[a] Data from Goren and Brennan (1979).
[b] Order of elution from diethylaminoethyl cellulose columns.

TABLE 23. Peptidolipids of mycolic acid-containing bacteria

Source	Structure	References
Nrda. asteroides	$\text{CH}_3\cdot(\text{CH}_2)_{16}\cdot\text{CH}\cdot\text{CH}_2\text{CO}\cdot\text{Thr}\cdot\text{Val}\cdot\text{Ala}$ (D, L, D) with Pro L; O-CO-Thr-[Ala or Val]-allo-Ileu (L, L, D); D L L L L	Guinand and Michel (1966); Guinand et al. (1966)
Mybc. paratuberculosis	$\text{CH}_3\cdot(\text{CH}_2)_{20}\cdot\text{CO}\cdot\text{Phe}\cdot\text{Ileu}\cdot\text{Ileu}\cdot\text{Phe}\cdot\text{Ala}\cdot\text{CH}_3$	Lanéelle (1969)
Mybc. fortuitum	$\text{CH}_3\cdot(\text{CH}_2)_{18-20}\cdot\text{CO}\cdot\text{Val}\cdot\text{MeLeu}\cdot\text{Val}_2\cdot\text{MeLeu}\cdot\text{AcThr}_2\cdot\text{Ala}\cdot\text{Pro}\cdot\text{CH}_3$	Barber et al. (1965)
'Mybc. rubrum'	Hydroxy fatty acid, Leu, Phe, Ala. Gly, Glu, Ser, Thr	Krasil'nikov and Koronelli (1971)

methyl-substituted portions so that the hydrophilic sugar sulphates may be projected outside the hydrophobic mycolic acid matrix (Minnikin, 1982). On the other hand, the shorter (6 to 8 carbons) methyl-branched chains of the mycocerosic acid substituents of the phthiocerol dimycocerosates (Table 21), phenol-phthiocerol mycosides and related lipids (Table 20) may be designed to locate the hydrophobic phthiocerol and phenol-phthiocerol units within the outer region of the mycolate matrix (Minnikin, 1982). Sugar substituents in the phenolphthiocerol mycosides (Table 20) could then project to the outside of the cell and express their specific effects on relations with host cells and colony morphology (Goren and Brennan, 1979; Minnikin, 1982). Waxes such as phthiocerol dimycocerosates (Table 21) are very inert and their main function may be principally to seal the cell surface and provide a resistant lipid barrier. The phenol-phthiocerol mycosides (Table 20) may be regarded as an 'activated wax' with the unusual sugar substituents expressing the characteristic biological activity (Hunter and Brennan, 1981; Minnikin, 1982). A similar relationship may exist between the 'apolar' and 'polar' varieties of C-mycosides (Table 19) found in other mycobacteria, the latter 'active' lipids being interspersed amidst a relatively inert basic region of apolar C-mycosides lacking characteristic sugars linked to *allo*-threonine-linked deoxytalose (Minnikin, 1982). These C-mycosides (Table 19) have single acyl chains available for interaction with a host mycolic acid matrix; the peptide units are relatively hydrophobic and may be associated with the terminal portions of the mycolate chains with the characteristic oligosaccharides protruding to the exterior (Minnikin, 1982). It has been proposed (Falk *et al.*, 1980) that glycolipid sulphates are implicated in selective uptake of potassium ions in halobacteria and such specific functions should also be considered for acylated trehalose sulphates (Table 22).

If the complex mycobacterial lipids interact with a bound mycolic acid matrix, in the manner indicated above (Fig. 2), to produce an effective outer membrane then it would be expected that the composition of the lipid constituents might vary systematically with changes in growth environment to maintain the integrity of this membrane. Indeed it has been shown that the length of the main chains of mycolic acids in *Mybc. phlei* increased with rises in growth temperature, it being notable that the size of the chain in 2-position remained relatively constant (Toriyama *et al.*, 1978, 1980). It would be of interest to study how the structures of both the complex free lipids such as the phthiocerol dimycocerosates (Table 21) or mycosides (Tables 19, 20) vary in relation to changes in mycolic acids with different growth environments.

The mycolic acids of mycobacteria (Tables 15, 16) are significantly different from those found in other taxa which are relatively simple mixtures having varying chain lengths and numbers of *cis* double bonds (Table 14).

It is reasonable to presume, however, that the bound mycolic acids in organisms other than mycobacteria interact together to produce an effective external membrane permeability barrier. Only limited information is available concerning the exact structures, particularly double bond position, of non-mycobacterial mycolic acids. The mycolic acids of *Cnbc. diphtheriae* PW8 are relatively simple being composed mainly of a 32-carbon saturated acid with lesser amounts of other homologous acids having a double bond in the mid-point of the main mycolate chain (Pudles and Lederer, 1954). The only detailed study of the position of the double bonds in the mycolic acids of *Nrda. asteroides* is that carried out by Bordet and Michel (1969). In Fig. 4 the mycolic acids of *Cnbc. diphtheriae* and *Nrda. asteroides* are arranged in a manner analogous to that proposed for mycobacterial mycolic acids (Fig. 2) (Minnikin, 1982).

It is notable that for the mycolic acids from *Nrda. asteroides* the length of the main chain as far as the first unsaturation is comparable with that of the branch in 2-position. This suggests that, as has been proposed for mycobacteria (Fig. 2) (Minnikin, 1982), the inner region is organized as a structural permeability barrier (Fig. 4). Differences in unsaturation between the various mycolic acid types in *Nrda. asteroides* appear to occur in the regions beyond the first unsaturation point (Bordet and Michel, 1969) and chain lengths increase with the number of double bonds for these types of mycolic acids (Maurice *et al.*, 1971; Minnikin *et al.*, 1974b; Yano *et al.*, 1978; Tomiyasu *et al.*, 1981). Discontinuities such as *cis* double bonds would give more disordered chains and the effective chain lengths might thereby be reduced; a triunsaturated mycolate might, therefore, protrude a similar distance to that for a saturated molecule. The outer regions of the mycolic acid matrix, in an organism such as *Nrda. asteroides*, might be relatively disordered and have a more fluid gel-like physical character in contrast to the more closely packed inner regions. Evidence for the importance of maintaining suitable physical properties of mycolate-based outer membranes in nocardiae and related organisms is provided by studies on the effect of growth temperature changes on the mycolic acid composition of '*Nocardia rubra*' (Tomiyasu *et al.*, 1981); an organism that should probably be placed in the genus *Rhodococcus* (Goodfellow and Alderson, 1977). It was found that for '*Nrda. rubra*' the degree of unsaturation increased dramatically as the temperature was decreased from 40 to 15°C.

The mycolic acids of *Cnbc. diphtheriae* have main chains comparable in size with the branch in 2-position so that both chains may be interacting together primarily to give an effective structural permeability barrier (Fig. 4). Though the main component of the mycolates of *Cnbc. diphtheriae* PW8 is a saturated 32-carbon acid, lesser amounts of other homologues and monounsaturated acids are also found (Pudles and Lederer, 1954; Minnikin and Goodfellow, 1980). The presence of these additional minor components

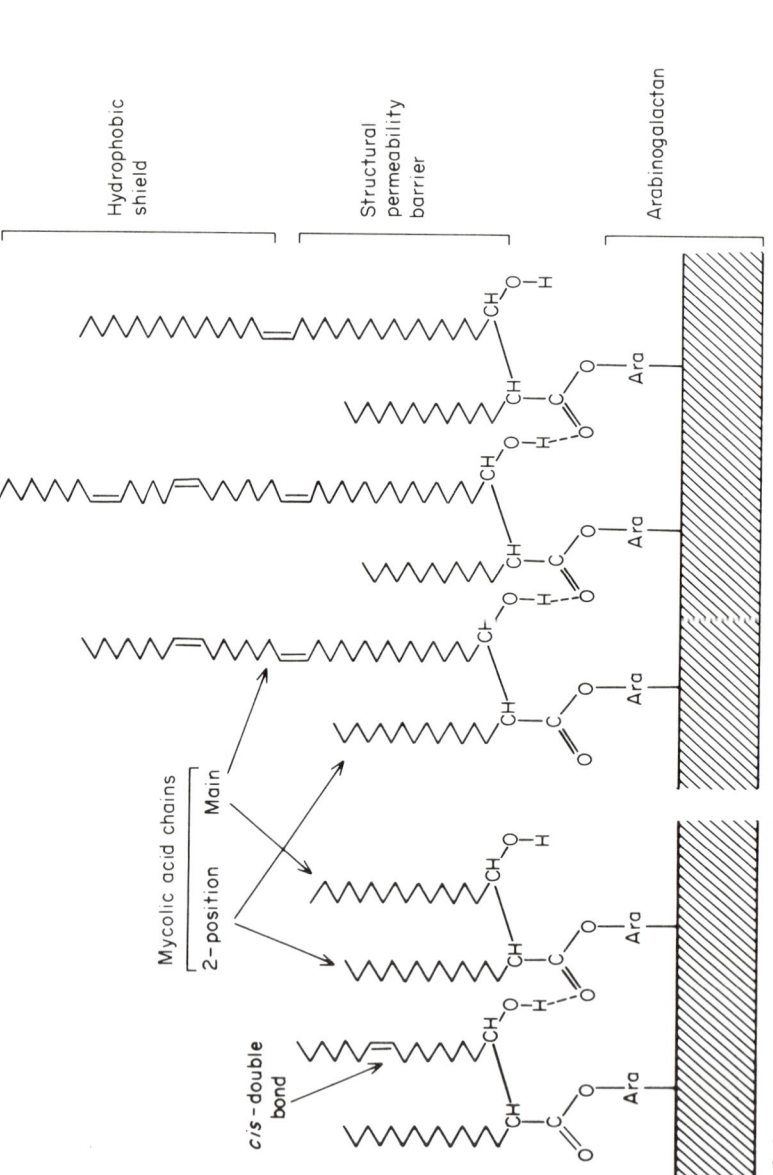

FIG. 4. Possible interactions of mycolic acids bound to arabinogalactan in envelopes of (a) *Corynebacterium diphtheriae* PW8 and (b) *Nocardia asteroides*. Mycolic acid chains are drawn to scale and are selected from natural chain lengths (Pudles and Lederer, 1954; Bordet and Michel, 1969). The overall arrangement of the mycolic acid matrix is similar to that proposed for mycobacteria (Minnikin, 1982) (see also Fig. 2).

will contribute to the production of suitable physical properties preventing over-condensation of mycolate chains. The mycolic acids of bacteria such as *Nrda. amarae* and '*aurantiaca*' strains have major proportions of compounds having a double bond in the chain in 2-position (Table 14). This suggests that in these organism the internal region of the mycolate matrix is less close-packed because of the presence of the double bonds.

The external membranes of non-mycobacterial mycolic acid containing actinomycetes appear to be essentially monolayers composed of balanced mixtures of saturated and unsaturated covalently bound mycolic acids. In contrast, many, but not necessarily all, mycobacteria have additional complex free lipids interacting with the mycolate chains to produce an effective bilayer. It has been suggested that nocardiae contain sulpholipids (Prabhudesai *et al.*, 1981), but no information is available concerning the acyl composition, so any relationship with a mycolic acid matrix cannot be assessed. It has been shown (Pommier and Michel, 1981) that the acidic glycolipid characterized from *Nrda. otitidis-caviarium* (neé *caviae*) (Table 5) is located in the wall rather than the plasma membrane. Amphipathic polar lipids in plasma membranes should have suitable mixtures of high and low melting acyl components (Tables 1, 2) so that the fact that only high-melting straight-chain acids are present in this acidic glycolipid (Table 5) supports an alternative location. The relatively short straight-chain acyl composition of the acidic glycolipid from '*Mybc. paraffinicum*' (Table 5) (Batrakov *et al.*, 1977) also points to a location away from the plasma membrane, perhaps in association with the wall mycolic acid matrix (Fig. 4). It is also possible that the peptidolipids from *Nrda. asteroides* and '*Mybc. rubrum*' (Table 23) are components of a wall membrane rather than a plasma membrane system.

5. Conclusions

The diversity of actinomycete lipid and peptidoglycan composition has been outlined in the preceding sections. The envelopes of these Gram-positive, morphologically diverse bacteria can be divided into those having a bound mycolic acid-based outer membrane and those apparently lacking such an organelle. Further clear subdivisions of the mycolic acid-containing organisms are possible according to the composition of characteristic lipid components such as mycolic acids (Tables 14–17), long-chain non-hydroxylated acids (Table 2) and menaquinones (Table 8).

Mycolic acid-containing actinomycetes can be separated into a number of envelope chemotypes as indicated in Table 24 and further detailed analyses may well enable further refinements of these groups to be made. Studies of the mycolic acids and complex outer membrane lipids allow

TABLE 24. Envelope chemotypes of mycolic acid-containing actinomycetes [a]

Peptido-glycan	Polar lipids		Menaquinones	Mycolic acids		Representative taxa
	Structural type [b]	Fatty acids [c]		No. of carbons	Acid released on pyrolysis	
A1γ {DPG, PI, PIM}	+PE	S,U,T	MK-9	68–74	20, 22[d]	'*aurantiaca*' taxon
			MK-8(H$_4$)	40–60	12–18	*Nocardia asteroides*
			MK-8(H$_2$)	34–52	12–16	*Rhodococcus rhodochrous*
				48–64	16, 18	*Rhodococcus bronchialis*
			MK-9(H$_2$)	60–90	22–26	*Mycobacterium*
				46–54	16, 18[d]	*Nocardia amarae*
				22–32	8, 10	*Corynebacterium bovis*
	−PE	S,U	MK-8(H$_2$)	28–36	12–18	*Corynebacterium diphtheriae*
			MK-9(H$_2$)	28–36	12–18	*Corynebacterium glutamicum*

[a] For source of data see Tables 2, 7, 13, 14, 17.
[b] For abbreviations see Table 7.
[c] Abbreviations: S, straight-chain; U, unsaturated; T, tuberculostearic (see Tables 1 and 2).
[d] Comparable proportions of monounsaturated components also present.

mycobacteria to be assigned to different envelope chemotypes as shown in Table 25. Substantiation of these mycobacterial chemotypes require the analysis of many more strains and the development of rapid small-scale analytical techniques.

The lack of comprehensive data also limits the reliable assignment of precise envelope chemotypes among actinomycetes lacking mycolic acids; a representative selection is shown, however, in Table 26. A particularly interesting group of bacteria are those exemplified by '*Nrda. autotrophica*'; these organisms apparently have polysaccharides containing arabinose and galactose but lack mycolic acids. Since an arabinogalactan is considered in

TABLE 25. Envelope chemotypes of mycobacteria[a]

Mycolic acid pattern[b]	Representative species	Characteristic outer membrane free lipids
A,B,C	*Mybc. tuberculosis*	Sulpholipids, methylphenolphthiocerol and phthiocerol dimycocerosates
	Mybc. bovis	
	Mybc. kansasii	Phenolphthiocerol mycosides
	Mybc. microti	?
	Mybc. gordonae	?
A,C	*Mybc. leprae*	Phenolphthiocerol mycosides, phthiocerol dimycocerosates
A,C,D	*Mybc. avium*	Glycopeptidolipid mycosides
	Mybc. intracellulare	
	Mybc. paratuberculosis	Peptidolipid
	Mybc. phlei	Trehalose phleates
A,E	*Mybc. fortuitum*	Peptidolipid[c]
	Mybc. farcinogenes	?
	Mybc. senegalense	Glycopeptidolipid mycosides[d]
	Mybc. smegmatis	
A,A$_1$	*Mybc. chelonei*	?

[a] For source of data see Tables 15, 16, 17, 19, 20, 21, 22, 23.
[b] Abbreviations: A, α-mycolate; A$_1$, α'-mycolate; B, methoxymycolate; C, ketomycolate; D, wax-ester mycolate; E, epoxymycolate (see Tables 15–17).
[c] Only in a single strain.
[d] Further studies needed to substantiate occurrence in more strains.

TABLE 26. Representative envelope chemotypes of actinomycetes lacking mycolic acids

Characteristic wall amino acids[a]	Peptidoglycan type[a]	Other wall characters[a]	Polar lipids		Menaquinones[d]	Representative taxa[b]
			Structural type[b]	Fatty acids[c]		
LL-DAP	A3γ	gly	PI, PIM, PE	S, I, A	MK-9(H_4, H_6, H_8)	*Streptomyces*
meso-DAP	A1γ	gly, OH-DAP	PI, PIM, PE	S, U, T, I, A	ND	*Micromonospora*
	A1γ	madurose	PI, PIM	S, U, T	MK-9(H_4, H_6, H_8)	*Actinomadura*
	?	arabinose, galactose	PI, PIM, PE, PME, PC	S, U, T, I, A	MK-8(H_4)	'*Nrda.*' *autotrophica*
Orn	A4β	asp or glu	PI, PIM?	S, I, A	MK-9(H_4)	*Cellulomonas*
Lys	A4α	asp or glu, ser or thr	PI, PIM?	S, I, A	MK-9(H_4)	*Oerskovia*

[a] See Table 13 for data and abbreviations.
[b] See Table 7 for data and abbreviations.
[c] See Table 2 for data. Abbreviations: S, straight chain; I, *iso*; A, *anteiso*; U, unsaturated; T, tuberculostearic.
[d] See Table 8 for data and abbreviations.

other bacteria to provide an anchorage for a mycolic acid-based outer membrane system (Fig. 2) the role of such polysaccharides in '*Nrda*'. *autotrophica* and related organisms should be clarified.

The genera *Cellulomonas* and *Oerskovia* (Table 26) provide an example of how morphologically very different organisms may have very closely related envelope chemical structure. The peptidoglycans of these bacteria have similar overall types (Table 26) but differ in the detailed amino acid composition (Table 13) (Seidl *et al.*, 1980; Stackebrandt *et al.*, 1980). A very close similarity between cellulomonads and oerskoviae was suggested by systematic lipid analysis (Minnikin *et al.*, 1979) (Tables 2, 7, 8), a conclusion reinforced by comparative analysis of ribosomal 16S RNA and DNA–DNA reassociation studies (Stackebrandt *et al.*, 1980).

Further systematic studies are necessary before a true understanding of the diversity and roles of the envelopes of actinomycetes and related bacteria can be approached. The chemical structures of many envelope polymers and lipids remain undetermined and their quantitative interrelationships are largely unknown. Studies on the effect of growth environment, including sporulation, would throw light on the role of particular components. Continuous culture systems, providing precisely controlled and systematically variable growth conditions have clarified the role of envelope components in bacilli and pseudomonads (Minnikin and Abdolrahimzadeh, 1974; Minnikin *et al.*, 1974a) and representative actinomycetes should be studied by such techniques. Another approach would be the selection and study of possible envelope mutants deficient in the synthesis of certain components. It would also be interesting to analyse the chemical composition of morphologically distinct parts of the same organisms, hyphae and substrate mycelia, for example.

Knowledge of how the envelopes of actinomycetes are synthesized, assembled and function is very limited at present. The data reviewed here, however, indicate that studies in this area would be very valuable in helping to understand the function of the envelopes of these taxa, which contain many industrially and clinically important bacteria.

6. Acknowledgements

Thanks are due to Dr. M. Goodfellow for his collaboration in recent studies on actinomycetes and to Doreen Moor for preparing the present manuscript. Recent studies on the chemistry of actinomycetes have been supported by grants from the Science and Engineering Research Council (GRA 88651 and GR/C/18830), Medical Research Council (G974/522/S and G7901343SB), World Health Organization (T16/181/L4/29), Royal Society and British Leprosy Relief Association (LEPRA).

REFERENCES

Adam, A., Petit, J. F. and Wietzerbin-Falszpan, J. (1969). L'acide N-glycolylmuramique, constituant des parois de *Mycobacterium smegmatis*: Identification par spectrometrie de masse. *FEBS Letters* **4**, 87–92.

Araki, Y., Nakatani, T., Hayashi, H. and Ito, E. (1971). Occurrence of non-N-substituted glucosamine residues in lysozyme-resistant peptidoglycan from *Bacillus cereus* cell walls. *Biochemical and Biophysical Research Communications* **42**, 691–697.

Araki, Y., Nakatani, T., Nakaoka, K. and Ito, E. (1972). Occurrence of N-non-substituted glucosamine residues in peptidoglycan of lysozyme-resistant cell walls from *Bacillus cereus*. *Journal of Biological Chemistry* **247**, 6312–6322.

Asselineau, J. (1966). *The Bacterial Lipids*. Hermann, Paris.

Asselineau, C. and Asselineau, J. (1978a). Lipides spécifiques des mycobactéries. *Annales de Microbiologie (Institut Pasteur)* **129A**, 49–69.

Asselineau, C. and Asselineau, J. (1978b). Trehalose-containing glycolipids. *Progress in the Chemistry of Fats and Other Lipids* **16**, 59–99.

Asselineau, C., Montrozier, H. and Promé, J. C. (1969). Présence d'acides polyinsatures dans une bactérie: isolement, à partir des lipides de *Mycobacterium phlei*, d'acide hexatriacontrapentaène-4,8,12,16,20-oique et d'acides analogues. *European Journal of Biochemistry* **10**, 580–584.

Asselineau, C. P., Montrozier, H. L., Promé, J. C., Savagnac, A. M. and Welby, M. (1972). Étude d'un glycolipide polyinsaturé synthétisé par *Mycobacterium phlei*. *European Journal of Biochemistry* **28**, 102–109.

Asselineau, C., Clavel, S., Clement, F., Daffé, M., David H., Lanéelle, M. A. and Promé, J. C. (1981). Constituants lipidiques de *Mycobacterium leprae* isolé de tatou infecté expérimentalement. *Annales de Microbiologie (Institut Pasteur)* **132A**, 19–30.

Azuma, I., Thomas, D. W., Adam, A., Ghuysen, J. M., Bonaly, R., Petit, J. F. and Lederer, E. (1970). Occurrence of N-glycolylmuramic acid in bacterial cell walls. *Biochimica et Biophysica Acta* **208**, 444–451.

Barber, M., Jolles, P., Vilkas, E. and Lederer, E. (1965). Determination of amino acid sequences in oligopeptides by mass spectrometry. I. The structure of fortuitine, an acyl nonapeptide methyl ester. *Biochemical and Biophysical Research Communications* **18**, 469–473.

Batrakov, S. G. and Bergelson, L. D. (1978). Lipids of the streptomycetes. Structural investigation and biological interrelation. *Chemistry and Physics of Lipids* **21**, 1–29.

Batrakov, S. C., Rosynow, B. V., Koronelli, T. V., Kozhuhova, R. A. and Bergelson, L. D. (1977). The lipids of mycobacteria. I. Unusual trehalose derivatives from *Mycobacterium paraffinicum*. *Bioorganicheskaya Khimiya* **3**, 55–67.

Beaman, B. L. (1975). Structural and biochemical alterations of *Nocardia asteroides* cell walls during its growth cycle. *Journal of Bacteriology* **123**, 1235–1253.

Becker, B., Lechevalier, M. P., Gordon, R. E. and Lechevalier, H. A. (1964). Rapid differentiation between *Nocardia* and *Streptomyces* by paper chromatography of whole-cell hydrolysates. *Applied Microbiology* **12**, 421–423.

Berd, D. (1972). Laboratory identification of clinically important aerobic actinomycetes. *Applied Microbiology* **25**, 665–681.

Boone, C. J. and Pine, L. (1968). Rapid method for characterisation of actinomycetes by cell wall composition. *Applied Microbiology* **16**, 279–284.

Bordet, C. and Michel, G. (1969). Structure et biogenèse des lipides à haut poids moléculaire de *Nocardia asteroides*. *Bulletin de la Société de chimie biologique* **51**, 527–548.

Bordet, C., Karahjoli, M., Gateau, O. and Michel, G. (1972). Cell walls of *Nocardia* and related actinomycetes: Identification of the genus *Nocardia* by cell wall analysis. *International Journal of Systematic Bacteriology* **22**, 251–259.

Brennan, P. J. (1981). Structures of the typing antigens of atypical antigens of atypical mycobacteria: A brief review of present knowledge. *Reviews of Infectious Diseases* **3**, 905–913.

Brennan, P. J., Souhrada, M., Ullom, B., McClatchy, J. K. and Goren, M. B. (1978). Identification of atypical mycobacteria by thin-layer chromatography of their surface antigens. *Journal of Clinical Microbiology* **8**, 374–379.

Brennan, P. J., Heifets, M. and Ullom, B. P. (1982). Thin-layer chromatography of lipid antigens as a means of identifying nontuberculous mycobacteria. *Journal of Clinical Microbiology* **15**, 447–455.

Brumfitt, W., Wardlaw, A. C. and Park, J. T. (1958). Development of lysozyme-resistance in *Micrococcus lysodiekticus* and its association with an increased *O*-acetyl content of the cell wall. *Nature, London* **181**, 1783–1784.

Collins, M. D. (1978). Lipids in coryneform taxonomy. Ph.D. Thesis University of Newcastle upon Tyne.

Collins, M. D. and Jones, D. (1981a). Distribution of isoprenoid quinone structural types in bacteria and their taxonomic implications. *Microbiological Reviews* **45**, 316–354.

Collins, M. D. and Jones, D. (1981b). Lipid composition of *Corynebacterium paurometabolum* (Steinhaus). *FEMS Microbiology Letters* **13**, 13–16.

Collins, M. D., Ross, H. N. M., Tindall, B. J. and Grant, W. D. (1981). Distribution of isoprenoid quinones in halophilic bacteria. *Journal of Applied Bacteriology* **50**, 559–565.

Collins, M. D., Goodfellow, M. and Minnikin, D. E. (1982). A survey of the structures of mycolic acids in *Corynebacterium* and related taxa. *Journal of General Microbiology* **128**, 129–149.

Cooper, D. G. and Zajic, J. E. (1980). Surface-active compounds from microorganisms. *Advances in Applied Microbiology* **26**, 229–253.

Cummins, C. S. and Harris, H. (1956). The chemical composition of the cell wall in some Gram-positive bacteria and its possible value as taxonomic character. *Journal of General Microbiology* **14**, 583–600.

Cummins, C. S. and Harris, H. (1958). Studies on the cell wall composition and taxonomy of Actinomycetales and related groups. *Journal of General Microbiology* **18**, 173–189.

Daffé, M., Lanéelle, M. A., Puzo, G. and Asselineau, C. (1981). Acide mycolique epoxydique: Un nouveau type d'acide mycolique. *Tetrahedron Letters* **22**, 4515–4516.

Draper, P., Dobson, G., Minnikin, D. E. and Minnikin, S. M. (1982). The mycolic acids of *Mycobacterium leprae* harvested from experimentally infected nine-banded armadillos. *Annales de Microbiolgie (Institut Pasteur)* **133B**, 39–47.

Durand, E., Welby, M., Lanéelle, G. and Tocanne, J. F. (1979). Phase behaviour of cord factor and related bacterial glycolipid toxins. *European Journal of Biochemistry* **93**, 103–112.

Falk, K. E., Karlsson, K. A. and Samuelsson, B. E. (1980). Structural analysis by mass spectrometry and NMR spectroscopy of the glycolipid sulphate from *Halobacterium salinarium* and a note on its possible function. *Chemistry and Physics of Lipids* **27**, 9–21.

Fleck, J., Mock, M., Minck, R. and Ghuysen, J. M. (1971). The cell envelope in *Proteus vulgaris* P-18. Isolation and characterization of the peptidoglycan component. *Biochimica et Biophysica Acta* **233**, 489–503.

Gastambide-Odier, M. (1973a). Variantes de mycosides caractérisées par des residus glycosidiques substitués par des chaines acyles. I: Spectres de masse des mycosides G' & A' peracetyles. *Organic Mass Spectrometry* **7**, 845–860.

Gastambide-Odier, M. (1973). Variantes de mycosides caractérisées par des residus glycosidiques substitués par des chaînes acyles. Nature mycolique des mycosides G'. *European Journal of Biochemistry* **33**, 81–86.

Gerber, N. N. (1975). Prodigiosin-like pigments. *Critical Reviews in Microbiology* **3**, 469–485.

Gerber, N. N. and Lechevalier, M. P. (1976). Prodiginine (prodigiosin-like) pigments from *Streptomyces* and other aerobic actinomycetes. *Canadian Journal of Microbiology* **22**, 658–667.

Ghuysen, J. M. (1968). Use of bacteriolytic enzymes in determination of wall structures and their role in cell metabolism. *Bacteriological Reviews* **32**, 425–464.

Goodfellow, M. and Alderson, G. (1977). The actinomycete genus *Rhodococcus*: a home for the '*rhodochrous*' complex. *Journal of General Microbiology* **100**, 99–122.

Goodfellow, M., Minnikin, D. E., Todd, C., Alderson, G., Minnikin, S. M. and Collins, M. D. (1982). Numerical and chemical classification of *Nocardia amarae*. *Journal of General Microbiology* **128**, 1283–1297.

Goren, M. B. and Brennan, P. J. (1979). Mycobacterial lipids: Chemistry and biologic activities. In: *Tuberculosis* (G. P. Youmans, ed.), pp. 63–193. W. B. Saunders Co., Philadelphia.

Guinand, M. and Michel, G. (1966). Structure d'un peptidolipide isolé de *Nocardia asteroides*, la peptidolipine NA. *Biochimica et Biophysica Acta* **125**, 75–91.

Guinand, M., Vacheron, M. J. and Michel, G. (1970). Structure des parois cellulaires des *Nocardia*. I. Isoelement et composition des parois de *Nocardia kirovani*. *FEBS Letters* **6**, 37–39.

Guinand, M., Vacheron, M. J., Michel, G., Das, B. C. and Lederer, E. (1966). Détermination de séquences d'acides aminés dans oligopeptides par la spectrométrie de masse − V. Structure de la val^6 − peptidolipine NA, nouveau peptidolipide de *Nocardia asteroides*. *Tetrahedron Supplement No. 7*, 271–276.

Harper, J. J. and Davis, G. H. G. (1979). Two-dimensional thin layer chromatography for amino acid analysis of bacterial cell walls. *International Journal of Systematic Bacteriology* **29**, 56–68.

Hayashi, H., Araki, Y. and Ito, E. (1973). Occurrence of glucosamine residues with free amino groups in cell wall peptidoglycan from bacilli as a factor responsible for resistance to lysozyme. *Journal of Bacteriology* **113**, 592–598.

Hoare, D. S. and Work, E. (1956). Distribution and metabolism of the stereoisomers of diaminopimelic acid in certain bacteria. *Journal of General Microbiology* **15**, xiii.

Hoare, D. S. and Work, E. (1957). The stereoisomers of α,ε-diaminopimelic acid. 2. Their distribution in the bacterial order Actinomycetales and in certain Eubacteriales. *Biochemical Journal* **65**, 441–447.

Hunter, S. W. and Brennan, P. J. (1981). A novel phenolic glycolipid from *Mycobacterium leprae* possibly involved in immunogenicity and pathogenicity. *Journal of Bacteriology* **147**, 728–735.

Ioneda, T. and Silva, C. L. (1979). Isolation and partial characterization of esters of trehalose from *Corynebacterium ovis* (*C. pseudotuberculosis*). *Chemistry and Physics of Lipids* **23**, 63–68.

Hůsek, P. and Macek, K. (1975). Gas chromatography of amino acids. *Journal of Chromatography* **113**, 139–230.

Johnson, J. L. and Cummins, C. S. (1971). Cell wall composition and deoxyribonucleic acid similarities among the anaerobic coryneforms, classical propionibacteria and strains of *Arachnia propionica*. *Journal of Bacteriology* **109**, 1047–1066.

Kato, M. and Maeda, J. (1974). Isolation and biochemical activities of trehalose-6-monomycolate of *Mycobacterium tuberculosis*. *Infection and Immunity* **9**, 8–14.

Kawamoto, I., Tetsuo, O. and Nara, T. (1981). Cell wall composition of *Micromonospora olivasterospora*, *Micromonospora sagamiensis*, and related organisms. *Journal of Bacteriology* **146**, 527–534.

Kawanami, J. (1971). Lipids of *Streptomyces toyocaensis*. On the structure of siolipin. *Chemistry and Physics of Lipids* **7**, 159–172.

Keddie, R. M. and Bousfield, I. J. (1980). Cell wall composition in the classification and identification of coryneform bacteria. In: *Microbiological Classification and Identification* (M. Goodfellow and R. G. Board, eds.), pp. 167–188. Academic Press, London.

Khuller, G. K. and Brennan, P. J. (1972). The polar lipids of some species of *Nocardia*. *Journal of General Microbiology* **73**, 409–412.

Koul, A. K. and Gastambide-Odier, M. (1977). Microanalyse rapide de dimycocérosate de phthiocérol, de mycosides et de glycérides dans les extraite à l'éther de pétrole de *Mycobacterium kansasii* et du BCG, souche Pasteur. *Biochimie* **59**, 535–538.

Krasil'nikov, N. A. and Koronelli, T. V. (1971). Nature of polar lipids from a paraffin-oxidizing culture of *Mycobacterium rubrum*. *Mikrobiologiya* **40**, 230–235 (English translation *Microbiology* **40**, 196–200).

Krinsky, N. I. (1976). Cellular damage initiated by visible light. In: *The Survival of Vegetative Microbes* (T. R. G. Gray and R. Postgate, eds.), pp. 209–239. University Press, Cambridge.

Kroppenstedt, R. M. and Kutzner, H. J. (1978). Biochemical taxonomy of some problem actinomycetes. *Zentralblatt für Bakteriologie, Parasitenkunde, Infektionskrankheiten und Hygiene. I. Abteilung, Supplement* **6**, 125–133.

Lanéelle, G. (1969). Mise en evidence d'une conformation stable d'un peptidolipide. *FEBS Letters* **4**, 210–212.

Lanéelle, G., Asselineau, J. and Chamoiseau, G. (1971). Présence de mycosides C' (formes simplifiées de mycoside C) dans les bactéries isolées de bovins atteints du farcin. *FEBS Letters* **19**, 109–111.

Lechevalier, H. A., Lechevalier, M. P. and Becker, B. (1966). Comparison of the chemical composition of cell walls of nocardiae with that of other aerobic actinomycetes. *International Journal of Systematic Bacteriology* **16**, 151–160.

Lechevalier, H. A., Lechevalier, M. P. and Gerber, N. N. (1971). Chemical composition as a criterion in the classification of actinomycetes. *Advances in Applied Microbiology* **14**, 47–72.

Lechevalier, M. P. (1976). The taxonomy of the genus *Nocardia*: Some light at the end of the tunnel. In: *The Biology of the Nocardiae* (M. Goodfellow, G. H. Brownell and J. Serrano, eds.), pp. 1–38. Academic Press, London.

Lechevalier, M. P. (1977). Lipids in bacterial taxonomy – a taxonomist's viewpoint. In: *CRC Critical Reviews in Microbiology* pp. 109–210. CRC Press, Ohio.

Lechevalier, M. P. and Lechevalier, H. (1970). Chemical composition as a criterion in the classification of aerobic actinomycetes. *International Journal of Systematic Bacteriology* **20**, 435–443.

Lechevalier, M. P., De Bièrve, C. and Lechevalier, H. (1977). Chemotaxonomy of aerobic actinomycetes: Phospholipid composition. *Biochemical Systematics and Ecology* **5**, 249–260.

Lederer, E., Adam, A., Ciorbaru, R., Petit, J-F. and Wietzerbin, J. (1975). Cell walls of mycobacteria and related organisms; chemistry and immunostimulant properties. *Molecular and Cellular Biochemistry* **7**, 87–104.

Liaaen-Jensen, S. and Andrewes, G. (1972). Microbial carotenoids. *Annual Review of Microbiology* **26**, 225–248.

Maurice, M. T., Vacheron, M. J. and Michel, G. (1971). Isolément d'acides nocardiques de plusieurs espèces de *Nocardia*. *Chemistry and Physics of Lipids* **7**, 9–18.

Minnikin, D. E. (1982). Lipids: Complex lipids, their chemistry, biosynthesis and roles. In *The Biology of the Mycobacteria* (C. Ratledge and J. L. Stanford, eds.), pp. 95–184. Academic Press, London.

Minnikin, D. E. and Abdolrahimzadeh, H. (1974). The replacements of phosphatidylethanolamine and acidic phospholipids by an ornithine-amide lipid and a minor phosphorus-free lipid in *Pseudomonas fluorescens* NCMB 129. *FEBS Letters* **43**, 257–260.

Minnikin, D. E. and Goodfellow, M. (1976). Lipid composition in the classification and identification of nocardiae and related taxa. In: *The Biology of the Nocardiae* (M. Goodfellow, G. H. Brownell and J. A. Serrano, eds.), pp. 160–219. Academic Press, London.

Minnikin, D. E. and Goodfellow, M. (1980). Lipid composition in the classification and identification of acid-fast bacteria. In: *Microbiological Classification and Identification* (M. Goodfellow and R. G. Board, eds.), pp. 189–256. Academic Press, London.

Minnikin, D. E. and Goodfellow, M. (1981). Lipids in the classification of actinomycetes. *Zentralblatt für Bakteriologie Parisitenkunde, Infektionskrankheiten und Hygiene. I. Abteilung, Supplement* **11**, 100–109.

Minnikin, D. E., Abdolrahimzadeh, H. and Baddiley, J. (1971). The interrelation of phosphatidylethanolamine and glycosyl diglycerides in bacterial membranes. *Biochemical Journal* **124**, 447–448.

Minnikin, D. E., Abdolrahimzadeh, H. and Baddiley, J. (1972). Variation of polar lipid composition of *Bacillus subtilis* (Marburg) with different growth conditions. *FEBS Letters* **27**, 16–18.

Minnikin, D. E., Abdolrahimzadeh, H. and Baddiley, J. (1974a). Replacement of acidic phospholipids by acidic glycolipids in *Pseudomonas diminuta*. *Nature, London* **249**, 268–269.

Minnikin, D. E., Patel, P. V. and Goodfellow, M. (1974b). Mycolic acids of representative strains of *Nocardia* and the "*rhodochrous*" complex. *FEBS Letters* **39**, 322–324.

Minnikin, D. E., Goodfellow, M. and Collins, M. D. (1978). Lipid composition in the classification and identification of coryneform and related taxa. In: *Coryneform Bacteria* (I. J. Bousfield and A. G. Callely, eds.), pp. 85–160. Academic Press, London.

Minnikin, D. E., Collins, M. D. and Goodfellow, M. (1979). Fatty acid and polar lipid composition in the classification of *Cellulomonas*, *Oerskovia* and related taxa. *Journal of Applied Bacteriology* **47**, 87–95.

Minnikin, D. E., Hutchinson, I. G., Caldicott, A. B. and Goodfellow, M. (1980). Thin-layer chromatography of methanolysates of mycolic acid-containing bacteria. *Journal of Chromatography* **188**, 221–233.

Minnikin, D. E., Minnikin, S. M. and Goodfellow, M. (1982). The oxygenated mycolic acids of *Mycobacterium fortuitum*, *M. farcinogenes* and *M. senegalense*. *Biochimica et Biophysica Acta* **712**, 616–620.

Misaki, A., Seto, N. and Azuma, I. (1974). Structure and immunological properties of D-arabino-D-galactans isolated from *Mycobacterium* species. *Journal of Biochemistry* **76**, 15–27.

Mompon, B., Federici, C., Toubiana, R. and Lederer, E. (1978). Isolation and structural determination of a "cord factor" (trehalose 6,6′-dimycolate) from *Mycobacterium smegmatis*. *Chemistry and Physics of Lipids* **21**, 97–101.

Moss, C. W., Diaz, F. J. and Lambert, M. A. (1971). Determination of diaminopimelic acid, ornithine, and muramic acid by gas chromatography. *Analytical Biochemistry* **44**, 458–461.

Navalkar, R. G., Wiegeshaus, E., Kondo, E., Kim, H. K. and Smith, D. W. (1965). Mycoside G, a specific glycolipid in *Mycobacterium marinum* (*balnei*). *Journal of Bacteriology* **90**, 262–265.

O'Donnell, A. G., Minnikin, D. E., Goodfellow, M. and Parlett, J. H. (1982). The analysis of actinomycete wall amino acids by gas chromatography. *FEMS Microbiology Letters* **15**, 75(E)–78(E).

Ogata, S., Tahara, Y. and Hongo, M. (1975). Chemical composition of cell wall peptidoglycan from *Clostridium saccharoperbutylacetonicum* studied with phage endolysins and gas chromatography. *Journal of General and Applied Microbiology* **21**, 65–74.

Pandhi, P. N. and Hammond, B. F. (1975). A glycolipid from *Rothia dentocariosa*. *Archives of Oral Biology* **20**, 399–401.

Pandhi, P. N. and Hammond, B. F. (1978). The polar lipids of *Actinomyces viscosus*. *Archives of Oral Biology* **23**, 17–21.

Perkins, H. R. (1965). 2,6-Diamino-3-hydroxypimelic acid in microbial cell wall mucopeptide. *Nature, London* **208**, 872–873.

Perkins, H. R. and Cummins, C. S. (1964). Ornithine and 2,4-diaminobutyric acid as components of the cell walls of plant pathogenic corynebacteria. *Nature, London* **201**, 1105–1107.

Petit, J. F., Adam, A., Wietzerbin-Falszpan, J., Lederer, E. and Ghuysen, J. M. (1969). Chemical structure of the cell wall of *Mycobacterium smegmatis*: I. Isolation and partial characterization of the peptidoglycan. *Biochemical and Biophysical Research Communications* **35**, 478–485.

Petit, J. F., Adam, A. and Wietzerbin-Falszpan, J. (1970). Isolation of UDP-N-glycoylmuramyl-(ALA, GLU, DAP) from *Mycobacterium phlei*. *FEBS Letters* **6**, 55–57.

Pommier, M. T. and Michel, G. (1979). Glyolipides des nocardiae. Isolement et caractérisation de mononocardomycolates et de dinocardomycolates de tréhalose dans *Nocardia caviae*. *Chemistry and Physics of Lipids* **24**, 149–155.

Pommier, M. T. and Michel, G. (1981). Structure of 2′, 3′-di-O-acyl-α-D-glucopyranosyl-(1 → 2)-D-glyceric acid, a new glycolipid from *Nocardia caviae*. *European Journal of Biochemistry* **118**, 329–333.

Prabhudesai, A. V., Kaur, S. and Khuller, G. K. (1981). Sulpholipids of *Nocardia* species – a preliminary report. *Indian Journal of Medical Research* **73**, 181–183.

Promé, J. C., Lacave, C. and Lanéelle, M. A. (1969). Sur les structures de lipides à ornithine de *Brucella melitensis* et de *Mycobacterium bovis*. *Compte rendu hebdomadaire des séances de l'Académie des sciences* **269C**, 1664–1667.

Promé, J. C., Lacave, C., Ahibo-Coffy, A. and Savagnac, A. (1976). Séparation et étude structurale des espèces moleculaires de monomycolates et de dimycolates de α-D-tréhalose présents chez *Mycobacterium phlei*. *European Journal of Biochemistry* **63**, 543–552.

Pudles, J. and Lederer, E. (1954). Sur l'isolement et la constitution chimique de l'acide coryno-mycolénique et de deux cetones de lipides de bacille diphtherique. *Bulletin de la Société de Chimie Biologique* **36**, 759–777.

Randall, H. M. and Smith, D. W. (1964). Characterisation of mycobacteria by infrared spectroscopic examination of their lipid fractions. *Zentralblatt für Bakteriologie, Parasitenkunde, Infektionskrankheiten und Hygiene. I. Abteilung* **194**, 190-201.

Rapp, P., Bock, H., Wray, V. and Wagner, F. (1979). Formation, isolation and characterization of trehalose dimycolates from *Rhodococcus erythropolis* grown on *n*-alkanes. *Journal of General Microbiology* **115**, 491–503.

Richter, G. (1977). Routine use of thin-layer chromatography for cell wall analysis of aerobic actinomycetes, including two strains from sediments of the North Sea. *Veroffentlichungen des Institut für Meeresforschung in Bremerhaven* **16**, 125–138.

Rogers, H. J., Perkins, H. R. and Ward, J. B. (1980). *Microbial Cell Walls and Membranes*. Chapman and Hall, London.

Rohmer, M., Bouvier, P. and Ourisson, G. (1979). Molecular evolution of biomembranes: Structural equivalents and phylogenetic precursors of sterols. *Proceedings of the National Academy of Sciences U.S.A.* **76**, 847–851.

Schleifer, K. H. and Kandler, O. (1972). Peptidoglycan types of bacterial cell walls and their taxonomic implication. *Bacteriological Reviews* **36**, 407–477.
Seidl, P. H., Faller, A. H. and Schleifer, K. H. (1980). Peptidoglycan types and cytochrome patterns of strains of *Oerskovia turbata* and *O. xanthineolytica*. *Archives of Microbiology* **127**, 173–178.
Silva, C. L., Gesztesi, J. L. and Ioneda, T. (1979). Trehalose mycolates from *Nocardia asteroides, Nocardia farcinica, Gordona lentifragmenta* and *Gordona bronchialis*. *Chemistry and Physics of Lipids* **24**, 17–25.
Stackebrandt, E., Häringer, M. and Schleifer, K. H. (1980). Molecular genetic evidence for the transfer of *Oerskovia* species into the genus *Cellulomonas*. *Archives of Microbiology* **127**, 179–185.
Staneck, J. L. and Roberts, G. D. (1974). Simplified approach to identification of aerobic actinomycetes by thin-layer chromatography. *Applied Microbiology* **28**, 226–231.
Strain, S. M., Toubiana, R., Ribi, E. and Parker, R. (1977). Separation of the mixture of trehalose 6,6'-dimycolates comprising the mycobacterial glycolipid fraction, 'P3'. *Biochemical and Biophysical Research Communications* **77**, 449–456.
Suzuki, T., Tanaka, K., Matsubara, I. and Kinoshita, S. (1969). Trehalose lipid and α-branched β-hydroxy fatty acid formed by bacteria grown on n-alkanes. *Agricultural and Biological Chemistry* **33**, 1619–1627.
Tabaud, H., Tisnovska, H. and Vilkas, E. (1971). Phospholipides et glycolipides d'une souche de *Micromonospora*. *Biochimie* **53**, 55–61.
Taber, H. (1980). Function of vitamin K_2 in microorganisms. In: *Vitamin K Metabolism and Vitamin K Dependant Proteins* (J. W. Suttie, ed.), pp. 177–187. University Park Press, Baltimore.
Thirkell, D. and Hunter, M. I. S. (1969). The polar carotenoid fraction from *Sarcina flava*. *Journal of General Microbiology* **58**, 293–299.
Tomiyasu, I., Toriyama, S., Yano, I. and Masui, M. (1981). Changes in molecular species composition of nocardomycolic acids in *Nocardia rubra* by the growth temperature. *Chemistry and Physics of Lipids* **28**, 41–54.
Toriyama, S., Yano, I., Masui, M., Kusunose, M. and Kusunose, E. (1978). Separation of C_{50-60} and C_{70-80} mycolic acid molecular species and their changes by growth temperature in *Mycobacterium phlei*. *FEBS Letters* **95**, 111–115.
Toriyama, S., Yano, I., Masui, M., Kusunose, E., Kusunose, M. and Akimori, N. (1980). Regulation of cell wall mycolic acid biosynthesis in acid-fast bacteria. I. Temperature-induced changes in mycolic acid molecular species and related compounds in *Mycobacterium phlei*. *Journal of Biochemistry* **88**, 211–221.
Turner, J. A. and Prebble, J. N. (1980). Protection of cell viability and respiratory quinone levels by carotenoid in *Micrococcus lysodeikticus* (*M. luteus*). *Journal of General Microbiology* **119**, 133–144.
Uchida, K. and Aida, K. (1977). Acyl type of bacterial cell wall: its simple identification by colorimetric method. *Journal of General and Applied Microbiology* **23**, 249–260.
Uchida, K. and Aida, K. (1979). Taxonomic significance of cell-wall type in *Corynebacterium-Mycobacterium-Nocardia* group by a glycolate test. *Journal of General and Applied Microbiology* **25**, 169–183.
Vacheron, M. J., Arpin, N. and Michel, G. (1970). Isolement d'esters de phleixanthophylle de *Nocardia kirovani*. *Compte rendu hebdomadaire des séances de l'Académie des sciences* **271C**, 881–884.
Vilkas, E. and Rojas, A. (1964). Sur les lipides de *Mycobacterium fortuitum*. *Bulletin de la Société de Chimie Biologique* **46**, 689–701.
Vilkas, E., Adam, A. and Senn, M. (1968). Isolement d'un nouveau type de diester de tréhalose à partir de *Mycobacterium fortuitum*. *Chemistry and Physics of Lipids* **2**, 11–16.
Wasserman, H. H., Keith, D. D. and Rodgers, G. C. (1976a). The structure of metacycloprodigiosin. *Tetrahedron* **32**, 1855–1861.
Wasserman, H. H., Rogers, G. C. and Keith, D. D. (1976b). Undecylprodigiosin. *Tetrahedron* **32**, 1851–1854.
Wheat, R. W. (1980). Composition, structure and biosynthesis of the bacterial cell envelope and energy storage polymers. In: *Zinsser Microbiology* (W. K. Joklik, H. P. Willet and D. B. Amos, eds.), pp. 106–134. Appleton Century Crofts, New York.

Wilkinson, S. G. (1968). Glycosyl diglycerides from *Pseudomonas rubescens*. *Biochimica et Biophysica Acta* **164**, 148–156.

Wilkinson, S. G. (1972). Composition and structure of the ornithine-containing lipid from *Pseudomonas rubescens*. *Biochimica et Biophysica Acta* **270**, 1–17.

Williams, S. T. and Wellington, E. M. H. (1980). Micromorphology and fine structure of actinomycetes. In: *Microbiological Classification and Identification* (M. Goodfellow and R. G. Board, eds.), pp. 139–165. Academic Press, London.

Work, E. and Dewey, K. L. (1953). The distribution of α,ε-diaminopimelic acid among various micro-organisms. *Journal of General Microbiology* **9**, 394–406.

Yamamoto, H. Y. and Bangham, A. D. (1978). Carotenoid organization in membranes. *Biochimica et Biophysica Acta* **507**, 119–127.

Yano, I., Kageyama, K., Ohno, Y., Masui, M., Kusunose, E. Kusunose, M. and Akimori, N. (1978). Separation and analysis of molecular species of mycolic acids in *Nocardia* and related taxa by gas chromatography mass spectrometry. *Biomedical Mass Spectrometry* **5**, 14–24.

Young, D. B. (1980). Identification of *Mycobacterium leprae*: use of wall-bound mycolic acids. *Journal of General Microbiology* **121**, 249–253.

Young, D. B. (1982). Mycobacterial lipids in infected tissue samples. *Annales de Microbiologie* (*Institut Pasteur*) **133B**, 53–58.

Yribarren, M., Vilkas, E. and Rozanis, J. (1974). Galactosyl diglyceride from *Actinomyces viscosus*. *Chemistry and Physics of Lipids* **12**, 172–175.

8
Clinical Significance of Actinomycetes

K. P. SCHAAL* and B. L. BEAMAN†

*Institute of Hygiene, University of Cologne, D-5000 Cologne 41, F.R.G. and
†Department of Medical Microbiology, School of Medicine, University of California, Davis, California 95616, U.S.A.*

1. Introduction	389
2. Endogenous actinomycete infections and impairments	390
A. Human actinomycoses	391
B. Lacrimal canaliculitis	400
C. Caries and periodontal disease	400
D. Animal actinomycoses	402
3. Exogenous actinomycete infections	403
A. Pulmonary and systemic nocardiosis	403
B. Localized cutaneous and subcutaneous nocardial infections	405
C. Actinomycetomas	406
D. Incidence and epidemiology of infections due to aerobic actinomycetes	407
E. Treatment of infections due to aerobic actinomycetes	411
F. Other 'opportunistic' pathogenic actinomycetes	412
G. Dermatophilosis or streptothrichosis	413
4. Actinomycete spores as allergens	414
A. Clinical picture	414
B. Aetiology	415
C. Incidence	415
D. Treatment and prophylaxis	415
References	416

1. Introduction

Gram-positive, filamentous, branching bacteria are known or suspected of being aetiologically involved in a variety of diseases and impairments affecting man and other homoeothermic animals. Although the medically or veterinary significant species constitute only a small part of these taxonomically heterogeneous actinomycetes, they are classified in a number of different genera and families. Thus, it is not surprising that the pathological lesions induced by members of the clinically significant taxa may differ greatly in clinical and histopathological appearance as well as in incidence,

THE BIOLOGY OF THE ACTINOMYCETES
ISBN 0-12-289670-X

Copyright © 1983 by Academic Press, London
All rights of reproduction in any form reserved

epidemiology and prognosis. On the other hand, marked resemblance may occur both between different actinomycete diseases and between their causative agents so that the proper clinical or microbiological diagnosis is often difficult to establish.

Notoriously difficult diagnostic problems have always hampered our understanding of actinomycete infections and have contributed to an almost unique accumulation of misunderstandings and poor deductions ever since the descriptions of bovine (Bollinger, 1877) and human (Israel, 1878) actinomycosis and bovine (Nocard, 1888) and human (Eppinger, 1891) nocardiosis, the first examples of actinomycete pathogenicity. As their designation indicates, actinomycetes ('ray fungi') were primarily confused with true fungi (eumycetes) and it was several decades before their bacterial nature was established. Further, the aetiological agents of bovine and human actinomycosis, which can now be readily distinguished, were long thought to be identical. Thus, human actinomycosis was mistaken for nocardiosis and free-living, non-pathogenic actinomycetes (streptomycetes) occurring as laboratory contaminants were wrongly considered to cause either disease (Bostroem, 1891).

These and other historical errors were only cleared up when improved methods of isolation and identification became available. However, the more reliable and feasible diagnostic techniques also led to the detection of new pathogenic species and new host–parasite interactions so that the spectrum of clinically significant actinomycetes and actinomycete-induced diseases has been and is still, gradually broadening. The discovery of new pathogens complicates the diagnostic work of clinicians and medical microbiologists who, in addition, have to face various nomenclatural changes brought about by recent advances in actinomycete systematics.

In the present chapter the actinomycetes are considered on their clinical and epidemiological importance and the reader is referred to Chapter 2 for a consideration of the current state of actinomycete systematics.

2. Endogenous Actinomycete Infections and Impairments

Actinomycetes have long been known to form an important component of the indigenous microflora of human and possibly animal mucous membranes, especially in the oral cavity (Bergey, 1907; Naeslund, 1925; Emmons, 1938; Lentze, 1948; Davis and Baird-Parker, 1959; Howell et al., 1959; Ritz, 1963; Blank and Georg, 1968; Collins et al., 1973). All of these commensal actinomycetes possess a fermentative carbohydrate metabolism, have comparatively exacting nutritional requirements, grow predominantly as facultative anaerobes and have not been reliably reported from sources outside the human and animal body (Slack, 1974; Slack and Gerencser, 1975).

Filamentous bacteria conforming to this description are currently classified in the genera *Actinomyces, Arachnia, Bifidobacterium, Bacterionema* and *Rothia* in the family *Actinomycetaceae* (Slack, 1974). However, the composition of this family has been increasingly questioned in light of data derived from recent taxonomic studies (Schaal and Pulverer, 1981; Schaal and Schofield, 1981a, b; Schofield and Schaal, 1981). Thus, the genus *Bifidobacterium* is no longer considered to be an actinomycete (Stackebrandt and Woese, 1981) and it has been proposed that *Bacterionema matruchotii* be transferred to the genus *Corynebacterium* as *Corynebacterium matruchotii* comb. nov. (see page 79). The genus *Actinomyces* appears to be heterogeneous and composed of at least two subgroups which might each deserve generic rank (Schofield and Schaal, 1981). Furthermore, two species of hitherto uncertain taxonomic affiliation, *Corynebacterium pyogenes* and *Actinobacterium meyerii* (Holmberg and Nord, 1975; Schofield and Schaal, 1981) share a high phenetic similarity to some *Actinomyces* species and it has been proposed that the former be transferred to the genus *Actinomyces* as *Actinomyces pyogenes* comb. nov. (see pages 78–79).

Irrespective of the latest taxonomic findings, however, several of the traditional members of the *Actinomycetaceae* are very similar to one another in terms of medical importance, pathogenesis and epidemiology. Indeed, diseases or impairments such as actinomycosis, lacrimal canaliculitis, periodontal disease and caries may be caused or initiated by *Actinomyces, Arachnia, Bifidobacterium, 'Bacterionema'* or *Rothia* strains. All of these taxa are commonly present on the mucosal surfaces of healthy individuals, but may also become invasive or harmful as endogenous pathogens.

A. HUMAN ACTINOMYCOSES

The name 'actinomycosis' dates back to the first extensive description of the bovine form of the disease by Otto Bollinger in 1877. Similar infections in man were first reported by James Israel in 1878. For many years *Actn. israelii*, which had been isolated and characterized by Bujwid (1889) and Wolff and Israel (1891), was thought to be the only causative agent of human actinomycosis. It is now known that additional *Actinomyces* species and other members of the *Actinomycetaceae* produce essentially identical syndromes. In order to prevent the introduction of further aetiological designations for the same clinical form of disease ('arachniosis', 'bifidobacteriosis') these actinomycete infections should be referred to as actinomycoses to accommodate their variable, but similar, aetiology.

(a) Clinical and Patho-anatomical Picture

As currently recognized, actinomycoses are subacute to chronic, granulomatous, inflammatory processes which usually give rise to suppuration and abscess formation and tend to develop draining sinus tracts

(Lentze, 1969; Slack and Gerencser, 1975; Pulverer and Schaal, 1978; Schaal, 1981; Schaal and Pulverer, 1981). A slowly emerging and progressing, painless infiltrate may be considered typical of the early stages of the disease, but actinomycoses may also manifest as acute abscesses or phlegmons. Advanced lesions show a characteristic and dangerous preference to spread *per continuitatem* irrespective of natural borders (Lentze, 1969) so that even deep-seated infections finally reach the common integument. The subcutaneous abscesses thus formed tend to rupture outwards spontaneously at multiple sites leaving a loculated system of draining sinuses which may temporarily heal and then reopen discharging a yellowish, purulent to serous exudate. Incision wounds made to drain the pus may stay open and continue to discharge the exudate.

In about 40% of cases (Pulverer and Schaal, 1978; Schaal, 1979, 1981; Schaal and Pulverer, 1981), the purulent discharge from actinomycotic lesions contains macroscopically visible ($\leqslant 1$ mm in diameter), yellowish to brownish particles which consist of a conglomerate of filamentous actinomycete microcolonies formed *in vivo*, various other bacteria, and tissue reaction material, especially polymorphonuclear granulocytes, surrounding the microbial centre. Typically, these granules are hard and require some pressure before they can be disintegrated. Under the microscope, they show a cauliflower-like appearance at low magnification (80–100$\times$, Fig. 1), and appear to be composed of roundish, densely packed structures with radially arranged peripheral filaments when viewed at higher magnification (400$\times$; Lentze, 1969; Slack and Gerencser, 1975; Schaal, 1979). Similar particles were described by Harz (1877) in his first microbiological report on bovine actinomycosis and were originally designated 'Drusen'; they are now usually referred to as 'sulphur granules'. Corresponding filamentous colonies can commonly be found in tissue sections but are often smaller than those from the discharge and show a club-shaped layer of hyaline material on the tips of peripheral hyphae. Such hyaline clubs, although characteristic diagnostically, are frequently missing in particles derived from pus (Schaal and Pulverer, 1981).

Given their endogenous origin, actinomycoses develop primarily in tissue adjacent to the mucous membranes, which are the natural habitat of the aetiological agents. Clinically, these predelection sites have been used to subdivide the disease into cervicofacial, thoracic and abdominal forms, respectively. In the extensive material collected at the Institute of Hygiene of the University of Cologne (Lentze, 1969; Pulverer and Schaal, 1978; Schaal, 1981), which comprises nearly 3,000 cases, involvement of the face and neck was encountered most frequently, and accounted for about 98% of the cases examined. Thoracic infections were observed in only 1.4% of cases and abdominal manifestations in 0.6% of the patients. These ratios differ greatly from those reported for the U.S.A. (Slack and Gerencser, 1975) where 40 to 50% of human actinomycoses were located in thoracic and abdominal sites. These regional differences in anatomical distribution are not easy to explain but they may be due to the fact that dental surgeons are most familiar with the disease in Germany where there is apparently also a higher overall incidence of actinomycoses compared with other countries (Lentze, 1970; Schaal, 1979).

The preferential involvement of the cervicofacial area may reflect the high populations of the actinomycetes in the oral cavity as well as predisposing conditions such as dental decay, tooth extraction and fracture of the jaw. Tissue surrounding the lower and upper jaw, cheek, neck, sinuses, parotid gland, tongue, lips and ears is primarily affected by actinomycotic infections. In 317 patients treated at the Department of Dental Surgery of the University of Cologne between 1952 and 1975 (Herzog, 1981) the anatomical distribution of the primary lesions was as follows: *corpus mandibulae*, 53.6%; cheek, 16.4%; chin, 13.3%; submaxillary ramus and angle, 10.7%; upper jaw, 5.7%; and mandibular joint, 0.3%. Direct invasion of the bone is rare, but a periostitis may develop and post-traumatic osteomyelitis is not uncommon (11.7% of the afore-mentioned cases). The regional lymphatics are usually not involved (Slack and Gerencser, 1975). The skin overlying the typical firm swellings often shows a reddish-purple cast and may be visibly impaired by scars and open sinus tracts (Fig. 2).

Thoracic actinomycoses may be due to inhalation or aspiration of the causative agents from the oral cavity, haematogenous extension, or descending cervicofacial infections. From a primary lesion in the lungs, the infection progresses directly from lobe to lobe forming multiple abscesses (Slack and Gerencser, 1975). The clinical and radiological picture is usually not characteristic and confusion with tumours, tuberculosis, or any non-specific type of infection may occur. The disease may spread to the heart, the pleura, or even to the chest wall producing draining sinus tracts which are then highly suggestive of actinomycosis (Fig. 3).

Abdominal actinomycotic processes have been associated with acute perforative gastrointestinal diseases and surgical or accidental trauma. Recently, an additional source of pelvic and abdominal actinomycete infections has been identified; in cervicovaginal secretions of women wearing intrauterine contraceptive devices or vaginal pessaries actinomycetes can frequently be demonstrated (Gupta *et al.*, 1976, 1978; Christ and Haja, 1978; K. P. Schaal, unpublished observations). The apparent change in the indigenous flora, together with the effect of a foreign body, can obviously favour the development of localized inflammations in adnexa and pelvis (Witwer *et al.*, 1977; Barnham *et al.*, 1978; Spence *et al.*, 1978; Luff *et al.*, 1978; Kohoutek and Nozicka, 1978) and even induce hepatic and intracranial actinomycotic abscesses (Gupta *et al.*, 1978). Initially, the symptoms are usually mild and not indicative of the type of disease. When an indurated mass has developed, this may simulate carcinoma until draining sinuses appear.

In addition to the three main forms of actinomycoses, infections of the central nervous system have been reported. These are rare and usually result from direct extension of cervicofacial lesions. Haematogenous spread is uncommon, but may occur from thoracic and abdominal primary foci. The clinical signs are those of other growing and displacing intracranial processes. Primary infections of the skin have been observed only exceptionally. They nearly always follow trauma resulting from human bites or fist fights stressing that the source of infection is the indigenous oral flora and not the environment (Lentze, 1938).

Without treatment actinomycoses show little tendency to heal spontaneously although abortive cases do occur. Usually the disease progresses slowly with partial remissions and subsequent relapses and the outcome is a major destruction of the organs affected or a life-threatening haematogenous generalization.

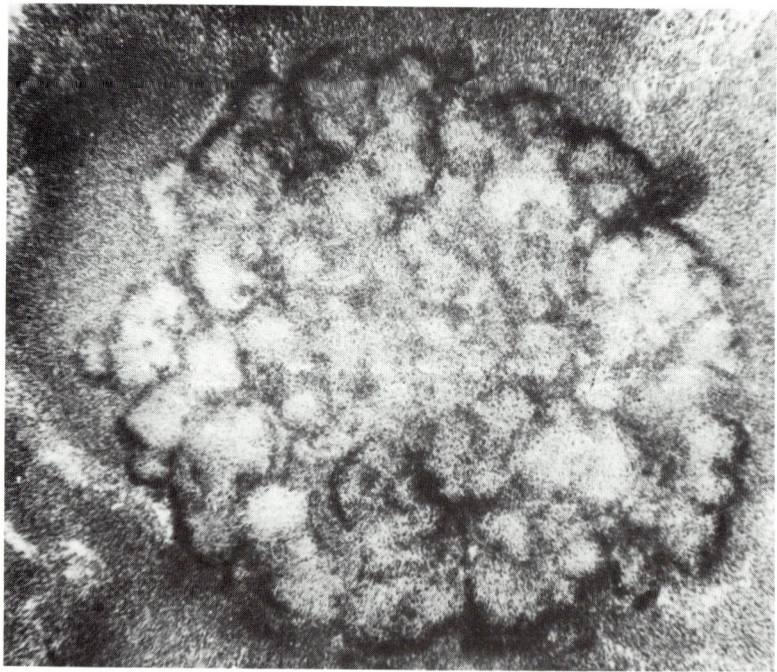

FIG. 1. Sulphur granule in pus taken from a chronic cervicofacial actinomycotic abscess. Note the lobulated, cauliflower-like appearance. (Granule embedded in 1% methylene blue solution under a cover slip; magnification about ×80; photographed by Professor F. Lentze, Cologne, F.R.G.)

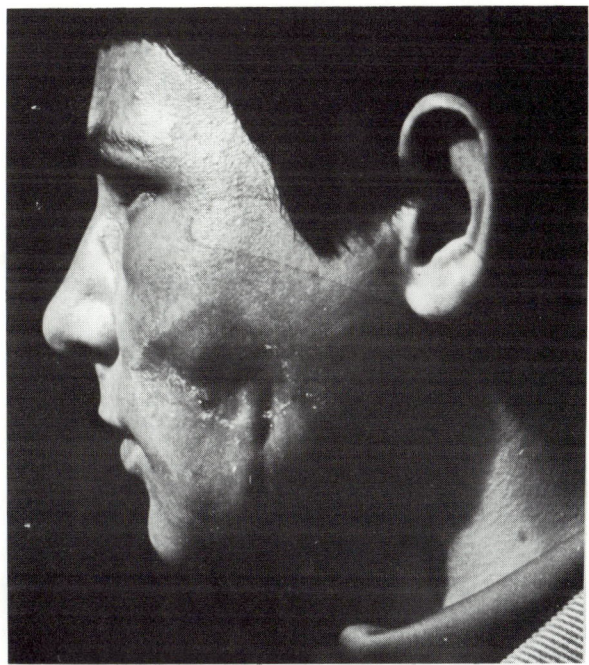

FIG. 2. Chronic cervicofacial actinomycosis extending to cheek and orbita. Central scarring and the open, discharging, old incision wound are very characteristic. (Photographed by S. Glanschneider, Cologne, F.R.G.)

(b) Aetiology

The aetiology of the human actinomycoses is complex in two respects; first, several actinomycete species, which belong to different genera of the family *Actinomycetaceae*, may incite clinically indistinguishable lesions and second, the microbial flora isolated from actinomycotic processes always contains one or more bacterial species in addition to the causative actinomycete.

In these multiple infections the fermentative actinomycetes act as primary pathogens which are basically responsible for the typical symptoms, the course and the prognosis of the disease. They have therefore been designated 'guiding organisms' by Lentze (1948, 1969, 1970). The so-called concomitant bacteria which are always present, but vary in number and species composition, are obviously necessary to strengthen the comparatively low invasive power of the actinomycetes (Holm, 1950, 1951; Lentze, 1948, 1953, 1969, 1970; Pulverer and Schaal, 1978; Schaal, 1979, 1981; Schaal and Pulverer, 1981), either by providing reduced conditions in the tissue or by supplementing the guiding organism with toxic or necrotizing extracellular products (Brede, 1959).

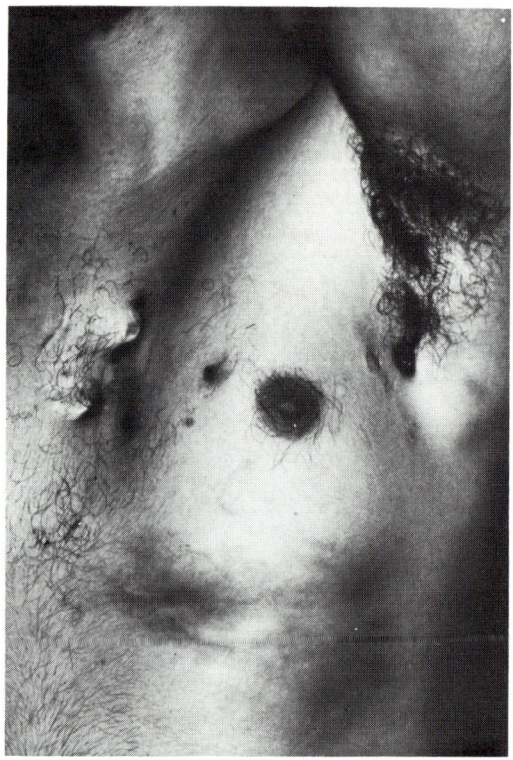

FIG. 3. Thoracic actinomycosis with multiple old and fresh sinus tracts and the beginning of abscess formation below the mamilla. (The patient was presented by Professor W. P. Herrmann, Bremen, F.R.G.)

In about one half of cases the mixed actinomycotic flora is composed of both aerobically growing bacteria and carboxyphils or strict anaerobes, the other half containing solely anaerobes (Lentze, 1969; Pulverer, 1974; Pulverer and Schaal, 1978; Schaal and Pulverer, 1981; Schaal, 1979, 1981). Among the aerobic concomitant bacteria *Staphylococcus aureus*, *Staphylococcus epidermidis* and alpha- and beta-haemolytic streptococci appear to be most prevalent whereas the *Enterobacteriaceae* are seldom encountered, at least as far as cervicofacial forms are concerned. Common anaerobic components of the mixed infections are carboxyphilic streptococci, *Peptococcaceae*, *Bacteroides asaccharolyticus*, *Bacteroides melaninogenicus* and other *Bacteroides* species, *Fusobacterium nucleatum*, *Leptotrichia buccalis* and '*Actinobacillus actinomycetem-comitans*'. In Cologne, the latter was found in about 25% of the cases examined and it seemed to be responsible for a particularly chronic and serious course of the disease (Pulverer and Schaal, 1978; Schaal and Pulverer, 1981).

The classical and most common pathogenic member of the *Actinomycetaceae* is *Actn. israelii* which has long been recognized as a human pathogen. In a recent evaluation of 943 actinomycoses studied between 1970 and 1979 (Schaal, 1981), this organism was isolated from 78% of the patients. Other actinomycete species such as *Actn. naeslundii, Arac. propionica, Actn. viscosus, Actn. odontolyticus, Bibc. (Actn.) eriksonii,* and *Cnbc. (Bact.) matruchotii* were encountered in 6.2%, 2.3%, 0.6%, 0.4%, 0.2% and 0.7% of cases, respectively. With the exception of *Cnbc. matruchotii*, pathogenicity has been claimed for all of these species and their isolation from typical actinomycotic lesions seems to support their aetiological role. However, in several instances, the purulent discharge has been found to contain two different actinomycete species; two cases with *Actn. israelii* and *Actn. naeslundii*, two with *Actn. israelii* and *Actn. viscosus*, and one with *Actn. israelii* and *Cnbc. matruchotii* (Schaal, 1979, 1981). These findings indicate that double actinomycete infections do occur and raise the question as to which species is the 'real' causative agent. Furthermore, only one of two different organisms present in a specimen may be identified by chance, and contaminations from mucosal secretions cannot always be excluded, so that it remains difficult to decide whether or not additional members of the family *Actinomycetaceae* can produce actinomycoses.

Nevertheless, at least *Arac. propionica* appears to show definite pathogenicity for humans. It is not infrequently isolated from actinomycotic lesions (Gerencser and Slack, 1967; Brock *et al.*, 1973; Conrad *et al.*, 1978; Pulverer and Schaal, 1978; Schaal, 1979, 1981) and F. Lentze (unpublished work) was able to cure a case of *Arachnia* infection, which did not respond to antibiotics and *Actn. israelii* heterovaccine, by administration of *Arac. propionica* autovaccine.

Among the other members of the *Actinomycetaceae* which have been incriminated as pathogens in invasive infections, diseases due to *Bibc. (Actn.) eriksonii* are comparatively well documented (Georg *et al.*, 1965; Green, 1978; Schaal and Pulverer, 1981). *Actinomyces naeslundii* and *Actn. viscosus* although not rare in pus specimens (Coleman and Georg, 1969; Gerencser and Slack, 1969; Lewis and Gorbach, 1972; Karetzky and Garvey, 1974; Scharfen, 1975a; Mosimann *et al.*, 1979; Thadepalli and Rao, 1979; Pulverer and Schaal, 1978; Schaal, 1979, 1981) often appear to be contaminants rather than the principal causative agents. In a few cases, however, such as an empyema of the knee-joint from which *Actn. naeslundii* was isolated in pure culture on several occasions over a period of 4 months (Schaal and Pulverer, 1981), a pathogenic role for these organisms is strongly favoured. Very rarely, *Actn. odontolyticus* may also be found in actinomycosis-like lesions (Guillou *et al.*, 1977; Mitchell *et al.*, 1977; Baron *et al.*, 1979; Schaal, 1981). *Corynebacterium matruchotii* (Wilhelmus *et al.*, 1979) and *Rothia dentocariosa* (Scharfen 1975b; Lutwick and Rockhill, 1978; Schaal, 1981) may occasionally cause abscess formation or other unspecific infections and *Rothia* has even been shown to produce infective endocarditis (Pape *et al.*, 1979; Schafer *et al.*, 1979).

(c) Incidence and Epidemiology

It is perhaps not surprizing that actinomycoses are world-wide in distribution and are not transmissible given the presence of the facultative pathogens on human mucosal surfaces. However, the prevalence of actinomycotic processes appears to vary regionally, especially as far as the cervicofacial forms are concerned, possibly reflecting the standard of dental care and use of antibiotics in dentistry. Such factors might explain the lower absolute and relative incidence of cervicofacial actinomycoses in the U.S.A. compared with Europe.

On the basis of histological findings, Hemmes (1963) calculated a morbidity rate of actinomycotic infections which amounted to one case per 119,000 inhabitants of the Netherlands per year. In the Cologne area of the Federal Republic of Germany, Lentze (1969) reported an incidence of 1:83,000 up to 1969. The morbidity rate calculated from material examined in Cologne during the past decade shows an average of one case per 40,000 Cologne inhabitants per year when both acute and chronic processes are considered (Schaal, 1979). This re-evaluation indicates that the incidence of actinomycoses has obviously been underestimated in the Cologne region and possibly also in other parts of Europe.

Another interesting finding relates to the uneven sex distribution of the disease. Most tabulations including sex ratios (Slack and Gerencser, 1975; Pulverer and Schaal, 1978; Schaal, 1981; Pulverer, 1974) indicate that males are involved about three to four times more often than females though these differences only apply to patients in sexual maturity. Before puberty and in the climacteric period, actinomycoses appear to be evenly distributed amongst patients of both sexes (Pulverer and Schaal, 1978; Schaal, 1981). Indeed, the disease can occur in persons of all age groups (Slack and Gerencser, 1975; Pulverer and Schaal, 1978; Schaal, 1981), the youngest patient in a recent series of cases (Schaal, 1981) was 1.5 months old, the oldest one 89 years of age. Nevertheless, the highest incidence has been observed in males aged 21 to 40 years and in females of 11 to 30 years (Pulverer and Schaal, 1978; Schaal and Pulverer, 1981; Schaal, 1981).

(d) Treatment

Before the introduction of sulphonamides and penicillins actinomycoses had to be considered infectious diseases of doubtful prognosis; many cervicofacial and nearly all of the thoracic and abdominal cases finally resulted in death or serious sequelae of the patients involved. Thymol (Myers, 1937), potassium iodide (Telford, 1915; Wangensteen, 1932), and X-ray irradiation (Wangensteen, 1932) although widely used have been shown to be completely ineffective. Only radical surgical measures provided some chance of success.

The first step towards a specific and effective treatment of actinomycotic lesions was the application of auto- and heterovaccines (Colebrook, 1921). A complex heterovaccine developed by Lentze (see Lentze, 1969) proved to be an efficient tool for treating cervicofacial infections and was used until recently in certain Departments of Dental Surgery in Germany.

However, the modern standard therapy for actinomycoses consists of a combination of surgical and chemotherapeutic measures. Facultatively pathogenic *Actinomycetaceae* are sensitive to a wide variety of antibacterial drugs (Schaal *et al.*, 1979; Schaal, 1979; Schaal and Pape, 1980; Lentze, 1969), especially to penicillins and cephalosporins (Clack and Gerencser, 1975). Nevertheless, not all of these substances are equally well suited for treating actinomycotic processes because both actinomycetes and the concomitant bacteria have to be considered when choosing the appropriate antibiotic. The latter may be resistant to β-lactam-antibiotics and/or may produce a β-lactamase thereby rendering the therapeutic measure insufficient. This is why benzyl penicillin, although usually recommended as the drug of choice (Slack and Gerencser, 1975), does not always cure the disease or prevent relapses. *Actinobacillus actinomycetem-comitans* and certain *Bacteroides* species, which are commonly resistant to penicillin G, may sustain the inflammation even after chemotherapeutic elimination of the actinomycetes or may protect the latter from the action of the drug.

According to our experience, aminopenicillins such as ampicillin or amoxycillin provide much better and more constant therapeutic results. However, in cases where *Bacteroides fragilis* or *Bacteroides thetaiotaomicron* are aetiologically involved (<5% of the cervicofacial forms) other modern antibiotics such as mezlocillin, cefoxitin, or possibly lamoxactam and carbenicillins may be more effective. Alternatively, ampicillin may be combined with metronidazole or clindamycin (Schaal and Pape, 1980). These combinations, possibly extended by an aminoglycoside, may be recommended especially for abdominal and thoracic infections which more commonly contain β-lactamase-producing *Bacteroidaceae*.

Whenever possible, chemotherapy should be complemented by surgery in order to remove the infected tissue and pus-filled sinuses. Heterovaccine, antibiotics, and surgery adequately provide good possibilities of safely curing even advanced cases of actinomycosis so that the prognosis of this disease must no longer be considered doubtful. However, its malignant potential remains and will become apparent when diagnosis and therapy are inadequate or delayed.

B. LACRIMAL CANALICULITIS

The presence of branching filamentous organisms in lacrimal concretions was described by Ferdinand Cohn in 1875. More recent examinations have shown that these filamentous organisms usually belong to the species *Actn. israelii*, *Actn. odontolyticus*, and *Arac. propionica* (Pine and Hardin, 1959; Pine *et al.*, 1960; Ellis *et al.*, 1961; Buchanan and Pine, 1962; Slack and Gerencser, 1975). According to our own experience *Arac. propionica* seems to be the most prevalent in both concretions and secretions from the lacrimal canaliculi.

The inflammatory reactions incited by these actinomycetes in the surroundings of the eye are usually mild and the infection is not invasive. In addition, concomitant bacteria are usually not present. Strictly speaking, therefore, the disease should not be termed actinomycosis, but lacrimal canaliculitis due to actinomycetes. In many patients there is a history of conjunctivitis followed by the development of an intermittent creamy discharge from the corner of one of the eyes (Slack and Gerencser, 1975). Yellowish concretions can be removed from the canaliculi by pressure or with a curette. The concretions contain one or two of the actinomycete species mentioned before, the latter can be cultured or demonstrated by immunofluorescence staining. Removal of the concretions together with local administration of antibiotics usually results in complete recovery.

The actinomycetes are most likely introduced into the eye from the oral cavity either by the patient's own fingers or by a reflux from the mouth into the tear duct. Lacrimal canaliculitis due to actinomycetes is not rare but, because the disease usually heals after removal of the concretions, there is often no need to establish an aetiological diagnosis.

C. CARIES AND PERIODONTAL DISEASE

From a practical and social point of view, caries and periodontal disease are probably the most important impairments in which *Actinomycetaceae* may be aetiologically involved. Undoubtedly, these diseases are very complex in both aetiology and pathogenesis. Microbes only constitute one link in a long chain of cause and effect and actinomycetes are certainly not the only, or even the most important, bacteria that contribute to the development of carious lesions and periodontitis.

Nevertheless, actinomycetes definitely play a part in the formation of dental plaque which is thought to be the initial step in the development of caries and periodontal disease. Filamentous bacteria have been found to form a major portion of the volume of plaque thereby supplying an enormous additional surface area for the attachment of other bacteria (Boyd and Williams, 1971). Furthermore, some oral actinomycetes possess mechanisms of adherence which enable them to colonize the smooth surface of teeth or to adhere to other microbes already· *in situ. Actinomyces naeslundii* (Slack and Gerencser, 1975), *Actn. viscosus* (Howell and Jordan, 1967; Miller *et al.*, 1978a; Pabst, 1977), and *Roth. dentocariosa* (Lesher and Gerencser, 1977) synthesize extracellular or cell-associated (Warner and Miller, 1978) levan which may prevent them from being washed off by salivary flow. In addition, levan may provide a reservoir of carbohydrate for growth because *Actn. viscosus* has been shown to have levan hydrolase activity (Miller and Somers, 1978). This observation may be an example of synergism in the oral cavity for other members of the mouth flora are also able to synthesize or degrade levan (DaCosta and Gibbons, 1968; van Houte and Jansen, 1968; Manly and Richardson, 1968). Adherence may also be mediated by surface fibrils which have been characterized in *Actn. viscosus* by means of electron microscopy and by the use of chemical and serological techniques (Cisar and Vatter, 1979). Interbacterial aggregation has been demonstrated between *Actn. naeslundii* and streptococci (Gibbons and Nygaard, 1970; Ellen and Balcerzak-Raczkowski, 1977; Miller *et al.*, 1978b) and between *Actn. viscosus* and veillonellae or streptococci (Gibbons and van Houte, 1973; McIntire *et al.*, 1978). In the process of calcification of plaque *Cnbc. matruchotii* may be important as it forms intracellular deposits of calcium phosphate which are indistinguishable from bone and tooth apatite (Ennever, 1960; Takazoe, 1961; Ennever *et al.*, 1971, 1978; Boyan-Salyers *et al.*, 1978).

Once plaque is formed, the underlying tooth is subjected to the action of various bacterial derived metabolic products. These include organic acids such as formic, acetic, propionic, lactic and succinic acids which are able to lower the pH to 5 or below thereby initiating demineralization of the enamel (Slack and Gerencser, 1975). The carbon sources necessary for the formation of acid end products are derived from either the host's diet or the metabolism of other microbes. Enzymes such as invertase and β-galactosidase have been identified from *Actn. viscosus* strains indicating the ability of the organism to utilize disaccharides which are usual components of food. Additionally, actinomycetes and streptococci on the one hand and *Veillonella* species on the other were found to form a food chain in which veillonellae degrade lactic and succinic acids produced by the actinomycetes and streptococci (Distler *et al.*, 1980; Distler and Kröncke, 1981).

In the complex development of caries, which finally leads to proteolytic degradation of the dentine and to extending cavities, the aetiological contribution of the actinomycetes is generally acknowledged. In detail, however, many questions remain open, as is exemplified by the finding that *Streptococcus mutans*, the best studied cariogenic *Streptococcus* species, other streptococci and lactobacilli predominantly induce enamel caries, whereas filamentous bacteria have been isolated from human root surface caries (Jordan and Hammond, 1972) but usually not from enamel lesions.

Dental plaque also represents the initial condition from which periodontal disease may develop (Hardie, 1974). Plaque and the gingival crevices expose the adjacent gingival epithelium to various potentially damaging metabolites of the microbes accumulating there. Although the intact epithelium may initially resist these metabolic products, an increase in bacterial numbers, a shift in species composition, an injury or irritation by dental calculus may induce an inflammatory reaction which progresses from gingivitis to periodontitis (Slack and Gerencser, 1975; Lesher et al., 1977). Factors such as bacterial enzymes, endotoxins, cytotoxins, hypersensitivity reactions and bacterial competition for nutrients could account for this epithelial damage. In actinomycetes, several of these factors have been detected: *Actn. viscosus* has been shown to produce chemotactic effects, to stimulate the release of mediators of inflammation from host immune cells, to mark fibroblasts for immune-mediated damage, and to possess amphipathic antigens (Engel et al., 1976, 1978; Burckhardt, 1978; Taichman et al., 1978; Wicken et al., 1978). *Corynebacterium matruchotii* enhances phagocytic and bactericidal functions (Nitta et al., 1977) and exhibits adjuvant activities (Nitta et al., 1978). Finally, *Actn. naeslundii* and *Actn. viscosus* have been shown to initiate periodontitis in hamsters and gnotobiotic rats (Socransky et al., 1970; Jordan et al., 1965, 1972; Llory et al., 1971; Crawford et al., 1978).

Although the carious cavity may be considered a wound, it does not show any tendency to heal. The only way of treating carious lesions is, therefore, to remove the decayed material and to close the defect by filling it with substances which are hard enough to withstand the pressure produced by the masticatory muscles and which can be fixed in place tightly and durably. Similarly, there is no selective treatment of periodontal disease unless the initiating conditions, especially plaque, have been cleared, although acute inflammatory reactions may improve after administration of chemotherapeutics such as penicillins or metronidazole (Ohkawa et al., 1969).

From a medical and social point of view, prevention is much more important than therapy. Prophylactic measures include efforts to prevent plaque from developing or to eradicate it after it has formed, or to make the enamel more resistant to decalcification by administering fluorides which bind with the hydroxyapatite. In addition, the diet should be controlled in respect to carbohydrates and there are still attempts to develop a vaccine.

D. ANIMAL ACTINOMYCOSES

Since Bollinger's first description of bovine actinomycosis in 1877, cattle have been shown to be the most frequently involved animals. The disease which is also called lumpy jaw resembles human actinomycoses in that it is a chronic suppurative process in which multiple abscesses and draining sinus tracts are formed. In contrast to human cases, bone is usually affected and slowly deformed by a destructive, and at the same time proliferating, osteitis (Slack and Gerencser, 1975). The principal causative agent, which most probably also invades the tissue endogenously, is *Actn. bovis* (Bollinger, 1877; Slack and Gerencser, 1975), but infections with *Actn. israelii* have been reported in cattle (King and Meyer, 1957; Cummins and Harris, 1958; Pine *et al.*, 1960).

Clinically somewhat different actinomycete diseases have been observed in swine (Magnusson, 1928; Franke, 1973), dogs (Georg *et al.*, 1972) and other domestic and wild animals. *Actinomyces bovis*, '*Actn. suis*', *Actn. israelii* and *Actn. viscosus* have been implicated as aetiological agents. However, some of these reports are not well documented so that detailed knowledge about pathogens and sources of infection is lacking.

3. Exogenous Actinomycete Infections

The soil and other natural habitats are important reservoirs of a diverse array of actinomycetes (Chapter 11). Most of these actinomycetes are obligate aerobes, many of which play a role in organic matter turnover although some occur in a dormant state. Few of these free-existing filamentous bacteria have gained medical significance, either as infective agents or as sources of potent allergens.

Three basic forms of infective disease due to aerobic actinomycetes have been recognized: (i) pulmonary and/or systemic nocardiosis; (ii) localized cutaneous or subcutaneous nocardiosis and (iii) actinomycetoma. The major pathogens responsible for these infections belong to the genera *Nocardia*, *Actinomadura* and *Streptomyces*; *Nocardia* species such as *Nrda. asteroides*, *Nrda. brasiliensis* or *Nrda. otitidis-caviarum* being most prevalent (Murray *et al.*, 1961 Gonzáles-Ochoa, 1962; Causey *et al.*, 1974; Beaman *et al.*, 1976; Barnetson and Milner, 1978; Satterwhite and Wallace, 1979; Harris, 1980).

A. PULMONARY AND SYSTEMIC NOCARDIOSIS

Pulmonary nocardiosis probably results from inhalation or aspiration of nocardial cells into the lungs leading to a primary infection which may either remain localized or may disseminate by way of the blood stream or lymphatics to other parts of the body. In addition, nocardiae may be introduced into the blood stream by trauma or accidental inoculation. The blood-borne organisms may then infect the lungs, brain, heart, kidneys or any other internal organ systems.

Nocardiosis may occur as an acute infection presenting as an abscess or as acute pneumonia with or without abscess formation. Disease may vary from a fulminating infection with aggressive tissue invasion and destruction to a benign self-limited or subclinical process that remains unrecognized in the patient. Further, nocardiosis may enter a progressive but chronic phase presenting as a chronic abscess, chronic fibronodular infection or occasionally as a granulomatous process. There is no specific clinical presentation that is diagnostic for nocardiosis (Fig. 4) and disease caused by

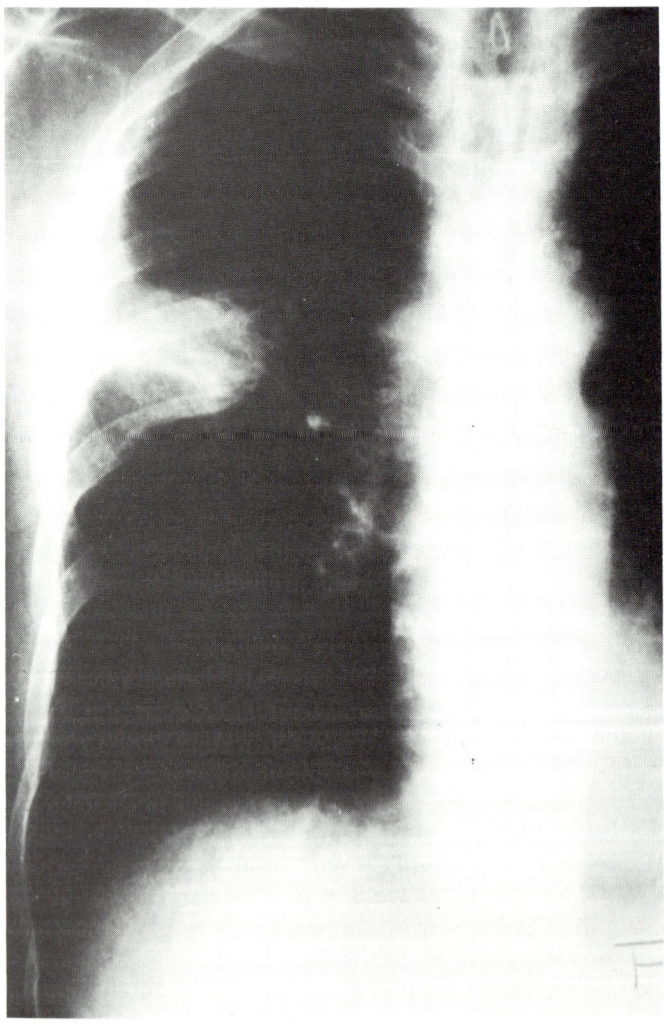

FIG. 4. X-ray photograph of a case of pulmonary nocardiosis caused by *Nocardia asteroides*, group A. Note the dense, roundish shadow in the right lung. (The patient was presented and discussed by Drs. M. F. Torres, L. Arango, H. de Toledo and I. Sabbaj, Guatemala.)

the nocardiae has often been misdiagnosed as pyogenic infections, tuberculosis, actinomycosis, mycoses of various aetiologies, tumours or various forms of cancer (Raich et al., 1961; Neu et al., 1967; Kurup et al., 1970; Presant et al., 1970; Palmer et al., 1974; Frazier et al., 1975; Rosett and Hodges, 1978; Curry, 1980; Scully et al., 1980).

Most cases of nocardiosis begin as a pulmonary disease. However, in approximately 30% of these infections, dissemination to the brain and central nervous system occurs. Further, disseminated nocardiosis involving the heart, kidneys, eyes, adrenals, spleen, liver, and bone is not infrequent (Freese et al., 1963; Susena et al., 1967; Presant et al., 1970; Chavez et al., 1972; Causey and Sieger, 1974; Frazier et al., 1975; Katz and Fauci, 1977; Sher et al., 1977; Cook and Farrar, 1978; Lissner et al., 1978; Byrne et al., 1979; Stuart, 1979).

There have been numerous reports in the literature of primary nocardiosis and Beaman et al. (1976) found that in between 15 and 40% of the cases studied, Nocardia was the primary pathogen and the patient had no known underlying or predisposing condition. However, it should be stressed that nocardiosis is most frequently recognized as an 'opportunistic' infection in patients with some underlying disease. The disorders that seem to predispose the individual to nocardiosis are broad in their spectrum and include leukemia, lymphoma and neoplasms in general, lupus erythematodes, Hodgkin's disease, traumatic wounds, tuberculosis, sarcoidosis, lipid pneumonia, asthma, emphysema, anthracosilicosis, Cushing's syndrome, dysproteinemia, haemolytic anaemia, nephrotic syndrome, Paget's disease, pulmonary alveolar proteinosis, and chronic granulomatous disease. Further, patients being treated with either corticosteroids or immunosuppressants during organ transplantation or antineoplastic therapy have a particularly high risk in developing nocardiosis (Andriole et al., 1964; Batshon et al., 1971; Presant et al., 1973; Causey and Sieger, 1974; Georg, 1974; Krick et al., 1975; Ralph et al., 1976; Rogers and Johnson, 1977; Balikian et al., 1978; Hoken et al., 1978; Kirmani et al., 1978; Peterson et al., 1978; Terezhalmy and Bottomley, 1978; Geiseler and Anderson, 1979; Hirst et al., 1979; Gorevic et al., 1980).

Approximately 80–90% of the cases of nocardiosis are caused by Nrda. asteroides while about 10–15% represent infections due to Nrda. brasiliensis. About 5% of the cases recognized were shown to be caused by Nrda. otitidis-caviarum (Beaman et al., 1976).

B. LOCALIZED CUTANEOUS AND SUBCUTANEOUS NOCARDIAL INFECTIONS

Primary cutaneous or subcutaneous nocardial infection usually depends upon traumatic inoculation of the organism into the skin. Following inoculation, frequently there is the development of a chancriform lesion that may persist and gradually enlarge. Regional lymph nodes become involved and in many instances the clinical picture appears to be similar to sporotrichosis (Rapaport, 1966; Moore and Conrad, 1967; Berd, 1973; Mitchell et al., 1975; Haim and Merzbach, 1979). However, in some cases only cellulitis, localized pustules or pyoderma occur and the lymphocutaneous syndrome may be absent. The lesions usually remain localized to the area of inoculation; however, dissemination can occur. The lesions may be self-limited and heal spontaneously or they may require surgical or therapeutic intervention.

When these infections occur as pustules, localized abscesses or pyoderma they are similar in appearance to staphylococcal infections and they will probably not be correctly identified (Rapaport, 1966; Haim and Merzbach, 1979; Satterwhite and Wallace, 1979). Further, the sporotrichoid form is clinically similar to sporotrichosis caused by the fungus *Sporothrix schenckii* and misdiagnosis is likely to occur (Moore and Conrad, 1967; Mitchell et al., 1975). Unlike nocardiosis, cutaneous infections do not frequently occur in compromised hosts but usually represent primary infection by either *Nrda. brasiliensis* or *Nrda. asteroides*. *Nocardia otitidis-caviarum* is less frequently recognized in cutaneous infections (Rapaport, 1966; Moore and Conrad, 1967; Berd, 1973; Causey and Sieger, 1974; Mitchell et al., 1975; Beaman et al., 1976; Haim and Merzbach, 1979).

C. ACTINOMYCETOMAS

Like the cutaneous infections described above, mycetomas usually begin at the site of a minor localized injury most frequently induced by a thorn or splinter. As a consequence, the disease is commonly recognized on the feet and hands, but it can occur anywhere on the body (Fig. 5). From 1 week to several months following inoculation a painless nodule may appear at the site of injury. This nodular mass may increase in size and become softened. It is filled with pus that will discharge through a small draining tract. This process is followed by cellular infiltration and chronic granulomatous inflammation with concomitant swelling and enlargement of the surrounding region. Multiple secondary nodules form and there is a purulent discharge through several sinus tracts (Fig. 5) which also exude serous and serosanguinous fluid.

Characteristic of this discharge is the presence of granules which are macroscopic colonies of the infecting organism surrounded by masses of inflammatory cells, especially polymorphonuclear phagocytes. The granules are pigmented and their colour varies depending upon the causative agent. Nocardial granules are usually small (<1 mm in diameter), soft, and lobulated and have a white to yellowish colour. *Actinomadura* species produce small or larger, soft granules which often show a cast of red colour (*Acmd. madurae*) or appear even deep red (Schaal, 1977). The particles found in *Streptomyces* infections have been reported to be large (1–2 mm in diameter), hard and yellow to brown to black (Gordon, 1974).

The mycetomatous infection usually remains localized with enlargement by direct extension of the organisms through the tissues. As the infection becomes more chronic, the sinus tracts extend more deeply into the body with involvement of both muscle and bone. Osteomyelitis, when it occurs, may be quite marked and detectable on X-ray. There is usually evidence of periostitis as well as cavities within the affected bone (Green and Adams, 1964; Whyte and Kaplan, 1969; Thammayya *et al.*, 1972; Greer, 1974; Barnetson and Milner, 1978; Sanyal *et al.*, 1978; Talwar and Sehgal, 1979).

More than 50% of the cases of mycetoma reported in the world literature were caused by *Nocardia*. *Nocardia brasiliensis* is most frequently recognized although *Nrda. asteroides* has also been identified as the cause of a significant number of cases. *Nocardia otitidis-caviarum* is frequently encountered in some regions. In addition to *Nocardia*, actinomycetomas may be caused by *Acmd. madurae, Acmd. pelletieri, Stmy. somaliensis,* and '*Stmy. paraguayensis*'. The generic designation of these latter agents is controversial (Green and Adams, 1964; Whyte and Kaplan, 1969; Thammayya *et al.*, 1972; Greer, 1974; Barnetson and Milner, 1978; Sanyal *et al.*, 1978; Talwar and Sehgal, 1979).

A further actinomycete, *Nrda.* (*Acmd.*) *dassonvillei*, which resembles some of the afore-mentioned species morphologically (Gordon and Horan, 1968) has been implicated in human infection, especially in cutaneous and subcutaneous forms and also in respiratory disease (Schaal, 1977).

D. INCIDENCE AND EPIDEMIOLOGY OF INFECTIONS DUE TO AEROBIC ACTINOMYCETES

Although *Nrda. asteroides, Nrda. brasiliensis* and *Nrda. otitidis-caviarum* have been found in almost every country of the world, the prevalence of these organisms within the environment appears to differ for each geographic location depending upon factors such as mean daily temperature, humidity and moisture, agricultural development, and soil composition. Thus, *Nrda. asteroides* is more readily isolated from the soils of the temperate regions of North America, Europe and other continents than is *Nrda. brasiliensis*. In contrast, *Nrda. brasiliensis* is most frequently found in tropical and subtropical regions of Central and South America. It is not surprising then, that the incidence of infection reflects this distribution of nocardiae within the environment (González-Ochoa, 1962; Georg, 1974; Pulverer and Schaal, 1978; Orchard, 1979).

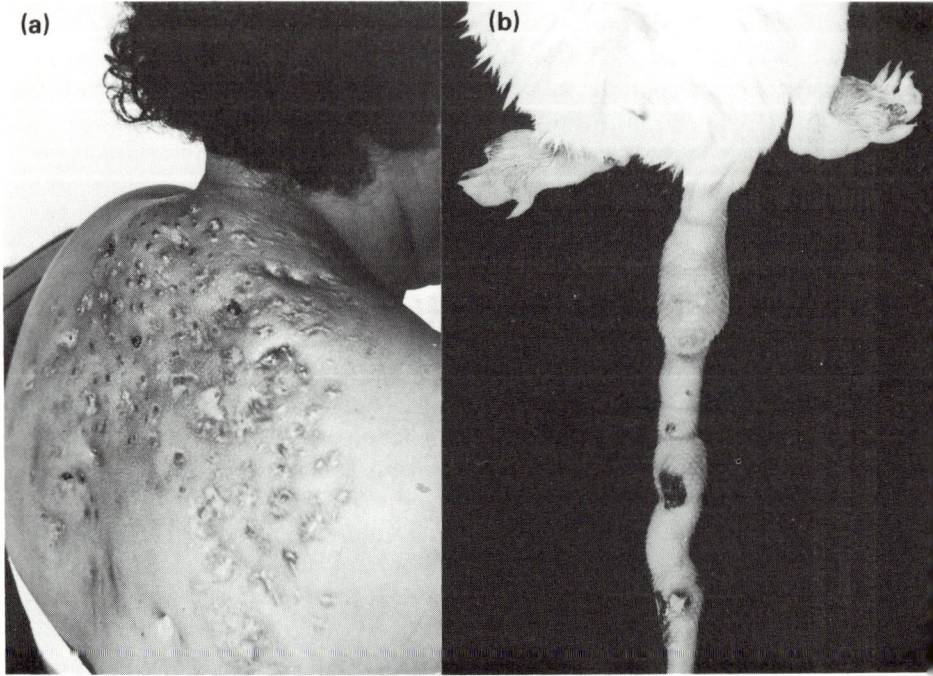

FIG. 5. (a) A mycetoma caused by *Nocardia brasiliensis* that has, over several years, extended to involve the entire back and shoulders and which has not completely responded to extensive chemotherapy. Note the typical multinodular lesions, massive enlargements and the numerous draining sinus tracts that are characteristic of mycetoma. (The patient was presented and discussed by Dr. R. Zamora and Dr. J. Serrano, Venezuela); (b) A mouse 4 months after intravenous injection of *Nocardia brasiliensis* isolated from a patient with mycetoma as shown in (a). Note the development of mycetomas in the cooler portions of the body, especially the feet and tail.

(a) *Infections due to* Nocardia asteroides

It is not possible to state with any certainty the prevalence of infections caused by *Nrda. asteroides* because there has been no active national or international surveillance effort to ascertain the incidence of disease. Therefore, our information must be dependent upon reports of cases within the world literature as well as a few studies that have been made by individual investigators.

There is general agreement that recognition of increased numbers of human infections has occurred in the world since 1960 (Murray *et al.*, 1961; Frazier *et al.*, 1975; Curry, 1980; Harris, 1980). Beaman *et al.* (1976) estimated that between 500 and 1000 cases of serious nocardial infections are diagnosed each year in the U.S.A. and that at least 82% of these infections are caused by *Nrda. asteroides*. Because of the difficulty encountered in the isolation and recognition of *Nocardia* and because most clinicians and hospitals are not specifically looking for these organisms, it is almost certain that the actual incidence of disease caused by *Nrda. asteroides* is significantly larger than the number reported by Beaman *et al.* (1976). Frazier *et al.* (1975) reported 25 patients with *Nocardia* at the Mayo Clinic during a 24 month period of time: 16 of these patients had evidence of clinical disease caused by the nocardiae while 9 of the patients had no definable disease caused by these organisms. In another study by Rosett and Hodges (1978), *Nocardia* was isolated from 42 patients during an 8 year period in two Kansas City hospitals. It was found that 47% of these cases had evidence of clinical disease while the remaining patients had no definable disease associated with the nocardial isolates. Georg (1974) discussed the prevalence of *Nrda. asteroides* in sputa of patients during a 2.5 year period at the Alabama State Department of Health and during a 5 year period at the South Carolina State Board of Health. A total of 349 isolates were obtained from 142,433 samples (0.24%).

The data from all of these studies suggest that infection of humans by *Nrda. asteroides* is not rare. Further, inapparent or subclinical infection without progressive disease may be more common than has been generally realized. From the literature it appears that, contrary to previous suggestions, *Nrda. asteroides* may be quite infectious but not highly invasive. Thus, subclinical or inapparent infections may be the rule rather than the exception (the same as with *Mybc. tuberculosis*) and only certain individuals, especially those that are immunocompromised, develop progressive disease. Since, in the U.S.A., many cases of nocardiosis are recognized in compromised patients receiving immunosuppressive therapy within the hospital, it is tempting to speculate that these patients acquired their infection at some earlier time and the immunosuppressive therapy resulted in a reactivation of the quiescent nidus of infection. This concept has been shown to be true with infections caused by *Mybc. intracellulare* and *Mybc. tuberculosis*.

It has been suggested that nocardial infections are not transmitted from human to human or from animal to human, but instead they are acquired only from the soil. It has now been shown that animal to animal transmission occurs in dairy cows (Bushnell *et al.*, 1979) and there is documented evidence of transmission from human to human in a hospital setting in England where seven patients placed in a room during a period of 6 months acquired nocardiosis. In fact, six of nine renal transplant patients developed disease caused by the same strain of *Nrda. asteroides* (Houang *et al.*, 1980). All of these patients were immunocompromised; nevertheless, it is clear that in certain circumstances nocardiosis should be considered a communicable disease.

(b) *Infections due to* Nocardia brasiliensis

At least 7% of the nocardial infections recognized in the U.S.A. each year are caused by *Nrda. brasiliensis*. More than half of these infections are limited to the skin and subcutaneous tissues; however, a third have been reported to cause systemic nocardiosis and in a significant number of cases the organisms remain localized within the lungs (Beaman *et al.*, 1976).

Even though *Nrda. brasiliensis* may be an opportunistic pathogen in compromised patients, it is most frequently recognized as a primary pathogen in otherwise healthy individuals. Further, infection usually results from direct inoculation into the skin by trauma, splinters or thorns from plants. *Nocardia brasiliensis* is not prominent in soils of temperate regions of the world while it tends to predominate in soils of tropical and subtropical regions. Thus, the incidence of infections, mostly as mycetomas, is relatively common in regions of Mexico, Central America, the northern portion of South America, and in regions of Africa, and India (González-Ochoa, 1962; Green and Adams, 1964; Rapaport, 1966; Moore and Conrad, 1967; Berd, 1973; Mitchell *et al.*, 1975; Talwar and Sehgal, 1979).

(c) *Infections due to* Nocardia otitidis-caviarum

Nocardia otitidis caviarum was first described as causing a middle-ear infection of guinea pigs (Snijders, 1924); however, in 1965 this organism was recognized as causing mycetomas in patients in Tunisia and Japan (Juminer *et al.*, 1965; Fukushiro and Mariat, 1965). Since these first reports, both systemic nocardiosis and mycetomas have been described in patients from India, Mexico and the U.S.A. (Thammayya *et al.*, 1972; Causey and Sieger, 1974; Causey *et al.*, 1974; Beaman *et al.*, 1976; Peterson *et al.*, 1978).

It is estimated that in patients in the U.S.A. recognized as having infections caused by *Nocardia*, at least 3% of these are caused by *Nrda. otitidis-caviarum* (Beaman *et al.*, 1976). Further, the prevalence of this organism in the soil appears to be less common than either *Nrda. asteroides* or *Nrda. brasiliensis* but of general world-wide distribution. Therefore, it does not appear to be any more common in the tropical and subtropical regions than in the temperate zones (Thammayya *et al.*, 1972; Causey and Sieger, 1974; Causey *et al.*, 1974; Beaman *et al.*, 1976; Peterson *et al.*, 1978).

(d) *Infections due to Other* Nocardia *Species*

A variety of species of *Nocardia* are occasionally encountered in clinical material obtained from humans and other animal species; these include '*Nrda*'. *autotrophica*, '*Nrda*'. *orientalis*, *Nrda. carnea*, '*Nrda*'. *aerocolonigenes*' and *Nrda. transvalensis* (Mishra *et al.*, 1980). However, their role as pathogens remains undefined. Nevertheless, it is likely that these organisms may be 'opportunistic' pathogens in the compromised host.

(e) Infections due to Actinomadura *and* Streptomyces

Mycetomas caused by *Actm. madurae* and *Actm. pelletieri* have been reported from Africa as well as the U.S.A., Mexico and other countries. At least *Actm. madurae* seems to be cosmopolitan. Similarly, *Stmy. somaliensis* infections are not restricted to one geographical area as its name implies. Disease caused by this organism has been encountered in Africa as well as in Mexico (Mackinnon and Artagaveytia-Allende, 1956; Mariat, 1963; Mahgoub, 1972).

E. TREATMENT OF INFECTIONS DUE TO AEROBIC ACTINOMYCETES

In contrast to human actinomycoses and many other bacterial infections nocardiosis and actinomycetoma have retained an unusually high degree of therapy resistance despite the availability of numerous potent antibacterial drugs. This somewhat surprising fact may be attributed to three major reasons, (i) clinical and microbiological diagnosis of both forms of disease has remained difficult to establish so that suitable therapeutic measures are often delayed, (ii) many patients, at least among those suffering from nocardiosis, are immunologically compromised so that natural defence mechanisms which are necessary to assist the action of antibiotics are not properly working and (iii) most of the aerobic actinomycetes in question exhibit a broad and pronounced natural resistance to many of the antimicrobial drugs currently in use. Methodological problems in susceptibility testing of nocardiae and actinomadurae further contribute to increased uncertainty so that the treatment is often empirical rather than based upon proven facts (Schaal and Leischik, 1969; Schaal and Heimerzheim, 1974; Schaal, 1977).

Besides surgical measures which may be helpful or even necessary in certain cases, antibacterial drugs form the therapeutic basis for treating nocardial infections. Since their introduction, sulphonamides have been known to be effective in nocardiosis (Benbow *et al.*, 1944). Especially *sulphadiazine*, which has proved to be active *in vitro* against most pathogenic *Nocardia* strains and has also shown comparatively favourable results *in vivo* (see Schaal and Heimerzheim, 1974). However, infections that do not respond to sulphadiazine or other sulphonamides are not rare.

Therefore, nearly all of the antimicrobial substances currently available have been tried *in vitro* and *in vivo* for their possible suitability in curing nocardial lesions. Penicillins and cephalosporins although occasionally applied are usually completely ineffective against *Nrda. asteroides* (Schaal and Leischik, 1969; Schaal and Heimerzheim, 1974; Schaal *et al.*, 1979) and should not be used. The efficacy of other antibiotics is variable depending on the subgroup of *Nrda. asteroides* to which an isolate belongs and on the derivative tested. Thus, certain strains are susceptible to aminoglycosides such as gentamicin, netilmicin, or amikacin (Schaal and Leischik, 1969;

Bach et al., 1973; Lerner and Baum, 1973; K. P. Schaal, unpublished data), to tetracyclines (minocycline) and to a combination of sulfamethoxazole and trimethoprim. In addition, fusidic-acid has been reported to be active *in vitro* against essentially all strains of *Nrda. asteroides* (Schaal and Leischik, 1969). The same is true for benzyl isothiocyanate which has not yet been clinically used, but would be limited to pulmonary infections because of its pharmacokinetics (Schaal and Heimerzheim, 1974). Nevertheless, sulphadiazine remains the drug of choice for treating *Nrda. asteroides* infections until sufficient clinical information has accumulated concerning the usefulness of other substances.

The considerations for treating *Nrda. brasiliensis* and *Nrda. otitidis-caviarum* infections are basically similar, but certain differences should be noted. In addition to sulphonamides, which may also be called the drugs of choice, sulfamethoxazole/trimethoprim and sulfones have been successfully used (Hogshead and Stein, 1970; Abbott et al., 1972; Mahgoub, 1972). Promising *in vitro* results were obtained with aminoglycosides. Fusidic acid is not active.

According to our present, rather limited knowledge, *Actinomadura* and *Streptomyces* infections should best respond to sulphadiazine or sulfamethoxazole plus trimethoprim. Alternatively, tetracyclines or sulfones (dapsone) could be tried.

F. OTHER 'OPPORTUNISTIC' PATHOGENIC ACTINOMYCETES

A variety of aerobic actinomycetes have been identified as the aetiology of serious, often fatal infections in the compromised host. In addition to the genera described above, *Oerskovia* and *Rhodococcus* are occasionally encountered but frequently misdiagnosed.

Sottnek et al. (1977) studied 35 isolates of *Oerskovia* obtained from clinical material. Nine of these isolates were identified as *Orsk. turbata* and 26 were *Orsk. xanthineolytica*. Five of the nine strains of *Orsk. turbata* were recovered from either heart valves or heart tissue, while only one of the *Orsk. xanthineolytica* isolates was associated with the heart. The remaining strains were isolated from either the blood, urine, sputum, lung, leg wounds, tear duct, eye, cerebrospinal fluid, liver or granuloma of the hand. The pathogenicity of these clinical isolates and their aetiological role were not evaluated. Reller et al. (1975) documented the first case of *Orsk. turbata* to cause endocarditis in a patient that had heart valve replacement while Cruickshank et al. (1979) documented a case of pyonephrosis caused by *Oerskovia* in a compromised patient.

Rhodococcus spp. have been isolated from a variety of environmental sources, however, they are rarely encountered as primary pathogens in healthy individuals. One of the first documented reports was a *Rhodococcus* infection in a young child who developed meningoencephalitis (Simon, 1962). Alture-Werber et al. (1968) reported two cases of severe systemic infections caused by *Rodc. rhodochrous* in 'compromised' patients. Since these early reports there have been several documented cases of serious infections in immunocompromised patients (Haburchak et al., 1978; Boughton and Atkin, 1980). *Rhodococcus* strains have been isolated from patients with chronic lung disease and in Japan 12 of 465 patients that were studied had *Rhodococcus* within their sputa (Tsukamura 1971, 1974). Corticosteroids significantly enhance host susceptibility to infection by this group of bacteria and the incidence of human infections caused by *Rhodococcus* is likely to be much more common than is generally recognized (Haburchak et al., 1978; Boughton and Atkin, 1980).

G. DERMATOPHILOSIS OR STREPTOTRICHOSIS

Dermatophilosis is an exudative, pustular dermatitis of world-wide distribution which primarily affects cattle, sheep and horses, but also many other domestic and feral animals (Gordon, 1964, 1974; Weber, 1978). A variety of names has been used for this disease before its aetiology was definitely clarified: streptotrichosis, cutaneous actinomycosis, mycotic dermatitis, nocardial dermatitis, lumpy wool and strawberry foot rot.

The causative agent of the infection is *Derm. congolensis*, the only species of the genus *Dermatophilus*. Dermatophili exhibit an unique life cycle leading to a characteristic morphology. In early growth stages *Dermatophilus* forms filamentous structures. These segment transversely as well as in at least two longitudinal planes to form packets of coccoid cells which become motile spores.

In animals, infections due to *Dermatophilus* first become apparent by the presence of tufted and incrusted hairs which is followed by scab formation. Later on, crusts can be easily removed leaving a reddish skin defect which tends to bleed (Gordon, 1974; Weber, 1978). Extensive infection often leads to death in lambs and cattle, either by concurrent infection or cachexia.

In humans, dermatophilosis is characterized by multiple pustules, furuncles, or desquamative eczema of the hands or forearms (Gordon, 1974). Crateriform keratolysis (pitted keratolysis) has also been observed (Weber, 1978). The human infection is usually self-limited; after 2 to 3 weeks, the disease heals spontaneously. Animal dermatophilosis has been successfully treated by simultaneous application of penicillin G and streptomycin (Weber, 1978).

Dermatophilus congolensis has hitherto solely been isolated from clinical specimens so that a direct transmission from diseased animals to other animals, including man, appears to be the usual route of infection. However, the contagiosity is apparently low between animals in temperate regions and the transmission requires injuries of the skin by thorns or splinters, high humidity or arthropods. The latter may not only produce lesions, but also carry the infective zoospores of the pathogen (Weber, 1978). Humans obviously acquire the disease through contact with infected animals so that dermatophilosis appears to be the only true anthropozoonosis among the actinomycete infections.

4. Actinomycete Spores as Allergens

Only recently, has it been realized that actinomycetes are important causes of respiratory allergy (Pepys *et al.*, 1963). Meanwhile, a good number of predominantly occupational diseases has been associated aetiologically with actinomycete spores which may be inhaled and induce a hypersensitivity reaction in the lungs. Such conditions are commonly referred to as farmer's lung disease, bagassosis, mushroom-worker's lung, humidifier fever, or byssinosis.

A. CLINICAL PICTURE

Allergic reactions to actinomycete particles are of the delayed type. Anatomically, they develop in the gas exchange tissue of the lung giving rise to a syndrome that is called allergic alveolitis or hypersensitivity pneumonitis.

Clinically, this syndrome is that of an interstitial pneumonitis which may have other causes as well. In its acute phase, the disease displays delayed-onset symptoms such as dyspnoea, fever, restrictive ventilatory defects, malaise, weight loss and radiographic changes (Berrens *et al.*, 1977; Burke *et al.*, 1977; Lacey, 1981). With repeated exposure, permanent lung damage with fibrosis can develop. Depending on the amount of antigen to which the patient is exposed, progressive destruction of the lungs may occur insidiously without expression of acute symptoms.

Histopathologically, allergic alveolitis is characterized by (i) an infiltration of the alveolar walls by mononucleated cells, (ii) high numbers of plasma cells in the infiltrate and (iii) a partly granulomatous bronchiolitis which can be distinguished from *Bronchiolitis obliterans* (Burke *et al.*, 1977).

In the lungs the size of the particles inhaled determines the type as well as the location of hypersensitivity reaction. Actinomycete spores, being mostly about 1 μm in diameter, are small enough to penetrate to the alveoli where they are deposited. To sensitize an individual constitutional predisposition is unimportant, but heavy exposure is necessary (Lacey, 1981). As in other delayed-type allergies, the onset of symptoms is delayed to at least four hours after exposure, the disease will fully develop within 12 hours and needs further 24 to 36 hours to resolve.

B. AETIOLOGY

Certain actinomycetes, especially thermophilic species, have been shown to multiply heavily in agricultural products such as hay, grain, cotton and sugar-cane bagasse and also in mushroom compost and humidifier water provided that they are supplied with an appropriate amount of humidity (Lacey, 1981; Burke et al., 1977). The spores produced by these organisms become easily detached and launched into the air when their substrate is disturbed. Under indoor conditions such air-borne spores may reach numbers which are sufficiently high to cause allergic symptoms.

Micropolyspora faeni, *Team. vulgaris*, *Sacs. viridis* and *Team. dichotomicus* have been implicated as aetiological agents of farmer's lung. *Thermoactinomyces sacchari* seems to be the most important cause of bagassosis and several thermophilic actinomycetes have been associated with mushroom-worker's lung, humidifier fever and byssinosis (for more details see Chapter 11).

C. INCIDENCE

Farmer's lung which is perhaps the best known of all forms of allergic alveolitis has been reported from most parts of Europe, from the U.S.A., and from Canada. Typically, it has a greater incidence in higher rainfall areas as has been demonstrated in detail for Great Britain (Lacey, 1981). Other forms of actinomycete-induced hypersensitivity pneumonitis are restricted to the environment that favours the growth of the respective organisms. This is also true for humidifier fever which requires a particular system for ventilation and heating to become prevalent (Burke et al., 1977).

D. TREATMENT AND PROPHYLAXIS

The most important means of treating allergic alveolitis is to identify the source and nature of the allergen and then prevent the patient from further exposure. However, this might be difficult in a given case so that antiallergic drugs such as corticosteroids may be needed to reduce the symptoms temporarily and improve the patient's condition.

To decrease the incidence of the diseases, moulding should be prevented. This can be achieved by drying to a low water content, by ensilage at high water content or by chemical preservatives such as propionic acid (Lacey, 1981). For artificial ventilation systems disinfectants are now available which are safe for man and animals and which inhibit growth of microbes in the humidifier water. Alternatively, steam should be used for humidification and care should be taken that the ventilating ducts remain dry.

REFERENCES

Abbott, L., Hunter, I. and Cutler, G. (1972). Mycetoma of the foot caused by *Nocardia brasiliensis*. *Medical Journal of Australia* **1**, 1136–1139.
Alture-Werber, E., O'Hare, D. and Louria, D. B. (1968). Infections caused by *Mycobacterium rhodochrous* and scotochromogens. *American Review of Respiratory Diseases* **97**, 694–698.
Andriole, V. T., Baldas, M. and Wilson, G. L. (1964). The association of nocardiosis and pulmonary alveolar proteinosis. *Annals of Internal Medicine* **60**, 266–274.
Bach, M. C., Sabath, L. D. and Findland, M. (1973). Susceptibility of *Nocardia asteroides* to 45 antimicrobial agents *in vitro*. *Antimicrobial Agents and Chemotherapy* **3**, 1–8.
Balikian, J. P., Herman, P. G. and Kopit, S. (1978). Pulmonary nocardiosis. *Radiology* **126**, 569–573.
Barnetson, R. S. and Milner, L. J. R. (1978). Mycetoma. *British Journal of Dermatology* **99**, 227–231.
Barnham, M., Burton, A. C. and Copland, P. (1978). Pelvic actinomycosis with IUCD. *British Medical Journal* **1**, 719–720.
Baron, E. J., Angevine, J. M. and Sundstrom, W. (1979). Actinomycotic pulmonary abscess in an immunosuppressed patient. *American Journal of Clinical Pathology* **72**, 637–639.
Batshon, B. A., Brosius, O. C. and Snyder, J. C. (1971). A case report of *Nocardia asteroides* of the eye. *Mycologia* **63**, 459–461.
Beaman, B. L., Burnside, J., Edwards, B. and Causey, W. (1976). Nocardial infections in the United States, 1972–1974. *Journal of Infectious Diseases* **134**, 286–289.
Benbow, E. P., Smith, D. T. and Grimson, K. S. (1944). Sulfonamide therapy in actinomycosis. *American Review of Tuberculosis* **49**, 395–407.
Berd, D. (1973). *Nocardia brasiliensis* infection in the United States: A report of nine cases and a review of the literature. *American Journal of Clinical Pathology* **60**, 254–258.
Bergey, D. H. (1907). *Actinomyces* der Mundhöhle. *Zentralblatt für Bakteriologie, Parasitenkunde und Infektionskrankheiten, Erste Abteilung: Referate* **40**, 361.
Berrens, L., De Ridder, G. and De Boer, F. (1977). Longitudinal studies of immunological parameters in farmer's lung. *Scandinavian Journal of Respiratory Diseases* **58**, 205–214.
Blank, C. H. and Georg, L. K. (1968). The use of fluorescent antibody methods for the detection and identification of *Actinomyces* species in clinical material. *Journal of Laboratory and Clinical Medicine* **71**, 283–293.
Bollinger, O. (1877). Über eine neue Pilzkrankheit beim Rinde. *Centralblatt für die medicinischen Wissenschaften* **15**, 481–485.
Bostroem, E. B. (1891). Untersuchungen über die Aktinomykose des Menschen. *Beiträge zur pathologischen Anatomie und zur allgemeinen Pathologie* **9**, 1–240.
Boughton, W. H. and Atkin, J. F. (1980). Ventricular peritoneal shunt infection caused by a member of the *Rhodococcus* complex. *Journal of Clinical Microbiology* **11**, 533–534.
Boyan-Salyers, B. D., Vogel, J. J. and Ennever, J. (1978). Basic biological sciences; pre-apatitic mineral deposition in *Bacterionema matruchotii*. *Journal of Dental Research* **57**, 291–295.
Boyd, A. and Williams, R. A. D. (1971). Estimation of the volumes of bacterial cells by scanning electron microscopy. *Archives of Oral Biology* **16**, 259–267.
Brede, H. D. (1959). Zur Ätiologie und Mikrobiologie der Aktinomykose. I. *In-vitro*-Versuche zur Frage der fermentativen Unterstützung des *A. israelii* durch Begleitbakterien. *Zentralblatt für Bakteriologie, Parasitenkunde und Infektionskrankheiten. I. Abteilung Originale* **174**, 110–122.
Brock, D. W., Georg, L. K., Brown, J. M. and Hicklin, M. D. (1973). Actinomycosis caused by *Arachnia propionica*. *American Journal of Clinical Pathology* **59**, 66–77.
Buchanan, B. B. and Pine, L. (1962). Characterization of a propionic acid producing actinomycete, *Actinomyces propionicus*, sp. nov. *Journal of General Microbiology* **28**, 305–323.
Bujwid, O. (1889). Über die Reinkultur des *Actinomyces*. *Zentralblatt für Bakteriologie, Parasitenkunde und Infektionskrankheiten* **6**, 630–633.
Burckhardt, J. J. (1978). Rat memory T lymphocytes: *in vitro* proliferation induced by antigens of *Actinomyces viscosus*. *Scandinavian Journal of Immunology* **7**, 167–172.
Burke, G. W., Carrington, C. B., Strauss, R., Fink, J. N. and Gaensler, E. A. (1977). Allergic alveolitis caused by home humidifiers. *Journal of the American Medical Association* **238**, 2705–2708.

Bushnell, R. B., Pier, A. C., Fichtner, R. E., Beaman, B. L., Boos, H. A. and Salman, M. D. (1979). Clinical and diagnostic aspects of herd problems with nocardial and mycobacterial mastitis. *American Association of Veterinary Laboratory Diagnosticians* **22**, 1–12.
Byrne, E., Brophy, B. P. and Perrett, L. V. (1979). Nocardia cerebral abscess: New concepts in diagnosis, management, and prognosis. *Journal of Neurology, Neurosurgery and Psychiatry* **42**, 1038–1045.
Causey, W. A. and Sieger, B. (1974). Systemic nocardiosis caused by *Nocardia brasiliensis. American Review of Respiratory Diseases* **109**, 134–137.
Causey, W. A., Arnell, P. and Brinker, J. (1974). Systemic *Nocardia caviae* infection. *Chest* **65**, 360–362.
Chavez, C. M., Causey, W. A. and Conn, J. H. (1972). Constrictive pericarditis due to infection with *Nocardia asteroides. Chest* **61**, 79–81.
Christ, M. L. and Haja, J. (1978). Cytologic changes associated with vaginal pessary use, with special reference to the presence of *Actinomyces. Acta Cytologica* **22**, 146–149.
Cisar, J. O. and Vatter, A. E. (1979). Surface fibrils (fimbriae) of *Actinomyces viscosus* T14V. *Infection and Immunity* **24**, 523–531.
Cohn, F. (1875). *Untersuchungen über Bacterien.* II. *Beiträge zur Biologie der Pflanzen* **III**, 141–207.
Colebrook, L. (1921). A report upon 25 cases of actinomycosis, with especial reference to vaccine therapy. *Lancet* **1**, 893–899.
Coleman, R. M. and Georg, L. K. (1969). Comparative pathogenicity of *Actinomyces naeslundii* and *Actinomyces israelii. Applied Microbiology* **18**, 427–432.
Collins, P. A., Gerencser, M. and Slack, J. M. (1973). Enumeration and identification of Actinomycetaceae in human dental calculus using the fluorescent antibody technique. *Archives of Oral Biology* **18**, 145–153.
Conrad, S. E., City, D., Breivis, J. and Fried, M. A. (1978). Vertebral osteomyelitis, caused by *Arachnia propionica* and resembling actinomycosis. *Journal of Bone and Joint Surgery* **60A**, 549–553.
Cook, F. V. and Farrar, W. E. (1978). Treatment of *Nocardia asteroides* infection with Trimethoprim-Sulfamethoxazole. *Southern Medical Journal* **71**, 512–515.
Crawford, J. M., Taubman, M. A. and Smith, D. J. (1978). The natural history of periodontal bone loss in germfree and gnotobiotic rats infected with periodontopathic microorganisms. *Journal of Periodontal Research* **13**, 316–325.
Cruickshank, J. G., Gawler, A. H. and Shaldon, C. (1979). *Oerskovia* species: Rare opportunistic pathogens. *Journal of Medical Microbiology* **12**, 513–515.
Cummins, C. S. and Harris, H. (1958). Studies on the cell-wall composition and taxonomy of Actinomycetales and related groups. *Journal of General Microbiology* **18**, 173–189.
Curry, W. A. (1980). Human nocardiosis: A clinical review with selected case reports. *Archives of Internal Medicine* **140**, 818–826.
DaCosta, T. and Gibbons, R. J. (1968). Hydrolysis of levan by human plaque streptococci. *Archives of Oral Biology* **13**, 609–617.
Davis, G. H. G. and Baird-Parker, A. G. (1959). *Leptotrichia buccalis. British Dental Journal* **106**, 70–73.
Distler, W. and Kröncke, A. (1981). Acid formation by mixed cultures of dental plaque bacteria *Actinomyces* and *Veillonella. Archives of Oral Biology* **26**, 123–126.
Distler, W., Ott, K. and Kröncke, A. (1980). Wechselwirkungen von *Streptococcus mutans, Actinomyces* und *Veillonella in vitro*—ein vereinfachtes Modell für den Kohlenhydratmetabolismus in der Plaque. *Deutsche zahnärztliche Zeitschrift* **35**, 548–553.
Ellen, R. P. and Balcerzak-Raczkowski, I. B. (1977). Interbacterial aggregation of *Actinomyces naeslundii* and dental plaque streptococci. *Journal of Periodontal Research* **12**, 11–20.
Ellis, P. P., Bausor, S. C. and Fulmer, J. M. (1961). *Streptothrix* canaliculitis. *American Journal of Ophthalmology* **52**, 36–43.
Emmons, C. W. (1938). The isolation of *Actinomyces bovis* from tonsillar granules. *Public Health Reports* **53**, 1967–1975.
Engel, D., Van Epps, D. and Clagett, J. (1976). *In vivo* and *in vitro* studies on possible pathogenic mechanisms of *Actinomyces viscosus. Infection and Immunity* **14**, 548–554.

Engel, D., Schroeder, H. E. and Page, R. C. (1978). Morphological features and functional properties of human fibroblasts exposed to *Actinomyces viscosus* substances. *Infection and Immunity* **19**, 287–295.

Ennever, J. (1960). Intracellular calcification by oral filamentous microorganisms. *Journal of Periodontology* **31**, 304–307.

Ennever, J., Vogel, J. J. and Streckfuss, J. L. (1971). Synthetic medium for calcification of *Bacterionema matruchotii*. *Journal of Dental Research* **50**, 1327–1330.

Ennever, J., Riggan, L. J., Vogel, J. J. and Boyan-Salyers, B. (1978). Characterization of *Bacterionema matruchotii* calcification nucleator. *Journal of Dental Research* **57**, 637–642.

Eppinger, J. (1891). Über eine neue pathogene *Cladothrix* und eine durch sie hervorgerufene Pseudotuberkulose. *Beiträge zur pathologischen Anatomie und zur allgemeinen Pathologie* **9**, 287–328.

Franke, F. (1973). Untersuchungen zur Ätiologie der Gesäugeaktinomykose des Schweines. *Zentralblatt für Bakteriologie, Parasitenkunde, Infektionskrankheiten und Hygiene*; *Erste Abteilung Originale A* **223**, 111–124.

Frazier, A. R., Rosenow, E. C. and Roberts, G. D. (1975). Nocardiosis: A review of 25 cases occurring during 24 months. *Mayo Clinic Proceedings* **50**, 657–663.

Freese, J. W., Young Jr., W. G. Sealy, W. C. and Conant, N. F. (1963). Pulmonary infection by *Nocardia asteroides*. *Journal of Thoracic and Cardiovascular Surgery* **46**, 537–547.

Fukushiro, R. and Mariat (1965). Note sur un mycétome à *Nocardia caviae* observé au Japan. *Bulletin de la Société de Pathologie Exotique* **58**, 185–188.

Geiseler, P. J. and Andersen, B. R. (1979). Results of therapy in systemic nocardiosis. *American Journal of Medical Sciences* **278**, 188–194.

Georg, L. K. (1974). *Nocardia* species as opportunists and current methods for their identification. In: *Opportunistic Pathogens* (J. E. Prier and H. Friedman, eds.), pp. 177–201. University Park Press, Baltimore.

Georg, L. K., Robertstad, G. W., Brinkmann, S. A. and Hicklin, M. D. (1965). A new pathogenic anaerobic *Actinomyces* species. *Journal of Infectious Diseases* **115**, 88–99.

Georg, L. K., Brown, J. M., Baker, H. J. and Cassell, G. H. (1972). *Actinomyces viscosus* as an agent of actinomycosis in the dog. *American Journal of Veterinary Research* **33**, 1457–1470.

Gerencser, M. A. and Slack, J. M. (1967). Isolation and characterization of *Actinomyces propionicus*. *Journal of Bacteriology* **94**, 109–115.

Gerencser, M. A. and Slack, J. M. (1969). Identification of human strains of *Actinomyces viscosus*. *Applied Microbiology* **18**, 80–87.

Gibbons, R. J. and Nygaard, M. (1970). Interbacterial aggregation of plaque bacteria. *Archives of Oral Biology* **15**, 1397–1400.

Gibbons, R. J. and Van Houte, J. (1973). On the formation of dental plaque. *Journal of Periodontology* **44**, 347–360.

González-Ochoa, A. (1962). Mycetomas caused by *Nocardia brasiliensis*: with a note on the isolation of the causative organism from the soil. *Laboratory Investigation* **11**, 1118–1123.

Gordon, M. A. (1964). The genus *Dermatophilus*. *Journal of Bacteriology* **88**, 509–522.

Gordon, M. A. (1974). Aerobic pathogenic Actinomycetaceae. In *Manual of Clinical Microbiology*, 2nd edition. (E. H. Lennette, E. H. Spaulding and J. P. Truant, eds.), pp. 175–188. American Society of Microbiology, Washington D.C.

Gordon, R. E. and Horan, A. C. (1968). *Nocardia dassonvillei*, a macroscopic replica of *Streptomyces griseus*. *Journal of General Microbiology* **50**, 235–240.

Gorevic, P. D., Katler, E. I. and Agus, B. (1980). Pulmonary nocardiosis: Occurrence in men with systemic lupus erythematosus. *Archives of Internal Medicine* **140**, 361–363.

Green, S. L. (1978). Case report. Fatal anaerobic pulmonary infection due to *Bifidobacterium eriksonii*. *Postgraduate Medicine* **63**, 187–192.

Green, W. O. and Adams, T. E. (1964). Mycetoma in the United States: A review and report of seven additional cases. *American Journal of Clinical Pathology* **42**, 75–91.

Greer, K. (1974). Nocardial mycetoma. *Virginia Medical Monthly* **101**, 193–195.

Guillou, J.-P., Durieux, R., Dublanchet, A. and Chevrier, L. (1977). *Actinomyces odontolyticus*, première étude realisée en France. *C.R. Academie des Sciences Paris* **285**, 1561–1564.

Gupta, P. K., Hollander, D. H. and Frost, J. K. (1976). Actinomycetes in cervico-vaginal smears. An association with IUD usage. *Acta Cytologica* **20**, 295–297.

Gupta, P. K., Erozan, Y. S. and Frost, J. K. (1978). Actinomycetes and the IUD: An update. *Acta Cytologica* **22**, 281–282.
Haburchack, D. R., Jeffery, B., Higbee, J. W. and Everett, E. D. (1978). Infections caused by *Rhodochrous*. *American Journal of Medicine* **65**, 298–302.
Haim, S. and Merzbach, D. (1979). Primary chancriform syndrome caused by *Nocardia asteroides*. *Mykosen* **22**, 360–363.
Hardie, J. (1974). Anaerobes in the mouth. In: *Infection with Non-sporing Anaerobic Bacteria* (I. Phillips and M. Sussman eds.), pp. 99–130. Churchill-Livingstone, Edinburgh.
Harris, G. K. (1980). Nocardiosis: A literature review and a case report of *Nocardia asteroides* infection. *American Journal of Medical Technology* **40**, 44–48.
Harz, C. O. (1877). *Actinomyces bovis*, ein neuer Schimmel in den Geweben des Rindes. *Jahresbericht der Königlichen Centralen Thierarzeneischule München für 1877/1878* **5**, 125–140.
Hemmes, G. D. (1963). Enige bevindingen over actinomycose. *Nederlands Tijdschrift voor Geneeskunde* **107**, 193.
Herzog, S. (1981). *Retrospektive Untersuchungen zur Klinik, Therapie und Bakteriologie der zerviko-fazialen Aktinomykosen an der Zahn- und Kieferklinik der Universität zu Köln von 1952 bis 1975–Auswertung der Aufzeichnungen von 317 Patienten*. MD Thesis, Köln.
Hirst, L. W., Harrison, G. K., Merg, W. G. and Stark, W. J. (1979). *Nocardia asteroides* keratitis. *British Journal of Ophthalmology* **63**, 449–454.
Hogshead, H. P. and Stein, G. H. (1970). Mycetoma due to *Nocardia brasiliensis*. *Journal of Bone Joint Surgery* **52**, 1229–1234.
Hoken, A. G., Smith, E. R. and Aarans, I. (1978). Goodpasture's syndrome complicated by *Nocardia asteroides* infection. *New Zealand Medical Journal* **87**, 38–41.
Holm, P. (1950). Studies on the aetiology of human actinomycosis. I. The 'other microbes' of actinomycosis and their importance. *Acta Pathologica et Microbiologica Scandinavica* **27**, 736–751.
Holm, P. (1951). Studies on the aetiology of human actinomycosis. II. Do the 'other microbes' of actinomycosis possess virulence? *Acta Pathologica et Microbiologica Scandinavica* **28**, 391–406.
Holmberg, K. and Nord, C.-E. (1975). Numerical taxonomy and laboratory identification of *Actinomyces* and *Arachnia* and some related bacteria. *Journal of General Microbiology* **91**, 17–44.
Houang, E. T., Lovett, I. S., Thompson, F. D., Harrison, A. R., Joekes, A. M. and Goodfellow, M. (1980). *Nocardia asteroides* infection—A transmissible disease. *Journal of Hospital Infection* **1**, 31–40.
Howell, A. and Jordan, H. V. (1967). Production of an extracellular levan by *Odontomyces viscosus*. *Archives of Oral Biology* **12**, 571–573.
Howell, A., Murphy, W. C., Paul, F. and Stephan, R. M. (1959). Oral strains of *Actinomyces*. *Journal of Bacteriology* **78**, 82–95.
van Houte, J. and Jansen, H. M. (1968). Levan degradation by streptococci isolated from human dental plaque. *Archives of Oral Biology* **13**, 827–830.
Israel, J. (1978). Neue Beobachtungen auf dem Gebiete der Mykosen des Menschen. *Archiv für Pathologie, Anatomie, Physiologie und Klinische Medizin* **74**, 15–53.
Jordan, H. V. and Hammond, B. F. (1972). Filamentous bacteria isolated from root surface caries. *Archives of Oral Biology* **17**, 1–12.
Jordan, H. V., Fitzgerald, R. J. and Stanley, H. R. (1965). Plaque formation and periodontal pathology in gnotobiotic rats infected with an oral actinomycete. *American Journal of Pathology* **47**, 1157–1167.
Jordan, H. V., Keyes, P. H. and Bellack, S. (1972). Periodontal lesions in hamsters and gnotobiotic rats infected with *Actinomyces* of human origin. *Journal of Periodontal Research* **7**, 21–28.
Juminer, B., Khalfat, A., Heldt, N. and Mariat, F. (1965). Deuxième cas tunisien de mycétome à *Nocardia*. *Bulletin de la Société de Pathologie Exotique* **58**, 177–185.
Karetzky, M. S. and Garvey, J. W. (1974). Empyema due to *Actinomyces naeslundii*. *Chest* **65**, 229–230.
Katz, P. and Fauci, A. S. (1977). *Nocardia asteroïdes* sinusitis: Presentation as a trimethoprim—sulfamethoxazole responsive fever of unknown origin. *Journal of the American Medical Association* **238**, 2391–2398.

King, S. and Meyer, E. (1957). Metabolic and serologic differentiation of *Actinomyces bovis* and anaerobic diphtheroids. *Journal of Bacteriology* **74**, 234–238.

Kirmani, N., Tuazon, C. U., Ocuin, J. A., Thompons, A. M., Kramer, N. C. and Geelhoed, G. W. (1978). Extensive cerebral nocardiosis cured with antibiotic therapy alone. *Journal of Neurosurgery* **49**, 924–928.

Kohoutek, M. and Nozicka, Z. (1978). Tubaraktinomykose also Komplikation der intrauterinen Antikonzeption. *Zentralblatt für Gynäkologie* **100**, 179–182.

Krick, J. A., Stinson, E. B. and Remington, J. S. (1975). *Nocardia* infection in heart transplant patients. *Annals of Internal Medicine* **82**, 18–26.

Kurup, P. V., Randhawa, H. S. and Gupta, N. P. (1970). Nocardiosis: A review. *Mycopathologia et Mycologia Applicata* **40**, 193–219.

Lacey, J. (1981). Airborne actinomycete spores as respiratory allergens. *Zentrablatt für Bakteriologie, Mikrobiologie und Hygiene. I. Abteilung, Supplement* **11**, 243–250.

Lentze, F. A. (1938). Die mikrobiologische Diagnostik der Aktinomykose. *Münchner medizinische Wochenschrift* **47**, 1826–1836.

Lentze, F. (1948). Die Ätiologie der Aktinomykose des Menschen. *Deutsche zahnärztliche Zeitschrift* **3**, 913–919.

Lentze, F. (1953). Zur Aetiologie und mikrobiologischen Diagnostik der Aktinomykose. *Estratto degli Atti del VI Congresso Internationale di Microbiologia, Roma.* 5 Sez. **XIV**, 145–148.

Lentze, F. (1969). Die Aktinomykose und die Nocardiosen. In: *Die Infektionskrankheiten des Menschen und ihre Erreger, Band I. Zweite Auflage* (A. Grumbach and O. Bonin, eds.), pp. 954–973. George Thieme-Verlag, Stuttgart.

Lentze, F. (1970). Klinik, Diagnostik und Therapie der Aktinomykosen. In: *Diagnostik und Therapie der Pilzkrankheiten und neuere Erkenntnisse in der Biochemie der pathogenen Pilze* (Kongressreferate, 6. Tagung der Deutschsprachigen Mykologischen Gesellschaft am 15. Juli 1966), pp. 83–92. Grosse Verlag, Berlin.

Lerner, P. I. and Baum, G. L. (1973). Antimicrobial susceptibility of *Nocardia* species. *Antimicrobial Agents and Chemotherapy* **4**, 85–93.

Lesher, R. J. and Gerencser, V. F. (1977). Levan production by a strain of *Rothia*: Activation of complement resulting in cytotoxicity for human gingival cells. *Journal of Dental Research* **56**, 1097–1105.

Lesher, R. J., Gerencser, V. F. and Morrison, D. J. (1977). Presence of *Rothia dentocariosa* strain 477 serotype 2 in gingiva of patients with inflammatory periodontal disease. *Journal of Dental Research* **56**, 189.

Lewis, R. and Gorbach, S. L. (1972). *Actinomyces viscosus* in man. *Lancet* **i**, 641–642.

Lissner, G. S., O'Grady, R. and Choromokos, E. (1978). Endogenous intraocular *Nocardia asteroides* in Hodgkin's disease. *American Journal of Ophthalmology* **86**, 388–394.

Llory, H., Guillo, B. and Frank, R. M. (1971). A cariogenic *Actinomyces viscosus* – a bacteriological and gnotobiotic study. *Helvetica Odontologica Acta* **15**, 134–138.

Luff, R. D., Gupta, P. K., Spence, M. R. and Frost, J. K. (1978). Pelvic actinomycosis and the intrauterine contraceptive device. *American Journal of Clinical Pathology* **69**, 581–586.

Lutwick, L. I. and Rockhill, R. C. (1978). Abscess associated with *Rothia dentocariosa*. *Journal of Clinical Microbiology* **8**, 612–613.

McIntire, F. C., Vatter, A. E., Baros, J. B. and Arnold, J. (1978). Mechanism of coaggregation between *Actinomyces viscosus* T14V and *Streptococcus sanguis* 34. *Infection and Immunity* **21**, 978–988.

Mackinnon, J. E. and Artagaveytia-Allende, R. C. (1956). The main species of pathogenic aerobic actinomycetes causing mycetomas. *Transactions of the Royal Society for Tropical Medicine and Hygiene* **50**, 31–40.

Magnusson, H. (1928). The commonest forms of a actinomycosis in domestic animals and their etiology. *Acta Pathologica et Microbiologica Scandinavia* **5**, 170–245.

Mahgoub, E. S. (1972). Treatment of actinomycetoma with sulfamethoxazole plus trimethoprim. *American Journal of Tropical Medicine and Hygiene* **21**, 332–335.

Manly, B. S. and Richardson, D. T. (1968). Metabolism of levan by oral samples. *Journal of Dental Research* **47**, 1080–1086.

Mariat, F. (1963). Sur la distribution géographique et la répartition des agents de mycétomes. *Bulletin de la Société de Pathologie Exotique* **56**, 35–45.

Miller, C. H. and Somers, P. J. (1978). Degradation of levan by *Actinomyces viscosus*. *Infection and Immunity* **22**, 266–274.

Miller, C. H., Warner, T. N., Palenik, C. J. and Somers, P. J. B. (1978a). Levan formation by whole cells of *Actinomyces viscosus* ATCC 15987. *Journal of Dental Research* **54**, 906.

Miller, C. H., Palenik, C. J. and Stamper, K. E. (1978b). Factors affecting the aggregation of *Actinomyces naeslundii* during growth and in washed cell suspensions. *Infection and Immunity* **21**, 1003–1009.

Mishra, S. K., Gordon, R. E. and Barnett, D. A. (1980). Identification of nocardiae and streptomycetes of medical importance. *Journal of Clinical Microbiology* **11**, 728–736.

Mitchell, G., Wells, G. M. and Goodman, J. S. (1975). Sporotrichoid *Nocardia brasiliensis* infection. *American Review of Respiratory Diseases* **112**, 721–723.

Mitchell, P. D., Hintz, C. S. and Haselby, R. C. (1977). Malar mass due to *Actinomyces odontolyticus*. *Journal of Clinical Microbiology* **5**, 658–660.

Moore, M. and Conrad, A. R. (1967). Sporotrichoid nocardiosis caused by *Nocardia brasiliensis*. *Archives of Dermatology* **95**, 390–393.

Mosimann, J., Hany, A. and Kayser, F. H. (1979). Pulmonale *Actinomyces-viscosus*-Infektion. *Schweizer medizinische Wochenschrift* **109**, 720–722.

Murray, J. F., Finegold, S. M., Froman, S. and Will, D. W. (1961). The changing spectrum of nocardiosis. *American Review of Respiratory Diseases* **83**, 315–330.

Myers, H. B. (1937). Thymol therapy in actinomycosis. *Journal of the American Medical Association* **108**, 1875.

Naeslund, C. (1925). Studies of *Actinomyces* from the oral cavity. *Acta Pathologica et Microbiologica Scandinavica* **2**, 110–140.

Neu, H. C., Silva, M., Hazen, E. and Rosenheim, S. H. (1967). Necroitizing nocardial pneumonitis. *Annals of Internal Medicine* **66**, 274–284.

Nitta, T., Okumura, S. and Nakano, M. (1977). Effect of *Bacterionema matruchotii* on immune response. I. Enhancement of the phagocytic and bactericidal functions. *Japanese Journal of Bacteriology* **32**, 691–696.

Nitta, T., Okumura, S., Tanabe, M. I. and Nakano, M. (1978). Water-soluble adjuvant obtained from *Bacterionema matruchotii*. *Infection and Immunity* **20**, 721–727.

Nocard, E. (1888). Note sur la maladie des boeufs de la Guadeloupe, connue sous le nom de farcin. *Annales de l'Institut Pasteur, Paris* **2**, 293–302.

Ohkawa, T., Kimura, K. and Sato, S. (1969). On the foul odour-removing effect of metronidazole in various branches of medicine. *Shindan to Chiryo*, 1278–1292.

Orchard, V. A. (1979). Nocardial infections of animals in New Zealand, 1976–1978. *New Zealand Veterinary Journal* **27**, 159–160.

Pabst, M. J. (1977). Levan and levansucrase of *Actinomyces viscosus*. *Infection and Immunity* **15**, 518–526.

Palmer, D. L., Harvey, R. L. and Wheeler, J. K. (1974). Diagnostic and therapeutic considerations in *Nocardia asteroides* infection. *Medicine* **53**, 397–401.

Pape, F., Singer, C., Kiehn, T. E., Lee, B. J. and Armstrong, D. (1979). Infective endocarditis caused by *Rothia dentocariosa*. *Annals of Internal Medicine* **91**, 746–747.

Pepys, J., Jenkins, P. A., Festenstein, G. N., Lacey, M. E., Gregory, P. H. and Skinner, F. A. (1963). Farmer's lung. Thermophilic actinomycetes as a source of 'farmer's lung hay' antigen. *Lancet* **ii**, 607–611.

Peterson, D. L., Hudson, L. D. and Sullivan, K. (1978). Disseminated *Nocardia caviae* with positive blood cultures. *Archives of Internal Medicine* **138**, 1164–1165.

Pine, L. and Hardin, H. (1959). *Actinomyces israelii*, a cause of lacrimal canaliculitis in man. *Journal of Bacteriology* **78**, 164–170.

Pine, L., Howell, A. and Watson, S. J. (1960). Studies of the morphological, physiological and biochemical characters of *Actinomyces bovis*. *Journal of General Microbiology* **23**, 403–424.

Presant, C. A., Wiernik, P. H. and Serpick, A. A. (1970). Disseminated extrapulmonary nocardiosis presenting as a renal abscess. *Archives of Pathology* **89**, 560–564.

Presant, C. A., Wiernik, P. H. and Serpick, A. A. (1973). Factors affecting survival in nocardiosis. *American Review of Respiratory Diseases* **108**, 1444–1448.

Pulverer, G. (1974). Problems of human actinomycosis. *Postepy Higienyi i Medycyny Doswiadczalnej* **28**, 253–260.

Pulverer, G. and Schaal, K. P. (1978). Pathogenicity and medical importance of aerobic and anaerobic actinomycetes. *Zentralblatt für Bakteriologie, Parasitenkunde, Infektionskrankheiten und Hygiene.* 1. Abteilung, Supplement **6**, 417–427.

Raich, R. A., Casey, F. and Hall, W. H. (1961). Pulmonary and cutaneous nocardiosis. *American Review of Respiratory Diseases* **83**, 505–509.

Ralph, R. A., Lemp, M. A. and Liss, G. (1976). *Nocardia asteroides* keratitis: A case report. *British Journal of Ophthalmology* **60**, 104–106.

Rapaport, J. (1966). Primary chancriform syndrome caused by *Nocardia brasiliensis. Archives of Dermatology* **93**, 62–64.

Reller, L. B., Maddoux, G. L., Eckman, M. R. and Pappas, G. (1975). Bacterial endocarditis caused by *Oerskovia turbata. Annals of Internal Medicine* **83**, 664–666.

Ritz, H. C. (1963). Localization of *Nocardia* in dental plaque by immunofluorescence. *Proceedings of the Society for Experimental Biology and Medicine* **113**, 925–929.

Rogers, S. J. and Johnson, B. L. (1977). Endogenous *Nocardia* endophthalmitis: Report of a case in a patient treated for lymphocytic lymphoma. *Annals of Ophthalmology* **9**, 1123–1131.

Rosett, W. and Hodges, G. R. (1978). Recent experiences with nocardial infections. *American Journal of the Medical Sciences* **276**, 279–285.

Sanyal, M., Thammayya, A. and Basu, N. (1978). Actinomycetoma caused by organisms of the *Nocardia asteroides* complex and closely related strains. *Mykosen* **21**, 109–121.

Satterwhite, T. K. and Wallace, R. J. (1979). Primary cutaneous nocardiosis. *Journal of the American Medical Association* **242**, 333–336.

Schaal, K. P. (1977). *Nocardia, Actinomadura* and *Streptomyces.* In: *CRC Handbook Series in Clinical Laboratory Sciences, Section E., Clinical Microbiology, Volume 1* (A. von Graevenitz, ed.), pp. 131–158. CRC Press, Cleveland.

Schaal, K. P. (1979). Die Aktinomykosen des Menschen – Diagnose und Therapie. *Deutsches Ärzteblatt* **76**, 1997–2006.

Schaal, K. P. (1981). Actinomycoses. *Revue de l'Institut Pasteur, Lyon* **14**, 279–288.

Schaal, K. P. and Heimerzheim, H. (1974). Mikrobiologische Diagnose und Therapie der Lungennocardiose. *Mykosen* **17**, 313–319.

Schaal, K. P. and Leischik, W. (1969). Zur Antibiotikatherapie der Nocardiose. *Deutsche medizinische Wochenschrift* **94**, 2505–2507.

Schaal, K. P. and Pape, W. (1980). Special methodological problems in antibiotic susceptibility testing of fermentative actinomycetes. *Infection* **8**, Supplement 2, 176–182.

Schaal, K. P. and Pulverer, G. (1981). The genera *Actinomyces, Agromyces, Arachnia, Bacterionema* and *Rothia.* In: *The Prokaryotes: A Handbook of Habitats, Isolation and Identification of Bacteria* (M. P. Starr, H. Stolp, H. G. Trüper, A. Balows and H. G. Schlegel, eds.), pp. 1921–1950. Springer, New York.

Schaal, K. P. and Schofield, G. M. (1981a). Taxonomy of Actinomycetaceae. *Revue de l'Institut Pasteur, Lyon* **14**, 27–39.

Schaal, K. P. and Schofield, G. M. (1981b). Current ideas on the taxonomic status of the Actinomycetaceae. *Zentralblatt für Bakteriologie, Mikrobiologie und Hygiene.* 1. Abteilung, Supplement **11**, 67–78.

Schaal, K. P., Schütt-Gerowitt, H. and Pape, W. (1979). Cefoxitin-Empfindlichkeit pathogener aerober and anaerober Aktinomyzeten. *Infection* **7**, Supplement 1, 47–51.

Schafer, F. J., Wing, E. J. and Norden, C. W. (1979). Infectious endocarditis caused by *Rothia dentocariosa. Annals of Internal Medicine* **91**, 747–748.

Scharfen, J. (1975a). Untraditional glucose fermenting actinomycetes as human pathogens. Part I. *Actinomyces naeslundii* as a cause of abdominal actinomycosis. *Zentralblatt für Bakteriologie, Parasitenkunde, Infektionskrankheiten und Hygiene.* 1. Abteilung Originale A **232**, 308–317.

Scharfen, J. (1975b). Untraditional glucose fermenting actinomycetes as human pathogens. II. *Rothia dentocariosa* as a cause of abdominal actinomycosis and a pathogen for mice. *Zentralblatt für Bakteriologie, Parasitenkunde, Infektionskrankheiten und Hygiene.* 1. Abteilung Originale A **233**, 80–92.

Schofield, G. M. and Schaal, K. P. (1981). A numerical taxonomic study of members of the Actinomycetaceae and related taxa. *Journal of General Microbiology* **127**, 237–259.

Scully, R. E., Galdabini, J. J. and McNeely, B. U. (1980). Case records of the Massachusetts General Hospital. *New England Journal of Medicine* **302**, 1194–1199.
Sher, N. A., Hill, C. W. Eifrig, D. E. (1977). Bilateral intraocular *Nocardia asteroides* infection. *Archives of Ophthalmology* **95**, 1415–1418.
Simon, C. (1962). Über eine Infektion mit *Mycobacterium rhodochrous* bei Pachymeningiosis haemorrhagica interna. *Tuberculosearzt* **16**, 152–158.
Slack, J. M. (1974). Family *Actinomycetaceae* and the genus *Actinomyces*. In: *Bergey's Manual of Determinative Bacteriology, Eighth edition*. (R. E. Buchanan and N. E. Gibbons, eds.), pp. 659–667. Williams and Wilkins, Baltimore.
Slack, J. M. and Gerencser, M. A. (1975). *Actinomyces, Filamentous Bacteria*. Burgess Publishing Company, Minneapolis.
Snijders, E. P. (1924). Verslaag van det weternschappenllijk gedeelte der vergaderingen van de afdelling Sumatra's Oostkust. *Geneeskundig tijdschrift voor Nederlandsch-Indie* **64**, 75–77.
Socransky, S. S., Hubersak, C. and Propas, D. (1970). Induction of periodontal destruction in gnotobiotic rats by a human oral strain of *Actinomyces naeslundii*. *Archives of Oral Biology* **15**, 993–995.
Sottnek, F. O., Brown, J. M., Weaver, R. E. and Carrol, G. F. (1977). Recognition of *Oerskovia* species in the clinical laboratory: Characterization of 35 isolates. *International Journal of Systematic Bacteriology* **27**, 263–270.
Spence, M. R., Gupta, P. K., Frost, J. K. and King, T. M. (1978). Cytologic detection and clinical significance of *Actinomyces israelii* in woman using intrauterine contraceptive devices. *American Journal of Obstetrics and Gynecology* **131**, 295–298.
Stackebrandt, E. and Woese, C. R. (1981). The evolution of prokaryotes. In: *Molecular and Cellular Aspects of Microbial Evolution* (M. J. Carlile, J. F. Collins and B. E. B. Moseley, eds.), pp. 1–31. University Press, Cambridge.
Stuart, M. (1979). Peritoneal nocardiosis. *Diseases of the Colon and Rectum* **22**, 183–184.
Susena, G. P., Al-Shamma, A., Rowe, J. C., Herbert, C. C., Bassis, M. L. and Coggs, G. C. (1967). Purulent constrictive pericarditis caused by *Nocardia asteroides*. *Annals of Internal Medicine* **67**, 1021–1032.
Taichman, N. S., Hammond, B. F., Tsai, Ch.-Ch., Baehni, P. C. and McArthur, W. P. (1978). Interaction of inflammatory cells and oral microorganisms. VII. *In vitro* polymorphonuclear responses to viable bacteria and to subcellular components of avirulent and virulent strains of *Actinomyces viscosus*. *Infection and Immunity* **21**, 594–604.
Takazoe, I. (1961). Study on the intracellular calcification of oral aerobic leptotrichia. *Shika Gakuho* **61**, 394–401.
Talwar, P. and Sehgal, S. C. (1979). Mycetomas in Northern India. *Sabouraudia* **17**, 287–291.
Telford, E. D. (1915). Acute actinomycosis of the parotid gland. *British Medical Journal* **2**, 534.
Terezhalmy, G. Y. and Bottomley, W. K. (1978). Pulmonary nocardiosis associated with primary nocardial infection of the oral cavity. *Oral Surgery* **45**, 200–206.
Thadepalli, H. and Rao, B. (1979). *Actinomyces viscosus* infections of the chest in humans. *American Review of Respiratory Diseases* **120**, 203–206.
Thammayya, A., Basu, N., Sur-Roy-Chowdhmy, D., Banerjee, A. K. and Sanyal, M. (1972). Actinomycetoma pedis caused by *Nocardia caviae* in India. *Sabouraudia* **10**, 19–23.
Tsukamura, M. (1971). Proposal of a new genus, *Gordona*, for slightly acid-fast organisms occurring in sputa of patients with pulmonary disease and in soil. *Journal of General Microbiology* **68**, 15–24.
Tsukamura, M. (1974). Necessity for differentiation of rhodochrous group from the tubercle bacilli. *Japanese Journal of Microbiology* **18**, 94–97.
Wangensteen, O. H. (1932). Actinomycosis of the thorax with report of a case successfully operated upon. *Journal of Thoracic Surgery* **1**, 612–631.
Warner, T. N. and Miller, C. H. (1978). Cell-associated levan of *Actinomyces viscosus*. *Infection and Immunity* **19**, 711–719.
Weber, A. (1978). Zur Dermatophilose bei Tier und Mensch. *Münchner tierärztliche Wochenschrift* **91**, 341–345.
Whyte, H. J. and Kaplan, W. (1969). Nocardial mycetoma resembling granuloma faciale. *Archives of Dermatology* **100**, 720–723.

Wicken, A. J., Broady, K. W., Evans, J. D. and Knox, K. W. (1978). New cellular and extracellular amphipathic antigen from *Actinomyces viscosus* NY1. *Infection and Immunity* **22**, 615–616.

Wilhelmus, K. R., Robinson, N. M. and Jones, D. B. (1979). *Bacterionema matruchotii* ocular infections. *American Journal of Ophthalmology* **87**, 143–147.

Witwer, M. W., Farmer, M. F., Wand, J. S. and Solomon, L. S. (1977). Extensive actinomycosis associated with an intrauterine contraceptive device. *American Journal of Obstetrics and Gynecology* **128**, 913–914.

Wolff, M. and Israel, J. (1891). Über die Reinkultur des *Actinomyces* und seine Übertragbarkeit auf Thiere. *Archiv für pathologische Anatomie, Physiologie und klinsche Medizin* **126**, 11–59.

9
Laboratory Diagnosis of Actinomycete Diseases

K. P. SCHAAL

Institute of Hygiene, University of Cologne, D-5000 Cologne 41, Federal Republic of Germany

1. Introduction	425
2. Laboratory diagnosis of diseases due to members of the family Actinomycetaceae	426
A. Collection and transport of clinical specimens	426
B. Direct examination of the specimens – demonstration and identification of sulphur granules	428
C. Isolation of *Actinomycetaceae* from clinical material	429
D. Identification of *Actinomycetaceae*	432
E. Serological diagnosis of actinomycoses	442
3. Laboratory diagnosis of diseases due to aerobic actinomycetes	442
A. Diagnosis of nocardiosis and actinomycetoma	442
B. Diagnosis of dermatophilosis	451
C. Diagnosis of allergic alveolitis due to actinomycetes	452
References	454

1. Introduction

Actinomycetes may be aetiologically involved in a number of diseases and impairments which differ greatly with respect to clinical manifestation, seriousness and prognosis (see Chapter 8). Invasive actinomycete infections such as actinomycoses, nocardiosis and actinomycetoma must, however, be considered potentially malignant conditions that can become dangerous to health, or even life, when treatment is inadequate or delayed. It follows, therefore, that specific therapeutic measures should be selected and applied as early as possible in the course of the disease. This practice usually presupposes a detailed microbiological diagnosis as actinomycete-induced lesions, unlike certain other bacterial infections, often lack pathognomonic clinical features and may not respond to antimicrobial drugs currently used as standard therapeutic agents in the management of common suppurative or septicaemic processes.

Unfortunately, the detection and identification of pathogenic actinomycetes in clinical microbiology laboratories may present problems which, in any given case, may be ascribed to one or more of the following reasons. (i) As actinomycete infections are still considered to be rare, most clinicians and medical microbiologists are not familiar with them or with their causative agents and are consequently not specifically looking for these pathogens, (ii) isolation of actinomycetes from clinical specimens although comparatively easy in principle often proves extremely difficult in practice, as standard cultural techniques may not be suitable, and the application of special media and growth conditions requires experience and a tentative clinical diagnosis which are commonly lacking, (iii) for the latter reasons other means of demonstrating actinomycetes in clinical material (e.g. immunofluorescence) are usually also not employed when necessary and (iv) reliable identification of actinomycetes often depends on the application of certain modern taxonomic techniques (see Chapter 2) which are not yet widely used in clinical laboratories.

The primary aim of this chapter is, therefore, to give a comprehensive survey of the media and methods currently available for the isolation and identification of clinically significant actinomycetes. Details on serological tests have also been included. Special emphasis has, however, been placed on techniques that are easily applicable in the clinical laboratory but which are sensitive and reliable enough to ensure that the diagnosis of the various actinomycete diseases can be established with some certainty.

2. Laboratory Diagnosis of Diseases due to Members of the Family *Actinomycetaceae*

Among the diseases and impairments that may be induced by fermentative actinomycetes of the family *Actinomycetaceae*, the microbiological diagnosis of actinomycoses is of the utmost importance. The other forms of disease in which these organisms may be aetiologically involved are less severe, can usually be cured without establishing an aetiological diagnosis or are predominantly amenable to symptomatic treatment.

A. COLLECTION AND TRANSPORT OF CLINICAL SPECIMENS

Depending on the location and stage of the disease it may be easy, difficult or nearly impossible to obtain suitable material for microscopic and cultural examination.

In principle, abscess content, sinus discharge, lacrimal concretions and secretions, bronchial secretions, cervicovaginal secretions, biopsy samples, plaque material, gingival crevice, saliva, sputum or even stool specimens may yield growth of fermentative actinomycetes or may be used to demonstrate the organisms microscopically. Few of these materials, however, provide sufficient evidence for diagnosing actinomycete infections. Detection of fermentative actinomycetes in plaque material, gingival crevice, saliva, sputum or cervicovaginal secretions or in stools primarily reflects the fact that these microorganisms are normal inhabitants of human mucous membranes and may contaminate any sample that has been in contact, directly or indirectly, with mucosal secretions. Such specimens can, therefore, only be used for the qualitative or quantitative assessment of the normal actinomycete flora in different associations involved in the pathogenesis of caries and periodontal disease, but not for diagnostic purposes in the strict sense (Schaal and Pulverer, 1981).

In order to prove that an infection is caused by an actinomycete, care must be taken to avoid contamination of specimens with bacteria from mucous membranes. Liquid pus, therefore, should preferably be collected by needle aspiration through the outer integument after careful disinfection of the skin. Needle aspiration should also be performed before an abscess is incised although pus may also be obtained through the incision wound. In the latter case, swabs, although widely employed by clinicians, are a most undesirable way of collecting samples when larger amounts of pus are present.

Percutaneous transtracheal aspiration, direct lung puncture or needle biopsy are the most reliable ways of obtaining suitable specimens in cases of pulmonary actinomycosis as they bypass areas with a normal actinomycete flora. Sputum generally does not yield results of diagnostic relevance and bronchoscopically drawn fluids or washings may also be contaminated.

In certain infections, especially in those located deep in the abdomen or brain, the infected material may not be accessible to puncture or may present as an infiltrate from which actinomycete-containing portions are difficult to remove. In such cases, the disease may remain unrecognized until abscesses or sinus tracts have formed. In the case of draining sinus tracts, the surface material is likely to contain contaminants. Material should therefore be aspirated by syringe and a plastic catheter placed as deeply as possible through the decontaminated skin opening into the fistula. Again, swabs will give less satisfactory results.

As facultative anaerobes, pathogenic *Actinomycetaceae* strains are usually quite resistant to oxygen but may be inactivated when specimens are dried in air, are cooled below 20°C or suffer pH changes. Thus, samples should be transported and examined as quickly as possible. In addition, transport measures that maintain anaerobic conditions should be used. This latter practice should not only increase the chances of isolating the actinomycetes, but will also allow the detection of oxygen-sensitive concomitant bacteria which, if not recognized, could account for therapeutic failures.

The easiest way of preventing oxygen contact for a limited period of time is to leave the aspirated material in the syringe, expel air bubbles and insert the needle into a sterile rubber stopper (Finegold et al., 1974). This simple technique is only recommended when large amounts of fluid pus and short transportation times are involved.

For small samples, swabs, biopsy material and/or prolonged transport, the application of special anaerobic transport systems is indispensable. Various systems and media have been devised and are commercially available (Finegold et al., 1974; Hallander et al., 1980; Suzuki and Ueno, 1981). In principle, most of these systems are suitable for actinomycetes as long as they do not contain a reduced salt solution which might be inhibitory to certain strains of Actinomycetaceae.

B. DIRECT EXAMINATION OF THE SPECIMENS – DEMONSTRATION AND IDENTIFICATION OF SULPHUR GRANULES

Sulphur granules, as defined earlier (see Chapter 8), can easily be detected in pus specimens with the unaided eye. Their recognition and collection for further study is facilitated by spreading out pus in a sterile Petri dish. Macroscopically, the granules appear as yellowish to brownish or reddish particles which are about 1 mm or less in diameter and they can be so numerous in pus from actinomycotic abscesses that the discharge may resemble semolina soup (Lentze, 1969). Similar particles, however, may be formed by other filamentous bacteria including *Leptotrichia buccalis* and certain aerobic actinomycetes or may be derived from necrotic tissue so that the actinomycotic nature of the granules has to be proven. First indication of a 'true' sulphur granule is that it is hard and requires some pressure to break it down whereas other particles, with the exception of *Streptomyces* granules (Schaal, 1977), usually disintegrate when they are transferred to slides or culture media using platinum loops.

Microscopy is necessary to confirm that particles are sulphur granules. Particles are placed on slides by means of a platinum loop or inoculation needle, embedded in a drop of 1% (w/v) Methylene Blue solution and covered by a cover slip under slight pressure. At low magnification (80–100×) sulphur granules exhibit a typical cauliflower-like appearance (Fig. 1) (see Chapter 8). At higher magnification, hyaline clubs may be visible at the tips of peripheral filaments, but these structures are often not present in wet mounts of 'free' granules from the discharge and should not be considered as an indispensable character. Increased pressure on top of cover slips often results in typical signs of disintegration; the granule is separated into a mass of blue-coloured leucocytes and several dense, roundish structures which show radially arranged filaments at the periphery.

Granules can also be examined further after Gram staining. Thus, particles are crushed between two slides, fixed and then stained. The characteristic structures observed at high magnification (800–1,000×) are Gram-positive branched filaments of bacterial size and possible diphtheroid rods in Y- and V-forms, various additional Gram-positive and Gram-negative rods and cocci, and numerous granulocytes (Fig. 1). Preparations of this nature are highly suggestive of sulphur granules and differentiate the latter from *Nocardia*, *Actinomadura* and *Streptomyces* granules which do not contain the concomitant bacteria. Particles of other origin, which lack the mycelial structures or are composed of much larger filaments (in case of fungal infections), are also easily distinguished from sulphur granules.

Alternatively, immunofluorescence may be applied directly to crushed granules, thereby allowing a possible immediate identification of the causative actinomycete species present (see below). When sulphur granules are not detected in specimens, i.e. in about 50% of cases (see Chapter 8), the microscopic picture of Gram-stained smears is much less clear; single, branched, Gram-positive filaments may be seen but usually only diphtheroid elements are found, the latter may be formed by propionibacteria, corynebacteria and other Gram-positive organisms as well as by actinomycetes.

In tissue section, the haematoxylin and eosin stain usually does not allow actinomycotic foci to be identified with any certainty. However, the application of the Gram-Weigert stain or the Brown and Brenn modification allows actinomycete filaments to be clearly distinguished.

C. ISOLATION OF *ACTINOMYCETACEAE* FROM CLINICAL MATERIAL

In many cases, actinomycoses cannot be diagnosed with sufficient certainty unless the aetiological agents have been cultured and identified. This is especially so when sulphur granules are not present in the discharge. Unfortunately, most members of the family *Actinomycetaceae* have quite exacting nutritional requirements, need a reduced oxygen tension for growth and are not easily detected among the concomitant flora. Because of these problems, cultural results from cases of suspected actinomycosis are commonly considered to be unreliable and much emphasis is placed on histological findings. In our experience, histology is much less sensitive and reliable than data derived from the application of suitable microbiological techniques.

(a) Methods for Obtaining Anaerobic Growth Conditions

Most *Actinomycetaceae*, except *Rothia* strains, require a reduction in atmospheric oxygen and the presence of carbon dioxide for optimum growth.

In contrast to many strict anaerobes, however, elaborate techniques for obtaining and maintaining an oxygen-free atmosphere are usually not necessary for the successful isolation of fermentative actinomycetes from clinical specimens. Thus, the actinomycete laboratory can usually do without pre-reduced media, the role tube system, glove boxes or similar complicated and expensive equipment. The latter may, however, be suitable for special research purposes, for instance in studies on the flora of dental plaque. In addition, certain oxygen-sensitive concomitant anaerobes may be missed when simple, semi-anaerobic techniques are used. Nevertheless, the easiest, least expensive and most widely applicable technique for producing reduced growth conditions for actinomycetes is that described by Fortner in 1928 (Lentze, 1938, 1969; Schaal and Pulverer, 1981). This technique is based upon the finding that certain members of the family *Enterobacteriaceae*, especially *Serratia marcescens*, consume oxygen and produce carbon dioxide when grown in a closed system (Fortner, 1928, 1929).

Briefly, the procedure operates as follows. One half of a suitable agar medium in a glass Petri dish is inoculated with the material to be examined or the strain to be subcultured. The other half of the medium is heavily seeded with *Serratia* using a spatula. The dish is then placed upside down upon a glass sheet and fixed and sealed with plasticine in order to make the system air-tight. Growth of the anaerobes usually starts within the first 4 to 8 hours of incubation. Leakage in the plasticine seal, which can occur when the plasticine is too brittle or the base plate damp, can easily be detected, as *Serratia* forms unpigmented colonies under reduced oxygen tension, but is red in the presence of air (Schaal and Pulverer, 1981). Practically all representatives of the family *Actinomycetaceae* can be cultured by this method, even the more aerophilic strains. Furthermore, the growing culture can be checked daily with a hand lens or under the microscope at low magnification, provided transparent culture media are used, without disturbing the semi-anaerobic atmosphere (Lentze, 1953).

Alternatively, anaerobic jars, such as Torbal or GasPak (BBL) jars, may be employed. These jars support the growth of many pathogenic *Actinomycetaceae* strains and of nearly all of their anaerobic companions. However, the more aerophilic strains, especially those of the catalase-positive species, may be suppressed. Anaerobic jars should therefore only be used in combination with techniques (candle jar, GasPak jar with carbon dioxide-generating unit) that allow for a parallel culture in air with increased carbon dioxide.

Anaerobiosis for fluid cultures may be obtained by placing tubes with suitable broth media into anaerobic jars. However, a simple 'biological' procedure gives the same or even better results in enrichment cultures for fermentative actinomycetes. Tarozzi and cooked meat media are modifications of the oxygen-binding principle using sterilized animal tissue.

Actinomycetes grow especially well on the piece of liver used for Tarozzi broth and can be seen macroscopically as white, cotton pad-like colonies adhering to the tissue. The colonies can be selectively aspirated from the tissue using a capillary pipette and examined further.

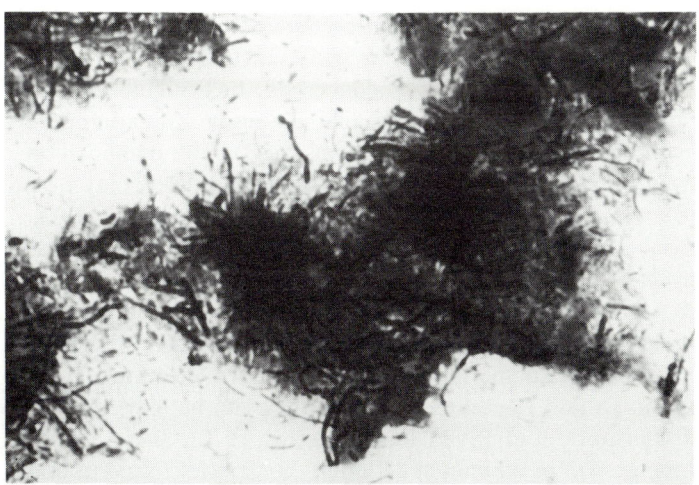

FIG. 1. Gram-stained smear from a sulphur granule (magnification about ×950). Note the mycelial structures and irregularly curved filaments, the other microbial particles represent the concomitant flora, and partially damaged leucocytes. (Figs. 1, 3, 4, 5, 6 photographed by S. Glanschneider.)

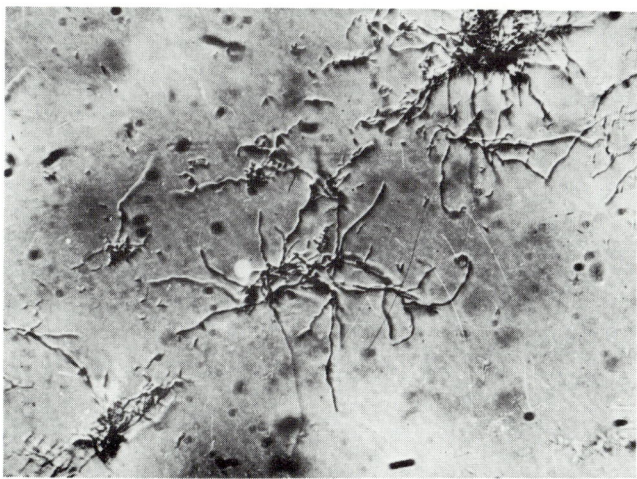

FIG. 2. Spider-like microcolonies of *Actinomyces israelii* grown on CC-medium for 48 hours at 37°C (magnification about ×160). (Photographed by Professor F. Lentze.)

(b) *Media and Isolation Procedures*

High-quality general-purpose media are usually required for the primary isolation and subsequent cultivation of *Actinomycetaceae* from clinical material. The media include Fluid Thioglycollate Broth, which may be supplemented with 0.1 to 0.2% (v/v) sterile rabbit serum, Brain Heart Infusion Agar, Trypticase Soy Agar and Schaedler Broth or Agar. Unfortunately, none of these media are completely satisfactory as far as the growth and detection of actinomycetes is concerned. In Cologne the best results have been obtained using a complex agar medium (CC-medium) developed some years ago by Heinrich and Korth (1967). The composition of this medium and instructions on how it should be prepared may be obtained from the original publication or from Schaal and Pulverer (1981).

There is clearly a need for reliable media to selectively isolate the actinomycetes given the obligate concomitant flora present in infected material. At present, no medium can be recommended to support the growth of all of the possible actinomycete pathogens. Selective media have been devised for studies of the mouth or plaque flora (Beighton and Colman, 1976; Kornman and Loesche, 1978) and they can be used together with a reduced transport fluid (Syed and Loesche, 1972); these media do not support the growth of all known actinomycete species found in this habitat.

Clinical specimens should be inoculated onto at least three to four different media including a fluid medium. Fortner plates can be checked daily for growth after incubation at 36°C. When anaerobic jars are preferred a duplicate or triplicate set of media should be inoculated and incubated at 36°C and examined after 3, 7 and 14 days, respectively. Surface cultures should be examined microscopically on transparent media at a magnification of 80 to 100× to detect the typical mycelial microcolonies of actinomycetes in the early stages of growth (Fig. 2). Fortner plates can be observed without opening if a long distance objective is employed (Lentze 1953; Schaal and Pulverer, 1981).

Sometimes it is difficult to obtain pure cultures of the actinomycete even after subculturing several times as certain concomitant bacteria repeatedly overgrow the plates. Depending on the kind of contaminant, commercially-available discs for antibiotic susceptibility testing may be used. Discs containing metronidazole, colistin, or nalidixic acid usually inhibit many of the contaminants but not the actinomycetes (Schaal and Pulverer, 1981).

D. IDENTIFICATION OF *ACTINOMYCETACEAE*

Although fermentative actinomycetes exhibit several characteristic properties potentially useful for identification, many clinical laboratories often encounter problems when trying to identify unknown isolates using morphology and standard laboratory tests. Thus, bacteria such as propionibacteria, eubacteria, lactobacilli, *Erysipelothrix* or *Arcanobacterium* can be difficult to distinguish from fermentative actinomycetes using such tests, and they may appear similar under the microscope. In addition, several actinomycete-like organisms in clinical specimens have yet to be characterized in detail and remain unnamed (Schofield and Schaal, 1981). Furthermore, there is no single simple test which can be used to differentiate actinomycete genera from one another or from a whole group of superficially related organisms.

(a) Serological Identification

Immunofluorescence is the technique most widely and successfully used in the serological identification of *Actinomycetaceae*. Compared with other diagnostic tests, immunofluorescence is reliable, comparatively easy, inexpensive and is not time-consuming. Direct (Slack *et al.*, 1961; Slack and Gerencser, 1975) and indirect (Schaal and Pulverer, 1973) modifications of the technique have been applied with good results.

For routine purposes, the procedure of Slack and Gerencser (1975) is readily applicable: smears from cultures or clinical material are air-dried, fixed by heating or flooding with methanol for 1 minute respectively, incubated with a drop of fluorescein isothiocyanate (FITC)-conjugated specific antiserum for 30 minutes in a moist chamber at room temperature, washed in two changes of pH 7.2 buffer (FTA haemagglutination buffer – BBL) for 5 minutes each, counterstained in 0.5% Evans Blue, and washed again in two changes of pH 9.0 buffer for 1 minute each. Air-dried smears are then embedded in a buffered glycerol mounting fluid (9 parts c.p. glycerol + 1 part pH 9 buffer) under a cover slip and examined at about 400× magnification using a microscope equipped for immunofluorescence work. Unfortunately, FITC-labelled specific antisera are not commercially available except for antisera against a few common *Actinomyces* species and serovars (Biological Reagent Section, CDC, Atlanta, Ga.).

As fluorescence serology can be applied to both clinical specimens and isolates from culture, some of the problems described above for the successful isolation of *Actinomycetaceae* can be avoided by using this technique. However, it must be noted that antigenically aberrant strains, hitherto unknown serovars or species, and undetected cross-reacting antibodies in the antiserum may cause misidentifications if immunofluorescence procedures are used as the sole means of identification.

(b) Morphological Characters

Until recently morphological characteristics were widely used in identification but they were of little value when an exact and reliable identification of an unknown isolate was required. Of course, branched filaments seen in clinical material or in smears from cultures (Fig. 3) are highly suggestive of actinomycetes. However, such observations do not allow isolates to be assigned to any of the broad groups of actinomycetes unless additional structures such as spores or signs of fragmentation are visible. A further difficulty is that various species of the *Propionibacteriaceae* may also exhibit filamentous forms under certain cultural conditions and may even appear branched.

Nevertheless, Gram-positive branching filaments which are non-acid-fast and irregularly curved and stained (Figs. 1 and 3) provide presumptive evidence of *Actinomyces* species although the latter may also appear as diphtheroidal rods in both pus and in culture. *Corynebacterium matruchotii* and *Rothia* often show characteristic morphological features that allow them to be presumptively identified (Schaal and Pulverer, 1981). The former is characterized by the formation of a thin filament attached to a large, bacillus-like body; this highly specific 'whip handle' morphology may also occur *in vivo* (Davis and Baird-Parker, 1959). Similarly, *Rothia* is the only fermentative actinomycete that produces numerous coccoid forms in addition to rods and filaments (Georg, 1974a).

From a diagnostic point of view, cultural morphology is more important than micromorphology. Typical filamentous spider-like colonies can be found (Fig. 4) especially when growing cultures are observed *in situ* after 24 to 48 hours of incubation (see above). Upon further incubation, the filaments usually fragment so that older colonies may be completely smooth and indistinguishable from those of other Gram-positive bacteria (Fig. 5). On the other hand, mature colonies may retain much of the mycelial appearance so that their actinomycete nature remains apparent even after prolonged incubation (Fig. 6).

Actinomyces bovis, *Actn. odontolyticus* and '*Actinobacterium meyerii*' usually form only a rudimentary mycelium in the very early stages of growth. In *Cnbc. matruchotii*, the filaments are visibly thicker and in *Arachnia* they often develop from the initial cell in one direction and not, as in *Actn. israelii*, with radial symmetry.

(c) Differentiation and Identification at the Generic Level

Because of pronounced morphological resemblances between the genera of the *Actinomycetaceae* and between them and other Gram-positive bacteria, it is not always clear whether clinical isolates are actinomycetes. Further, simple physiological tests, suitable for differentiation at the species level, often fail to answer this basic question. Thus, in addition to the traditional methods, modern and more elaborate techniques may be necessary to achieve the reliable identification of actinomycetes. Some of the newer techniques can be readily used in clinical microbiological laboratories whereas others should remain the domain of reference laboratories.

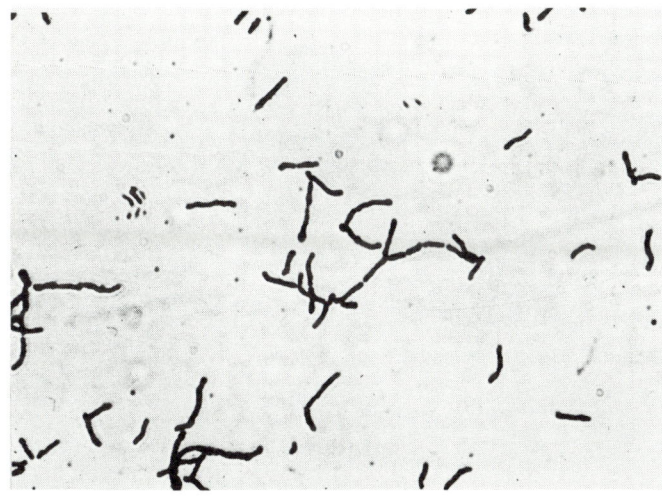

FIG. 3. Gram-stained smear from a pure culture of *Actinomyces israelii* (magnification about ×950), grown on BHI-agar for 7 days at 37°C.

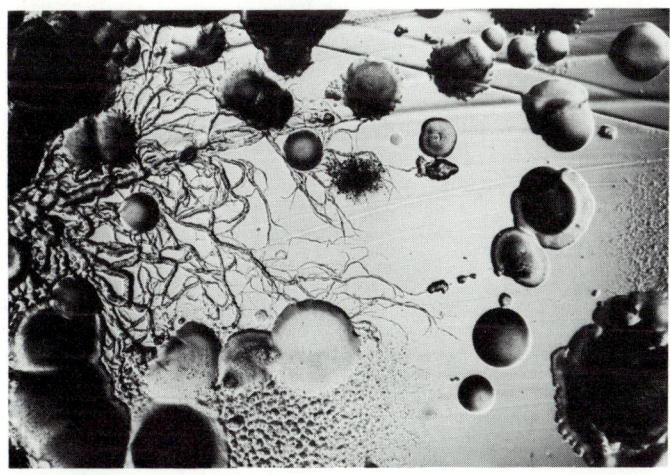

FIG. 4. Mixed actinomycotic flora from a case of pelvic actinomycosis (magnification about ×30), grown for 14 days at 37°C on CC-medium using the Fortner's method. Note the small, dense, filamentous actinomycete colony (slightly, above the centre of the photograph) among various colony forms of concomitant bacteria.

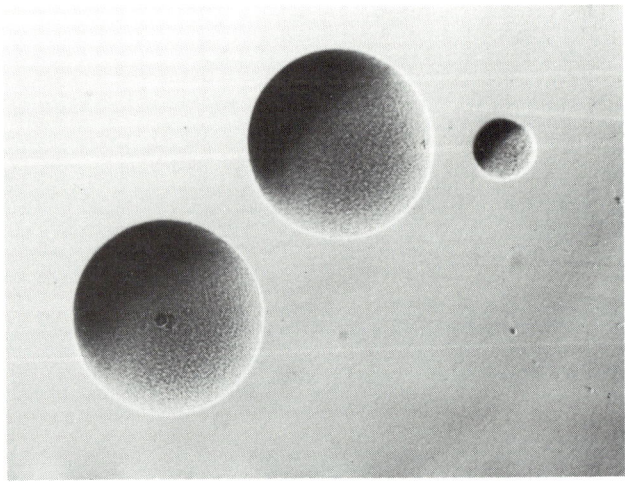

FIG. 5. Smooth mature colonies of *Actinomyces viscosus* grown on BHI-agar for 7 days at 37°C (magnification about ×100).

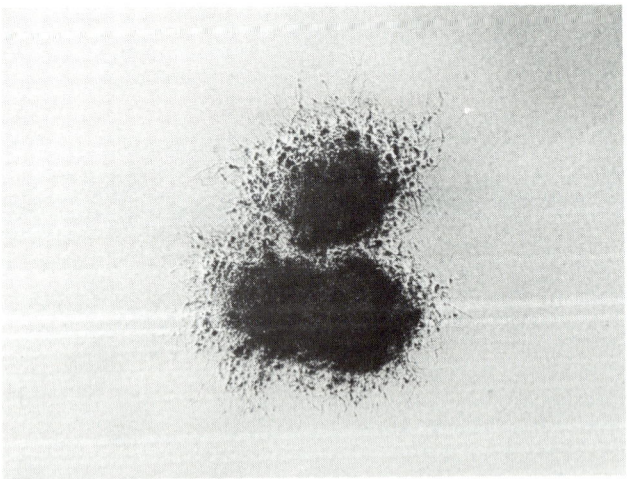

FIG. 6. Filamentous mature colonies of *Actinomyces naeslundii* with dense, granular centre grown on CC-medium for 7 days at 37°C (magnification about ×100).

Gas–liquid chromatography (GLC) is a useful, practicable and widely used technique that can be applied in clinical laboratories for the detection of acid fermentation end products. The GLC procedures described in the VPI Anaerobic Laboratory Manual (Holdeman et al., 1977) are well suited to the actinomycetes. In addition, analysis of whole-organism hydrolysates for characteristic cell-wall and lipid markers can provide valuable data. Technically simple procedures (see below) are available which make these chemical tests suitable for routine purposes. Most of the *Actinomycetaceae* and superficially similar organisms can be differentiated at the genus level using these chemical techniques and two physiological tests (Table 1).

(d) *Identification at the Specific Level*

Physiological markers such as growth on different media, atmospheric requirements and acid production from carbohydrates have long been used for the reliable identification of clinically significant *Actinomycetaceae* to the species level. Once the genus of an unknown isolate has been determined using the methods outlined above, these more conventional tests can be applied (Slack and Gerencser, 1975) in order to differentiate the species. It is, however, very important to carry out the physiological tests with great care as they are sensitive to small changes in methodology. Improved results may be obtained by using micromethods which provide a more precise relationship between inoculum and medium, and are more easily standardized (Schofield and Schaal, 1979a, b, 1980a, b, 1981). Similarly, commercial test kits (e.g., Minitek System – BBL) are available, which to some extent solve the problem of quality standardization.

The diagnostic tables shown in Tables 2 and 4 are predominantly based upon such micromethods (Schofield and Schaal, 1981). Conventional tests may give both identical and divergent results. Further differentiation within established species (Table 3) requires especially careful adaptation of the test conditions. A number of species whose taxonomic affiliation is still under discussion have been included in Tables 2 and 4 as such strains can be encountered in clinical material and may be difficult to differentiate from 'true' actinomycetes.

Table 3 has been included to facilitate the identification of the heterogeneous species *Actn. naeslundii* and *Actn. viscosus* and to demonstrate that the serovars of *Actn. israelii, Actn. naeslundii, Actn. viscosus* and *Arac. propionica* can be recognized using physiological properties (see Chapter 2; Schofield and Schaal, 1981).

Strains that cannot be assigned to any of the established species are not infrequently isolated from clinical specimens. Some of these organisms require further study but are listed in Tables 2 and 3 to aid in their recognition and to prevent them from being overlooked. *Arcanobacterium haemolyticum* and *Erysipelothrix rhusiopathiae* are included in Tables 1 and 4 as they may be confused with pathogenic actinomycetes.

TABLE 1. Differentiation of the genera of the family *Actinomycetaceae* from one another and from related taxa[a]

Genus	Physiological characters		Chemotaxonomic characters			Major acid end products from glucose fermentation[d]										
	Catalase	Nitrate reduction	DL-DAP[b]	LL-DAP[b]	Mycolic acids[c]	Acetic	Propionic	iso-Butyric	n-Butyric	iso-Valeric	n-Valeric	iso-Caproic	n-Caproic	Pyruvic	Lactic	Succinic
Actinomyces	-[e]	d	-	-	-	+	-	-	-	-	-	-	-	-	+	+
Arachnia	-	+	-	+	-	+	+	-	-	-	-	-	-	-	(+)	(+)
Bifidobacterium	-	-	-	-	-	+	-	-	-	-	-	-	-	-	+	(+)
Bacterionema (*Cnbc. matruchotii*)	+	+	+	-	+[f]	+	d	-	-	-	-	-	-	d	d	-
Rothia	+	+	-	-	-	+	-	-	-	-	-	-	-	(+)	+	(+)
Propionibacterium	d	d	(+)	-	-	+	+	-	-	d	-	-	-	-	(+)	(+)
Eubacterium	-	d	d	-	-	(+)	(+)	-	+	-	-	-	d	-	d	(+)
Lactobacillus	-	-	-	-	-	(+)	-	-	-	-	-	-	-	-	+	+
Erysipelothrix	-	-	-	-	-	+	-	-	-	-	-	-	-	-	+	+
Arcanobacterium (*Cnbc. haemolyticum*)	-	d	-	-	-	+	-	-	-	-	-	-	-	-	+	d

+, present; -, absent; d, species or type differences, and (+) present in small amounts or absent.
[a] Data from Slack and Gerencser (1975); Holmberg and Nord (1975); Holdeman et al. (1977); Schaal et al. (1980); Schaal and Schofield (1981a, b).
[b] DAP, *meso*- or LL-diaminopimelic acid in whole-cell hydrolysates.
[c] Mycolic acids in whole-cell methanolysates.
[d] Produced in PYG medium (Holdeman et al., 1977).
[e] *Actinomyces viscosus* is catalase-positive.
[f] *Propionibacterium lymphophilum* does not contain DAP.

Test	Actin. israelii	Actin. naeslundii	Actin. viscosus	Actin. bovis	Actin. odontolyticus	Arac. propionica	'Bibe. eriksonii'	'Actinobacterium meyerii'	Actin. (Cnbc.) pyogenes	Unnamed oral isolates (cluster 4)[b]
Aerobic growth	−	d	d	−	−	−	−	−	−	−
Colony rough	+	d	d	−	−	+	−	−	−	−
Catalase	−	−	+	−	−	−	−	−	−	d
Nitrite reduction	d	+	−	−	−	−	ND	−	−	d
Nitrate reduction	+	+	d	−	+	+	−	+	−	+
Aesculin hydrolysis	−	+	−	+	−	−	+	−	d	d
Starch hydrolysis	−	−	d	+	d	−	+	d	−	−
Acid from:										
Arabinose	d	−	−	−	d	d	+	d	−	d
Cellobiose	+	d	−	−	−	+	+	−	−	d
Glucose	+	+	+	+	+	+	+	+	+	+
Glycerol	−	−	−	−	−	d	−	−	−	−
meso-Inositol	+	d	+	+	−	+	−	−	−	−
Mannitol	d	d	d	−	−	d	+	−	+	d
Raffinose	+	+	−	−	−	+	+	−	−	−
Ribose	−	−	+	+	+	+	+	−	−	+
Sorbitol	d	d	d	−	−	d	+	−	−	d
Sucrose	+	+	+	+	+	+	+	+	+	+
Trehalose	+	+	d	−	d	+	+	+	+	+
Xylose	−	−	−	−	−	−	−	−	−	d
Ammonia from:										
Arginine	+	d	−	−	−	−	ND	−	−	−
Urea	−	+	d	−	−	−	ND	−	−	d

+, positive; −, negative; d, strain differences; and ND, not determined.

[a] Data from Holmberg and Nord (1975); Holdeman *et al.* (1977); Schofield and Schaal (1980a, b; 1981) and Schaal and Schofield (1981a, b), and [b] after Schofield and Schaal (1981).

TABLE 3. Differentiation of heterogeneous *Actinomyces* and *Arachnia* species[a]

Test	*Actn. israelii* I[b]	*Actn. israelii* II[b]	*Actn. naeslundii* I+II	*Actn. naeslundii* III[b]	Unnamed oral strains (cluster 3b)[c]	*Actn. viscosus* I	*Actn. viscosus* II	'atypical'	*Arac. propionica* I	*Arac. propionica* II
Catalase	−	−	−	−	−	+	+	+	−	−
Nitrite reduction	−	−	+	−	+	−	−	d	−	−
Nitrate reduction	+	d	+	+	+	d	d	+	+	+
Aesculin hydrolysis	+	+	+	+	+	−	−	−	−	−
Acid from:										
Adonitol	−	−	−	−	−	−	−	−	−	−
Arabinose	+	−	−	−	−	−	−	−	d	−
Cellobiose	+	+	d	−	−	−	d	d	d	−
Glycerol	−	−	−	d	d	−	−	+	−	−
Ribose	+	+	d	+	d	−	d	+	d	d
Sorbitol	d	−	−	−	−	−	−	−	+	−
Ammonia from:										
Arginine	+	+	d	−	+	−	d	−	−	−
Ornithine	−	−	−	−	−	−	−	−	+	−
Serine	−	−	−	−	−	−	−	−	+	−
Urea	−	−	+	−	d	d	d	−	−	−

+, positive; −, negative and d, strain differences.
[a] Data from Schofield and Schaal (1980a, b, 1981) and Schaal and Schofield (1981a, b).
[b] Serovars and [c] after Schofield and Schaal (1981).

TABLE 4. Physiological characteristics of Corynebacterium, Rothia, Arcanobacterium and Erysipelothrix[a]

Test	Corynebacterium (Bacterionema) matruchotii	Rothia dentocariosa	Arcanobacterium (Corynebacterium) haemolyticum	Erysipelothrix rhusiopathiae
Aerobic growth	+	+	−	−
Colony rough	+	+	−	−
Catalase	+	+	−	−
Nitrite reduction	−	+	−	−
Nitrate reduction	+	+	d	−
Aesculin hydrolysis	−	+	−	−
Starch hydrolysis	+	−	d	−
Acid from:				
Arabinose	−	−	−	−
Cellobiose	−	−	−	−
Glucose	+	+	+	+
Glycerol	−	−	−	−
meso-Inositol	−	−	−	−
Mannitol	d	−	−	−
Raffinose	−	−	−	−
Ribose	d	−	d	+
Sorbitol	−	−	−	−
Sucrose	+	+	d	−
Trehalose	−	+	d	−
Xylose	−	−	−	−
Ammonia from:				
Arginine	−	−	−	+
Urea	d	−	−	−

+, positive; +, negative and d, strain differences
[a] Data from Schofield and Schaal (1981).

E. SEROLOGICAL DIAGNOSIS OF ACTINOMYCOSES

Many attempts have been made, and various techniques have been tried, to diagnose actinomycoses serologically. Agglutination, complement fixation and immunofluorescence tests have all been found to be diagnostically irrelevant. However, Holmberg and his colleagues (Holmberg et al., 1975; Holmberg, 1981) have recently reported promising results with a crossed-immunoelectrophoresis technique which gave results that appeared to correlate well with the clinical findings. If these results can be confirmed by other investigators, an additional diagnostic tool might become available.

3. Laboratory Diagnosis of Diseases due to Aerobic Actinomycetes

The various actinomycete diseases require knowledge and experience of a number of procedures and instruments if they are to be diagnosed with the same degree of accuracy and reliability. Understandably, clinical microbiology laboratories often specialize in fields that reflect the local epidemiological situation and prefer to ask a reference laboratory for assistance when a rare disease is encountered. In Europe, infections due to aerobic actinomycetes are undoubtedly rare so that few clinicians and microbiologists are familiar with the diseases or with their causative agents. However, early recognition of *Actinomadura*, *Nocardia* and *Streptomyces* infections is highly dependent on at least a tentative diagnosis derived from microbiological tests since the clinical symptoms may be completely misleading. Thus, methods for the isolation and tentative identification of aerobic actinomycetes should be available in as many clinical laboratories as possible.

A. DIAGNOSIS OF NOCARDIOSIS AND ACTINOMYCETOMA

Similar diagnostic procedures may be applied to both forms of disease. However, cultivation and identification of actinomadurae usually present more problems than are encountered with the nocardiae.

(a) Collection and Transport of Clinical Specimens

Nocardia, *Actinomadura* and *Streptomyces* species have been successfully isolated from sputum, bronchial washings, exudate, pus, cerebrospinal fluid, biopsy and autopsy materials, sinus discharge and urine.

In pulmonary nocardiosis, sputum is the most easily available material. However, since each specimen does not necessarily contain nocardiae, several fresh, single samples should be collected and processed. On the other hand, potentially pathogenic actinomycetes may be isolated from sputum as contaminants or in cases of benign subclinical infection (see

Chapter 8). It is therefore necessary to culture the same organism from several independent samples to be sure of its aetiological role. From time to time, nocardiae do not appear at all in sputum or bronchial washings. In such cases, material can be obtained by direct lung puncture or transthoracic needle biopsy.

In cases of systemic nocardiosis, abscesses frequently develop and these can be punctured or incised in order to collect pus. Involvement of the central nervous system is diagnosed from brain abscess material or cerebrospinal fluid. The latter may give negative results in culture, even though a serious infection is present, as the pathogens may only be present in an abscess or L-forms may have formed (Beaman, 1981) which cannot be detected by conventional techniques.

Suitable specimens from cutaneous infections and actinomycetomas are primarily sinus discharge and biopsy material. The technique of collecting fluid from sinus tracts is the same as described above. Swabs are not recommended but, when they are used, they should be premoistened with sterile tap water without further additives. If granules are present, they should be collected as they are of the utmost diagnostic importance (Gordon, 1974; Schaal, 1977).

Although the aerobic actinomycetes are not very sensitive to adverse environmental conditions samples should be transported as fast as possible in order to prevent overgrowth by contaminants. This is especially important for samples of sputa and sinus discharge. If a delay in transit or storage is inevitable, the material should be kept at about +4°C without the addition of preservatives (Schaal, 1977).

(b) *Direct Examination of the Specimens*

The presence of granules may be checked in the same way as described for actinomycotic sulphur granules. Usually, material from pulmonary, cutaneous and systemic nocardioses does not contain these structures. In actinomycetomas, however, granules are commonly found and may aid in an immediate tentative identification of the causative agent as they differ in size, consistency and colour from species to species (see Chapter 8).

Fluid material may be examined in wet mounts under the microscope without staining. Spinal fluid, various exudates, or urine are centrifuged (4,000 r.p.m.) and a drop of the sediment is observed, between slide and cover slip, for the presence of branching filaments using greatly reduced light or phase contrast (Gordon, 1974).

Stained smears are prepared from fresh material, from sediments after centrifugation, or from granules crushed between two slides. In positive cases, the Gram-stain will reveal Gram-positive, branching filaments and possibly also rod- and coccoid-like elements. In contrast to actinomycotic granules, a marked concomitant flora will not be present in specimens without an indigenous flora.

The well-known acid-fastness of *Nocardia* species is often more pronounced in clinical than in cultured material. Nevertheless, acid-fastness cannot always be demonstrated with the conventional Ziehl-Neelsen technique; better results can be obtained using the 'modified Kinyoun acid-fast stain' (Georg, 1974b). Even with the latter technique, nocardiae may only be partially acid-fast, i.e. they show both acid-fast and non-acid-fast rods and filaments.

In tissue sections, the haematoxylin and eosin stain usually fails to demonstrate the actinomycetes although typical necrotic areas surrounded by leucocytes may be found. With the Gram-Weigert technique, the Brown and Brenn stain or acid-fast stains, however, the organisms can be demonstrated both in granules and in tissue.

(c) *Isolation of* Nocardia, Actinomadura *and* Streptomyces *Species*

All of these organisms usually grow well on most general-purpose media, though actinomadurae are sometimes difficult to cultivate. For non-selective isolation from non-contaminated samples, such as pus, spinal fluid and biopsy material, the following media have been successfully used: Sabouraud Dextrose Agar (Gordon, 1974); Brain Heart Infusion Agar (Oxoid, CM 261; Schaal, 1972a, b). These media should be transparent to facilitate direct microscopic examination as outlined for the fermentative actinomycetes.

In cultures from samples that contain large numbers of other microbes, actinomycetes will frequently be overgrown and consequently will not be detected. Selective media should, therefore, be simultaneously employed. Several selective media have been recommended but it seems that all of them can be inhibitory to some of the actinomycetes sought.

For nocardiae, the following media and techniques may provide satisfactory results: the paraffin-bait technique (Gordon and Hagan, 1936); *Nocardia* selective agar (Schaal and Heimerzheim, 1974); *Nocardia* selective medium (Orchard and Goodfellow, 1974) and Löwenstein-Jensen medium (Georg, 1974b). Pretreatment of material using the concentration and digestion-decontamination procedures applied in the isolation of mycobacteria cannot be recommended as they considerably reduce the isolation rate of nocardiae (Schaal, 1977).

Actinomadurae and streptomycetes will not grow on all of the media outlined above but they may form small colonies on *Nocardia* selective agar. In general, a set of different media should be simultaneously applied and they may be supplemented by other routinely used culture systems. All cultures should be incubated aerobically at 25 to 27°C and at 36°C for up to 3 weeks and examined both macroscopically and microscopically for growth every 2 days. In contrast to sulphur granules, actinomycetoma granules can be washed in sterile tap water, before they are crushed to obtain material for inoculation.

Subcultures for isolating the actinomycetes in pure culture should be performed using selective and non-selective agars. Growth may be scraped from paraffin baits and streaked onto the appropriate media; alternatively, paraffin-coated glass rods may be firmly rolled over the agar media. Fungi which frequently contaminate paraffin baits can be kept in check by adding amphotericin B (50 μg ml^{-1}) to one of the selective substrates.

Actinomycete colonies can be distinguished from other bacteria by their filamentous appearance (similar to that in Fig. 6) especially in the early stages of growth. They can also be recognized by their aerial mycelium and colour (Tables 5 and 6). Occasionally, Gram and acid-fast stains and preliminary physiological tests are necessary to decide whether pure cultures of unknown isolates are actinomycetes.

(d) Identification

Although the actinomycete nature of an aerobic isolate is often immediately obvious, identification to the species level can be difficult. Physiological tests are not always reliable, especially when the generic position of the isolate has not been determined. However, presumptive evidence of the species to which an aerobic isolate may belong (Table 5) can be derived using a few so-called degradation tests (Gordon *et al.*, 1974) and from simple agar diffusion tests that employ antibiotic-containing discs on DST-agar (Oxoid, CM 261) (Schaal, 1977; Alderson and Goodfellow, 1979). It should be noted that variation in single test results may occur so that definite diagnosis requires confirmation using additional criteria.

(i) *Identification at the generic level.* Reliable differentiation at the generic level is usually only possible when modern chemotaxonomic techniques are applied. Simplified procedures have been devised for a number of chemical tests and are applicable to clinical microbiology laboratories.

The determination of certain diagnostic cell wall components provides especially useful data. Some of these chemical markers are only present in the walls so that whole-organism hydrolysis can be used to detect them. It is not, therefore, necessary to prepare purified cell walls, a procedure which would in any case be beyond the scope of routine laboratories. In whole-organism hydrolysates, the presence or absence of 2,6-diaminopimelic acid (DAP) and the differentiation of its isomers are of great diagnostic relevance. In addition, such hydrolysates may contain diagnostic sugars which are also of value in both classification and identification (Table 6). In routine work the simplified techniques of Becker *et al.* (1964), Lechevalier (1968),

TABLE 5. Presumptive identification of selected *Nocardia*, *Rhodococcus*, *Actinomadura*, *Nocardiopsis* and *Streptomyces* species[a]

Species	Decomposition of:						Sensitivity to:						Substrate mycelium colour	Aerial mycelium colour	Soluble pigment	Colony margin highly filamentous
	Casein	Tyrosine	Xanthine	Hypoxanthine	Adenine	Testosterone	Lysozyme	Penicillin G (10 i.u.)	Gentamicin (10 μg)	Vancomycin (5 μg)	Fusidic acid (10 μg)	Acid-fastness				
Nocardia asteroides subgroup A[b]	−	−	−	−	−	+	−	+/−	+	−	+	+/−	C, Y	W, P	B/−	+
subgroup B[b]	−	−	−	−	−	+	−	−	−	−	−	+/−	O	W	−	+/−
Nrda. brasiliensis	+	+	−	+	−	+	+	−	+	−	−	+/−	Y, O, R	W	−	−
Nrda. otitidis-caviarum	−	−	+	+	−	+	+	−	+	−	+	+/−	C, O, BR	W	B/−	+/−
Rhodococcus rhodochrous	−	+	−	d	−	−	+	+	+	−	+	+/−	W, C, P	W	−	−
Rodc. bronchialis	−	−	−	−	−	−	+	+	+	−	+	+/−	O, R P, R	−	−	+/−
Actinomadura madurae	+	+	−	+	−	−	+	−	ND	ND	ND	−	W, Y, R	W	Y, B, Pu	−
Actm. pelletieri	+	+	−	+	+	−	+	+	ND	ND	ND	−	R	−	R, B	+/−
Nocardiopsis dassonvillei	+	+	+	+	−	+	+	−	ND	ND	ND	−	Y, O, B RB, Gy	W, Gy Bl, BlG	Y, B Gy, Pu	+
Streptomyces somaliensis	+	+	−	−	−	d	+	+	ND	ND	ND	−	Cr, B	W, Y, BlBk	Bl, Y	+
'*Stmy. paraguayensis*'	+	+	+	+	ND	ND	−	ND	ND	ND	ND	−	Cr, B, Gy	W	B/−	+

+, positive; −, negative; d, strain differences; +/− = weakly positive or partially acid-fast; ND, not determined; B, brown; Bk, black; Bl, blue; C, colourless; Cr, cream; G, green; Gy, grey; O, orange; P, pink; Pu, purple; R, red; W, white and Y, yellow.
[a] Data from Schaal (1977) and Goodfellow and Schaal (1979) and [b] after Schaal (1974); Schaal and Heimerzheim (1974) and Schaal and Reutersberg (1978).

TABLE 6. Differentiation of selected genera of aerobic actinomycetes and related taxa[a]

Taxon	Whole-organism hydrolysate analysis:					Mycolic acids		Morphology			Acid-fastness
	DL-DAP[d]	LL-DAP[d]	Arabinose	Galactose	Madurose	Single spot pattern	Multiple spot pattern	Substrate mycelium	Aerial mycelium	Fragmentation of substrate mycelium	
Corynebacterium	+	−	+	+	−	+	−	−	−	−	−
Mycobacterium	+	−	+	+	−	−	+	+/−	−	+	+
Nocardia	+	−	+	+	−	+	−	+/−	+/−	+	+/−
Rhodococcus	+	−	+	+	−	+	−	+/−	−	−	+/−
Streptomyces	−	+	−	d	−	−	−	+	+	−	−
Actinomadura	+	−	−	+	+	−	−	+	+/−	−	−
Nocardiopsis	+	−	−	d	−	−	−	+	+	−	−
Dermatophilus	+	−	−	ND	+	−	−	+	−	O[b]	−
Oerskovia	−	−	−	+	−	−	−	+	−	+[c]	−

+, present; −, absent; +/−, rudimentary or partially present; d, strain or species differences; and ND, not determined.
[a] Data from Cross and Goodfellow (1973); Schaal (1977); Goodfellow and Schaal (1979); and Berd (1973).
[b] Filaments divide by longitudinal and transverse septa to form motile spores.
[c] Fragmentation into motile rod-like elements and [d] meso- and LL-diaminopimelic acid.

Berd (1973) and Staneck and Roberts (1974) are recommended for the detection of diagnostic amino acids and sugars. The introduction of thin-layer chromatography made these techniques applicable to the routine laboratory as results could be read much more quickly than was possible using the old paper chromatography procedure (Becker et al., 1964).

It is often useful, if not necessary, to determine whether or not mycolic acids are present in whole-organism methanolysates. Qualitative evaluation of the mycolic acids present can be done easily and quickly using a thin-layer chromatography technique devised by Minnikin et al. (1975). Methanolysates of most mycobacteria give a multispot pattern of mycolates whereas corynebacteria, nocardiae, and rhodococci only produce a single characteristic spot (Table 6).

The diagnostic table shown in Table 6, which is based on the two chemical methods outlined above and a few morphological characters, can be used to identify most clinical isolates of aerobic actinomycetes to the generic level.

(ii) *Identification at the specific level.* Definite identification of clinical isolates to the species level is still not without problems. Quite a number of physiological tests still have to be applied when a reliable diagnosis is required.

Besides the degradation tests already mentioned, the ability of strains to use a variety of organic compounds provide data of value in the identification of aerobic actinomycetes (Tsukamura, 1969; Goodfellow, 1971; Schaal, 1972a, b, 1974; Schaal and Heimerzheim, 1974; Schaal and Reutersberg, 1978). In the Cologne laboratory, such tests seem to be superior to the 'fermentation' tests for carbohydrates devised by Gordon et al. (1974) as most aerobic actinomycetes usually produce small amounts of acid end products which are difficult to detect under routine conditions; test error is consequently a problem. Methodologically, the media and methods described by Goodfellow (1971) are recommended. The techniques used by Schaal (1972a) may be slightly more reproducible, but have not been published in detail elsewhere. A detailed description of the methods may, however, be obtained from the author upon request.

As can be seen from Table 7, some of the species give variable responses with some of the tests. This is the main reason why so many tests are required to identify these organisms with sufficient certainty. The subdivision of the heterogeneous *Nrda. asteroides* is still under discussion but it is clear that this taxon accommodates several subgroups (Tsukamura, 1969; Goodfellow, 1971; Schaal, 1972a, b; Schaal and Heimerzheim, 1974; Schaal and Reutersberg, 1978; Orchard and Goodfellow, 1980; Goodfellow and Minnikin, 1981). The properties of two of the major subgroups A and B, which contain clinically significant isolates (Pulverer and Schaal, 1978), are shown in Table 7. Reliable serological procedures are neither available for the

TABLE 7. Additional physiological tests for the identification of species of aerobic actinomycetes which are clinically significant or may cause problems in differential diagnosis[a]

Test	Nrda. asteroides A[b]	Nrda. asteroides B[b]	Nrda. brasiliensis	Nrda. otitidis-caviarum	Nrda. transvalensis	Nrda. vaccinii	Actm. madurae	Actm. pelletieri	Nrdp. dassonvillei	Stmy. somaliensis	Rodc. rhodochrous	Rodc. bronchialis	Rodc. corallinus	Rodc. erythropolis	Rodc. ruber	Rodc. equi	Orsk. turbata
Utilization as sole C-source:																	
Adonitol	−	−	−	−	+	−	+	−	−	−	−	−	−	−	−	−	−
L-Arabinose	−	−	−	−	−	+	+	−	−	−	−	−	−	−	−	−	d
Benzoate	−	+	−	−	−	−	−	−	−	−	+	+	+	d	+	ND	−
2,3-Butylene glycol	d	−	+	d	ND	+	−	−	ND	ND	ND	ND	+	ND	ND	ND	ND
Citrate	d	−	+	−	+	d	−	d	+	d	ND	ND	+	ND	+	−	d
Gluconate	−	−	+	+	d	+	d	−	−	−	ND	ND	ND	+	ND	ND	+
meso-Inositol	−	−	−	+	d	ND	+	−	ND	ND	ND	+	−	−	+	−	−
D-Mannitol	−	+	−	+	d	−	−	−	+	−	+	+	+	ND	+	−	−
1,2-Propylene glycol	−	+	−	−	−	+	−	−	d	−	−	+	−	+	+	−	−
L-Rhamnose	−	−	d	−	ND	ND	−	−	ND	ND	ND	−	−	ND	−	ND	ND
D-Sorbitol	−	−	−	−	d	d	−	−	−	−	+	+	+	+	+	−	d
D-Xylose	−	−	−	−	−	d	+	−	d	−	−	−	−	−	−	−	d

TABLE 7 (cont.)

Test	Nrda. asteroides A[a]	Nrda. asteroides B[b]	Nrda. brasiliensis	Nrda. otitidis-caviarum	Nrda. transvalensis	Nrda. vaccinii	Actm. madurae	Actm. pelletieri	Nrda. dassonvillei	Smy. somaliensis	Rodc. rhodochrous	Rodc. bronchialis	Rodc. corallinus	Rodc. erythropolis	Rodc. ruber	Rodc. equi	Orsk. turbata
Utilization as N+C source:																	
Gelatin	−	−	+	−	ND	ND	+	d	ND	+	−	−	−	−	−	+	ND
L-Serine	−	−	+	d	ND	ND	−	−	ND	ND	−	−	−	−	−	−	−
Decomposition of:																	
Adenine	−	−	−	−	d	−	−	−	+	−	d	−	−	+	d	+	−
Aesculin	+	+	+	+	+	+	+	−	d	−	ND	ND	ND	ND	ND	ND	+
Casein	−	−	+	−	−	−	+	+	+	+	−	−	−	−	−	−	−
Elastin	−	−	+	−	−	−	+	+	+	−	+	−	−	d	−	−	+
Tyrosine	−	−	+	−	−	−	+	+	+	+	+	−	−	+	+	−	+
Urea	+	+	+	+	+	+	−	−	d	−	ND	ND	+	+	+	+	ND

+, positive; −, negative; d, strain differences; and ND, not determined or without diagnostic value.
[a] Data from Prauser et al. (1970); Goodfellow (1971); Berd (1973); Schaal and Heimerzheim (1974); Goodfellow and Alderson (1977); Schaal (1977); Gordon et al. (1978); Goodfellow and Schaa (1979) and Mishra et al. (1980), and [b] Subgroups after Schaal (1974) and Schaal and Reutersberg (1978).

identification of aerobic actinomycetes nor for the detection of antibodies in diseased patients.

B. DIAGNOSIS OF DERMATOPHILOSIS

The procedures recommended for the diagnosis of *Dermatophilus* infections differ from those used for other aerobic actinomycetes, and consequently are dealt with separately.

(a) Collection and Transport of Specimens

The most suitable specimens come from unopened pustules, but exudate, scrapings, crusts, scabs or biopsy samples may also allow the demonstration and identification of *Derm. congolensis.* Materials should be submitted to laboratories in sterile, dry tubes or on swabs moistened with sterile tap water. Specimens should be kept at ambient temperature and transported without preservatives (Gordon, 1974).

(b) Direct Examination

Smears or tissue sections may be stained using the Giemsa, Grocott and Gram stains or with Methylene Blue. The Giemsa method has been found to be especially useful (Gordon, 1974). Under the microscope, *Derm. congolensis* may be seen in any stage of its life cycle. Typical diagnostic structures are branched filaments, 2–5 μm in diameter, which divide both transversely and longitudinally to form packets of coccoid cells. These structures may also be demonstrated in wet mounts.

(c) Isolation

Clinical material should be inoculated onto beef infusion-blood agar plates which are then incubated aerobically at 36°C. Highly contaminated samples may require an animal passage: crusts and scabs are ground and applied to the shaved and scarified skin of a rabbit. Lesions develop within 2 to 7 days and can be used to isolate the organism in pure culture.

Alternatively, the chemotactic response of zoospores to carbon dioxide may be used for separating the pathogen from other microbes. Scab material is immersed in a bottle of distilled water for 3 to 4 hours at room temperature, then exposed to carbon dioxide in a candle jar for 15 minutes. Thereafter, the surface film is streaked onto isolation media (Gordon, 1974) which may be supplemented with polymyxin B $(1,000 \text{ i.u. ml}^{-1})$ as selective principle (Weber, 1978).

Blood-agar plates may additionally be incubated in air with 5–10% carbon dioxide added.

(d) Identification

After 24 hours of incubation at 36°C tiny colonies appear which typically cause pitting in the medium (Gordon, 1974). After 2 to 5 days the colonies develop an orange pigmentation and are 4 to 6 mm in diameter. β-Haemolysis slowly becomes apparent in areas of heavy growth.

The microscopic picture from culture material is the same as that from clinical specimens although morphological strain variation may be pronounced.

Additional criteria which may aid in identification when the unique morphology cannot clearly be detected are: the presence of DL-DAP and madurose in whole-organism hydrolysates (Table 6), hydrolysis of casein and starch but not of tyrosine and xanthine, liquefaction of gelatin and Loeffler's coagulated serum, inability to reduce nitrate or produce indole and acid production from glucose and fructose in proteose-peptone broth, but not from sucrose, lactose, ocylose, dulcitol, mannitol, sorbitol and salicin (Gordon, 1964, 1974).

C. DIAGNOSIS OF ALLERGIC ALVEOLITIS DUE TO ACTINOMYCETES

As in other allergic diseases, serological procedures form the basis for diagnosing allergic alveolitis or hypersensitivity pneumonitis. In these delayed-type hypersensitivity reactions, circulating IgG and IgM antibodies against antigens from the causative actinomycete spores (see Chapter 8) are detectable in the sera of the patients involved. With soluble antigen preparations, these specific immunoglobulins can be demonstrated using precipitation techniques.

The practical diagnostic value of these tests mainly depends upon the availability of suitable antigen preparations which, because of their comparatively high specificity, have to cover the whole range of possible organisms responsible for allergic alveolitis. However, only some of the forms of hypersensitivity pneumonitis are as yet aetiologically well characterized (Lacey, 1981) so that diagnostic antigens are not always available. Furthermore, commercial antigen preparations are practically limited to extracts of *Micropolyspora faeni* and *Thermoactinomyces vulgaris* the latter often being impure as it contains components of *Team. vulgaris* (= *Team. candidus*) and *Team. thalpophilus* (Lacey, 1981).

To obtain a panel of specific diagnostic antigens which cover all of the important known actinomycete species inducing hypersensitivity reactions in the lungs, antigenic extracts have to be produced individually in the laboratory performing these tests.

(a) Preparation of Antigenic Extracts

Two comparatively simple and reliable techniques for preparing antigens for precipitin testing in allergic alveolitis are commonly used. One has been described in detail in the laboratory manual 'Serology of Fungal Infections and Farmer's Lung Disease' (Evans, 1976).

Briefly, in this method the respective organisms are incubated in two litres of 3% casein hydrolysate broth (pH 7.2) at a suitable temperature (45 to 55°C for *Mips. faeni*) for up to 14 days; the whole culture is concentrated by pouring it into sterile dialysis tubing and blowing air over the tubing; when the volume is reduced to approximately 50 ml, the content of the tubing is dialysed against running tap water for 2 days and against two changes of distilled water for another 12 hours; then, the contents are centrifuged at 10,000 g for 10 minutes at 4°C; the supernatant is filtered through a membrane filter (0.45 μm pore size) and used as antigen. The latter can be stored at −20°C or freeze-dried.

A second method for the preparation of antigens has been devised by Edwards (1972). In this double-dialysis technique, nutrients (e.g. nutrient broth) are enclosed in dialysis tubing (internal phase) and equilibrated with an aqueous solution of dialysable low molecular-weight compounds (e.g. sodium chloride) in a suitable container (external phase). The external phase is seeded with the actinomycete and the whole system is incubated. The external fluid is then placed into another sterile dialysis tubing and concentrated and dialysed as described above. This method may be facilitated by reversing the two phases and inoculating the internal phase so that the latter can be used directly for concentration and dialysis (T. Cross, personal communication).

(b) Antiserum Production

For control purposes antisera can be raised in rabbits to the antigens although animals do not produce antibody against the complete spectrum of antigens (Evans, 1976).

(c) Serological Tests

Several techniques including the double diffusion method of Ouchterlony, counter immunoelectrophoresis and other modifications of the immunoelectrophoretic principle have been used. For routine purposes, the simple double immunodiffusion procedure yields satisfactory results.

As with other serological tests, a positive reaction in this system only indicates that the patient has been in contact immunologically with the organisms involved, but not that he necessarily suffers from a disease due to hypersensitivity response. In addition, false positive reactions, although very infrequent, may occur (Evans, 1976).

REFERENCES

Alderson, G. and Goodfellow, M. (1979). Classification and identification of actinomycetes causing mycetoma. *Postepy Higieny i Medycyny Doswiadczalnej* **33**, 109–124.

Beaman, B. L. (1981). The possible role of L-phase variants of *Nocardia* in chronic infections. *Zentralblatt für Bakteriologie, Mikrobiologie und Hygiene.* I. *Abteilung, Supplement* **11**, 221–227.

Becker, B., Lechevalier, M. P., Gordon, R. E. and Lechevalier, H. A. (1964). Rapid differentiation between *Nocardia* and *Streptomyces* by paper chromatography of whole-cell hydrolysates. *Applied Microbiology* **12**, 421–423.

Beighton, D. and Colman, G. (1976). A medium for the isolation and enumeration of oral *Actinomycetaceae* from dental plaque. *Journal of Dental Research* **55**, 875–878.

Berd, D. (1973). Laboratory identification of clinically important aerobic actinomycetes. *Applied Microbiology* **25**, 665–681.

Cross, T. and Goodfellow, M. (1973). Taxonomy and classification of the actinomycetes. In: *Actinomycetales: Characteristics and Practical Importance* (G. Sykes and F. A. Skinner, eds.), pp. 11–112. Academic Press, London.

Davis, G. H. G. and Baird-Parker, A. G. (1959). *Leptotrichia buccalis. British Dental Journal* **106**, 70–73.

Edwards, J. H. (1972). The double dialysis method of producing farmer's lung antigens. *Journal of Laboratory and Clinical Medicine* **79**, 683–688.

Evans, E. G. V. (ed.) (1976). *Serology of Fungal Infection and Farmer's Lung Disease. A Laboratory Manual.* University Printing Service, Leeds.

Finegold, S. M., Sutter, V. L., Attebery, H. R. and Rosenblatt, J. E. (1974). Isolation of anaerobic bacteria. In: *Manual of Clinical Microbiology*, 2nd Edition (E. H. Lennette, E. H. Spaulding and J. P. Truant, eds.), pp. 365–375. American Society for Microbiology, Washington D.C.

Fortner, J. (1928). Ein einfaches Plattenverfahren zur Züchtung strenger Anaerobier. *Zentralblatt für Bakteriologie, Parasitenkunde, Infektionskrankheiten und Hygiene.* I. *Abteilung Originale* **108**, 155–159.

Fortner, J. (1929). Zur Technik der anaeroben Züchtung. *Zentralblatt für Bakteriologie, Parasitenkunde, Infektionskrankheiten und Hygiene.* I. *Abteilung Originale* **110**, 233–256.

Georg, L. K. (1974a). Genus *Rothia*. In: *Bergey's Manual of Determinative Bacteriology*, Eighth Edition (R. E. Buchanan and N. E. Gibbons, eds.), pp. 679–681. Williams and Wilkins, Baltimore.

Georg, L. K. (1974b). *Nocardia* species as opportunists and current methods for their identification. In: *Opportunistic Pathogens* (J. E. Prier and H. Friedman, eds.), pp. 177–201. University Park Press, Baltimore.

Goodfellow, M. (1971). Numerical taxonomy of some nocardioform bacteria. *Journal of General Microbiology* **69**, 33–80.

Goodfellow, M. and Alderson, G. (1977). The actinomycete-genus *Rhodococcus*: A home for the 'rhodochrous' complex. *Journal of General Microbiology* **100**, 99–122.

Goodfellow, M. and Minnikin, D. E. (1981). Classification of nocardioform bacteria. *Zentralblatt für Bakteriologie, Mikrobiologie und Hygiene.* I. *Abteilung, Supplement* **11**, 7–16.

Goodfellow, M. and Schaal, K. P. (1979). Identification methods for *Nocardia, Actinomadura* and *Rhodococcus*. In: *Identification Methods for Microbiologists*, Second Edition (F. A. Skinner and D. W. Lovelock, eds.), pp. 261–276. Society for Applied Bacteriology, Technical Series 14. Academic Press, London.

Gordon, M. A. (1964). The genus *Dermatophilus*. *Journal of Bacteriology* **88**, 509–522.

Gordon, M. A. (1974). Aerobic pathogenic *Actinomycetaceae*. In: *Manual of Clinical Microbiology*, 2nd Edition (E. H. Lennette, E. H. Spaulding and J. P. Truant, eds.), pp. 175–188. American Society for Microbiology, Washington D.C.

Gordon, R. E. and Hagan, W. A. (1936). A study of some acid-fast actinomycetes from soil with special reference to pathogenicity for animals. *Journal of Infectious Diseases* **59**, 200–206.

Gordon, R. E., Barnett, D. A., Handerhan, J. E. and Pang, C. H. N. (1974). *Nocardia coeliaca*, *Nocardia autotrophica* and the nocardin strain. *International Journal of Systematic Bacteriology* **24**, 54–63.

Gordon, R. E., Mishra, S. K. and Barnett, D. A. (1978). Some bits and pieces of the genus *Nocardia: N. carnea, N. vaccinii, N. transvalensis, N. orientalis* and *N. aerocolonigenes*. *Journal of General Microbiology* **109**, 69–78.

Hallander, H. O., Flodström, A. and Aberg, C. (1980). Collection and transport of specimens for anaerobic culture. *Infection* **8**, Supplement 2, 147–150.

Heinrich, S. and Korth, H. (1967). Zur Nährbodenfrage in der Routinediagnostik der Aktinomykose: Ersatz unsicherer biologischer Substrate durch ein standardisiertes Medium. In: *Krankheiten durch Aktinomyceten und verwandte Erreger* (H. J. Heite, ed.), pp. 16–20. Springer-Verlag, Berlin.

Holdeman, L. V., Cato, E. P. and Moore, W. E. C. (1977). *V.P.I. Anaerobe Laboratory Manual*, 4th Edition. Southern Printing Co., Blacksburg.

Holmberg, K. (1981). Immunodiagnosis of human actinomycosis. *Zentralblatt für Bakteriologie, Mikrobiologie und Hygiene*. I. *Abteilung, Supplement* **11**, 259–261.

Holmberg, K. and Nord, C.-E. (1975). Numerical taxonomy and laboratory identification of *Actinomyces* and *Arachnia* and some related bacteria. *Journal of General Microbiology* **91**, 17–44.

Holmberg, K., Nord, C.-E. and Wadström, T. (1975). Serological studies of *Actinomyces israelii* by crossed immunoelectrophoresis. *Infection and Immunity* **12**, 387–397.

Kornman, K. S. and Loesche, W. J. (1978). New medium for isolation of *Actinomyces viscosus* and *Actinomyces naeslundii* from dental plaque. *Journal of Clinical Microbiology* **7**, 514–518.

Lacey, J. (1981). Airborne actinomycete spores as respiratory allergens. *Zentralblatt für Bakteriologie, Mikrobiologie und Hygiene*. I. *Abteilung, Supplement* **11**, 243–250.

Lechevalier, M. P. (1968). Identification of aerobic actinomycetes of clinical importance. *Journal of Laboratory and Clinical Medicine* **71**, 934–944.

Lentze, F. A. (1938). Die mikrobiologische Diagnostik der Aktinomykose. *Münchner medizinische Wochenschrift* **47**, 1826–1836.

Lentze, F. (1953). Zur Aetiologie und mikrobiologischen Diagnostik der Aktinomykose. *Estratto degli Atti del VI Congresso Internationale de Microbiologia, Roma 5*, Sez. XIV, 145–148.

Lentze, F. (1969). Die Aktinomykose und die Nocardiosen. In: *Die Infektionskrankeiten des Menschen und ihre Erreger, Band I, Zweite Auflage* (A. Grumbach and O. Bonin, eds.), pp. 954–973. Georg Thieme-Verlag, Stuttgart.

Minnikin, D. E., Alshamaony, L. and Goodfellow, M. (1975). Differentiation of *Mycobacterium, Nocardia* and related taxa by thin-layer chromatographic analysis of whole-organism methanolysates. *Journal of General Microbiology* **88**, 200–204.

Mishra, S. K., Gordon, R. E. and Barnett, D. A. (1980). Identification of nocardiae and streptomycetes of medical importance. *Journal of Clinical Microbiology* **11**, 728–736.

Orchard, V. A. and Goodfellow, M. (1974). The selective isolation of *Nocardia* from soil using antibiotics. *Journal of General Microbiology* **85**, 160–162.

Orchard, V. A. and Goodfellow, M. (1980). Numerical classification of some named strains of *Nocardia asteroides* and related isolates from soil. *Journal of General Microbiology* **118**, 295–312.

Prauser, H., Lechevalier, M. P. and Lechevalier, H. (1970). Description of *Oerskovia* gen. n. to harbor Ørskov's motile *Nocardia*. *Applied Microbiology* **19**, 534.

Pulverer, G. and Schaal, K. P. (1978). Pathogenicity and medical importance of aerobic and anaerobic actinomycetes. *Zentralblatt für Bakteriologie, Mikrobiologie und Hygiene*. I. *Abteilung, Supplement* **11**, 417–427.

Schaal, K. P. (1972a). *Die Nocardiosen des Menschen. Untersuchungen zur Taxonomie, Ökologie und Pathogenität von Nocardien*. Habilitationsschrift, Köln.

Schaal, K. P. (1972b). Zur mikrobiologischen Diagnostik der Nocardiose. *Zentralblatt für Bakteriologie, Parasitenkunde, Infektionskrankheiten und Hygiene*. I. *Abteilung Originale A* **220**, 242–246.

Schaal, K. P. (1974). Differentiation of strains thought to belong to the species *Nocardia asteroides* by biochemical and serological methods. In: *Proceedings of the First International Conference on the Biology of the Nocardiae*. (G. Brownell, ed.), pp. 12–13. McGowen Printing Co., Augusta, Georgia.

Schaal, K. P. (1977). *Nocardia, Actinomadura* and *Streptomyces*. In: *CRC Handbook Series in Clinical Laboratory Sciences, Section E., Clinical Microbiology,* Volume 1 (A. von Graevenitz, ed.), pp. 131–144. CRC Press, Cleveland.
Schaal, K. P. and Heimerzheim, H. (1974). Mikrobiologische Diagnose und Therapie der Lungennocardiose. *Mykosen* 17, 313–319.
Schaal, K. P. and Pulverer, G. (1973). Fluoreszenzserologische Differenzierung von fakultativ anaeroben Aktinomyzeten. *Zentralblatt für Bakteriologie, Parasitenkunde, Infektionskrankheiten und Hygiene.* I. *Abteilung Originale,* A 225, 424–430.
Schaal, K. P. and Pulverer, G. (1981). The genera *Actinomyces, Agromyces, Arachnia, Bacterionema* and *Rothia.* In: *The Prokaryotes: A Handbook of Habitats, Isolation and Identification of Bacteria* (M. P. Starr, H. Stolp, H. G. Trüper, A. Balows and H. G. Schlegel, eds.), pp. 1923–1950. Springer Verlag, New York.
Schaal, K. P. and Reutersberg, H. (1978). Numerical taxonomy of *Nocardia asteroides. Zentralblatt für Bakteriologie, Parasitenkunde, Infektionskrankheiten und Hygiene.* I. *Abteilung, Supplement* 6, 53–62.
Schaal, K. P. and Schofield, G. M. (1981a). Taxonomy of *Actinomycetaceae. Revue de l'Institut Pasteur, Lyon* 14, 27–39.
Schaal, K. P. and Schofield, G. M. (1981b). Current ideas on the taxonomic status of the *Actinomycetaceae. Zentralblatt für Bakteriologie, Mikrobiologie und Hygiene.* I. *Abteilung, Supplement* 11, 67–78.
Schaal, K. P., Schofield, G. M. and Pulverer, G. (1980). Taxonomy and clinical significance of *Actinomycetaceae* and *Propionibacteriaceae. Infection* 8, *Supplement* 2, 122–130.
Schofield, G. M. and Schaal, K. P. (1979a). Application of the Minitek differentiation system in the classification and identification of *Actinomycetaceae. FEMS Microbiology Letters* 5, 311–313.
Schofield, G. M. and Schaal, K. P. (1979b). A simple basal medium for carbon source utilization tests with the anaerobic actinomycetes. *FEMS Microbiology Letters* 5, 309–310.
Schofield, G. M. and Schaal, K. P. (1980a). Carbohydrate fermentation patterns of facultatively anaerobic actinomycetes using micromethods. *FEMS Microbiology Letters* 8, 67–69.
Schofield, G. M. and Schaal, K. P. (1980b). Rapid micromethods for detecting deamination of decarboxylation of amino acids, indole production and reduction of nitrate and nitrite by facultatively anaerobic actinomycetes. *Zentralblatt für Bakteriologie, Parasitenkunde, Infektionskrankheiten und Hygiene.* I. *Abteilung Originale* A 247, 383–391.
Schofield, G. M. and Schaal, K. P. (1981). A numerical taxonomic study of members of the *Actinomycetaceae* and related taxa. *Journal of General Microbiology* 127, 237–259.
Slack, J. M. and Gerencser, M. A. (1975). *Actinomyces, Filamentous Bacteria.* Burgess Publishing Company, Minneapolis.
Slack, J. M., Winger, A. and Moore, D. W. Jr. (1961). Serological grouping of *Actinomyces* by means of fluorescent antibodies. *Journal of Bacteriology* 82, 54–65.
Staneck, J. L. and Roberts, G. D. (1974). Simplified approach to identification of aerobic actinomycetes by thin-layer chromatography. *Applied Microbiology* 28, 226–231.
Suzuki, S. and Ueno, K. (1981). *Anaerobic Bacteria. Illustrated Laboratory Techniques Series.* Volume I (N. Kosakai, ed.), pp. 32–36. Igaku-Shoin, Tokyo.
Syed, S. A. and Loesche, W. J. (1972). Survival of human dental plaque flora in various transport media. *Applied Microbiology* 24, 638–644.
Tsukamura, M. (1969). Numerical taxonomy of the genus *Nocardia. Journal of General Microbiology* 56, 265–287.
Weber, A. (1978). Zur Dermatophilose bei Tier und Mensch. *Münchner tierärztliche Wochenschrift* 91, 341–345.

10
Actinomycete Pathogenesis

B. L. BEAMAN

Department of Medical Microbiology, School of Medicine, University of California, Davis, California 95616, USA

1. Experimental Infections	457
2. Host–Parasite Interactions	462
3. Chemical Determinants of Pathogenicity	467
4. The Cell Wall as a Determinant of Pathogenesis	472
5. Acknowledgements	474
References	474

1. Experimental Infections

Various actinomycetes have been found to be pathogenic for humans and for a large variety of animals. The genera most frequently associated with disease have been described in Chapter 8. Mice, guinea pigs, hamsters and rabbits have all served as model systems for studying the mechanisms of pathogenesis of many of these bacteria (Slack and Gerencser, 1975; Beaman, 1976; Youmans, 1979; Fig. 1).

In 1891, Wolff and Israel were among the first to infect rabbits and guinea pigs experimentally with *Actinomyces israelii* strains isolated from human infections. Later, Naeslund (1929) infected rabbits with pure cultures of *Actn. israelii* administered by various routes. Since these early studies it has been shown that both mice and hamsters can be infected with *Actn. israelii* and that mucin enhances the infectivity of the organism though many reports are quite variable concerning the infectivity of *Actn. israelii* for laboratory animals (Slack and Gerencser, 1975). It appears, however, that typical, progressive lesions containing sulphur granules can be induced, to a variable degree, in laboratory animals using pure cultures of *Actn. israelii* while the use of mucin or large clumps of bacteria enhance the ability to establish progressive disease. The clinical picture of actinomycosis is almost always one of a mixed microbial infection; only rarely is *Actn. israelii* isolated from infected patients in pure culture. Although a variety of bacteria appear to be associated with *Actn. israelii* in humans, strains of *Actinobacillus actinomycetemcomitans*, *Bacteroides* spp. and anaerobic cocci are the most frequently recognized partners (Pulverer and Schaal, 1978). All of these organisms can also induce infections, especially abscesses, in humans. The significance of these mixed microflora in the development of actinomycosis has still to be adequately studied in experimental models.

Infections have been established in mice with *Actn. bovis*, '*Actn. (Bifidobacterium) eriksonii*', *Actn. naeslundii*, *Actn. odontolyticus* and *Actn. viscosus* (Slack and Gerencser, 1975). In this model it appears that infections are most readily induced using *Actn. israelii* followed by strains of *Actn. viscosus*, *Actn. bovis*, *Actn. naeslundii* and '*Actn. eriksonii*'; *Actn. odontolyticus* appears to be the least able to induce progressive disease in mice (Fig. 1A).

Actinomyces viscosus is a major component of the bacterial population of subgingival plaque in humans and probably plays a significant role in periodontal disease, root surface and fissure caries (Slack and Gerencser, 1975; Brecher et al., 1978; Brecher and van Houte, 1979). The organism can also cause periodontal disease in laboratory animals (Brecher and van Houte, 1979) and can effectively colonize the fissures of molars in rats (Brecher and van Houte, 1979). The ability of *Actn. viscosus* strains to colonize the pits and fissures of molar teeth is related to the age of the animal with differences noted in the organism's ability to attach to tooth surfaces and colonize tooth fissures. These differences might be related to the reduced amount of enamel epithelium and connective tissue associated with molar teeth in animals of different age groups (Brecher and van Houte, 1979). In contrast, it appears that host factors such as antibodies, indigenous microflora and saliva have little or no influence on the interactions of *Actn. viscosus* with the surface of the teeth of rats; the diet of the animals does not alter the age dependence of these interactions (Brecher and van Houte, 1979).

Arachnia propionica strains can cause infection in humans that resemble actinomycosis caused by *Actinomyces* spp. (Conrad et al., 1978). It has also been shown that arachniae can establish slow progressive infections when injected into the peritoneal cavities of mice (Buchanan and Pine, 1962; Georg and Coleman, 1970). The specific mechanisms of pathogenesis of this organism have not been established but it has been suggested that bacterial aggregates are important. *Arachnia propionica* strains form a gelatinous matrix when grown *in vitro* and this material enhances the formation of bacterial aggregates. It has been suggested that this formation of aggregates may represent one of the mechanisms of pathogenesis (Abe et al., 1978).

There have been many reports on the relative pathogenecity of different nocardiae for laboratory animals and it appears that not all strains of animals are equally susceptible to infection to *Nrda. asteroides*, *Nrda. brasiliensis* and *Nrda. otitidis-caviarum* (Drake and Henrici, 1943; Runyon, 1951; Brizin and Lenhart, 1957; Macotella-Ruiz and Mariat, 1963;

Mohapatra and Pine, 1963; Mason and Hathaway, 1969; Kurup et al., 1970; Smith and Hayward, 1971; González-Ochoa, 1973; Folb et al., 1976, 1977; González-Ochoa and Sandoval-Cuellar, 1976). It is important, therefore, to recognize differences in host susceptibility when studying different strains of *Nocardia* and in establishing experimental models (Beaman, 1976). Other important factors that influence interactions of nocardiae with experimental animals include (a) the route of exposure, (b) the state of the animal host, (c) the specific strain involved, (d) the methods employed to prepare the inoculum, (e) the culture medium used to grow the test strain, (f) the age and stage of development of the inoculum, and (g) the methods used to assess infectivity (Beaman, 1973; Beaman and Maslan, 1977, 1978; Beaman et al., 1978a,c, 1980a,b; Fig. 1B, C, D).

There are marked differences in the abilities of different strains of *Nrda. asteroides* to cause both acute and chronic progressive infections in mice. Further, depending upon the route of inoculation, many strains exhibit a predilection for specific internal organs (Beaman and Maslan, 1977; Beaman et al., 1980a, b; Fig. 1). The cerebrum is a frequent secondary site of infection in humans following haematogenous dissemination during pulmonary nocardiosis (Beaman et al., 1976; Saenz Lope and Guiterrez, 1977; Harris, 1980). Many strains of *Nrda. asteroides* cause brain infections when injected intravenously into mice (Beaman et al., 1980a, b) and histological studies demonstrated that the infectious foci are mostly confined to the cerebrum. The mechanisms of this association may reflect the large content of high molecular-weight lipid and polysaccharide compounds bound to the surface of the nocardiae which, in turn, could specifically interact with the large amounts of sphingomyelin within the brain.

When nocardiae become established within host tissues, certain properties of the organisms are altered. It has been observed that strains which are least virulent are altered the most during adaptation to the parasitic state while the more aggressive ones appear to be changed the least within the tissues (Beaman, 1973, 1976, 1980a). Ultrastructural, tinctorial and histological properties give some indication as to the types of changes that occur during growth within the host. Thus, for example, the acid-fastness of these organisms is enhanced during growth *in vivo* (Beaman, 1973). This particular change suggests that the accumulation of high molecular-weight lipids such as mycolic acids is enhanced during the growth of nocardiae within the host's cells. Recently, the results of *in vitro* experiments have shown that strains of *Nrda. asteroides* grown in highly activated mouse peritoneal macrophages are strongly acid-alcohol-fast, whereas the same organisms grown in normal peritoneal macrophages gave a negative response to the stain. The enhanced ability of strains to evade destruction within the activated macrophages may reflect the altered composition of the bacterial envelope which, in turn, is a consequence of intracellular growth within a severely restrictive environment. These observations also imply that it is the interaction of nocardiae with activated macrophages that results in the increased acid-fastness of nocardiae in host tissues (Filice *et al.*, 1980a).

Germ-free mice offer a unique tool for studying host–*Nocardia* relationships as such animals are free from the nonspecific influences of exogenous microbial interactions (Beaman *et al.*, 1980b). *Nocardia asteroides* strains

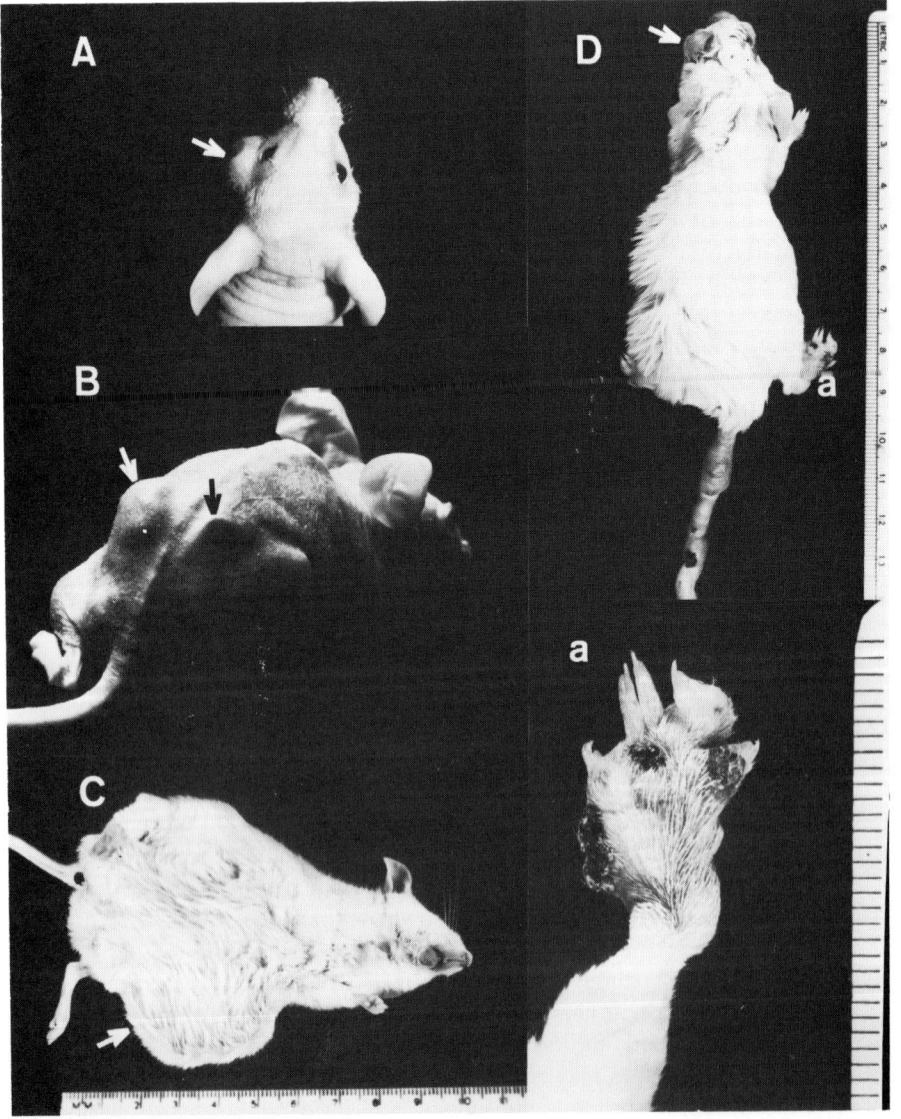

were therefore administered to germ-free and lipopolysaccharide (LPS)-treated germ-free mice by either intranasal inoculation or by intravenous injection. At specific time intervals the interactions of the nocardiae within the lungs, brain, adrenals, kidneys, liver and spleen were determined. It was shown that germ-free mice were significantly more susceptible to acute infection than were the conventionally grown animals, whereas, in contrast, the LPS-treated germ-free mice were more resistant to infection. The host response in the lungs and brain was affected most dramatically whereas the interactions within the liver and spleen were essentially unaltered in these groups of mice. These results indicate the importance of nonspecifically activated macrophages and the development of cell-mediated immunity in host resistance to nocardial infection. Further, these data establish the importance of a resident microflora in non-specific host resistance to either systemic or pulmonary nocardiosis (Beaman et al., 1980a).

Host–parasite interactions with nocardial strains are complex. However, it was shown that, by using different routes of infection with various nocardiae under controlled conditions of growth, the specific host–parasite interactions at the cellular level can be studied using specifically manipulated mouse models that have genetically controlled and well characterized immunological defects (Beaman and Maslan, 1977, 1978; Beaman et al., 1978a,c, 1980a,b).

Rhodococcus strains have been isolated from pulmonary infections and shown to be the cause of serious disease in immunosuppressed patients (Haburchack et al., 1978). Rhodococci appear to be true 'opportunists' in

FIG. 1. Experimental murine infections caused by members of the *Actinomycetales* (A). An athymic (nude) mouse 3 months after intranasal administration of 1.5×10^7 c.f.u. of *Actinomyces israelii* serotype I. This T-cell deficient animal developed a cervicofacial lesion that is characteristic of disease caused by this organism (arrow) (Beaman et al., 1979). (B) An athymic mouse 3 months after intravenous injection of 2.6×10^5 c.f.u. of *Nocardia asteroides* GUH-2. Note the bilateral infection originating from the kidneys and extending through the back of this mouse (arrows) (Beaman et al., 1978a). (C) An immunologically intact mouse 1 year after intravenous injection of approximately 1×10^8 L-forms of *Nocardia otitidis-caviarum* 112. The animal remained asymptomatic for about 8 months followed by a dramatic emergence of a lesion originating from the base of the spine. This chronic lesion enlarged during the following months to involve the entire abdominal region of the mouse and at the time of necropsy it was found to weigh 50 g (arrow). Histology of this lesion showed granules surrounded by inflammatory cells characteristic of chronic mycetomatous lesions. Both L-phase variants and normal cells of *Nrda. otitidis-caviarum* were isolated from this lesion (B. L. Beaman, unpublished work). (D) A mouse 4 months after intravenous injection of 2.5×10^6 c.f.u. of *Nocardia brasiliensis* 17E which was originally isolated from a patient with extensive mycetomatous lesions on the back. This organism causes extensive lesions involving the cooler parts of the body (such as the feet, tail and nose; arrow). These lesions were chronic and progressive and did not appear until about 2.5 months after injection (18 out of 20 mice receiving intravenous injections of this organism became infected as shown in this figure within 6 months). Inset (a) is a view of one of the feet of the mouse shown in (D). Note the large tumorous lesions with draining sinus tracts which are characterisitic of the chronic lesions seen in humans and caused by this organism (B. L. Beaman, unpublished observations).

so far as they seem unable to cause disease in 'normal' hosts. Experiments using guinea pigs have shown that steroid treatment is necessary before disease can be induced by *Rhodococcus* strains. In the author's laboratory mice have been infected with strains of either *Rodc. ruber* or *Rodc. rhodochrous* by both intravenous and intraperitoneal routes. Even when 1×10^9 colony forming units (c.f.u.) were administered to each mouse, progressive disease was not induced in any of the animals with any of the test strains. When, however, inoculum supplemented with mucin or mineral oil was administered to the animals intraperitoneally, some mice developed lesions. Organ clearance studies have shown that 100% of the cells of *Rodc. ruber* are killed within 4 days after intravenous injection of 1×10^8 c.f.u. per mouse (J. Rosen and B. L. Beaman, unpublished observations).

Animal models of infection with strains of *Actinomadura*, *Bacterionema* (*Cnbc.*) *matruchotii*, *Oerskovia*, *Rothia* and *Streptomyces* have not been reported. The rabbit does, however, appear to be a useful model for studying experimental infections with *Dermatophilus congolensis* though on these organisms several studies have also been carried out using various domestic livestock and mice as experimental animals (Pullman et al., 1967; Abu-Samra, 1978; Makinde and Wilkie, 1979). Experimental infections using mycobacteria and corynebacteria have been studied extensively but are beyond the purview of this chapter. Please refer to the reviews presented by Youmans (1979) and Barksdale (1970).

2. Host–Parasite Interactions

The resistance of humans and other animals to infection by actinomycetes involves a variety of mechanisms that include mechanical, physical, chemical and cellular processes (Beaman, 1976; Youmans, 1979). Many of these components are non-specific and they function equally against most microorganisms (i.e. the ciliary action of the epithelial cells of the upper respiratory tract, the mucous membranes, and the integrity of the skin) but, once the bacterial cell has crossed these barriers and gained entrance into the host, cellular and humoral processes function to eliminate the invading microorganism (Youmans, 1979). These processes involve complex interactions of T-lymphocytes, B-lymphocytes, macrophages and polymorphonuclear neutrophils (PMNs) as well as other components of the reticuloendothelial system (Davis et al., 1973).

Macrophages and PMN leukocytes eliminate strains of most actinomycete taxa from host tissues either by a non-selective phagocytosis of the invading organisms or by specific mechanisms under the control of humoral and cell-mediated immune processes (Slack and Gerencser, 1975; Beaman, 1976; Youmans, 1979). Several factors influence the fate of the ingested actinomycetes. These include the metabolic state of the phagocyte, the availability, types and concentrations of lysosomal enzymes within macrophages and PMN leukocytes, the presence or absence of specific opsonins, the presence of non-specific factors such as complement, the condition of the host, and the state of the actinomycete (Mackaness, 1969; Slack and Gerencser, 1975; Beaman, 1976; Barksdale and Kim, 1977; Youmans, 1979).

The specific mechanisms of host resistance to infections caused by actinomycetes are not clearly understood. There have been many studies dealing with resistance to strains of *Mycobacterium* and *Nocardia*, and a few reports have dealt with host–parasite interactions with *Actinomyces*. Almost nothing, however, is known concerning the mechanisms of host–parasite interactions with other pathogenic actinomycetes.

Strains of *Nocardia* and *Mybc. tuberculosis*, together with most other pathogenic mycobacteria, can grow as facultative intracellular parasites within macrophages (Beaman and Smathers, 1976; Beaman, 1977, 1979; Youmans, 1979; Davis-Scibienski and Beaman, 1980a; Filice *et al.*, 1980a, b; Fig. 2a). In contrast, most of the current information available suggests that the anaerobic actinomycetes may not be facultative intracellular parasites and, once these organisms are phagocytosed, they are probably destroyed by both macrophages and PMN leukocytes (Slack and Gerencser, 1975; Beaman *et al.*, 1979).

Facultative intracellular parasites such as *Mycobacterium* and *Nocardia* are usually limited to macrophages because these phagocytic cells live long enough to permit growth and persistence of the parasite. Polymorphonuclear leukocytes, on the other hand, are short lived and therefore intracellular parasites taken into these cells generally do not have adequate time to replicate. Further, the PMN leukocyte appears to be more bactericidal for some microorganisms, such as *Listeria monocytogenes*, than macrophages (Filice *et al.*, 1980b). The interactions of cells of *Nrda. asteroides* with human PMN leukocytes and monocytes in both the presence and absence of antibody obtained from the sera of patients with nocardiosis has been studied in detail (Filice *et al.*, 1980b). The monocytes and neutrophils effectively killed *Listeria monocytogenes* and *Staphylococcus aureus* but had no effect on the viability of *Nrda. asteroides* strains. Filice *et al.* (1980b) concluded that normal human neutrophils and monocytes did not have a major role in host resistance to infections by *Nrda. asteroides*. In addition, specific agglutinating antibody has not been found to have any effect on the leukocyte–*Nocardia* interaction.

If peripheral blood monocytes obtained from humans are permitted to attach to a glass surface for several hours they acquire the characteristics of maturing macrophages. Human macrophages obtained in this manner do not kill *Nrda. asteroides*, but instead nocardial cells are phagocytosed by these macrophages. Subsequently, the nocardial cells grow out of the human macrophages (G. A. Filice, personal communication; B. L. Beaman, unpublished work). It is not clear from these studies which cells are responsible for host resistance to infection by *Nrda. asteroides* but it has been shown that macrophage activation, and interaction of these macrophages with specifically sensitized lymphocytes, are important in host resistance to both *Mybc. tuberculosis* and *Listeria monocytogenes* strains (Mackaness, 1969; Youmans, 1979; Filice *et al.*, 1980a).

In pulmonary nocardiosis, it is generally believed that *Nrda. asteroides* enters the host by way of inhalation. The primary mechanisms of eliminating these bacteria from the lungs are by ciliary activity of the bronchial epithelium and by phagocytosis by alveolar macrophages. The initial events, therefore, must involve the interaction of *Nrda. asteroides* with alveolar macrophages (Beaman *et al.*, 1978c).

The role of alveolar macrophages in host defense against infection with virulent strains of *Nrda. asteroides* has been determined by using macrophages obtained from the lungs of both normal and immunized rabbits (Beaman, 1979). It was shown that cells of *Nrda. asteroides* in logarithmic growth differed from organisms in the stationary phase, but the nocardial cells in all phases of growth were able to survive and grow within normal macrophages. However, specifically activated macrophages obtained from immunized rabbits inhibited the growth of nocardial cells in stationary phase much more effectively than logarithmic phase cells. Further, when these bacterial cells were preincubated with sera from immunized rabbits it was found that there was some enhancement of phagocytosis and inactivation of the filamentous form (logarithmic phase) of *Nrda. asteroides* by presensitized alveolar macrophages, but there was little or no effect on uptake and killing of the stationary phase of these organisms (Beaman, 1979). Alveolar macrophages are very effective in killing both logarithmic and stationary phase cells of *Nrda. asteroides* when preincubated with specifically primed lymphocytes from lymph nodes combined with immune serum with complement activity in the presence of surfactant obtained from the lungs. All of these components are required for effective destruction of *Nrda. asteroides*. A deficiency or loss of any one of these factors could, therefore, sufficiently compromise the host and enhance its susceptibility to pulmonary nocardiosis (Davis-Scibienski and Beaman, 1980c; Fig. 2A, B).

Immunodeficient murine systems have been developed which allow more specific dissection of host response to microbial infections. These unique systems have recently been used to study host response to some actinomycetes (Beaman and Maslan, 1977; Folb *et al.*, 1977; Beaman *et al.*, 1978a, c, 1979, 1980a, b). For example, congenitally athymic (Nu/Nu) and hereditarily asplenic (Dh/+) mice have been given suspensions of *Nrda. asteroides* and *Actn. israelii* by several different inoculation routes. These studies revealed the following: (1) T-cell lymphocytes are essential for an effective host clearance of *Nrda. asteroides*; (2) T-cells are not required for complete clearance of *Actn. israelii* from the lungs of mice; (3) T-cells appear to be important for the prevention of localized cervico-facial actinomycosis but other factors appear to be involved; and (4) alveolar macrophages by themselves are not sufficient to destroy *Nrda. asteroides* upon initial contact but, in sharp contrast, these macrophages are very effective against *Actn. israelii* (Beaman 1978a, c, 1979).

Additional information concerning the role of macrophages in host resistance to *Nrda. asteroides* has been obtained from manipulations of these murine systems. Data obtained from germ-free animals either treated with lipopolysaccharide or untreated clearly indicate that *in vivo* the non-specific activation of macrophages is important in host resistance to *Nrda. asteroides* (Beaman *et al.*, 1980a, b). These observations are strongly supported by *in vitro* studies on the interaction of either '*Corynebacterium parvum*' or *Toxoplasma gondii* activated macrophages with several strains of *Nrda. asteroides* (Filice *et al.*, 1980a, b). All of these studies indicate that the activated state of the macrophage is an essential component in determining the host's resistance to *Nrda. asteroides*. These and other studies also demonstrate that macrophages alone are not enough for effective protection of the host against *Nrda. asteroides* (Beaman *et al.*, 1978a, c, 1980a, b; Davis-Scibienski and Beaman, 1980c; Filice *et al.*, 1980a).

Host–parasite interactions with *Actinomyces* strains appear to be different from those involved with *Nrda. asteroides* (Engel *et al.*, 1976; Brecher *et al.*, 1978; Beaman *et al.*, 1979; Brecher and van Houte, 1979; Shull, 1979). This is not surprising since *Actinomyces* differ substantially from *Nrda. asteroides* strains in respect to several basic biological properties (Slack and Gerencser, 1975). Host–parasite interaction with strains of *Actn. viscosus* has been studied most extensively (Slack and Gerencser, 1975; Engel *et al.*, 1976; Brecher *et al.*, 1978; Brecher and van Houte, 1979; Shull, 1979). Since these organisms are mostly anaerobic or microaerophilic it is not unreasonable to suggest that PMN leukocytes and possibly macrophages can kill *Actinomyces* strains by production of superoxide radicals. It has been shown that, when some microorganisms interact with phagocytes, there is an induction of a strong burst of oxidative metabolism within the phagocyte. This process results in the production of superoxide radicals that are very toxic to most microorganisms, especially to anaerobic organisms that may lack or have reduced amounts of superoxide dismutase and catalase (Drivalle and Gregory, 1979; Welch *et al.*, 1979).

Unlike most other Gram-positive bacteria, *Actn. viscosus* and *Actn. naeslundii* strains have been shown to produce pili (Cisar *et al.*, 1978; Ellen *et al.*, 1978; Cisar and Vatter, 1979). These structures may be important in the attachment of the bacterial cells to the surface of teeth or to gingival epithelium and they may have some antiphagocytic function. It has also

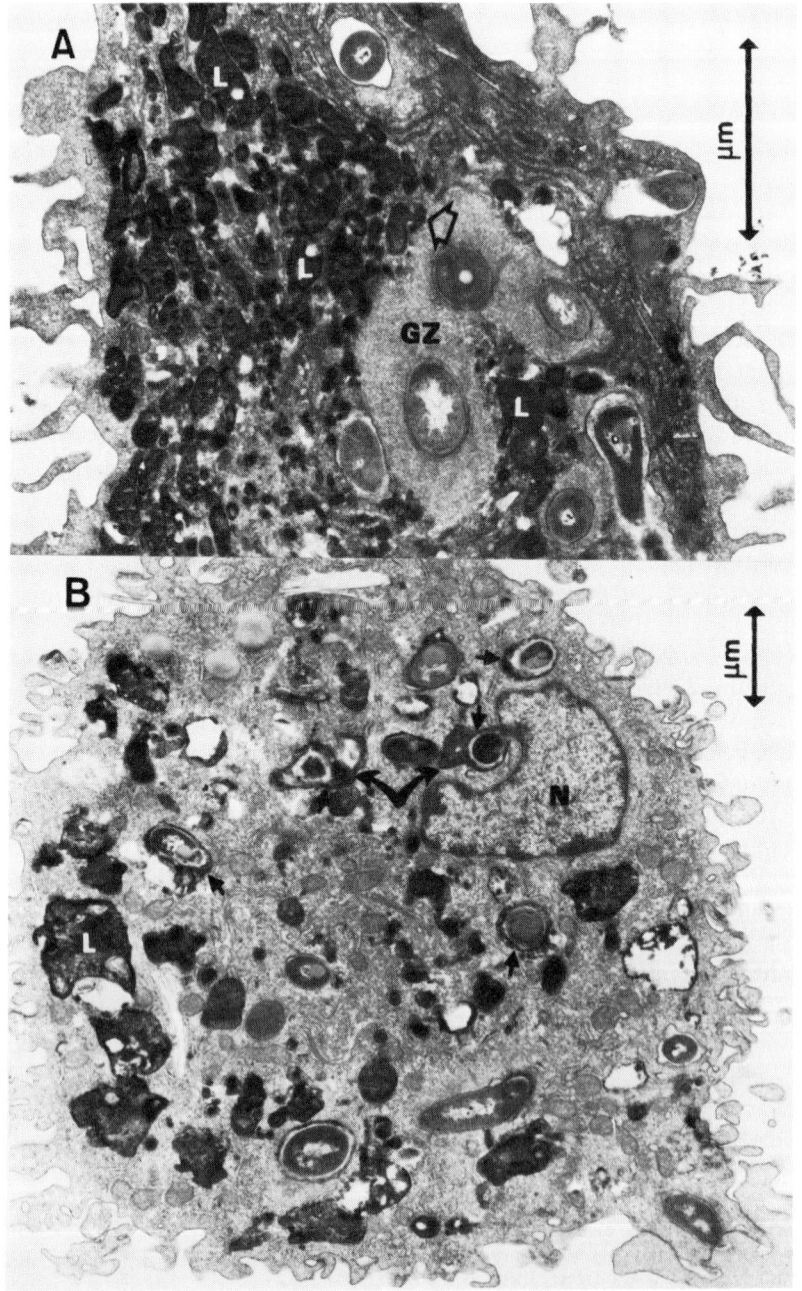

been shown that strains of *Actn. viscosus* induce a chemotactic effect on PMN leukocytes and stimulate the release of lysosomal enzymes from these leukocytes which can induce a severe localized inflammatory response. In addition, the organisms contain mitogenic activity for B-cell lymphocytes as well as being able to stimulate host cells to release mediators of inflammation (Engel *et al.*, 1976). It has been suggested, therefore, that *Actinomyces* strains may be pathogenic and induce host tissue damage by inducing a strong host response with the specific accumulation of PMN leukocytes and macrophages. These phagocytic cells, of course, contribute to the defence of the host against *Actinomyces* strains but, in so doing, they also contribute to tissue destruction through the release of their hydrolytic enzymes. Such tissue damage would contribute further to the anaerobic milieu that would promote additional growth of strains of *Actinomyces*. Thus, the long-term inflammatory lesions characteristic of actinomycosis, sulphur granules surrounded by PMN leukocytes, macrophages, plasma cells and soft tissue destruction, would be the result (Slack and Gerencser, 1975; Engel *et al.*, 1976; Brecher *et al.*, 1978; Cisar *et al.*, 1978; Ellen *et al.*, 1978; Beaman *et al.*, 1979; Brecher and van Houte, 1979; Cisar and Vatter, 1979; Shull, 1979).

3. Chemical Determinants of Pathogenicity

Actinomycetes do not appear to synthesize specific exotoxins related to pathogenesis and disease production, and capsules that prevent phagocytosis are not generally recognized in pathogenic strains. Indeed, most actinomycetes that have been studied are readily ingested by the host's phagocytes. The capacity of these organisms to invade the host and cause disease must, therefore, be controlled by other components associated with the bacterial cell. Several substances have been studied most extensively in mycobacteria and, although the exact mechanisms of pathogenesis are not understood, a great deal of insight has been gained (Barksdale and Kim, 1977; Youmans, 1979).

FIG. 2. Electron micrographs of rabbit alveolar macrophages infected with *Nocardia asteroides* GUH-2. The lysosomes are labelled with horseradish peroxidase to permit the demonstration of phagosome-lysosome fusion (Davis-Schibienski and Beaman, 1980c). (A) A normal alveolar macrophage. Note the large granular zone (GZ) surrounding the organisms (open arrow). The lysosomes (L) are not able to fuse with the phagocytic vacuole containing the nocardiae. This virulent strain of *Nocardia asteroides* inhibits normal phagosome-lysosome fusion in these macrophages. (B) An alveolar macrophage from an immunized rabbit. This macrophage was incubated with lymph node lymphocytes, alveolar lining material and antibody from the immunized rabbits and infected with *Nrda. asteroides*. These immunological components have significantly enhanced the ability of the macrophage's lysosomes to fuse with the phagosome and destroy the cells of *Nocardia* (arrows). (N = macrophage nucleus).

Corynebacteria, mycobacteria, nocardiae and some related actinomycetes have a complex cell envelope composed of several classes of free and bound lipids, peptides and polysaccharides. In *Mycobacterium* strains many of these wall components have been implicated as factors contributing to the organisms pathogenesis and host response to mycobacterial infection. Some of these compounds such as 'cord factor' (trehalose-6,6'-dimycolate) are toxic and are able to induce pathology within the host whereas others such as 'wax D' (peptidoglycomycolate) play a role in the host's development of delayed-type hypersensitivity (Goren and Brennan, 1979). Many of these compounds may also play a vital part in protecting the bacterial cell from host defence mechanisms. It is not possible here to review the complex chemical composition of these organisms, therefore the excellent reviews of Barksdale and Kim (1977) and Goren and Brennan (1979) should be consulted.

The walls of *Nrda. asteroides* strains undergo significant structural and chemical changes during the different phases of growth in Brain Heart Infusion (BHI) broth (Beaman, 1975; Beaman and Maslan, 1978); the extent and nature of these changes vary considerably depending upon the strain (B. L. Beaman, unpublished observations). Thus, many of the compounds such as trehalose dimycolate, peptidolipids, other glycolipids, peptidoglycolipids, carbohydrates, fatty acids and peptidoglycan change in amount relative to each other depending upon the conditions of growth and culture strain (Beaman, 1975; Beaman and Maslan, 1978). Which of these compounds affects nocardial virulence has not been determined, but it seems likely that many of these substances would have a major effect on host–parasite interactions. Because of the change in the relative amounts of these compounds during the growth phase one would predict that the same organism at different stages of growth (i.e. logarithmic phase versus stationary phase) would differ significantly in both its virulence and host–parasite interactions. This was found to be true (Beaman and Maslan, 1978). In all strains of *Nocardia* studied, it was observed that the cells from the logarithmic phase of growth were significantly more virulent than those from the same culture during stationary phase. In fact, with *Nrda. asteroides* GUH-2, the logarithmic phase cells grown in BHI broth were more than one thousand times more pathogenic for mice than were stationary phase cells from the same culture flask (Beaman and Maslan, 1978). It should be possible to define these differences chemically.

Beaman (1979) showed that logarithmic phase cells of *Nrda. asteroides* GUH-2 are more toxic for macrophages than stationary phase cells. Recently, it was found that the logarithmic phase cells appeared to contain significantly more trehalose dimycolate (cord factor) than the stationary phase organisms from the same culture (T. Ioneda and B. L. Beaman, personal communication). Since this toxic glycolipid was first discovered by Bloch in 1950 and chemically characterized by Noll *et al.* (1956), a great deal of research has been done on its role in the virulence of *Mybc. tuberculosis* and its affect on the host (Block, 1950; Noll *et al.*, 1956; Bekierkunst *et al.*, 1969; Kato, 1970; Moore *et al.*, 1972). Cord factor has both granuloma-inducing and adjuvant properties. It also has antitumour activity and non-specifically activates macrophages. In addition, it is toxic for mice and greatly affects mitochondrial structure and physiological activities and appears to be related, in some way, to the virulence of *Mybc. tuberculosis.* It seems reasonable to assume, therefore, that trehalose dimycolates from strains of *Nrda. asteroides* may be toxic and might play some role in nocardial pathogenicity. For a more complete discussion of this topic the reviews of Barksdale and Kim (1977) and Goren and Brennan (1979) are recommended.

Cord factors alone cannot explain the mechanisms of pathogenesis and virulence for *Corynebacterium, Mycobacterium* and *Nocardia* because it has been shown that members of these three genera may possess trehalose dimycolate, yet be incapable of establishing infection within a host (Barksdale and Kim, 1977; Goren and Brennan, 1979). Another group of compounds found in the walls of *Mybc. tuberculosis* has been implicated in the virulence of this organism. These are the sulpholipids which, like the cord factors, also contain trehalose esters (Middlebrook *et al.*, 1959; Goren and Brennan, 1979). Unlike cord factor, the sulpholipids alone do not appear to be significantly toxic though they affect mitochondrial membranes and alter macrophage functions. Generally, there is a direct correlation between the ability of *Mybc. tuberculosis* to bind the dye Neutral Red and the virulence of the organism. This Neutral Red binding capacity is due to the presence primarily of the sulpholipids and, to a lesser extent, the phospholipids within the wall (Goren and Brennan, 1979). The purified sulpholipids can be taken up by macrophages and become associated with the lysosomes within the macrophage and so inhibit their ability to fuse with the phagocytic vacuole. Sulpholipid on the surface of the mycobacteria appears, therefore, to be responsible for the specific inhibition of phagosome–lysosome fusion during ingestion of the mycobacterial cells by macrophages. Since the lysosomes cannot fuse, the degradative enzymes cannot be discharged into the phagosome containing the mycobacteria. This could have a major effect on the macrophage's ability to kill or inhibit mycobacteria and might permit the bacterial cell to grow within the phagocyte (D'Arcy Hart and Armstrong, 1974; Goren *et al.*, 1976; Goren and Brennan, 1979). Virulent strains of *Nrda. asteroides* inhibit phagosome–lysosome fusion within rabbit alveolar macrophages whereas less virulent strains do not. Strains of *Nrda. asteroides* that are of intermediate virulence possess an intermediate capacity to inhibit phagosome–lysosome fusion. Thus, there appears to be a direct correlation between the ability of *Nrda.*

asteroides to inhibit phagosome–lysosome fusion and nocardial virulence (Davis-Scibienski and Beaman, 1980a). In addition, the logarithmic phase cells of *Nrda. asteroides* inhibit phagosome–lysosome fusion more effectively than stationary phase cells of the same culture (Davis-Scibienski and Beaman, 1980b). Interestingly, it was observed that these logarithmic phase cells of *Nrda. asteroides* bound Neutral Red much more intensely than did the stationary phase cells (C. Davis-Scibienski and B. L. Beaman, personal communication). These observations indicate that a mechanism similar to that observed with the sulpholipids in mycobacteria may be involved in the virulence of *Nrda. asteroides* (Fig. 2A, B).

There are many other components of the walls of *Corynebacterium*, *Mycobacterium* and *Nocardia* that elicit specific host responses, or may be involved in the pathogenesis of these organisms. Thus, for example, the wax D fraction of *Mybc. tuberculosis* has powerful capabilities as an immunoadjuvant and strongly affects host responsiveness (Barksdale and Kim, 1977; Goren and Brennan, 1979). The wall wax D residues that have the strongest adjuvant activity to stimulate both immunoglobulin production and delayed hypersensitivity contain arabinogalactan-mycolate linked to a peptidoglycan subunit. It has been found that the peptidoglycan subunit is essential for the maximal amount of activity. The smallest unit of this complex found to have adjuvant activity is *N*-acylmuramyl-L-alanyl-D-isoglutamine which can be referred to as muramyldipeptide (MDP). To date this peptidoglycolipid has only been found in strains of *Corynebacterium*, *Mycobacterium*, *Nocardia* and *Rhodococcus* but the muramyldipeptide is a common peptidoglycan fragment that can be derived from a variety of diverse bacteria (Barksdale and Kim, 1977; Goren and Brennan, 1979). Even though muramyldipeptide is an immunoadjuvant, it is not very effective in stimulating a macrophage response and it does not enhance macrophage activity against *Listeria monocytogenes* (Finger and von Konig, 1980). The specific addition of glycerol-mycolate to MDP enhances its adjuvant activity and its ability to induce delayed-type hypersensitivity and host resistance to microbial infections presumably by non-specific activation of macrophages. Elimination of muramic acid from this polymer renders the complex inactive as an adjuvant but the complex retains the capacity to stimulate host resistance to microbial infection (Parant *et al.*, 1980).

Wall subunits obtained from nocardiae induce a mitogenic response in B-cell lymphocytes and these B-cell mitogens have been characterized and shown to be independent of T-cell function. The mitogen activity occurs for B-cells obtained from mice, rabbits and humans (Ciorbaru *et al.*, 1976; Ortiz-Ortiz *et al.*, 1979). The latter workers demonstrated that extracts of *Nrda. brasiliensis* are strong activators of B-cells even in T-deficient hosts. The B-cell mitogen from *Nrda. brasiliensis* strains is not only a potent stimulator of B-cells but results in polyclonal B-cell activation of spleen cells. The mitogen can overcome the need for helper T-cells in an *in vitro* antibody response. The B-cell mitogens could, therefore, play an important role in host responsiveness to infection with nocardiae.

In order for an organism to be a pathogen it must be able to grow within the host. One of the many limiting factors within the host appears to be the lack of available iron since most iron is firmly bound within the host by iron chelators (transferrin and lactoferrin). Bacteria cannot grow without adequate levels of iron, therefore the availability of iron determines the fate of the microorganism within the host (Kochan, 1973). Virulent strains of *Mycobacterium* and *Nocardia* possess within their walls complex secondary hydroxamates that are very strong iron-chelating agents and appear to be involved in iron transport into the organism. These lipid-associated compounds from mycobacteria are called mycobactins and those from nocardiae are referred to as nocobactins (Snow, 1970; Ratledge and Snow, 1974; Ratledge and Patel, 1976; Ratledge and Ewig, 1978). These iron-binding compounds appear to provide a mechanism by which the organism can obtain adequate amounts of iron from the host thereby allowing it to grow successfully within host tissues. Mycobactin and nocobactin may therefore be important in the overall host–parasite interaction and they may be necessary for the virulence of these organisms (Barksdale and Kim, 1977; Goren and Brennan, 1979).

The basal cell wall component of *Corynebacterium*, *Mycobacterium* and *Nocardia* strains is composed of peptidoglycan covalently linked to polymers of arabinogalactan-mycolates. As mentioned above, many additional classes of compounds unique to these and related organisms reside within the wall. In sharp contrast, other actinomycetes do not have this wall organization. The chemical basis for the pathogenesis of *Actinomyces* must, therefore, be quite different from the morphologically similar nocardiae since cells of the former lack mycolic acids, arabinogalactan, and do not contain any of the other complex glycolipids found in the nocardial wall. Very little information is available concerning the mechanisms of pathogenesis of *Actn. israelii* though several studies have yielded possible mechanisms for the virulence of *Actn. viscosus* (Slack and Gerencser, 1975; Engel *et al.*, 1976; Cisar *et al.*, 1978; Ellen *et al.*, 1978; Brecher *et al.*, 1978; Beaman *et al.*, 1979; Brecher and van Houte, 1979; Cisar and Vatter, 1979; Shull, 1979). The virulent *Actn. viscosus* T14-Vi has a wall antigen with increased amounts of 6-deoxytalose whereas the avirulent strain T14-Av has only trace amounts of this compound (Cisar and Vatter, 1979). In addition, other wall differences have been noted between these two strains (Cisar and Vatter, 1979). The avirulent strain produces large amounts of extracellular polysaccharide material that heavily coats the cell whereas the virulent T14-Vi strain produce little of this material. Data suggest that this polysaccharide 'capsule' is important in the determination of virulence because in the avirulent T14-Av strain this material buries the pili; these protein structures remain exposed in the unencapsulated T14-Vi strain. Thus, these observations suggest that the availability of the pili to interact with the host is important in the determination of virulence. The differences between these two organisms appear to reside in their ability to colonize the teeth (Brecher *et al.*, 1978; Cisar *et al.*, 1978; Brecher and van Houte, 1979; Cisar and Vatter, 1979).

Engel *et al.* (1976) found that cell homogenates of *Actn. viscosus* strains induce an acute inflammatory response with a direct chemotactic effect on polymorphonuclear phagocytes. The substance responsible for this chemotactic effect appears to be of low molecular weight and is active only in the presence of human serum. Further, complement appears to be involved in this process. In addition to the chemotactic factor, the cell homogenates stimulate the host cells to release mediators of inflammation. These investigators believe that these components in the cell homogenates may explain some of the pathogenic capabilities of *Actn. viscosus* (Engel *et al.*, 1976).

4. The Cell Wall as a Determinant of Pathogenesis

A large variety of complex compounds are present in the walls of actinomycetes. Many of these substances have been implicated in the virulence of these microorganisms and some have been shown to be toxic to host cells or elicit a specific type of host response. It is possible to remove the wall from these bacteria within the laboratory by a variety of methods, such as treatment with lysozyme, or by incubating the organisms with wall inhibitors like penicillin or D-cycloserine. Also, host cells such as macrophages can strip off the wall of actinomycetes both *in vitro* and *in vivo* (Bourgeois and Beaman 1974, 1976; Beaman and Smather, 1976; Beaman *et al.*, 1978b; Beaman, 1980a, b). If an osmotically suitable environment is provided, these organisms do not lyse but instead form either sphaeroplasts or protoplasts. These wall-deficient forms may be able to grow and replicate and thereby become L-phase or transitional phase variants (Mattman, 1974). If these wall-less forms continue to grow in a suitable environment and if they can be serially transferred in the absence of the inducing agent without reversion to their parental form, then they are referred to as L-forms (Mattman, 1974).

Protoplasts, sphaeroplasts, L-phase variants and L-forms have been described in strains of several genera of the *Actinomycetales* including *Corynebacterium* (Kagan and Savenkova, 1960; Savenkova, 1962; Kanei *et al.*, 1978), *Mycobacterium* (Thacore and Willett, 1966; Imaeda, 1975; Mattam, 1970; Imaeda, 1975; Moscovic, 1978), *Nocardia* (Bourgeois and Beaman, 1976; Beaman *et al.*, 1978b), *Streptomyces* (Bradley, 1959; Sagara *et al.*, 1971; Okanishi *et al.*, 1974; Gumpert, 1978), *Agromyces* (Horwitz and

Casida, 1975), *Bacterionema* (Takazoe, 1979) and *Actinomyces* (Merline and Mattman, 1971). The pathogenic capacities of wall-less organisms have been extensively studied but reports are inconclusive and often contradictory. There is, however, a growing amount of data supporting the view that L-phase variants and L-forms are involved in the latency of certain infections and some L-forms appear to be pathogenic (Mattman, 1974; Beaman, 1980a, b).

L-phase variants and sphaeroplasts of *Mybc. tuberculosis, Mybc. avium,* and other mycobacteria have been isolated from infected individuals (Mattman, 1970, 1974). Attempts to study the pathogenicity of L-phase variants of *Mybc. tuberculosis* were reported by Ratnam and Chandrasekhar (1976). These investigators found that relatively low numbers of L-phase cells did not induce pathological changes in the lungs of guinea pigs but, after several weeks, some of the animals developed disease and acquired tuberculin hypersensitivity as the result of reversion of the sphaeroplasts to the parental acid-fast bacillary form. These results demonstrate that sphaeroplasts of *Mybc. tuberculosis* can remain viable within the lung of the host for a relatively long period of time and suggest that L-phase variants may play a significant role in latency which is characteristic of mycobacterial disease (Ratnam and Chandrasekhar, 1976).

Peritoneal macrophages obtained from mice and maintained *in vitro* can induce L-phase variants of *Nrda. asteroides* 10905. Further, these L-phase organisms persist for several days within the macrophages and then increase in numbers (Bourgeois and Beaman, 1974). Similar results are obtained when this strain of *Nrda. asteroides* is incubated with alveolar macrophages from rabbits (Beaman and Smathers, 1976). More recently, L-phase variants of *Nrda. otitidis-cavarium* 112 were shown to be induced by alveolar macrophages within the intact murine lung (Beaman, 1980a). These wall-deficient organisms contributed to the pathological development of alveolar consolidation resulting in animal death. It was shown conclusively that the L-phase of *Nrda. otitidis-caviarum* 112 was pathogenic when induced in the lungs of mice (Beaman, 1980a). Further, considerable evidence has been obtained demonstrating that L-phase variants of both *Nrda. otitidis-caviarum* 112 and *Nrda. asteroides* GUH-5 can persist within host tissues for more than 1 year. The reversion of L-phase colonies to normal organisms within the tissues resulted in granule formation and a chronic inflammatory response characteristic of a mycetomatous lesion (Beaman, 1980b, unpublished work). Thus, L-phase variants of a few strains of *Nocardia* were shown to be pathogenic when the wall-less bacterial forms were injected either intravenously or intraperitoneally into normal mice (Beaman, 1980b, unpublished work; Fig. 1C).

These data support the view that L-phase variants of several actinomycetes play a significant role in disease in humans. For example, L-phase variants of *Nrda. asteroides* were repeatedly isolated from the spinal fluid of a patient suffering from pulmonary nocardiosis. Normal nocardial cells were never isolated from the central nervous system of this patient yet, after several months, the patient died of central nervous system infection and brain abscess (Beaman, 1980b). Dr. S. R. Pal (personal communication) reported a case of an L-form of an unknown aerobic actinomycete associated with chronic infection of the central nervous system. During the course of this infection normal organisms were not isolated from the patient, but L-forms were repeatedly recovered and some of these were observed to revert to an aerobic actinomycete with physiological properties more closely resembling *Actinomadura* than *Nocardia* (S. R. Pal, personal communication). In addition, L-phase variants of *Actn. naeslundii* and *Actn. odontolyticus* have been recovered from the blood in patients suffering from leukemia (Merline and Mattman, 1971).

The fact that some of the wall-deficient variants of actinomycetes may be pathogenic or persist within the host in a latent form for extended periods of time demonstrate that the integrity of the complex wall structure possessed by these organisms is not essential for pathogenicity. These observations do not, however, imply that the various components of the wall are not essential for the organism's pathogenesis. It is well established that many L-forms continue to synthesize and secrete toxins and wall components even though they lack a wall. Thus, some L-forms from Gram-negative bacteria can still synthesize endotoxin (Dasinger and Suter, 1962; Weibull *et al.*, 1967; Kalmanson *et al.*, 1968) and L-forms of *Streptococcus pyogenes* may continue to synthesize M-protein even though the cells do not possess a wall (Freimer *et al.*, 1959; Widdowson *et al.*, 1970). Whether or not pathogenic L-forms of the actinomycetes continue to synthesize and secrete the various complex wall components such as cord factor, sulpholipids, wax-D, mycobactin or nocobactin and other chemicals remains to be determined.

5. Acknowledgements

I am grateful for the help and comments of the individuals who have discussed various aspects of this manuscript during its preparation. Much of the work discussed in the review was supported by Public Health Service grants AI-13167 and AI-15114 from the National Institutes of Allergy and Infectious Diseases. I thank Marilyn Wheeler for her expert typing of this manuscript.

REFERENCES

Abe, P. M., Majeski, J. A. and Stauffer, L. R. (1978). Histological changes observed in an *Arachnia propionica* infection of mice. *Journal of Surgical Research* **25**, 174–179.

Abu-Samra, M. T. (1978). The effect of prednisolone trimethylacetate on the pathogenicity of *Dermatophilus congolensis* to white mice. *Mycopathologia* **66**, 1–9.
Barksdale, L. (1970). *Corynebacterium diphtheriae* and its relatives. *Bacteriological Reviews* **34**, 378–422.
Barksdale, L. and Kim, K. S. (1977). *Mycobacterium*. *Bacteriological Reviews* **41**, 217–372.
Beaman, B. L. (1973). An ultrastructural analysis of *Nocardia* during experimental infections in mice. *Infection and Immunity* **8**, 828–840.
Beaman, B. L. (1975). Structural and biochemical alterations of *Nocardia asteroides* cell walls during its growth cycle. *Journal of Bacteriology* **123**, 1235–1253.
Beaman, B. L. (1976). Possible mechanisms of nocardial pathogenesis. In: *The Biology of the Nocardiae*, (M. Goodfellow, G. H. Brownell, and J. A. Serrano, eds.), pp. 386–417. Academic Press, London.
Beaman, B. L. (1977). The *in vitro* response of rabbit alveolar macrophages to infection with *Nocardia asteroides*. *Infection and Immunity* **15**, 925–937.
Beaman, B. L. (1979). The interactions of *Nocardia asteroides* obtained from the lungs of "normal" and immunized rabbits. *Infection and Immunity* **26**, 355–361.
Beaman, B. L. (1980a). The induction of L-phase variants of *Nocardia caviae* within the intact murine lung. *Infection and Immunity* **29**, 244–251.
Beaman, B. L. (1980b). The possible role of L-phase variants of *Nocardia* in chronic infections. *Zentralblatt für Bakteriologie, Parasitenkunde, Infektionskrankheiten und Hygiene*. 1. *Abteilung, Supplement* **11**, 222–227.
Beaman, B. L. and Smathers, M. (1976). Interaction of *Nocardia asteroides* with cultured rabbit alveolar macrophages. *Infection and Immunity* **13**, 1126–1131.
Beaman, B. L. and Maslan, S. (1977). The effect of cyclophosphamide on experimental *Nocardia asteroides* infection in mice. *Infection and Immunity* **16**, 995–1004.
Beaman, B. L. and Maslan, S. (1978). Virulence of *Nocardia asteroides* during its growth cycle. *Infection and Immunity* **20**, 290–295.
Beaman, B. L., Burnside, J., Edwards, B. and Causey, W. (1976). Nocardial infections in the United States, 1972–1974. *Journal of Infectious Diseases* **134**, 286–289.
Beaman, B. L., Gershwin, M. E. and Maslan, S. (1978a). Infectious agents in immunodeficient murine models: Pathogenicity of *Nocardia asteroides* in congenitally athymic (nude) and hereditarily asplenic (Dh/ +) mice. *Infection and Immunity* **20**, 381–387.
Beaman, B. L., Serrano, J. A. and Serrano, A. A. (1978b). Comparative ultra-structure within the *Nocardia*. *Zentralblatt für Bakteriologie, Parasitenkunde, Infektionskrankheiten und Hygiene*. 1. *Abteilung, Supplement* **6**, 201–220.
Beaman, B. L., Goldstein, E., Gershwin, M. E., Maslan, S. and Lippert, W. (1978c). Lung response of congenitally athymic (nude), heterozygous, and Swiss Webster mice to aerogenic and intranasal infection by *Nocardia asteroides*. *Infection and Immunity* **22**, 867–877.
Beaman, B. L., Gershwin, M. E. and Maslan, S. (1979). Infectious agents in immunodeficient murine models: Pathogenicity of *Actinomyces israelii* Serotype 1 in congenitally athymic (nude) mice. *Infection and Immunity* **24**, 583–585.
Beaman, B. L., Gershwin, M. E., Scates, S. and Ohsugi, Y. (1980a). The immunobiology of germfree mice infected with *Nocardia asteroides*. *Infection and Immunity* **29**, 733–793.
Beaman, B. L., Maslan, S., Scates, S. and Rosen, J. (1980b). The effect of route of inoculation on host resistance to *Nocardia*. *Infection and Immunity* **28**, 185–189.
Bekierkunst, A., Levy, I. S., Yarkoni, E., Vilkas, E., Adam, A. and Lederer, E. (1969). Granuloma formation induced in mice by chemically defined mycobacterial fractions. *Journal of Bacteriology* **100**, 95–102.
Block, H. (1950). Studies on the virulence of tubercle bacilli: Isolation and biological properties of a constituent of virulent organisms. *Journal of Experimental Medicine* **91**, 197–217.
Bourgeois, L. and Beaman, B. L. (1974). Probable L-forms of *Nocardia asteroides* induced in cultured mouse peritoneal macrophages. *Infection and Immunity* **9**, 576–590.
Bourgeois, L. and Beaman, B. L. (1976). *In vitro* spheroplast and L-form induction within the pathogenic nocardiae. *Journal of Bacteriology* **127**, 584–594.
Bradley, S. G. (1959). Protoplasts of *Streptomyces griseus* and *Nocardia paraguayensis*. *Journal of Bacteriology* **77**, 115–116.

Brecher, S. M. and van Houte, J. (1979). Relationship between host age and susceptibility to oral colonization by *Actinomyces viscosus* in Sprague-Dawley rats. *Infection and Immunity* **26**, 1137–1145.

Brecher, S. M., van Houte, J. and Hammond, B. F. (1978). Role of colonization in the virulence of *Actinomyces viscosus* strains T14-Vi and T14-Av. *Infection and Immunity* **22**, 603–614.

Brizin, B. and Lenart, J. (1957). Experimental nocardiosis in white mice. *Acta Medica (Yugoslavia)* **11**, 292–297.

Buchanan, B. B. and Pine, L. (1962). Characterization of a propionic acid producing actinomycete, *Actinomyces propionicus*, sp. nov. *Journal of General Microbiology* **28**, 305–323.

Ciorbaru, R., Pettit, J. F., Lederer, E., Zissman, E., Bona, C. and Chedid, L. (1976). Presence and subcellular localization of two distinct mitogenic fractions in the cells of *Nocardia rubra* and *Nocardia opaca*: Preparation of soluble mitogenic peptidoglycan fractions. *Infection and Immunity* **13**, 1084–1090.

Cisar, J. O. and Vatter, A. E. (1979). Surface fibrils (fimbrae) of *Actinomyces viscosus* T14V. *Infection and Immunity* **24**, 523–531.

Cisar, J. O., Vatter, A. E. and McIntire, F. C. (1978). Identification of the virulence-associated antigen on the surface fibrils of *Actinomyces viscosus* T14. *Infection and Immunity* **19**, 312–319.

Conrad, S. E., Breivis, J. and Fried, M. A. (1978). Vertebral osteomyelitis caused by *Arachnia propionica* and resembling actinomycosis. *American Journal of Bone Joint Surgery* **60**, 549–553.

D'Arcy Hart, P. and Armstrong, J. A. (1974). Strain virulence and the lysosomal response in macrophages infected with *Mycobacterium tuberculosis*. *Infection and Immunity* **10**, 742–746.

Dasinger, B. L. and Suter, E. (1962). Endotoxic activity of L-forms derived from *Salmonella paratyphi* B. *Proceedings of the Society of Experimental Biology and Medicine* **111**, 399–400.

Davis, B. D., Dulbecco, R., Eisen, H. N., Ginsberg, H. S., Wood, W. B. and McCarty, M. (1973). *Microbiology*, 2nd Edition, pp. 512–595. Harper and Row, New York.

Davis-Scibienski, C. and Beaman, B. L. (1980a). The interaction of *Nocardia asteroides* with rabbit alveolar macrophages: The association of virulence, viability, ultrastructural damage and phagosome-lysosome fusion. *Infection and Immunity* **28**, 610–619.

Davis-Scibienski, C. and Beaman, B. L. (1980b). The interaction of *Nocardia asteroides* with rabbit alveolar macrophages: The effect of phase of growth and viability on phagosome-lysosome fusion. *Infection and Immunity* **29**, 24–29.

Davis-Scibienski, C. and Beaman, B. L. (1980c). Interaction of alveolar macrophages with *Nocardia asteroides*: Immunologic enhancement of phagocytosis, phagosome–lysosome fusion and microbicidal activity. *Infection and Immunity* **30**, 578–587.

Drake, C. H. and Henrici, A. T. (1943). *Nocardia asteroides*. Its pathogencity and allergic properties. *American Review of Tuberculosis* **48**, 184–198.

Drivalle, C. T. and Gregory, E. M. (1979). Superoxide dismutase and O_2 lethality in *Bacteroides fragilis*. *Journal of Bacteriology* **138**, 139–145.

Ellen, R. P., Walker, D. L. and Chan, K. H. (1978). Association of long surface appendages with adherence-related functions of the Gram-positive species *Actinomyces naeslundii*. *Journal of Bacteriology* **134**, 1171–1175.

Engel, D., van Epps. D. and Clagett, J. (1976). *In vivo* and *in vitro* studies on possible pathogenic mechanisms of *Actinomyces viscosus*. *Infection and Immunity* **14**, 548–554.

Filice, G. A., Beaman, B. L. and Remington, J. S. (1980a). Effects of activated macrophages on *Nocardia asteroides*. *Infection and Immunity* **27**, 643–649.

Filice, G. A., Beaman, B. L., Krick, J. A. and Remington, J. S. (1980b). Effects of human neutrophils and monocytes on *Nocardia asteroides*: Failure of killing despite occurrence of the oxidative metabolic burst. *Journal of Infectious Diseases* **142**, 432–438.

Finger, H. and Wirsing von Konig, C. H. (1980). Failure of synthetic muranyl dipeptide to increase antibacterial resistance. *Infection and Immunity* **27**, 288–291.

Folb, P. I., Jaffe, R. and Altman, G. (1976). *Nocardia asteroides* and *Nocardia brasiliensis* infections in mice. *Infection and Immunity* **13**, 1490–1496.

Folb, P. I., Timme, A. and Horowitz, A. (1977). *Nocardia* infections in congenitally athymic (nude) mice and other inbred mouse strains. *Infection and Immunity* **18**, 459–466.

Freimer, E. H., Krause, R. M. and McCarty, M. (1959). Studies of L-forms and protoplasts of Group A streptococci. I. Isolation, growth and bacteriologic characteristics. *Journal of Experimental Medicine* **110**, 853–874.

Georg, L. K. and Coleman, R. M. (1970). Comparative pathogenicity of various *Actinomyces* species. In: *The Actinomycetales* (H. Prauser, ed.,), pp. 35–45. VEB Gustav Fischer, Jena.

González-Ochoa, A. (1973). Virulence of nocardiae. *Canadian Journal of Microbiology* **19**, 901–904.

González-Ochoa, A. and Sandoval-Cuellar, A. (1976). Different degrees of morbidity in white mice induced by *Nocardia brasiliensis, Nocardia asteroides,* and *Nocardia caviae. Sabouraudia* **14**, 255–259.

Goren, M. B. and Brennan, P. J. (1979). Mycobacterial lipids: Chemistry and biologic activities. In: *Tuberculosis*, (Youmans, G. P., ed.) pp. 63–193. W. B. Saunders, Philadelphia.

Goren, M. B., Hart, P. D., Young, M. R. and Armstrong, J. A. (1976). Prevention of phagosome-lysosome fusion in cultured macrophages by sulfatides of *Mycobacterium tuberculosis. Proceedings of the National Academy of Science* **73**, 2510–2514.

Gumpert, J. (1978). Ultrastructure and modifiability of the cell envelope in *Streptomyces hygroscopicus. Zentralblatt für Bakteriologie, Parasitenkunde, Infektionskrankheiten und Hygiene.* I. *Abteilung, Supplement* **6**, 221–233.

Haburchack, D. R., Jeffrey, B., Higbee, J. W. and Everett, E. E. (1978). Infections caused by rhodochrous. *American Journal of Medicine* **65**, 298–302.

Harris, G. K. (1980). Nocardiosis: A literature review and a case report of *Nocardia asteroides* infection. American Journal of Medical Technology **46**, 44–48.

Horwitz, A. H. and Casida, L. E. (1975). L-phase variants of *Agromyces ramosus. Antonie van Leeuwenhoek* **41**, 153–171.

Imaeda, T. (1975). Ultrastructure of L-phase variants isolated from a culture of *Mycobacterium phlei. Journal of Medical Microbiology* **8**, 389–395.

Kagan, G. I. and Savenkova, V. T. (1960). The technique of obtaining L-forms of *C. diphtheriae* and some morphological peculiarities. *Zhurnal Mikrobiologii, Epidemiologii I Immunobiologii* **31**, 55–58.

Kalmanson, G. M., Kubota, M. Y. and Guze, L. B. (1968). Production of the Schwartzman reaction with microbial L-forms. *Journal of Bacteriology* **96**, 646–651.

Kanei, C., Uchida, T. and Yoneda, M. (1978). Isolation of the L-phase variant from toxigenic *Corynebacterium diphtheriae* C7 (B). *Infection and Immunity* **20**, 167–172.

Kato, M. (1970). Molecular configuration and toxicity of cord factor. *Japanese Journal of Medical Science Biology* **23**, 267–271.

Kochan, I. (1973). The role of iron in bacterial infections with specific considerations of host-tubercle bacillus interaction. *Current Topics in Microbiology and Immunobiology* **60**, 1–30.

Kurup, P. V., Randhawa, H. S., Sandu, R. S. and Abraham, S. (1970). Pathogenicity of *Nocardia caviae, N. asteroides* and *N. brasilensis. Mycopathologia Mycologia Applicata* **40**, 113–130.

Mackaness, G. B. (1969). The influence of immunologically committed lymphoid cells on macrophage activity *in vivo. Journal of Experimental Medicine* **129**, 973–992.

Macotella-Ruiz, E. and Mariat, F. (1963). Sur la production de mycetoma experimentaux par *Nocardia brasiliensis* et *Nocardia asteroides. Bulletin of Society of Pathology Exotica* **89**, 426–431.

Makinde, A. A. and Wilkie, B. N. (1979). Humoral and cell mediated immune response to crude antigens of *Dermatophilus congolensis* during experimental infection in rabbits. *Canadian Journal of Comparative Medicine* **43**, 68–77.

Mason, K. N. and Hathaway, B. M. (1969). A study of *Nocardia asteroides.* White mice used as test animals. *Archives of Pathology* **87**, 389–392.

Mattman, L. H. (1970). Cell wall deficient forms of mycobacteria. *Annals of New York Academy of Science* **174**, 852–861.

Mattman, L. (1974). Cell wall deficient forms. CRC Press, Cleveland.

Merline, J. R. and Mattman, L. H. (1971). Cell wall deficient forms of actinomyces complicating two cases of leukemia. *Michigan Academy* **3**, 113–121.

Middlebrook, G., Coleman, C. M. and Schaefer, W. B. (1959). Sulfolipid from virulent tubercle bacilli. *Proceedings of National Academy of Science* **45**, 1801–1804.

Mohapatra, L. N. and Pine, L. (1963). Studies on the pathogenicity of aerobic actinomycetes inoculated into mice intravenously. *Sabouraudia* **2**, 176–184.

Moore, V. L., Myrvik, Q. N. and Kato, M. (1972). Role of cord factor (trehalose-6,6'-dimycolate) in allergic granuloma formation in rabbits. *Infection and Immunity* **6**, 5–8.
Moscovic, E. (1978). Sarcoidosis and mycobacterial L-forms. In: *Pathology Annual, Part 2* (S. C. Sommers and P. P. Rosen, eds.), pp. 69–164. New York: Appleton-Century-Crofts.
Naeslund, C. (1929). Experimentelle Uebertragung der Aktinomykoses von Menschen und versuchstiere Mittels Reinkulturen der Actinomycesform. Wolff-Israel (*Act. israeli*). *Acta Pathologica et Microbiologica Scandinavica* **6**, 66–77.
Noll, H., Block, H., Asselineau, J. and Lederer, E. (1956). The chemical structure of cord factor of *Mycobacterium tuberculosis*. *Biochemistry Biophysics Acta* **20**, 299–309.
Okanishi, M., Suzuki, K. and Umezawa, H. (1974). Formation and reversion of streptomycete protoplasts: Cultural condition and morphology. *Journal of General Microbiology* **80**, 389–400.
Ortiz-Ortiz, L., Parks, D. E., Lopez, J. S. and Weigle, W. O. (1979). B lymphocyte activation with an extract of *Nocardia brasiliensis*. *Infection and Immunity* **25**, 627–634.
Parant, M. A., Audibent, F. M., Chedid, L. A., Level, M. R., Lefrancier, P. L., Choay, J. P. and Lederer, E. (1980). Immunostimulant activities of a lipophilic muramyl dipeptide derivative and of desmuramyl peptidolipid analogs. *Infection and Immunity* **27**, 826–831.
Pullman, J. D., Kelley, D. C. and Coles, F. H. (1967). Immunologic studies of natural and experimental cutaneous streptothricosis infections in cattle. *American Journal of Veterinary Research* **28**, 447–455.
Pulverer, G. and Schaal, K. P. (1978). Pathogenicity and medical importance of aerobic and anaerobic actinomycetes. *Zentralblatt für Bakteriologie, Parasitenkunde, Infektionskrankheiten und Hygiene. I. Abteilung, Supplement* **6**, 417–427.
Ratnam, S. and Chandrasekhar, S. (1976). The pathogenicity of spheroplasts of *Mycobacterium tuberculosis*. *American Review of Respiratory Disease* **114**, 549–554.
Ratledge, C. and Ewig, D. (1978). The separation of the mycobactins from *Mycobacterium smegmatis* by using high pressure liquid chromatography. *Journal of Biochemistry* **175**, 853–857.
Ratledge, C. and Patel, P. V. (1976). The isolation, properties and taxonomic relevance of lipid-soluble, iron-binding compounds (the nocobactins) from *Nocardia*. *Journal of General Microbiology* **93**, 141–152.
Ratledge, C. and Snow, G. A. (1974). Isolation and structure of nocobactin NA, a lipid-soluble iron bind compound from *Nocardia asteroides*. *Journal of Biochemistry* **139**, 407–413.
Runyon, E. H. (1951). *Nocardia asteroides*: Studies of its pathogenicity and drug sensitivities. *Journal of Laboratory Clinical Medicine* **37**, 713–720.
Saenz Lope, E. and Guiterrez, D. C. (1977). *Nocardia asteroides* primary cerebral abscess and secondary meningitis. *Acta Neurochirurgica* **37**, 139–145.
Sagara, Y. K., Fukui, F., Ota, F., Yoshida, N., Kashiyama, T. and Fujimoto, M. (1971). Rapid formation of protoplasts of *Streptomyces griseoflavus* and their fine structure. *Japanese Journal of Microbiology* **15**, 73–84.
Savenkova, V. T. (1962). Some peculiarities in the formation of L-forms of *C. diphtheriae*. *Zhurnal Mikrobiologii, Epidemiologii I Immunobiologii* **33**, 91–93.
Shull, J. D. (1979). Induction of cell mediated immunity to *Actinomyces viscosus* in mice. *Archives Oral Biology* **24**, 421–425.
Slack, J. M. and Gerencser, M. A. (1975). *Actinomyces, Filamentous Bacteria: Biology and Pathogenicity*. Burgess Publishing Co., Minneapolis.
Smith, I. M. and Hayward, A. H. S. (1971). *Nocardia caviae* and *Nocardia asteroides*: Comparative bacteriological and mouse pathogenicity studies. *Journal of Comparative Pathology* **81**, 79–88.
Snow, G. A. (1970). Mycobactins: Iron-chelating growth factors from mycobacteria. *Bacteriological Review* **34**, 99–125.
Takazoe, I. (1979). Formation of *Bacterionema matruchotii* protoplasts. *Journal of Dental Research* **58**, 1709–1713.
Thacore, H. and Willett, H. P. (1966). The formation of spheroplasts of *Mycobacterium tuberculosis* in tissue culture cells. *American Review of Respiratory Disease* **93**, 786–796.
Weibull, C., Bickel, W. D., Haskins, W. T., Milner, K. C. and Ribi, E. (1967). Chemical, biological and structural properties of stable *Proteus* L-forms and their parent bacteria. *Journal of Bacteriology* **93**, 1143–1159.

Welch, D. F., Sword, C. P., Brehms, S. and Dusanic, D. (1979). Relationship between superoxide dismutase and pathogenic mechanisms of *Listeria monocytogenes*. *Infection and Immunity* **23**, 863–872.

Widdowson, J. P., Maxted, W. R. and Grant, D. L. (1970). The production of opacity in serum by Group A streptococci and its relationship with the presence of M protein. *Journal of General Microbiology* **61**, 343–353.

Wolff, M. and Israel, I. (1981). Ueber Reincultur des *Actinomyces* und sein Uebertragbarkeit auf Thiere. *Archives of Pathology. Anatomy, Physiology, Klinical Medicine* **126**, 11–59.

Youmans, G. P. (1979). *Tuberculosis*. W. B. Saunders Co., Philadelphia.

11
Ecology of Actinomycetes

S. T. WILLIAMS,* S. LANNING*
and E. M. H. WELLINGTON†

*Department of Botany, University of Liverpool, Liverpool, UK
and † Department of Biology, Liverpool Polytechnic, Liverpool, UK

1. Introduction	481
2. Detection of actinomycetes in natural habitats	482
3. Actinomycetes in soil	486
A. Form, growth and dispersal	486
B. Influence of environmental factors	487
C. Roles in decomposition processes	491
D. The rhizosphere	492
E. Antibiotics and biological control of root pathogens	493
4. Actinomycetes in compost, fodder and related substrates	494
A. Manures and composts	495
B. Fodders, grain and related substrates	496
C. Airborne actinomycete spores as respiratory allergens	498
5. Actinomycetes as plant pathogens	499
A. Coryneform pathogens	499
B. Streptomycete pathogens	500
6. Nitrogen fixation by endophytic actinomycetes	501
A. Occurrence of endophytes	502
B. Specificity of endophyte–host associations	503
C. Infection, nodule development and structure	504
D. Nitrogen fixation	504
7. Actinomycetes in fresh water and marine environments	504
A. Actinomycetes in fresh water	505
B. Actinomycetes as a cause of tastes and odours in potable water	506
C. Marine actinomycetes	507
8. Actinomycetes as symbionts of animals	509
9. Ecology of actinophage	510
A. Occurrence	511
B. Influence of environmental factors	512
10. Conclusions	513
References	513

1. Introduction

Actinomycetes occur in a wide variety of natural and man-made habitats, growing on a vast range of substrates within them (Table 1). Although most are strict saprophytes, some form parasitic or symbiotic associations with plants or animals. Evidence of their occurrence and numbers in habitats is extensive. However, such information alone is of limited ecological value and leaves open to speculation what roles actinomycetes play in their environments. Here we have attempted to answer this question by assessing current knowledge of the activities of actinomycetes in a range of diverse environments.

It is first useful to consider briefly some of the general approaches made to the study of actinomycete ecology.

2. Detection of Actinomycetes in Natural Habitats

In order to study the ecology of most microbes, it is first necessary to isolate them. Procedures for the selective isolation and enumeration of actinomycetes have been recently reviewed (Williams and Wellington, 1982a, b). Therefore, details of the considerable variety of methods used will not be given here.

Aims of isolation vary, but the ecologist is usually interested initially in discerning the qualitative and quantitative nature of a whole population or one or more of its specific components. Although some genera are relatively easy to isolate (e.g. *Micromonospora*, *Streptomyces*), others have been isolated on only one or two occasions (e.g. *Planobispora*, Thiemann and Beretta, 1968; *Planomonospora*, Thiemann *et al.*, 1967). The plant endophyte, *Frankia*, eluded isolation until quite recently (Callaham *et al.*, 1978). Novel genera, such as *Actinosynnema* (Hasegawa *et al.*, 1978), are still occasionally detected.

The four major stages of an isolation procedure are (i) selection of material containing the microbes, (ii) pretreatment of the material, (iii) growth on laboratory media and (iv) incubation and colony selection. Selectivity may be introduced at any of these stages, though not always intentionally.

Selection of material is obviously determined by the ecological aims. These can range from the copious general surveys of populations in different soil types, to isolation of a single species from a root nodule.

Various pretreatments have been used to increase the proportion of actinomycetes to other bacteria on isolation plates. Heat treatments are quite often used. Examples are the heating of soil up to 100°C for 1 hour to isolate *Microbispora* and other rare genera (Nonomura and Ohara, 1969, 1971a, b, c, d), keeping soil at 40°C for 2–16 hours to isolate streptomycetes from roots and soil (Williams *et al.*, 1972), heating water, soil and dung extracts at 55°C for 6 minutes to isolate *Rodc. coprophilus* and other actinomycetes (Rowbotham and Cross, 1977) and keeping soil at 100°C for 15 minutes to isolate *Actinomadura* and other genera (Athalye *et al.*, 1981). Membrane filtration may be used to concentrate actinomycete propagules in water (Burman *et al.*, 1969; Al-Diwany. *et al.*, 1978), while agitation of dry, deteriorated plant materials in a sedimentation chamber has been used to isolate thermophiles (Lacey and Dutkiewicz, 1976). Recently, Palleroni (1980) applied a chemotactic method to isolate motile spores of the *Actinoplanaceae* from soil.

Selectivity of the isolation medium is influenced primarily by its nutrient composition, pH and addition of selective inhibitors. A wide range of formulations have been recommended for isolation of actinomycetes. Components have usually been chosen by trial and error, therefore the basis for their selectivity is not always clear. Thus, examples of three agar media commonly used to isolate actinomycetes from soil and water are starch–casein (Küster and Williams, 1964), colloidal chitin (Hsu and Lockwood, 1975) and M3 medium (Rowbotham and Cross, 1977), yet these differ markedly in composition. As most actinomycetes are neutrophilic, media usually approximate to pH 7.0. However, if the reaction is lowered to pH 4.5–5.0, acidophilic and acidoduric isolates are obtained from acidic soils and wastes (Khan and Williams, 1975; Hagedorn, 1976). Incorporation of antibiotics into isolation media is now a widespread and effective means of increasing selectivity. Antifungal antibiotics, such as cycloheximide and mycostatin, do not inhibit actinomycetes and are useful when dealing with many soils and plant materials (e.g. Gregory and Lacey, 1963; Williams and Davies, 1965; Cross, 1968). Antibacterial antibiotics have to be selected with care as many actinomycetes are susceptible to them. Nevertheless, there are an increasing number of reports of their successful use in selective isolation of particular genera or species. Some good examples are the use of novobiocin to isolate *Thermoactinomyces vulgaris* (Cross, 1968; Cross and Johnston, 1972), chlortetracycline and methacycline for *Nocardia* spp. (Orchard *et al.*, 1977), kanamycin for '*Thermomonospora chromogena*' (McCarthy and Cross, 1981) and rifampicin for *Actinomadura* strains (Athalye *et al.*, 1981).

Isolation plates for most actinomycetes are incubated at 25–30°C, apart from thermophiles, which are kept at 45–55°C. The major variable at this stage is the length of the incubation period. If this is prolonged beyond the normal 1–2 week period, slower growing strains, often overlooked, may be detected. Thus, the most likely reason for the many previously unsuccessful attempts to isolate the root endophyte, *Frankia*, is that it grows slowly on isolation media (Lechevalier, 1981).

Isolation and also enumeration of actinomycetes is almost invariably achieved by a dilution plate procedure. This does not provide information on the location and form of growth in the natural habitat, nor is it an accurate measurement of activity. Direct observation and homogenization experiments have indicated that most colonies on soil dilution plates originate from spores or other resting propagules (Lloyd, 1969a; Mayfield *et al.*,

TABLE 1. Characteristic habitats of some actinomycete genera

Genera	Saprophytic					Parasitic			Symbiotic	
	Soil and plant material	Fresh water	Salt water	Manure and composts	Man and animals	Man	Animals	Plants	Animals	Plants
Actinomadura	+			?		+				
Actinomyces	+				+	+	+			
Actinoplanes	+	+								
Actinosynemma	+									
Arachnia						+				
'*Coryneform*' bacteria[a]	+	+	+	?	?	+	+	+	?	
Dactylosporangium	+	+								
Dermatophilus						+	+			
Frankia	?									+
Geodermatophilus	+	+								
Microbispora	+		+							
Microellobosporia	+			+						
Micropolyspora	+					+				
Micromonospora	+	+	+						?	

Genus							
Microtetraspora	+						
Mycobacterium	+		+				
Nocardia	+		+		+	+	
Nocardioides	+				+		?
Nocardiopsis	+				+		
Oerskovia	+						
Planobispora	+						
Planomonospora	+						
Pseudonocardia	+						
Rhodococcus	+		+	+	+	+	+
Saccharomonospora				+			
Saccharopolyspora				+			
Spirillospora		+					
Streptomyces	+	+		+	+	+	
Streptosporangium	+	+					
Thermoactinomyces						+	?
Thermomonospora						+	

[a] Coryneform bacteria as defined by Keddie (1978), which include *Arthrobacter, Brevibacterium, Cellulomonas, Corynebacterium, Curtobacterium* and *Microbacterium*.

1972). The same is probably true for many other habitats. Homogenization and ultrasonication of lake muds indicated that, while *Micromonospora* occurred as mycelium and spores, streptomycetes and nocardioforms existed as spores or hyphal fragments (Johnston and Cross, 1976b).

Probably the best technique for detecting growth sites of actinomycetes on natural substrates is scanning electron microsopy (Mayfield *et al.*, 1972; Williams and Wellington, 1982b). However, it is not always possible to identify the growth observed with any degree of accuracy.

Thus, much of the information we have about the roles of actinomycetes in natural habitats, with the exception of some parasitic and mutualistic associations, is based on extrapolation from the behaviour of isolates in the laboratory.

3. Actinomycetes in Soil

The vast majority of cultures of aerobic actinomycetes have originated from soil. Viable counts of over four millions per gram may be obtained from fertile soils (Flaig and Kutzner, 1960) and even these are probably underestimates as no single isolation procedure allows detection of all genera. Over 20 genera have been isolated from soil (Williams and Wellington, 1982b) and many species lists produced. Streptomycetes are ubiquitous. Lechevalier and Lechevalier (1967) found 95% of over 5000 isolates were streptomycetes. Many studies on 'actinomycetes' in soil have, in effect, been on streptomycetes. Although much attention will naturally be paid to the activities of streptomycetes in soil, the literature is too extensive to cover fully. Reviews on soil streptomycetes have been made by Lacey (1973), Küster (1976), Williams (1978) and Kutzner (1981). Although *Streptomyces* is undoubtedly the most important genus in soil, the frequency and importance of some other genera may have been underestimated. Application of more efficient isolation procedures can demonstrate that a genus such as *Actinomadura*, once thought to be a rare alien, is quite widespread and numerous (Nonomura and Ohara, 1971d; Preobrazhenskaya *et al.*, 1975; Athalye *et al.*, 1981). Viable counts of actinomycetes are generally lowest in water-logged, anaerobic soils and in acidic soils (Waksman, 1959). However, counts are significantly increased if suitable selective procedures are used for microaerophiles (Casida, 1965; Gledhill and Casida, 1969a, b) or acidophiles (Williams *et al.*, 1971; Hagedorn, 1976).

A. FORM, GROWTH AND DISPERSAL

These aspects have received little attention, most ecological studies being based on dilution plate counts. The limited information available is almost entirely restricted to streptomycetes. Although many genera have originated from soil, in most cases, no attempt has been made to ascertain how or if they grow and survive in this habitat. Soil may often contain pathogenic actinomycetes, such as nocardiae, the status of which in this habitat was discussed by Schaal and Bickenbach (1978) and Orchard (1981).

The behaviour of streptomycetes in unsterile soil has been studied by light microscopy of buried slides (Pfennig, 1958), transmission electron microscopy and incident light microscopy (Szabo et al., 1964), scanning electron microscopy (Williams and Mayfield, 1971; Mayfield et al., 1972) and immunofluorescence microscopy (Efremenkova et al., 1978). Results of these studies were summarized by Williams (1978). The general picture which emerges is that streptomycetes exist for long periods as arthrospores or chlamydospores. They germinate in the presence of exogenous nutrients, lack of which prevents germination of most or all spores added to unsterile soil. When localized organic substrates (e.g. roots, fungi) are available they are rapidly colonized by mycelial growth. This soon results in production of spores when nutrients are exhausted. The mean doubling time of streptomycetes in soil was estimated to be 1.7 days (Mayfield et al., 1972) which probably reflects their short intermittent periods of active growth, rather than supporting the widely held conception that they are 'slow-growing' microbes. Specific growth rates and doubling times for streptomycetes in laboratory culture were found to be approximately intermediate between those of other bacteria and fungi (Flowers and Williams, 1977a). Non-sporing actinobacteria, such as *Arthrobacter*, probably survive in soil as coccoid forms (Luscombe and Gray, 1974; Gray, 1976).

Streptomycete spores can be released above the soil when particles are disturbed by wind or rain (Lloyd, 1969b). Within the soil, dispersal of spores is affected by movement of water and arthropods (Ruddick and Williams, 1972). Detachment of spores by water is influenced by the form of the sporing apparatus and surface wettability. Thus the single, hydrophilic conidia of *Micromonospora* were more readily detached and dispersed in water than were the hydrophobic spore chains of streptomycetes. In contrast, hydrophobic spores readily adhered to arthropod exocuticles and were dispersed by these animals in soil and litter.

B. INFLUENCE OF ENVIRONMENTAL FACTORS

Temperature is an obvious factor determining activity in soil. However, there have been surprisingly few studies of its effect in the field, with the exception of thermophiles, which will be discussed later. Temperature can be indirectly implicated in the numerous examples of the influences of seasons or climatic regions on population size and composition. As examples of seasonal influence, numbers of streptomycetes in grassland were highest in summer (Küster, 1976) and nocardiae were most numerous in pasture in winter (Orchard, 1981). Climatic influence can be illustrated by the claim that '*Stmy. malachiticus*' is confined to sub-tropical and tropical soils (Küster, 1970), as is *Nrda. otitidis-caviarum* (Schaal and Bickenbach, 1978). As most actinomycetes are mesophilic, it must be assumed that in many soils they do not experience their optimum growth temperature. A notable exception was the isolation of almost pure cultures of nocardiae from soils at 30–40°C in a thermal region of New Zealand (Orchard, 1981).

Most actinomycete isolates behave as neutrophiles in culture, with a growth range from about pH 5.0 to 9.0 and an optimum around pH 7.0. Therefore pH is a major environmental factor determining the distribution and activity of soil actinomycetes. Nevertheless, neutrophiles do occur in low numbers in soils below pH 5.0. The early work of Jensen (1928) detected a group of streptomycetes (termed *Actinomyces acidophilus*) in acidic soils which required a low pH for growth in culture, i.e. they were acidophilic.

More recent work (Williams *et al.*, 1971; Khan and Williams, 1975; Hagedorn, 1976) has shown that acidophilic and acidoduric streptomycetes are numerous and widespread in natural and man-made acidic soils. The former grow between about pH 3.5 and 6.5; thus they are excluded when neutral isolation media are used. They are presumably active in decomposition processes in such soils and their exo-enzymes, diastases and chitinases, are adapted to function at a lower pH than those from neutrophiles (Williams and Flowers, 1978; Williams and Robinson, 1981). Most acidophilic isolates to date have been regarded as streptomycetes, although they are clearly separated from their neutrophilic counterparts in numerical classification (Williams and Khan, 1974). Further studies of existing and future isolates may well reveal acidophilism in other genera.

The existence of neutrophilic streptomycetes in acidic soils may be explained by the periodic occurrence of micro-sites of higher pH, produced by ammonification of substrates such as amino acids or chitin, initiated by acidophilic or acidoduric soil microbes (Williams and Mayfield, 1971). Coupled with this, spores of neutrophiles are tolerant to acidity (Flowers and Williams, 1977b). An interesting succession from acidophilic to neutrophilic streptomycetes occurs when chitin or dead fungal mycelium is added to an acidic soil (Williams and Robinson, 1981). Initially, acidophiles were active but, as ammonification proceeded and the pH rose, neutrophile numbers increased significantly whereas acidophile numbers dropped; this was most clearly demonstrated in a poorly buffered mineral soil horizon.

There are few reports of basophilic actinomycetes. A species (*Stmy. caeruleus*) which grew from pH 6.5 to 9.5 was isolated from soil near to a salt lake (Taber, 1960). It seems likely that others remain to be detected.

Moisture tension and aeration of soils are intimately related, as rates of gaseous diffusion are drastically reduced at low tensions, i.e. when soil pores are filled with water. Therefore, the most extreme effects that can be observed are those of aeration in water-logged soil, and of reduced water availability on microbial growth and survival.

Soil actinomycetes are generally considered to be strict aerobes, although *Oerskovia* species will grow anaerobically in some laboratory media (Lechevalier, 1972) and the microaerophilic *Actn. humiferus* (Gledhill and Casida, 1969a) and *Agromyces ramosus* (Gledhill and Casida, 1969b) were isolated from soil. Radial growth of streptomycetes in soil was lowered at low moisture tension (pF 1.0) when pores were almost water-filled (Williams *et al.*, 1972). In contrast, a *Micromonospora sp.* showed maximum growth at pF 1.0. The characteristic habitat of this genus is lake muds (Waksman, 1959), but its predominance in wet soils has also been noted (Singal *et al.*, 1978; Cross, 1981), thus suggesting that *Micromonospora* also is possibly capable of microaerophilic growth. Stotzky and Goos (1965) reported that soil actinomycetes were particularly sensitive to high carbon dioxide and low oxygen concentrations, while streptomycetes isolated from various depths in soil were able to grow in sterile soil at low oxygen concentrations but not when carbon dioxide concentrations exceeded 10% (Williams *et al.*, 1972). Such levels of carbon dioxide occur in water-filled pores (Griffin, 1966).

As with many soil microbes, optimum growth of actinomycetes occurs when pores are partially filled with water. Thus, optimum radial growth of streptomycetes in sterile soil was between pF 1.5–2.5 (Williams *et al.*, 1972). As the soil becomes drier (>pF 4.0), growth is severely limited and eventually ceases. In these conditions, survival becomes ecologically significant. Streptomycete spores survived desiccation in soil more successfully than vegetative hyphae or *Pseudomonas* sp. cells (Williams *et al.*, 1972). As a result, streptomycetes often constitute a greater proportion of the colonies on isolation plates from dry soil, leading to the misconception that they are more abundant in dry soils than wet ones. The sclerotia of *Chainia olivacea* were found to be even more resistant than streptomycete spores to desiccation (Lechevalier, 1981). Some streptomycetes from arid Australian soils grew on laboratory media at high osmotic potentials (-100 to -150 bars) (Wong and Griffin, 1974) and halophilic strains have also been isolated (Kayamura and Takada, 1970). The most strikingly halophilic actinomycete is *Actinopolyspora halophila* (Gochnauer *et al.*, 1975), but its natural habitat is unknown.

Almost all soils contain a significant proportion of clay and humic colloidal material. Adsorption between colloids, microbial cells and their extracellular products occurs, and thus such materials can markedly affect microbial activity at the microenvironmental level. Addition of calcium montmorillonite to cultures of *Streptomyces, Micromonospora* and *Nocardia* species accelerated growth, glucose utilization and carbon dioxide evolution (Martin *et al.*, 1976). Calcium humate also stimulated growth of some strains. Streptomycete spores appeared to adsorb to kaolin but not to montmorillonite except at low pH (Ruddick and Williams, 1972). Sites of ammonia adsorption, leading to micro-sites of increased pH in acidic soil, were associated with humic material (Williams and Mayfield, 1971).

The soil microbial environment is altered by man, more or less intentionally, when land is cultivated. However, less predictable effects occur when various alien materials and substances reach the soil by accident or design.

The effect of application of sewage sludge to pasture soils on the frequency of some actinomycete genera has been studied by Orchard (1979, 1980, 1981). Numbers of nocardiae (mainly *Nrda. asteroides*) and micromonosporae increased significantly over 14 months after application of dried sludge. However, with wet sludge, *Thermoactinomyces* was most noticeably increased.

Heavy metals are now a common form of soil pollutant, leading to retardation of many important microbial processes. Babich and Stotzky (1977) found that, in general, actinomycetes were more tolerant to cadmium than were other bacteria; toxicity was pH dependent, usually being greatest at pH 8 or 9 and was lowered by clay minerals. Jordan and Lechevalier (1975) reported increased zinc tolerance among actinomycetes isolated from a forest soil near to a smelter. However, actinomycetes were generally less tolerant than other bacteria or fungi. A similar observation was made for isolates from lead mine waste, but no evidence for increased lead tolerance was obtained (Williams *et al.*, 1977).

Actinomycetes isolated from contaminated soil grew well on paraffin; except for three mycobacteria strains oxidizing nonane, no actinomycetes oxidized this or octane (Krassilnikov *et al.*, 1969). *Nocardia asteroides* and *Rhodococcus* spp. were particularly abundant in chernozem soils polluted with petroleum and were isolated on media containing n-alkanes (Nesterenko *et al.*, 1978). Jensen (1975) noted that coryneforms had been most often reported to be stimulated by addition of hydrocarbons to soil. When the effect of mineral oils and oily waste on the bacterial flora of soils was studied, *Arthrobacter* strains were among those stimulated whereas *Streptomyces* strains were repressed. It was concluded that the latter were unable to compete with other bacteria at least in the early stages of hydrocarbon degradation. '*Nocardia salmonicolor*' and an *Actinoplanes* sp. from soil were able to oxidize carbon monoxide to carbon dioxide (Bartholomew and Alexander, 1979).

There are an increasing number of studies on the effects of pesticides and herbicides on actinomycetes and other soil microbes. Thus, for example, soil actinomycetes were inhibited in soil and culture by the fungicide PCNB (*p*-chloronitrobenzene) (Katan and Lockwood, 1970), a range of other fungicides (Corden and Young 1965), heptachlor (Shamiyeh and Johnson, 1973) and 2,4-D (2,4-dichlorophenoxyacetic acid (Ou *et al.*, 1978). In contrast, actinomycetes were not significantly reduced in soil by DDT (1,1,1-trichloro-2,2-bis(*p*-chlorophenylethane) and mycelia added to soil accumulated it and also dieldrin and PCNB (Ko and Lockwood, 1968). Atrazine was also strongly accumulated and it increased numbers of actinomycetes (Percich and Lockwood, 1978). Chlordane-treated soil yielded a *Nocardiopsis* sp. able to degrade the insecticide in culture (Beeman and Matsumura, 1981) and soil enriched with carboxanilide fungicides provided a *Nocardia*-like isolate which could use various fungicides as sole sources of carbon and nitrogen (Bachofer *et al.*, 1973).

C. ROLES IN DECOMPOSITION PROCESSES

There is a vast amount of literature on the enzyme-producing and degradative abilities of soil actinomycetes *in vitro*. It is beyond the scope of this chapter to review these data and, indeed, the ecological relevance of much of this information is not clear. Actinomycetes are not found in large numbers on live or newly-dead green plants (Lechevalier, 1981) although *Corynebacterium* can occur in the phylloplane (Dickinson *et al.*, 1975). They are usually considered to be most active in the later stages of decomposition of plant and other materials, playing an important part in the degradation of complex and relatively recalcitrant polymers (Lacey 1973; Lechevalier, 1981). Therefore, the evidence for involvement of actinomycetes in the decomposition of some of these complex polymers which are abundant in soil will be considered here.

Despite the long-held belief that actinomycetes decompose lignin and cellulose in soil, it is only recently that we have obtained convincing evidence for their capacity to degrade lignocellulose, which is a major, naturally occurring, component of recalcitrant plant residues (Küster, 1979). The results of Crawford and his co-workers, using ^{14}C-labelled substrates have shown that some streptomycetes can degrade both the cellulosic and lignin components, the latter involving oxidation of both aromatic ring and side chain carbon to carbon dioxide (Crawford, 1978; Crawford and Sutherland, 1979; Phelan *et al.*, 1979). *Streptomyces flavovirens* decomposed intact cell walls of phloem from Douglas fir, although the bulk of degraded material was carbohydrate (Sutherland *et al.*, 1979). *Streptomyces* strains degraded lignocelluloses of grass, softwoods and hardwoods, the former being most readily attacked (Antai and Crawford, 1981). *Streptomyces badius* differed from white-rot fungi in being able to degrade lignin in the absence of a readily available carbon source and in its tolerance of levels of organic or inorganic nitrogen, which suppress lignin degradation by these fungi (Barder and Crawford, 1981). Other actinomycetes are also able to participate in lignin decomposition. *Nocardia* spp. from soil released $^{14}CO_2$ from labelled coniferyl alcohol, the main precursor of lignin biosynthesis, and from plant lignins (Trojanowski *et al.*, 1977). Degradation of lignin-related compounds by *Nocardia* and *Rhodococcus* strains has also been shown (Eggeling and Sahm, 1980, 1981; Rast *et al.*, 1980).

Actinomycetes have been implicated in the decomposition of humus for some time (Waksman, 1959), but the ill-defined chemical nature of humus and its fractions, humic acid and fulvic acid, make it difficult to obtain definitive evidence. It has been claimed that streptomycetes can use humic substances in the presence of an available carbon source (Szegi and Gulyas, 1968). Nocardiae have also been implicated (Küster, 1979); *Nocardia* isolates liberated carbon dioxide from humic acid, fulvic acid and total humus extracts over a period of 280 days (Steinbrenner and Mundstock, 1975). Similarly, it has been claimed that *Rhodococcus* (*Nocardia*) spp. were able to use humic acid *in vitro* over a period of 4 months incubation (Cross *et al.*, 1976).

It has also been claimed for many years that streptomycetes, like some fungi, can synthesize humic compounds in culture (see review by Kutzner, 1968). Although it is difficult to make exact chemical comparisons, it has been claimed that the 'melanins' produced by many streptomycetes are similar to humic acids (Matschke 1970c, d, e; Huntjens, 1972). '*Streptomyces aureus*' produced tyrosinase and laccase and synthesized tricyclic hydrocarbons which then oxidized and polymerized with other amino acids to form humic acids in 14 days (Matschke, 1970a). However, maximum concentration of humic acids was obtained after 2 year's incubation (Matschke, 1970b).

There is some evidence that actinomycetes are involved in the degradation of many other naturally occurring polymers in soil. These include hemicelluloses (Waksman and Diehm, 1931; Iizuka and Kawaminami, 1965), pectin (Kaiser, 1971), keratin (Noval and Nickerson, 1959; Mostafa and Hussein, 1974; Young and Smith, 1975), chitin (Okafor, 1967; Gray and Baxby, 1968; Hsu and Lockwood, 1975; Williams and Robinson, 1981), and fungal cell wall material (Mitchell, 1963; Mann *et al.*, 1978; Williams and Robinson, 1981). The last of these probably provides one of the major substrates for soil streptomycetes (Williams, 1978). Arthropod exoskeletons provide another potential source of chitin (Okafor, 1966).

D. THE RHIZOSPHERE

There is little evidence that actinomycetes play a major role in the rhizosphere, apart from their possible role in suppression of root pathogens (see next section).

Positive rhizosphere effects (based on R:S ratios) reported include those for nocardiae and *Calotropsis gigantea* (Agate and Bhat, 1964), streptomycetes and *Juncus* (Dickinson and Dooley, 1967), nocardiae and soyabeans (Thomas and Khurana, 1980), while Bernhard (1967) found an increase in actinomycetes in the rhizospheres of only 4 of 14 plants studied. Vruggink (1976) concluded that soil type had more effect on actinomycete populations than did various crop plants. R:S ratios ranging from 16 to 50 were obtained for actinomycetes in the rhizosphere of the sand dune plants, *Ammophila arenaria* and *Agropyron junceiforme* (Watson and Williams, 1974). However, actinomycetes were unable to respond to exudates from young roots. Old, dead roots of *Ammophila* were, however, readily colonized by *Streptomyces, Micromonospora, Nocardia* and *Rhodococcus* strains. There was little qualitative difference between soil and rhizosphere populations, but it has been claimed that differences occur between rhizoplane populations and those in the rhizosphere (Vruggink, 1976) and in the soil (Buti, 1978). Greater ability among rhizosphere isolates to hydrolyse starch (Abraham and Herr, 1964) has been reported. Some rhizosphere isolates can synthesize gibberellin-like substances *in vitro* (Katznelson and Cole, 1965) and some actinomycetes can produce indole acetic acid in culture and when inoculated into soil (Purushothaman *et al.*, 1974).

E. ANTIBIOTICS AND BIOLOGICAL CONTROL OF ROOT PATHOGENS

It is often assumed that actinomycetes, particularly streptomycetes, play a major role in antagonistic interactions in soil because of their greater capacity for antibiotic production *in vitro*. Although it is over 40 years since the isolation of the first antibiotic from an actinomycete, there is still doubt about the occurrence of these substances in natural soil (Williams and Khan, 1974; Gottlieb, 1976; Williams, 1982). Several workers have argued from various viewpoints that antibiotics are natural products (Williams and Khan, 1974; Hopwood and Merrick, 1977; Martin and Demain, 1980). However, direct evidence for their presence in natural soil is still lacking. This may indicate that they are not produced or that they are present but are not detectable due to their short-term stability, adsorption to soil colloids, lack of sensitive detection methods or the spatially and temporally restricted growth of their producers (Williams, 1982).

Despite the lack of direct evidence, rhizosphere streptomycetes have often been invoked as agents for control of fungal root pathogens. However, many other antagonistic interactions, other than antibiosis, may occur between streptomycetes and fungi (see Williams, 1978) and the significance of antibiosis in biological control of root pathogens has been questioned (Baker, 1968). Typically, reduced incidence of root infection has been correlated with an increase in numbers of streptomycetes in the rhizosphere which inhibit the pathogen *in vitro*, usually by production of antifungal antibiotics (Table 2). In addition, control of aspergilli in soil of pot plants (Staib *et al.*, 1980) and of wood-infecting fungi (Blanchette *et al.*, 1981) by streptomycetes has been claimed.

TABLE 2. Examples of 'biological control' of root-infecting fungi by actinomycetes

Actinomycete antagonists	Pathogens controlled	Reference
Streptomyces spp.	*Fusarium oxysporum*	Whaley and Boyle
	Phymatotrichum omnivorum	(1967)
	Rhizoctonia solani	
	Verticillium albo-atrum	
Streptomyces spp.	*Rhizoctonia solani*	Sneh *et al.* (1971)
Stmy. diastochromogenes	*Pythium debaryanum*	Kaspari (1973)
		Kaspari and Schönbeck (1973)
Stmy. griseus	*Rhizoctonia solani*	Merriman *et al.* (1974a, b)
Streptomyces sp.	*Gaeumannomyces graminis*	Smiley (1978a, b)
Stmy. griseoalbus	*Phellinus weirii*	Rose *et al.* (1980)
	Fomes annosus	
	Phytophthora cinnamoni	
Streptomyces spp.	*Rhizoctonia bataticola*	Sing and Mehrotra (1980)

Inoculation of seeds or plants with antagonistic actinomycetes to reduce disease severity has been attempted. These include inoculation of *Stmy. griseus* to potato cuttings (Chambers and Millington, 1974) and to wheat and carrot seeds (Merriman *et al.*, 1974a, b), and various streptomycetes to wheat seedlings (Sivasithamparam and Parker, 1978) and grain (Sing and Mehrotra, 1980). Some evidence for lowered disease incidence was obtained.

Amendment of soil with organic materials may also control pathogens, apparently by stimulating saprophytic antagonists such as streptomycetes. Chitin amendment has been shown to be particularly successful, control of *Fusarium oxysporum* (Buxton *et al.*, 1965), *Rhizoctonia solani* (Henis *et al.*, 1967) and other fungi being achieved. Control of '*Stmy. scabies*' on potato tubers was also obtained (Vruggink, 1970). The antifungal factor in chitin-amended soil was extracted (Sneh and Henis, 1972) but it was a lipid-containing complex, not the polyene antibiotic, which most of the isolates produced in culture.

4. Actinomycetes in Compost, Fodder and Related Substrates

Many mesophilic actinomycetes are active on such substrates in the initial stages of decomposition. However, the capacity for self-heating during decomposition often provides ideal conditions for obligate or facultative thermophilic actinomycetes able to grow at temperatures above about 40°C. Some genera (e.g. *Thermoactinomyces, Saccharomonospora*) are strictly thermophilic, while other genera (e.g. *Microbispora, Micropolyspora, Pseudonocardia, Streptomyces* and *Thermomonospora*) contain thermophilic species. *Thermoactinomyces* is notable for its production of resistant endospores similar to those of *Bacillus* and certain other bacterial genera (Cross, 1968; Cross *et al.*, 1968b). This may explain the occurrence of these obligate thermophiles in relatively low temperature environments, such as soil, lake muds and marine sediments (Cross and Johnston, 1972) and in archaeological excavations (Unsworth *et al.*, 1977). The occurrence of actinomycetes in self-heating materials is summarized in Table 3. The ecology of actinomycetes in such habitats has been reviewed by Lacey (1973, 1978).

A. MANURES AND COMPOSTS

Thermophilic actinomycetes grow well on animal manures (Waksman *et al.*, 1939) although temperatures inside piles during the initial phases of composting are probably too high for their activity (Lacey, 1973). It seems likely that at least one mesophilic actinomycete, *Rodc. coprophilus*, is truly coprophilous (Rowbotham and Cross, 1977).

Thermophilic fermentation of pig faeces and straw by *Thermomonospora* spp. and *Psnc. thermophila* at 55°C resulted in cellulose breakdown and increase in digestible protein (Ginnivan *et al.*, 1977). The latter species was initially isolated from rotting manure using anaerobic techniques (Henssen, 1957). Deodorization of pig faeces by actinomycetes has also been reported (Ohta and Ikeda, 1978). When manure is used for preparation of mushroom composts, thermophilic actinomycetes (particularly *Thermomonospora* spp.) grow well during the second, indoor phase of preparation, when conditions are humid and temperatures reach about 60°C (Fergus, 1964; Lacey, 1973).

With the exception of mushroom compost, there have been few detailed studies of the succession and activity of thermophilic actinomycetes in composted materials. *Thermoactinomyces* spp. were found to be active in self-heating of peat piles (Küster and Locci, 1963) and in grass compost (Erikson, 1952). Spontaneous heating of wheat straw compost was accompanied by an increase in actinomycete numbers (Chang and Hudson, 1967) and they were involved in changes during composting of spruce-bark in the laboratory (Bågstam, 1978). Actinomycetes increased during the later stages of decomposition of municipal waste (Filip and Küster, 1979). *Thermomonospora curvata* produces thermostable C_1 and C_x cellulases and is active in decomposition of municipal waste compost (Stutzenberger *et al.*, 1970; Stutzenberger, 1971; 1972a, b); heavy metals in such composts were shown to inhibit cellulase production (Stutzenberger and Sterpu, 1978).

Mesophilic actinomycetes may be present in sewage sludge. Activated sludge from a dairy contained *Corynebacterium, Microbacterium* and *Rhodococcus* strains, while in municipal sludge the former two genera and *Arthrobacter* sp. predominated (Seiler *et al.*, 1980). A strain of *Rodc. (Nrda.) erythropolis* was used to remove phthalate esters from activated sludge and to prevent deflocculation (Kurane *et al.*, 1979). A novel species, *Nrda. amarae*, was detected in foam of activated sludge where it grew in large quantities (Lechevalier and Lechevalier, 1974). In some cases, growth at the surface of treatment tanks was dense enough to impair functioning of the plant (Lechevalier, 1981).

There is considerable interest in the commercial exploitation of the many enzymes that have been isolated from thermophiles in culture, but it is beyond the scope of this chapter to review this topic. Examples that may be of ecological relevance include amylases from *Team. vulgaris* (Kuo and Hartman, 1967) and *Temo. curvata* (Glymph and Stutzenberger, 1977), amylase and protease from *Sacs. (Temo.) viridis* (Upton and Fogarty, 1977), β-1,3-glucanase from a thermophilic streptomycete (Lilley and Bull, 1974) and cellulases from '*Temo. fusca*' and *Stmy. thermodiastaticus* (Crawford and McCoy, 1972). The former species was also involved in decomposition of lignocellulose pulps (Crawford, 1974). A thermoactinomycete degraded up to 70% of a 1% cellulose substrate in 24 hours at 55°C and the use of these microbes commercially for the rapid saccharification of cellulosic materials has been suggested (Hagerdal *et al.*, 1978).

B. FODDERS, GRAIN AND RELATED SUBSTRATES

A range of mesophilic and thermophilic actinomycetes plays a part in the deterioration of such substrates (Table 3). The most prevalent in almost all substrates is *Team. vulgaris*, with *Streptomyces* spp. and *Mips. faeni* also very widespread. One of the commonest streptomycetes isolated is *Stmy. albus* (Lacey, 1981) which, though mesophilic, can grow at 45°C.

The succession and biochemical changes in mouldy hay have been most clearly defined (Gregory *et al.*, 1963; Festenstein *et al.*, 1965; Lacey, 1978). The maximum temperature attained and the succession of actinomycetes are both closely correlated with the water content of the hay. At 29% water content, 33°C was reached, whereas at 42%, the temperature rose to 60°C. As heating increased from ambient temperature to 65°C, predominant actinomycetes changed first from *Stmy. griseus* to *Stmy. albus*, then to *Mips. faeni*, *Team. vulgaris*, and *Sacs. viridis* (Lacey, 1978). Maximum numbers of most species occurred after 11 days. Growth of the thermophilic actinomycetes was favoured in localized areas of increased pH and possibly occurred on cell walls of earlier fungal colonizers. Some evidence for protein breakdown by *Mips. faeni* was obtained.

TABLE 3. Predominant actinomycetes in heated materials (from Lacey, 1978, 1981)

Species	Hay	Cereal grain	Sugar-cane bagasse	Cotton	Mushroom compost	Sewage sludge	Straw
Actinomadura-like species							+
Micropolyspora faeni	+	+					+
Nocardia spp.						+	
Nocardiopsis dassonvillei				+			
Pseudonocardia sp.			+				
Saccharomonospora viridis	+	+		+			+
Saccharopolyspora hirsuta			+				
Streptomyces spp.	+	+	+	+	+	+	+
Thermoactinomyces sacchari				+			
Tcam. thalpophilus	+	+		+			+
Tcam. vulgaris	+	+					+
Thermomonospora spp.					+		

A similar succession occurred on moist barley grain stored in unsealed silos where the heating was related to rate of removal of grain from the surface as well as water content (Lacey, 1971a). *Micropolyspora faeni* occurred only in grain containing 35–40% water, heated to 50–60°C, whereas *Team. vulgaris* was present over a wider temperature range than in hay. Streptomycetes predominated in high-moisture (27%) field corn and silo-stored corn which was untreated or treated with ammonia, ammonium isobutyrate or propionic-acetic acid (Lyons et al., 1975).

Sugar-cane bagasse leaves the mill containing about 50% water and 3–6% sugar, providing ideal conditions for microbial growth (Lacey, 1978). When baled, maximum temperatures greater than 50°C were common and sometimes remained above 40°C for long periods. After initial fungal growth, actinomycetes predominated, as in hay, although different types occurred; these included two novel species, *Team. sacchari* (Lacey, 1971b) and *Saccharopolyspora hirsuta* (Lacey and Goodfellow, 1975).

C. AIRBORNE ACTINOMYCETE SPORES AS RESPIRATORY ALLERGENS

This topic has been recently reviewed by Lacey (1981). Actinomycete spores are generally ideally suited for aerial dispersal; they are usually produced in the aerial environment above the substrata, have dry, hydrophobic surfaces and are easily detached by disturbance of the substrate or air movement. Their numbers in outdoor air are soon diluted but very large numbers can occur indoors when materials such as deteriorated hay, grain, cotton and sugar-cane bagasse are handled. Up to 1.6×10^9 spores m^{-3} occurred with mouldy hay, of which 98% were 'actinomycetes + bacteria', while 2.9×10^9 spores m^{-3}, half of which were 'actinomycetes + bacteria', occurred when moist barley was unloaded from a silo (Gregory et al., 1963; Gregory and Lacey, 1963; Lacey, 1971a). Even undisturbed grain produced 10^6 to 10^7 spores m^{-3} air (Lacey and Lacey, 1964; Lacey, 1971a). On mushroom farms when spawn was mixed with compost, 7.4×10^8 spores m^{-3} air were detected, with *Thermomonospora* spp. predominating (Lacey, 1974). In cotton mills, up to 2.3×10^6 m^{-3} 'actinomycetes + bacteria' occurred with *Nrdo. dassonvillei* and *Stmy. griseus* being most common (Lacey, 1977).

Actinomycete spores have been shown to be involved in the allergies known as farmer's lung, bagassosis and mushroom-worker's lung. Farmer's lung is associated with the development of thermophilic actinomycetes in self-heated hay. Extracts of mouldy hay give precipitin reactions against sera from patients with farmer's lung (Gregory et al., 1964) and three groups of antigens with differing physical and chemical properties have been recognized. *Micropolyspora faeni* gave all three antigens, while *Team. vulgaris* gave one (Pepys et al., 1963; Cross et al., 1968a). Other actinomycetes, including *Sacs. viridis* (Barrowcliff and Arblaster, 1968; Wenzel *et al.*, 1974) and *Team. dichotomica* (Molina, 1974) have also been implicated as causal agents. Since the original work on farmer's lung, it has been recognized that the thermoactinomycetes involved fall into two distinct species, *Team. vulgaris* and *Team. thalpophilus* (Cross and Unsworth, 1976). As the latter grows more rapidly, most antigens prepared are for this species, including the one used by Pepys *et al.* (1963), yet more farmers have precipitin to the former species. Thus it is essential that the species used is identified correctly (Lacey, 1981).

Bagassosis is caused by inhalation of dust when stored bagasse is handled. Actinomycetes from bagasse have been tested immunologically against sera from patients. It was concluded by Lacey (1981) that *Team. sacchari* spores are likely to be an important cause of this disease, as this species was usually more abundant in bagasse than *Team. vulgaris* (Seabury *et al.*, 1968; Salvaggio *et al.*, 1969; Lacey, 1971b, 1974) and possessed common as well as specific antigens (Lacey, 1971b). Respiratory allergy in mushroom workers may occur immediately during work in the cropping houses or as a delayed reaction in those spawning the compost. The latter is probably caused by spores of actinomycetes which grow during the second phase of composting. However, it has not yet been possible to demonstrate precipitins to the predominant *Thermomonospora* spp.

5. Actinomycetes as Plant Pathogens

Compared to fungi and Gram-negative bacteria, relatively few actinomycetes are phytopathogens. Two broad groups can be recognized, the coryneform actinomycetes and streptomycetes (Table 4) although a few other genera have been occasionally implicated in minor diseases.

A. CORYNEFORM ACTINOMYCETES

This topic has been recently reviewed by Vidaver and Starr (1981). Actinomycetes which they assigned to *Corynebacterium* have long been recognized as causal agents of a variety of diseases, most being associated with a single plant genus. Little is known of their general ecology; their enumeration, extent of spread and existence outide the host would be better understood if selective isolation media were available (Vidaver and Starr, 1981).

The taxonomy of this group is somewhat confused. At present the identity of most coryneform pathogens is determined by inoculation tests in host plants, although serodiagnosis has been used to detect and identify '*Cnbc*'. *insidiosum* (Hale, 1972) and '*Cnbc*'. *sepedonicum* (Slack *et al.*, 1979). The generic placement of coryneform phytopathogens is also still in dispute (Chapter 2; Jones, 1975; Dye and Kemp, 1977).

TABLE 4. Examples of phytopathogenic actinomycetes

Causal agents	Disease	References
'Corynebacterium' fascians	Leaf gall of many plants Fasciation of sweet pea	Tilford (1936)
'Cnbc. tritici' 'Cnbc. iranicum,' 'Cnbc.' rathayi	Gumming of cereals	Scharif (1961)
'Cnbc.' insidiosum	Alfalfa wilt	McCulloch (1925)
'Cnbc.' nebraskense	Wilt and blight of corn	Schuster (1975)
'Cnbc.' sepedonicum	Potato ring rot	Skaptason and Burkholder (1942)
'Streptomyces scabies'	Common scab of potato, sugar beet etc.	Hoffmann (1958); Lapwood (1973)
Stmy. griseus, Stmy. aureofaciens, Stmy. flaveolus	Common scab of potato	Hoffmann (1958); Corbaz (1964)
Streptomyces spp. 'Stmy. scabies'	Russet scab of potato	Bang (1979a, b); Harrison (1962); Vruggink and Maat (1968)
Stmy. ipomoeae	Rot of sweet potato	Person and Martin (1940)

B. STREPTOMYCETE PATHOGENS

The major disease caused by streptomycetes is scab of potatoes and sugar beet (Table 4). Reviews of this topic have been made by Lapwood (1973) and Kutzner (1981).

Common scab causes disfiguration of potato tubers, the scab lesions resulting from formation of wound tissues in response to penetration by the streptomycete. Degrees of blemishing were defined by McKee (1958) but two major categories can be recognized, normal scab which involves deep or superficial discrete lesions, and russet scab which is a brown, superficial roughening of most of the tuber skin (Harrison, 1962; Bang, 1979a). Although scab is confined to the skin, the market value of potatoes is decreased and 'table grade' in Britain excludes tubers with scab (Lapwood, 1973). Yield is not normally affected (Wenzl and Denel, 1971) but some reduction with severe russet scab was reported by Bang (1979b).

Considerable confusion surrounds the specific identity of the causal agents of common and russet scab, due to the lack of reliable criteria for recognition of streptomycete species. It is not apposite to discuss details of the taxonomic problems here. Some idea of the current position is given by the fact that the description of 'Stmy. scabies' (Shirling and Gottlieb,

1968) does not conform to that of earlier workers (e.g. Waksman, 1961). Also, isolates shown to be pathogenic by Hoffmann (1958) conformed to Waksman's description, while those isolated by Corbaz (1964) did not. Involvement of other streptomycete species has also been claimed (Table 4). Attempts to develop specific antisera for pathogenic streptomycetes failed as the serological reactions did not correlate with pathogenicity (Bowman and Weinhold, 1963; Vruggink and Maat, 1968). It has also been claimed that streptomycetes causing russet scab are distinct from '*Stmy. scabies*' (Harrison, 1962; Bang, 1979a). It seems likely that more than one species is involved in production of scab, but their taxonomy needs to be more clearly defined, their pathogenicity proven by host reinoculation, and their behaviour in soil clarified.

Infection occurs only in young tubers through stomata or newly formed lenticels (Fellows, 1926) and the tubers are most susceptible when stomata are changing into lenticles (Lapwood and Adams, 1973). Disease incidence of common scab is greatest in dry, neutral to alkaline soils (Labruyere, 1971; Lapwood, 1973) and is most severe in Britain when June and July are dry (Lapwood, 1966). In contrast, incidence of russet scab increases in wet soil (Harrison, 1962; Labruyere, 1971). For common scab, dry weather during the month after tuber initiation is critical for infection (Lapwood, 1973). The reasons for this are not clear but it was suggested that, in wet conditions, '*Stmy. scabies*' infection was prevented by enhanced growth of bacterial competitors or antagonists (Adams and Lapwood, 1978). A practical consequence is that incidence of common scab can be significantly reduced by irrigation for the first 3 weeks after tuber initiation (Lapwood, 1966; Lapwood and Hering, 1970; Lapwood *et al.*, 1970).

6. Nitrogen Fixation by Endophytic Actinomycetes

As yet there is no convincing evidence for nitrogen fixation by free-living actinomycetes. Recent surveys of a range of species from various habitats (Pridham and Kroppenstedt, 1979; Pearson *et al.*, 1982) failed to detect nitrogenase activity; the latter workers did obtain evidence for nitrogen fixation by *Mycoplana* spp., but the inclusion of this genus in the actinomycetes is dubious. Earlier claims for nitrogen fixation by *Rodc. erythropolis* (*Nrda. calcarea*) and *Orsk.* (*Nrda.*) *cellulans* (Metcalfe and Brown, 1957) have been repudiated (Hill and Postgate, 1969) and '*Mybc.*' *flavum*, which does have nitrogenase (Biggins and Postgate, 1969), was re-classified as a new species of *Xanthobacter* (Malik and Claus, 1979).

In contrast, it has been clear for many years that actinomycetes living as endophytes in root nodules of certain shrubs and trees could fix nitrogen *in situ* (see reviews by Becking, 1970a, 1975, 1977; Bond, 1967, 1976). These organisms were regarded as actinomycetes on the basis of their morphology and ultrastructure in the nodule (e.g. Becking, 1975) and were placed in the genus *Frankia* (Becking, 1970b). The eventual isolation of the endophyte was first achieved by Callaham *et al.* (1978). This, together with results from the International Biological Programme, has stimulated interest in these microbes.

A. OCCURRENCE OF ENDOPHYTES

Actinomycetes are known to form symbiotic associations in the root nodules of almost 170 species of woody, dicotyledonous plants from 15 genera (Torrey, 1978). Some examples, including those species from which the endophyte has been isolated (most of which produced effective nodules on re-inoculation), are given in Table 5. The plants are typically pioneer colonizers of nutrient-poor soils and inhabit a wide range of habitats, including forests, bogs, sand dunes, arid soils and mine wastes. Although many of the plants themselves are not of great economic importance, there is increasing evidence that they are of value in promoting growth of more useful plants, particularly in forestry plantations (Silvester, 1976). For example, mixed stands of black cottonwood and red alder gave higher dry matter production and 9% more soil nitrogen than did pure cottonwood plantings (Bell and Radwan, 1979). Nodulated plants also have great potential in reclamation of waste lands, such as those produced by mining operations (Torrey and Tjepkema, 1979).

Isolation of the endophyte from several plant species (Table 5) has confirmed that they are actinomycetes. It has been stressed, however, that the genus *Frankia* should not necessarily be restricted to endophytes but

TABLE 5. Some examples of plants forming root nodules containing actinomycetes

Genus	No. of species[a] nodulated	Isolations of the endophyte	Reference
Alnus	33	*A. glutinosa*	Quispel and Tak (1978)
		A. incana	Lechevalier (1981)
		A. rubra	Berry and Torrey (1979)
		A. viridis	Baker *et al.* (1979)
Casuarina	24	*C. equisetifolia*	Baker (1982)
Ceanothus	31	*C. americanus*	Baker (1982)
Comptonia	1	*C. peregrina*	Callaham *et al.* (1978)
Elaeagnus	16	*E. umbellata*	Baker *et al.* (1979)
Hippophaë	1		
Myrica	26	*M. pensylvanica*	Lechevalier and Lechevalier (1979)

[a] After Torrey (1978).

should include any free-living actinomycetes which may be found to have the unique capacity to produce non-motile spore-containing sporangia in liquid culture (Lechevalier and Lechevalier, 1979). They also urged that speciation of *Frankia* should not be based on host specificity as it was conceived before isolation of any endophytes (Becking, 1970b). Nodules of *Alnus glutinosa* contained large quantities of *meso*-diaminopimelic acid but it was absent from nodules of *Hippophaë rhamnoides* and *Myrica gale* (van Dijk, 1978, 1979). This suggests that the endophytes are by no means taxonomically homogeneous.

Evidence for the occurrence of these actinomycetes in soil has been obtained by studying infection rates of potential host plants. Van Dijk (1979) concluded that the presence of a suitable host was the prime factor but that extranodular growth or long-term survivial of the endophytes could not be ruled out. Endophytes of *Alnus glutinosa* and *Myrica gale* were detected in soils that had been free from these plants for many years (Rodriguez-Barrueco, 1968). In view of the capacity of at least some endophytes to grow *in vitro* on relatively simple media, the possibility of their growth in the soil or the rhizosphere cannot be discounted (Lechevalier, 1981).

B. SPECIFICITY OF ENDOPHYTE–HOST ASSOCIATIONS

There are dangers in drawing conclusions from previous experiments carried out under non-sterile conditions using crushed nodules as inoculum. As pure cultures of endophytes have only recently become available, it is premature to draw general conclusions on the occurrence of cross-inoculation barriers.

Alnus glutinosa was infected by at least two endophyte strains, one of which spored in the nodules and one that did not; the former was almost 1000 times more infective, but acetylene reduction rates did not differ (van Dijk, 1978, 1979). Seedlings of *Myrica gale*, grown xeroponically, were inoculated with suspensions of nodules from the same plant and with pure endophyte cultures from *Comptonia peregrina*, a closely related species. Both treatments induced nodule formation (Torrey and Callaham, 1979). *Alnus glutinosa* was successfully inoculated with the *Comptonia* endophyte and its presence in the nodule was confirmed by fluorescein-labelled antibodies (Lalonde, 1979). Although sporing did not occur in the nodules and nodule formation took longer than with the usual *Alnus* endophyte, nitrogen was fixed. A pure culture of a *Frankia* endophyte was shown to form effective nodules with species of *Alnus*, *Comptonia* and *Myrica* (Baker and Torrey, 1980). In contrast, another *Frankia* strain appeared to be restricted to the *Elaeagnaceae* and formed ineffective nodules (Baker *et al.*, 1980). So far, two host compatibility groups have been recognized; *Alnus*, *Myrica* and *Comptonia*; and *Eleagnus*, *Hippophaë* and *Shepherdia* (Baker, 1982).

C. INFECTION, NODULE DEVELOPMENT AND STRUCTURE

Five stages were recognized in the development of induced nodules in *Myrica gale* (Torrey and Callaham, 1978). These were root hair infection, pre-nodule formation, nodule lobe formation, a transition period and nodule root development. Infection through root hairs occurs in all genera studied so far (Torrey, 1978; Callaham *et al.*, 1979). This leads to root hair deformation and growth of hyphae through the hair, resulting in penetration of an outer cortical cell by dissolution of the wall. The infection spreads to other cortical cells, stimulating hypertrophy and division of the cells to form a pre-nodule. This is associated with stimulation of multiple, modified lateral-root-like primordia which ultimately form a condensed, highly branched system.

Micromorphology and ultrastructure of nodules have been studied in detail (see reviews of Becking, 1975; Gardner, 1976). Generally, three phases are recognized, invasive hyphae, vesicles formed within cells at hyphal tips and thicker-walled spore-like bodies which are released upon decay of the host root. Recent morphological studies on isolated strains include those of Newcomb *et al.* (1979) and Lechevalier and Lechevalier (1979).

D. NITROGEN FIXATION

Rates of fixation by nodulated plants are very similar to those of legumes (Torrey, 1978). Because the hosts are perennials, fixed nitrogen is made available to the soil on a long-term basis, although nitrogenase activity decreases as plants and nodules get older. This is probably due to a decrease in the proportion of metabolically active vesicle clusters, which are thought to be the sites of nitrogen fixation, in nodule tissue. Most information on the physiology and biochemistry of nitrogen fixation has been obtained using nodule homogenates (e.g. Benson and Eveleigh, 1979) or cell-free extracts of nodules (e.g. Benson *et al.*, 1979). However, recently nitrogenase activity has also been detected in pure cultures of *Frankia* (Tjepkema *et al.*, 1980; Gauthier *et al.*, 1981). This will no doubt facilitate more detailed studies of the biochemistry of nitrogen fixation by these actinomycetes. The endophytes have many similarities to other nitrogen-fixing systems, with respect to the activities and function of their nitrogenase, hydrogenase and ammonia-assimilating enzymes.

7. Actinomycetes in Fresh Water and Marine Environments

The ecology of actinomycetes in water has received comparatively little attention. Consequently, many basic questions about their growth and activities remain to be answered. This topic has been recently most adequately reviewed by Cross (1981).

A. ACTINOMYCETES IN FRESH WATER

Actinomycetes can be readily isolated from fresh water, though usually in low numbers. Genera most frequently present are *Actinoplanes, Micromonospora, Nocardia, Rhodococcus, Streptomyces* and *Thermoactinomyces*. The mere isolation of actinomycetes from water does not prove that they are truly aquatic (i.e. that they grow in this environment), as the possibility of wash-in from surrounding terrestrial environments must always be considered. Cross (1981) evaluated the evidence for the existence of aquatic actinomycetes.

The resistant endospores of *Thermoactinomyces* are produced in self-heating composts, fodders, etc. but can be washed into streams, rivers and lakes; some of these spores have also been found in muds and sediments (Cross and Johnston, 1972; Cross and Attwell, 1974). It seems safe to assume that these thermophiles are unable to grow at the ambient temperatures in most fresh water environments. A survey of actinomycetes in streams (Rowbotham and Cross, 1977) showed that *Rodc. coprophilus* predominated in those polluted by effluent from dairy farms. This species is coprophilous and its resting coccal stage passes into water where it can survive but probably does not grow. Its presence in water could serve as a useful indicator of water pollution by cattle (Al-Diwany and Cross, 1978). It has been claimed that aquatic streptomycetes exist and that their primary mycelium is facultatively anaerobic, while their secondary growth is aerobic (Francisco and Silvey, 1971). Growth of streptomycetes on the chitinous exoskeletons of *Procambarus versutus* in a woodland stream was observed by scanning electron microscopy (Aumen, 1980). Numbers of streptomycetes and other genera were substantially increased in foam on river water; it was suggested that this was due to the concentration of hydrophobic spores and cells at the water/air interface (Al-Diwany and Cross, 1978). Cross (1981) concluded that streptomycetes are terrigenous species which are occasionally washed into water. It has been known for some time that *Micromonospora* strains occur frequently in streams, rivers, lakes and sediments (Erikson, 1941; Umbreit and McCoy, 1941). These workers regarded them as indigenous inhabitants of such habitats and it was claimed by Erikson that they participated in decomposition of cellulose, chitin and lignin.

Collins and Willoughby (1962) concluded that micromonosporae isolated from a eutrophic lake originated from lake mud, while Willoughby (1969b) and Willoughby and Baker (1969) regarded this genus as being truly aquatic. *Micromonospora* was the dominant actinomycete genus in a range of lakes

(Johnston and Cross, 1979b), being particularly prevalent in deeper mud layers; spores were recovered from sediments deposited at least 100 years previously (Cross and Attwell, 1974). The presence of both spores and mycelium in lake muds was detected by Johnston and Cross (1976b) and micromonosporae were also able to grow in natural or amended muds in the laboratory. Cross (1981) concluded that while spores could survive for long periods in muds, and that therefore wash-in could not be discounted, it is possible that micromonosporae grow in sediments or on organic debris in fresh water. *Actinoplanes*, which produces motile spores, can be isolated from soils, rivers and lakes (Willoughby, 1969b, 1971). Willoughby (1969a) and Willoughby and Baker (1969) found the typical habitat of the *Actinoplanaceae* to be decomposing vegetation (especially twigs and leaves) at the edges of lakes or in shallow streams. The wet–dry regime in such habitats was apparently most suitable for growth, spore production (which was shown to be stimulated by humic and fulvic acids) and spore dispersal. An explanation for the occurrence of *Actinoplanes* in wet and dry habitats was provided by Cross (1981) who reported that the spore vesicles withstood prolonged desiccation and released their motile spores for dispersal when rehydrated. Other actinomycetes which have been claimed to be inhabitants of fresh water include *Mybc. kansasii* (Joynson, 1979), *Actn. madurae* (Lawson and Davey, 1972) and *Arthrobacter, Corynebacterium* and *Nocardia* species (Donderski and Strzelcyzk, 1974). The former species has also been detected in water supplies (Powell and Steadham, 1981).

B. ACTINOMYCETES AS A CAUSE OF TASTES AND ODOURS IN POTABLE WATER

Earthy tastes and odours occasionally occur in drinking water rendering it unpalatable. For many years, actinomycetes, particularly streptomycetes, have been prime suspects as the causal agents. However, it is only in the last 15 years that the compounds concerned have been isolated and characterized, thus facilitating a more rational approach to this problem. The chemistry and distribution of these compounds were reviewed by Gerber (1979a, b), who concluded that geosmin and methyl*iso*-borneol were the most frequently produced. They have been isolated from cultures of streptomycetes and occasionally from other genera. It has also been possible to detect them in soil (Buttery and Garibaldi, 1976) and in odour-polluted waters (Rosen *et al.*, 1970; Gerber, 1979a, b). It must be emphasized that cyanobacteria and fungi can also produce such compounds and that they can occur in foodstuffs (see Gerber and Lechevalier, 1977). Other compounds have also been implicated, for example butyric acid (Weete *et al.*, 1979), and more probably remain to be detected. Standards of the compounds are essential for analysis of culture extracts or water samples; these are now available for geosmin (Gerber and Lechevalier, 1977).

There has been, and still is, much interest in the possibility of preventing the development of taints in reservoirs and water supply systems. The major problem has been detection of the sites of production. This topic has been reviewed by Cross (1981) who emphasized that these compounds are not produced by streptomycete spores but arise as secondary metabolites subsequent to hyphal growth. Unfortunately, many workers have attempted to correlate the occurrence of taints in water with increases in plate counts of actinomycetes; as most colonies on such plates arise from spores, this approach has met with limited success. It is therefore important to consider all the possible growth sites of odour-producing actinomycetes.

Production of odoriferous compounds, such as geosmin, undoubtedly occurs in soil and there is always the likelihood that they will be washed into water. Land frequently fertilized with organic manures and effluents contains more actively growing actinomycetes (Al-Diwany and Cross, 1978), thus taints are more common in eutrophic waters than in oligotrophic lakes and reservoirs.

Plant and animal debris at water margins can provide substrates for actinomycete growth and geosmin production (Silvey and Roach, 1975; S. T. Williams and S. Wood, unpublished observations). This will be particularly important when water levels drop during long dry periods. Under such conditions, sediments in lakes and reservoirs can also be exposed to aerobic warm conditions and provide a substrate for growth.

The water mass of most lakes and reservoirs in temperate regions is unlikely to provide a suitable environment for production of odoriferous compounds. As stated previously, it is arguable that streptomycetes are able to grow in water (Cross, 1981) and anyway, nutrient levels in water of reservoirs susceptible to taints are insufficient to support geosmin production even at optimum temperatures (S. T. Williams and S. Wood, unpublished observations).

Actinomycete spores are relatively resistant to chlorination treatments used to disinfect water supplies (Burman, 1973), and therefore can be detected in distribution systems (Silvey and Roach, 1975; Lechevalier *et al.*, 1980). As pipes tend to contain organic debris and oxygenated water, actinomycete growth and production of taints can occur. Actinomycetes can also cause other problems in water and sewage pipes due to the capacity of streptomycetes and *Nrda. asteroides* to degrade vulcanized rubber pipe-joint rings (Cundell and Mulcock, 1973, 1975; Hutchinson *et al.*, 1975).

C. MARINE ACTINOMYCETES

A variety of genera have been isolated from sea water and marine sediments, including *Streptomyces, Micromonospora, Microbispora, Nocardia, Thermoactinomyces* and coryneforms (Grein and Meyers, 1958; Weyland, 1969, 1981a; Cross and Johnston, 1972; Bousfield, 1978). Numbers are generally higher in shallow coastal areas and lowest in deep-sea sediments, but seldom exceed a few thousand per ml of water; relatively few studies of ocean sediments have been made. As with fresh water, one is again faced with the problem of distinguishing indigenous marine actinomycetes from those washed in from terrestrial environments. Qualitative differences between actinomycete populations close to, and distant from, land have been made. Streptomycetes predominated in shallow areas of the Pacific (Okami and Okazaki, 1978) and Atlantic oceans (Weyland, 1981a), while micromonosporae and nocardioforms predominated in deep sea sediments (Weyland, 1981a). Actinomycetes were present in samples taken from 7790m depth in the Puerto Rican Trench, 60 miles offshore (Walker and Colwell, 1975). However, such results do not necessarily prove that the isolates were indigenous marine microbes.

Marine Gram-negative bacteria can be distinguished by their specific ion requirements (although they are not generally tolerant to salt concentrations higher than those in sea water), their psychrophilism and, in some cases, their barotolerance or barophilism (MacLeod, 1965). There have been relatively few attempts to apply these criteria to 'marine' actinomycetes. Okazaki and Okami (1972, 1975) found that marine isolates were generally more salt-tolerant than terrestrial strains, but greater tolerance was found among sand-dune isolates than those from sea water (Watson and Williams, 1974). The main coryneform genera isolated from the North Sea were *Corynebacterium, Curtobacterium* and *Brevibacterium* but it was suggested that they had an insignificant role in this environment as they were all mesophilic (Bousfield, 1978). Helmke (1981) found that some actinomycetes could grow at hydrostatic pressures equivalent to those in deep-sea sediments. While there was some evidence of greater tolerance in marine isolates, some terrestrial strains (e.g. *Rodc. erythropolis*) also showed marked resistance to high pressures. Nocardioform isolates, which predominated in open sea areas of the North Atlantic, were regarded as indigenous marine organisms as they were salt-dependent, psychrotrophic and barotolerant (Weyland, 1981b). This probably represents the clearest evidence to date for the existence of truly marine actinomycetes.

Not surprisingly, little is known of the roles of actinomycetes in the sea. Alginate and lamarinarin decomposers were isolated from sea weeds (Chester *et al.*, 1956) and actinomycetes were involved in decay of wood in sea water (Eaton and Dickinson, 1976). Willingham *et al.* (1966) showed that actinomycetes could degrade starch, cellulose and lignin in sea water under laboratory conditions. *Nocardia* strains from marine environments predominated among isolates oxidizing petroleum hydrocarbons (Zobell *et al.*, 1943; Mulkins-Phillips and Stewart, 1974). The same genus together with 'actinomycetes' and coryneforms was among petroleum-degrading actinomycetes isolated from Chesapeake Bay (Austin *et al.*, 1977). In a recent review of this important aspect of marine pollution, Atlas (1981) included the genera *Arthrobacter, Brevibacterium, Corynebacterium* and *Nocardia* among the microbes important in degradation of petroleum hydrocarbons in aquatic environments.

8. Actinomycetes as Symbionts of Animals

Actinomycetes can play a role as symbionts in the intestinal tract of animals that feed on the organic substances present in the soil. Streptomycetes constituted 15 to 50% of the total gut microflora of larvae from different *Bibio marci* populations (Szabo et al., 1967). *Streptomyces finlayi* was usually the predominant species comprising over 80% of the total streptomycetes. This contrasts with the high diversity of actinomycete species in the soil and suggests that soil actinomycetes are subjected to considerable selection pressure in the gut. The predominant species are well adapted to multiply in the gut environment and play an important role in the metabolism of the larvae, particularly in protein decomposition, and in the elimination of other microorganisms ingested with the food. However, no *Streptomyces* species have been found that can live and propagate exclusively in the gut (Szabo and Marton, 1966).

Actinomycetes may also play a significant role in the digestive processes of termites. The lower termites are largely xylophagous and a primary mutualism exists with intestinal flagellates which digest cellulose anaerobically to form a range of organic acids, some of which are absorbed by the host. Hungate (1946) isolated an anaerobic actinomycete, '*Mims. propionici*', capable of decomposing cellulose from the gut of *Amitermes minimus*. He also observed a morphologically identical strain in a culture of protozoa from the rumen of cattle. Sebald and Prevot (1962) isolated an anaerobic species of *Micromonospora* ('*Mims. acetiformici*') from the posterior gut of *Reticulitermes lucifugus*, but this differed from '*Mims. propionici*' in cultural properties, fermentative pattern and the absence of cellulolytic capacity. It is possible that this organism had progressed from a saprophytic to a parasitic existence by loss of some enzymes associated with respiration and nutrition.

The higher termites utilize most plant, wood and soil organic matter as food sources. Direct observation suggested that actinomycetes might form a significant component of the gut microflora of *Procubitermes aburiensis*; they were also more abundant relative to non-filamentous forms than in freshly ingested soil (Bignell et al., 1979, 1980, 1981). Their specific attachment to retaining structures in the mesenteron, mixed segment and colon suggested a possible role in digestion. Many soil actinomycetes secrete phenoloxidases and may assist in humus formation from phenolic complexes derived from plant tissues (see page 492); thus the presence of actinomycetes in the gut of soil-feeding termites suggests that their nutrition may include a significant component from the soil organic matter fraction (Bignell *et al.*, 1979).

A mutualistic symbiotic relationship has been demonstrated between the blood-sucking arthropod, *Rhodnius prolixus* and *Rodc.* (*Nrda.*) *rhodnii* (Baines, 1956; Harington, 1960; Gumpert and Schwartz. 1962; Cavanagh and Marsden, 1969). Normal development to the adult stage was inhibited in the absence of the *Nocardia* symbiont (Brecher and Wrigglesworth, 1944; Baines, 1956; Harington, 1960; Lake and Friend, 1968; Nyirady, 1973; Auden, 1974). However, the inhibition could be removed and sexual maturity attained by reinfection of *Rhodnius prolixus* with *Rodc. rhodnii* at any stage of development (Baines, 1956). A number of bacteria not normally associated with *Rhodnius prolixus* were also able to remove the inhibition. These included species of *Corynebacterium, Nocardia* and *Rhodococcus*, together with a strain of *Pseudomonas* (Gumpert, 1962a, b). In contrast, normal development to maturity occurred when *Rhodnius prolixus* was reared germ-free and fed on germ-free mice (Nyirady, 1973; Auden, 1974). It is thought that normal development is inhibited in *Rhodnius prolixus* nymphs due to an insufficient concentration of certain required nutrients in the blood of some host animals and that the critical metabolites are supplied by the symbiont, *Rodc. rhodnii*. These metabolites may include folic acid, pantothenic acid, pyridoxine, thiamine, nicotinic acid and riboflavin (Harington, 1960; Lake and Friend, 1968).

It is possible that *Rodc. rhodnii* is an aggregate of strains capable of forming a symbiotic and stable relationship with one particular insect and that selection pressure within the insect gut favours those strains capable of supplying the necessary nutrients (Gumpert, 1962a, b).

9. Ecology of Actinophage

Little is known of the significance of actinophage in natural environments, although it is clear that they are of widespread occurrence and there is information on their reactions to certain environmental factors. Anderson (1957) suggested that bacteriophage may influence natural populations in three ways. Firstly, virulent phage might eliminate sensitive hosts, leading to their replacement by resistant mutants with possibly different ecological potential. Secondly, temperate phage might eliminate part of a population and lysogenize the rest, the prophage causing changes in the survivors. Finally, temperate or virulent phage might be agents of transduction between some members of the population. The numerous studies of actinophage *in vitro* suggest that there is no *a priori* reason why these interactions should not occur in nature, but direct evidence remains to be obtained.

A. OCCURRENCE

Qualitative studies using enrichment techniques have shown that actinophage are present in most natural habitats that favour actinomycete activity such as soils, composts, straw, freshwater, human and animal faeces. Phage active against many actinomycete genera have been described.

Lack of information on the frequency, distribution and ecological significance of actinophage in natural systems is primarily due to the lack of efficient methods for direct isolation. Titres of actinophage obtained by direct methods are extremely low; typical counts in soils are 40 p.f.u. g^{-1} for *Atbc. globiformis* (Casida and Liu, 1974) and 33 to 270 p.f.u. g^{-1} for *Stmy. chrysomallus* (Collard, 1970). Such low counts may be due to the inactivity of the host, the failure to concentrate phage from natural systems and losses of phage incurred during isolation.

Lanning and Williams (1982) determined some important stages at which actinophage may be lost during isolation from soil by measuring the percentage recovery of actinophage added to sterile soils. Between 30 and 100% of added phage were lost when soil extracts were shaken, 56% during filtration and up to 20% during centrifugation. Complex eluents, such as nutrient broth, containing a high concentration of protein enhanced the recovery of actinophage from soil.

The occurrence of actinophage in natural habitats is obviously dependent on the presence of actively growing susceptible hosts. Corke *et al.* (1964) noted a seasonal fluctuation in number of phage for '*Stmy. scabies*' and suggested that this might be related to host numbers. Most phage isolated from natural habitats are polyvalent, i.e. capable of attacking a wide range of host organisms, thus increasing their chances of survival (Gilmour and Buthala, 1950; Carvajal, 1955; Bradley *et al.*, 1961), while others may be quite specific (Khavina and Rautenstein, 1958; Willoughby *et al.*, 1972). Sykes *et al.* (1981) showed that, although phage could not be isolated from acid soils (pH < 5.5), some phage isolated from neutral soils were capable of attacking both neutrophilic and acidophilic streptomycetes. Willoughby *et al.* (1972), on the other hand, demonstrated a highly specific phage for *Actinoplanes*; both phage and host were readily isolated from allochthonous leaves at a lake margin, but neither were detected in surface benthic lake mud. It was suggested that the presence of specific actinophage in fresh water might be a guide to the activity of their host strains.

The possible existence of actinophage ecotypes in soil and water has been suggested by Willoughby (1976), who failed to isolate a phage from local soils which was active against a *Streptomyces* strain from another area. Similarly, Welsch *et al.* (1963) were unable to recover phage from soils enriched with *Streptomyces* strains taken from other regions, but phage were obtained from *Streptomyces* strains isolated from the same soils. Robinson and Corke (1959) also showed that the extract from any given soil contained more phage for actinomycetes isolated from the same soil than from other sources.

B. INFLUENCE OF ENVIRONMENTAL FACTORS

Relatively little is known about the factors that influence the ecology of actinophage. Some of the parameters which might influence phage–host interaction in the natural environment include concentration of susceptible host cells, pH, temperature, ionic concentration, soil colloids and phage-inactivating agents.

The role of pH and ions has been studied in the laboratory, for phage acting upon *Stmy. griseus* (Walton, 1951; Hurd and Gilmour, 1951; Gold, 1959) and upon *Stmy. chrysomallus* (Welsch, 1956). Khavina and Rautenstein (1959) failed to detect actinophage in soils below pH 5.0 despite the presence of susceptible hosts. They suggested that this was due to inactivation of the phage at low pH during isolation. Sykes *et al.* (1981) were also unable to isolate phage from soils below pH 6.0 despite the presence of acidophilic streptomycetes in these soils. They showed that selected actinophage were stable between pH 5.5 and 9.0 in sterile soils and broth, but were rapidly inactivated at lower or higher pH values. Survival of these phage was good in neutral soils, but negligible in acidic soils. They concluded that free phage were unable to remain infective in acidic soils, although acidophilic streptomycetes were susceptible to phage *in vitro*. Acidity was shown to have variable effects on stages in phage replication, including adsorption, penetration and length of the latent period. The latter effect was directly related to the metabolic activity of the host.

Laboratory studies have shown that the limits of temperature for reproduction and survival vary for different actinophage (Welsch, 1969) and are positively correlated with the limits of tolerance of the host (Kuroda and Bradley, 1967). Little is known of the effect of temperature on phage in the natural environment. However, Patel (1969a) demonstrated the presence of phage for *Team. vulgaris* in compost and showed it to be capable of survival and activity at elevated temperatures. Phage for the thermophilic *Thermomonospora* sp. (Patel, 1969b), *Team. candidus* and *Mips. faeni* (Kurup and Heinzen, 1978) have also been reported. Sykes and Williams (1978) suggested that soil colloids may have a major influence on actinophage in soil. Adsorbed phage remained infective, possibly protected and in close proximity to adsorbed host spores, thus increasing the chances of phage–host interaction. Adsorbed phage, however, were shown to be more susceptible to inactivation by low pH values than free phage in moderately acid soils.

10. Conclusions

Current evidence for the activities of actinomycetes in natural environments is not always unequivocal but it is clear that they are involved in important processes in a wide range of habitats. These include the following.

(i) Degradation of plant, animal and microbial polymers in soil and litter.
(ii) Degradation of petroleum hydrocarbons and other man-made products in terrestrial and aquatic environments.
(iii) Degradation and self-heating of composts, fodders and related materials.
(iv) Induction of respiratory allergies, particularly among agricultural workers.
(v) Causing plant diseases.
(vi) Biological control of root-infecting pathogens.
(vii) Nitrogen fixation in nodules of non-leguminous plants.
(viii) Production of tastes and odours in natural potable waters.

REFERENCES

Abraham, T. A. and Herr, L. J. (1964). Activity of actinomycetes from rhizosphere and non-rhizosphere soils of corn and soybean in four physiological tests. *Canadian Journal of Microbiology* **10**, 281–285.

Adams, M. J. and Lapwood, D. H. (1978). Studies on the lenticel development, surface microflora and infection by common scab (*Streptomyces scabies*) of potato tubers growing in wet and dry soils. *Annals of Applied Biology* **90**, 335–343.

Agate, A. E. and Bhat, J. Y. (1964). Microflora associated with the rhizosphere of *Calotropis gigantea*. *Journal of the Indian Institute of Science* **46**, 1–10.

Al-Diwany, L. J. and Cross, T. (1978). Ecological studies on nocardioforms and other actinomycetes in aquatic habitats. *Zentralblatt für Bakteriologie, Parasitenkunde, Infektionskrankheiten und Hygiene. I. Abteilung Supplement* **6**, 153–160.

Al-Diwany, L. J., Unsworth, B. A. and Cross, T. (1978). A comparison of membrane filters for counting *Thermoactinomyces* endospores in spore suspensions and river water. *Journal of Applied Bacteriology* **45**, 249–258.

Anderson, E. S. (1957). Bacteriophages and bacterial ecology. In: *Microbial Ecology* (R. E. O. Williams and C. C. Spicer, eds.), pp. 189–217. Cambridge University Press, Cambridge.

Antai, S. P. and Crawford, D. L. (1981). Degradation of softwood, hardwood and grass lignocelluloses by two *Streptomyces* strains. *Applied and Environmental Microbiology* **42**, 378–380.

Athalye, M., Lacey, J. and Goodfellow, M. (1981). Selective isolation and enumeration of actinomycetes using rifampicin. *Journal of Applied Bacteriology* **51**, 289–298.

Atlas, R. M. (1981). Microbial degradation of petroleum hydrocarbons: an environmental perspective. *Microbiological Reviews* **45**, 180–209.

Auden, D. T. (1974). Studies on the development of *Rhodnius prolixus* and the effects of its symbiote *Nocardia rhodnii*. *Journal of Medical Entomology* **11**, 63–71.

Aumen, N. G. (1980). Microbial succession on a chitinous substrate in a woodland stream. *Microbial Ecology* **6**, 317–327.

Austin, B., Calomiris, J. J., Walker, J. D. and Colwell, R. R. (1977). Numerical taxonomy and ecology of petroleum degrading bacteria. *Applied and Environmental Microbiology* **34**, 60-68.

Babich, H. and Stotzky, G. (1977). Sensitivity of various bacteria, including actinomycetes, and fungi to cadmium and the influence of pH on sensitivity. *Applied and Environmental Microbiology* **33**, 681–695.

Bachofer, R., Oltmanns, O. and Lingens, F. (1973). Isolation and characterization of a Nocardia-like soil bacterium growing on carboxanilde fungicides. *Archiv für Mikrobiologie*, **90**, 141–149.

Bågstam, G. (1978). Population changes in microorganisms during composting of spruce-bark. 1. Influence of temperature control. *European Journal of Applied Microbiology and Biotechnology* **5**, 315–330.

Baines, S. (1956). The role of the symbiotic bacteria in the nutrition of *Rhodnius prolixus* (Hemiptera). *Journal of Experimental Biology* **33**, 533–541.

Baker, D. (1982). A cumulative listing of isolated frankiae, the symbiotic, nitrogen-fixing actinomycetes. *The Actinomycetes* **17**, 35–42.

Baker, D. and Torrey, J. G. (1980). Characterization of an effective actinorhizal microsymbiont *Frankia*, sp. Avc 11 (*Actinomycetales*). *Canadian Journal of Microbiology* **26**, 1066–1071.

Baker, D., Torrey, J. G. and Kidd, G. H. (1979). Isolation by sucrose-density fractionation and cultivation *in vitro* of actinomycetes from nitrogen-fixing root nodules. *Nature, London* **281**, 76–78.

Baker, D., Newcomb, W. and Torrey, J. G. (1980). Characterization of an ineffective actinorhizal microsymbiont, *Frankia* sp. Eu 11 (*Actinomycetales*). *Canadian Journal of Microbiology* **26**, 1072–1089.

Baker, R. (1968). Mechanisms of biological control of soil-borne pathogens. *Annual Review of Phytopathology* **6**, 263–294.

Bang, H. O. (1979a). Studies on potato russet scab. 1. A characterization of different isolates from northern Sweden. *Acta Agriculturae Scandinavica* **29**, 145–150.

Bang, H. O. (1979b). Studies on potato russet scab. 2. Influence of infection on the production capacity of seed. *Potato Research* **23**, 203–208.

Barder, M. J. and Crawford, D. L. (1981). Effects of carbon and nitrogen supplementation on lignin and cellulose decomposition by a *Streptomyces*. *Canadian Journal of Microbiology* **27**, 859–63.

Barrowcliffe, D. E. and Arblaster, P. G. (1968). Farmer's lung: a study of an early acute fatal case. *Thorax* **23**, 490–500.

Bartholomew, G. W. and Alexander, M. (1979). Microbial metabolism of carbon monoxide in culture and in soil. *Applied and Environmental Microbiology* **37**, 932–937.

Becking, J. H. (1970a). Plant- endophyte symbiosis in non-leguminous plants. *Plant and Soil* **32**, 611–654.

Becking, J. H. (1970b). Frankiaceae fam. nov. (*Actinomycetales*) with one new combination and six new species of the genus *Frankia* Brunchorst 1866. *International Journal of Systematic Bacteriology* **20**, 201–220.

Becking, J. H. (1975). Root nodules in non-legumes. In: *The Development and Function of Roots* (J. G. Torrey and D. T. Clarkson, eds.), pp. 507–566. Academic Press, London.

Becking, J. H. (1977). Dinitrogen-fixing associations in higher plants other than legumes. In: *A Treatise on Dinitrogen Fixation. Section III. Biology* (R. W. F. Hardy and W. S. Silver, eds.), pp. 185–275. J. Wiley, New York.

Beeman, R. W. and Matsumura, F. (1981). Metabolism of *cis*- and *trans*-chlordane by soil micro-organisms. *Journal of Agricultural and Food Chemistry* **29**, 84–89.

Bell, D. E. and Radwan, M. A. (1979). Growth and nitrogen relations of coppiced black cottonwood and red alder in pure and mixed plantings. *Botanical Gazette* **140** Supplement, S97–S101.

Benson, D. R. and Eveleigh, D. E. (1979). Ultrastructure of the nitrogen-fixing symbiont of *Myrica pensylvanica*. L. (Bayberry) root nodules. *Botanical Gazette* **140** Supplement, S15–S21.

Benson, D. R., Arp., D. J. and Burris, R. H. (1979). Cell-free nitrogenase and hydrogenase from actinorhizal root nodules. *Science* **205**, 688–689.

Bernhard, K. (1967). Zahl und Artenspektrum der Actinomyceten in der engen Rhizosphäre von Kultur- und Wildpflanzen. *Zentralblatt für Bakteriologie, Parasitenkunde, Infektionskrankheiten und Hygiene*. II. Abteilung **121**, 353–361.

Berry, A. and Torrey, J. G. (1979). Isolation and characterization *in vivo* and *in vitro* of an actinomycetous endophyte from *Alnus rubra*. Bong. In: *Symbiotic Nitrogen Fixation in the Management of Temperate Forests* (J. C. Gordon, C. T. Wheeler and D. A. Perry, eds.), pp. 69–83. Oregon State University Press, Corvallis.

Biggins, D. R. and Postgate, J. R. (1969). Nitrogen fixation by cultures and cell-free extracts of *Mycobacterium flavum* 301. *Journal of General Microbiology* **56**, 181–193.

Bignell, D. E., Oskarsson, H. and Anderson, J. M. (1979). Association of actinomycete-like bacteria with soil-feeding termites (Termitidae, Termitinae). *Applied and Environmental Microbiology* **37**, 339–342.

Bignell, D. E., Oskarsson, H. and Anderson, J. M. (1980). Distribution and abundance of bacteria in the gut of a soil-feeding termite *Procubitermes aburiensis* (*Termitidae, Termitinae*). *Journal of General Microbiology* **117**, 393–403.

Bignell, D. E., Oskarsson, H. and Anderson, J. H. (1981). Association of actinomycetes with soil-feeding termite; a novel symbiotic relationship? *Zentralblatt für Bakteriologie, Mikrobiologie und Hygiene.* I. *Abteilung, Supplement* **11**, 201–206.

Blanchette, R. A., Sutherland, J. B. and Crawford, D. L. (1981). Actinomycetes in discoloured wood of living silver maple. *Canadian Journal of Botany* **59**, 1–7.

Bond, G. (1967). Fixation of nitrogen by higher plants other than legumes. *Annual Review of Plant Physiology* **18**, 107–126.

Bond, G. (1976). The results of the IBP survey of root-nodule formation in nonleguminous angiosperms. In: *Symbiotic Nitrogen Fixation in Plants* (P. S. Nutman, ed.), pp. 443–474. Cambridge University Press, Cambridge.

Bousfield, I. J. (1978). The taxonomy of coryneform bacteria from the marine environment. In: *Coryneform Bacteria* (I. J. Bousfield and A. G. Callely, eds.), pp. 217–233. Academic Press, London.

Bowman, T. and Weinhold, A. R. (1963). Serological relationship of the potato scab organism and other species of *Streptomyces*. *Nature, London* **200**, 599–600.

Bradley, S. G., Anderson, D. L. and Jones, L. A. (1961). Phylogeny of actinomycetes as revealed by susceptibility to actinophage. *Development in Industrial Microbiology* **2**, 223–237.

Brecher, G. and Wrigglesworth, V. B. (1944). The transmission of *Actinomyces rhodnii* and its influence on the growth of the host. *Parasitology* **35**, 220–224.

Burman, N. P. (1973). Occurrence and significance of actinomycetes in water supply. In: *Actinomycetales: Characteristics and Practical Importance* (G. Sykes and F. A. Skinner, eds.), pp. 219–230. Academic Press, London.

Burman, N. P., Oliver, C. W. and Stevens, J. K. (1969). Membrane filtration techniques for the isolation from water of coliaerogenes, *Escherichia coli*, faecal streptococci, *Clostridium perfringens*, actinomycetes and microfungi. In: *Isolation Methods for Microbiologists* (D. A. Shapton and G. W. Gould, eds.), pp. 127–134. Academic Press, London.

Buti, I. (1978). On the streptomycetes flora of the rhizoplanes of *Medicago sativa* and *Trifolium alevandrium*. I. Composition of the species of *Streptomyces* populations. *Agrokemika Talajtan* **27**, 151–161.

Buttery, R. G. and Garibaldi, J. A. (1976). Geosmin and methyl *iso*-borneol in garden soil. *Journal of Agriculture and Food Chemistry* **24**, 1246–1247.

Buxton, E. W., Khalifa, O. and Ward, V. (1965). Effect of soil amendment with chitin on pea wilt caused by *Fusarium oxysporum* f. *pisi*. *Annals of Applied Biology* **55**, 83–88.

Callaham, D., Del-Tredici, P. and Torrey, J. G. (1978). Isolation and cultivation *in vitro* of the actinomycete causing root nodulation in *Comptonia*. *Science* **199**, 899–902.

Callaham, D., Newcomb, W., Torrey, J. G. and Peterson, R. L. (1979). Root hair infection in actinomycete-induced root nodule initiation in *Casuarina*, *Myrica* and *Comptonia*. *Botanical Gazette* **140** Supplement, S51–S59.

Carvajal, F. (1955). Host-parasite relations with a polyvalent steptomycophage from *Streptomyces griseus*. *Antibiotics and Chemotherapy* **5**, 28–37.

Casida, L. E. (1965). An abundant micro-organism in soil. *Applied Microbiology* **13**, 327–334.

Casida, L. A. and Liu, K. C. (1974). *Arthrobacter globiformis* and its bacteriophage in soil. *Applied Microbiology* **28**, 951–959.

Cavanagh, P. and Marsden, P. D. (1969). Bacteria isolated from the gut of some reduviid bugs. *Transactions of the Royal Society of Tropical Medicine and Hygiene* **63**, 415–416.

Chambers, S. C. and Millington, J. R. (1974). Studies on *Fusarium* species associated with a field planting of 'pathogen-tested' potatoes. *Australian Journal of Agricultural Research* **25**, 293–297.

Chang, Y. and Hudson, H. J. (1967). The fungi of wheat straw compost. I. Ecological studies. *Transactions of the British Mycological Society* **50**, 649–666.
Chester, C. G. C., Apinis, A. and Turner, M. (1956). Studies of the decomposition of seaweeds and seaweed products by microorganisms. *Proceedings of the Linnean Society, London* **166**, 87–97.
Collard, C. A. (1970). Comparative studies of methods of isolation of actinophage starting with their natural habitat. *Comptes rendus des Séances de la Société de Biologie Filiales* **164**, 465–468.
Collins, V. G. and Willoughby, L. G. (1962). The distribution of bacteria and fungal spores in Blelham Tarn with particular reference to an experimental overturn. *Archiv für Mikrobiologie* **43**, 294–307.
Corbaz, R. (1964). Étude des streptomycètes provoquat la gale commune de la pomme de terre. *Phytopathologische Zeitschrift* **51**, 351–360.
Corden, M. E. and Young, R. A. (1965). Changes in the soil miroflora following fungicide treatments. *Soil Science* **99**, 272–277.
Corke, C. T., Robinson, J. B. and Douglas, R. J. (1964). Serological relationships of soil actinophages. *Canadian Journal of Microbiology* **10**, 897–903.
Crawford, D. L. (1974). Growth of *Thermomonospora fusca* on lignocellulosic pulps of varying lignin content. *Canadian Journal of Microbiology* **20**, 1069–1072.
Crawford, D. L. (1978). Lignocellulose decomposition by selected *Streptomyces* strains. *Applied and Environmental Microbiology* **35**, 1041–1045.
Crawford, D. L. and McCoy, E. (1972). Cellulases of *Thermomonospora fusca* and *Streptomyces thermodiastaticus*. *Applied Microbiology* **24**, 150–152.
Crawford, D. L. and Sutherland, J. B. (1979). The role of actinomycetes in the decomposition of lignocellulose. *Developments in Industrial Microbiology* **20**, 143–151.
Cross, T. (1968). Thermophilic actinomycetes. *Journal of Applied Bacteriology* **31**, 36–53.
Cross, T. (1981). Aquatic actinomycetes: a critical survey of the occurrence, growth and role of actinomycetes in aquatic habitats. *Journal of Applied Bacteriology* **50**, 391–424.
Cross, T. and Attwell, R. W. (1974). Recovery of viable thermoactinomycete spores from deep mud cores. In: *Spore Research 1973* (A. M. Barker, G. W. Gould and J. Wolf, eds.), pp. 11–20. Academic Press, London.
Cross, T. and Johnston, D. W. (1972). *Thermoactinomyces vulgaris*. II. Distribution in natural habitats. In: *Spore Research* (A. M. Barker, G. W. Gould and J. Wolf, eds.), pp. 315–330. Academic Press, London.
Cross, T. and Unsworth, B. A. (1976). Farmer's lung: a neglected antigen. *Lancet* i 958–959.
Cross, T., Maciver, A. M. and Lacey, J. (1968a). The thermophilic actinomycetes in mouldy hay: *Micropolyspora faeni* sp. nov. *Journal of General Microbiology* **50**, 351–359.
Cross, T., Walker, P. D. and Gould, G. W. (1968b). Thermophilic actinomycetes producing resistant endospores. *Nature, London* **220**, 352–354.
Cross, T., Rowbotham, T. J., Mishustin, E. N., Tepper, E. Z., Antoine-Portaels, F., Schaal, K. P. and Bickenbach, H. (1976). The ecology of nocardioform actinomycetes. In: *The Biology of the Nocardiae*, (M. Goodfellow, G. H. Brownell and J. A. Serrano, eds.), pp. 337–371. Academic Press, London.
Cundell, A. M. and Mulcock, A. P. (1973). Microbiological deterioration of natural rubber pipe-joint rings. *Material und Organismen* **8**, 165–177.
Cundell, A. M. and Mulcock, A. P. (1975). The biodegradation of vulcanized rubber. *Developments in Industrial Microbiology* **16**, 88–96.
Dickinson, C. H. and Dooley, M. J. (1967). The microbiology of cut-away peat. I. Descriptive ecology. *Plant and Soil* **27**, 172–186.
Dickinson, C. H., Austin, B. and Goodfellow, M. (1975). Quantitative and qualitative studies of phylloplane bacteria from *Lolium perenne*. *Journal of General Microbiology* **91**, 157–166.
Dijk, Van C. (1978). Spore formation and endophyte diversity in root nodules of *Alnus glutinosa* (L) Vill. *New Phytologist* **81**, 601–615.
Dijk, Van, C. (1979). Endophyte distribution in the soil. In: *Symbiotic Nitrogen Fixation in the Management of Temperate Forests* (J. C. Gordon, C. T. Wheeler and D. A. Perry. eds.), pp. 84–94. Oregon State University Press, Corvallis.
Donderski, W. and Strzelczyk, E. (1974). Generic composition and nutritional requirements of bacteria isolated from three lakes. *Acta Microbiologica Polonica Series B* **6**, 67–74.

Dye, D. W. and Kemp, W. J. (1977). A taxonomic study of plant pathogenic *Corynebacterium* species. *New Zealand Journal of Agricultural Research* **20**, 563–582.
Eaton, R. A. and Dickinson, D. J. (1976). The peformance of copper–chrome-arsenic treated wood in the marine environment. *Material und Organismen* **11** Supplement, 521–529.
Efremenkova, L. M., Kozhevin, P. A. and Zviagintzev, D. G. (1978). Application of an indirect immunofluorescence technique for studying soil actinomycetes *Streptomyces olivocinereus. Mikrobiologiya* **47**, 1122–1124.
Eggeling, L. and Sahm, H. (1980). Degradation of coniferyl alcohol and other lignin-related aromatic compounds by *Nocardia* sp. DSM 1069. *Archiv für Mikrobiologie* **126**, 141–148.
Eggeling, L. and Sahm, H. (1981). Degradation of lignin-related aromatic compounds by *Nocardia* spec. DSM1069 and specificity of demethylation. *Zentralblatt für Bakteriologie, Mikrobiologie und Hygiene.* 1. Abteilung, Supplement **11**, 361–366.
Erikson, D. (1941). Studies on some lake-mud strains of *Micromonospora. Journal of Bacteriology* **41**, 277–300.
Erikson, D. (1952). Temperature/growth relationships of a thermophilic actinomycete *Micromonospora vulgaris. Journal of General Microbiology* **6**, 286–294.
Fellows, H. (1926). Relation of growth in the potato tuber to the potato scab disease. *Journal of Agricultural Research* **32**, 757–781.
Fergus, C. L. (1964). Thermophilic and thermotolerant molds and actinomycetes of mushroom compost during peak heating. *Mycologia* **56**, 267–284.
Festenstein, G. N., Lacey, J., Skinner, F. A., Jenkins, P. A. and Pepys, J. (1965). Self-heating of hay and grain in dewar flasks and the development of farmer's lung antigens. *Journal of General Microbiology* **41**, 389–407.
Filip, Z. and Küster, E. (1979). Microbial activity and the turnover of organic matter in a municipal refuse disposed of in a landfill. *European Journal of Applied Microbiology* **7**, 371–379.
Flaig, W. and Kutzner, H. J. (960). Beitrag zur Ökologie der Gattung *Streptomyces* Waksman et Henrici. *Archiv für Mikrobiologie* **35**, 207–228.
Flowers, H. F. and Williams, S. T. (1977a). Measurement of growth rates of streptomycetes: comparison of turbidimetric and gravimetric techniques. *Journal of General Microbiology* **98**, 285–289.
Flowers, T. H. and Williams, S. T. (1977b). The influence of pH on the growth rate and viability of neutrophilic and acidophilic streptomycetes. *Microbios* **18**, 223–228.
Francisco, D. E. and Silvey, J. K. G. (1971). The effect of CO inhibition on the growth of an aquatic streptomycete. *Canadian Journal of Microbiology* **17**, 347–351.
Gardner, I. C. (1976). Ultrastructural studies of non-leguminous root nodules. In: *Symbiotic Nitrogen Fixation in Plants* (P. S. Nutman, ed.), pp. 485–496. Cambridge University Press, Cambridge.
Gauthier, D., Diem, H. G. and Dommergues, Y. (1981). *In vitro* nitrogen fixation by two actinomycete strains isolated from *Casuarina* nodules. *Applied and Environmental Microbiology* **41**, 306–308.
Gerber, N. N. (1979a). Odorous substances from actinomycetes. *Developments in Industrial Microbiology* **20**, 225–238.
Gerber, N. N. (1979b). Volatile substances from actinomycetes: their role in the odor pollution of water. *Critical Reviews in Microbiology* **7**, 191–214.
Gerber, N. N. and Lechevalier, H. A. (1977). Production of geosmin in fermentors and extraction with an ion-exchange resin. *Applied and Environmental Microbiology* **34**, 857–858.
Gilmour, C. H. and Buthala, A. (1950). The isolation and study of actinophages from the soil. *Bacteriological Proceedings* 1950, 17.
Ginnivan, M. J., Woods, J. L. and O'Callaghan, J. R. (1977). Thermophilic fermentation of pig faeces and straw by actinomycetes. *Journal of Applied Bacteriology* **43**, 231–238.
Gledhill, W. E. and Casida, L. E. (1969a). Predominant catalase-negative soil bacteria. II. Occurrence and characterization of *Actinomyces humiferus* sp. n. *Applied Microbiology* **18**, 114–121.
Gledhill, W. E. and Casida, L. E. (1969b). Predominant catalase-negative soil bacteria. III. *Agromyces*, gen. n., micro-organisms intermediary to *Actinomyces* and *Nocardia. Applied Microbiology* **18**, 340–349.

Glymph, J. L. and Stutzenberger, F. J. (1977). Production, purification and character of α-amylase from *Thermomonospora curvata*. *Applied and Environmental Microbiology* **34**, 391.
Gochnauer, M. B., Leppard, G. G., Kommaratat, P., Kates, M., Novitsky, T. and Kushner, D. J. (1975). Isolation and characterization of *Actinopolyspora halophila* gen. et sp. nov., an extremely halophilic actinomycete. *Canadian Journal of Microbiology* **21**, 1500–1511.
Gold, W. (1959). Effects of medium on actinophage. *Annals of the New York Academy of Science* **81**, 994–1002.
Gottlieb, D. (1976). The production and role of antibiotics in soil. *Journal of Antibiotics* **29**, 987–1000.
Gray, T. R. G. (1976). Survival of vegetative microbes in soil. In: *The Survival of Vegetative Microbes* (T. R. G. Gray and J. R, Postgate, eds.), pp. 327–364. Cambridge University Press, Cambridge.
Gray, T. R. G. and Baxby, P. (1968). Chitin decomposition in soil. II. The ecology of chitinoclastic micro-organisms in forest soil. *Transactions of the British Mycological Society* **51**, 293–309.
Gregory, P. H. and Lacey, M. E. (1963). Mycological examination of dust from mouldy hay associated with farmer's lung disease. *Journal of General Microbiology* **30**, 75–88.
Gregory, P. H., Lacey, M. E., Festenstein, G. N. and Skinner, F. A. (1963). Microbial and biochemical changes during the moulding of hay. *Journal of General Microbiology* **33**, 147–174.
Gregory, P. H., Festenstein, G. N., Lacey, M. E. and Skinner, F. A. (1964). Farmer's lung disease; the development of antigens in moulding hay. *Journal of General Microbiology* **36**, 429–440.
Grein, A. and Meyers, S. P. (1958). Growth characteristics and antibiotic production of actinomycetes isolated from littoral sediments and materials suspended in sea water. *Journal of Bacteriology* **76**, 457–463.
Griffin, D. M. (1966). Soil physical factors and the ecology of fungi. IV. Influence of the soil atmosphere. *Transactions of the British Mycological Society* **49**, 115–119.
Gumpert, J. (1962a). Die Funktion der symbiontischen Bakterien in dem Triatomien. *Zentralblatt für Bakteriologie, Parasitenkunde, Infektionskrankheiten und Hygiene*. I. Abteilung **184**, 315–318.
Gumpert, J. (1962b). Untersuchungen über die Symbiose von Tieren mit Pilzen und Bakterien. X. Die Symbiose der Triatominen. 2. Infektion symbiotenfreier Triatominen mit symbiontischen und Mikroorganismen und gemeinsame eigen schaften der symbiontische Stamme. *Zeitschrift für Allgemeine Mikrobiologie* **2**, 290–302.
Gumpert, J. and Schwartz, W. (1962). Untersuchungen über die symbiose von Tieren mit Pilzen und Bakterien. X. Die Symbiose der Triatominen. 1. Aufzucht symbiotenhaltiger und symbiotenfreier Triatominen und eigenschaften der bei Triatominen verkommenden Mikroorganismen. *Zeitschrift für Allgemeine Mikrobiologie* **2**, 209–225.
Hagedorn, C. (1976). Influences of soil acidity on *Streptomyces* populations inhabiting forest soils. *Applied and Environmental Microbiology* **32**, 368–375.
Hagerdal, B. G. R., Ferchak, J. D. and Kendall Pye, E. (1978). Cellulolytic enzyme system of *Thermoactinomyces* sp. grown on microcrystalline cellulose. *Applied and Environmental Microbiology* **36**, 606–612.
Hale, C. M. (1972). Rapid identification methods for *Corynebacterium insidiosum* (McCulloch, 1925) Jensen, 1934. *New Zealand Journal of Agricultural Research* **15**, 149–154.
Harington, J. S. (1960). Synthesis of thiamine and folic acid by *Nocardia rhodnii*. *Nature, London* **188**, 1027–1028.
Harrison, M. D. (1962). Potato russet scab, its cause and factors affecting its development. *American Potato Journal* **39**, 368–387.
Hasegawa, T., Lechevalier, M. P. and Lechevalier, H. (1978). New genus of the *Actinomycetales: Actinosynnema* gen. nov. *International Journal of Systematic Bacteriology* **28**, 304–310.
Helmke, E. (1981). Growth of actinomycetes from marine and terrestrial origin under increased hydrostatic pressure. *Zentralblatt für Bakteriologie, Mikrobiologie und Hygiene*. I. Abteilung Supplement, **11**, 321–327.
Henis, Y., Sneh, B. and Katan, J. (1967). Effect of organic amendments on *Rhizoctonia* and accompanying microflora in soil. *Canadian Journal of Microbiology* **13**, 643–650.

Henssen, A. (1957). Beiträge zur Morphologie und Systematik der thermophilen Actinomyceten. *Archiv für Mikrobiologie* **26**, 373–414.
Hill, S. and Postgate, J. R. (1969). Failure of putative nitrogen-fixing bacteria to fix nitrogen. *Journal of General Mirobiology* **56**, 277–285.
Hoffmann, G. M. (1958). Untersuchungen zur Aetiologie pflanzlicher Actinomycosen. *Phytopathologische Zeitschrift* **34**, 1–56.
Hopwood, D. A. and Merrick, M. J. (1977). Genetics of antibiotic production. *Bacteriological Reviews* **41**, 595–635.
Hsu, S. C. and Lockwood, J. L. (1975). Powdered chitin as a selective medium for enumeration of actinomycetes in water and soil. *Applied Microbiology* **29**, 422–426.
Hungate, R. E. (1946). Studies on cellulose fermentation. II. An anaerobic cellulose decomposing actinomycete, *Micromonospora propionici* n. sp. *Journal of Bacteriology* **51**, 51–56.
Huntjens, J. L. M. (1972). Amino acid composition of humic acid-like polymers produced by streptomycetes and of humic acid from pasture and arable land. *Soil Biology and Biochemistry* **4**, 379–345.
Hurd, W. L. and Gilmour, C. M. (1951). Factors influencing the multiplication of actinophage. *Bacteriological Proceedings* 1951, 48–49.
Hutchinson, M., Ridgway, J. W. and Cross, T. (1975). Biodeterioration of rubber in contact with water, sewage and soil. In: *Microbial Aspects of the Deterioration of Materials* (D. W. Lovelock and R. A. Gilbert, eds.), pp. 187–202. Academic Press, London.
Iizuka, H. and Kawaminami, T. (1965). Studies on the xylanase from *Streptomyces*. I. Purification and some properties of xylanase from *Streptomyces xylophagus* nov. sp. *Agricultural and Biological Chemistry* **29**, 520–524.
Jensen, H. L. (1928). *Actinomyces acidophilus* n. sp – a group of acidophilus actinomycetes isolated from the soil. *Soil Science* **25**, 225–236.
Jensen, V. (1975). Bacterial flora of soil after application of oily waste. *Oikos* **26**, 152–158.
Johnston, D. W. and Cross, T. (1976a). The occurrence and distribution of actinomycetes in lakes of the English Lake District. *Freshwater Biology* **6**, 457–463.
Johnston, D. W. and Cross, T. (1976b). Actinomycetes in lake muds: dormant spores or metabolically active mycelium? *Freshwater Biology* **6**, 464–469.
Jones, D. (1975). A numerical taxonomy study of coryneform and related bacteria. *Journal of General Microbiology* **87**, 52–96.
Jordan, M. J. and Lechevalier, M. P. (1975). Effects of zinc smelter emissions on forest soil microflora. *Canadian Journal of Microbiology* **21**, 1855–1865.
Joynson, D. H. M. (1979). Water: the natural habitat of *Mycobacterium kansasii*. *Tubercle* **60**, 77–81.
Kaiser, P. (1971). L'activaté pectinolytique des actinomycetes. *Annales de l'Institut Pasteur, Paris* **121**, 389–404.
Kaspari, H. (1973). Untersuchungen über Bildung und Aktavität von Streptomyceten-Antibiotika im Boden. I. Bildung von Anthrachinon-Antibiotika im Boden. *Zentralblatt für Bakteriologie, Parasitenkunde, Infectionskrankheiten und Hygiene.* II. *Abteilung* **128**, 764–771.
Kaspari, H. and Schonbeck, F. (1973). Untersuchungen über Bildung und Aktivität von Streptomyceten-Antibiotikum im Boden. II. Aktivität von Anthrachion-Antibiotika im Boden. *Zentralblatt für Bakteriologie, Parasitenkunde, Infectionskrankheiten und Hygiene. II. Abteilung* **128**, 772–779.
Katan, J. and Lockwood, J. L. (1970). Effect of pentachloronitrobenzene on colonization of alfalfa residues by fungi and streptomycetes in soil. *Phytopathology* **60**, 1578–1582.
Katznelson, H. and Cole, S. E. (1965). Production of gibberellin-like substances by bacteria and actinomycetes. *Canadian Journal of Microbiology* **11**, 733–742.
Kayamura, Y. and Takada, H. (1970). Some characteristics of a *Streptomyces* KY67, isolated from salt farm in Japan. *Transactions of the Mycological Society of Japan* **11**, 7–10.
Keddie, R. M. (1978). What do we mean by coryneform bacteria? In: *Coryneform Bacteria* (I. J. Bousfield and A. G. Calleley, eds.), pp. 2–12. Academic Press, London.
Khan, M. R. and Williams, S. T. (1975). Studies on the ecology of actinomycetes in soil. VIII. Distribution and characteristics of acidophilic actinomycetes. *Soil Biology and Biochemistry* **7**, 345–348.
Khavina, E. S. and Rautenstein, Y. I. (1958). *Actinomyces olivaceus* actinophages and lysogenicity among the cultures of this species. *Microbiologiya* **27**, 433–439.

Khavina, E. S. and Rautenstein, Y. I. (1959). Effect of pH of the medium on the isolation of actinophages from podzol soil. *Microbiologiya* **28**, 685–691.
Ko, W. H. and Lockwood, J. L. (1968). Accumulation and concentration of chlorinated hydrocarbon pesticides by micro-organisms in soil. *Canadian Journal of Microbiology* **14**, 1075–1078.
Krasilnikov, N. A., Zenova, G. M. and Stepanova, L. N. (1969). Utilization of hydrocarbons by actinomycetes. *Mikrobiologiya* **38**, 962–967.
Kuo, M. J. and Hartman, P. A. (1967). Purification and partial characterization of *Thermoactinomyces vulgaris* amylases. *Canadian Journal of Microbiology* **13**, 1157–1163.
Kurane, R., Suzuki, T. and Takahara, Y. (1979). Removal of phthalate esters by activated sludge inoculated with a strain of *Nocardia erythropolis*. *Agricultural and Biological Chemistry* **43**, 421–427.
Kuroda, S. and Bradley, S. G. (1967). Temperature sensitive replication of an actinophage for *Streptomyces aureofaciens*. *Canadian Journal of Microbiology* **13**, 1569–1575.
Kurup, V. P. and Heinzen, R. J. (1978). Isolation and characterization of actinophages of *Thermoactinomyces* and *Micropolyspora*. *Canadian Journal of Microbiology* **24**, 794–797.
Küster, E. (1970). Note on the taxonomy and ecology of *Streptomyces malachiticus* and related species. In: *The Actinomycetales* (H. Prauser, ed.), pp. 169–172. VEB Gustav Fischer Verlag, Jena.
Küster, E. (1976). Ecology and predominance of soil streptomycetes. In: *Actinomycetes – The Boundary Microorganisms* (T. Arai, ed.), pp. 109–121. Toppan Co., Tokyo.
Küster, E. (1979). Bedeutung der Aktinomyceten für den Abbau von Cellulose, Lignin und Huminstoffen im Boden. *Zeitschrift für Pflanzenernährung Dungung und Bodenkunde* **142**, 365–374.
Küster, E. and Locci, R. (1963). Studies on peat and peat microorganisms. I. Taxonomic studies on thermophilic actinomycetes isolated from peat. *Archiv für Mikrobiologie* **45**, 188–197.
Küster, E. and Williams, S. T. (1964). Selection of media for isolation of streptomycetes. *Nature, London* **202**, 928–929.
Kutzner, H. J. (1968). Über die Bildung von Huminstoffen durch Streptomyceten. *Landwirtschaftliche Forschung* **21**, 48–61.
Kutzner, H. J. (1981). The family *Streptomycetaceae*. In: *The Prokaryotes: A Handbook on Habitats, Isolation and Identification of Bacteria. Volume II.* (M. P. Starr, H. Stolp, H. G. Trüper, A. Balows and H. G. Schlegel, eds.), pp. 2028–2090. Springer-Verlag, Berlin, Heidelberg and New York.
Labruyere, R. E. (1971). Common scab and its control in seed potato crops. *Agricultural Research Reports* **767**, 72 pp, Wageningen.
Lacey, J. (1971a). The microbiology of moist barley storage in unsealed silos. *Annals of Applied Biology* **69**, 187–212.
Lacey, J. (1971b). *Thermoactinomyces sacchari* sp. nov., a thermophilic actinomycete causing bagassosis. *Journal of General Microbiology* **66**, 327–338.
Lacey, J. (1973). Actinomycetes in soils, composts and fodders. In: *Actinomycetales, Characteristics and Practical Importance*. (G. Sykes and F. A. Skinner, eds.), pp. 231–251. Academic Press, London and New York.
Lacey, J. (1974). Allergy in mushroom workers. *Lancet* **i**, 366.
Lacey, J. (1977). Microorganisms in air of cotton mills. *Lancet* **ii**, 455–456.
Lacey, J. (1978). Ecology of actinomycetes in fodders and related substrates. *Zentralblatt für Bakteriologie, Parasitenkunde, Infektionskrankheiten und Hygiene. I. Abteilung, Supplement* **6**, 162–170.
Lacey, J. (1981). Airborne actinomycete spores as respiratory allergens. *Zentralblatt für Bakteriologie, Mikrobiologie und Hygiene. I. Abteilung, Supplement* **11**, 243–250.
Lacey, J. and Dutkiewicz, J. (1976). Isolation of actinomycetes and fungi from mouldy hay using a sedimentation chamber. *Journal of Applied Bacteriology* **41**, 315–319.
Lacey, J. and Goodfellow, M. (1975). A novel actinomycete from sugar-cane bagasse: *Saccharopolyspora hirsuta* gen. et sp. nov. *Journal of General Microbiology* **88**, 75–85.
Lacey, J. and Lacey, M. E. (1964). Spore concentrations in the air from farm buildings. *Transactions of the British Mycological Society* **47**, 547–552.

Lake, P. and Friend, W. G. (1968). The use of artificial diets to determine some of the effects of *Nocardia rhodnii* on the development of *Rhodnius prolixus*. *Journal of Insect Physiology* **14**, 543–562.

Lalonde, M. (1979). Immunological and ultrastructural demonstration of nodulation in the European *Alnus glutinosa* (L) Gaertn. host plant by an actinomycetal isolate from the North American *Comptonia peregrina* (L) Coult. root nodule. *Botanical Gazette* **140** Supplement, S35–S43.

Lanning, S. and Williams, S. T. (1982). Methods for the direct isolation and enumeration of actinophages from soil. *Journal of General Microbiology* **128**, 2063–2071.

Lapwood, D. H. (1966). The effects of soil moisture at the time potato tubers are forming on the incidence of common scab (*Streptomyces scabies*). *Annals of Applied Biology* **58**, 447–456.

Lapwood, D. H. (1973). *Streptomyces scabies* and potato scab disease. In: *Actinomycetales: Characteristics and Practical Importance*. (G. Sykes and F. A. Skinner, eds.), pp. 253–260. Academic Press, London and New York.

Lapwood, D. H. and Adams, M. J. (1973). The effect of a few days rain on the distribution of common scab (*Streptomyces scabies*) on young potato tubers. *Annals of Applied Biology* **73**, 277–283.

Lapwood, D. H. and Hering, T. F. (1970). Soil moisture and the infection of young potato tubers by *Streptomyces scabies* (Common scab). *Potato Research* **13**, 296–304.

Lapwood, D. H., Wellings, L. W. and Rosser, W. R. (1970). The control of common scab of potatoes by irrigation. *Annals of Applied Biology* **66**, 397–405.

Lawson, E. N. and Davey, L. M. (1972). A waterborne actinomycete resembling strains causing mycetoma. *Journal of Applied Bacteriology* **35**, 389–394.

Lechevalier, H. A. and Lechevalier, M. P. (1967). Biology of actinomycetes. *Annual Review of Microbiology* **21**, 71–100.

Lechevalier, M. P. (1972). Description of a new species, *Oerskovia xanthineolytica* and emendation of *Oerskovia* Prauser et al. *International Journal of Systematic Bacteriology* **22**, 260–264.

Lechevalier, M. P. (1981). Ecological associations involving actinomycetes. *Zentralblatt für Bakteriologie, Mikrobiologie und Hygiene*. I. Abteilung, Supplement **11**, 159–166.

Lechevalier, M. P. and Lechevalier, H. A. (1974). *Nocardia amarae* sp. nov. an actinomycete common in foaming activated sludge. *International Journal of Systematic Bacteriology* **24**, 278–288.

Lechevalier, M. P. and Lechevalier, H. A. (1979). The taxonomic position of the actinomycetic endophytes. In: *Symbiotic Nitrogen Fixation in the Management of Temperate Forests* (J. C. Gordon, C. T. Wheeler and D. A. Perry, eds.), pp. 111–122. Oregon State University Press, Corvallis.

Lechevalier, M. P., Seidler, R. J. and Evans, T. M. (1980). Enumeration and characterization of standard plate count bacteria in chlorinated and raw water supplies. *Applied and Environmental Microbiology* **40**, 922–930.

Lilley, G. and Bull, A. T. (1974). The production of β1,3-glucanase by a thermophilic species of *Streptomyces*. *Journal of General Microbiology* **83**, 123–133.

Lloyd, A. B. (1969a). Behaviour of streptomycetes in soil. *Journal of General Microbiology* **56**, 165–170.

Lloyd, A. B. (1969b). Dispersal of streptomycetes in air. *Journal of General Microbiology* **57**, 35–40.

Luscombe, B. M. and Gray, T. R. G. (1973). Characteristics of *Arthrobacter* grown in continuous culture. *Journal of General Microbiology* **82**, 213–222.

Lyons, A. J., Pridham, T. G. and Rogers, R. F. (1975). Actinomycetales from corn. *Applied Microbiology* **29**, 246–249.

McCarthy, A. J. and Cross, T. (1981). A note on a selective isolation medium for the thermophilic actinomycete *Thermomonospora chromogena*. *Journal of Applied Bacteriology* **51**, 299–302.

McCulloch, L. (1925). *Aplanobacter insidiosum* n. sp., the cause of an alfalfa disease. *Phytopathology* **15**, 496–497.

McKee, R. K. (1958). Assessment of the resistance of potato varieties to common scab. *European Potato Journal* **1**, 65–80.

Macleod, R. A. (1965). The question of the existence of specific marine bacteria. *Bacteriological Reviews* **29**, 9–25.

Malik, K. A. and Claus, D. (1979). *Xanthobacter flavus*, a new species of nitrogen-fixing hydrogen bacteria. *International Journal of Systematic Bacteriology* **29**, 283–287.

Mann, J. W., Jeffries, T. W. and Macmillan, J. D. (1978). Production and ecological significance of yeast cell wall-degrading enzymes from *Oerskovia*. *Applied and Environmental Microbiology* **36**, 594–605.

Martin, J. F. and Demain, A. L. (1980). Control of antibiotic synthesis. *Microbiological Reviews* **44**, 230–251.

Martin, J. P., Filip, Z. and Haider, K. (1976). Effect of montmorillonite and humate on growth and metabolic activity of some actinomycetes. *Soil Biology and Biochemistry* **8**, 409–413.

Matschke, J. (1970a). Ein Beitrag zur Huminstoff-synthese durch *Streptomyces aureus*. 1. Stoffwechselvorgänge von *S. aureus*. *Zentralblatt für Bakteriologie und Parasitenkunde*. II. *Abteilung* **125**, 85–99.

Matschke, J. (1970b). Ein Beitrag zur Huminstoff-synthese durch *Streptomyces aureus*. 2. Absorption-spectrographisches Verhalten der isolierten Huminstoff-Fraktionen. *Zentralblatt für Bakteriologie und Parasitenkunde*. II. *Abteilung* **125**, 150–161.

Matschke, J. (1970c). Ein Beitrag zur Huminstoff-synthese durch *Streptomyces aureus*. 3. Chemische Untersuchungen an den Organismen-Huminstoffen. *Zentralblatt für Bakteriologie und Parasitenkunde*. II. *Abteilung* **125**, 162–169.

Matschke, J. (1970d). Ein Beitrag zur Huminstoff-synthese durch *Streptomyces aureus*. 4. Infrarotspektoskopischer Vergleich. *Zentralblatt für Bakteriologie und Parasitenkunde*. II. *Abteilung* **125**, 438–447.

Matschke, J. (1970e). Ein Beitrag zur Huminstoff-synthese durch *Streptomyces aureus*. 5. Untersuchugen an Sephadex fraktionerten Streptomyceten-Kulturfiltraten. *Zentralblatt für Bakteriologie und Parasitenkunde*. II. *Abteilung* **125**, 448–457.

Mayfield, C. I., Williams, S. T., Ruddick, S. M. and Hatfield, H. L. (1972). Studies on the ecology of actinomycetes in soil. IV. Observations on the form and growth of streptomycetes in soil. *Soil Biology and Biochemistry* **4**, 79–91.

Merriman, P. R., Price, R. D. and Baker, K. F. (1974a). The effect of inoculation of seed with antagonists of *Rhizoctonia solani* on the growth of wheat. *Australian Journal of Agricultural Research* **25**, 213–218.

Merriman, P. R., Price, R. D., Kollmorgen, J. F. Piggott, T. and Ridge, E. H. (1974b). Effect of seed inoculation with *Bacillus subtilis* and *Streptomyces griseus* on the growth of cereals and carrots. *Australian Journal of Agricultural Research* **25**, 219–226.

Metcalfe, G. and Brown, M. E. (1957). Nitrogen fixation by new species of *Nocardia*. *Journal of General Microbiology* **17**, 567–572.

Mitchell, R. (1963). Addition of fungal cell-wall components of soil for biological disease control. *Phytopathology* **53**, 1068–1071.

Molina, C. (1974). Farmer's lung in France. In: *Aspergillosis and Farmer's Lung in Man and Animals* (R. de Haller and F. Suter, eds.) Hans Huber Publishers, Bern.

Mostafa, S. A. and Hussein, A. M. (1974). Biological and biochemical studies on a keratinolytic thermophilic actinomycete isolated from Egyptian soil. *Zentralblatt für Bakteriologie, Parasitenkunde, Infektionskrankheiten und Hygiene*. II. *Abteilung* **129**, 571–579.

Mulkins-Phillips, G. J. and Stewart, J. E. (1974). Distribution of hydrocarbon-utilizing bacteria in northwestern Atlantic waters and coastal sediments. *Canadian Journal of Microbiology* **20**, 955–962.

Nesterenko, P. A., Kasumova, S. A. and Kvasnikov, E. I. (1978). Microorganisms of the *Nocardia* genus and the 'rhodochrous' group in soils of the Ukranian S.S.R. *Mikrobiologiya* **47**, 866–870.

Newcomb, W., Callaham, D., Torrey, J. G. and Peterson, R. L. (1979). Morphogenesis and fine structure of the actinomycetous endophyte of nitrogen-fixing root nodules of *Comptonia peregrina*. *Botanical Gazette* **140** (*Supplement*), 522–534.

Nonomura, H. and Ohara, Y. (1969). Distribution of actinomycetes in soil. VI. A culture method effective for both preferential isolation and enumeration of *Microbispora* and *Streptosporangium* strains in soil (Part 1). *Journal of Fermentation Technology* **47**, 463–469.

Nonomura, H. and Ohara, Y. (1971a). Distribution of actinomycetes in soil. VIII. Green-spore group of *Microtetraspora*, its preferential isolation and taxonomic characteristics. *Journal of Fermentation Technology* **49**, 1–7.
Nonomura, H. and Ohara, Y. (1971b). Distribution of actinomycetes in soil. IX. New species of the genera *Microbispora* and *Microtetraspora* and their isolation method. *Journal of Fermentation Technology* **49**, 887–894.
Nonomura, H. and Ohara, Y. (1971c). Distribution of actinomycetes in soil. X. New genus and species of monosporic actinomycetes. *Journal of Fermentation Technology* **49**, 895–903.
Nonomura, H. and Ohara, Y. (1971d). Distribution of actinomycetes in soil. XI. Some new species of the genus *Actinomadura* Lechevalier *et al. Journal of Fermentation Technology* **49**, 904–912.
Noval, J. J. and Nickerson, W. J. (1959). Decomposition of native keratin by *Streptomyces fradiae*. *Journal of Bacteriology* **77**, 251–263.
Nyirady, S. A. (1973). The germfree culture of three species of *Triatominae: Triatoma protracta* (Uhler), *Triatoma rubida* (Uhler) and *Rhodnius prolixus* (Stal). *Journal of Medical Entomology* **10**, 417–448.
Ohta, Y. and Ikeda, M. (1978). Deodorization of pig faeces by actinomycetes. *Applied and Environmental Microbiology* **36**, 487–491.
Okafor, N. (1966). Ecology of micro-organisms on chitin buried in soil. *Journal of General Microbiology* **44**, 311–327.
Okafor, N. (1967). Decomposition of chitin by micro-organisms isolated from a temperate and a tropical soil. *Nova Hedwigia* **13**, 209–226.
Okami, Y. and Okazaki, T. (1978). Actinomycetes in marine environments. *Zentralblatt für Bakteriologie, Parasitenkunde, Infektionskrankheiten und Hygiene. I. Abteilung Supplement* **6**, 145–152.
Okazaki, T. and Okami, Y. (1972). Studies on marine micro-organisms. II. Actinomycetes in Sagami Bay and their antibiotic substances. *Journal of Antiobiotics* **25**, 461–466.
Okazaki, T. and Okami, Y. (1975). Actinomycetes tolerant to increased NaCl concentration and their metabolites. *Journal of Fermentation Technology* **53**, 833–840.
Orchard, V. A. (1979). Effect of sewage sludge additions on *Nocardia* in soil. *Soil Biology and Biochemistry* **11**, 217–220.
Orchard, V. A. (1980). Long term effect of sludge additions on populations of *Nocardia asteroides*, *Micromonospora* and *Thermoactinomyces* in soil. *Soil Biology and Biochemistry* **12**, 477–481.
Orchard, V. A. (1981). The ecology of *Nocardia* and related taxa. *Zentralblatt für Bakteriologie, Mikrobiologie und Hygiene, Supplement* **11**, 167–180.
Orchard, V. A., Goodfellow, M. and Williams, S. T. (1977). Selective isolation and occurrence of nocardiae in soil. *Soil Biology and Biochemistry* **9**, 233–238.
Ou, L.-T., Davidson, J. M. and Rothwell, D. F. (1978). Responses of soil microflora to high 2,4-D applications. *Soil Biology and Biochemistrry* **10**, 443–445.
Palleroni, N. J. (1980). A chemostatic method for the isolation of *Actinoplanaceae*. *Archiv für Mikrobiologie* **128**, 53–55.
Patel, J. J. (1969a). Phages of lysogenic *Thermoactinomyces vulgaris*. *Archiv für Mikrobiologie* **69**, 294–300.
Patel, J. J. (1969b). Phage-like particles from a lysogenic *Thermomonospora*. *Archiv für Mikrobiologie* **65**, 401–402.
Pearson, H. W., Howsley, R. and Williams, S. T. (1982). A study of nitrogenase activity in *Mycoplana* species and free-living actinomycetes. *Journal of General Microbiology* **128**, 2073–2080.
Pepys, J., Jenkins, P. A., Festenstein, G. N., Gregory, P. H., Lacey, M. E. and Skinner, F. A. (1963). Farmer's lung: Thermophilic actinomycetes as a source of 'farmer's lung hay' antigen. *Lancet* **ii**, 607–611.
Percich, J. A. and Lockwood, J. L. (1978). Interaction of atrazine with soil micro-organisms: population changes and accumulation. *Canadian Journal of Microbiology* **24**, 1145–1152.
Person, L. H. and Martin, W. J. (1940). Soil rot of sweet potatoes in Louisiana. *Phytopathology* **30**, 913–926.

Pfennig, N. (1958). Beobachtungen des Wachstumverhaltens von Streptomyceten auf Rossicholodny-Aufwuchsplatten im Boden. *Archiv für Mikrobiologie* **31**, 206–216.

Phelan, M. B., Crawford, D. L. and Pometto III, A. L. (1979). Isolation of lignocellulose-decomposing actinomycetes and degradation of specifically ^{14}C-labelled lignocelluloses by six selected *Streptomyces* strains. *Canadian Journal of Microbiology* **25**, 1270–1276.

Powell, B. L. and Steadham, J. E. (1981). Improved technique for isolation of *Mycobacterium kansasii* from water. *Journal of Clinical Microbiology* **13**, 969–975.

Preobrazhenskaya, T. P., Lavrova, M. V., Ukholina, R. S. and Nechaeva, N. P. (1975). Isolation of new species of *Actinomadura* on selective media with streptomycin and bruneomycin. *Antibiotiki* **20**, 404–409.

Pridham, T. G. and Kroppenstedt, R. M. (1979). Acetylene reduction tests with streptomycetes, streptoverticillia and other micro-organisms. *Actinomycetes and Related Organisms* **14**, 28–35.

Purushothaman, D., Marimuthu, T., Venkataramanan, C. V. and Kesavan, R. (1974). Role of actinomycetes in the biosynthesis of indole acetic acid in soil. *Current Science* **43**, 413–414.

Quispel, A. and Tak, T. (1978). Studies on the growth of the endophyte of *Alnus glutinosa* (L.) Vill. in nutrient solutions. *New Phytologist* **81**, 587–600.

Rast, H. G., Engelhardt, G., Diegler, W. and Wallhoffer, P. R. (1980). Bacterial degradation of model compounds for lignin and chlorophenol derived lignin bound residues. *FEMS Microbiology Letters* **8**, 259–63.

Robinson, J. B. and Corke, C. T. (1959). Preliminary studies on the distribution of actinophages in soil. *Canadian Journal of Microbiology* **5**, 479–484.

Rodriguez-Barrueco, C. (1968). The occurrences of the root-nodule endophytes of *Alnus glutinosa* and *Myrica gale* in soil. *Journal of General Microbiology* **52**, 189–194.

Rose, S. L., Li, C.-Y. and Hutchins, S. (1982). A streptomycete antagonistic to *Phellinus weirii*, *Fomes annosus* and *Phytophthora cinnamoni*. *Canadian Journal of Microbiology* **26**, 583–587.

Rosen, A. A., Mashni, C. I. and Safferman, R. S. (1970). Recent developments in the chemistry of odour in water: the cause of earthy/musty odour. *Water Treatment and Examination* **19**, 106–119.

Rowbotham, T. J. and Cross, T. (1977). Ecology of *Rhodococcus coprophilus* and associated actinomycetes in fresh water and agricultural habitats. *Journal of General Microbiology* **100**, 231–240.

Ruddick, S. M. and Williams, S. T. (1972). Studies on the ecology of actinomycetes in soil. V. Some factors influencing the dispersal and adsorption of spores. *Soil Biology and Biochemistry* **4**, 93–103.

Salvaggio, J., Arquembourg, P., Seabury, J. and Buechner, H. (1969). Bagassosis. IV. Precipitins against extracts of thermophilic actinomycetes in patients with bagassosis. *American Journal of Medicine* **46**, 538–44.

Schaal, K. P. and Bickenbach, H. (1978). Soil occurrence of pathogenic nocardiae. *Zentralblatt für Bakteriologie Parasitenkunde, Infectionskrankheiten und Hygiene. I. Abteilung, Supplement* **6**, 429–434.

Scharif, G. (1961). *Corynebacterium iranicum* sp. nov. on wheat (*Triticum vulgare* L.) in Iran and a comparative study of it with *C. tritici* and *C. rathayi*. *Entomologie et Phytopathologie Appliqués (Téheran)* **19**, 1–24.

Schuster, M. L. (1975). Leaf freckles and wilt of corn incited by *Corynebacterium nebraskense* Schuster, Hoff, Mandel, Lazar, 1972. *Research Bulletin* **270**. The Agricultural Experiment Station, University of Nebraska.

Seabury, J., Salvaggio, J., Buechner, H. and Kundur, V. G. (1968). Bagassosis. III. Isolation of thermophilic and mesophilic actinomycetes and fungi from mouldy bagasse. *Proceedings of the Society for Experimental Biology and Medicine* **129**, 351–360.

Sebald, H. and Prevot, A. R. (1962). Étude d'une nouvelle espèce anaérobic stricte *Micromonospora acetiformici* n. sp. isolée de l'intestin postérieur de *Reticulitermes lucifugus* var. Saintonnensis. *Annales de l'Institut Pasteur, Paris* **102**, 199–214.

Seiler, H., Braatz, R. and Ohmayer, G. (1980). Numerical cluster analysis of the coryneform bacteria from activated sludge. *Zentralblatt für Bakteriologie, Mikrobiologie und Hygiene. I. Abteilung C* **1**, 357–375.

Shamiyeh, M. B. and Johnson, L. F. (1973). Effect of heptachlor on numbers of bacteria, actinomycetes and fungi in soil. *Soil Biology and Biochemistry* **5**, 309–314.

Shirling, E. B. and Gottlieb, D. (1968). Co-operative descriptions of type cultures of *Streptomyces*. III. additional species descriptions from first an second studies. *International Journal of Systemic Bacteriology* **18**, 279–391.

Silvester, W. B. (1976). Ecological and economic significance of the non-legume symbioses. In: *Proceedings of the 1st International Symposium of Nitrogen Fixation* (W. E. Newton and C. J. Hyman, eds.), pp. 489–506. Washington State University Press, Pullman.

Silvey, J. K. G. and Roach, A. W. (1975). The taste and odor producing aquatic actinomycetes. *Critical Reviews in Environmental Control* **5**, 233–273.

Sing, P. J. and Mehrotra, R. S. (1980). Biological control of *Rhizoctonia bataticola* on grain by coating seed with *Bacillus* and *Streptomyces* spp. and their influence on plant growth. *Plant and Soil* **56**, 475–483.

Singal, E. M., Ivanitskaya, L. P., Smirnova, M. P. and Navashin, S. M. (1978). Direct isolation of *Micromonospora* cultures from moist soils and silts. *Antibiotiki* **23**, 693–696.

Sivasithamparam, K. and Parker, C. A. (1978). Effects of certain isolates of bacteria and actinomycetes on *Gaeumannomyces graminis*. var. *tritici* and take-all of wheat. *Australian Journal of Botany* **26**, 773–782.

Skaptason, J. B. and Burkholder, W. H. (1942). Classification and nomenclature of the pathogen causing bacterial ring rot of potatoes. *Phytopathology* **32**, 439–441.

Slack, S. A., Kelman, A. and Perry, J. B. (1979). Comparison of three serodiagnostic assays for detection of *Corynebacterium sepedonicum*. *Phytopathology* **69**, 186–189.

Smiley, R. W. (1978a). Antagonists of *Gaeumannomyces graminis* from the rhizoplane of wheat in soils fertilised with ammonium- or nitrate-nitrogen. *Soil Biology and Biochemistry* **10**, 169–174.

Smiley, R. W. (1978b). Colonization of wheat roots by *Gaeumannomyces graminis* inhibited by specific soils, microorganisms and ammonium-nitrogen. *Soil Biology and Biochemistry* **10**, 175–179.

Sneh, B. and Henis, Y. (1972). Production of antifungal substances active against *Rhizoctonia solani* in chitin-amended soil. *Phytopathology* **62**, 595–599.

Sneh, B., Katan, H. and Henis, Y. (1971). Mode of inhibition of *Rhizoctonia solani* in chitin-amended soil. *Phytopathology* **61**, 1113–1117.

Staib, F., Mishra, S. K. and Blisse, A. (1980). Interaction between aspergilli and streptomycetes in the soil of potted indoor plants: a preliminary report (contibution to the epidemiology of aspergillosis). *Mycopathologia* **70**, 9–12.

Steinbrenner, K. and Mundstock, I. (1975). The decomposition of humus by *Nocardia*. *Archiv für Acker and Pflanzenbau and Bodenkunde* **19**, 243–250.

Stotzky, G. and Goos, R. D. (1965). Effect of high CO_2 and low O_2 tensions on the soil microbiota. *Canadian Journal of Microbiology* **11**, 853–868.

Stutzenberger, F. J. (1971). Cellulase production by *Thermomonospora curvata* isolated from municipal solid waste compost. *Applied Microbiology* **22**, 147–152.

Stutzenberger, F. J. (1972a). Cellulolytic activity of *Thermomonospora curvata*. 1. Nutritional requirements for cellulase production. *Applied Microbiology* **24**, 77–82.

Stutzenberger, F. J. (1972b). Cellulolytic activity of *Thermomonospora curvata*. 2. Optimal assay conditions, partial purification and product of the cellulase. *Applied Microbiology* **24**, 83–90.

Stutzenberger, F. and Sterpu, I. (1978). Effect of municipal refuse metals in cellulase production by *Thermomonospora curvata*. *Applied and Environmental Microbiology* **36**, 201–204.

Stutzenberger, F. J., Kaufmann, A. J. and Lossin, R. D. (1970). Cellulolytic activity in municipal solid waste composting. *Canadian Journal of Microbiology* **16**, 553–560.

Sutherland, J. B., Blanchette, R. A., Crawford, D. L. and Pometto, A. L. (1979). Breakdown of Douglas fir phloem by a lignocellulose-degrading *Streptomyces*. *Current Microbiology* **2**, 123–126.

Szabo, I. and Marton, M. (1966). Selection and succession of microbial associations in different soil types on the root surface of plants and in the intestine of soil animals. *Annales de l'Institut Pasteur, Paris, Supplement* **III**, 178–196.

Szabo, I., Marton, M. and Partai, G. (1964). Micro-milieu studies in the A-horizon of a mull-like rendsina. In: *Soil Micromorphology* (J. Jongerius, ed.), pp. 33–45. Elsevier, London and New York.

Szabo, I., Marton, M., Ferenczy, L. and Buti, I. (1967). Intestinal microflora of the larvae of St. Mark's fly. II. Computer analysis of intestinal actinomycetes from the larvae of a bibio population. *Acta Microbiologia Academiae Scientarum Hungaricae* **14**, 239–249.

Sykes, I. K. and Williams, S. T. (1978). Interactions of actinophage and clays. *Journal of General Microbiology* **108**, 97–102.

Sykes, I. K., Lanning, S. and Williams, S. T. (1981). The effect of pH on soil actinophage. *Journal of General Micribiology* **122**, 271–280.

Szegi, J. and Gulyas, F. (1968). Data on the humus-decomposing activity of some streptomycetes and microscopic fungi. *Agrokemika Talajtan* **17**, 109–119.

Taber, W. A. (1960). Evidence for the existence of acid-sensitive actinomycetes in soil. *Canadian Journal of Microbiology* **6**, 503–514.

Thiemann, J. E. and Beretta, G. (1068). A new genus of the *Actinoplanaceae*; *Planobispora* gen. nov. *Archiv für Mikrobiologie* **62**, 157–166.

Thiemann, J. E., Pagani, H. and Beretta, G. (1967). A new genus of the *Actinoplanaceae*; *Planomonospora* gen. nov. *Giornale de Microbiologia* **15**, 27–38.

Thomas, G. V. and Khurana, A. S. (1980). Rhizosphere fungal and actinomycetal flora of soybean. *Indian Journal of Ecology* **7**, 281–287.

Tilford, P. E. (1936). Fasciation of sweet peas caused by *Phytomonas facians* n. sp. *Journal of Agricultural Research* **53**, 383–394.

Tjepkema, J. D., Ormerod, W. and Torrey, J. G. (1980). Vesicle-formation and acetylene reduction in *Frankia* sp. CP.11 cultured in defined media. *Nature, London* **287**, 633–635.

Torrey, J. G. (1978). Nitrogen fixation by actinomycete-nodulated angiosperms. *Bioscience* **28**, 586–592.

Torrey, J. G. and Callaham, D. (1978). Determinate development of nodule roots in actinomycete-induced root nodules of *Myrica gale*. *Canadian Journal of Botany* **56**, 1357–1364.

Torrey, J. G. and Callaham, D. (1979). Early nodule development in *Myrica gale*. *Botanical Gazette* **140** *Supplement*, S10–S14,

Torrey, J. G. and Tjepkema, J. D. (1979). Symbiotic nitrogen fixation in actinomycete-nodulated plants. *Botanical Gazette* **140** *Supplement*, i-ii.

Trojanowski, J., Haider, K. and Sundman, V. (1977). Decomposition of ^{14}C-labelled lignin and phenols by a *Nocardia* sp. *Archiv für Mikrobiologie* **114**, 149–153.

Umbreit, W. W. and McCoy, E. (1941). The occurrence of actinomycetes of the genus *Micromonospora* in inland lakes. *Symposium on Hydrobiology*, University of Wisconsin 1940, pp. 106–114.

Unsworth, B. A., Cross, T., Seaward, M. R. D. and Sims, R. E. (197). The longevity of thermoactinomycete endospores in natural substrates. *Journal of Applied Bacteriology* **42**, 45–52.

Upton, M. E. and Fogarty, W. M. (1977). Production and purification of thermostable amylase and protease of *Thermomonospora viridis*. *Applied and Environmental Microbiology* **33**, 59.

Vidaver, A. K. and Starr, M. P. (1981). Phytopathogenic coryneform and related bacteria. In: *The Prokaryotes: A Handbook on Habitats, Isolation, and Identification of Bacteria. Volume 2* (M. P. Starr, H. Stolp, H. G. Trüper, A. Balows and H. G. Schlegel, eds.), pp. 1879–1887. Springer-Verlag, Berlin, Heidelberg and New York.

Vruggink, H. (1970). The effect of chitin amendment on actinomycetes in soil and on the infection of potato tubers by *Streptomyces scabies*. *Netherlands Journal of Plant Pathology* **76**, 293–95.

Vruggink, H. (1976). Influence of agricultural crops on the actinomycetes flora in soil. *Plant and Soil* **44**, 639–654.

Vruggink, H. and Maat, D. Z. (1968). Serological recognition of *Streptomyces* species causing scab on potato tubers. *Netherlands Journal of Plant Pathology* **74**, 35–43.

Waksman, S. A. (1959). *The Actinomycetes, Volume 1. Nature, Occurrence and Activities.* Williams and Wilkins, Baltimore.

Waksman, S. A. (1961). *The Actinomycetes, Volume 2. Classification, Identification and Description of Genera and Species.* Baltimore: Williams and Wilkins.

Waksman, S. A. and Diehm, R. A. (1931). On the decomposition of hemicelluloses by microorganisms. II. Decomposition of hemicelluloses by fungi and actinomyces. *Soil Science* **32**, 97–117.

Waksman, S. A., Umbreit, W. W. and Cordon, T. C. (1939). Thermophilic actinomycetes and fungi in soils and in composts. *Soil Science* **47**, 37–61.
Walker, J. D. and Colwell, R. R. (1975). Factors affecting enumeration and isolation of actinomycetes from Chesapeake Bay and southeastern Atlantic Ocean sediments. *Marine Biology, Berlin* **30**, 193–201.
Walton, R. B. (1951). Effect of cations upon multiplication of actinophage for *Streptomyces griseus*. *Antibiotics and Chemotherapy* **1**, 518–522.
Watson, E. T. and Williams, S. T. (1974). Studies on the ecology of actinomycetes in soil. VII. Actinomycetes in a coastal sand belt. *Soil Biology and Biochemistry* **6**, 43–52.
Weete, J. D., Huane, W. Y. and Laseter, J. L. (1979). *Streptomyces* sp. a source of odorous substances in potable water. *Water Air Soil Pollution* **11**, 217–223.
Welsch, M. (1956). Influence de l'age d'un *Streptomyces* ser ses réactions à un actinophage. *Comptes rendus de la Société de Biologie* **150**, 609–613.
Welsch, M. (1969). Biology of actinophages. In: *Genetics and Breeding in Streptomycetes* (G. Sermonti and M. Alacevic, eds.), pp. 43–62. Yugoslav Academy of Science and Arts, Zagreb.
Welsch, M., Rutten-Pinckaers, A. and Selman, M. (1963). Recherches sur des *Streptomyces* d'Afrique Centrale. IV. Iso-antibiose, lysogénie et actinophages libres. *Extrait du Bulletin de la Société Royale des Sciences de Liège* **7–8**, 529–573.
Wenzl, H. and Denel, J. (1971). Schorfbefall und Pflanzgutwert. Report of the meeting of section pathology (Munich, West Germany, 1971) *Potato Research* **14**, 334.
Wenzel, F. J., Gray, R. L., Roberts, R. C. and Emanuel, D. A. (1974). Serological studies in farmer's lung. Precipitins to the thermophilic actinomycetes. *American Review of Respiratory Diseases* **109**, 464–468.
Weyland, H. (1969). Actinomycetes in North Sea and Atlantic Ocean sediments. *Nature, London* **223**, 858.
Weyland, H. (1981a). Distribution of actinomycetes on the sea floor. *Zentralblatt für Bakteriologie, Mikrobiologie und Hygiene. I. Abteilung, Supplement* **11**, 185–193.
Weyland, H. (1981b). Characteristics of actinomycetes isolated from marine sediments. *Zentralblatt für Bakteriologie, Mikrobiologie und Hygiene. I. Abteilung, Supplement* **11**, 309–314.
Whaley, J. W. and Boyle, A. M. (1967). Antibiotic production by *Streptomyces* species from the rhizosphere of desert plants. *Phytopathology* **57**, 347–351.
Williams, S. T. (1978). Streptomycetes in the soil ecosystem. *Zentralblatt für Bakteriologie, Parasitenkunde, Infektionskrankheiten und Hygiene. I. Abteilung, Supplement* **6**, 137–144.
Williams, S. T. (1982). Are antibiotics produced in soil? *Pedobiologia* **23**, 427–435.
Williams, S. T. and Davies, F. L. (1965). Use of antibiotics for selective isolation and enumeration of actinomycetes in soil. *Journal of General Microbiology* **38**, 251–261.
Williams, S. T. and Flowers, T. H. (1978). The influence of pH on starch hydrolysis by neutrophilic an acidophilic streptomycetes. *Microbios* **20**, 99–106.
Williams, S. T. and Khan, M. R. (1974). Antibiotics – a soil microbiologist's viewpoint. *Postepy Higieny i Medycyny Doswiadczalnej* **28**, 295–400.
Williams, S. T. and Mayfield, C. I. (1971). Studies on the ecology of actinomycetes in soil. III. The behaviour of streptomycetes in acid soil. *Soil Biology and Biochemistry* **3**, 197–208.
Williams, S. T. and Robinson, C. S. (1981). The role of streptomycetes in decomposition of chitin in acidic soils. *Journal of General Microbiology* **127**, 55–63.
Williams, S. T. and Wellington, E. M. H. (1982a). Principles and problems of selective isolation of microbes. In: *Bioactive Microbial Products: Search and Discovery* (J. D. Bu'lock, L. J. Nisbet and D. J. Winstanley, eds.), pp. 9–26. Academic Press, London.
Williams, S. T. and Wellington, E. M. H. (1982b). Actinomycetes. In: *Methods of Soil Analysis. Part 2. Chemical and Microbiological Properties.* 2nd Edn. (A. L. Page, R. H. Miller and D. R. Keeney, eds.), pp. 969–987. American Society of Agronomy and Soil Science Society of America, Madison, Wisconsin.
Williams, S. T., Davies, F. L. and Mayfield, C. I. and Khan, M. R. (1971). Studies on the ecology of actinomycetes in soil. II. The pH requirements of streptomycetes from two acid soils. *Soil Biology and Biochemistry* **3**, 187–195.
Williams, S. T., Shameemullah, M., Watson, E. T. and Mayfield, C. I. (1972). Studies on the ecology of actinomycetes in soil. VI. The influence of moisture tension on growth and survival. *Soil Biology and Biochemistry* **4**, 215–225.

Williams, S. T., McNeilly, T. and Wellington, E. M. H. (1977). The decomposition of vegetation growing on metal mine waste. *Soil Biology and Biochemistry* **9**, 271–275.

Willingham, C. A., Roach, A. W. and Silvey, J. K. G. (1966). Comparative studies of substrate degradation by marine-occurring actinomycetes. *American Midland Naturalist* **75**, 232–241.

Willoughby, L. G. (1969a). A study on aquatic actinomycetes: the allochthouous leaf component. *Nova Hedwigia* **18**, 45–113.

Willoughby, L. G. (1969b). A study of the aquatic actinomycetes of Blelham Tarn. *Hydrobiologia* **34**, 465–483.

Willoughby, L. G. (1971). Observations on some aquatic actinomycetes of streams and rivers. *Freshwater Biology* **1**, 23–27.

Willoughby, L. G. (1976). The activity of *Streptomyces* phage-virus in fresh water. *Hydrobiologia* **49**, 215–228.

Willoughby, L. G. and Baker, C. D. (1969). Humic and fulvic acids and their derivatives as growth and sporulation media for aquatic actinomycetes. *Verhandlungen International Verein Limnologie* **17**, 795–801.

Willoughby, L. G., Smith, S. M. and Bradshaw, R. M. (1972). Actinomycete virus in fresh water. *Freshwater Biology* **2**, 19–26.

Wong, P. T. W. and Griffin, D. M. (1974). Effect of osmotic potential on streptomycete growth, antibiotic production and antagonism to fungi. *Soil Biology and Biochemistry* **6**, 319–325.

Young, R. A. and Smith, R. E. (1975). Degradation of feather keratin by culture filtrates of *Streptomyces fradiae*. *Canadian Journal of Microbiology* **21**, 583.

Zobell, C. E., Grant, C. W. and Haas, H. F. (1943). Marine micro-organisms which oxidize petroleum hydrocarbons. *Bulletin of the American Association of Petroleum Geologists* **27**, 1175–1193.

Index

A

Actinobacillus actinomycetemcomitans, 33, 396, 399, 457
Actinobacteria, 12, 13, 14, *see also* Individual genera, 29–59
Actinobacterium meyerii, 35, 391, 434, 439
Actinobifida chromogena, 120
Actinomadura, 4, 16, 21, 105, 108–109, 118, 128, 289, 406
    classification and general properties, 106–107
    diagnosis of disease, 429, 442–447
    ecology, 482–487
    infection due to, 411
    lipid and peptidoglycan composition, 340, 346, 353–355, 380
    numerical taxonomy, 23, 24, 26
    pathogenesis, 462, 474
Actinomadura africana, 106, 118
Acmd. coeruleofusca, 106, 118
Acmd. dassonvillei, 21, 107, 118, 407
Acmd. fastidiosa, 106
Acmd. flava, 118
Acmd. flexuosa, 106, 109
Acmd. helvata, 107
Acmd. longispora, 106, 118
Acmd. madurae, 69, 106–107, 407, 506
    diagnosis of disease, 446–450
    lipid and peptidoglycan composition, 348, 351, 356
Acmd. pelletieri, 106–107, 348, 351, 407, 411
    diagnosis of disease, 446–450
Acmd. pusilla, 107
Acmd. roseoviolacea, 107
Acmd. verrusocospora, 106
Actinomyces, 3, 166, 391, 472
    actinobacteria, 29–31, 38, 57
    chemotaxonomy, 12–14, 21
    classification and general properties, 32–35
    ecology, 484
    identification, 432–442
    numerical taxonomy, 26
    pathogenesis, 458, 465–466, 471
    plasma membrane lipids, 339–340
Actinomyces acidophilus, 488
Actm. bovis, 2, 25–26, 33–34, 170, 403, 434, 439, 458
Actm. dentocariosus, 57
Actm. eriksonii, 34–35, 397, 458
Actm. humeriferus, 33–34, 489
Actm. israelii, 24, 348, 356, 391
    classification of *Actinomyces*, 33–35
    classification of *Arachnia*, 37–38
    clinical significance of actinomycetes, 397–403
    diagnosis of actinomycete disease, 431–440
    pathogenesis, 457, 464–465, 471
Actm. naeslundii, 27, 33–34, 170, 458, 465, 474
    clinical significance of actinomycetes, 397, 401–402
    diagnosis of actinomycete disease, 436–440
Actm. odontolyticus, 33–34, 397, 400, 434, 439, 458, 474
Actm. pyogenes, 33, 35, 439
Actm. suis, 33–35, 403
Actm. viscosus, 33–34, 170, 342, 438
    clinical significance of actinomycetes, 397, 401–403
    diagnosis of actinomycete disease, 436–440
    pathogenesis, 458, 465–467, 471–472
Actinomycetaceae, 26, 29–30, 397, 399–400
Actinomycetales, 1, 29, 124, 131, 472
Actinomycetes, 463 *see also* Individual genera
    cell structure, 172–180
    eubacteria, 173–178

Actinomycetes (*cont.*)
  importance of, 1–5
  with multilocular sporangia 69–74
  in soil, 486–494
  spores of, 180–188
Actinomycetoma, 403, 406–407, 411, 425
Actinomycosis, 425–427, 465, 467
  animal, 402–403
  human, 391–400
Actinophage, 216–218, *see also* Phage
Actinoplanaceae, 26, 97, 99, 105, 183, 483, 506
Actinoplanes, 4, 10, 13, 15, 24, 59, 66, 115, 353, 356
  classification and general properties, 60–66
  ecology, 484, 490, 505–506, 511
Actinoplanes armeniacus, 63
Actp. coeruleus, 61
Actp. ferrugineus, 61
Actp. italicus, 61
Actp. missouriensis, 62, 326
Actp. philippinensis, 61–62
Actp. rectilineatus, 61
Actp. teichomyceticus, 61
Actp. utahensis, 62
Actinoplanetes, 13, 15, 99, *see also* Individual genera, 59–69
Actinopolyspora, 17, 22, 121–122, 355
  classification and general properties, 122–123
  plasma membrane lipids, 339–340, 346
Actinopolyspora halophila, 122–123, 348, 489
Actinopycnidium, 26, 96, 103
Actinosporangium, 26, 96, 103
Actinosynnema, 17, 24, 482, 484
  classification and general properties, 116–117
Actinosynnema mirum, 116–117
Adonitol, in diagnosis of actinomycete disease, 440, 449
Aerial mycelium, 169–170
Agglutinins, 153, 156, 161, 162
Agromyces, 12, 14, 25, 30–31, 59, 356, 472
  classification and general properties, 35–36
Agromyces ramosus, 36, 59, 489

Agropyron junceiforme, 493
Alanine, 11, 40, *see also* Walls
Alkaloids, 314–322
Amino acid amides, 345–346
Amino acids in cell walls, *see* Walls
Ammophila arenaria, 493
Amorphosporangium, 10, 15, 24, 59, 66, 353
  classification and general properties, 63
Amorphosporangium auranticolor, 63
Amorphosporangium globisporus, 63
Ampullariella, 13, 15, 24, 59, 60, 66, 353
  classification and general properties, 64
Ampullariella regularis, 64
Anthracyclines, 326
Antibiotics, 322–330, 399, 411, 483, 493–494
  aminoglycoside, 263
  fluorescein-labelled, 503
  genetic analysis, 246–267
  production, 3–4, 259, 260–262
  resistance, 259–264, 273
Arabinogalactan, 18, 20, 355, 358
Arabinose, 18, 257–258, 355, *see also* Individual genera, classification and general properties
  diagnosis of actinomycete disease, 439–449
Arachnia, 12, 14, 24–26
  actinobacteria, 29–31
  classification and general properties, 37–38
  diagnosis of actinomycete disease, 434, 438, 440
  ecology, 484
  infection and impairment, 391
  lipid and peptidoglycan composition, 339–340
Arachnia propionica, 37–38, 397, 400, 437–440, 458
Arcanobacterium, 14, 21, 25, 31, 433, 438, 441
  classification and general properties, 38–39
Arcanobacterium haemolyticum, 39, 437, 441
Arthrobacter, 9, 12, 169
  actinobacteria, 30–31, 45
  chemotaxonomy, 13–14, 19, 21

classification and general properties, 40–44
defining actinomycetes, 29
ecology, 485–487, 490, 496, 506, 509
nocardioform actinomycete, 76, 78
numerical taxonomy, 23–25
Arthrobacter albidus, 78
Atbc. atrocyaneus, 43
Atbc. aurescens, 43
Atbc. citreus, 41
Atbc. crystallopoites, 43
Atbc. duodecadis, 41–44
Atbc. flavescens, 41–43
Atbc. globiformis, 41–44, 511
Atbc. histidinolovorans, 43
Atbc. ilicis, 43
Atbc. nicotianae, 42–43
Atbc. pascens, 43
Atbc. polychromogenes, 43
Atbc. ramosus, 43
Atbc. sensu lato, 41–42
Atbc. sensu stricto, 41, 43
Atbc. simplex, 13, 42–43, 101, 303–304, 340
Atbc. terregens, 36, 42–44
Atbc. tumescens, 42–43
Atbc. ureafasciens, 43
'*Aurantiaca*' taxon, 377–378
chemotaxonomy, 15–20
nocardioform actinomycetes, 74, 78
numerical taxonomy, 22–23
peptidoglycan, 355
plasma membrane lipid, 340, 348
wall membrane lipid, 358

B

Bacillaceae, 11–12, 29, 131
Bacillus, 10–13, 68, 182, 495
Bacillus cereus, 178, 312
B. subtilis, 165, 175, 231
Bacterioidaceae, 399
Bacterioides, 399, 457
Bacterioides asaccharolyticus, 396
Bacterioides fragilis, 399
Bacterioides melaninogenicus, 396
Bacterioides thetaiotaomicron, 399
Bacterionema, 29, 391, 438, 462
Bacterionema matruchotii, 30, 78, 391, 397, 441
Bagassosis, 499

Bifidobacterium, 13, 26, 29–30, 35, 371, 438
Bifidobacterium adolescentis, 35
Bibc. dentium, 35
Bibc. eriksonii, 397, 439, 458
Bifunctional replicons, 276–277
Brevibacterium, 12, 14
actinobacteria, 30–31, 41, 57
classification and general properties, 44–46
ecology, 484, 508–509
nocardioform actinomycetes, 76, 78
numerical taxonomy, 23–24, 29
Brevibacterium albidum, 49
Brev. ammoniagenes, 78
Brev. citreum, 49
Brev. divaricatum, 78
Brev. fermentans, 53
Brev. flavum, 78
Brev. helvolum, 36
Brev. immariophilum, 78
Brev. imperiale, 36, 52
Brev. insectiphilum, 36, 58
Brev. iodinum, 46
Brev. lactofermentum, 78
Brev. linens, 13, 45–46
Brev. lipolyticum, 43
Brev. liquefaciens, 78
Brev. luteum, 49
Brev. lyticum, 53
Brev. protophormiae, 43, 58
Brev. pusillum, 49
Brev. roseum, 78
Brev. saperdae, 49
Brev. sulphureum, 43
Brev. testaceum, 49
Brev. vitarumen, 78
Brochothrix, 51
Brochothrix thermosphacta, 51
Bronchiolitis obliterans, 414

C

Calcium, 187
Cannabinoids, 312–314
Caries and peridontal disease, 400–402
Carotenoids, 347–350
Caseobacter, 8, 15, 20, 22, 74
classification and general properties, 75–76
Caseobacter polymorphus, 76, 358
Causal agents of disease, 2–3

Cell wall, *see* Wall
Cellobiose, 261
  diagnosis of actinomycete disease, 439–442
Cellulomonas, 9, 12, 380
  actinobacteria, 30–31, 53
  chemotaxonomy, 13–14, 19
  classification and general properties, 46–48
  ecology, 485
  numerical taxonomy, 21–25, 29
  plasma membrane lipid, 339–340, 346, 348
Cellulomonas biazotea, 48
Celm. carta, 53
Celm. cartalyticum, 48
Celm. cellasea, 48, 356
Celm. flavigena, 47–48, 356
Celm. fimi, 48
Celm. gelida, 48, 356
Celm. uda, 48
Chainia, 10, 13, 16, 172, 185
  numerical taxonomy, 24, 26
  streptomycetes, 96, 103
Chainia olivacea, 189
Chemical determinants of pathogenicity, 467–472
Chemotaxonomy, 12–21, 438, *see also* Individual genera, classification and general properties
Chloramphenicol, 263–264
Chromobacterium iodinum, 46
Chromosomal inheritance, 251–256
Chromosomal mapping, 250–251
Classification, 7–131
Clinical significance of actinomycetes, 389–415
Clostridium, 10–11, 13, 182
Coenocytic structures, 202
Conjugation,
  Mycobacterium, 220–222
  natural gene exchange, 233–244
Cord factor, 358, 468–469, 474, *see also* Trehalose mycolates
Corynebacteria, 19, 21, 355, 365
  actinobacteria, 30–31, 35–36
Corynebacterium, 9, 12, 28, 309
  actinomycetes, 35, 39, 45, 55, 57
  chemotaxonomy, 13–15, 20–21
  classification and general properties, 77–80
  diagnosis of actinomycete disease, 441
  ecology, 485, 491, 496, 499–500, 506–510
  nocardioform actinomycetes, 74–76, 92, 94
  numerical taxonomy, 22–23
  pathogenesis, 469–472
  peptidoglycans, 353–355
  plasma membrane lipid, 340, 346
  wall membrane lipids, 356–358
Corynebacterium acetoacidophilum, 78
Cnbc. alkanum, 78
Cnbc. aquaticum, 36, 58, 59, 78
Cnbc. barkeri, 58, 78
Cnbc. betae, 49
Cnbc. beticola, 57, 78
Cnbc. bovis, 21, 79, 358, 378
  plasma membrane lipids, 340–348
Cnbc. callunae, 78
Cnbc. diphtheriae, 2, 18, 375–378
  nocardioform actinomycetes, 77, 79
  plasma membrane lipids, 348–349
Cnbc. fascians, 94, 349, 500
Cnbc. flaccumfaciens, 49
Cnbc. flavescens, 51, 79
Cnbc. glutamicum, 18, 348, 378
  nocardioform actinomycetes, 78–79
Cnbc. haemolyticum, 39, 78, 438, 441
Cnbc. herculis, 78
Cnbc. hoagii, 78
Cnbc. hydrocarboclastus, 78, 209
Cnbc. insidiosum, 57, 58, 499–500
Cnbc. iranicum, 58, 59, 500
Cnbc. kutscheri, 79
Cnbc. laevaniformis, 36, 52, 78
Cnbc. lilium, 78
Cnbc. liquefaciens, 79
Cnbc. manihot, 53, 78
Cnbc. matruchotii, 30, 79, 462
  diagnosis of actinomycete disease, 434, 438, 441
  infection and impairment, 391–397, 401–402
Cnbc. mediolanum, 58, 59, 78
Cnbc. melassecola, 78
Cnbc. michiganense, 49, 57, 58
Cnbc. mycetoides, 79
Cnbc. nebraskense, 49, 57, 58, 500
Cnbc. nephridii, 57, 78
Cnbc. okanaganae, 58, 78

Cnbc. oortii, 49
Cnbc. parvum, 465
Cnbc. paurometabolum, 78, 94, 348
Cnbc. poinsettiae, 49
Cnbc. pseudodiphtheriticum, 77
Cnbc. pseudotuberculosis, 77, 79
Cnbc. pyogenes, 26, 35, 39, 77–78, 391, 439
Cnbc. rathayi, 500
Cnbc. renale, 77, 79
Cnbc. rubrum, 78
Cnbc. sensu lato, 21
Cnbc. sensu stricto, 18–19, 26, 30
Cnbc. sepedonicum, 49, 51, 58, 499–500
Cnbc. tritici, 59, 500
Cnbc. vitarumen, 79
Cnbc. xerosis, 77
Coryneforms, 169, 507
Coryneform actinomycetes, 499–500
Coryneform bacteria, 9, 338, 485
    actinobacteria, 30
    chemotaxonomy, 18–20
    numerical taxonomy, 23, 28
Coryneform taxa, 12, 19, 27
Curtobacterium, 12, 25, 29, 78
    actinobacteria, 31, 36, 44–45, 51, 58
    chemotaxonomy, 13–14, 19, 21
    classification and general properties, 48–50
    ecology, 485, 508
Curtobacterium, betae, 50
Curt. citreum, 49
Curt. flaccumfaciens, 49, 50
Curt. luteum, 49
Curt. oortii, 50
Curt. poinsettiae, 50
Curt. pusillum, 49
Curt. saperdae, 49
Curt. testaceum, 49
Cyclopropane fatty acids, 339–340
Cycloserine, 219
Cytoplasm, 172, 183, 188, 245, 251
Cytoplasmic membrane, 173–175, 182

D

Dactylosporangium, 66–67
Dactylosporangium aurantiacum, 67
Dcso. matsukiense, 67
Dcso. thailandense, 67
Definition of actinomycetes, 28–29
Deoxyhexose, 62–69

Deoxyribonucleic acid (DNA), *see also* Individual genera, classification and general properties
    base composition, 10–11, 14–20, 29–30, 220, 231–232
    chromosomal, 231–232
    genetics of nocardioform actinomycetes, 216–221
    homology, 27, 34, 214, 381
    streptomyces genetics, 230–233, 242–255, 258–268, 273–277
    plasmid, 222, 232, 240, 252
Deoxyribonucleic acid–ribonucleic acid (DNA–RNA) homology, 10, 36, 38, 69
Dermatophilaceae, 26
Dermatophilosis (streptotrichosis), 413–414
    diagnosis, 451–452
Dermatophilus, 8, 15, 69, 105, 168, 356, 413, 484
    classification and general properties, 70–71
    numerical taxonomy, 24, 26
Dermatophilus congolensis, 70–71, 413, 451, 462
Detection of actinomycete, 482–486
Diagnosis of allergic alveolitis, 452–453
Diagnosis of disease, 426–453
Diamino acids, 353, *see also* Walls
2,4-Diaminobutyric acid (DAB), 19
Diaminopimelic acid (DAP), 352–353, 438, 445
LL-Diaminopimelic acid, 31, 37
meso-Diaminopimelic acid (*meso*-DAP), 11, 18–20, 26, 311, 355, *see also* Walls
Diphosphatidylglycerol, 20
DNA *see* Deoxyribonucleic acid

E

Ecology, 481–513
Ecology of actinophage, 510–512
Elytrosporangium, 10, 13, 16, 26, 97, 103
Endogenous infection and impairment, 390–403
Endonuclease digestion, 10
Endonucleases, 217–218, 274
Endophytes, 502–503
Endospores, 182–183, 188

Enterobacteriaceae, 396, 430
Enterothallic spores, 182, 185
Envelope lipid and peptidoglycan composition, 337–381
Erysipelothrix, 433, 438, 441
Erysipelothrix rhusiopathiae, 437, 441
Erythromycin, 221, 263
Escherichia coli, 202, 231
　genetic analysis, 248, 255, 257, 265–266
　gene cloning, 267–268, 274, 276–277
　natural gene exchange, 237–238, 242
Ethambutol, 219
Eubacteriales, 202, 210
Eubacterium, 10, 26, 438
Excellospora, 16, 105, 124
　classification and general properties, 108–109
Excellospora viridilutea, 108
Exoenzymes, 266–267
Exogenous infection, 403–414
Experimental infections, 457–462
Extrachromosomal elements, *see also* Plasmids
　in streptomyces genetics, 231–256
　of *Mycobacterium*, 222–223

F

Fatty acid, 19–25, 187, 223, 290, 338–340, 378, 380, 468, *see also* Individual genera, classification and general properties
Frankia, 9, 15, 69, 105, 483–484, 502–504
　classification and general properties, 71–73
Frankia alni, 73
Fresh water and marine environments, 504–509
Fructose, 260–261, 452, *see also* Individual genera, classification and general properties
Fusion inhibition, 262
Fusobacterium nucleatum, 396

G

G+C base composition, *see* Deoxyribonucleic acid (DNA) base composition

Galactose, 257–258, 355, *see also* Individual genera, classification and general properties
Gene cloning in *Streptomyces*, 267–277
Gene cloning system, 211, 230
Genetics, 4, 5, 201–223, 229–278, *see also* Conjugation, Lysogeny, Phage, Protoplast and Transduction
Genetic analysis, 246–259
　in *Streptomyces* biology, 256–267
Genome complexity, 231
Geodermatophilus, 8, 12, 69, 168, 484
　chemotaxonomy, 13, 15
　classification and general properties, 73–74
　numerical taxonomy, 24, 26
Geodermatophilus obscurus, 74
Geosmin, 506–507
Gluconate, 261, 449
Glucosamine, 350
Glucose, 257–258, *see also* Individual genera, classification and general properties
　diagnosis of actinomycete disease, 439, 442, 452
Glycerol, 257–258, *see also* Individual genera, classification and general properties
　diagnosis of actinomycete disease, 439–442
Glycine, 18, *see also* Walls
Glycolipids, 342–345, 468, 471
Gordona, 92, 94, 353
Gordona aurantiaca, 94, 190
Gram-negative bacteria, 233, 243, 429, 499, 508
Gram-positive bacteria, 201, 377, 429, 434, *see also* Individual genera, classification and general properties
Growth of actinomycete cells, 188–191
Guanine + cytosine (G+C mol %), *see* Deoxyribonucleic acid (DNA) base composition

H

Holothallic spores, 182
Host–parasite interactions, 462–467
Hydrocarbons, 289–296

INDEX

Hypersensitivity, delayed-type, 468, 470, 473

I

Identification of *Actinomycetaceae*, 432–442
Immunoelectrophoresis, 453
Immunofluorescence, 189, 429, 433, 487
Infection due to aerobic actinomycetes, 407–411
  treatment, 411–412
Inositol, 439
  diagnosis of actinomycete disease, 442, 449
Intrasporangium, 16, 24, 26, 115, 356
  classification and general properties, 98–99
Intrasporangium calvum, 99
Isolation of *Actinomycetaceae* from clinical material, 429–432
Isoprenoid quinones, 11, 19–20, 347

J

Jensenia, 92
Jensenia canicruria, 210

K

Kanamycin, 255
Kineosporia, 16, 24
  classification and general properties, 100
Kineosporia aurantiaca, 100
Kitasatoa, 10, 13, 16, 26
  classification and general properties, 97–98
Kitasatoa displospora, 98
Kits. kauaiensis, 98
Kits. nagasakiensis, 98
Kits. purpurea, 98
Kurthia, 11, 29
  eubacteria, 433
  laboratory diagnosis of disease, 425–453
  lactobacilli, 433

L

Lacrimal canaliculitis, 400
Lactobacillaceae, 30
Lactobacillus, 30, 438
Leprosy, 211

Leptotrichia buccalis, 396, 428
Lethal zygosis, 239–241
Leucocytes, 462–467
Leucomycin, 255, 262
Lipids, 172–174, 216, *see also* Individual genera, classification and general properties
  actinobacteria, 29–30
  chemotaxonomy, 19, 21
  mycobacteria, 19
  numerical taxonomy, 22–25
Liposome-mediated transformation, 245–246
Listeria denitrificans, 55
Listeria monocytogenes, 463–464, 470
Lymphocytes, 462, 464–467, 470
Lysine, 31–33, 40, 42, 43, 341, 352, *see also* Walls
Lysogeny,
  Rhodococcus, 215–216
  Mycobacterium, 223
Lysosomes, 469–470
Lysozyme, 189, 191, 216, 472

M

Macrolides, 325
Macrophage, 459–469, 472
Maduromycetes, 105–115, *see also* Individual genera
Maduromycetes, 16, *see also* Individual genera, 105–115
Madurose, 447, 452
Magnesium, 187
Malate dehydrogenase, 11
Mannitol, 261, 439
  diagnosis of actinomycete disease, 442, 449–452
Mavioquinone, 347
Melanin, 255
Menaquinones, 11, 20–21, 24–25, 31, 347–350, *see also* Individual genera, classification and general properties
meso-Diaminopimelic acid, *see* Diaminopimelic acid
Mesosomes, 175–176
Methylenomycin A, 263
Microaerophilic actinomycete, 188
Microbacterium, 12, 14, 59, 78
  actinobacteria, 31, 36

Microbacterium (cont.)
   classification and general properties, 50–52
   ecology, 484, 496
   nocardioform bacteria, 218–223
   numerical taxonomy, 23, 25
Microbacterium ammoniophilum, 78
Miba. flavum, 51, 78
Miba. lacticum, 13, 51–52
Miba. liquefaciens, 51
Miba. mesentericum, 51
Miba. thermosphactum, 51
Microbispora
   actinomycete spores, 185, 188
   chemotaxonomy, 16
   classification and general properties, 109–110
   ecology, 482, 484, 495
   maduromycetes, 105–106
   numerical taxonomy, 24, 26
   plasma membrane lipid, 340
   thermomonosporas, 115
   wall membrane lipid, 356
Microbispora aerata, 109
Mibs. amethystogenes, 109
Mibs. echinospora, 110
Mibs. parva, 109
Mibs. rosea, 109–110
Mibs. thermodiastatica, 110
Micrococcus, 12–14, 29–31, 44
Micrococcus luteus, 348
Microellobospora, 10, 168, 188
   chemotaxonomy, 13, 16
   ecology, 484
   numerical taxonomy, 24, 26–27
   streptomycetes, 97, 103
Microellobosporia flavea, 174
Micromonospora, 4, 10–11, 169, 353, 356, 380
   actinomycete spores, 185, 188
   actinoplanetes, 59–60, 66
   chemotaxonomy, 13, 15
   classification and general properties, 67–69
   ecology, fresh water and marine environment, 505, 507
      natural habitats, 482, 484, 486
      soil, 487, 489–490, 493
      as symbionts of animals, 509
   numerical taxonomy, 24
   plasma membrane lipids, 340, 342, 346

Micromonospora acetoformici, 68, 509
Mims. caballi, 68
Mims. gallica, 68
Mims. melanosporea, 68, 181
Mims. propionici, 68, 509
Mims. ruminantium, 68
Micropolyspora, 12, 169, 182, 188, 340, 355, see also Individual genera, 121–130
   chemotaxonomy, 16–18
   classification and general properties, 123–124
   ecology, 484, 495
   maduromycetes, 108
   numerical taxonomy, 22
Micropolyspora angiospora, 124, 187
Mips. brevicatena, 26, 89, 121, 123–124
Mips. caesia, 127
Mips. coerulea, 127
Mips. faeni, 124, 415, 452, 512
   actinomycete spores, 181
   ecology, 496–498
   plasma membrane lipids, 339, 346, 348
Mips. flexuosa, 108
Mips. internatus, 124, 127
Mips. rectivirgula, 124, 186
Mips. rubrobrunea, 108, 124
Mips. viridinigra, 108, 124
   unassigned species, classification and general properties, 128–130
Microtetraspora, 16, 24, 26
   classification and general properties, 110–111
   ecology, 484
   maduromycetes, 105–106
   plasma membrane lipids, 340, 346
Microtetraspora caesia, 111
Mits. fusca, 111
Mits. glauca, 111
Mits. niveoalba, 111
Mits. viridis, 111
Mits. viridis var *intermedia*, 111
Miscellaneous compounds of synthetic origin, 296–299
Mitogens, 470
Monocytes, 463
Morphology, 165–191
Multilocular sporangia, 15
Muramic acid, 14–17, 19, 31, 350, 352–353, see also Walls

Mycelial organization of
   actinomycetes, 166–172
Mycelium, 12, 170, 202, 235–237, 243,
   242–243, 255, 258, 340, 488
Mycobacteriophage, 216, 220, 222–223
Mycobacterium, 8, 9, 28, 128, 170, 378
   chemotaxonomy, 13, 15, 18–19
   classification and general properties,
      80–88
   ecology, 485
   nocardioform actinomycetes, 74–75,
      78, 94
   nocardioform bacteria, 202, 204, 212
   numerical taxonomy, 22–23, 27
   pathogenesis, 463, 468–472
   peptidoglycans, 353–356
   plasma membrane lipids, 340, 346–
      349
   wall membrane lipids, 365, 368
   xenobiotics, 307, 309
Mycobacterium abscessus, 85
M. africanum, 82, 83
M. agri, 86–87
M. album, 94
M. aichiense, 87
M. asiaticum. 83, 84
M. aurum, 87, 88, 94, 345
M. avium, 83, 85, 347, 473
   nocardioform actinomycetes, 82–84
   wall membrane lipids, 364–366
M. avium-intracellulare, 222
M. borstelense, 86
M. bovis, 28, 83
   nocardioform actinomycetes, 81–82
   wall membrane lipids, 364, 366, 370
M. bovis BCG, 219, 345
M. chelonea, 86, 87, 223, 364
   nocardioform actinomycetes, 85–87
M. chitae, 86, 87
M. chubuense, 87
M. diernhoferi, 87, 88
M. duvalii, 87, 88
M. farcinogenes, 83, 84, 86, 364
   nocardioform actinomycetes, 83–84,
      86
M. flavescens, 87–88
M. flavum, 501
M. fortuitum, 86–87, 223, 342
   nocardioform actinomycetes, 85–87
   wall membrane lipids, 356, 363–364,
      373
M. gadium, 87–88

M. gastri, 83
M. giae, 86
M. gilvum, 87–88
M. gordonae, 83, 84, 364
M. habana, 83
M. haemophilum, 83, 85
M. intracellulare, 83, 85, 409
   nocardioform actinomycetes, 82–84
   wall membrane lipids, 364–365,
      369
M. kansasii, 83, 202, 223, 349, 506
   wall membrane lipids, 364–365, 370
M. komossense, 87–88
M. leprae, 3, 81–82, 83, 85, 202, 364,
   370
M. lepraemurium, 85–86, 202
M. malmoense, 83, 84
M. marinum, 83, 84, 366, 370
M. microti, 82, 83, 364
M. minetti, 86
M. neoaurum, 87, 88
M. nonchromogenicum, 83
M. obuense, 87
M. paraffinicum, 342, 377
M. parafortuitum, 87, 88
M. paratuberculosis, 83, 85, 364, 373
M. peregrinum, 86
M. phlei, 86–88, 219
   peptidoglycans, 361–367, 374
   plasma membrane lipids, 348–349
M. rhodesiae, 87
M. rhodochrous, 92, 312
M. rubrum, 367, 373, 377
M. runyonii, 86
M. scrofulaceum, 83, 202, 365, 368
   nocardioform actinomycetes, 82–83
M. senegalense, 86, 87, 171, 364, 368
   nocardioform actinomycetes, 84–87
M. shimoidei, 85
M. simiae, 83
M. smegmatis, 86–87, 357, 360, 364–368
   nocardioform bacteria, 202, 219–222
   wall membrane lipids, 357, 360,
      364–368
M. sphagni, 87–88
M. szulgae, 83, 84
M. terrae, 83
M. thermoresistibile, 87–88
M. tokaiense, 87
M. triviale, 83
M. tuberculosis, 3, 20, 28, 83, 85, 202,
   219, 409

M. tuberculosis (cont.)
  actinomycete pathogenesis, 463–464, 469, 473
  nocardioform actinomycetes, 80–82, 84
  wall membrane lipids, 361–372
M. ulcerans, 83, 84
M. vaccae, 87, 88
M. xenopi, 82, 83
Mycocerosic acids, 367, 374
Mycolic acid, 20–23, 438, 448, 459, 471, *see also* Individual genera, classification and general properties, 91–129
  lipid and peptidoglycan composition, 338–339, 348, 355–358, 364–381
Mycolipenic acid, 366
Mycoplana, 501
Mycosides, 370, 374
Myrica gale, 504

N

Natural gene exchange, 232–244
Neomycin, 263, 276
  kanamycin, 221
  phosphotransferase, 256–260
Neutrophils, 462–463
Nitrogen fixation, 501–504
Nocardia, 5, 9, 168, 202, 353–356, *see also* Individual genera, 203–210
  actinobacteria, 31, 53, 57
  cell structure, 175–177
  chemotaxonomy, 13, 15, 18–21
  classification and general properties, 89–91
  defining actinomycetes, 28–29
  diagnosis of actinomycete disease, 429, 442–447
  ecology, 485, 490–493, 497, 504, 506–510
  infection, 403–412
  maduromycetes, 107
  micropolysporas, 121, 124, 128
  nocardioform actinomycetes, 74–75, 78, 92, 94
  numerical taxonomy, 22, 26–27
  pathogenesis, 459–474
  peptidoglycan, 353, 355
  plasma membrane lipid, 340, 346, 348
  thermomonosporas, 116
  wall chemotype, 356
  xenobiotics, 289–296, 301–311, 332
Nocardia aerocolonigenes, 91, 129, 410
Nrda. amarae, 20, 89–90, 348, 358, 377–378, 496
Nrda. asteroides, 27, 89–91, 191, 348, 358, 378
  cell structure, 173, 177
  diagnosis of actinomycete disease, 446–450
  ecology, 490, 507
  infection, 403–407, 411–412
  infection due to, 408–409
  mycelial organization, 167
  nocardioform bacteria, 202–206
  pathogenesis, 458–473
  wall membrane lipid, 373–377
Nrda. autotrophica, 90, 129, 346, 348, 378, 380, 410
Nrda. blackwellii, 305
Nrda. brasiliensis, 90–91, 348, 358, 458, 470
  diagnosis of actinomycete disease, 446–450
  infection, 403–407, 410, 412
Nrda. brevicatena, 26, 90, 124
Nrda. calcarea, 215, 501
Nrda. canicruria, 210, 211, 213
Nrda. capreola, 129
Nrda. carnea, 90, 410
Nrda. caviae, 90
Nrda. cellulans, 53, 501
Nrda. coeliaca, 90
Nrda. corallina, 292–293
Nrda. dassonvillei, 407, 498
Nrda. dentocariosus, 57
Nrda. erythropolis, 210–211, 301, 496
Nrda. farcinica, 90
Nrda. globerula, 210, 215
Nrda. kirovani, 347, 349
Nrda. leishmanii, 129
Nrda. mediterranea, 19, 90, 129, 202–207, 246, 328
Nrda. minima, 292
Nrda. opaca, 202–203, 207–210
Nrda. orientalis, 90, 129, 410
Nrda. otitidis-caviarum, 90–91, 174, 358, 377, 473
  diagnosis of actinomycete disease, 446–450

ecology, 488
infection, 403–407, 410, 412
pathogenesis, 458
plasma membrane lipid, 342–348
Nrda. polychromogenes, 342
Nrda. restricta, 203, 207–210, 304
Nrda. rhodnii, 510
Nrda. rubra, 375
Nrda. rugosa, 129
Nrda. salivae, 57
Nrda. salmonicolor, 290, 292, 314, 490
Nrda. sensu Wakeman, 21, 89
Nrda. tartaricans, 295
Nrda. transvalensis, 90, 410, 449–450
Nrda. turbata, 53
Nrda. vaccinii, 90–91, 449–450
Other nocardial species, 207–210
Nocardia and nocardioform bacteria, 203–210
Nocardial infections, 405–406
Nocardioform actinomycetes, 8–9, 13, 15, 18, 19, 23, 28, 177, see also Individual genera, classification and general properties, 74–94
  genetics of nocardioform bacteria, 201, 206–207, 210, 215, 223
  morphology, 168, 172, 177
  genetics, 201–224
Nocardioforms, 169, 188, 216
Nocardioides, 16, 24, 26, 43, 340, 485
  classification and general properties, 100–101
Nocardioides albus, 27, 43, 101
Nrdo. simplex, 43, 101
Nocardiopsis,
  actinomycete spores, 182
  chemotaxonomy, 13, 17, 21
  classification and general properties, 117–118
  diagnosis of actinomycete disease, 446–447
  ecology, 485, 491
  lipid and peptidoglycan composition, 340, 346
  maduromycetes, 106–107
  numerical taxonomy, 23–24, 26
  thermomonosporas, 115
Nocardiopsis dassonvillei, 118, 348, 446–450, 497
Nrdp. syringae, 118

Nocardiosis, 211, 403, 409, 411, 425
  diagnosis of actinomycete disease, 442–451
  pathogenesis, 459, 461–464, 474
Nucleic acids, see Deoxyribonucleic acid (DNA) and ribonucleic acid (RNA)
Nucleotides, 10, 214
Numerical phenetic analysis, 203, see also Individual genera, classification and general properties
Numerical taxonomy, 21–28

O

Oerskovia, 12, 29, 31, 168, 289, 380, 412
  actinobacteria, 245, 248, 254
  chemotaxonomy, 13–14, 19, 21
  classification and general properties, 52–53
  ecology, 485, 489
  numerical taxonomy, 23, 25
  pathogenesis, 462
  peptidoglycan, 353, 355
  plasma membrane lipid, 339–340, 346, 348
Oerskovia cellulans, 501
Orsk. turbata, 53, 356, 412, 449–450
Orsk. xanthineolytica, 53, 356, 412
Oligonucleotides, 10
Oligonucleotide sequencing, 29
Ornithine, 31, 33, see also Walls

P

Pathogenic actinomycetes, 412–413
Pathogenesis, 457–474
Peptidoglycans, 11, 177, 337–381, 468, 470–471, see also Walls
  actinobacteria, 30–31
  chemotaxonomy, 14–17
  numerical taxonomy, 18–20
Peptidolipid, 468
Peptococcaceae, 396
Pesticides, 300–302
Phage,
  gene cloning in *Streptomyces*, 268, 272–276
  genetics of *Streptomyces*, 232, 256
  restriction modification (RM) system, 265
Rhodococcus, 216–218, 223

Phage vector molecules, 273–276
Phagocytes, 463, 465–467
Phagocytosis, 462, 464
Phagosome, 469–470
Phleic acids, 367
Phosphatidylcholine, 20
Phosphatidylethanolamine, 20, see also Individual genera, classification and general properties
Phosphatidylinositol, 20, see also Individual genera, classification and general properties
Phosphatidylinositol mannosides, 20, see also Individual genera, classification and general properties
Phosphatidyl-monomethylethanolamine, 20
Phospholipids, 22, 24–25, 341, 469
Phthioceranic acids, 367
Phthiocerol waxes, 374
Phylogenetic relationship, 13
Pilimelia, 15, 24, 59–60
　classification and general properties, 64–66
Pilimelia columellifera, 65
Pilm. terevasa, 65
Planobispora, 10, 16, 59, 105, 340
　classification and general properties, 111–112
　ecology, 482, 485
　numerical taxonomy, 24, 26
Planobispora longispora, 112
Plob. rosea, 112
Planomonospora, 10, 16, 59, 105, 112, 340
　actinomycete spores, 182, 185
　classification and general properties, 112–113
　ecology, 482, 485
　numerical taxonomy, 24, 26
Planomonospora parontospora, 112–113
Plants and fungi (in ecology)
　Amitermes minimus, 509
　Bibio marci, 509
　Calotropsis gigantea, 493
　Fomes annusus, 494
　Fusarium oxysporum, 494
　Juncus, 493
　Gaemannomyces graminis, 494
　Phellinus weirii, 494
　Phymototrichum omnivorum, 494
　Phytophthora cinnamoni, 494
　Procambarus vensutus, 505
　Procubitermes aburiensis, 509
　Pythium debaryanum, 494
　Reticulitermes fucifugus, 509
　Rhizoctonia bataticola, 494
　Rhizoctonia solari, 494
　Rhodnius prolixus, 510
　Verticillium albo-atrum, 494
Plant pathogens, 499–501
Plasma membrane lipids, 338–350
Plasmids, see also Deoxyribonucleic acid
　genetics of nocardioform bacteria, 208–222
　Streptomyces genetics, 230–244, 252–253, 256–266, 276–277
　vector molecules, 269–273
Plom. venezuela, 112
Polar lipid, 19–21, 346, see also Phospholipids
Polyene antibiotic, 187
Polymorphonuclear granulocytes, 392
Polysaccharide, 173–174, 177, 187
Prodiginine, 19, 347, 350
Prokaryotes, 2
Promicromonospora, 13, 15, 21, 25, 29, 31, 48, 346
　classification and general properties, 53–54
Promicromonospora citrea, 54
Propionibacterium, 13, 26, 29–30, 38
Propionibacterium terium, 438
Protoplast,
　formation, 216
　fusion, 244–245, 250–252
　Streptomyces, 268–269
Pseudomonas, 307, 489, 510
Pseudomonas diminuta, 341
Pseudonocardia, 17, 22, 121, 182, 340, 346, 355
　classification and general properties, 125–126
　ecology, 485, 495, 497
Pseudonocardia spinosa, 125–126
Psnc. thermophila, 124, 348, 495
Pulmonary and systemic nocardiosis, 403–405

R

Raffinose, in diagnosis of actinomycete disease, 439, 442

Renibacterium, 12, 14, 25, 31
  classification and general properties, 54–55
Renibacterium salmoninarum, 55
Restriction,
  enzyme analysis of chromosomal DNA, 231–232, 275
  modification (RM) systems, 264–266
Reticuloendothelial system, 462
Rhizopus arrhizus, 287
Rhizospheres, 492–493
Rhodococcus, 5, 8–9, 128, 169, 202, 220, 289, 307, 375, 412
  actinobacteria, 41, 45
  chemotaxonomy, 13, 15, 18–21
  classification and general properties, 91–94
  diagnosis of actinomycete disease, 446–447
  ecology, 485, 490–493, 496, 504, 510
  Nocardia, 203–204
  nocardioform actinomycetes, 74–76, 78
  numerical taxonomy, 22–27
  pathogenesis, 461, 470
  peptidoglycans, 353, 355
  plasma membrane lipids, 340, 346
Rhodococcus bronchialis, 93–94, 169
  diagnosis of actinomycete disease, 446–450
  lipid and peptidoglycan composition, 348, 358, 378
  nocardioform actinomycetes, 91–94
Rodc. coprophilus, 93–94, 169, 482, 495, 505
  lipid and peptidoglycan composition, 348, 358
  nocardioform actinomycetes, 92–93
Rodc. corallinus, 449–450
Rodc. equi, 92–94, 169, 348, 358, 449–450
Rodc. erythropolis, 93–94, 169, 348, 358
  diagnosis of actinomycete disease, 449–450
  ecology, 496, 501, 508
  nocardioform actinomycetes, 93–94
  nocardioform bacteria, 203, 206–217
Rodc. globerulus, 93, 167, 169
Rodc. luteus, 93–94
Rodc. maris, 93–94
Rodc. rhodnii, 93–94, 169, 171, 348, 358, 510

Rodc. rhodochrous, 92–94, 169, 190, 213, 412, 462
  diagnosis of actinomycete disease, 446–450
  lipid and peptidoglycan composition, 348, 358, 378
  nocardioform actinomycetes, 93–94
Rodc. ruber, 93–94, 348, 358, 449–450, 462
Rodc. rubropertinctus, 93–94, 169, 348, 358
Rodc. sputi, 169
Rodc. terrae, 93–94, 169, 348, 358
Rhodococcus and nocardioform bacteria, 210–218
Ribonuclease, 10
Ribonucleic acid (RNA), 268, 381
Ribose, in diagnosis of actinomycete disease, 439–442
Ribosomal ribonucleic acid (rRNA), 10–13, 26–30, 38, 44
Ribosomes, 172, 221, 263
Rifamycin, 19, 206–207, 219, 260, 328
Root nodules, 502–504
Root pathogens, 493–494
Rothia, 12, 15, 21, 355, 391
  actinobacteria, 29–31
  classification and general properties, 56–57
  diagnosis of actinomycete disease, 434, 438, 441
  numerical taxonomy, 23–26
  pathogenesis, 462
Rothia dentocariosa, 56, 342, 397, 401, 441

S

Saccharomonospora, 11, 17, 22, 121, 355, 485, 495
  classification and general properties, 126–127
Saccharomonospora viridis, 415, 496–497, 499
Saccharopolyspora, 17, 22, 121–122, 355, 485
  classification and general properties, 128
Saccharapolyspora hirsuta, 128, 497–498
Salmonella typhimurium, 249
Saprophytes, 3, 481
Schizosaccharomyces pombe, 189

Sclerota, 172
Serratia marcescens, 430
Silicon, 187
Sorbitol, in diagnosis of actinomycete disease, 439–440, 453
Spirillospora, 16, 24, 59, 66, 105, 340, 485
    classification and general properties, 113
Spirillospora albida, 105, 113
Spores as allergens, 414–415
Sporichthya, 16, 170
    classification and general properties, 101–102
Sporichthya polymorpha, 101–102
Sporoactinomycetes, 8–10, 18–19, 26, 169, 177, 186, 202
Sporogenous actinomycetes, 168
Staphylococcus aureus, 165, 396, 463
Staphylococcus epidermidis, 396
Steroids, 302–310
Streptoalloteichus, 17, 111
    classification and general properties, 118–119
Streptoalloteichus hindustanus, 119
Streptococcus, 10, 401
Streptococcus faecalis, 176, 178
Streptococcus mutans, 401
Streptococcus pyogenes, 474
Streptomyces, 5, 10, 29
    actinomycete infection, 407
    actinomycete spores, 182–188
    actinoplanes, 60, 63
    alkaloids, 319–320
    antibiotics, 327
Streptomyces
    biology, 256
    genetic analysis, 267
    genetics, 230–278
        and eubacteria, 253, 263
        differentiation, 258–259
    as symbionts of animals, 509
    cannabinoids, 312
    cell structure, 173, 177, 180
    chemotaxonomy, 13, 16, 18, 21
    classification and general properties, 102–103
    diagnosis of actinomycete disease, 428–429, 442–447
    ecology of actinophage, 511
    in compost and related substrates, 495–497
    in fresh water and marine environments, 505, 507
    in natural habitats, 482, 485
    in plant pathogens, 500
    in soil, 486, 490–494
    envelope chemotypes, 380
    gene analysis, 258–263, 265, 267
    gene cloning, 273, 275–277
    infection due to, 411
    maduromycetes, 115
    micropolysporas, 121, 128
    mycelial organization, 168–170
    Nocardia, 203
    numerical taxonomy, 24, 26–27
    pathogenesis, 462, 472
    peptidoglycans, 351, 353
    plasma membrane lipids, 339–340, 343–348
    Rhodococcus, 210, 216
    steroids, 306
    streptomycetes, 94–100, 104
    thermomonosporas, 118
    wall membrane lipid, 356
    xenobiotics, 289, 296
Streptomyces acrimycini, 244, 246, 263–264
Stmy. africanus, 129
Stmy. albovinaceus, 129
Stmy. albus, 103, 231, 250, 496
Stmy. ambofaciens, 262, 325
Stmy. antibioticus, 244, 262, 273
Stmy. armeniacus, 63
Stmy. aureofaciens, 174, 181, 298, 301, 311, 324, 500
Stmy. aureoverticillus, 345
Stmy. aureus, 492
Stmy. azureus, 263
Stmy. badius, 491
Stmy. bikiniensis var. *zorbonensis*, 260
Stmy. caeruleus, 488
Stmy. chrysomallus, 511–512
Stmy. cinnamoneous, 301
Stmy. cinnamonesis, 317–318
Stmy. clavuligerus, 4, 261
Stmy. coelicolor, 4, 185, 187, 205–206
    genetics, 230–278
    sex plasmids in, 233–239, 253
Stmy. diastochromogenes, 494
Stmy. erythaeus, 307

Stmy. finlayi, 509
Stmy. flaveolus, 500
Stmy. flavovirens, 491
Stmy. foersteri, 2
Stmy. fradiae, 244, 251, 263, 276
Stmy. fulvissimus, 319
Stmy. glaucescens, 246, 250, 255, 266
Stmy. globisporus, 345
Stmy. griseoalbus, 494
Stmy. griseofuscus, 244
Stmy. griseus, 27, 187, 244
   ecology, 494–498, 500, 512
   xenobiotics, 317–321
Stmy. ipomoeae, 500
Stmy. kasugaensis, 261–262
Stmy. kitasatoensis, 326
Stmy. lavendulae, 105
Stmy. lividans
   gene cloning, 269–276
   genetic analysis, 256, 265
   natural gene exchange, 233–234, 240–243
   protoplast fusion, 244
Stmy. longisporuber, 351
Stmy. malachiticus, 488
Stmy. mediterranea, 203, 256, 326, 329–330
Stmy. olivaceus, 345, 348
Stmy. ostreogriseus, 174, 181
Stmy. paraguayensis, 407, 446
Stmy. parvulus, 240–241, 244, 262, 265, 270
Stmy. peuceticus var. caesius, 69
Stmy. platensis, 319–320
Stmy. punipalus, 320
Stmy. reticuli, 244, 255, 262, 272
Stmy. rimosus, 233, 243, 246, 250, 255–256, 319
Stmy. roseochromogenes, 303–304
Stmy. roseoflavus var. rosefungini, 168, 187
Stmy. scabies, 250, 494, 500–501, 511
Stmy. (née Oospora) scabies, 2
Stmy. sioyaensis, 345
Stmy. somaliensis, 407, 411, 446–450
Stmy. steffiburgensis, 327
Stmy. stretomycini, 191
Stmy. tenebrarius, 119
Stmy. tenjimariensis, 262
Stmy. tetranda, 320
Stmy. thermodiastatious, 496

Stmy. tolypophorus, 129
Stmy. toyocaensis, 345
Stmy. venezuelae, 187, 232–233, 259, 261
Stmy. vinaceus, 276
Stmy. violaceoruber, 240, 242
Stmy. viridochromogenes, 173, 190, 301
Streptomycetaceae, 101
Streptomycetes, 13, 16, 26, 106, 296, see also Individual genera, 94–105
   morphology, 170, 177, 186–188
   pathogens, 500–501
   Streptomyces genetics, 230, 233, 239, 249–273, 347, 351
Streptomycin, 177, 218–222
Streptosporangiaceae, 99
Streptosporangium, 10, 13, 16, 24, 59, 105, 340, 485
   classification and general properties, 114–115
Streptosporangium alba var. thermophilum, 114
Stso. bovinum, 114–115
Stso. corrugatum, 114
Stso. indianesis, 115
Stso. nondiastaticum, 114
Stso. pseudovulgare, 114
Stso. roseum, 114
Stso. viridogriseum, 114
Streptothrix chromogens, 3
Streptotrichosis (see Dermatophilosis)
Streptoverticillia, 170
Streptoverticillium, 10, 170, 340, 346, 351
   chemotaxonomy, 13, 16, 21
   classification and general properties, 104–105
   numerical taxonomy, 24, 26, 29
Streptoverticilium baldacci, 104
Substrate mycelium, 166–169
Sucrose, in diagnosis of actinomycete disease, 439, 442, 452
Sulpholipid, 367, 469–470, 474
Symbionts of animals, 509–510

T

Taxonomic methods, 12
Teichoic acid, 177, see also Individual genera, classification and general properties
Terpenes, 310–312

Tetracycline, 177, 276, 324
Thermoactinomyces, 130–131, see also Individual genera, classification and general properties
Thermoactinomyces, 11, 25, 29, 114, 169
  actinomycete spores, 182–183, 188
  chemotaxonomy, 13, 17–18
  ecology, 485, 490, 495, 504, 507
  thermomonosporas, 120–121
Thermoactinomyces candidus, 131, 512
Team. cellulosae, 121
Team. dichotomica, 166
Team. dichotomicus, 130, 415, 499
Team. monospora, 127
Team. peptonophilus, 130
Team. sacchari, 11, 131, 415, 497–499
Team. thalpophilus, 130, 452, 497, 499
Team. viridis, 127, 496
Team. vulgaris, 131, 188, 233, 415, 452
  ecology, 483, 496–499, 512
Thermoactinomycetes, 11, 16, 18
Thermomonosporas, 115–121, see also Individual genera, classification and general properties
Thermomonospora, 11–12, 340
  chemotaxonomy, 13, 16–17
  classification and general properties, 119–121
  compost and related substrates, 495, 497–499
  ecology, 485, 512
  numerical taxonomy, 24, 26
Thermomonospora alba. 120
Temo. chromogena, 120, 483
Temo. curvata, 120, 495–496
Temo. falcata, 120
Temo. fusca, 11, 120–121, 496
Temo. galeriensis, 121
Temo. mesophila, 120
Temo. mesouviformis, 120
Temo. viridis, 127, 496
Thermopolyspora flexuosa, 108
Thermopolyspora glauca, 127
Thermopolyspora polyspora, 124

Thermopolyspora rectivirgula, 124
Thiostrepton, 263
Toxoplasma gondii, 465
Transduction
  in Mycobacterium, 219–220
  natural gene exchange, 233
Transformation, of xenobiotics, 287–332
Transposable genetic elements, 253–256
Trehalose, in diagnosis of actinomycete disease, 439–449
Trehalose mycolates, 19, 357–358, 365–366, see also Cord Factor
Tuberculosis, 211
Tuberculostearic acid, 21, 339–340

U

Unassigned actinobacteria, 57–59

V

Veillonella, 401
Viomycin, 263

W

Wall analyses, 11, 18
Wall chemotypes, 18–26, 335
Wall constituents, 14–17
Wall membrane lipids, 355–377
Wall structures, see also Individual genera, classification and general properties
  as determinant of pathogenicity, 472–474
  use in classification, 29–30

X

Xanthobacter, 501
Xylose, 219–220, see also, Individual genera, classification and general properties
  in diagnosis of actinomycete disease, 439–453